LINEAR ALGEBRA

WOO, SUNG SIK

KYOWOO

Woo, Sung Sik
Department of Mathematics
Ewha Women's University

Kyowoo Co., Ltd.
8, Yangnyeongsi-ro, Dongdaemun-gu,
Seoul, Korea 130-861,

Phone: 82-2-925-2861
F a x: 82-2-925-2860
e-mail: kyowoo@kyowoo.co.kr
http://www.kyowoo.co.kr

Printed in the Republic of Korea.
ISBN 979-11-251-0262-5 (93410)

Preface

Linear algebra is one of the foundational subject not only for the mathematics major student but also for the students majoring the area of natural science and social science. Primary purpose of this text is to provide a solid mathematical foundation of the theory of linear algebra for a student. Therefore, this text is written from the view point of a supplier of linear algebra not as an applier of linear algebra. And the readers of this text I have in mind are the students of mathematics major or the area close to it. I believe that the number of texts written from the view point of the latter is strictly bigger than the number of the texts written from the view point of the former. We tried to make every assertion in this text is provable. If this is not possible, we refer to a proper text.

We will be generous for the logical independence among the theorems and definitions. A result in a theorem may be stated in another theorem in different form or terminology and we repeated the definitions whenever we think it is necessary.

The text consists of eight chapters and four parts of appendixes. The chapters 1–5 contains the basics of linear algebra which can be described as a linear algebra of $\mathbb{R}^n$. Those chapters contain an ample amount of materials for a semester course most of which are computational in nature. The key words here are system of linear equations, Gauss elimination method, basis, left and right inverse of a matrix, invertible matrix, elementary matrix, determinant, change of coordinate matrix, eigenvector, diagonalization, Gram Schmidt process, Cauchy Schwartz inequality, least square problem.

The chapters 6–8 are more theoretical and aimed for another semester course. The key words there are abstract vector space, Jordan canonical forms, rational canonical forms, inner product space, normal operators, Hermitian matrix, unitary matrix, spectral theorem, polar decomposition, bilinear forms and Sylvester's Theorem.

In Chapter 1, we start with simple examples of linear equations. And then proceed to find a general method – Gauss elimination method – to solve a system of linear equations. We interpret a system of linear equations as an equation of matrices. We use the notion of vector space $\mathbb{R}^n$ to reinterpret a matrix equation as an equation of vectors. We then introduce the concept of subspace and we show that the set of solution of a system of homogeneous equations becomes a subspace. We define the concept of linear independence of vectors and spanning set of vectors of a subspace. We then introduce an important concept dimension in an early stage so that we can define the rank and nullity of a matrix.

In Chapter 2, we introduce linear maps between $\mathbb{R}^n$'s and we show that a linear map between $\mathbb{R}^n$'s can be represented by a matrix. And we study the relationship between the linear maps and the matrices corresponding to the linear maps. We prove an important rank theorem which relates between the nullity and the rank of a matrix. Elementary matrices are defined and show that the elementary row operations are the same as multiplying the elementary matrices. And we show that an invertible matrices are a product of elementary matrices. We also characterize the invertible matrices in various way.

In Chapter 3, we study one of the most important subject in linear algebra, the determinant which has many applications as well. We study various properties of determinant and we show how the determinant is applied – Cramer's rule, Lagrange interpolation and so on.

In Chapter 4, we study bases of a subspace of $\mathbb{R}^n$ and study basic properties of bases. We then define the linear maps between the subspaces and we represent a linear map in a matrix using the bases of subspaces. We also study the relationship between the matrix representations of the linear maps in different bases.

In Chapter 5, we learn two subjects – eigenvectors for matrices and the inner product structure on $\mathbb{R}^n$. We define diagonalizability of a matrix and see that not all square matrices are diagonalizable. And using the standard inner product structure on $\mathbb{R}^n$ we prove the standard inequalities like Cauchy Schwartz inequality, triangle inequality. Also we study the least square problems.

In Chapter 6, we give a formal definition of 'abstract' vector space over a field F. It will turn out that many mathematical objects can be realized as a vector space and we can analyze them by studying the vector space. We introduce the concepts of basis, dimension and direct sum of vector spaces. Also the notions of linear map, nullity and rank are treated in an abstract way in contrast to the treatments in Chapter 2 in which the treatments were more concrete. In §6.5, we define the dual space of a vector space and we introduce briefly multilinear algebra in §6.6 which can be skipped in the first reading. In §6.7 we repeat the diagonalization of linear maps of §5.1 in more abstract way. In §6.8 we define the cyclic subspaces and prove Cayley-Hamilton Theorem.

In Chapter 7, we consider a vector space together with a fixed linear operator T on a vector space V over a field F. We prove that V is a direct sum of T-cyclic subspaces which is a special case of structure theorem of finitely generated modules over a principal ideal domain. Our proof of decomposition theorem into cyclic subspaces follows closely the proof of the general theorem over a principal ideal domain. Such a decomposition immediately leads to the rational canonical form for T. When the characteristic polynomial of a linear operator splits a slight modification leads to the Jordan canonical form of the linear operator T. Jordan canonical form has wider applications than the rational canonical form and we treat Jordan canonical form in more detail than the rational canonical form.

Chapter 8 is about the inner product which we called, in Chapter 5, a dot product of $\mathbb{R}^n$. We define the inner product structure abstractly on a vector space over the field $\mathbb{C}$ of complex numbers. Throughout the text we write a vector of $\mathbb{C}^n$ as a column, not as a row, and as a consequence we need to require conjugate linearity in the first variable instead of conjugate linearity of the inner product in the second variable. As we did in §5.2, we study orthogonality of the inner product spaces and Gram Schmidt process again.

Next, we study about the adjoint of a linear operator on an inner product space which corresponds to the adjoint of a complex matrix. We also observe that the adjoint is closely related with the dual of a linear map. Then we define the normal operators which plays an important role in the inner product spaces. We prove that a normal operator is diagonalizabe if its characteristic polynomial splits. As consequences of the theorem we draw conclusions for Hermitian, real symmetric, positive matrices and unitary, orthogonal matrices. Also we study the canonical forms for the normal operators whose characteristic polynomial does not necessarily split. Another consequence of diagonalizablity of the normal operators we prove Spectral Theorem. And using the results of positive matrices we prove singular and polar decomposition theorems.

The last section §8.6 is a brief introduction to the theory of bilinear forms. One of the fundamental theorem of this section is the theorem of Sylvester's Law of Inertia which is about the signature of the real symmetric matrices.

We collect the standard basic mathematical facts we need in the appendixes. The first appendix is about set, functions and equivalence relations. The second appendix deal with basic algebraic materials – integers, fields and complex numbers. Next we discuss about polynomials. And finally we discuss about maximal principle and cardinality in set theory. We prove that the cardinality of bases of a vector space are the same. We also prove that the cardinality of a basis of the dual space of an infinite dimensional vector space V is strictly bigger than the cardinality of a basis of V.

Throughout the text, vectors in F^n are denoted by the columns. That way we can realize naturally a matrix multiplication by an $m \times n$ matrix A as a linear map $T_A : F^n \to F^m$. Another effect, as we already pointed out, is that the standard inner product in the vector space $\mathbb{C}^n$ is conjugate linear in the first variable instead of the second variable. We thought consistency in the notation of the vectors in F^n is more important.

Woo, Sung Sik

Notations

$\mathbb{N} = \{1, 2, 3, \dots\}$, the natural numbers

$\mathbb{Z} = \{0, \pm 1, \pm 2, \dots\}$, the integers

$\mathbb{Q} = \{\frac{n}{m} \mid m, n \in \mathbb{Z}\}$, the rational numbers

$\mathbb{R} =$ the real numbers

$\mathbb{C} = \{a + bi \mid a, b \in \mathbb{R}\}$, the complex numbers.

$X \setminus Y = \{x \in X \mid x \notin Y\}$, the set difference

$X \subsetneq Y$ means $X \subseteq Y$ but $X \neq Y$

'$A \Rightarrow B$' stands for 'A implies B'

'$A \Leftrightarrow B$' stands for 'A implies B and B implies A'

'$A := @^9!s * \&$뻬' will mean 'we define A to be $@^9!s * \&$뻬'

'$@^9!s * \&$뻬 $:= A$' will mean '$@^9!s * \&$뻬' will be denoted by A

$A \smile B$, A and B are row equivalent, §1.1.1

$A^{(i)}, A_{(j)}, A^{(I)}, A_{(J)}, A^{(I)}_{(J)}$, §1.2.2

$C(A)$, the set of columns of the matrix A, §1.2.2, §2.1.1

$\mathrm{Span}(S)$, the subset spanned by S, §1.2.3

$\mathbb{R}^n, \mathbb{R}_n$, the column vectors, §1.2.1

$\mathrm{Col}(A), \mathrm{Row}(A)$, the column and the row space of A, §1.3.1

$\mathcal{E}^n = \{\mathbf{e}^i\}, \mathcal{E}_n = \{\mathbf{e}_i\}$, the standard basis, §1.3.4

$\rho(A), \nu(A)$ (resp. $\rho(T), \nu(T)$), the rank and the nullity of A (resp. T), §1.3.4 (resp. §4.2)

$\mathrm{diag}[\lambda_1, \lambda_2, ..., \lambda_n]$, the diagonal matrix, §2.1.1

$M(m \times n), L(m, n)$, the matrices and the linear maps, §2.1.1

A^k, the k-th power of A, §2.2.2

A^t, the transpose of A, §2.2.2

$\tilde{A}_{ij}, \tilde{A}_{IJ}$, §3.2

$\mathrm{adj}(A)$, the adjoint (adjugate) of A, §3.4.3

$\mathrm{sign}(\sigma)$, §3.3.4

$\det(A)($ or $|A|), \mathrm{Tr}(A)$, the determinant and the trace of A, §3.1, §3.2, §3.3.4

$C(f)$, the companion matrix, §3.4.5, §7,1,2

$[T]^{\mathcal{C}}_{\mathcal{B}}$, the matrix of T in the bases $\mathcal{B}, \mathcal{C}$, §4.2.1

$T(S) = \{T(\mathbf{v}_1), ..., T(\mathbf{v}_n)\}$, for a map $T : V \to W$ and $S = \{\mathbf{v}_1, ..., \mathbf{v}_n\}$, §4.2.2

$\det(T), \mathrm{Tr}(T)$, the determinant and the trace of a linear map T, §4.3

$A \sim B$, A and B are similar, §4.3, §6.4.2

$g|f$, f is divisible by g for the polynomials g, f, §5.1.2, §7.1.1

$f_A(t)$, the characteristic polynomial of A, §5.1.2

$\mathbf{u} \cdot \mathbf{v}$, the dot product in $\mathbb{R}^n$, §5.2.1

$\|\mathbf{u}\|$, the norm of $\mathbf{u}$, §5.2.1, §8.1.1

$\mathbf{u} \perp \mathbf{v}$, $\mathbf{u}$ and $\mathbf{v}$ are orthogonal, §5.2.1, §8.1.1

$S^\perp$, the orthogonal complement, §5.2.1

$P_W(\mathbf{y}), \hat{\mathbf{y}}$, the orthogonal projection, §5.2.2, §8.1.1

$P_n(F)$, the polynomials of deg $\leq n$ over F, §6.1.1

U/V, the quotient space, Exercise 6.1.9

A^*, the adjoint of a complex matrix, §6.2.2

$W_1 \oplus W_2$, the direct sum, §6.2.3

$V^\#$, the dual space of V, §6.5.1

S^0, the annihilator of S, Exercise 6.5.17

$U \otimes V$, the tensor product of vector spaces, §6.6.1

$A \otimes B$, the Kronecker product of matrices, §6.6.1

$\bigwedge^r(V), \mathbf{v}_1 \wedge \mathbf{v}_2 \wedge \cdots \wedge \mathbf{v}_r$, the r-th exterior product, §6.6.2

$A \oplus B$, the direct sum of square matrices A, B, §6.8

$C_T(\mathbf{v})$, the cyclic subspace spanned by $\mathbf{v}$, §6.8

$f_T(t), q_T(t)$, the characteristic and minimal polynomial, §7.1.2

$N(g), L(g)$ for a polynomial g, §7.2.2

$J(\lambda, m)$, the Jordan block, §7.3.1

$C_\lambda(\mathbf{v}), \mathcal{B}_\mathbf{v}, \mathcal{B}_\mathbf{v}(\lambda)$, the Jordan cyclic basis,... §7.3.1

$\check{\mathcal{B}}, \check{\beta}_i$, the bases with reversed order, §7.3.1

$\langle\ ,\ \rangle$, the inner product, §8.1.1

$V = W \bigoplus W'$, the orthogonal sum, §8.1.3

$B(\mathbf{u}, \mathbf{v})$, the bilinear form, §8.5.1

Contents

Chapter 1

Linear Equations and Subspaces

Simultaneous polynomial equations are of the form

$$\begin{cases} f_1(x_1, x_2, ..., x_n) = b_1 \\ f_2(x_1, x_2, ..., x_n) = b_2 \\ \quad \cdots\cdots \\ f_m(x_1, x_2, ..., x_n) = b_m \end{cases} \tag{1.1}$$

where f_i is a polynomial in the variables $x_1, x_2, ..., x_n$. A **system of linear equations** refer to the equations (1.1) where the degree of f_i is one. Linear equations are the only equations we can really solve to our satisfaction. As the degree of f_i grows the degree of difficulty in solving the equations grows much faster than the exponential function. Therefore, in many areas of mathematics people try to consider the linear part of mathematical objects, like the tangent line of a curve which already contains a good deal of information about the curve.

In this chapter, we study a method to solve the systems of linear equations systematically by using, so called the Gauss elimination method. We then interpret a system of linear equations in terms of a linear map of $\mathbb{R}^n$'s and the set of solutions as the preimage of the column vector comprised of the b_i's in (1.1) of the linear map. Also, we introduce the concept of subspace, linear independence and spanning set of vectors.

1.1 Solving Linear Equations by Gauss Elimination

1.1.1 Solving Explicit Equations

A **linear equation** in 2 variables is of the form

$$2x + 3y = 1$$

The solutions for the equation is, geometrically, represented by a straight line passing through the points $(0, \frac{1}{3}), (\frac{1}{2}, 0)$ namely, the line with the solpe $-\frac{2}{3}$ and y-intercept $\frac{1}{3}$. A linear equation in 3 variables is of the form

$$2x + y + 4z = 2$$

The solutions to the equation above is represented by a plane in 3 space passing through the points $(1, 0, 0), (0, 2, 0), (0, 0, \frac{1}{2})$. A **system of linear equations** in three variables is of the form

$$\begin{cases} 2x + 3y \qquad = 1 \\ 2x + \ y + 4z = 2 \end{cases}$$

Each equation represents a plane in 3 space and the solutions to the system is represented by the intersection of two planes which is in general a line in 3 space and so on.

Let's solve a system of linear equations explicitly:

Example 1.1.1. *Let's solve a system of linear equations in four variables:*

$$\begin{cases} \quad -x_2 -3x_3 +x_4 = -1 \\ x_1 + x_2 - x_3 + 2x_4 = 2 \\ \qquad 2x_3 + 4x_4 = 2 \end{cases} \tag{1.2}$$

First interchange the first two equations.

$$\begin{cases} x_1 + x_2 - x_3 + 2x_4 = 2 \\ \quad -x_2 -3x_3 +x_4 = -1 \\ \qquad 2x_3 + 4x_4 = 2 \end{cases}$$

Add the second equation to the first one:

$$\begin{cases} x_1 \qquad -4x_3 + 3x_4 = 1 \\ \quad -x_2 -3x_3 +x_4 = -1 \\ \qquad 2x_3 + 4x_4 = 2 \end{cases}$$

Divide the third equation by 2:

$$\begin{cases} x_1 \qquad -4x_3 + 3x_4 = 1 \\ \quad -x_2 -3x_3 +x_4 = -1 \\ \qquad x_3 + 2x_4 = 1 \end{cases}$$

Add 4 times the third equation to the first one; add 3 times the third equation to the second one;

$$\begin{cases} x_1 \qquad\qquad +11x_4 = 5 \\ \quad -x_2 \qquad +7x_4 = 2 \\ \qquad\quad x_3 \ +2x_4 = 1 \end{cases}$$

Finally multiply the second equation by -1;

$$\begin{cases} x_1 \qquad\qquad +11x_4 = 5 \\ \quad x_2 \qquad -7x_4 = -2 \\ \qquad\quad x_3 \ +2x_4 = 1 \end{cases} \tag{1.3}$$

We obtain

$$\begin{cases} x_1 + 11x_4 = 5 \\ x_2 - 7x_4 = -2 \\ x_3 + 2x_4 = 1 \\ x_4 = \text{ arbitrary number} \end{cases} \quad \text{or} \quad \begin{cases} x_1 = -11x_4 + 5 \\ x_2 = 7x_4 - 2 \\ x_3 = -2x_4 + 1 \\ x_4 = \text{ arbitrary number} \end{cases} \tag{1.4}$$

In doing these operations to solve the equations (1.2), *the solutions of the equation remain unchanged. Hence the solutions are given by*

$$\begin{bmatrix} x_1 \\ x_2 \\ x_3 \\ x_4 \end{bmatrix} = \begin{bmatrix} -11x_4 + 5 \\ 7x_4 - 2 \\ -2x_4 + 1 \\ x_4 \end{bmatrix} = x_4 \begin{bmatrix} -11 \\ 7 \\ -2 \\ 1 \end{bmatrix} + \begin{bmatrix} 5 \\ -2 \\ 1 \\ 0 \end{bmatrix} \tag{1.5}$$

When we solve the equation (1.2) we applied the following operations:

(1) Add a constant multiple of an equation to another equation.

(2) Interchange two equations.

(3) Multiply an equation by a <u>nonzero</u> constant.

While we perform these operations the variables $x_1, x_2, \ldots$ plays really no essential role – they only indicate the positions of the coefficients. So we omit the variables $x_1, x_2, \ldots$ and write the coefficients in an array of numbers so called a **matrix** as in (1.6) below. The number in the i-th row and j-th column will be called the (i, j)-th entry. A matrix with m rows and n columns will be called an $m \times n$ ('m by n') matrix.

For example, we write the equation (1.2) in the following matrix

$$M_0 = \left[\begin{array}{cccc|c} 0 & -1 & -3 & 1 & -1 \\ 1 & 1 & -1 & 2 & 2 \\ 0 & 0 & 2 & 4 & 2 \end{array}\right] \tag{1.6}$$

where the vertical line, which we may omit, is inserted to distinguish between the coefficients of the unknowns and the constant terms of the equation. And then we perform the operations of matrices corresponding to (1), (2) and (3) on M_0 to obtain the matrix

$$M_\infty = \left[\begin{array}{cccc|c} 1 & 0 & 0 & 11 & 5 \\ 0 & 1 & 0 & -7 & -2 \\ 0 & 0 & 1 & 2 & 1 \end{array}\right]$$

corresponding to (1.3). In short, what we did is:

$$M_0 = \left[\begin{array}{cccc|c} 0 & -1 & -3 & 1 & -1 \\ 1 & 1 & -1 & 2 & 2 \\ 0 & 0 & 2 & 1 & 2 \end{array}\right] \xrightarrow[\text{corresponding to (1),(2),(3)}]{\text{perform matrix operations}} M_\infty = \left[\begin{array}{cccc|c} 1 & 0 & 0 & 11 & 5 \\ 0 & 1 & 0 & -7 & -2 \\ 0 & 0 & 1 & 2 & 1 \end{array}\right] \tag{1.7}$$

From M_∞ we can read off the solutions.

Definition 1.1.2. *The row operations on matrices corresponding to the operations* (1),(2),(3) *on the equations which we call the* **elementary row operations** *are:*

(1)$'$ *Add c times the i-th row to the j-th row which we denote by $cR_i + R_j$* (**the i-th row remain unchanged**)

(2)$'$ *Interchange the i-th row and the j-th row which we denote by R_{ij}*

(3)$'$ *Multiply a <u>nonzero</u> constant c to the i-th row: which we denote by cR_i*

Elementary column operations *are defined by replacing 'row' by 'column'.*

To solve the system of equations (1.2) apply elementary row operations to M_0 repeatedly to get M_∞. Let's find a systematic way to get M_∞ from M_0.

(1) We need 1 in (1,1) position. So create 1 in (1,1) position by using R_{ij}, cR_i

(2) Using 1 in (1,1) position of (1) produce 0's below (1,1) position by using $cR_i + R_j$.

(3) Move the row having the first nonzero entry in its row to the 2nd row by using the operation R_{ij} and make it to be 1 by using cR_i.

(4) Produce 0's below and above 1 in (3) by using $cR_i + R_j$.

(5) Repeat (1) – (4) to obtain the matrix M_∞.

The resulting matrix will be of the form

$$\left[\begin{array}{cccc|c} 1 & 1 & 0 & 0 & 1 \\ 0 & 0 & 1 & 0 & 2 \\ 0 & 0 & 0 & 1 & 1 \end{array}\right]$$

The process (1) – (5) to find the solutions of a system of linear equation is called **Gauss elimination method.**

Now we follow these steps to translate the computations of Example 1.1.1 in terms of matrices:

$$M_0 = \left[\begin{array}{cccc|c} 0 & -1 & -3 & 1 & -1 \\ 1 & 1 & -1 & 2 & 2 \\ 0 & 0 & 2 & 4 & 2 \end{array}\right] \xrightarrow{R_{12}} \left[\begin{array}{cccc|c} 1 & 1 & -1 & 2 & 2 \\ 0 & -1 & -3 & 1 & -1 \\ 0 & 0 & 2 & 4 & 2 \end{array}\right] \xrightarrow[\frac{1}{2}R_3]{1R_2+R_1}$$

$$\left[\begin{array}{cccc|c} 1 & 0 & -4 & 3 & 1 \\ 0 & -1 & -3 & 1 & -1 \\ 0 & 0 & 1 & 2 & 1 \end{array}\right] \xrightarrow[3R_3+R_2;-1R_2]{4R_3+R_1} \left[\begin{array}{cccc|c} 1 & 0 & 0 & 11 & 5 \\ 0 & 1 & 0 & -7 & -2 \\ 0 & 0 & 1 & 2 & 1 \end{array}\right] = M_\infty$$

Now substituting back the variables x_1, x_2, x_3, x_4 we get (1.4).

The 3×4 matrix A on the left side of M_0, formed by the coefficients of the variables

$$A = \begin{bmatrix} 0 & -1 & -3 & 1 \\ 1 & 1 & -1 & 2 \\ 0 & 0 & 2 & 4 \end{bmatrix}$$

is called the **coefficient matrix** and the matrix M_0 is called the **augmented matrix** corresponding to the system of linear equation (1.2). If $\mathbf{b} = \mathbf{0}$ we may omit the last column so that the coefficient matrix is the same as the augmented matrix. The matrix of the form M_∞ is called the **reduced echelon form** (see Definition 1.1.12) of M_0 which we will discuss soon.

The following fact is obvious.

Theorem 1.1.3. *There is an one-to-one correspondence between the family of system of linear equations and the family of augmented matrices.*

A system of equations may have no solutions in which case we say the system is **inconsistent**. For example, if M_∞ is of the form

$$\left[\begin{array}{ccc|c} 1 & 2 & -1 & 1 \\ 0 & 0 & 0 & 2 \end{array}\right]$$

then the system is inconsistent. Because the second row, when substituting back in variables, will become $0 = 2$ which is absurd. If a system of linear equations has at least one solution then we say the system is **consistent**. We will also say an augmented matrix is consistent if the corresponding system has a solution.

Definition 1.1.4. *We say two systems of linear equations having the same number of variables and equations are* **equivalent** *if they have the same solutions.*

Let A, B be matrices of the same size. We say A is **row equivalent** *to B if B is obtained from A by performing elementary row operations. In this case, we write $A \smile B$.*

Note that the definition of row equivalence is not symmetric – if A is row equivalent to B then we don't know yet if B is row equivalent to A. For this we observe that the elementary row operations are always reversible. See also Exercise 5.

Theorem 1.1.5. *The elementary row operations are reversible. The reversing (elementary) row operations for $cR_i + R_j, R_{ij}, cR_i$ are $-cR_i + R_j, R_{ij}, \frac{1}{c}R_i$ respectively. Hence, if $A \smile B$ then $B \smile A$.*

Proof. Obvious. □

Corollary 1.1.6. *Let A, B be $m \times n$ matrices. Suppose A, B are row equivalent. If there is $\mathbf{b} \in \mathbb{R}^m$ such that $[B|\mathbf{b}]$ is inconsistent then there is $\mathbf{b}'$ such that $[A|\mathbf{b}']$ is inconsistent.*

Proof. Let $R_1, R_2, ..., R_k$ be the elementary row operations applied to A to get B and let $R'_1, R'_2, ..., R'_k$ be their reversing elementary row operations. Then the operations $R'_1, R'_2, ..., R'_k$ will reduce $[B|\mathbf{b}]$ to a matrix of the form $[A|\mathbf{b}']$. Then clearly the system of linear equations corresponding to $[A|\mathbf{b}']$ is inconsistent. □

By considering the equations corresponding to the matrix the following fact is obvious.

Theorem 1.1.7. *Let A, B be row equivalent augmented matrices. Then the corresponding system of linear equations have the same solutions. In particular, the system of linear equations corresponding to A is consistent if and only if the system of linear equations corresponding to B is consistent.*

It would be nice to record each elementary operation $cR_i + R_j, R_{ij}, cR_i$ we perform to check the possible error of our computations.

Example 1.1.8. *We want to determine whether the following system is consistent.*

$$\begin{cases} \quad\quad\; x_2 - 4x_3 = 8 \\ 2x_1 - 3x_2 + 2x_3 = 1 \\ 5x_1 - 8x_2 + 7x_3 = 1 \end{cases} \tag{1.8}$$

We perform the elementary row operations to the augmented matrix:

$$\left[\begin{array}{ccc|c} 0 & 1 & -4 & 8 \\ 2 & -3 & 2 & 1 \\ 5 & -8 & 7 & 1 \end{array}\right] \xrightarrow[\frac{1}{2}R_1]{R_{12}} \left[\begin{array}{ccc|c} 1 & -\frac{3}{2} & 1 & \frac{1}{2} \\ 0 & 1 & -4 & 8 \\ 5 & -8 & 7 & 1 \end{array}\right] \xrightarrow{-5R_1+R_3} \left[\begin{array}{ccc|c} 1 & -\frac{3}{2} & 1 & \frac{1}{2} \\ 0 & 1 & -4 & 8 \\ 0 & -\frac{1}{2} & 2 & -\frac{3}{2} \end{array}\right] \xrightarrow{\frac{1}{2}R_2+R_3}$$

$$\left[\begin{array}{ccc|c} 1 & -\frac{3}{2} & 1 & \frac{1}{2} \\ 0 & 1 & -4 & 8 \\ 0 & 0 & 0 & \frac{5}{2} \end{array}\right]$$

The last row, when substituted back the variables, becomes $0x_1 + 0x_2 + 0x_3 = \frac{2}{5}$ which is absurd. Hence the equation (1.8) is inconsistent.

Example 1.1.9. *To determine whether the following system is consistent*

$$\begin{cases} \phantom{2x_1 - {}} x_2 - 4x_3 = -3 \\ 2x_1 - 3x_2 + 2x_3 = -1 \\ 5x_1 - 8x_2 + 7x_3 = -1 \end{cases}$$

we perform the elementary row operations to the augmented matrix:

$$\left[\begin{array}{ccc|c} 0 & 1 & -4 & -3 \\ 2 & -3 & 2 & -1 \\ 5 & -8 & 7 & -1 \end{array}\right] \xrightarrow[\frac{1}{2}R_1]{R_{12}} \left[\begin{array}{ccc|c} 1 & -\frac{3}{2} & 1 & -\frac{1}{2} \\ 0 & 1 & -4 & -3 \\ 5 & -8 & 7 & -1 \end{array}\right] \xrightarrow{-5R_1+R_3} \left[\begin{array}{ccc|c} 1 & -\frac{3}{2} & 1 & -\frac{1}{2} \\ 0 & 1 & -4 & -3 \\ 0 & -\frac{1}{2} & 2 & \frac{3}{2} \end{array}\right] \xrightarrow{\frac{1}{2}R_2+R_3}$$

$$\left[\begin{array}{ccc|c} 1 & -\frac{3}{2} & 1 & -\frac{1}{2} \\ 0 & 1 & -4 & -3 \\ 0 & 0 & 0 & 0 \end{array}\right] \xrightarrow{\frac{3}{5}R_2+R_1} \left[\begin{array}{ccc|c} 1 & 0 & -5 & -5 \\ 0 & 1 & -4 & -3 \\ 0 & 0 & 0 & 0 \end{array}\right]$$

Hence the system of equations are consistent and we have the solution $x_1 = -5 + 5x_3, x_2 = -3 + 4x_3, x_3 =$ *arbitrary number. Or, the solutions are*

$$\begin{bmatrix} x_1 \\ x_2 \\ x_3 \end{bmatrix} = \begin{bmatrix} -5 + 5x_3 \\ -3 + 4x_3 \\ x_3 \end{bmatrix} = \begin{bmatrix} -5 \\ -3 \\ 0 \end{bmatrix} + x_3 \begin{bmatrix} 5 \\ 4 \\ 1 \end{bmatrix}$$

Observe that the two examples above have the same coefficient matrix. The first equation has no solution whereas the second equation does have solutions. Why this happens? We will figure it out in Chapter 2.

Practice Problem 1.1.1. *Solve the system of linear equations by reducing the augmented matrix to a reduced echelon form.*

$$\begin{cases} \phantom{2x_1 - {}} x_2 - 4x_3 = -1 \\ 2x_1 - 3x_2 + 2x_3 = 0 \\ x_1 - 2x_2 \phantom{{}+{}} - x_3 = -1 \end{cases}$$

Example 1.1.10. *We want to find the solution of the following system.*

$$\begin{cases} x + y + z + w = 1 \\ \phantom{x + {}} y \phantom{{}+ z} + w = -1 \\ x \phantom{{}+ y + z} - w = 1 \end{cases}$$

Form an augmented matrix corresponding to the system and perform elementary row operations:

$$\left[\begin{array}{cccc|c} 1 & 1 & 1 & 1 & 1 \\ 0 & 1 & 0 & 1 & -1 \\ 1 & 0 & 0 & -1 & 1 \end{array}\right] \xrightarrow{-R_1+R_3} \left[\begin{array}{cccc|c} 1 & 1 & 1 & 1 & 1 \\ 0 & 1 & 0 & 1 & -1 \\ 0 & -1 & -1 & -2 & 0 \end{array}\right] \xrightarrow{1R_2+R_3}$$

$$\left[\begin{array}{cccc|c} 1 & 1 & 1 & 1 & 1 \\ 0 & 1 & 0 & 1 & -1 \\ 0 & 0 & -1 & -1 & -1 \end{array}\right] \xrightarrow{-1R_3} \left[\begin{array}{cccc|c} 1 & 1 & 1 & 1 & 1 \\ 0 & 1 & 0 & 1 & -1 \\ 0 & 0 & 1 & 1 & 1 \end{array}\right]$$

At this point we could go back to the equation by back substitution and solve the equation:

$$\begin{cases} x + y + z + w = 1 \\ \phantom{x + {}} y \phantom{{}+ z} + w = -1 \\ \phantom{x + y + {}} z + w = 1 \end{cases}$$

Hence back substituting the equality $w = 1 - z$ obtained from the last equation to the second equation, we get $y = -2 + z$ and $x = 2 - z$. Therefore, the solution will be

$$\begin{bmatrix} x \\ y \\ z \\ w \end{bmatrix} = \begin{bmatrix} 2-z \\ -2+z \\ z \\ 1-z \end{bmatrix} = \begin{bmatrix} 2 \\ -2 \\ 0 \\ 1 \end{bmatrix} + z \begin{bmatrix} -1 \\ 1 \\ 1 \\ -1 \end{bmatrix}$$

Or we can proceed to get the reduced echelon form:

$$\xrightarrow[-R_3+R_1]{-R_2+R_1} \left[\begin{array}{cccc|c} 1 & 0 & 0 & -1 & 1 \\ 0 & 1 & 0 & 1 & -1 \\ 0 & 0 & 1 & 1 & 1 \end{array}\right]$$

from which we get the solution

$$\begin{bmatrix} x \\ y \\ z \\ w \end{bmatrix} = \begin{bmatrix} 1+w \\ -1-w \\ 1-w \\ w \end{bmatrix} = \begin{bmatrix} 1 \\ -1 \\ 1 \\ 0 \end{bmatrix} + w \begin{bmatrix} 1 \\ -1 \\ -1 \\ 1 \end{bmatrix}$$

Observe that we may get different expression of the solution still the two expression represent the same set of solutions. We can check this, for example

$$\begin{bmatrix} 2 \\ -2 \\ 0 \\ 1 \end{bmatrix} + z \begin{bmatrix} -1 \\ 1 \\ 1 \\ -1 \end{bmatrix} = \left(\begin{bmatrix} 2 \\ -2 \\ 0 \\ 1 \end{bmatrix} + \begin{bmatrix} -1 \\ 1 \\ 1 \\ -1 \end{bmatrix} \right) + (z-1) \begin{bmatrix} -1 \\ 1 \\ 1 \\ -1 \end{bmatrix} = \begin{bmatrix} 1 \\ -1 \\ 1 \\ 0 \end{bmatrix} - t \begin{bmatrix} 1 \\ -1 \\ -1 \\ 1 \end{bmatrix}$$

where we set $t = z - 1$ and vice versa.

Example 1.1.11. *A necessary and sufficient condition for the system*

$$\begin{cases} ax + by = c \\ a'x + b'y = c' \end{cases}$$

to have a unique solution is

$$ab' - a'b \neq 0 \tag{1.9}$$

In this case, the solution is

$$x = \frac{b'c - bc'}{ab' - a'b}, \quad y = \frac{ac' - a'c}{ab' - a'b} \tag{1.10}$$

In fact, if $a = a' = 0$ then either the equation is inconsistent ($b'c \neq bc'$) or has infinitely many solutions ($b'c = bc'$). Hence we may assume $a \neq 0$ or $a' \neq 0$ say, $a \neq 0$. Form the augmented matrix and use Gauss elimination:

$$\left[\begin{array}{cc|c} a & b & c \\ a' & b' & c' \end{array}\right] \xrightarrow[-\frac{a'}{a}R_1+R_2]{} \left[\begin{array}{cc|c} a & b & c \\ 0 & \frac{ab'-a'b}{a} & \frac{ac'-a'c}{a} \end{array}\right]$$

If the second row of the last matrix consists of zeros then the system has infinitely many solutions. Hence for the equation to have a unique solution it is necessary and sufficient that

$\frac{ab'-a'b}{a} \neq 0$ *which is equivalent to* $ab' - a'b \neq 0$. *To find the solution proceed to apply* $a^{-1}R_1$ *and* $\frac{a}{ab'-a'b}R_2$ *to get*

$$\xrightarrow[\frac{a}{ab'-a'b}R_2]{\frac{1}{a}R_1} \left[\begin{array}{cc|c} 1 & \frac{b}{a} & \frac{c}{a} \\ 0 & 1 & \frac{ac'-a'c}{ab'-a'b} \end{array}\right] \xrightarrow[-\frac{b}{a}R_2+R_1]{} \left[\begin{array}{cc|c} 1 & 0 & \frac{b'c-bc'}{ab'-a'b} \\ 0 & 1 & \frac{ac'-a'c}{ab'-a'b} \end{array}\right]$$

If the denominator of the right hand side column of the last matrix is nonzero then we get the desired solution.

1.1.2 Linear Equations in n Variables

So far we considered systems of linear equations in small number of variables. Now we consider a general system of linear equations which is of the form

$$\begin{cases} a_{11}x_1 + a_{12}x_2 + \cdots + a_{1n}x_n = b_1 \\ a_{21}x_1 + a_{22}x_2 + \cdots + a_{2n}x_n = b_2 \\ \quad \cdots\cdots \\ a_{m1}x_1 + a_{m2}x_2 + \cdots + a_{mn}x_n = b_m \end{cases} \tag{1.11}$$

If $b_i = 0$ for all i then the linear system of equations (1.11) is said to be **homogeneous**. A homogeneous equation always has a solution $x_1 = x_2 = \cdots = x_n = 0$ which we call the **trivial solution**. Hence a homogeneous system of linear equation is never inconsistent. If there is nonzero b_i then we say the system is **nonhomogeneous**.

The steps to solve a system of linear equations of the previous subsection can be applied to solve a system of linear equations (1.11) in n variables.

(1) Write the augmented matrix for (1.11)

$$M_0 := \left[\begin{array}{cccc|c} a_{11} & a_{12} & \cdots & a_{1n} & b_1 \\ a_{21} & a_{22} & \cdots & a_{2n} & b_2 \\ & \cdots & \cdots & & \vdots \\ a_{m1} & a_{m2} & \cdots & a_{mn} & b_m \end{array}\right] = [A|\mathbf{b}] \tag{1.12}$$

where

$$A = \begin{bmatrix} a_{11} & a_{12} & \cdots & a_{1n} \\ a_{21} & a_{22} & \cdots & a_{2n} \\ & \cdots & \cdots & \\ a_{m1} & a_{m2} & \cdots & a_{mn} \end{bmatrix}, \quad \mathbf{b} = \begin{bmatrix} b_1 \\ b_2 \\ \vdots \\ b_n \end{bmatrix}$$

(2) Create 1 in (1,1) position of M_0 by using R_{ij}, cR_i

(3) Using 1 in (1,1) position from (2) produce 0's below (1,1) position by using $cR_i + R_j$.

(4) Move the row having the first nonzero entry to the second row by using R_{ij} and make it to be 1 by using cR_i.

(5) Produce 0's below and above 1 from (4) by using $cR_i + R_j$.

(6) Repeat (1)–(5) to obtain the matrix, for example, of the form

$$M_\infty = \left[\begin{array}{ccccccc|c} \boxed{1} & 2 & 0 & 0 & 0 & 1 & 0 & -1 \\ 0 & 0 & \boxed{1} & 0 & 1 & 2 & 0 & 2 \\ 0 & 0 & 0 & \boxed{1} & 0 & 1 & 0 & -1 \\ 0 & 0 & 0 & 0 & 0 & 0 & \boxed{1} & 3 \\ 0 & 0 & 0 & 0 & 0 & 0 & 0 & 0 \end{array}\right] \tag{1.13}$$

where the **leading 1's** of each row is indicated by $\boxed{1}$. The the solution to the equation is then, by back substituting the vaiables

$$\begin{cases} x_1 = -2x_2 - x_6 - 1 \\ x_2 = \text{ arbitrary number} \\ x_3 = -x_5 - 2x_6 + 2 \\ x_4 = -x_6 - 1 \\ x_5 = \text{ arbitrary number} \\ x_6 = \text{ arbitrary number} \\ x_7 = 3 \end{cases} \tag{1.14}$$

In this case we can write the solutions as:

$$\begin{bmatrix} x_1 \\ x_2 \\ x_3 \\ x_4 \\ x_5 \\ x_6 \\ x_7 \end{bmatrix} = \begin{bmatrix} -1 - 2x_2 - x_6 \\ x_2 \\ 2 - x_5 - 2x_6 \\ -1 - x_6 \\ x_5 \\ x_6 \\ 3 \end{bmatrix} = \begin{bmatrix} -1 \\ 0 \\ 2 \\ -1 \\ 0 \\ 0 \\ 3 \end{bmatrix} + x_2 \begin{bmatrix} -1 \\ 1 \\ 0 \\ 0 \\ 0 \\ 0 \\ 0 \end{bmatrix} + x_5 \begin{bmatrix} 0 \\ 0 \\ -1 \\ 0 \\ 1 \\ 0 \\ 0 \end{bmatrix} + x_6 \begin{bmatrix} -1 \\ 0 \\ -2 \\ -1 \\ 0 \\ 1 \\ 0 \end{bmatrix} \tag{1.15}$$

Here, the variables x_2, x_5, x_6 are called the **free variables**. A **particular solution** is obtained by taking some special values for the free variables; e.g. $x_1 = -1, x_2 = 0, x_3 = 2, x_4 = -1, x_5 = 0, x_6 = 0, x_7 = 3$ is a particular solution obtained by letting $x_2 = x_5 = x_6 = 0$.

The process of row operations to solve a system of linear equations is called the **Gauss elimination method**†. And the position where $\boxed{1}$ is located in M_∞ will be called a **pivot position** of M_0. So (1,1), (2,3),... positions of M_0 are the pivot positions of M_0 of (1.12). The column containing the pivot position will be called the **pivot column** say 1st, 3rd, 4th and 7th columns of M_∞. The other columns say the 2nd, 5th and 6th columns corresponding to the free variables x_2, x_5, x_6 are called the **nonpivot columns**.

The matrices of the form M_∞ will be called a reduced echelon form. Formally we define:

Definition 1.1.12. *A matrix satisfying*

(1) *the first nonzero entry of each row is* $\boxed{1}$

(2) *the entries below and above* $\boxed{1}$ *are* 0

(3) *the position of* $\boxed{1}$ *of a row precede the position of* $\boxed{1}$ *of the next row*

is called the **(row) reduced echelon form**.

†Johann Carl Friedrich Gauss (1777 – 1855) was a German mathematician and physicist who made significant contributions to many fields, including algebra, analysis, astronomy, differential geometry, electrostatics, geodesy, geophysics, magnetic fields, matrix theory, mechanics, number theory, optics and statistics.

For example, reduced echelon form looks like

$$\begin{bmatrix} \boxed{1} & * & 0 & 0 & * & * & 0 \\ 0 & 0 & \boxed{1} & 0 & * & * & 0 \\ 0 & 0 & 0 & \boxed{1} & * & * & 0 \\ 0 & 0 & 0 & 0 & 0 & 0 & \boxed{1} \\ 0 & 0 & 0 & 0 & 0 & 0 & 0 \end{bmatrix} \tag{1.16}$$

Example 1.1.13. *To solve the system of linear equations*

$$\begin{cases} x_1 - x_2 + 3x_3 - x_4 + x_5 = 0 \\ x_2 - x_3 + 2x_4 - 2x_5 = -1 \\ 3x_1 - x_2 + x_3 - x_4 + x_5 = 1 \end{cases}$$

by reducing the augmented matrix to the reduced echelon form we follow the steps above

$$\left[\begin{array}{ccccc|c} 1 & -1 & 3 & -1 & 1 & 0 \\ 0 & 1 & -1 & 2 & -2 & -1 \\ 3 & -1 & 1 & -1 & 1 & 1 \end{array}\right] \xrightarrow{-3R_1+R_3} \left[\begin{array}{ccccc|c} 1 & -1 & 3 & -1 & 1 & 0 \\ 0 & 1 & -1 & 2 & -2 & -1 \\ 0 & 2 & -8 & 2 & -2 & 1 \end{array}\right] \xrightarrow{-2R_2+R_3}$$

$$\left[\begin{array}{ccccc|c} 1 & -1 & 3 & -1 & 1 & 0 \\ 0 & 1 & -1 & 2 & -2 & -1 \\ 0 & 0 & -6 & -2 & 2 & 3 \end{array}\right] \xrightarrow{-\frac{1}{6}R_3} \left[\begin{array}{ccccc|c} 1 & -1 & 3 & -1 & 1 & 0 \\ 0 & 1 & -1 & 2 & -2 & -1 \\ 0 & 0 & 1 & \frac{1}{3} & -\frac{1}{3} & -\frac{1}{2} \end{array}\right] \xrightarrow{R_3+R_2}$$

$$\left[\begin{array}{ccccc|c} 1 & -1 & 3 & -1 & 1 & 0 \\ 0 & 1 & 0 & \frac{7}{3} & -\frac{7}{3} & -\frac{3}{2} \\ 0 & 0 & 1 & \frac{1}{3} & -\frac{1}{3} & 3 \end{array}\right] \xrightarrow{-3R_3+R_1} \left[\begin{array}{ccccc|c} 1 & -1 & 0 & -2 & 2 & \frac{3}{2} \\ 0 & 1 & 0 & \frac{7}{3} & -\frac{7}{3} & -\frac{3}{2} \\ 0 & 0 & 1 & \frac{1}{3} & -\frac{1}{3} & -\frac{1}{2} \end{array}\right] \xrightarrow{1R_2+R_1}$$

$$\left[\begin{array}{ccccc|c} 1 & 0 & 0 & \frac{1}{3} & -\frac{1}{3} & 0 \\ 0 & 1 & 0 & \frac{7}{3} & -\frac{7}{3} & -\frac{3}{2} \\ 0 & 0 & 1 & \frac{1}{3} & -\frac{1}{3} & -\frac{1}{2} \end{array}\right]$$

Therefore, $x_1 = -\frac{1}{3}x_4 + \frac{1}{3}x_5, x_2 = -\frac{3}{2} - \frac{7}{3}x_4 + \frac{7}{3}x_5, x_3 = -\frac{1}{2} - \frac{1}{3}x_4 + \frac{1}{3}x_5$. *Hence the general solution is*

$$\mathbf{x} = \begin{bmatrix} 0 \\ -\frac{3}{2} \\ -\frac{1}{2} \\ 0 \\ 0 \end{bmatrix} + s \begin{bmatrix} -1 \\ -7 \\ -1 \\ 3 \\ 0 \end{bmatrix} + t \begin{bmatrix} 1 \\ 7 \\ 1 \\ 0 \\ 3 \end{bmatrix}$$

Practice Problem 1.1.2. *Reduce the following matrices to the reduced echelon form*

(1) $\begin{bmatrix} 2 & 3 & 0 & 1 & 0 \\ 1 & 1 & 2 & -1 & -1 \\ 1 & -1 & -1 & 0 & 1 \end{bmatrix}$

(2) $\begin{bmatrix} -1 & -1 & 1 & 0 & 1 & 0 \\ 0 & -1 & 1 & -1 & 0 & 1 \\ 1 & 0 & 1 & 0 & 1 & -1 \end{bmatrix}$.

Starting from a matrix A there are many different ways to reach a reduced echelon form. Then we may get different reduced echelon form by taking different row operations which means we get 'different' solutions. We will show that the reduced echelon form is unique whose proof will be deferred until Chapter 2.

Corollary 2.3.8. *The reduced echelon form of a matrix is unique.*

To solve a system of linear equations one may not go this far, namely to annihilate all entries above the pivot position as in Example 1.1.10. Consider an example:

$$\begin{cases} \quad\; 2x_2 - 2x_3 + x_4 + 3x_5 = -1 \\ x_1 - x_2 + 3x_3 - 2x_4 \qquad\;\; = 0 \\ x_1 + 2x_2 \qquad\; - 2x_4 + 2x_5 = -2 \end{cases} \tag{1.17}$$

Write the corresponding augmented matrix and apply the elementary row operations

$$\left[\begin{array}{ccccc|c} 0 & 2 & -2 & 1 & 3 & -1 \\ 1 & -1 & 3 & -2 & 0 & 0 \\ 1 & 2 & 0 & -2 & 2 & -2 \end{array}\right] \xrightarrow[-R_1+R_3]{R_{12}} \left[\begin{array}{ccccc|c} 1 & -1 & 3 & -2 & 0 & 0 \\ 0 & 2 & -2 & 1 & 3 & -1 \\ 0 & 3 & -3 & 0 & 2 & -2 \end{array}\right] \xrightarrow[-2R_3]{-\frac{3}{2}R_2+R_3}$$
$$\left[\begin{array}{ccccc|c} 1 & -1 & 3 & -2 & 0 & 0 \\ 0 & 2 & -2 & 1 & 3 & -1 \\ 0 & 0 & 0 & 3 & 5 & 1 \end{array}\right] \tag{1.18}$$

Now by substituting back the variables we get

$$\begin{cases} x_1 - x_2 + 3x_3 - 2x_4 \qquad\; = 0 \\ \quad\; 2x_2 - 2x_3 + x_4 + 3x_5 = -1 \\ \qquad\qquad\quad 3x_4 + 5x_5 = 1 \end{cases} \tag{1.19}$$

Solve for x_4 in the third equation (the first variable with nonzero coefficient of the last equation) and then substitute the expression for $x_4 = \frac{1}{3} - \frac{5}{3}x_5$ into the second equation and solve for x_2 (the first variable with nonzero coefficient in the second to the last equation)

$$x_2 = -\frac{2}{3} + x_3 - \frac{2}{3}x_5$$

Now substitute this into the first equation to get

$$x_1 = -2x_3 - 4x_5$$

Hence we get the solution

$$\begin{cases} x_1 = -2x_3 - 4x_5 \\ x_2 = -\frac{2}{3} + x_3 - \frac{2}{3}x_5 \\ x_3 = \text{arbitrary number} \\ x_4 = \frac{1}{3} - \frac{5}{3}x_5 \\ x_5 = \text{arbitrary number} \end{cases}$$

and x_3, x_5 are the free variables.

We could have solve for x_5 instead of x_4 at the beginning and substitute $x_5 = \frac{1}{5} - \frac{3}{5}x_4$. Then we have

$$\begin{cases} x_1 = -\frac{4}{5} - 2x_3 + \frac{12}{5}x_4 \\ x_2 = -\frac{4}{5} + x_3 + \frac{2}{5}x_4 \\ x_3 = \text{arbitrary number} \\ x_4 = \text{arbitrary number} \\ x_5 = \frac{1}{5} - \frac{3}{5}x_4 \end{cases}$$

In this case, x_4, x_3 will be free variables. Hence **free variables are not uniquely determined** in general.

Now, we can go further to carry on the row operations on (1.18) to get the reduced echelon form

$$\left[\begin{array}{ccccc|c} 1 & 0 & 2 & 0 & 4 & 0 \\ 0 & 1 & -1 & 0 & \frac{2}{3} & -\frac{2}{3} \\ 0 & 0 & 0 & 1 & \frac{5}{3} & \frac{1}{3} \end{array}\right]$$

The first nonzero entry of each row of the last matrix in (1.18) need not be 1 and the entries below it are 0's but the entries above it need not be 0. This is a matrix satisfying a weaker condition than (3) for reduced echelon form, namely

Definition 1.1.14. *A matrix satisfying the condition,*

(3′) *the first nonzero entry of a nonzero row proceed the first nonzero entry of the next row*

is called an **echelon form.**

It looks like

$$\begin{bmatrix} * & * & * & * & * & * & * \\ 0 & 0 & * & * & * & * & * \\ 0 & 0 & 0 & 0 & 0 & * & * \\ 0 & 0 & 0 & 0 & 0 & 0 & 0 \end{bmatrix} \tag{1.20}$$

and *the first nonzero entries of each row are the pivot positions.* For example,

$$\begin{bmatrix} \boxed{3} & 2 & 1 & 0 & 1 & 2 & 1 \\ 0 & 0 & \boxed{3} & 1 & 2 & 0 & 1 \\ 0 & 0 & 0 & 0 & 0 & \boxed{5} & 2 \\ 0 & 0 & 0 & 0 & 0 & 0 & 0 \end{bmatrix}$$

The columns containing the boxed numbers are the pivot columns.

Theorem 1.1.15. *A system of linear equation is consistent if and only if an echelon form of the augmented matrix does not contain a row of the form*

$$[\,0 \ \cdots \ 0 \mid b] \quad \text{where} \quad b \neq 0$$

If a linear system is consistent then either

$$\begin{cases} \textit{the system has a unique solution or} \\ \textit{the system has infinitely many solutions.} \end{cases}$$

depending on whether the solution has a free variable or not.

Proof. The first statement is obvious. For the second statement, if the solution has a free variable i.e., the reduced echelon form has a nonpivot column, then there are infinitely many solutions. Otherwise every column of the reduced echelon form is a pivot column

$$\left[\begin{array}{ccccc|c} 1 & 0 & 0 & \cdots & 0 & b_1' \\ 0 & 1 & 0 & \cdots & 0 & b_2' \\ \vdots & & \ddots & & \vdots & \vdots \\ 0 & & \cdots & & 1 & b_n' \end{array}\right]$$

and hence, by substituting back the variables we see that the solution is unique. □

Example 1.1.16. *Let us solve the following system of equations by reducing the augmented matrix into an echelon form:*

$$\begin{cases} x_1 + 2x_2 + x_4 + 2x_5 = 1 \\ 2x_1 - x_2 + x_3 - x_4 = 1 \\ -x_1 + 3x_2 - x_3 + 2x_4 + 2x_5 = 0 \\ -3x_1 + x_2 - x_3 - 2x_5 = -2 \end{cases}$$

First, write the augmented matrix for the equation and perform the row operations (here we omit the vertical line separating the coefficients and the constants)

$$\begin{bmatrix} 1 & 2 & 0 & 1 & 2 & 1 \\ 2 & -1 & 1 & -1 & 0 & 1 \\ -1 & 3 & -1 & 2 & 2 & 0 \\ -3 & 1 & -1 & 0 & -2 & -2 \end{bmatrix} \xrightarrow{-2R_1+R_2} \begin{bmatrix} 1 & 2 & 0 & 1 & 2 & 1 \\ 0 & -5 & 1 & -3 & -4 & -1 \\ -1 & 3 & -1 & 2 & 2 & 0 \\ -3 & 1 & -1 & 0 & -2 & -2 \end{bmatrix} \xrightarrow{1R_1+R_3}$$

$$\begin{bmatrix} 1 & 2 & 0 & 1 & 2 & 1 \\ 0 & -5 & 1 & -3 & -4 & -1 \\ 0 & 5 & -1 & 3 & 4 & 1 \\ -3 & 1 & -1 & 0 & -2 & -2 \end{bmatrix} \xrightarrow{3R_1+R_4} \begin{bmatrix} 1 & 2 & 0 & 1 & 2 & 1 \\ 0 & -5 & 1 & -3 & -4 & -1 \\ 0 & 5 & -1 & 3 & 4 & 1 \\ 0 & 7 & -1 & 3 & 4 & 1 \end{bmatrix} \xrightarrow{R_2+R_3}$$

$$\begin{bmatrix} 1 & 2 & 0 & 1 & 2 & 1 \\ 0 & 5 & -1 & 3 & 4 & 1 \\ 0 & 0 & 0 & 0 & 0 & 0 \\ 0 & 7 & -1 & 3 & 4 & 1 \end{bmatrix} \xrightarrow[R_{34}]{-\frac{7}{5}R_2+R_4} \begin{bmatrix} 1 & 2 & 0 & 1 & 2 & 1 \\ 0 & 5 & -1 & 3 & 4 & 1 \\ 0 & 0 & \frac{2}{5} & -\frac{6}{5} & -\frac{8}{5} & -\frac{2}{5} \\ 0 & 0 & 0 & 0 & 0 & 0 \end{bmatrix} \xrightarrow{\frac{5}{2}R_3}$$

$$\begin{bmatrix} 1 & 2 & 0 & 1 & 2 & 1 \\ 0 & 5 & -1 & 3 & 4 & 1 \\ 0 & 0 & 1 & -3 & -4 & -1 \\ 0 & 0 & 0 & 0 & 0 & 0 \end{bmatrix}$$

Substituting back the variables we get

$$\begin{cases} x_1 + 2x_2 + x_4 + 2x_5 = 1 \\ 5x_2 - x_3 + 3x_4 + 4x_5 = 1 \\ x_3 - 3x_4 - 4x_5 = -1 \end{cases}$$

From the third equation we get

$$x_3 = -1 + 3x_4 + 4x_5$$

Next, we do the same for the second equation:

$$5x_2 = 1 + x_3 - 3x_4 - 4x_5 = 1 + (-1 + 3x_4 + 4x_5) - 3x_4 - 4x_5 = 0$$

And then

$$x_1 = 1 - 2x_2 - x_4 - 2x_5 = 1 - x_4 - 2x_5$$

Hence the solution to the equation is

$$\begin{cases} x_1 = 1 - x_4 - 2x_5 \\ x_2 = 0 \\ x_3 = -1 + 3x_4 + 4x_5 \\ x_4 = \text{arbitrary number} \\ x_5 = \text{arbitrary number} \end{cases}$$

The reduced echelon form will then be

$$\begin{bmatrix} 1 & 0 & 0 & 1 & 2 & 1 \\ 0 & 1 & 0 & 0 & 0 & 0 \\ 0 & 0 & 1 & -3 & -4 & -1 \\ 0 & 0 & 0 & 0 & 0 & 0 \end{bmatrix}$$

from which we get the same solutions.

Practice Problem 1.1.3. *Find the solutions of the following system of linear equations by reducing the augmented matrix to an echelon form.*

$$(1) \begin{cases} 2x_1 + 3x_2 + x_4 = 1 \\ x_1 + x_3 - x_4 = 0 \\ x_2 - x_3 + 2x_4 = -1 \end{cases}$$

$$(2) \begin{cases} -x_1 - x_2 + x_3 + x_5 = 0 \\ - x_2 + x_3 - x_4 = 1 \\ x_1 + x_3 + x_5 = -1 \end{cases}$$

Now we have an obvious but important fact which is a primitive form of Theorem 2.3.17.

Theorem 1.1.17. *Let A be an $m \times n$ matrix. Then*

$$\begin{pmatrix} \text{the number of} \\ \text{pivot columns} \end{pmatrix} + \begin{pmatrix} \text{the number of} \\ \text{nonpivot columns} \end{pmatrix} = n \qquad (1.21)$$

Proof. Instead of giving a formal proof just look at (1.16). □

In terms of equations the nonpivot columns corresponds to the free variables.

Answers for Practice Problems 1.1

(1.1.1) The reduced echelon form is $\left[\begin{array}{ccc|c} 1 & 0 & 0 & \frac{3}{8} \\ 0 & 1 & 0 & \frac{1}{2} \\ 0 & 0 & 1 & \frac{3}{8} \end{array}\right]$. Hence the solution is $x_1 = \frac{3}{8}, x_2 = \frac{1}{2}, x_3 = \frac{3}{8}$.

(1.1.2) (1) $\begin{bmatrix} 1 & 0 & 0 & -\frac{3}{4} & \frac{1}{2} \\ 0 & 1 & 0 & -\frac{5}{4} & \frac{1}{2} \\ 0 & 0 & 1 & \frac{1}{2} & -1 \end{bmatrix}$ (2) $\begin{bmatrix} 1 & 0 & 0 & -1 & -1 & 1 \\ 0 & 1 & 0 & 2 & 2 & -3 \\ 0 & 0 & 1 & 1 & 2 & -2 \end{bmatrix}$

(1.1.3) (1) $x_1 = -4 - 2x_4,\ x_2 = 3 + x_4,\ x_3 = 4 + 3x_4,\ x_4 =$arbitrary number
(2) $x_1 = 1 + x_4 + x_5,\ x_2 = -3 - 2x_4 - 2x_5,\ x_3 = -2 - x_4 - 2x_5$, x_4 and x_5 are free

1.1 Exercises

1. Determine whether the following matrices are echelon form or reduced echelon form or none of these. Indicate the pivot positions when they make sense.

(a) $\begin{bmatrix} 1 & 0 & 1 & 1 & 0 \\ 0 & 1 & 0 & 0 & 0 \\ 0 & 0 & 1 & 0 & 0 \\ 0 & 0 & 0 & 0 & 1 \end{bmatrix}$ (b) $\begin{bmatrix} 1 & 0 & 1 & 1 & 0 \\ 0 & 2 & 0 & 0 & 0 \\ 0 & 0 & 3 & 0 & 0 \\ 0 & 0 & 0 & 0 & 1 \end{bmatrix}$

(c) $\begin{bmatrix} 1 & 0 & 1 & 1 & 0 \\ 0 & 2 & 0 & 0 & 0 \\ 0 & 1 & 3 & 0 & 0 \\ 0 & -1 & 0 & 0 & 1 \end{bmatrix}$ (d) $\begin{bmatrix} 1 & 0 & 0 & 1 & 0 \\ 0 & 1 & 0 & 0 & 0 \\ 0 & 0 & 0 & 1 & 0 \\ 0 & 0 & 0 & 0 & 0 \end{bmatrix}$

(e) $\begin{bmatrix} 1 & 0 & 0 & 0 & 1 \\ 0 & 0 & 1 & 0 & 0 \\ 0 & 1 & 0 & 1 & 1 \end{bmatrix}$ (f) $\begin{bmatrix} 1 & 0 & 0 & 0 & 2 \\ 0 & 2 & 1 & 1 & 0 \\ 0 & 0 & 0 & 1 & 1 \end{bmatrix}$

2. Determine whether the system is consistent. If consistent find the solutions.

(a) $\begin{cases} x_2 - 4x_3 = 8 \\ 2x_1 - 3x_2 + 2x_3 = 1 \\ 5x_1 - 8x_2 + 7x_3 = 1 \end{cases}$ (b) $\begin{cases} -x_2 - 3x_3 + x_4 = -1 \\ x_1 + x_2 - x_3 + 2x_4 = 2 \\ 2x_3 + 2x_4 = 1 \end{cases}$

(c) $\begin{cases} x_1 + x_2 = -1 \\ x_1 - x_2 = 2 \\ x_1 + 2x_2 = -1 \end{cases}$ (d) $\begin{cases} x_1 + x_2 - x_3 = 1 \\ x_1 - x_2 + 2x_3 = 1 \\ x_1 + 2x_2 - x_3 = 2 \end{cases}$

(e) $\begin{cases} x_1 + x_2 + x_3 = 1 \\ -x_2 - x_3 + 2x_4 = 1 \\ x_1 + x_2 + x_4 = 2 \\ -x_1 + x_2 + 2x_3 - x_4 = 0 \end{cases}$ (f) $\begin{cases} x_1 + x_2 = 1 \\ 3x_1 - x_2 = -1 \\ 6x_1 + 2x_2 = 2 \\ 5x_1 + x_2 = 1 \end{cases}$

3. Find the solutions of the following homogeneous equations.

(a) $\begin{cases} x_2 - 4x_3 = 0 \\ 2x_1 - x_2 + 2x_3 = 0 \\ 5x_1 - 2x_2 + x_3 = 0 \end{cases}$ (b) $\begin{cases} -x_2 - 3x_3 + x_4 = 0 \\ x_1 + x_2 - x_3 + 2x_4 = 0 \\ x_3 + 2x_4 = 0 \end{cases}$

(c) $\begin{cases} x_1 + x_2 = 0 \\ x_1 - x_2 = 0 \\ x_1 + 2x_2 = 0 \end{cases}$ (d) $\begin{cases} x_1 + x_2 - 2x_3 = 0 \\ x_1 - x_2 + 2x_3 = 0 \\ -2x_2 + 4x_3 = 0 \end{cases}$

(e) $\begin{cases} x_1 - x_3 = 0 \\ x_1 - x_2 + x_4 = 0 \\ x_2 + x_3 + 2x_5 = 0 \end{cases}$ (f) $\begin{cases} 2x_2 + x_3 = 0 \\ x_1 + x_4 = 0 \\ -2x_2 + x_3 + x_5 = 0 \end{cases}$

4. Find the solutions of the following nonhomogeneous equations.

(a) $\begin{cases} x_2 - x_3 = 1 \\ x_2 + 2x_3 = 1 \\ 5x_1 + x_3 = 1 \end{cases}$ (b) $\begin{cases} -x_2 - x_3 + x_4 = 1 \\ x_1 + x_2 + 2x_4 = 0 \\ x_3 + 2x_4 = 2 \end{cases}$

(c) $\begin{cases} 2x_1 + x_2 = 0 \\ x_1 - x_2 = -1 \\ x_1 + 2x_2 = 1 \end{cases}$ (d) $\begin{cases} x_1 + x_2 - x_3 = 1 \\ x_1 - x_2 + x_3 + x_4 = 1 \\ -2x_2 + x_3 + x_5 = 1 \end{cases}$

(e) $\begin{cases} x_1 - x_3 = 0 \\ x_1 - x_2 + x_4 = 1 \\ x_2 + x_3 + 2x_5 = -1 \end{cases}$ (f) $\begin{cases} 2x_2 + x_3 = 0 \\ x_1 + x_4 = 1 \\ -2x_2 + x_3 + x_5 = -2 \end{cases}$

5. Prove that row equivalence is an equivalence relation. That is, for the $m \times n$ matrices A, B, C we have

   (a) $A \smile A$

   (b) If $A \smile B$ then $B \smile A$.

   (c) If $A \smile B$ and $B \smile C$ then $A \smile C$.

6. Let A, B be $m \times n$ matrices. Prove that A, B are row equivalent if and only if their reduced echelon forms are the same.

7. Provide a proof for the true statement(s) and provide a counterexample for the false statement(s).

   (a) If the number of variables of a system of linear equations is bigger than the number of equations then the system always has a solution.

   (b) If a system of linear equations has more equations than the variables then the system has no solution.

   (c) A system of linear equations having more than one solution has infinitely many solutions.

   (d) A system of homogeneous equation may have no solution.

   (e) If a system of linear equations has only one equation then the system is consistent.

   (f) There is a system of linear equations having exactly five solutions.

   (g) The system of linear equations $\begin{cases} ax + by = c \\ a'x + b'y = c' \end{cases}$ has no solution or infinite solution if $ab' = a'b$.

   (h) If a system of linear equations is inconsistent then the number of equations is bigger than the number of variables.

1.2 Vector Equations and Subspaces

In this section, we interpret a system of linear equations in terms of vectors and matrices. We introduce the concept of subspace and show that a subspace of $\mathbb{R}^n$ is a special kind of subsets of $\mathbb{R}^n$. An important notion 'spanning set' of a subspace is also introduced.

1.2.1 Vector Space $\mathbb{R}^n$ and Subspaces

We denote the vectors in $\mathbb{R}^2$ by $\mathbf{u} = \begin{bmatrix} u_1 \\ u_2 \end{bmatrix}, \mathbf{v} = \begin{bmatrix} v_1 \\ v_2 \end{bmatrix}$. We add these vectors using the parallelogram law

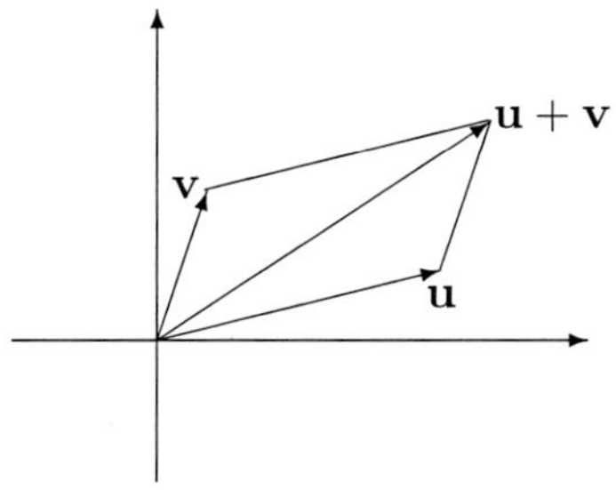

which should be familiar. Algebraically the addition can be defined by

$$\mathbf{u} + \mathbf{v} = \begin{bmatrix} u_1 \\ u_2 \end{bmatrix} + \begin{bmatrix} v_1 \\ v_2 \end{bmatrix} = \begin{bmatrix} u_1 + v_1 \\ u_2 + v_2 \end{bmatrix}$$

and the constant multiple of a vector $\mathbf{u}$ is defined by

$$c\mathbf{u} = \begin{bmatrix} cu_1 \\ cu_2 \end{bmatrix}$$

In general, the set of $n \times 1$ column vectors called the **space of columns vectors** is denoted by $\mathbb{R}^n$

$$\mathbb{R}^n = \left\{ \begin{bmatrix} u_1 \\ u_2 \\ \vdots \\ u_n \end{bmatrix} \middle| \; u_i \in \mathbb{R} \right\} \tag{1.22}$$

Often, we call a column vector simply a **vector**. The vector with $u_1 = u_2 = \cdots = u_n = 0$ is denoted by $\mathbf{0} = \begin{bmatrix} 0 \\ 0 \\ \vdots \\ 0 \end{bmatrix}$ which we call the **zero vector**.

If $\mathbf{u} = \begin{bmatrix} u_1 \\ u_2 \\ \vdots \\ u_n \end{bmatrix}$, $\mathbf{v} = \begin{bmatrix} v_1 \\ v_2 \\ \vdots \\ v_n \end{bmatrix}$ are two vectors and $c \in \mathbb{R}$ is a constant which we also call a **scalar**, then we define the **addition** and the **scalar multiplication** by

$$\mathbf{u} + \mathbf{v} = \begin{bmatrix} u_1 + v_1 \\ u_2 + v_2 \\ \vdots \\ u_n + v_n \end{bmatrix}, \quad c\mathbf{u} = \begin{bmatrix} cu_1 \\ cu_2 \\ \vdots \\ cu_n \end{bmatrix}$$

And we write $-\mathbf{u} = (-1)\mathbf{u}$ namely,

$$-\mathbf{u} = \begin{bmatrix} -u_1 \\ -u_2 \\ \vdots \\ -u_n \end{bmatrix}$$

Theorem 1.2.1. *For* $\mathbf{u}, \mathbf{v}, \mathbf{w} \in \mathbb{R}^n$ *and* $c, d \in \mathbb{R}$ *the following properties hold*

(1) $\mathbf{u} + \mathbf{v} = \mathbf{v} + \mathbf{u}$ (2) $(\mathbf{u} + \mathbf{v}) + \mathbf{w} = \mathbf{u} + (\mathbf{v} + \mathbf{w})$

(3) $\mathbf{u} + \mathbf{0} = \mathbf{0} + \mathbf{u} = \mathbf{u}$ (4) $\mathbf{u} + (-\mathbf{u}) = \mathbf{0}$

(5) $c(\mathbf{u} + \mathbf{v}) = c\mathbf{u} + c\mathbf{v}$ (6) $(c + d)\mathbf{u} = c\mathbf{u} + d\mathbf{u}$

(7) $c(d\mathbf{u}) = (cd)\mathbf{u}$ (8) $1\mathbf{u} = \mathbf{u}$

Proof. These properties follows from the corresponding properties of real numbers. For example, we prove (1) for $n = 2$. Let $\mathbf{u} = \begin{bmatrix} u_1 \\ u_2 \end{bmatrix}, \mathbf{v} = \begin{bmatrix} v_1 \\ v_2 \end{bmatrix}$. Then

$$\begin{aligned} \mathbf{u} + \mathbf{v} &= \begin{bmatrix} u_1 + v_1 \\ u_2 + v_2 \end{bmatrix} \qquad \text{(by commutativity of addition of real numbers)} \\ &= \begin{bmatrix} v_1 + u_1 \\ v_2 + u_2 \end{bmatrix} = \mathbf{v} + \mathbf{u} \end{aligned}$$

To prove (8) for $n = 2$, we compute

$$1\mathbf{u} = 1\begin{bmatrix} u_1 \\ u_2 \end{bmatrix} = \begin{bmatrix} 1u_1 \\ 1u_2 \end{bmatrix} = \begin{bmatrix} u_1 \\ u_2 \end{bmatrix} = \mathbf{u}$$

sicne $1u_i = u_i$ for $u_i \in \mathbb{R}$ $(i = 1, 2)$. □

Practice Problem 1.2.1. *Prove* (6) *and* (7) *of* Theorem 1.2.1 *for* $n = 3$.

The set of $1{\times}n$ row vectors $[u_1, u_2, ..., u_n]$ called the **space of row vectors** is denoted by $\mathbb{R}_n$.

$$\mathbb{R}_n = \Big\{ [u_1, u_2, ..., u_n] \;\Big|\; u_i \in \mathbb{R} \Big\} \tag{1.23}$$

We can define the addition and scalar multiplication for the row vectors in the same manner: If $\mathbf{u} = [u_1, u_2, ..., u_n], \mathbf{v} = [v_1, v_2, ..., v_n], c \in \mathbb{R}$ then we define

$$\begin{aligned} \mathbf{u} + \mathbf{v} &= [u_1 + v_1, u_2 + v_2, ..., u_n + v_n] \\ c\mathbf{u} &= [cu_1, c_2u_2, ..., cu_n] \end{aligned}$$

All the properties of Theorem 1.2.1 are true for the row vectors.

1.2.2 Vector Equation and Matrix Equation

Definition 1.2.2. *An* $m{\times}n$ **matrix** A *is an array of* mn *numbers which consists of* m *rows and* n *columns.*

$$A = \begin{bmatrix} a_{11} & a_{12} & \cdots & a_{1n} \\ a_{21} & a_{22} & \cdots & a_{2n} \\ & \cdots & \cdots & \\ a_{m1} & a_{m2} & \cdots & a_{mn} \end{bmatrix}$$

The set of all $m\times n$ matrices will be denoted by $M(m\times n)$. The (i,j)-th position of A is denoted by A_{ij} namely, $A_{ij} = a_{ij}$. We also write $A = [a_{ij}]$.

Two $m\times n$ matrices A, B are defined to be equal if $A_{ij} = B_{ij}$ i.e., their (i,j)-th positions are the same. An $n\times n$ matrix a **square matrix** *of size n. A square matrix of size n with 1's on the diagonal and zero elsewhere is called the* **identity matrix** *and is denoted by I_n.*

$$I_n = \begin{bmatrix} 1 & 0 & \cdots & 0 \\ 0 & 1 & \cdots & 0 \\ & & \ddots & \\ 0 & \cdots & 0 & 1 \end{bmatrix}$$

We write $A^{(i)}, A_{(i)}$ for the i-th **column** *and the j-th* **row** *of A respectively and $C(A)$ for the set of columns of A:*

$$A^{(i)} = \begin{bmatrix} a_{1i} \\ a_{2i} \\ \vdots \\ a_{mi} \end{bmatrix}, \ A_{(j)} = [a_{i1}, a_{i2}, ..., a_{in}], \ C(A) = \{A^{(1)}, A^{(2)}, ..., A^{(n)}\} \tag{1.24}$$

If $I = \{i_1, i_2, ..., i_k\}, J = \{j_1, j_2, ..., j_l\}$ then we write $A^{(I)}$ and $A_{(J)}$ for the matrix formed by the $I = \{i_1, i_2, ..., i_k\}$ columns and $J = \{j_1, j_2, ..., j_l\}$ rows:

$$A^{(I)} = [A^{(i_1)} \ A^{(i_2)} \cdots A^{(i_k)}], \quad A_{(J)} = \begin{bmatrix} A_{(j_1)} \\ A_{(j_2)} \\ \vdots \\ A_{(j_l)} \end{bmatrix} \tag{1.25}$$

Hence an $m\times n$ matrix can be viewed as n column vectors $A^{(i)}$ $(i = 1, 2, ..., n)$ or m row vectors $A_{(j)}$ $(j = 1, 2, ..., m)$. A vector in $\mathbb{R}^n$ is an $n\times 1$ matrix and a 1×1 matrix is just a scalar or a constant.

Example 1.2.3. *Let*

$$\begin{bmatrix} 1 & 2 & 3 & 4 & 5 \\ a & b & c & d & e \\ A & B & C & D & E \\ ! & \$ & \# & \% & \& \end{bmatrix}$$

Then

$$A^{(2)} = \begin{bmatrix} 2 \\ b \\ B \\ \$ \end{bmatrix}, \quad A_{(3)} = [A \ B \ C \ D \ E]$$

If $I = \{2, 3, 5\}, J = \{2, 4\}$ and $K = \{1, 3\}$ then

$$A^{(I)} = \begin{bmatrix} 2 & 3 & 5 \\ b & c & e \\ B & C & E \\ \$ & \# & \& \end{bmatrix}, \quad A_{(J)} = \begin{bmatrix} a & b & c & d & e \\ ! & \$ & \# & \% & \& \end{bmatrix}, \quad (A_{(J)})^{(K)} = A_{(J)}^{(K)} = \begin{bmatrix} a & c \\ ! & \# \end{bmatrix}$$

Definition 1.2.4. *Let* $\mathbf{v}_1, \mathbf{v}_2, ..., \mathbf{v}_p \in \mathbb{R}^n$ *and* $c_1, c_2, ..., c_p \in \mathbb{R}$ *then the vector* $\mathbf{y}$ *defined by*

$$\mathbf{y} = c_1\mathbf{v}_1 + c_2\mathbf{v}_2 + \cdots + c_p\mathbf{v}_p$$

is called a **linear combination** *of* $\{\mathbf{v}_1, \mathbf{v}_2, ..., \mathbf{v}_p\}$.

For example, if $\mathbf{v}_1 = \begin{bmatrix} 1 \\ 2 \\ -1 \end{bmatrix}$, $\mathbf{v}_2 = \begin{bmatrix} 1 \\ 4 \\ 5 \end{bmatrix}$ and $\mathbf{b} = \begin{bmatrix} 5 \\ 14 \\ 7 \end{bmatrix}$ then

$$3\mathbf{v}_1 + 2\mathbf{v}_2 = \mathbf{b} \tag{1.26}$$

Hence $\mathbf{b}$ is a linear combination of $\{\mathbf{v}_1, \mathbf{v}_2\}$.

Using the notation of vectors the equation (1.11) of §1.1 can be expressed as a **vector equation**:

$$x_1 \begin{bmatrix} a_{11} \\ a_{21} \\ \vdots \\ a_{m1} \end{bmatrix} + x_2 \begin{bmatrix} a_{12} \\ a_{22} \\ \vdots \\ a_{m2} \end{bmatrix} + \cdots + x_n \begin{bmatrix} a_{1n} \\ a_{2n} \\ \vdots \\ a_{mn} \end{bmatrix} = \begin{bmatrix} b_1 \\ b_2 \\ \vdots \\ b_m \end{bmatrix} \tag{1.27}$$

If we let

$$\mathbf{a}_1 = \begin{bmatrix} a_{11} \\ a_{21} \\ \vdots \\ a_{m1} \end{bmatrix}, \mathbf{a}_2 = \begin{bmatrix} a_{12} \\ a_{22} \\ \vdots \\ a_{m2} \end{bmatrix}, \ldots, \mathbf{a}_n = \begin{bmatrix} a_{1n} \\ a_{2n} \\ \vdots \\ a_{mn} \end{bmatrix} \text{ and } \mathbf{b} = \begin{bmatrix} b_1 \\ b_2 \\ \vdots \\ b_m \end{bmatrix}$$

then (1.27) becomes

$$x_1\mathbf{a}_1 + x_2\mathbf{a}_2 + \cdots + x_n\mathbf{a}_n = \mathbf{b} \tag{1.28}$$

i.e., $\mathbf{b}$ is a linear combination of $\{\mathbf{a}_1, \mathbf{a}_2, ..., \mathbf{a}_n\}$.

Let A be an $m \times n$ matrix and let $\mathbf{a}_i$ be the i-th column $A^{(i)}$. Let $\mathbf{x} = \begin{bmatrix} x_1 \\ \vdots \\ x_n \end{bmatrix} \in \mathbb{R}^n$.
View A as n column vectors $A = [\mathbf{a}_1\ \mathbf{a}_2\ \cdots\ \mathbf{a}_n]$, the **product** $A\mathbf{x}$ is define to be the linear combination of column vectors,

$$A\mathbf{x} = \begin{bmatrix} \mathbf{a}_1 & \mathbf{a}_2 & \cdots & \mathbf{a}_n \end{bmatrix} \begin{bmatrix} x_1 \\ \vdots \\ x_n \end{bmatrix} = x_1\mathbf{a}_1 + x_2\mathbf{a}_2 + \cdots + x_n\mathbf{a}_n \tag{1.29}$$

which is the same as the left hand side of (1.11) of §1.1. Hence if $\mathbf{b} \in \mathbb{R}^n$ then the **matrix equation**

$$A\mathbf{x} = \mathbf{b} \tag{1.30}$$

is the same as the system of linear equation (1.11).

On the other hand, as we observed before, the systems of linear equations (1.11) are in one-to-one correspondence with the augmented matrices

$$[A|\mathbf{b}] = \begin{bmatrix} \mathbf{a}_1 & \mathbf{a}_2 & \cdots & \mathbf{a}_n & | & \mathbf{b} \end{bmatrix} \tag{1.31}$$

From our discussion the following theorem is obvious.

Theorem 1.2.5. *Let A be an $m \times n$ matrix with $A^{(i)} = \mathbf{a}_i$ and let $\mathbf{b} \in \mathbb{R}^m$. Then the the following data are in one-to-one correspond to each other*

(1) *The system of linear equations* (1.11).

(2) *The vector equation $x_1\mathbf{a}_1 + \cdots + x_n\mathbf{a}_n = \mathbf{b}$ of* (1.28).

(3) *The matrix equation $A\mathbf{x} = \mathbf{b}$ of* (1.30).

(4) *The augmented matrix $[A \mid \mathbf{b}]$ of* (1.31).

1.2.3 Spanning Vectors

Definition 1.2.6. *For $S = \{\mathbf{v}_1, ..., \mathbf{v}_p\} \subseteq \mathbb{R}^n$ we define the* **subset spanned (generated) by** *S to be the set of all linear combinations of vectors in S:*

$$\mathrm{Span}\,(S) := \{c_1\mathbf{v}_1 + \cdots + c_p\mathbf{v}_p \in \mathbb{R}^n \mid c_1, ..., c_p \in \mathbb{R}\}$$

If S any subset of $\mathbb{R}^n$ (not necessarily finite) then we define Span$\,(S)$ *in the same way:*

$$\mathrm{Span}\,(S) = \{c_1\mathbf{v}_1 + \cdots + c_k\mathbf{v}_k \in \mathbb{R}^n \mid c_1, ..., c_k \in \mathbb{R}, \mathbf{v}_1, ..., \mathbf{v}_k \in S\}$$

If S is empty then we define Span$\,(\emptyset)$ *to be the the set consisting only of the zero vector:*

$$\mathrm{Span}\,(\emptyset) = \{\mathbf{0}\}$$

For example, if $S = \left\{\begin{bmatrix}1\\1\end{bmatrix}\right\}$ then Span$\,(S)$ consists of the vectors lying on the line $y = x$ in the plane. If $S = \left\{\begin{bmatrix}1\\1\\0\end{bmatrix}, \begin{bmatrix}1\\0\\0\end{bmatrix}\right\}$ then Span$\,(S)$ will be the vectors consisting of the vectors lying on the xy plane. In general, for $\mathbf{v} \in \mathbb{R}^3$ the subset spanned by $\mathbf{v}$, Span$\,\{\mathbf{v}\}$ is the line through the origin parallel to the vector $\mathbf{v}$. And for $\mathbf{v}_1, \mathbf{v}_2 \in \mathbb{R}^3$ the subset Span$\,\{\mathbf{v}_1, \mathbf{v}_2\}$ is the plane through the origin containing $\mathbf{v}_1$ and $\mathbf{v}_2$.

If $S \subseteq T$ are sets of vectors then it is obvious that

$$S \subseteq \mathrm{Span}\,(S) \subseteq \mathrm{Span}\,(T) \tag{1.32}$$

Let A be the matrix with the i-th column $\mathbf{v}_i$

$$A = [\mathbf{v}_1\ \mathbf{v}_2\ \cdots\ \mathbf{v}_p] \text{ and } \mathbf{c} = \begin{bmatrix}c_1\\ \vdots\\ c_p\end{bmatrix} \in \mathbb{R}^p$$

then by (1.29) we can write

$$A\mathbf{c} = c_1\mathbf{v}_1 + \cdots + c_p\mathbf{v}_p \tag{1.33}$$

Hence

$$\mathrm{Span}\,(S) = \{A\mathbf{c} \mid \mathbf{c} \in \mathbb{R}^p\}$$

Theorem 1.2.7. (1) *The vector equation $x_1\mathbf{a}_1 + x_2\mathbf{a}_2 + \cdots + x_n\mathbf{a}_n = \mathbf{b}$ has a solution if and only if $\mathbf{b} \in$* Span$\,\{\mathbf{a}_1, \mathbf{a}_2 \cdots \mathbf{a}_n\}$.

(2) *Let $A = [\mathbf{a}_1\ \mathbf{a}_2 \cdots \mathbf{a}_n]$. Then the equation $A\mathbf{x} = \mathbf{b}$ has a solution for any $\mathbf{b} \in \mathbb{R}^m$ if and only if every row of the reduced echelon form of A is nonzero.*

Proof. (1) Obvious.

(2) If every row of the reduced echelon form of A is nonzero then the reduced echelon form of $[A|\mathbf{b}]$ will not contain a row of the form $[\,0\ 0 \cdots\ 0 \mid b\,]$. Hence by Theorem 1.1.15 the equation has a solution. Conversely, suppose the reduced echelon form A' of A has a row, say the last row, consisting only of 0's. Choose $\mathbf{b}'$ whose last coordinate is 1. Then $[A'|\mathbf{b}']$ is inconsistent. Since A and A' are row equivalent, by Corollary 1.1.6, there is $\mathbf{b}$ such that $[A|\mathbf{b}]$ is inconsistent. □

The equality $3\mathbf{v}_1 + 2\mathbf{v}_2 = \mathbf{b}$ of (1.26) says $\mathbf{b} \in \operatorname{Span}\{\mathbf{v}_1, \mathbf{v}_2\}$. Hence the system of equations

$$\begin{cases} x_1 + x_2 = 5 \\ 2x_1 + 4x_2 = 14 \\ -x_1 + 5x_2 = 7 \end{cases}$$

has a solution, namely $x_1 = 3, x_2 = 2$.

In Example 1.1.8, the system of equations (1.8) has no solution which says that $\mathbf{b} = \begin{bmatrix} 8 \\ 1 \\ 1 \end{bmatrix} \notin \operatorname{Span}(S)$ where $S = \left\{ \mathbf{v}_1 = \begin{bmatrix} 0 \\ 2 \\ 5 \end{bmatrix}, \mathbf{v}_2 = \begin{bmatrix} 1 \\ -3 \\ 8 \end{bmatrix}, \mathbf{v}_3 = \begin{bmatrix} -4 \\ 2 \\ 7 \end{bmatrix} \right\}$. And Example 1.1.9 shows that $\begin{bmatrix} -3 \\ -1 \\ -1 \end{bmatrix} \in \operatorname{Span}(S)$.

Example 1.2.8. *Let* $\mathbf{a}_1 = \begin{bmatrix} 1 \\ 2 \\ -1 \end{bmatrix}$, $\mathbf{a}_2 = \begin{bmatrix} 1 \\ 4 \\ 5 \end{bmatrix}$, $\mathbf{a}_3 = \begin{bmatrix} 0 \\ -2 \\ 1 \end{bmatrix}$, $\mathbf{a}_4 = \begin{bmatrix} 1 \\ 0 \\ 1 \end{bmatrix}$ *and* $\mathbf{b} = \begin{bmatrix} 1 \\ 2 \\ 0 \end{bmatrix}$.

Then $\mathbf{b} \in \operatorname{Span}\{\mathbf{a}_1, \mathbf{a}_2, \mathbf{a}_3, \mathbf{a}_4\}$ *if and only if the system of equations corresponding to the augmented matrix*

$$\begin{bmatrix} \mathbf{a}_1 & \mathbf{a}_2 & \mathbf{a}_3 & \mathbf{a}_4 & | & \mathbf{b} \end{bmatrix} = \left[\begin{array}{cccc|c} 1 & 1 & 0 & 1 & 1 \\ 2 & 4 & -2 & 0 & 2 \\ -1 & 5 & 1 & 1 & 0 \end{array}\right]$$

has a solution. So let's execute our computation:

$$\left[\begin{array}{cccc|c} 1 & 1 & 0 & 1 & 1 \\ 2 & 4 & -2 & 0 & 2 \\ -1 & 5 & 1 & 1 & 0 \end{array}\right] \xrightarrow{-2R_1+R_2} \left[\begin{array}{cccc|c} 1 & 1 & 0 & 1 & 1 \\ 0 & 2 & -2 & -2 & 0 \\ -1 & 5 & 1 & 1 & 0 \end{array}\right] \xrightarrow{1R_1+R_3}$$

$$\left[\begin{array}{cccc|c} 1 & 1 & 0 & 1 & 1 \\ 0 & 2 & -2 & -2 & 0 \\ 0 & 6 & 1 & 2 & 1 \end{array}\right] \xrightarrow{\frac{1}{2}R_3} \left[\begin{array}{cccc|c} 1 & 1 & 0 & 1 & 1 \\ 0 & 1 & -1 & -1 & 0 \\ 0 & 6 & 1 & 2 & 1 \end{array}\right] \xrightarrow{-6R_2+R_3}$$

$$\left[\begin{array}{cccc|c} 1 & 1 & 0 & 1 & 1 \\ 0 & 1 & -1 & -1 & 0 \\ 0 & 0 & 7 & 8 & 1 \end{array}\right] \xrightarrow{\frac{1}{7}R_3} \left[\begin{array}{cccc|c} 1 & 1 & 0 & 1 & 1 \\ 0 & 1 & -1 & -1 & 0 \\ 0 & 0 & 1 & \frac{8}{7} & \frac{1}{7} \end{array}\right] \xrightarrow{1R_1+R_2}$$

$$\left[\begin{array}{cccc|c} 1 & 1 & 0 & 1 & 1 \\ 0 & 1 & 0 & \frac{1}{7} & \frac{1}{7} \\ 0 & 0 & 1 & \frac{8}{7} & \frac{1}{7} \end{array}\right] \xrightarrow{-1R_2+R_1} \left[\begin{array}{cccc|c} 1 & 0 & 0 & \frac{6}{7} & \frac{6}{7} \\ 0 & 1 & 0 & \frac{1}{7} & \frac{1}{7} \\ 0 & 0 & 1 & \frac{8}{7} & \frac{1}{7} \end{array}\right]$$

Hence the corresponding system of equations are consistent and the solutions are given by $x_1 = \frac{6}{7} - \frac{6}{7}x_4, x_2 = \frac{1}{7} - \frac{1}{7}x_4, x_3 = \frac{1}{7} - \frac{8}{7}x_4$ *and* x_4 *is arbitrary. If we take* $x_4 = 1$ *then we get a simple solution* $x_1 = 0, x_2 = 0, x_3 = -1, x_4 = 1$. *Corresponding to this solution we have*

$$-\mathbf{a}_3 + \mathbf{a}_4 = \mathbf{b}$$

If we choose $x_4 = -1$ *then* $x_1 = \frac{12}{7}, x_2 = \frac{2}{7}, x_3 = \frac{9}{7}, x_4 = -1$. *Hence*

$$\frac{12}{7}\mathbf{a}_1 + \frac{2}{7}\mathbf{a}_2 + \frac{9}{7}\mathbf{a}_3 - \mathbf{a}_4 = \mathbf{b}$$

Hence writing a vector as a linear combination of vectors amounts to solving a system of linear equation. In this case, there are infinitely many different ways of writing $\mathbf{b}$ as linear combinations of $\{\mathbf{a}_1, \mathbf{a}_2, \mathbf{a}_3, \mathbf{a}_4\}$.

Is this always the case? Is there a case when a vector is a linear combination of other vectors in a unique way? We will find the answers to these questions.

Practice Problem 1.2.2. *Let* $\mathbf{v}_1 = \begin{bmatrix} 1 \\ 2 \\ 1 \end{bmatrix}, \mathbf{v}_2 = \begin{bmatrix} 1 \\ 1 \\ 0 \end{bmatrix}, \mathbf{v}_3 = \begin{bmatrix} 0 \\ -1 \\ 1 \end{bmatrix}, \mathbf{v}_4 = \begin{bmatrix} 1 \\ 0 \\ 1 \end{bmatrix}$ *and* $\mathbf{b} = \begin{bmatrix} 2 \\ 0 \\ 1 \end{bmatrix}$. *Decide whether* $\mathbf{b}$ *is a linear combination of* $S = \{\mathbf{v}_1, \mathbf{v}_2, \mathbf{v}_3, \mathbf{v}_4\}$. *If so, is the expression of* $\mathbf{b}$ *as a linear combination of* S *unique? If it is not the case, find two different ways to express* $\mathbf{b}$ *as linear combinations of* S.

Example 1.2.9. *Let* $S = \left\{ \begin{bmatrix} x \\ y \end{bmatrix} \middle| \; |x| \le 1, |y| \le 1 \right\}$. *We claim that* $\text{Span}\,(S) = \mathbb{R}^2$. *For if* $\mathbf{v} = \begin{bmatrix} a \\ b \end{bmatrix} \in \mathbb{R}^2$ *then* $\mathbf{v} = a\begin{bmatrix} 1 \\ 0 \end{bmatrix} + b\begin{bmatrix} 0 \\ 1 \end{bmatrix}$ *with* $\begin{bmatrix} 1 \\ 0 \end{bmatrix}, \begin{bmatrix} 0 \\ 1 \end{bmatrix} \in S$. *The set* S *is an infinite set. We only need two vectors* $\mathbf{v}_1 = \begin{bmatrix} 1 \\ 0 \end{bmatrix}, \mathbf{v}_2 = \begin{bmatrix} 0 \\ 1 \end{bmatrix}$ *to span the same space* $\text{Span}\,\{\mathbf{v}_1, \mathbf{v}_2\} = \text{Span}\,(S) = \mathbb{R}^2$.

1.2.4 Subspace

Consider a line V through the origin in $\mathbb{R}^2$ which consists of the scalar multiples of a nonzero vector $\mathbf{v}$; $V = \{c\mathbf{v} | c \in \mathbb{R}\}$. Then a sum of two vectors and a scalar multiple of vectors in V are again a vector in V since they are all constant multiples of $\mathbf{v}$. Such a subset is called a subspace. We formally define:

Definition 1.2.10. *A nonempty subset* V *of* $\mathbb{R}^n$ *is a* **subspace** *if* $\mathbf{v}, \mathbf{v}_1, \mathbf{v}_2 \in V, c \in \mathbb{R}$ *then*

$$\mathbf{v}_1 + \mathbf{v}_2 \in V, \;\; c\mathbf{v} \in V \tag{1.34}$$

For example, if $\mathbf{v} \in \mathbb{R}^n$ then $V = \text{Span}\,\{\mathbf{v}\} = \{c\mathbf{v} \mid c \in \mathbb{R}\}$ is a subspace. In fact, if $\mathbf{v}_1 = c_1\mathbf{v}, \mathbf{v}_2 = c_2\mathbf{v}$ and $c \in \mathbb{R}$ then $\mathbf{v}_1 + \mathbf{v}_2 = (c_1 + c_2)\mathbf{v} \in V$ and $c\mathbf{v}_1 = (cc_1)\mathbf{v} \in V$. We will show that $\text{Span}\,(S)$ is a subspace for any subset S (Theorem 1.2.11).

If we take $c = 0$ then the second condition of (1.34) guarantees that any subspace contains the zero vector:

$$\mathbf{0} \in V$$

since $0 \cdot \mathbf{v} = \mathbf{0}$ for any $\mathbf{v} \in V$. Also note that operations of vectors in a subspace enjoy the properties of Theorem 1.2.1. By (1.34), if $\mathbf{v}_1, \mathbf{v}_2, ..., \mathbf{v}_p$ are vectors in a subspace V then for any scalars $c_1, c_2, ..., c_p$ we have

$$c_1\mathbf{v}_1 + c_2\mathbf{v}_2 + \cdots + c_p\mathbf{v}_p \in V$$

The subset $\{\mathbf{0}\}$ consisting only of the zero vector is a subspace, called the **trivial subspace**. And $\mathbb{R}^n$ itself is also a subspace. A subspace V of $\mathbb{R}^n$ such that $V \neq \mathbf{0}$ and $V \neq \mathbb{R}^n$ will be called a **proper subspace**.

A comment on the notation for the trivial subspace: If V is a trivial subspace then we should write $V = \{\mathbf{0}\}$. However, for simplicity, for convenience, we often write $V = \mathbf{0}$ for the trivial subspace. If V, W are subspaces with $W \subseteq V$ then we say W is a subspace of V.

Since every subspace contains $\mathbf{0}$ we see that the subset

$$V_1 = \left\{ \begin{bmatrix} x_1 \\ x_2 \end{bmatrix} \middle| \, x_1 \in \mathbb{R}, x_2 = 1 \right\}$$

is not a subspace of $\mathbb{R}^2$. And the subset

$$V_2 = \left\{ \begin{bmatrix} x_1 \\ x_2 \end{bmatrix} \middle| \, x_i \in \mathbb{R}, x_2 \geq 0 \right\}$$

is not a subspace of $\mathbb{R}^2$ either. For $\begin{bmatrix} 1 \\ 2 \end{bmatrix} \in V$ but $(-2)\begin{bmatrix} 1 \\ 2 \end{bmatrix} = \begin{bmatrix} -2 \\ -4 \end{bmatrix} \notin V_2$.

Theorem 1.2.11. *Let S be a subset of $\mathbb{R}^n$ then* Span (S) *is a subspace of $\mathbb{R}^n$ containing S. The corresponding statement for the row space $\mathbb{R}_n$ is true.*

Proof. Elements of Span (S) can be written as $\mathbf{u} = \sum c_i\mathbf{v}_i, \mathbf{v} = \sum d_i\mathbf{v}_i$ $(\mathbf{v}_i \in S)$ where some c_i, d_i may be zero. Then $\mathbf{u} + \mathbf{v} = \sum(c_i + d_i)\mathbf{v}_i$. Hence $\mathbf{u} + \mathbf{v} \in$ Span (S). Further, if c is a scalar then $c\mathbf{v} = c\sum c_i\mathbf{v}_i = \sum c\,c_i\mathbf{v}_i$ belongs to Span (S). Hence Span (S) is a subspace. The second statement is similarly proved. □

Let V be a subspace of $\mathbb{R}^n$ and S be a subset of V. If Span $(S) = V$ then we say S **spans** V or V is a subspace **spanned by** of S.

Theorem 1.2.12. *Let A be an $m\times n$ matrix and $\mathbf{u}, \mathbf{v} \in \mathbb{R}^n$ and c is a scalar. Then*

$$A(\mathbf{u} + \mathbf{v}) = A\mathbf{u} + A\mathbf{v}\,, \qquad A(c\mathbf{u}) = cA\mathbf{u}$$

Proof. Let $A^{(i)} = \mathbf{a}_i$ and $\mathbf{u} = \begin{bmatrix} u_1 \\ \vdots \\ u_n \end{bmatrix}, \mathbf{v} = \begin{bmatrix} v_1 \\ \vdots \\ v_n \end{bmatrix}$. Then

$$\begin{aligned} A(\mathbf{u} + \mathbf{v}) &= \textstyle\sum(u_i + v_i)\mathbf{a}_i && \text{by (1.29)} \\ &= \textstyle\sum u_i\mathbf{a}_i + \sum v_i\mathbf{a}_i && \text{by Theorem 1.2.1 (6)} \\ &= A\mathbf{u} + A\mathbf{v} && \text{by (1.29)} \end{aligned}$$

The second equality is similarly proved. □

Corollary 1.2.13. *Let A be an $m\times n$ matrix. Then $W = \{A\mathbf{u} \mid \mathbf{u} \in \mathbb{R}^n\}$ is a subspace of $\mathbb{R}^m$ spanned by the columns $\{A^{(1)}, A^{(2)}, ..., A^{(n)}\}$.*

Proof. By Theorem 1.2.11, it suffices to show that W is spanned by the set of columns of A. For this we need to show that every vector in W is a linear combination of the columns of A. In fact, if $A\mathbf{u} \in W$ and $\mathbf{u} = \begin{bmatrix} u_1 \\ \vdots \\ u_n \end{bmatrix}$ then $A\mathbf{u} = u_1 A^{(1)} + u_2 A^{(2)} + \cdots + u_n A^{(n)}$. Hence every vector in W is a linear combination of the columns of A. □

Corollary 1.2.14. *The set of solutions of a system of homogeneous equations in n variable is a subspace of* $\mathbb{R}^n$.

Proof. Let A be the coefficient matrix for a system of homogeneous equations and let $\mathbf{u}, \mathbf{v}$ be its solutions; $A\mathbf{u} = \mathbf{0}, A\mathbf{v} = \mathbf{0}$. Then

$$\begin{aligned} A(\mathbf{u}+\mathbf{v}) &= A\mathbf{u} + A\mathbf{v} && \text{Theorem 1.2.12} \\ &= \mathbf{0} + \mathbf{0} = \mathbf{0} && \text{Theorem 1.2.1(3) with } \mathbf{u} = \mathbf{0} \end{aligned}$$

Hence $\mathbf{u}+\mathbf{v}$ is also a solution. If $c \in \mathbb{R}$ then, by Theorem 1.2.12 and Theorem 1.2.1, we have $A(c\mathbf{u}) = cA(\mathbf{u}) = c\mathbf{0} = \mathbf{0}$. Hence $c\mathbf{u}$ is also a solution. Therefore the set of solutions of a homogeneous equations form a subspace. □

The set of solutions of a system of homogeneous equations and the set of linear combinations of the columns of a matrix are subspaces. We define:

Definition 1.2.15. *Let A be an $m \times n$ matrix. The subspace spanned by the columns of A is called the* **column space** *of A which we denote by* Col(A). *The set of solutions to the homogeneous equation $A\mathbf{x} = \mathbf{0}$ is called the* **solution space** *or* **null space** *of A and we denote it by $N(A)$:*

$$\text{Col}(A) = \{A\mathbf{u} \mid \mathbf{u} \in \mathbb{R}^n\}, \quad N(A) = \{\mathbf{x} \mid A\mathbf{x} = \mathbf{0}\}$$

The subspaces of $\mathbb{R}^n$ are rather special kinds of subsets. For example, the only subspace of $\mathbb{R}$ are the trivial subspace and $\mathbb{R}$ itself (Exercise 6). Now we claim that the subspaces of $\mathbb{R}^2$ are the trivial subspace, the lines through the origin and $\mathbb{R}^2$ itself. In fact,

Example 1.2.16. *We will show that the subspaces of $\mathbb{R}^2$ are $\{\mathbf{0}\}$, the lines through the origin and $\mathbb{R}^2$ itself.*

If $H = \{\mathbf{0}\}$, the trivial subspace, then it is one of our subspace. Now suppose $H \neq \{\mathbf{0}\}$ then there is $\mathbf{x} = \begin{bmatrix} x_1 \\ x_2 \end{bmatrix} \neq \mathbf{0}$ in H. Consider $H' := \{c\mathbf{x} \mid c \in \mathbb{R}\} \subseteq H$.

If $H' = H$ then H is a subspace which is a line through the origin. Now suppose $H' \subsetneq H$. We claim that $H = \mathbb{R}^2$. To show this first, notice that $H' \subsetneq H$ means there is $\mathbf{y} = \begin{bmatrix} y_1 \\ y_2 \end{bmatrix}$ in $H \setminus H'$. Now such $\mathbf{y}$ exists if and only if there is no $0 \neq c \in \mathbb{R}$ with $\mathbf{y} = c\mathbf{x}$. On the other hand, there is nonzero such $c \in \mathbb{R}$ if and only if $y_1 = cx_1$ and $y_2 = cx_2$; that is $x_1 y_2 = \frac{y_1}{c} c x_2 = x_2 y_1$. Hence we have shown that there is $\mathbf{y} = \begin{bmatrix} y_1 \\ y_2 \end{bmatrix} \in H \setminus H'$ if and only if $x_1 y_2 \neq x_2 y_1$.

To show $H = \mathbb{R}^2$ we need to show that any $\mathbf{z} \in \mathbb{R}^2$ is a linear combination of $\{\mathbf{x}, \mathbf{y}\}$. Let $\mathbf{z} = \begin{bmatrix} z_1 \\ z_2 \end{bmatrix} \in \mathbb{R}^2$. We need to find $c, d \in \mathbb{R}$ such that $\mathbf{z} = c\mathbf{x} + d\mathbf{y}$. This is equivalent to the existence of solution of the equation (in the variables c, d)

$$\begin{cases} cx_1 + dy_1 = z_1 \\ cx_2 + dy_2 = z_2 \end{cases} \tag{1.35}$$

in c, d. Now by Example 1.1.11, *the equation* (1.35) *has a unique solution if $x_1y_2 \neq x_2y_1$ which is guaranteed by the condition $H' \subsetneq H$ as we saw above. Then the solution is given by*

$$c = \frac{y_2z_1 - y_1z_2}{x_1y_2 - x_2y_1}, \qquad d = \frac{x_1z_2 - x_2z_1}{x_1y_2 - x_2y_1}$$

Hence $H = \mathbb{R}^2$. Later we will show this easily using the concept 'dimension'.

From this example, we can infer that the subspaces of $\mathbb{R}^3$ are $\{\mathbf{0}\}$, the lines through the origin, the planes through the origin and $\mathbb{R}^3$ itself. Soon we will prove this easily by using 'dimension' (Example 1.3.31).

Theorem 1.2.17. *Let S be a subset of $\mathbb{R}^n$ and V be a subspace of $\mathbb{R}^n$. Then $S \subseteq V$ if and only if* $\text{Span}(S) \subseteq V$. *Namely,* $\text{Span}(S)$ *is the smallest subspace containing S.*

Proof. Suppose $S \subseteq V$. Let $\mathbf{v} = \sum c_i\mathbf{v}_i \in \text{Span}(S)$ with $\mathbf{v}_i \in S$. Then since $\mathbf{v}_i \in V$ we have $c_i\mathbf{v}_i \in V$ and also $\sum c_i\mathbf{v}_i \in V$ since V is a subspace i.e., $\text{Span}(S) \subseteq V$. Converse is obvious since $S \subseteq \text{Span}(S)$.

For the last statement, we know that $\text{Span}(S)$ is a subspace containing S (Theorem 1.2.11). Since every subspace contain S also contains $\text{Span}(S)$ we see that $\text{Span}(S)$ is the smallest subspace containing S. □

Definition 1.2.18. *Let V_1, V_2 be subspaces of $\mathbb{R}^n$. We define the* **sum** *of V_1 and V_2 by*

$$V_1 + V_2 = \{\mathbf{v}_1 + \mathbf{v}_2 \mid \mathbf{v}_i \in V_i \ (i = 1, 2)\}$$

Theorem 1.2.19. *Let V_1, V_2 be subspaces of $\mathbb{R}^n$.*

(1) *The intersection $V_1 \cap V_2$ is a subspace.*

(2) *The sum $V_1 + V_2$ is a subspace. Further $V_1 + V_2$ is the smallest subspace of $\mathbb{R}^n$ containing $V_1 \cup V_2$.*

Proof. Let $\mathbf{u}, \mathbf{v} \in V_1 \cap V_2$ and $c \in \mathbb{R}$. Then $\mathbf{u} + \mathbf{v} \in V_1$ since $\mathbf{u}, \mathbf{v} \in V_1$ and V_1 is a subspace. By the same token we have $\mathbf{u} + \mathbf{v} \in V_2$. Hence $\mathbf{u} + \mathbf{v} \in V_1 \cap V_2$. Similarly $c\mathbf{u} \in V_1$ and $c\mathbf{u} \in V_2$. Hence $c\mathbf{u} \in V_1 \cap V_2$.

Proof for the sum is Exercise 7. □

For example, if $V_i = \text{Span}\{\mathbf{v}_i\}$ $(i = 1, 2)$ then $V_1 + V_2 = \text{Span}\{\mathbf{v}_1, \mathbf{v}_2\}$.

Answers for Practice Problems 1.2

(1.2.1) (6) $(c + d)\mathbf{u} = \begin{bmatrix} u_1 \\ u_2 \\ u_3 \end{bmatrix} = \begin{bmatrix} (c+d)u_1 \\ (c+d)u_2 \\ (c+d)u_3 \end{bmatrix} = \begin{bmatrix} cu_1 + du_1 \\ cu_2 + du_2 \\ cu_3 + du_3 \end{bmatrix} = \begin{bmatrix} cu_1 \\ cu_2 \\ cu_3 \end{bmatrix} + \begin{bmatrix} du_1 \\ du_2 \\ du_3 \end{bmatrix} =$
$c\begin{bmatrix} u_1 \\ u_2 \\ u_3 \end{bmatrix} + d\begin{bmatrix} u_1 \\ u_2 \\ u_3 \end{bmatrix} = c\mathbf{u} + d\mathbf{u}$

(7) $c(d\mathbf{u}) = c\begin{bmatrix} du_1 \\ du_2 \\ du_3 \end{bmatrix} = \begin{bmatrix} c(du_1) \\ c(du_2) \\ c(du_3) \end{bmatrix} = \begin{bmatrix} (cd)u_1 \\ (cd)u_2 \\ (cd)u_3 \end{bmatrix} = (cd)\begin{bmatrix} u_1 \\ u_2 \\ u_3 \end{bmatrix} = (cd)\mathbf{u}$

(1.2.2) The vector equation $x_1\mathbf{v}_1 + x_2\mathbf{v}_2 + x_3\mathbf{v}_3 + x_4\mathbf{v}_4 = \mathbf{b}$ has the solution $x_1 = -\frac{1}{2}, x_2 = \frac{5}{2} - x_4, x_3 = \frac{3}{2} - x_4, x_4 =$ arbitrary number. Hence $\mathbf{b}$ is a linear combination of S

and there are infinitely many different ways to write $\mathbf{b}$ as a linear combination of S. And we get different expressions of $\mathbf{b}$ as linear combinations of S by taking different values of x_4.

1.2 Exercises

1. Describe Span (S).

   (a) Let $S = \left\{ \mathbf{v}_1 = \begin{bmatrix} 1 \\ -1 \\ 0 \end{bmatrix}, \mathbf{v}_2 = \begin{bmatrix} -1 \\ -1 \\ 0 \end{bmatrix} \right\}$. (b) Let $S = \left\{ \mathbf{v}_{s,t} = \begin{bmatrix} s \\ t \\ 0 \end{bmatrix} \middle| s < 0, t > 0 \right\}$.

   (c) Let $S = \left\{ \begin{bmatrix} x \\ y \\ z \end{bmatrix} \middle| x > 0, y > 0, z = 1 \right\}$. (d) Let $S = \left\{ \mathbf{v} = \begin{bmatrix} 0 \\ 1 \\ 0 \end{bmatrix} \right\}$.

2. Prove the true statement(s) and provide a counter example for the false statement(s).

   (a) The empty set is a subspace of $\mathbb{R}^n$ for any n.

   (b) The span of the empty set, Span $(\emptyset)$ is also empty set.

   (c) If V is a subspace of $\mathbb{R}^n$ then there is a subset (finite or not) S of V such that Span $(S) = V$.

   (d) If V_1 and V_2 are subspaces then $V_1 \cup V_2$ is also a subspace.

   (e) The subset $S = \left\{ \begin{bmatrix} x \\ y \end{bmatrix} \in \mathbb{R}^2 \middle| x \geq 0 \right\}$ is a subspace of $\mathbb{R}^2$.

   (f) The subset $S = \left\{ \begin{bmatrix} x \\ y \end{bmatrix} \in \mathbb{R}^2 \middle| 2x + y = 0 \text{ and } x - y = 1 \right\}$ is a subspace of $\mathbb{R}^2$.

   (g) The subset $S = \left\{ \begin{bmatrix} x \\ y \end{bmatrix} \in \mathbb{R}^2 \middle| 2x + y = 0 \right\}$ is a subspace of $\mathbb{R}^2$.

3. Prove that V is a subspace of $\mathbb{R}^3$ and find a finite spanning set of vectors for V.

   (a) $V = \left\{ \begin{bmatrix} x \\ y \\ z \end{bmatrix} \middle| -x + 2y + z = 0 \right\}$. (b) $V = \left\{ \begin{bmatrix} x \\ y \\ z \end{bmatrix} \middle| x = y = z \right\}$.

   (c) $V = \left\{ \begin{bmatrix} x \\ y \\ z \end{bmatrix} \middle| x = z \right\}$. (d) $V = \left\{ \begin{bmatrix} x \\ y \\ z \end{bmatrix} \middle| x^2 + y^2 + z^2 = 0 \right\}$.

4. Prove that V is a subspace of $\mathbb{R}^4$ and find a finite spanning set of vectors for V.

   (a) $V = \left\{ \begin{bmatrix} x \\ y \\ z \\ w \end{bmatrix} \middle| x + y + z + w = 0 \right\}$. (b) $V = \left\{ \begin{bmatrix} x \\ y \\ z \\ w \end{bmatrix} \middle| y = z \right\}$.

   (c) $V = \left\{ \begin{bmatrix} x \\ y \\ z \\ w \end{bmatrix} \middle| x = z, y = w \right\}$. (d) $V = \left\{ \begin{bmatrix} x \\ y \\ z \\ w \end{bmatrix} \middle| \begin{matrix} -x + 2y - z = 0 \\ y + w = 0 \end{matrix} \right\}$.

5. Let $S = \left\{ \mathbf{v}_1 = \begin{bmatrix} 1 \\ 1 \\ 0 \\ 1 \\ -1 \end{bmatrix}, \mathbf{v}_2 = \begin{bmatrix} -1 \\ 0 \\ 0 \\ 1 \\ 0 \end{bmatrix}, \mathbf{v}_3 = \begin{bmatrix} 1 \\ 0 \\ 1 \\ 0 \\ -1 \end{bmatrix} \right\}$. Decide whether the following vectors belong to Span (S) or not. If it does write it as a linear combination of S.

   (a) $\begin{bmatrix} 1 \\ 1 \\ 1 \\ 1 \\ 1 \end{bmatrix}$ (b) $\begin{bmatrix} 1 \\ -1 \\ 2 \\ 1 \\ 0 \end{bmatrix}$ (c) $\begin{bmatrix} 1 \\ 0 \\ 0 \\ 0 \\ 0 \end{bmatrix}$ (d) $\begin{bmatrix} -1 \\ 1 \\ -1 \\ 2 \\ 0 \end{bmatrix}$ (e) $\begin{bmatrix} 1 \\ 1 \\ 1 \\ 2 \\ -2 \end{bmatrix}$

6. Prove that the subspaces of $\mathbb{R}$ are either the trivial subspace or $\mathbb{R}$ itself.

7. Prove Theorem 1.2.19 (2).

8. Let $\mathbf{v}_i \neq \mathbf{0}$ and let $V_i = \text{Span}\{\mathbf{v}_i\}$ $(i = 1, 2)$.

   (a) Prove that $V_1 + V_2 = \text{Span}\{\mathbf{v}_1, \mathbf{v}_2\}$.

   (b) Prove that $V_1 \cap V_2$ contains a nonzero vector if and only if there is a nonzero scalar c such that $\mathbf{v}_2 = c\mathbf{v}_1$. In this case, $V_1 = V_2 = V_1 \cap V_2 = V_1 + V_2$.

9. Let $S = \{\mathbf{v}_1, \mathbf{v}_2, ..., \mathbf{v}_n\}$ be nonzero vectors and let $V_i = \text{Span}(\mathbf{v}_i)$. Prove that

$$\text{Span}(S) = V_1 + V_2 + \cdots + V_n$$

10. Let V_1, V_2 be subspaces of $\mathbb{R}^n$. Prove that $V_1 \cup V_2$ is a subspace if and only if either $V_1 \subseteq V_2$ or $V_1 \supseteq V_2$.

1.3 Bases of Subspaces

1.3.1 Spanning vectors for Subspaces

Definition 1.3.1. *Let V be a subspace of $\mathbb{R}^n$ and let $S = \{\mathbf{v}_1, \mathbf{v}_2, ..., \mathbf{v}_p\}$ be a subset of V. If* Span $(S) = V$ *then we say V is* **spanned** *(or* **generated***) by S. And S is called a* **set of generators** *or* **spanning set** *of V.*

Theorem 1.3.2. *Let S_1, S_2 be two sets of vectors in $\mathbb{R}^n$. Let $V_1 = \text{Span}(S_1), V_2 = \text{Span}(S_2)$.*

(1) *We have $V_1 \subseteq V_2$ if and only if $S_1 \subseteq V_2$.*

(2) *We have $V_1 = V_2$ if and only if $S_1 \subseteq V_2$ and $S_2 \subseteq V_1$.*

Proof. (1) If $V_1 \subseteq V_2$ then since $S_1 \subseteq \text{Span}(S_1) = V_1$ we have $S_1 \subseteq V_2$. Conversely, by Theorem 1.2.17, the condition $S_1 \subseteq V_2$ implies $V_1 \subseteq V_2$.

(2) Follows from (1) since $V_1 = V_2$ is equivalent to $V_1 \subseteq V_2$ and $V_1 \supseteq V_2$. □

Definition 1.3.3. *Consider an $m \times n$ matrix*

$$A = \begin{bmatrix} a_{11} & a_{12} & \cdots & a_{1n} \\ a_{21} & a_{22} & \cdots & a_{2n} \\ & \cdots & \cdots & \\ a_{m1} & a_{m2} & \cdots & a_{mn} \end{bmatrix}$$

The subspace spanned by the columns $\{A^{(1)}, A^{(2)}, ..., A^{(n)}\}$ (resp.*rows* $\{A_{(1)}, A_{(2)}, ..., A_{(m)}\}$) *of* $\mathbb{R}^m$ (resp. $\mathbb{R}_n$) *is called the* **column space** (resp. **row space**) *of* A *and we will denote it by* $\mathrm{Col}(A)$ (resp. $\mathrm{Row}(A)$)

$$\mathrm{Col}(A) = \mathrm{Span}\,\{A^{(1)}, A^{(2)}, ..., A^{(n)}\} \tag{1.36}$$

$$\mathrm{Row}(A) = \mathrm{Span}\,\{A_{(1)}, A_{(2)}, ..., A_{(m)}\} \tag{1.37}$$

By Corollary 1.2.13, the column space $\mathrm{Col}(A)$ consists of the vectors $A\mathbf{u}$ ($\mathbf{u} \in \mathbb{R}^n$).

Theorem 1.3.4. *Let* A, B *be* $m \times n$ *matrices. If* A *and* B *are row equivalent then* $\mathrm{Row}(A) = \mathrm{Row}(B)$.

Proof. We need to show that each elementary row operation on A does not change its row space. Let $S = \{A_{(1)}, A_{(2)}, ..., A_{(m)}\}$ be the set of rows of A and $V = \mathrm{Row}(A)$. Let S' be the rows obtained by performing an elementary operation on A and $V' = \mathrm{Span}\,(S')$. By Theorem 1.3.2, we need to show that $S \subseteq V'$ and $S' \subseteq V$. First consider the elementary operation $R_i + cR_j$. Then $S' = \{A_{(1)}, ..., A_{(i)} + cA_{(j)}, ..., A_{(m)}\}$. Obviously $S' \subseteq V$. To see $S \subseteq V'$ we only need to check $A_{(i)} \in V'$. But this follows from the equality $A_{(i)} = (A_{(i)} + cA_{(j)}) - cA_{(j)} \in V'$. Similarly the other elementary row operations do not change the row space of A. □

The converse of Theorem 1.3.4 is true (Exercise 9).

Example 1.3.5. *Let* $A = \begin{bmatrix} 1 & 0 & -1 \\ 0 & 1 & 2 \end{bmatrix}$. *We add the second row to the first row of* A. *And then add* -1 *times the first row to the second row to get* $B = \begin{bmatrix} 1 & 1 & 1 \\ -1 & 0 & 1 \end{bmatrix}$. *Hence* A *and* B *are row equivalent and hence, by* Theorem 1.3.4, $\mathrm{Row}(A) = \mathrm{Row}(B)$. *To apply* Theorem 1.3.2, *we need to show that* $\{A_{(1)}, A_{(2)}\} \subseteq \mathrm{Span}\,\{B_{(1)}, B_{(2)}\}$ *and* $\{B_{(1)}, B_{(2)}\} \subseteq \mathrm{Span}\,\{A_{(1)}, A_{(2)}\}$. *In fact,* $A_{(1)} = -B_{(2)}$ *and* $A_{(2)} = B_{(1)} + B_{(2)}$; *and therefore* $\{A_{(1)}, A_{(2)}\} \subseteq \mathrm{Span}\,\{B_{(1)}, B_{(2)}\}$. *The opposite inclusion is similarly proved. Thus* $\mathrm{Row}(A) = \mathrm{Row}(B)$.

Let A be an $m \times n$ matrix with $A^{(i)} = \mathbf{a}_i \in \mathbb{R}^m$ and let $\mathbf{x} = \begin{bmatrix} x_1 \\ \vdots \\ x_n \end{bmatrix} \in \mathbb{R}^n$. View A as n column vectors $A = [\mathbf{a}_1, \mathbf{a}_2, ..., \mathbf{a}_n]$. The column space $\mathrm{Col}(A)$ consists of the **product** $A\mathbf{x}$ defined as in (1.29)

$$A\mathbf{x} = x_1\mathbf{a}_1 + x_2\mathbf{a}_2 + \cdots + x_n\mathbf{a}_n$$

Hence, as we already observed (Corollary 1.2.13)

$$\mathrm{Col}(A) = \{A\mathbf{x} \mid \mathbf{x} \in \mathbb{R}^n\}$$

Hence $\mathbf{b} \in \mathrm{Col}(A)$ if and only if there is $\mathbf{c} \in \mathbb{R}^n$ with

$$A\mathbf{c} = \mathbf{b}$$

namely, the equation $A\mathbf{x} = \mathbf{b}$ has a solution $\mathbf{c}$.

We state this fact formally which is Theorem 1.2.7 (1) in slightly different language.

Theorem 1.3.6. *Let A be an $m \times n$ matrix. Then the equation $A\mathbf{x} = \mathbf{b}$ has a solution if and only if $\mathbf{b} \in \mathrm{Col}(A)$.*

Example 1.3.7. *Let* $A = \begin{bmatrix} 1 & 4 & -1 \\ 0 & 3 & 3 \end{bmatrix}$ *and* $\mathbf{x} = \begin{bmatrix} x_1 \\ x_2 \\ x_3 \end{bmatrix}$. *Then*

$$A\mathbf{x} = \begin{bmatrix} 1 & 4 & -1 \\ 0 & 3 & 3 \end{bmatrix} \begin{bmatrix} x_1 \\ x_2 \\ x_3 \end{bmatrix} = x_1 \begin{bmatrix} 1 \\ 0 \end{bmatrix} + x_2 \begin{bmatrix} 4 \\ 3 \end{bmatrix} + x_3 \begin{bmatrix} -1 \\ 3 \end{bmatrix} = \begin{bmatrix} x_1 + 4x_2 - x_3 \\ 3x_2 + 3x_3 \end{bmatrix}$$

If we let

$$\mathbf{a}_1 = \begin{bmatrix} 1 \\ 0 \end{bmatrix}, \ \mathbf{a}_2 = \begin{bmatrix} 4 \\ 3 \end{bmatrix}, \ \mathbf{a}_3 = \begin{bmatrix} -1 \\ 3 \end{bmatrix} \ \text{and } \mathbf{b} = \begin{bmatrix} 3 \\ 6 \end{bmatrix}$$

then the matrix equation $A\mathbf{x} = \mathbf{b}$ is the same as the vector equation

$$x_1\mathbf{a}_1 + x_2\mathbf{a}_2 + x_3\mathbf{a}_3 = \mathbf{b} \tag{1.38}$$

which in turn is equivalent to the system of linear equations

$$\begin{cases} x_1 + 4x_2 - x_3 = 3 \\ \quad\quad 3x_2 + 3x_3 = 6 \end{cases} \tag{1.39}$$

The corresponding augmented matrix has the reduced echelon form

$$\left[\begin{array}{ccc|c} 1 & 0 & -5 & -5 \\ 0 & 1 & 1 & 2 \end{array}\right] \tag{1.40}$$

Hence the system has the solution

$$\begin{cases} x_1 = -5 + 5x_3 \\ x_2 = 2 - x_3 \\ x_3 = \text{ arbitrary number} \end{cases}$$

Since the system (1.39) *has a solution we conclude that $\mathbf{b} \in \mathrm{Col}(A)$. For example, if we take $x_3 = 0$ then we have*

$$-5\mathbf{a}_1 + 2\mathbf{a}_2 = \mathbf{b}$$

or of we take $x_3 = \frac{1}{2}$ then we have

$$-\frac{5}{2}\mathbf{a}_1 - \frac{3}{2}\mathbf{a}_2 + \frac{1}{2}\mathbf{a}_3 = \mathbf{b}$$

Hence there are infinitely many different ways writing $\mathbf{b}$ as linear combinations.

Practice Problem 1.3.1. *Let* $\mathbf{v}_1 = \begin{bmatrix} 1 \\ 2 \\ -1 \end{bmatrix}, \mathbf{v}_2 = \begin{bmatrix} 0 \\ 1 \\ 0 \end{bmatrix}, \mathbf{v}_3 = \begin{bmatrix} 2 \\ -1 \\ 1 \end{bmatrix}, \mathbf{v}_4 = \begin{bmatrix} 1 \\ 0 \\ 1 \end{bmatrix}$. *Find two expressions of* $\mathbf{v} = \begin{bmatrix} 3 \\ 1 \\ 0 \end{bmatrix}$ *as linear combinations of $\mathbf{v}_1, \mathbf{v}_2, \mathbf{v}_3, \mathbf{v}_4$.*

Theorem 1.3.8. *Let A be an $m\times n$ matrix with the columns $A^{(i)} = \mathbf{v}_i$ $(i = 1, 2, ..., n)$. The following conditions are equivalent:*

(1) *For any $\mathbf{b} \in \mathbb{R}^m$ the equation $A\mathbf{x} = \mathbf{b}$ has a solution.*

(2) *Every $\mathbf{b} \in \mathbb{R}^m$ is a linear combination of columns $\{\mathbf{v}_1, \mathbf{v}_2, ..., \mathbf{v}_n\}$ of A.*

(3) *The columns $\{\mathbf{v}_1, \mathbf{v}_2, ..., \mathbf{v}_n\}$ of A span $\mathbb{R}^m$.*

(4) *Every row of A has a pivot position i.e, each row of the reduced echelon form B of A is nonzero. Namely, B is of the form*

$$B = \begin{bmatrix} 1 & 0 & * & 0 & 0 & * & * \\ & 1 & * & 0 & 0 & * & * \\ & & & 1 & \vdots & \vdots & \\ & & & & 1 & * & * \end{bmatrix}$$

In particular, if $m = n$ then the conditions are equivalent to that $B = I_n$.

Proof. Equivalence of (1), (2) and (3) are immediate. The equivalence of (1) and (4) is Theorem 1.2.7. □

Notice that the coefficient matrix of Example 1.3.7 satisfies the condition of Theorem 1.2.7 (4) by (1.40). Hence for any $\mathbf{b} \in \mathbb{R}^2$ the vector equation (1.38) has a solution.

On the other hand, consider the row operations on the following matrix A:

$$A := \begin{bmatrix} 2 & 0 & 2 & 1 \\ 0 & 1 & -1 & 0 \\ 2 & 1 & 1 & 1 \end{bmatrix} \xrightarrow{-R_1+R_3} \begin{bmatrix} 2 & 0 & 2 & 1 \\ 0 & 1 & -1 & 0 \\ 0 & 1 & -1 & 0 \end{bmatrix} \xrightarrow{-R_2+R_3} \begin{bmatrix} 2 & 0 & 2 & 1 \\ 0 & 1 & -1 & 0 \\ 0 & 0 & 0 & 0 \end{bmatrix} \tag{1.41}$$

Let C be the last matrix of (1.41) and let $\mathbf{c} = \begin{bmatrix} 0 \\ -1 \\ 1 \end{bmatrix}$. Then the augmented matrix $[C \mid \mathbf{c}]$ is inconsistent. Under the reversed row operations of (1.41), $\mathbf{c}$ becomes $\mathbf{b} = \begin{bmatrix} 0 \\ -1 \\ 0 \end{bmatrix}$. Hence the augmented matrix $[A|\mathbf{b}]$ is also inconsistent and hence the corresponding system of equations

$$\begin{cases} 2x_1 \quad\quad + 2x_3 + x_4 = 0 \\ \quad\quad x_2 + x_3 \quad\quad = -1 \\ 2x_1 + x_2 + x_3 + x_4 = 0 \end{cases}$$

corresponding to $[A|\mathbf{b}]$ is inconsistent. Therefore the equation $A\mathbf{x} = \mathbf{b}$ has no solution.

Corollary 1.3.9. *If any one of the conditions of* Theorem 1.3.8 *holds then $n \geq m$. Further, if $m = n$ then the reduced echelon form of A is the identity matrix I_n. In this case, the system of equations corresponding to $A\mathbf{x} = \mathbf{b}$ has a unique solution for any $\mathbf{b} \in \mathbb{R}^n$.*

Proof. The condition (4) implies that every row contains a pivot position i.e., there are m pivot columns. Since the number of columns is the sum of the numbers of pivot and the number of nonpivot columns we see that $n \geq m$. The last statement is obvious. □

1.3.2 General Solutions of Linear Equations

We sum up the methods of solving the systems of linear equations we have treated. We first consider examples of homogeneous equations.

Example 1.3.10. *To solve*

$$\begin{cases} x_1 + x_2 + 2x_3 + x_4 = 0 \\ x_1 + 2x_2 - x_3 + x_4 = 0 \end{cases} \tag{1.42}$$

write the coefficient matrix $\begin{bmatrix} 1 & 1 & 2 & 1 \\ 1 & 2 & -1 & 1 \end{bmatrix}$. *Its reduced echelon form is* $\begin{bmatrix} 1 & 0 & 5 & 1 \\ 0 & 1 & -3 & 0 \end{bmatrix}$. *Hence the solution is given by*

$$\begin{cases} x_1 = -5x_3 - x_4 \\ x_2 = 3x_3 \\ x_3 = \text{any} \\ x_4 = \text{any} \end{cases}$$

Therefore we can write any solution as

$$\begin{bmatrix} -5x_3 - x_4 \\ 3x_3 \\ x_3 \\ x_4 \end{bmatrix} = x_3 \begin{bmatrix} -5 \\ 3 \\ 1 \\ 0 \end{bmatrix} + x_4 \begin{bmatrix} -1 \\ 0 \\ 0 \\ 1 \end{bmatrix} \tag{1.43}$$

where x_3, x_4 are free variables. Hence the solution space is spanned by

$$\left\{ \mathbf{v}_1 = \begin{bmatrix} -5 \\ 3 \\ 1 \\ 0 \end{bmatrix}, \mathbf{v}_2 = \begin{bmatrix} -1 \\ 0 \\ 0 \\ 1 \end{bmatrix} \right\} \tag{1.44}$$

That is $N(A) = \text{Span}\{\mathbf{v}_1, \mathbf{v}_2\}$.

Example 1.3.11. *To decide whether the following system has a nontrivial solution or not*

$$\begin{cases} 2x_1 + x_2 + x_3 = 0 \\ -3x_1 - 2x_2 + x_3 = 0 \\ x_1 + x_2 - 2x_3 = 0 \end{cases} \tag{1.45}$$

we form the coefficient matrix and perform elementary row operations:

$$\begin{bmatrix} 2 & 1 & 1 \\ -3 & -2 & 1 \\ 1 & 1 & -2 \end{bmatrix} \xrightarrow[3R_1+R_2]{\frac{1}{2}R_1} \begin{bmatrix} 1 & \frac{1}{2} & \frac{1}{2} \\ 0 & -\frac{1}{2} & \frac{5}{2} \\ 1 & 1 & -2 \end{bmatrix} \xrightarrow{-1R_1+R_3} \begin{bmatrix} 1 & \frac{1}{2} & \frac{1}{2} \\ 0 & -\frac{1}{2} & \frac{5}{2} \\ 0 & \frac{1}{2} & -\frac{5}{2} \end{bmatrix} \xrightarrow{-2R_2}$$

$$\begin{bmatrix} 1 & \frac{1}{2} & \frac{1}{2} \\ 0 & 1 & -5 \\ 0 & \frac{1}{2} & -\frac{5}{2} \end{bmatrix} \xrightarrow[-\frac{1}{2}R_2+R_1]{-\frac{1}{2}R_2+R_3} \begin{bmatrix} 1 & 0 & 3 \\ 0 & 1 & -5 \\ 0 & 0 & 0 \end{bmatrix}$$

Hence the solutions to (1.45) *are given by* $\begin{bmatrix} -3 \\ 5 \\ 1 \end{bmatrix} t$ $(t \in \mathbb{R})$. *Or* $N(A) = \text{Span}\left\{ \begin{bmatrix} -3 \\ 5 \\ 1 \end{bmatrix} \right\}$.

We will show that the solutions of a homogeneous equations is a unique linear combination of a finite vectors.

Practice Problem 1.3.2. *Solve the following systems of linear equations:*

$$\begin{cases} \quad\; 2x_2 + x_3 - x_4 = 0 \\ x_1 \qquad\; + x_3 + x_4 = 0 \\ 2x_1 + 2x_2 \qquad - x_4 = 0 \end{cases}$$

Now consider a system of nonhomogenous linear equation $A\mathbf{x} = \mathbf{b}$ with $\mathbf{b} \neq \mathbf{0}$. Unlike the solutions of homogeneous equation, a sum or a scalar multiple of solutions of a nonhomogenous equation is *not* a solution to the nonhomgenous equation as we can see from one of the previous examples.

We will show that the solutions of a system of nonhomogenous equations are given by the sum of a system of homogeneous equations and a particular solution.

Theorem 1.3.12. *Let A be an $m \times n$ matrix. Let $A\mathbf{x} = \mathbf{b}$ be a consistent system of linear equations. If $\mathbf{p}$ is a particular solution for the equation $A\mathbf{x} = \mathbf{b}$ then the general solution to the equation $A\mathbf{x} = \mathbf{b}$ is given by $\mathbf{p} + \mathbf{v}_h$ where $\mathbf{v}_h$ is a solution to the homogeneous equation $A\mathbf{x} = \mathbf{0}$.*

Proof. Suppose $\mathbf{v}$ is a solution of $A\mathbf{x} = \mathbf{b}$. Then $A(\mathbf{v} - \mathbf{p}) = A\mathbf{v} - A\mathbf{p} = \mathbf{b} - \mathbf{b} = \mathbf{0}$ by Theorem 1.2.12. Hence $\mathbf{v} - \mathbf{p}$ is a solution of $A\mathbf{x} = \mathbf{0}$. If we let $\mathbf{v}_h = \mathbf{v} - \mathbf{p}$ then $\mathbf{v} = \mathbf{v}_h + \mathbf{p}$.

Conversely, we have $A(\mathbf{p} + \mathbf{v}_h) = A\mathbf{p} + A\mathbf{v}_h = \mathbf{b} + \mathbf{0} = \mathbf{b}$. ☐

Example 1.3.13. (1) *Let $A = \begin{bmatrix} 1 & 2 & -1 \\ -2 & -3 & 2 \\ 1 & 1 & -1 \end{bmatrix}$ and $\mathbf{b} = \begin{bmatrix} 2 \\ -1 \\ -1 \end{bmatrix}$. We want to find the general solutions of the equation $A\mathbf{x} = \mathbf{b}$ as a sum of particular solution and the general solutions of a homogeneous equation.*

For this form augmented matrix and perform the row operations:

$$\left[\begin{array}{ccc|c} 1 & -2 & -1 & 2 \\ -2 & -3 & 2 & -1 \\ 1 & 1 & -1 & -1 \end{array}\right] \xrightarrow[-1R_1+R_3]{2R_1+R_2} \left[\begin{array}{ccc|c} 1 & 2 & -1 & 2 \\ 0 & 1 & 0 & 3 \\ 0 & -1 & 0 & -3 \end{array}\right] \xrightarrow[2R_2+R_1]{1R_2+R_3} \left[\begin{array}{ccc|c} 1 & 0 & -1 & -4 \\ 0 & 1 & 0 & 3 \\ 0 & 0 & 0 & 0 \end{array}\right]$$

The general solutions are given by $x_1 = -4 - 3x_3$, $x_2 = 3$, $x_3 =$arbitrary number. If we choose $x_3 = 1$ then we get a particular solution $\mathbf{p} = \begin{bmatrix} -7 \\ 3 \\ 1 \end{bmatrix}$ and the general solution for the homogeneous equation $A\mathbf{x} = \mathbf{0}$ is $\mathbf{v}_h = \begin{bmatrix} 1 \\ 0 \\ 1 \end{bmatrix} t$ $(t \in \mathbb{R})$. Therefore, the solutions for the equation is given by $\mathbf{p} + \mathbf{v}_h t$.

(2) *We want to find the solutions for the system of equations:*

$$\begin{cases} 2x - y + z \qquad\; = 1 \\ \qquad\; y + z - 2w = 0 \\ x \qquad\; - z \; + w = 2 \end{cases}$$

For this we form the augmented matrix and perform row operations:

$$\begin{bmatrix} 2 & -1 & 1 & 0 & 1 \\ 0 & 1 & 1 & -2 & 0 \\ 1 & 0 & -1 & 1 & 2 \end{bmatrix} \xrightarrow{-\frac{1}{2}R_1+R_3} \begin{bmatrix} 2 & -1 & 1 & 0 & 1 \\ 0 & 1 & 1 & -2 & 0 \\ 0 & \frac{1}{2} & -\frac{3}{2} & 1 & \frac{3}{2} \end{bmatrix} \xrightarrow{-\frac{1}{2}R_2+R_3} \begin{bmatrix} 2 & -1 & 1 & 0 & 1 \\ 0 & 1 & 1 & -2 & 0 \\ 0 & 0 & -2 & 2 & \frac{3}{2} \end{bmatrix} \xrightarrow[\frac{1}{2}R_3+R_2]{\frac{1}{2}R_3+R_1}$$

$$\begin{bmatrix} 2 & -1 & 0 & 1 & \frac{7}{4} \\ 0 & 1 & 0 & -1 & \frac{3}{4} \\ 0 & 0 & -2 & 2 & \frac{3}{2} \end{bmatrix} \xrightarrow{R_2+R_1} \begin{bmatrix} 2 & 0 & 0 & 0 & \frac{10}{4} \\ 0 & 1 & 0 & -1 & \frac{3}{4} \\ 0 & 0 & -2 & 2 & \frac{3}{2} \end{bmatrix} \xrightarrow[\frac{1}{2}R_3]{\frac{1}{2}R_1} \begin{bmatrix} 1 & 0 & 0 & 0 & \frac{5}{4} \\ 0 & 1 & 0 & -1 & \frac{3}{4} \\ 0 & 0 & 1 & -1 & -\frac{3}{4} \end{bmatrix}$$

Hence the solutions are given by $x = \frac{5}{4}$, $y = \frac{3}{4} + w$, $z = -\frac{3}{4} + w$. *If we choose* $w = 0$ *then the particular solution is* $\mathbf{p} = \begin{bmatrix} \frac{5}{4} \\ \frac{3}{4} \\ -\frac{3}{4} \\ 0 \end{bmatrix}$ *and the general solution for the corresponding homogeneous equation is* $t\mathbf{v}_h$ $(t \in \mathbb{R})$ *where* $\mathbf{v}_h = \begin{bmatrix} 0 \\ 1 \\ 1 \\ 1 \end{bmatrix}$; *the general solutions are* $\mathbf{p} + t\mathbf{v}_h$.

Practice Problem 1.3.3. *Find the general solution for the equation*

$$\begin{cases} x_1 + 3x_2 + x_3 \quad - x_5 = 1 \\ x_1 + x_2 + x_3 + x_4 = 0 \\ 2x_1 \quad + x_3 + 2x_4 - x_5 = 3 \end{cases}$$

1.3.3 Linear Independence

Consider Example 1.3.10. Notice that any solution vectors can be written in a unique way as a linear combination of $\{\mathbf{v}_1, \mathbf{v}_2\}$. In fact, suppose $s\mathbf{v}_1 + t\mathbf{v}_2 = s'\mathbf{v}_1 + t'\mathbf{v}_2$ then $(s - s')\mathbf{v}_1 + (t - t')\mathbf{v}_2 = \mathbf{0}$;

$$(s - s') \begin{bmatrix} -5 \\ -3 \\ 1 \\ 0 \end{bmatrix} + (t - t') \begin{bmatrix} -1 \\ 0 \\ 0 \\ 1 \end{bmatrix} = \begin{bmatrix} 0 \\ 0 \\ 0 \\ 0 \end{bmatrix}$$

Comparing the third and fourth position of the column vectors in the equality, we see that $(s - s') = 0, (t - t') = 0$. Therefore $s = s', t = t'$. Thus the two solution vectors $\left\{ \begin{bmatrix} -5 \\ -3 \\ 1 \\ 0 \end{bmatrix}, \begin{bmatrix} -1 \\ 0 \\ 0 \\ 1 \end{bmatrix} \right\}$ satisfies the following property, by letting $c_1 = s - s', c_2 = t - t'$:

$$\text{Whenever } c_1 \begin{bmatrix} -5 \\ -3 \\ 1 \\ 0 \end{bmatrix} + c_2 \begin{bmatrix} -1 \\ 0 \\ 0 \\ 1 \end{bmatrix} = \begin{bmatrix} 0 \\ 0 \\ 0 \\ 0 \end{bmatrix} \text{ we have } c_1 = c_2 = 0. \tag{1.46}$$

which is the same thing to say that the vector equation $x_1\mathbf{v}_1 + x_2\mathbf{v}_2 = \mathbf{0}$ has only the trivial solution. In fat, let $A = [\mathbf{v}_1\ \mathbf{v}_2]$ and $\mathbf{x} = \begin{bmatrix} x_1 \\ x_2 \end{bmatrix}$ then the equation $A\mathbf{x} = \mathbf{0}$ has only the trivial solution since the reduced echelon form of A is $\begin{bmatrix} 1 & 0 \\ 0 & 1 \\ 0 & 0 \\ 0 & 0 \end{bmatrix}$.

The set of vectors $\{\mathbf{v}_1, \mathbf{v}_2\}$ with such a property will be said to be 'linearly independent'. Formally we define;

Definition 1.3.14. *A set of vectors* $S = \{\mathbf{v}_1, \mathbf{v}_2, ..., \mathbf{v}_p\}$ *in* $\mathbb{R}^n$ *is said to be* **linearly independent** *if the vector equation* $x_1\mathbf{v}_1 + x_2\mathbf{v}_2 + \cdots + x_p\mathbf{v}_p = \mathbf{0}$ *has only the trivial solution. Otherwise, we say* $\{\mathbf{v}_1, \mathbf{v}_2, ..., \mathbf{v}_p\}$ *is* **linearly dependent** *i.e., the vector equation has a nontrivial solution. That is, there are* $c_1, c_2, ..., c_p$*, not all zero, such that*

$$c_1\mathbf{v}_1 + c_2\mathbf{v}_2 + \cdots + c_p\mathbf{v}_p = \mathbf{0} \tag{1.47}$$

The equality of the form (1.47) *is called a* (*nontrivial*) **linear relation** *among* S*. The empty set* $\emptyset$ *is defined to be linearly independent.*

If $\mathbf{v} \neq \mathbf{0}$ then the set $\{\mathbf{v}\}$ consisting of a single nonzero vector is linearly independent. If one of $S = \{\mathbf{v}_1, ..., \mathbf{v}_p\}$ is a zero vector or any two of them are equal then S is linearly dependent.

Theorem 1.3.15. *Let* $S \subseteq S'$ *be finite sets of vectors. If* S *is linearly dependent then so is* S'*. If* S' *is linearly independent then so is* S*.*

Proof. The second assertion is the contrapositive of the first assertion. Hence we only need to show the first statement which is obvious. □

As we observed above, if $\mathbf{v}_1$ is a nonzero vector then $\{\mathbf{v}_1\}$ is linearly independent. If $\mathbf{v}_2$ is another nonzero vector then $\{\mathbf{v}_1, \mathbf{v}_2\}$ is linearly dependent if and only if there are constants c_1, c_2 such that $c_1\mathbf{v}_1 + c_2\mathbf{v}_2 = \mathbf{0}$. We claim that both c_1, c_2 are nonzero. In fact, if one of c_1, c_2 is 0 say $c_1 = 0$ then $c_2\mathbf{v}_2 = \mathbf{0}$. Hence $\mathbf{v}_2 = \mathbf{0}$ which contradicts to that they are nonzero. Hence both c_1, c_2 are nonzero; $\mathbf{v}_1 = -\frac{c_2}{c_1}\mathbf{v}_2$. Therefore two nonzero vectors $\{\mathbf{v}_1, \mathbf{v}_2\}$ are linearly dependent if and only if one is a scalar multiple of the other. In this case, we say the vectors $\{\mathbf{v}_1, \mathbf{v}_2\}$ are **collinear**. For example, two vectors $\left\{\begin{bmatrix}1\\1\\1\end{bmatrix}, \begin{bmatrix}1\\2\\3\end{bmatrix}\right\}$ are linearly independent since they are not collinear. This result will be generalized in Theorem 1.3.19.

The following theorem is a counter part of Theorem 1.3.8.

Theorem 1.3.16. *If* $A = [\mathbf{v}_1\ \mathbf{v}_2 \cdots \mathbf{v}_n]$ *is an* $m \times n$ *matrix whose i-th column is* $\mathbf{v}_i \in \mathbb{R}^m$*. The following conditions are equivalent.*

(1) $\{\mathbf{v}_1, \mathbf{v}_2, ..., \mathbf{v}_n\}$ *is linearly independent.*

(2) *The equation* $A\mathbf{x} = \mathbf{0}$ *has only the trivial solution.*

(3) *Every column of* A *is a pivot column i.e., the reduced echelon form* B *of* A *is of the form*

$$B := \begin{bmatrix} 1 & 0 & 0 & 0 \\ 0 & 1 & 0 & 0 \\ & & \ddots & \\ 0 & 0 & 0 & 1 \\ 0 & 0 & 0 & 0 \\ & \cdots & \cdots & \\ 0 & 0 & 0 & 0 \end{bmatrix} \tag{1.48}$$

In particular, if $m = n$ *then the conditions are equivalent to that* $B = I_n$*.*

Proof. (1) $\Leftrightarrow$ (2) Obvious from the definition of linear dependence.

(2) $\Leftrightarrow$ (3) The equation $A\mathbf{x} = \mathbf{0}$ has a unique solution namely $x_1 = x_2 = \cdots = x_n = 0$ if and only if the solution has no free variable i.e., every column is a pivot column. And this is the case when the diagonals of the reduced echelon form B are 1's. $\square$

Example 1.3.17. (1) *Let* $\mathbf{v}_1 = \begin{bmatrix} 1 \\ 2 \\ 3 \end{bmatrix}, \mathbf{v}_2 = \begin{bmatrix} 4 \\ 5 \\ 6 \end{bmatrix}, \mathbf{v}_3 = \begin{bmatrix} 2 \\ 1 \\ 0 \end{bmatrix}$. *We want to decide whether* $S = \{\mathbf{v}_1, \mathbf{v}_2, \mathbf{v}_3\}$ *is linearly dependent. The set of vectors* $\{\mathbf{v}_1, \mathbf{v}_2, \mathbf{v}_3\}$ *is linearly dependent if and only if the vector equation* $x_1\mathbf{v}_1 + x_2\mathbf{v}_2 + x_3\mathbf{v}_3 = \mathbf{0}$ *has a nontrivial solution. We solve the vector equation which amounts to solve the homogeneous equation*

$$\begin{cases} x_1 + 4x_2 + 2x_3 = 0 \\ 2x_1 + 5x_2 + x_3 = 0 \\ 3x_1 + 6x_2 \qquad = 0 \end{cases}$$

Solving this we get $x_1 = 2x_3, x_2 = -x_3, x_3 =$ *arbitrary number. Hence if we let* $x_3 = 1$ *then we get a nontrivial solution* $x_1 = 2, x_2 = -1, x_3 = 1$; $2\mathbf{v}_2 - \mathbf{v}_2 + \mathbf{v}_3 = \mathbf{0}$. *Hence* S *is linearly dependent.*

(2) *We want to determine whether the set of columns of the matrix* $A = \begin{bmatrix} 1 & -2 & -1 & 2 \\ -2 & 1 & -3 & 1 \\ 1 & 0 & -1 & 2 \end{bmatrix}$ *is linearly independent or not. This time we use* Theorem 1.3.16 (3): *The reduced echelon form of* A *is* $\begin{bmatrix} 1 & 0 & 0 & 1 \\ 0 & 1 & 0 & 0 \\ 0 & 0 & 1 & -1 \end{bmatrix}$. *This not of the form* (1.48). *Hence the columns of* A *are linearly dependent.*

Corollary 1.3.18. *Let* $S = \{\mathbf{v}_1, \mathbf{v}_2, ..., \mathbf{v}_p\}$ *be a set of vectors in* $\mathbb{R}^n$. *If* $p > n$ *then* S *is linearly dependent. If* S *is linearly independent then* $p \leq n$.

Proof. Let A be the $n \times p$ matrix consisting of the columns $S = \{\mathbf{v}_1, ..., \mathbf{v}_p\}$. Then, as $p > n$, the reduced echelon form of A is not of the form (1.48). Hence by Theorem 1.3.16 (3), S is linearly dependent. The second statement is the contrapositive of the second statement. $\square$

Note that in Example 1.3.17 (2), we could have conclude that the columns are linearly dependent without computation by Corollary 1.3.18.

The remark following Theorem 1.3.15 shows $\{\mathbf{v}_1, \mathbf{v}_2\}$ is linearly dependent if and only if $\mathbf{v}_2 \in \text{Span}\{\mathbf{v}_1\}$. In general we have;

Theorem 1.3.19. *Let* $S = \{\mathbf{v}_1, \mathbf{v}_2, ..., \mathbf{v}_p\}$ *be a linearly independent set of nonzero vectors in a subspace* V *of* $\mathbb{R}^n$. *Let* $\mathbf{0} \neq \mathbf{v}_{p+1} \in V$. *Then* $S \cup \{\mathbf{v}_{p+1}\}$ *is linearly dependent if and only if* $\mathbf{v}_{p+1} \in \text{Span}(S)$. *In this case,*

$$\text{Span}(S) = \text{Span}(S \cup \{\mathbf{v}_{p+1}\}) \tag{1.49}$$

Proof. Assume $S \cup \{\mathbf{v}_{p+1}\}$ is linearly dependent i.e., $c_1\mathbf{v}_1 + \cdots + c_{p+1}\mathbf{v}_{p+1} = \mathbf{0}$ not all c_i's are zero. First we claim that $c_{p+1} \neq 0$. If $c_{p+1} = 0$ then $c_1\mathbf{v}_1 + \cdots + c_p\mathbf{v}_p = \mathbf{0}$ and hence $c_i = 0$ $(i = 1, 2, ..., p)$ since S is linearly independent. This is a contradiction because not all c_i's are zero. Hence $c_{p+1} \neq 0$. Thus

$$\mathbf{v}_{p+1} = -\frac{1}{c_{p+1}}(c_1\mathbf{v}_1 + c_2\mathbf{v}_2 + \cdots + c_p\mathbf{v}_p) \tag{1.50}$$

Therefore $\mathbf{v}_{p+1} \in \text{Span}\{\mathbf{v}_1, \mathbf{v}_2, ..., \mathbf{v}_p\}$.

Conversely, let $\mathbf{v}_{p+1} \in \text{Span}\{\mathbf{v}_1, \mathbf{v}_2, ..., \mathbf{v}_p\}$. Then $\mathbf{v}_{p+1}$ is a linear combination of the vectors $\mathbf{v}_1, \mathbf{v}_2, ..., \mathbf{v}_p$ say, $\mathbf{v}_{p+1} = \sum_{i=1}^{p} c_i \mathbf{v}_i$ then $\sum_{i=1}^{p} c_j \mathbf{v}_j - \mathbf{v}_{p+1} = \mathbf{0}$. Hence $S \cup \{\mathbf{v}_{p+1}\}$ is linearly dependent.

To show the equality (1.49), the inclusion $\subseteq$ is obvious by (1.32). For the opposite inclusion, notice that $\mathbf{v}_{p+1} \in \text{Span}\{\mathbf{v}_1, ..., \mathbf{v}_p\}$ by (1.50) and hence $\{\mathbf{v}_1, ..., \mathbf{v}_{p+1}\} \subseteq \text{Span}\{\mathbf{v}_1, ..., \mathbf{v}_p\}$. By Theorem 1.3.2 we have the desired inclusion. □

Hence we can say that the more vectors the more likely the vectors are linearly dependent and the less vectors the more likely the vectors are linearly independent.

1.3.4 Bases of Subspaces

Now we introduce an important concept called a basis. First consider an example.

Example 1.3.20. *Let* $\mathbf{v}_1 = \begin{bmatrix} 1 \\ 2 \\ 0 \end{bmatrix}, \mathbf{v}_2 = \begin{bmatrix} -1 \\ 0 \\ 1 \end{bmatrix}, \mathbf{v}_3 = \begin{bmatrix} 3 \\ 4 \\ -1 \end{bmatrix}$. *Let* $S = \{\mathbf{v}_1, \mathbf{v}_2, \mathbf{v}_3\}$ *and* $V =$ Span (S). *We see, by* Theorem 1.3.19, Span $\{\mathbf{v}_1\} \neq$ Span $\{\mathbf{v}_1, \mathbf{v}_2\}$ *since* $\{\mathbf{v}_1, \mathbf{v}_2\}$ *is linearly independent for* $\mathbf{v}_2$ *is not a constant multiple of* $\mathbf{v}_1$. *And* Span $\{\mathbf{v}_1, \mathbf{v}_2\} =$ Span $\{\mathbf{v}_1, \mathbf{v}_2, \mathbf{v}_3\} = V$ *since* $\mathbf{v}_3 = 2\mathbf{v}_1 - \mathbf{v}_2$ (Theorem 1.3.19).

Hence the single vector $\{\mathbf{v}_1\}$ *is not enough to span* V. *The two vectors* $\{\mathbf{v}_1, \mathbf{v}_2\}$ *are good enough to span* V. *The three vectors* $\{\mathbf{v}_1, \mathbf{v}_2, \mathbf{v}_3\}$ *are more than enough to span* V*;* $\mathbf{v}_3$ *is redundant. Therefore two vectors* $\{\mathbf{v}_1, \mathbf{v}_2\}$ *are 'optimal' subset of* S *to span* V. *We will call such an optimal set of vectors to span a subsapce* V *a 'basis' of* V.

We formalize the notion 'optimal spanning set' in the following definition:

Definition 1.3.21. *Let* $V \subseteq \mathbb{R}^m$ *be a subspace. Let* $\mathcal{B} = \{\mathbf{v}_1, \mathbf{v}_2, ..., \mathbf{v}_n\}$ *be an ordered subset of* V. *We say* $\mathcal{B}$ *is a* **basis** *of* V *if* $\mathcal{B}$ *spans* V *and is linearly independent.*

In Example 1.3.10, $S = \{\mathbf{v}_1, \mathbf{v}_2\}$ is a basis for the null space $N(A)$. In fact, it is linearly independent by (1.46). And any solution to the equation (1.42) is a linear combination of S.

A basis is not just a subset of a subspace V. *It comes with a specific ordering.* Hence if $\mathcal{B} = \{\mathbf{v}_1, \mathbf{v}_2\}$ is a basis of V then $\mathcal{B}' = \{\mathbf{v}_2, \mathbf{v}_1\}$ is also a basis. But $\mathcal{B}$ is not the same as $\mathcal{B}'$.

Explicit conditions for $\mathcal{B} = \{\mathbf{v}_1, \mathbf{v}_2, ..., \mathbf{v}_n\} \subseteq V$ to be a basis of a subspace V are:

(1) if $\mathbf{b} \in V$ then we can find constants $c_1, c_2, \ldots, c_n$ such that $\mathbf{b} = \sum c_i \mathbf{v}_i$. In other words, the vector equation $x_1\mathbf{v}_1 + x_2\mathbf{v}_2 + \cdots + x_n\mathbf{v}_n = \mathbf{b}$ has a solution for any $\mathbf{b} \in V$.

(2) if $c_1\mathbf{v}_1 + c_2\mathbf{v}_2 + \cdots + c_n\mathbf{v}_n = \mathbf{0}$ then $c_1 = c_2 = \cdots = c_n = 0$. In other words, the equation $x_1\mathbf{v}_1 + x_2\mathbf{v}_2 + \cdots + x_n\mathbf{v}_n = \mathbf{0}$ has the only trivial solution.

Example 1.3.22. *Consider a reduced echelon form*

$$B = \begin{bmatrix} \boxed{1} & 2 & 0 & 0 & 0 & 1 \\ 0 & 0 & \boxed{1} & 0 & 1 & 2 \\ 0 & 0 & 0 & \boxed{1} & 0 & 1 \\ 0 & 0 & 0 & 0 & 0 & 0 \end{bmatrix}$$

Then $S = \{B_{(1)}, B_{(2)}, B_{(3)}\}$ *is a basis for* Row(B). *In fact, obviously* S *span* Row(B). *To see the linear independence of* S *let* $c_1B_{(1)} + c_2B_{(2)} + c_3B_{(3)} = \mathbf{0}$. *Then since* $c_1B_{(1)} + c_2B_{(2)} +$

$c_3B_{(3)} = [c_1, 2c_1, c_2, c_3, c_2, c_1 + 2c_2 + c_3]$ *the first, third and fourth places which corresponds to the pivot columns are* c_1, c_2, c_3*, they are all* 0.

Example 1.3.23. (1) *Let* $V = \{\mathbf{0}\}$ *be the trivial subspace of* $\mathbb{R}^n$. *Then since* Span $(\emptyset) = \{\mathbf{0}\}$ (Definition 1.2.6) *we see that* $\emptyset$ *span* V. *Further, the empty set is linearly independent* (Definition 1.3.14). *Hence the empty set is a basis for the trivial subspace.*

(2) *For* $n \geq 1$, *let* $\mathbf{e}^i = \begin{bmatrix} 0 \\ \vdots \\ 1 \\ \vdots \\ 0 \end{bmatrix} \in \mathbb{R}^n$ *be the column vector with* 1 *in the i-th place and zero elsewhere. Then it is easy to show that* $\mathcal{E}^n = \{\mathbf{e}^1, \mathbf{e}^2, ..., \mathbf{e}^n\}$ *is a basis of* $\mathbb{R}^n$ *which we call the* **standard basis** *of* $\mathbb{R}^n$.

Similarly, let $\mathbf{e}_i = [0, ..., 1, ..., 0]$ *be the row vector with* 1 *in the i-th place and zero elsewhere. Then* $\mathcal{E}_n = \{\mathbf{e}_1, \mathbf{e}_2, ..., \mathbf{e}_n\}$ *is a basis of* $\mathbb{R}_n$ *which we call the* **standard basis for the row space** $\mathbb{R}_n$.

Theorem 1.3.24. *A basis of* $\mathbb{R}^n$ *consists exactly n vectors. Let* $S = \{\mathbf{v}_1, \mathbf{v}_2, ..., \mathbf{v}_n\} \subseteq \mathbb{R}^n$ *and let* $A = [\mathbf{v}_1\ \mathbf{v}_2 \cdots \mathbf{v}_n]$. *Then the following conditions are equivalent.*

(1) *The reduced echelon form of* A *is* I_n.

(2) *The set of vectors* S *span* $\mathbb{R}^n$.

(3) *The set of vectors* S *is linearly independent.*

(4) *The set of vectors* S *is a basis of* $\mathbb{R}^n$.

Proof. The equivalence (1) and (2) is Theorem 1.3.8. The equivalence (1) and (3) is Theorem 1.3.16 and obviously (4) implies (2). But we know (2) $\Rightarrow$ (1) $\Rightarrow$ (3) and hence (2) $\Rightarrow$ (4). □

Example 1.3.25. *Let* $\mathcal{B} = \left\{ \begin{bmatrix} 1 \\ 1 \\ -1 \end{bmatrix}, \begin{bmatrix} 1 \\ -1 \\ 2 \end{bmatrix}, \begin{bmatrix} 1 \\ 1 \\ 1 \end{bmatrix} \right\}$. *To see if* $\mathcal{B}$ *is a basis of* $\mathbb{R}^3$ *consider the matrix* $A = \begin{bmatrix} 1 & 1 & 1 \\ 1 & -1 & 1 \\ -1 & 2 & 1 \end{bmatrix}$. *Now, the reduced echelon form for* A *is the identity matrix* $\begin{bmatrix} 1 & 0 & 0 \\ 0 & 1 & 0 \\ 0 & 0 & 1 \end{bmatrix}$. *Hence* $\mathcal{B}$ *is a basis for* $\mathbb{R}^3$.

Theorem 1.3.26. *Let* A *be an* $m \times n$ *matrix and* B *be the row reduced echelon form of* A.

(1) *The nonzero rows of* B *form a basis of* Row(A).

(2) *The pivot columns of* B *form a basis for* Col(B) (not for Col(A)).

Proof. (1) Since A, B are row equivalent we have Row(A) = Row(B) (Theorem 1.3.4). Hence the nonzero rows of B span Row(A). We need to prove that the nonzero rows $\{B_{(1)}, B_{(2)}, ..., B_{(k)}\}$ of B are linearly independent. Suppose $\sum_{i=1}^k c_i B_{(i)} = \mathbf{0}$ in $\mathbb{R}_m$ and $B_{(i)}$ contains the pivot position $B_{ij} = 1$. Then the j-th place of $\sum c_i B_{(i)}$ is c_i for $i = 1, 2, ..., k$ as in Example 1.3.22. And hence $c_i = 0$ for all i.

(2) The pivot columns of B are of the form $\mathbf{e}^i$. Hence the pivot columns of B are linearly independent. And the nonpivot columns of B are linear combinations of pivot columns of B. Hence the pivot columns of B span Col(B). Therefore the pivot columns form a basis for Col(B). □

For a basis of Col(A), instead of Col(B), see Theorem 2.3.6.

Theorem 1.3.27. *Let V be a subspace of $\mathbb{R}^m$. Let $\mathcal{B} = \{\mathbf{v}_1, \mathbf{v}_2, ..., \mathbf{v}_n\} \subseteq V$ and let $A = [\mathbf{v}_1\ \mathbf{v}_2 \cdots \mathbf{v}_n]$ be the matrix whose i-th column is $\mathbf{v}_i$. Then $\mathcal{B}$ is a basis of V if and only if for any $\mathbf{b} \in V$ the vector equation*

$$A\mathbf{x} = \mathbf{b}, \quad \mathbf{x} = \begin{bmatrix} x_1 \\ \vdots \\ x_n \end{bmatrix}$$

has a unique solution. Namely, every $\mathbf{b} \in V$ can be written uniquely as a linear combination of $\mathcal{B}$.

Proof. Suppose $\mathcal{B}$ is a basis of V. Let $\mathbf{b} \in V$ and consider the matrix equation $A\mathbf{x} = \mathbf{b}$. Then since $\mathcal{B}$ spans V we can write $\mathbf{b} = \sum c_i\mathbf{v}_i$. Hence $x_i = c_i$ $(i = 1, 2, ..., n)$ is a solution for the equation $A\mathbf{x} = \mathbf{b}$. Now suppose $\mathbf{c} = \begin{bmatrix} c_1 \\ \vdots \\ c_n \end{bmatrix}, \mathbf{c}' = \begin{bmatrix} c'_1 \\ \vdots \\ c'_n \end{bmatrix}$ are solutions of the equation $A\mathbf{x} = \mathbf{b}$. Then $A\mathbf{c} = \mathbf{b} = A\mathbf{c}'$ and hence $A(\mathbf{c} - \mathbf{c}') = \mathbf{0}$ by Theorem 1.2.12; $\sum(c_i - c'_i)\mathbf{v}_i = \mathbf{0}$. Since $\mathcal{B}$ is linearly independent we have $c_i = c'_i$; $\mathbf{c} = \mathbf{c}'$. Hence the solution is unique.

Conversely, assume that the equation $A\mathbf{x} = \mathbf{b}$ has a unique solution for any $\mathbf{b} \in V$. We have to show that the columns $\mathcal{B} = \{\mathbf{v}_i\}$ of A span V and linearly independent. To see $\mathcal{B}$ is linearly independent, by our assumption, $A\mathbf{x} = \mathbf{0}$ has a unique solution. Hence, by Theorem 1.3.16, the columns of A are linearly independent. To see $\mathcal{B}$ span V, let $\mathbf{b} \in V$. By our assumption $A\mathbf{x} = \mathbf{b}$ has a solution, say $\mathbf{x} = \mathbf{c} = \begin{bmatrix} c_1 \\ \vdots \\ c_n \end{bmatrix}$. Since $\mathcal{B}$ consists of the columns of A and $A\mathbf{c} = \mathbf{b}$, we conclude that $\sum c_i\mathbf{v}_i = \mathbf{b}$. Hence $\mathcal{B}$ span V. □

In Example 1.3.20, the subset $\{\mathbf{v}_1, \mathbf{v}_2\}$ of S is a basis of V. And the subset $\{\mathbf{v}_1, \mathbf{v}_3\}$ of S is also a basis. How about $S = \{\mathbf{v}_2, \mathbf{v}_3\}$?

Theorem 1.3.27, without mentioning the equations, can be rephrased as:

Theorem 1.3.28. *Let $V \subseteq \mathbb{R}^m$ be a subspace and $\mathcal{B} = \{\mathbf{v}_1, \mathbf{v}_2, ..., \mathbf{v}_n\} \subseteq V$. Then $\mathcal{B}$ is a basis of V if and only if every $\mathbf{v} \in V$ can be written uniquely as a linear combination of $\mathcal{B}$.*

Proof. Assume $\mathcal{B}$ is a basis. By definition, $\mathcal{B}$ spans V; for any $\mathbf{v} \in V$ there are constants $c_1, ..., c_n$ such that $\mathbf{v} = c_1\mathbf{v}_1 + \cdots + c_n\mathbf{v}_n$. To prove uniqueness, we need to show if $\mathbf{v} = d_1\mathbf{v}_1 + \cdots + d_n\mathbf{v}_n$ as well then $c_i = d_i$. In fact, $\mathbf{0} = \sum c_i\mathbf{v}_i - \sum d_i\mathbf{v}_i = \sum(c_i - d_i)\mathbf{v}_i$. Since $\mathcal{B}$ is linearly independent, $c_i - d_i = 0$; $c_i = d_i$.

Conversely, assume the condition holds. Since every $\mathbf{v} \in V$ is of the form $\mathbf{v} = c_1\mathbf{v}_1 + \cdots + c_n\mathbf{v}_n$, we see that $\mathcal{B}$ span V. To prove linear independence suppose $c_1\mathbf{v}_1 + \cdots + c_n\mathbf{v}_n = \mathbf{0}$. Then we can also write $\mathbf{0} = 0\mathbf{v}_1 + \cdots + 0\mathbf{v}_n$. By uniqueness, $c_i = 0$. □

Example 1.3.29. *Consider, again, the equation of* Example 1.3.10

$$\begin{cases} x_1 + x_2 + 2x_3 + x_4 = 0 \\ x_1 + 2x_2 - x_3 + x_4 = 0 \end{cases} \tag{1.51}$$

In Example 1.3.10, *we solved the equation for x_1, x_2 in terms of x_3, x_4 to show that $N(A) =$* Span $\{\mathbf{v}_1, \mathbf{v}_2\}$. *And* (1.46) *shows that $\{\mathbf{v}_1, \mathbf{v}_2\}$ is linearly independent. Hence $\{\mathbf{v}_1, \mathbf{v}_2\}$ is a basis of $N(A)$.*

Now we can solve for the other variables, say in the variables x_3, x_4 for instance: Subtract the second equation from the first equation and then add $\frac{1}{3}$ times the first equation to the second equation we get

$$\begin{cases} \quad -x_2 + 3x_3 \quad = 0 \\ x_1 + \frac{5}{3}x_2 \quad + x_4 = 0 \end{cases}$$

Hence the solution is

$$\begin{bmatrix} x_1 \\ x_2 \\ \frac{1}{3}x_2 \\ -x_1 - \frac{5}{3}x_2 \end{bmatrix} = x_1 \begin{bmatrix} 1 \\ 0 \\ 0 \\ -1 \end{bmatrix} + \frac{x_2}{3} \begin{bmatrix} 0 \\ 3 \\ 1 \\ -5 \end{bmatrix} \tag{1.52}$$

Hence any solution of the equation can be written as a linear combination of

$$\left\{ \mathbf{v}_1' = \begin{bmatrix} 1 \\ 0 \\ 0 \\ -1 \end{bmatrix}, \mathbf{v}_2' = \begin{bmatrix} 0 \\ 3 \\ 1 \\ -5 \end{bmatrix} \right\} \tag{1.53}$$

in a unique fashion. Therefore, $\{\mathbf{v}_1', \mathbf{v}_2'\}$ is also a basis of $N(A)$.

We have two different bases depending on our choice of free variables. However, what is common in the two bases are their numbers of elements. We will prove in Chapter 4 that every subspace has a basis (Theorem 4.1.2) and any bases contain the same number of elements. Henceforth, we will assume the following fact from Chapter 4.

Corollary 4.1.4. *Let $\mathcal{B}, \mathcal{B}'$ be bases of V consisting of finite vectors. Then $\mathcal{B}$ and $\mathcal{B}'$ have the same number of elements.*

Definition 1.3.30. *The number of elements of a basis of a subspace of V is called the* **dimension** *of V and we denote it by $\dim(V)$. The dimension of the null space $N(A)$ is called the* **nullity** *of A and is denoted by $\nu(A)$. The dimension of the column space $\mathrm{Col}(A)$ is called the* **rank** *of A and will be denoted by $\rho(A)$.*

$$\nu(A) = \dim N(A), \quad \rho(A) = \dim \mathrm{Col}(A)$$

We will also use the following facts deferring the proof until Chapter 4.

Theorem 4.1.12. *Let $V_1 \subseteq V_2$ be subspaces of $\mathbb{R}^n$. Then*

(1) $\dim(V_1) \leq \dim(V_2)$

(2) *If* $\dim(V_1) = \dim(V_2)$ *then* $V_1 = V_2$.

Dimension is very useful tool. Here is an illustration of usefulness of the concept dimension. In Example 1.2.16, we showed that the subspaces of $\mathbb{R}^2$ are the trivial subspace, line through the origin and $\mathbb{R}^2$. For this, we made some computations. Here is an easy way to see this using dimension.

Example 1.3.31. *Let V be a subspace of $\mathbb{R}^2$. Then the dimension of V is 0 or 1 or 2. If $\dim V = 0$ then $V = \{\mathbf{0}\}$. If $\dim V = 1$ and if $\{\mathbf{v}\}$ is a basis then V is a line containing $\mathbf{v}$ passing through the origin. If $\dim V = 2$ then $V = \mathbb{R}^2$ by* Theorem 4.1.12 *since $\dim \mathbb{R}^2 = 2$.*

Now, we will show that the nullity of a matrix A is the same as the number of nonpivot columns A which corresponds to the free variables for the system of homogeneous system of linear equations $A\mathbf{x} = \mathbf{0}$. First, we illustrate this before we give a formal proof.

Example 1.3.32. *Let A be a 3×5 matrix and consider a homogeneous equation $A\mathbf{x} = \mathbf{0}$. Suppose the reduced echelon form A' of A is*

$$A' := \begin{bmatrix} 1 & 2 & 0 & -1 & 4 \\ 0 & 0 & 1 & 2 & -5 \\ 0 & 0 & 0 & 0 & 0 \end{bmatrix}$$

The nonpivot columns are second, fourth and fifth columns and the solutions are given by

$$x_1 = -2x_2 + x_4 - 4x_5, \quad x_3 = -2x_4 + 5x_5 \tag{1.54}$$

and x_2, x_3, x_5 are free variables corresponding to the nonpivot columns. Hence the general solutions will be

$$\begin{bmatrix} x_1 \\ x_2 \\ x_3 \\ x_4 \\ x_5 \end{bmatrix} = \begin{bmatrix} -2x_2 + x_4 - 4x_5 \\ x_2 \\ -2x_4 + 5x_5 \\ x_4 \\ x_5 \end{bmatrix} = x_2 \begin{bmatrix} -2 \\ 1 \\ 0 \\ 0 \\ 0 \end{bmatrix} + x_4 \begin{bmatrix} 1 \\ 0 \\ 0 \\ 1 \\ 0 \end{bmatrix} + x_5 \begin{bmatrix} -4 \\ 0 \\ 5 \\ 0 \\ 1 \end{bmatrix} \tag{1.55}$$

If we let

$$S = \left\{ \mathbf{v}_2 = \begin{bmatrix} -2 \\ 1 \\ 0 \\ 0 \\ 0 \end{bmatrix}, \mathbf{v}_4 = \begin{bmatrix} 1 \\ 0 \\ 0 \\ 1 \\ 0 \end{bmatrix}, \mathbf{v}_5 = \begin{bmatrix} -4 \\ 0 \\ 5 \\ 0 \\ 1 \end{bmatrix} \right\} \tag{1.56}$$

then the equality (1.55) shows that S span $N(A)$. To see linear independence of S suppose $c_2\mathbf{v}_2 + c_4\mathbf{v}_4 + c_5\mathbf{v}_5 = \mathbf{0}$. Then the second, fourth and fifth place (corresponding to the free variables x_2, x_4, x_5) of $c_2\mathbf{v}_2 + c_4\mathbf{v}_4 + c_5\mathbf{v}_5$ are c_2, c_4 and c_5 respectively. Hence they are equal to 0. Therefore, S is linearly independent.

Theorem 1.3.33. *Let A be an $m\times n$ matrix. The nullity of A is the same as the number of nonpivot columns and it is the same as the number of free variables of the solutions of the homogeneous equation $A\mathbf{x} = \mathbf{0}$.*

Proof. For the proof bear in mind Example 1.3.32. Let $\mathbf{x} = \begin{bmatrix} x_1 \\ \vdots \\ x_n \end{bmatrix}$ be a solution for $A\mathbf{x} = \mathbf{0}$ and let

$$I = \{i \mid x_i \text{ is a free variable}\}, \quad J = \{j \mid x_j \text{ is a nonfree variable}\}$$

Hence if $j \in J$ then, by considering the reduced echelon form of A, we can write $x_j = \sum_{i\in I} a_{ij}x_i$ as in (1.54). For each $i \in I$, define $\mathbf{v}_i \in \mathbb{R}^n$ by

$$(k\text{-th place of } \mathbf{v}_i) = \begin{cases} a_{ik} & (k \in J) \\ 1 & (i = k) \\ 0 & \text{otherwise} \end{cases} \tag{1.57}$$

as in (1.56). It is obvious that $S = \{\mathbf{v}_i \mid i \in I\}$ spans $N(A)$ as in (1.55). We want to show that S is also linearly independent. Suppose $\sum a_i\mathbf{v}_i = \mathbf{0}$. Then the k-th place of $\sum a_i\mathbf{v}_i$ is a_i by (1.57). Hence $a_i = 0$ for all i. Thus S is a basis of $N(A)$. □

Corollary 1.3.34. *With the same notations of the theorem, let I and J be the indices of free and nonfree variables. Then we can choose a basis $\{\mathbf{v}_i \mid i \in I\}$ for $N(A)$ by* (1.57)

We will show that the rank of A is the same as the number of pivot columns of the (reduced) echelon form (Theorem 2.3.6).

Example 1.3.35. *We look for a basis for the null space $N(A)$ of $A = \begin{bmatrix} 2 & -1 & 1 & 0 & 1 \\ 0 & 1 & 1 & -2 & 0 \\ 1 & 0 & -1 & 1 & 2 \end{bmatrix}$. To ease the computation we have chosen the matrix we have used in* Example 1.3.13 (2). *The reduced echelon form of A is $\begin{bmatrix} 1 & 0 & 0 & 0 & \frac{5}{4} \\ 0 & 1 & 0 & -1 & \frac{3}{4} \\ 0 & 0 & 1 & -1 & -\frac{3}{4} \end{bmatrix}$. Hence the solutions for the homogeneous equation $A\mathbf{x} = \mathbf{0}$ is $s \begin{bmatrix} 0 \\ 1 \\ 1 \\ 1 \\ 0 \end{bmatrix} + t \begin{bmatrix} -5 \\ -3 \\ 3 \\ 0 \\ 4 \end{bmatrix}$. Thus if we let $\mathbf{v}_1 = \begin{bmatrix} 0 \\ 1 \\ 1 \\ 1 \\ 0 \end{bmatrix}$, $\mathbf{v}_2 = \begin{bmatrix} -5 \\ -3 \\ 3 \\ 0 \\ 4 \end{bmatrix}$ then $N(A)$ is spanned by $S = \{\mathbf{v}_1, \mathbf{v}_2\}$. To see S is linearly independent, observe that the 4-th and 5-th positions, which corresponds to the nonpivot columns of A, of $s\mathbf{v}_1 + t\mathbf{v}_2$ are $s, 4t$. Hence $s = t = 0$ if $s\mathbf{v}_1 + t\mathbf{v}_2 = \mathbf{0}$.*

We summarize the conditions for a set of vectors to be linearly independent, span $\mathbb{R}^m$ and a basis.

Theorem 1.3.36. *Let $\{\mathbf{v}_1, \mathbf{v}_2, ..., \mathbf{v}_n\}$ be a set of vectors in $\mathbb{R}^m$. Let A be the $m \times n$ matrix $A = [\mathbf{v}_1\ \mathbf{v}_2 \cdots\ \mathbf{v}_n]$. Let B be the reduced echelon form of A.*

(1) *The following conditions are equivalent.*

  (a) *The columns of A span $\mathbb{R}^m$.*

  (b) *For any $\mathbf{b} \in \mathbb{R}^m$ the equation $A\mathbf{x} = \mathbf{b}$ has a solution.*

  (c) *Every row of A has a pivot position, namely B is of the form*

$$B = \begin{bmatrix} 1 & * & 0 & * & 0 \\ 0 & 0 & 1 & * & 0 \\ 0 & 0 & 0 & 0 & 1 \end{bmatrix} \tag{1.58}$$

Explicitly, $m \leq n$ and B has no row entirely comprised of 0's.

(2) *The following conditions are equivalent.*

  (a) *The columns of A are linearly independent.*

  (b) *The equation $A\mathbf{x} = \mathbf{0}$ has only the trivial solution.*

  (c) *Every column of A is a pivot column, namely B is of the form*

$$B = \begin{bmatrix} 1 & 0 & 0 & 0 \\ 0 & 1 & 0 & 0 \\ 0 & 0 & \ddots & 0 \\ 0 & 0 & 0 & 1 \\ 0 & 0 & 0 & 0 \\ & \cdots & \cdots & \\ 0 & 0 & 0 & 0 \end{bmatrix} \tag{1.59}$$

Explicitly, $m \geq n$ *and* $B_{ij} = \begin{cases} 0 & (i > j) \\ 0 & (i = j) \end{cases}$.

(3) *If* $m = n$ *then following conditions are equivalent.*

(a) *The columns of* A *span* $\mathbb{R}^n$ *i.e., the equation* $A\mathbf{x} = \mathbf{b}$ *has a solution for any* $\mathbf{b} \in \mathbb{R}^n$.

(b) *The columns of* A *are linearly independent i.e.,* $A\mathbf{x} = \mathbf{0}$ *has only the trivial solution.*

(c) *The columns of* A *form a basis of* $\mathbb{R}^n$ *i.e., the equation* $A\mathbf{x} = \mathbf{b}$ *has a unique solution for any* $\mathbf{b} \in \mathbb{R}^n$.

(d) *The reduced echelon form* B *of* A *is* I_n.

$$B = \begin{bmatrix} 1 & 0 & 0 & 0 \\ 0 & 1 & 0 & 0 \\ 0 & 0 & \ddots & 0 \\ 0 & 0 & 0 & 1 \end{bmatrix}$$

Proof. (1) Theorem 1.3.8
(2) Theorem 1.3.16
(3) Theorem 1.3.24. □

Practice Problem 1.3.4. *Decide whether the columns of the following matrices are linearly independent or span* $\mathbb{R}^m$ *or form a basis for* $\mathbb{R}^m$ *where* m *is the number of rows.*

(1) $A = \begin{bmatrix} 2 & -1 & 1 & 0 & 1 \\ 0 & 1 & 1 & -2 & 0 \\ 1 & 0 & -1 & 1 & 2 \end{bmatrix}$ (2) $A = \begin{bmatrix} 1 & 0 & 1 \\ 0 & 1 & 0 \\ 1 & 0 & -1 \end{bmatrix}$ (3) $A = \begin{bmatrix} 1 & 0 & 1 \\ 0 & 1 & 1 \\ 1 & 0 & 1 \\ 1 & 1 & 2 \end{bmatrix}$

Answers for Practice Problems 1.3

(1.3.1) $\mathbf{v} = (1 + \frac{c}{3})\mathbf{v}_1 - \frac{4c}{3}\mathbf{v}_2 + (1 - \frac{2c}{3})\mathbf{v}_3 + c\mathbf{v}_4$ for any scalar c. If we take $c = 0$ then $\mathbf{v} = \mathbf{v}_1 + \mathbf{v}_3$ and of we take $c = 1$ then $\mathbf{v} = \frac{4}{3}\mathbf{v}_1 - \frac{4}{3}\mathbf{v}_2 + \frac{1}{3}\mathbf{v}_3 + \mathbf{v}_4$.

(1.3.2) The solutions for the equations are $t \begin{bmatrix} -2 \\ 5 \\ -4 \\ 6 \end{bmatrix}$ $(t \in \mathbb{R})$.

(1.3.3) A particular solution is $\mathbf{p} = \begin{bmatrix} 3 \\ 1 \\ -5 \\ 1 \\ 0 \end{bmatrix}$; $s \begin{bmatrix} -1 \\ 1 \\ -2 \\ 2 \\ 0 \end{bmatrix} + t \begin{bmatrix} 3 \\ 1 \\ -4 \\ 0 \\ 2 \end{bmatrix}$ $(s, t \in \mathbb{R})$ are the solutions for homogeneous equation.

(1.3.4) (1) The reduced echelon form of A is $\begin{bmatrix} 1 & 0 & 0 & 0 & \frac{5}{4} \\ 0 & 1 & 0 & -1 & \frac{3}{4} \\ 0 & 0 & 1 & -1 & -\frac{3}{4} \end{bmatrix}$. Hence the columns of A span $\mathbb{R}^3$ but they are linearly dependent by Theorem 1.3.36.

(2) The reduced echelon form of A is $\begin{bmatrix} 1 & 0 & 0 \\ 0 & 1 & 0 \\ 0 & 0 & 1 \end{bmatrix}$. By Theorem 1.3.36, the columns of A is a basis of $\mathbb{R}^3$.

(3) The reduced echelon form of A is $\begin{bmatrix} 1 & 0 & 1 \\ 0 & 1 & 1 \\ 0 & 0 & 0 \\ 0 & 0 & 0 \end{bmatrix}$. By Theorem 1.3.36, the columns of A is not linearly independent nor span $\mathbb{R}^4$ neither.

1.3 Exercises

1. Let $A = \begin{bmatrix} 1 & -1 & 0 \\ 0 & 2 & 3 \\ 2 & 0 & 3 \end{bmatrix}$. Decide whether the following vectors belong to the column space $\mathrm{Col}(A)$ of A. If so write it as a linear combination of the columns of A.

   (a) $\mathbf{b}_1 = \begin{bmatrix} 1 \\ 1 \\ 0 \end{bmatrix}$ (b) $\mathbf{b}_2 = \begin{bmatrix} 0 \\ -1 \\ -1 \end{bmatrix}$ (c) $\mathbf{b}_3 = \begin{bmatrix} 1 \\ 1 \\ 1 \end{bmatrix}$ (d) $\mathbf{b}_4 = \begin{bmatrix} 1 \\ 5 \\ 7 \end{bmatrix}$ (e) $\mathbf{b}_5 = \begin{bmatrix} 1 \\ 0 \\ 0 \end{bmatrix}$

2. Prove that $\left\{ \mathbf{v}_1 = \begin{bmatrix} a \\ b \end{bmatrix}, \mathbf{v}_2 = \begin{bmatrix} c \\ d \end{bmatrix} \right\}$ is a basis of $\mathbb{R}^2$ if and only if $ad - bc \neq 0$.

3. Let $\mathcal{B} = \{\mathbf{v}_1, \mathbf{v}_2\}$ be a basis of a subspace V of $\mathbb{R}^m$. Prove that

   (a) $\mathcal{B}' = \{\mathbf{v}_1 - \mathbf{v}_2, \mathbf{v}_1 + 2\mathbf{v}_2\}$ is a basis of V.

   (b) Prove that $\{a\mathbf{v}_1, c\mathbf{v}_1 + d\mathbf{v}_2\}$ is a basis of V if and only if $ad \neq 0$ i.e., $a \neq 0$ and $d \neq 0$.

   (c) Find a necessary and sufficient condition on a, b, c, d for $\{a\mathbf{v}_1 + b\mathbf{v}_2, c\mathbf{v}_1 + d\mathbf{v}_2\}$ to be a basis of V.

4. Decide whether the following sets S is a basis for the given subspace V.

   (a) $S = \left\{ \begin{bmatrix} 1 \\ 1 \\ 1 \end{bmatrix}, \begin{bmatrix} 1 \\ 0 \\ -1 \end{bmatrix}, \begin{bmatrix} 0 \\ 1 \\ 2 \end{bmatrix} \right\}, V = \mathbb{R}^3$

   (b) $S = \left\{ \begin{bmatrix} 2 \\ 3 \\ 1 \\ 2 \end{bmatrix}, \begin{bmatrix} 1 \\ 0 \\ 0 \\ 2 \end{bmatrix}, \begin{bmatrix} 3 \\ 1 \\ 2 \\ 6 \end{bmatrix} \right\}, V = \mathbb{R}^4$

   (c) $S = \left\{ \begin{bmatrix} 2 \\ 3 \\ 1 \end{bmatrix}, \begin{bmatrix} 1 \\ 0 \\ 2 \end{bmatrix}, \begin{bmatrix} 3 \\ 1 \\ 2 \end{bmatrix} \right\}, V = \mathbb{R}^3$

   (d) $S = \left\{ \begin{bmatrix} 1 \\ 1 \\ 1 \\ 1 \end{bmatrix}, \begin{bmatrix} 0 \\ 1 \\ 0 \\ -1 \end{bmatrix}, \begin{bmatrix} -1 \\ 0 \\ 1 \\ 2 \end{bmatrix}, \begin{bmatrix} 0 \\ 1 \\ 1 \\ 2 \end{bmatrix} \right\}, V = \mathbb{R}^4$

(e) $S = \left\{ \begin{bmatrix} 1 \\ 0 \\ 1 \\ 0 \end{bmatrix}, \begin{bmatrix} 1 \\ 1 \\ 3 \\ 1 \end{bmatrix}, \begin{bmatrix} 0 \\ 1 \\ 2 \\ 1 \end{bmatrix} \right\}, V = \left\{ \begin{bmatrix} x \\ y \\ z \\ w \end{bmatrix} \middle| \; x + y - z - w = 0 \right\}$

(f) $S = \left\{ \begin{bmatrix} 1 \\ 1 \\ 0 \\ 3 \end{bmatrix}, \begin{bmatrix} 0 \\ 1 \\ 1 \\ 0 \end{bmatrix}, \begin{bmatrix} 1 \\ 0 \\ 1 \\ 1 \end{bmatrix} \right\}, V = \left\{ \begin{bmatrix} x \\ y \\ z \\ w \end{bmatrix} \middle| \; 2x + y - z = w \right\}$

5. Find a basis for the following subspaces:

(a) $V = \left\{ \begin{bmatrix} x \\ y \\ z \end{bmatrix} \middle| \; 2x - y + 3z = 0 \right\}$

(b) $V = \left\{ \begin{bmatrix} x \\ y \\ z \end{bmatrix} \middle| \; \begin{array}{l} 2x - y = 0 \\ x - 2z = 0 \end{array} \right\}$

(c) $V = \left\{ \begin{bmatrix} x \\ y \\ z \\ w \end{bmatrix} \middle| \; -2x + y + 3z = w \right\}$

(d) $V = \left\{ \begin{bmatrix} x \\ y \\ z \\ w \end{bmatrix} \middle| \; \begin{array}{lllll} 2x & -y & +z & -2w & = 0 \\ -x & +2y & +z & -w & = 0 \end{array} \right\}$

(e) $V = \left\{ \begin{bmatrix} x \\ y \\ z \\ w \end{bmatrix} \middle| \; \begin{array}{llll} x & +y & +z & = 0 \\ y & +z & +w & = 0 \\ z & -w & & = 0 \end{array} \right\}$

(f) $V = \left\{ \begin{bmatrix} x \\ y \\ z \\ w \end{bmatrix} \middle| \; \begin{array}{lll} x & +y & = 0 \\ y & +z & = 0 \\ z & -w & = 0 \end{array} \right\}$

6. Prove the true statement(s) and provide a counter example for the false statement(s).

(a) If V is a subspace of $\mathbb{R}^5$ then the number of vectors of a basis of V does not exceed 5.

(b) If S is a linearly dependent set of nonzero vectors in $\mathbb{R}^4$ then the number of elements of S is bigger than 4.

(c) If S is a set of vectors which spans $\mathbb{R}^3$ then S contains at least three vectors.

(d) Let A be an 3×5 matrix with two pivot positions. Then $\nu(A) = 3$.

(e) The set of solutions of a system of linear equations, homogeneous or not, in n variables is a subspace.

(f) A system of homogeneous equation may have no solution.

(g) There is a system of linear equations having exactly two solutions.

(h) If the number of equations of a system of linear equation is less than the number of variables then the system has at least one solution.

(i) If A and B are row equivalent then $\mathrm{Col}(A) = \mathrm{Col}(B)$.

7. Decide whether the following column vectors of A are linearly independent or not.

(a) $\begin{bmatrix} -1 & 1 \\ 1 & -1 \end{bmatrix}$ (b) $\begin{bmatrix} 1 & -1 & 0 \\ 0 & 0 & 1 \\ 1 & 1 & 1 \end{bmatrix}$ (c) $\begin{bmatrix} -1 & -1 & 0 \\ 0 & 0 & 1 \\ 1 & 1 & 1 \end{bmatrix}$

(d) $\begin{bmatrix} 1 & 0 & 1 & 1 \\ 0 & 1 & 1 & 1 \\ 1 & 1 & 1 & 0 \\ 1 & 1 & 0 & 1 \end{bmatrix}$ (e) $\begin{bmatrix} 1 & 0 & 1 & 1 \\ 1 & 1 & 1 & 0 \\ 1 & 1 & 0 & 1 \\ -1 & 0 & 0 & 1 \end{bmatrix}$ (f) $\begin{bmatrix} 1 & 0 & 1 & 1 \\ 0 & 1 & 1 & 1 \\ 1 & 1 & 1 & 0 \\ -1 & 0 & 0 & 1 \end{bmatrix}$

8. Decide whether the following column vectors of A are linearly independent or spans $\mathbb{R}^m$ or a basis of $\mathbb{R}^m$ where m is the number of rows.

(a) $\begin{bmatrix} -1 & 1 \\ 1 & 1 \end{bmatrix}$ (b) $\begin{bmatrix} 1 & 1 & 0 \\ 0 & 0 & 1 \\ 1 & 1 & 1 \end{bmatrix}$ (c) $\begin{bmatrix} -1 & -1 & 0 \\ 0 & 0 & 1 \\ 1 & -1 & 1 \end{bmatrix}$

(d) $A = \begin{bmatrix} 3 & 2 & 2 & 1 & -2 \\ 1 & 2 & 0 & 1 & 0 \\ 1 & 0 & 1 & 0 & -1 \\ 2 & 0 & 2 & 0 & -2 \end{bmatrix}$ (e) $A = \begin{bmatrix} 1 & 1 & 0 & 1 \\ 1 & 0 & 1 & 1 \\ 0 & -1 & 1 & 0 \\ 0 & 1 & 1 & 1 \end{bmatrix}$ (f) $A = \begin{bmatrix} 1 & 0 & 1 & 1 \\ 0 & 1 & 1 & 1 \\ 1 & 1 & 1 & 0 \\ 1 & 1 & 0 & 1 \\ -1 & 0 & 0 & 1 \end{bmatrix}$

9. Let A, B be $m \times n$ matrices. Prove that if $\mathrm{Row}(A) = \mathrm{Row}(B)$ then A and B are row equivalent.

10. Let $\mathbf{v}_i \in \mathbb{R}^n$ and $V_i = \mathrm{Span}\,\{\mathbf{v}_i\}$ $(i = 1, 2)$. Prove that $\mathrm{Span}\,\{\mathbf{v}_1, \mathbf{v}_2\} = \mathrm{Span}\,(V_1 \cup V_2)$.

Chapter 2

Matrices and Linear Maps on $\mathbb{R}^n$

In Chapter 1, we saw that the matrices are useful tool to solve linear equations. In this chapter, we study various aspects of matrices: We associate a matrix a function which we call a linear map and we define operations of matrices and they form a 'vector space'.

To each matrix we associate the subspaces called, the column space and the null space and we relate the dimensions of these two subspaces.

And we study the matrices with special properties namely invertible matrices. For this purpose, we apply the method we used to solve a system of equations.

2.1 Matrices and Linear Maps on $\mathbb{R}^n$

2.1.1 Operations of Matrices

Recall some notations from Chapter 1 (Definition 1.2.2). Let A be an $m{\times}n$ matrix

$$A = \begin{bmatrix} a_{11} & a_{12} & \cdots & a_{1n} \\ a_{21} & a_{22} & \cdots & a_{2n} \\ & \cdots & \cdots & \\ a_{m1} & a_{m2} & \cdots & a_{mn} \end{bmatrix} = [a_{ij}] \tag{2.1}$$

The set of all $m{\times}n$ matrices will be denoted by $M(m{\times}n)$. If $m = n$ then we will simply write $M(n)$ for $M(n{\times}n)$. The (i, j)-th position of A is denoted by A_{ij}. If A, B are $m{\times}n$ matrices then $A = B$ if and only if $A_{ij} = B_{ij}$ for all i, j. The i-th column and j-th row of A are denoted by $A^{(i)}, A_{(j)}$

$$A^{(i)} = \begin{bmatrix} a_{1i} \\ a_{2i} \\ \vdots \\ a_{mi} \end{bmatrix}, \quad A_{(j)} = [\, a_{j1}, a_{j2}, ..., a_{jm} \,] \quad \begin{pmatrix} i = 1, 2, ..., n \\ j = 1, 2, ..., m \end{pmatrix} \tag{2.2}$$

We denote the set of columns by $C(A)$

$$C(A) = \{A^{(1)}, A^{(2)}, ..., A^{(n)}\}$$

With this notation, we can write $\mathrm{Col}(A) = \mathrm{Span}\ C(A)$. The columns $A^{(i)}$ of A can be viewed as matrices of size $m{\times}1$ and the row vectors $A_{(j)}$ as $1{\times}n$ matrices. If $\mathbf{a}_i$ is an $n{\times}1$ column vector and $A^{(i)} = \mathbf{v}_i$ then we also write $A = [\mathbf{v}_1\ \mathbf{v}_2 \cdots\ \mathbf{v}_n]$.

Definition 2.1.1. *An $n \times n$ matrix A is said to be* **upper triangular** (*resp.* **strictly upper triangular**) *if $A_{ij} = 0$ for $i > j$ (resp. $i \geq j$). We say A is* **lower triangular** (*resp.* **strictly lower triangular**) *if $A_{ij} = 0$ for $i < j$ (resp. $i \leq j$). An upper or lower (strict or not) triangular matrix will be called a* **triangular matrix**. *A* **diagonal matrix** *is a square matrix with only possible nonzero entries are the diagonal entries which we also denote by* $\mathrm{diag}[a_{11}, a_{22}, ..., a_{nn}]$.

$$\begin{bmatrix} a_{11} & 0 & \cdots & 0 \\ 0 & a_{22} & \cdots & 0 \\ \vdots & \vdots & \ddots & \vdots \\ 0 & \cdots & 0 & a_{nn} \end{bmatrix} = \mathrm{diag}[a_{11}, a_{22}, ..., a_{nn}] \tag{2.3}$$

The **identity matrix**, *denoted by I_n, is an $n \times n$ diagonal matrix with 1's on the diagonal and 0 elsewhere.*

$$I_n = \begin{bmatrix} 1 & 0 & \cdots & 0 \\ 0 & 1 & \cdots & 0 \\ \vdots & \vdots & \ddots & \vdots \\ 0 & \cdots & 0 & 1 \end{bmatrix} \tag{2.4}$$

A matrix with zeros everywhere is called the **zero matrix** *which we denote by* 0.

$$0 = \begin{bmatrix} 0 & 0 & \cdots & 0 \\ 0 & 0 & \cdots & 0 \\ \vdots & \vdots & \ddots & \vdots \\ 0 & \cdots & 0 & 0 \end{bmatrix} \tag{2.5}$$

The set of all $m \times n$ matrices will be denoted by $M(m \times n)$. When $n = m$, the square matrices of size n, will be denoted by $M(n)$ for $M(n \times n)$.

$$M(m \times n) = \{A \mid A \text{ is an } m \times n \text{ matrix}\}$$
$$M(n) = \{A \mid A \text{ is an } n \times n \text{ matrix}\}$$

Now, we define the addition and scalar multiplication on the set of $m \times n$ matrices.

Definition 2.1.2. *Let $A = [a_{ij}], B = [b_{ij}]$ be $m \times n$ matrices and c be a scalar. We define the* **addition** $A + B$ *and* **scalar multiplication** cA *to be the $m \times n$ matrices defined by*

$$(A + B)_{ij} = a_{ij} + b_{ij} \text{ and } \quad (cA)_{ij} = ca_{ij} \tag{2.6}$$

Example 2.1.3. *Let* $A = \begin{bmatrix} 2 & -3 & -1 \\ 0 & 1 & -1 \end{bmatrix}, B = \begin{bmatrix} 3 & 1 & 0 \\ 1 & 0 & 1 \end{bmatrix}$. *Find $A + B, 3A$.*

Theorem 2.1.4. *Let A, B, C be $m \times n$ matrices and r, s be scalars. Then we have*

(1) $A + B = B + A$ (2) $(A + B) + C = A + (B + C)$

(3) $A + 0 = A$ (4) $r(A + B) = rA + rB$

(5) $(r + s)A = rA + sA$ (6) $r(sA) = (rs)A$

Proof. To prove (a), for instance, we need to show that both $A + B$ and $B + A$ have the same (i, j) entries. Now $(A + B)_{ij} = A_{ij} + B_{ij}$ and $(B + A)_{ij} = B_{ij} + A_{ij}$. These are equal since addition of real numbers is commutative. The proofs of the others are exercises. □

Practice Problem 2.1.1. *Prove $(A + B) + C = A + (B + C)$.*

2.1.2 Matrices and Linear Maps on $\mathbb{R}^n$

A function f from $\mathbb{R}^n$ to $\mathbb{R}^m$ is a rule assigning a vector $f(\mathbf{x}) = \mathbf{y} \in \mathbb{R}^m$ to each $\mathbf{x} \in \mathbb{R}^n$. In symbol, we write $f : \mathbb{R}^n \to \mathbb{R}^m$. For general definition of the functions see Definition A.1.1. We will study some special kind of functions so called the 'linear functions'.

Definition 2.1.5. *A function $T : \mathbb{R}^n \to \mathbb{R}^m$ is called a* **linear map (transformation, mapping, function)** *if T satisfies, for $\mathbf{u}, \mathbf{v} \in \mathbb{R}^n$ and $c \in \mathbb{R}$*

$$T(\mathbf{u} + \mathbf{v}) = T(\mathbf{u}) + T(\mathbf{v}) \quad \text{and} \quad T(c\mathbf{u}) = cT(\mathbf{u}) \tag{2.7}$$

The set of all linear maps from $\mathbb{R}^n$ to $\mathbb{R}^m$ will be denoted by $L(n, m)$.

$$L(n, m) = \{T \mid T : \mathbb{R}^n \to \mathbb{R}^m \text{ linear}\}$$

The set of all images $\{T(\mathbf{x}) \in \mathbb{R}^m \mid \mathbf{x} \in \mathbb{R}^n\}$ is called the **range space** *or* **image** *of T and we denote by*

$$R(T) = \{T(\mathbf{x}) \in \mathbb{R}^m \mid \mathbf{x} \in \mathbb{R}^n\}$$

The **null space** *$N(T)$ of T is the defined by*

$$N(T) = \{\mathbf{x} \in \mathbb{R}^n \mid T(\mathbf{x}) = \mathbf{0}\}$$

A bijective linear map $T : \mathbb{R}^n \to \mathbb{R}^n$ is called an **isomorphism**.

If $T : \mathbb{R}^n \to \mathbb{R}^m$ is a linear map then

$$T(\mathbf{0}) = \mathbf{0}$$

In fact, $T(\mathbf{0}) = T(\mathbf{0} + \mathbf{0}) = T(\mathbf{0}) + T(\mathbf{0})$. Letting $T(\mathbf{0}) = \begin{bmatrix} u_1 \\ \vdots \\ u_m \end{bmatrix}$ we get $\begin{bmatrix} u_1 \\ \vdots \\ u_m \end{bmatrix} = \begin{bmatrix} 2u_1 \\ \vdots \\ 2u_m \end{bmatrix}$.

Hence $u_i = 0$ for all i. Therefore, $T(\mathbf{0}) = \mathbf{0}$.

By repeated application of (2.7), we have

$$T(c_1\mathbf{v}_1 + \cdots + c_p\mathbf{v}_p) = c_1T(\mathbf{v}_1) + \cdots + c_pT(\mathbf{v}_p)$$

Now, we consider geometric picture of linear maps. The line through the origin containing $\mathbf{v}$ is defined by

$$L(\mathbf{v}) = \{c\mathbf{v} \mid c \in \mathbb{R}\}$$

A linear map $T : \mathbb{R}^3 \to \mathbb{R}^3$ transforms the line $L(\mathbf{v})$ to the same kind

$$T(L(\mathbf{v})) = \{cT(\mathbf{v}) \mid c \in \mathbb{R}\}$$

namely the line passing through the origin containing $T(\mathbf{v})$.

The plane through the origin containing $\mathbf{u}, \mathbf{v} \in \mathbb{R}^3$ can be described by

$$P(\mathbf{u}, \mathbf{v}) := \{c\mathbf{u} + d\mathbf{v} \mid c, d \in \mathbb{R}\}$$

and a linear map T transforms $P(\mathbf{u}, \mathbf{v})$ into the same kind $P(T(\mathbf{u}), T(\mathbf{v}))$.

A linear map T transforms a line segment $\{t\mathbf{u} + (1-t)\mathbf{v} \mid 0 \le t \le 1\}$ joining two points $\mathbf{u}, \mathbf{v} \in \mathbb{R}^3$ into the line $tT(\mathbf{u}) + (1-t)T(\mathbf{v})$ joining $T(\mathbf{u}), T(\mathbf{v})$.

The **identity map** $I_{\mathbb{R}^n} : \mathbb{R}^n \to \mathbb{R}^n$ defined by $I_{\mathbb{R}^n}(\mathbf{v}) = \mathbf{v}$ is a linear map. The **zero map** $0 : \mathbb{R}^n \to \mathbb{R}^m$ defined by $0(\mathbf{v}) = \mathbf{0}$ is a linear map. The map $\tau : \mathbb{R}^n \to \mathbb{R}_n$ assigning the transpose $\mathbf{u}^t \in \mathbb{R}_n$ to each column vector $\mathbf{u} \in \mathbb{R}^n$ is a bijective liner map, an 'isomorphism'.

It will turn out that the linear maps are rather special kinds of functions.

Example 2.1.6. (1) *A linear map from $\mathbb{R}$ into itself is a multiplication by some scalar. In fact, if $T : \mathbb{R} \to \mathbb{R}$ is a linear map and let $T(1) = a \in \mathbb{R}$. Then $T(x) = T(x \cdot 1) = xT(1) = xa$. Hence T is the multiplication-by-a map:*

$$T(x) = ax \tag{2.8}$$

Now consider a linear map $T : \mathbb{R}^2 \to \mathbb{R}$. Let $\mathbf{e}^1 = \begin{bmatrix} 1 \\ 0 \end{bmatrix}, \mathbf{e}^2 = \begin{bmatrix} 0 \\ 1 \end{bmatrix}$ and $T(\mathbf{e}^1) = a$, $T(\mathbf{e}^2) = b$. If $\mathbf{x} = \begin{bmatrix} x \\ y \end{bmatrix}$ then $\mathbf{x} = x\mathbf{e}^1 + y\mathbf{e}^2$. Hence, by linearity of T, we have $T(\mathbf{x}) = T(x\mathbf{e}^1 + y\mathbf{e}^2) = xT(\mathbf{e}^1) + yT(\mathbf{e}^2) = ax + by$. Hence

$$T(\mathbf{x}) = ax + by \tag{2.9}$$

where $a = T(\mathbf{e}^1), b = T(\mathbf{e}^2)$. If we let $A = [\,a\ b\,]$ be a $1{\times}2$ matrix then $T(\mathbf{x}) = A\mathbf{x}$.

(2) *Let $A \in M(m{\times}n)$. Then the map $T_A : \mathbb{R}^n \to \mathbb{R}^m$ defined by $T_A(\mathbf{x}) = A\mathbf{x}$ is a linear map by* Theorem 1.2.12.

Recall (Definiton 1.2.10) that a nonempty subset V of $\mathbb{R}^n$ is a subspace if $\mathbf{v}_1 + \mathbf{v}_2 \in V$ and $c\mathbf{v} \in V$ whenever $\mathbf{v}, \mathbf{v}_1, \mathbf{v}_2 \in V,\ c \in \mathbb{R}$.

Now we justify the names 'range space' and 'null space'.

Theorem 2.1.7. *Let $T : \mathbb{R}^n \to \mathbb{R}^m$ be a linear map. Then the range space $R(T)$ and the null space $N(T)$ are subspaces of $\mathbb{R}^m$ and $\mathbb{R}^n$ respectively.*

Proof. To show $R(T)$ is a subspace let $\mathbf{w}_1, \mathbf{w}_2 \in R(T)$ and $c \in \mathbb{R}$. Then there are $\mathbf{v}_1, \mathbf{v}_2 \in \mathbb{R}^n$ such that $T(\mathbf{v}_1) = \mathbf{w}_1, T(\mathbf{v}_2) = \mathbf{w}_2$. Hence $\mathbf{w}_1 + \mathbf{w}_2 = T(\mathbf{v}_1) + T(\mathbf{v}_2) = T(\mathbf{v}_1 + \mathbf{v}_2) \in R(T)$; $\mathbf{w}_1 + \mathbf{w}_2 \in R(T)$. Now $c\mathbf{w}_1 = cT(\mathbf{v}_1) = T(c\mathbf{v}_1) \in R(T)$; $c\mathbf{w}_1 \in R(T)$. Therefore, $R(T)$ is a subspace of R^m.

Proof for $N(T)$ is an exercise (Exercise 4). □

Theorem 2.1.8. *Let $T : \mathbb{R}^n \to \mathbb{R}^m$ be a linear map. If $S = \{\mathbf{v}_1, \mathbf{v}_2, ..., \mathbf{v}_p\}$ is linearly dependent then so is $T(S) = \{T(\mathbf{v}_1), T(\mathbf{v}_2), ..., T(\mathbf{v}_p)\}$.*

Proof. Exercise 8 □

We saw in Example 2.1.6 (1) that $T : \mathbb{R}^2 \to \mathbb{R}$ is determined by the values $T(\mathbf{e}^1), T(\mathbf{e}^2)$. More generally, we show that a linear map is determined by the values on a basis.

Theorem 2.1.9. *Let $\mathcal{B} = \{\mathbf{v}_1, \mathbf{v}_2, ..., \mathbf{v}_n\}$ be a basis of $\mathbb{R}^n$ and let $\{\mathbf{w}_1, \mathbf{w}_2, ..., \mathbf{w}_n\}$ be a set of n vectors in $\mathbb{R}^m$. Then there is a unique linear map $T : \mathbb{R}^n \to \mathbb{R}^m$ such that $T(\mathbf{v}_i) = \mathbf{w}_i$ $(i = 1, 2, ..., n)$.*

Proof. Let $\mathbf{v} \in \mathbb{R}^n$ and let $\mathbf{v} = \sum a_i\mathbf{v}_i$. We define $T(\mathbf{v}) = \sum a_i\mathbf{w}_i$. We need to show that T is linear and it is unique. For linearity, let $\mathbf{v}' = \sum a_i'\mathbf{v}_i$ then $T(\mathbf{v}') = \sum a_i'\mathbf{w}_i$. And

$$\begin{aligned} T(\mathbf{v} + \mathbf{v}') &= T\left(\sum a_i\mathbf{v}_i + \sum a_i'\mathbf{v}_i\right) = T\left(\sum (a_i + a_i')\mathbf{v}_i\right) \\ &= \sum (a_i + a_i')\mathbf{w}_i = \sum a_i\mathbf{w}_i + \sum a_i'\mathbf{w}_i = T(\mathbf{v}) + T(\mathbf{v}') \end{aligned}$$

Also it is easy to check that $T(c\mathbf{v}) = cT(\mathbf{v})$ for a constant c.

For uniqueness, suppose $T' : \mathbb{R}^n \to \mathbb{R}^m$ is also a linear map such that $T'(\mathbf{v}_i) = \mathbf{w}_i$. Let $\mathbf{v} = \sum a_i\mathbf{v}_i$. Then $T'(\mathbf{v}) = \sum a_i T'(\mathbf{v}_i) = \sum a_i\mathbf{w}_i = T(\mathbf{v})$. Hence $T = T'$. □

Probably the most important linear maps are the 'matrix linear maps' which are associated with matrices.

Theorem 2.1.10. *Let A be an $m\times n$ matrix. The map $T_A : \mathbb{R}^n \to \mathbb{R}^m$ defined by*

$$T_A(\mathbf{x}) = A\mathbf{x} \tag{2.10}$$

is a linear map. We call the map T_A the **matrix linear map** *associated to A.*

The range of T_A is the same as the column space of A and the null space of T_A is the same as the solution space of A:

$$R(T_A) = \mathrm{Col}(A)$$
$$N(T_A) = N(A)$$

Proof. Linearity of T_A follows from Theorem 1.2.12.

To see the equality $R(T_A) = \mathrm{Col}(A)$, observe that $T_A(\mathbf{e}^i) = A^{(i)}$. Hence $C(A) \subseteq R(T_A)$ and hence $\mathrm{Span}\, C(A) = \mathrm{Col}(A) \subseteq R(T_A)$. To show the inclusion $R(T_A) \subseteq \mathrm{Col}(A)$ let $\mathbf{w} \in R(T_A)$. Then there is $\mathbf{v} \in \mathbb{R}^n$ such that $\mathbf{w} = T_A(\mathbf{v})$. Let $\mathcal{E} = \{\mathbf{e}^i\}$ be the standard basis of $\mathbb{R}^n$ and let $\mathbf{v} = \sum a_i\mathbf{e}^i$. Then $T_A(\mathbf{v}) = T_A(\sum a_i\mathbf{e}^i) = \sum a_i T_A\mathbf{e}^i = \sum a_i A^{(i)}$ which belongs to $\mathrm{Col}(A)$.

The equality $N(T_A) = N(A)$ is an Exercise 4. □

Example 2.1.11. (1) *Let $A_\theta = \begin{bmatrix} \cos\theta & -\sin\theta \\ \sin\theta & \cos\theta \end{bmatrix}$. Then the matrix linear map T_θ for the matrix A_θ is the* **rotation by the angle** *θ which is an isomorphism.*

(2) *If $A = \begin{bmatrix} 1 & 0 \\ 0 & -1 \end{bmatrix}$ then T_A represent the reflection with respect to x-axis.*

(3) *Let $\mathcal{E} = \{\mathbf{e}^1, \mathbf{e}^2\}$ be the standard basis of $\mathbb{R}^2$. By* Theorem 2.1.9, $T(\mathbf{e}^1) = \begin{bmatrix} 2 \\ 1 \end{bmatrix}$ *and* $T(\mathbf{e}^2) = \begin{bmatrix} 3 \\ -1 \end{bmatrix}$ *determines a linear map $T : \mathbb{R}^2 \to \mathbb{R}^2$. Let $\mathbf{x} = \begin{bmatrix} x \\ y \end{bmatrix}$. Then* $T(\mathbf{x}) = T(x\mathbf{e}^1 + y\mathbf{e}^2) = xT(\mathbf{e}^1) + yT(\mathbf{e}^2) = x\begin{bmatrix} 2 \\ 1 \end{bmatrix} + y\begin{bmatrix} 3 \\ -1 \end{bmatrix} = \begin{bmatrix} 2 & 3 \\ -1 & 1 \end{bmatrix}\begin{bmatrix} x \\ y \end{bmatrix}$. *Hence $T = T_A$ where* $A = \begin{bmatrix} 2 & 3 \\ -1 & 1 \end{bmatrix}$.

A comment on notation: A usual notation of a function $f : \mathbb{R}^3 \to \mathbb{R}^2$ in calculus looks like

$$f(x, y, z) = (x^2 \sin y, y \ln z)$$

which becomes

$$f\left(\begin{bmatrix} x \\ y \\ z \end{bmatrix}\right) = \begin{bmatrix} x^2 \sin y \\ y \ln z \end{bmatrix}$$

in our vector notation. This notation is rather grotesque looking. So we will often write

$$f(x, y, z) = \begin{bmatrix} x^2 \sin y \\ y \ln z \end{bmatrix} \tag{2.11}$$

instead. Of course the function f is not linear.

Remark. *A function* $f : \mathbb{R}^n \to \mathbb{R}^m$ *can be written as* $f(\mathbf{x}) = \begin{bmatrix} f_1(\mathbf{x}) \\ \vdots \\ f_m(\mathbf{x}) \end{bmatrix}$ $(\mathbf{x} \in \mathbb{R}^n)$. *In analysis,* f *is defined to be* **continuous** *(resp.* **differentiable**) *if each* $f_i : \mathbb{R}^n \to \mathbb{R}$ *is continuous (resp. differentiable).*

For example, let $f_1(x, y, z) = x^2 \sin y$, $f_2(x, y, z) = y \ln z$ *in* (2.11). *We know, from calculus, that* f_1, f_2 *are continuous at the points on which the function* f *is defined. Hence the function* f *in* (2.11) *is continuous and differentiable.*

When $A = [a_{ij}]$ *is an* $m \times n$ *matrix how does the matrix linear map* $T_A : \mathbb{R}^n \to \mathbb{R}^m$ *can be represented in this form? Let* $\mathbf{x} = \begin{bmatrix} x_1 \\ \vdots \\ x_n \end{bmatrix}$. *Then* $T_A(\mathbf{x})$, *in vector notation, becomes*

$$T_A(\mathbf{x}) = \begin{bmatrix} T_1(\mathbf{x}) \\ T_2(\mathbf{x}) \\ \vdots \\ T_m(\mathbf{x}) \end{bmatrix} = \begin{bmatrix} A_{(1)}\mathbf{x} \\ A_{(2)}\mathbf{x} \\ \vdots \\ A_{(m)}\mathbf{x} \end{bmatrix} \quad \text{where } A_{(i)}\mathbf{x} = \sum_{k=1}^{n} a_{ik} x_k \tag{2.12}$$

Notice that each $T_i : \mathbb{R}^n \to \mathbb{R}$ *is a polynomial of degree one in* $x_1, x_2, ..., x_n$ *viz* T_i *is a polynomial function in the variables* $x_1, x_2, ..., x_n$. *Also notice that each component function* T_i, *being a polynomial, is continuous as well as differentiable. Therefore, we conclude that a matrix linear transformation is continuous and differentiable.*

For example, if $A = \begin{bmatrix} 1 & -3 \\ 3 & 5 \\ -1 & 7 \end{bmatrix}$ then the expression in (2.12) is

$$T_A(\mathbf{x}) = \begin{bmatrix} x_1 - 3x_2 \\ 3x_1 + 5x_2 \\ -x_1 + 7x_2 \end{bmatrix}$$

where $\mathbf{x} = \begin{bmatrix} x_1 \\ x_2 \end{bmatrix}$. Then, for example, $T_1(\mathbf{x}) = x_1 - 3x_2$. And if $\mathbf{x} = \begin{bmatrix} 2 \\ -1 \end{bmatrix}$ then

$$T_A(\mathbf{x}) = A\mathbf{x} = \begin{bmatrix} 1 & -3 \\ 3 & 5 \\ -1 & 7 \end{bmatrix} \begin{bmatrix} 2 \\ -1 \end{bmatrix} = \begin{bmatrix} 5 \\ 1 \\ -9 \end{bmatrix}$$

Practice Problem 2.1.2. *Let* $A = \begin{bmatrix} 1 & 2 \\ 3 & 4 \end{bmatrix}$. *Find* $T_A(\mathbf{v})$ *where* $\mathbf{v} = \begin{bmatrix} a \\ b \end{bmatrix}$.

Answers for Practice Problems 2.1

(2.1.1) We need to show that the (i, j)-th positions of the both sides coincide. The (i, j)-th position of the lefty side is $(A + B)_{ij} + C_{ij} = (A_{ij} + B_{ij}) + C_{ij}$. Similarly, the (i, j)-th position of the right hand side is $A_{ij} + (B_{ij} + C_{ij})$. Now $(A_{ij} + B_{ij}) + C_{ij} = A_{ij} + (B_{ij} + C_{ij})$ by the properties of real numbers.

(2.1.2) $\begin{bmatrix} a + 2b \\ 3a + 4b \end{bmatrix}$

2.1 Exercises

1. Find $T(\mathbf{v})$ for the given T and $\mathbf{v}$.

   (a) $T(x_1, x_2) = \begin{bmatrix} x_1 - x_2 \\ -2x_1 + 3x_2 \end{bmatrix}$, $\mathbf{v} = \begin{bmatrix} 1 \\ -2 \end{bmatrix}$

   (b) $T(x_1, x_2, x_3) = \begin{bmatrix} x_1 + x_3 \\ x_2 - x_3 \end{bmatrix}$, $\mathbf{v} = \begin{bmatrix} 1 \\ 2 \\ -1 \end{bmatrix}$

   (c) $T(x_1, x_2, x_3, x_4) = \begin{bmatrix} x_1 + x_3 \\ x_2 - x_4 \\ x_1 - x_3 + x_4 \end{bmatrix}$, $\mathbf{v} = \begin{bmatrix} -1 \\ 2 \\ 0 \\ 1 \end{bmatrix}$

   (d) $T = T_{A_\theta}$ with $A_\theta = \begin{bmatrix} \cos\theta & -\sin\theta \\ \sin\theta & \cos\theta \end{bmatrix}$, and $\mathbf{v} = \begin{bmatrix} \cos\tau \\ \sin\tau \end{bmatrix}$.

   (e) $T = T_A$ with $A = \begin{bmatrix} 1 & 0 \\ 0 & -1 \end{bmatrix}$ and $\mathbf{v} = \begin{bmatrix} \cos\tau \\ \sin\tau \end{bmatrix}$.

2. Find $T(\mathbf{v})$ for the given T and $\mathbf{v}$.

   (a) $T = T_A$ with $A = [1 \;\; 2]$ and $\mathbf{v} = \begin{bmatrix} -1 \\ 1 \end{bmatrix}$

   (b) $T = T_A$ with $A = \begin{bmatrix} 1 & 2 \\ 3 & 4 \end{bmatrix}$ and $\mathbf{v} = \begin{bmatrix} -1 \\ 1 \end{bmatrix}$

   (c) $T = T_A$ with $A = \begin{bmatrix} 1 & 0 & 2 \\ 1 & 1 & -1 \\ 0 & -1 & 1 \end{bmatrix}$ and $\mathbf{v} = \begin{bmatrix} 0 \\ -1 \\ 2 \end{bmatrix}$

   (d) $T = T_A$ with $A = \begin{bmatrix} 1 & 1 & 1 & 0 \\ 0 & -1 & -1 & 1 \\ -1 & 0 & 0 & -2 \end{bmatrix}$ and $\mathbf{v} = \begin{bmatrix} 1 \\ 0 \\ 1 \\ 2 \end{bmatrix}$

   (e) $T = T_A$ with $A = \begin{bmatrix} 1 & 1 & 0 \\ 0 & -1 & 1 \\ 0 & 0 & -2 \\ -1 & -1 & 3 \end{bmatrix}$ and $\mathbf{v} = \begin{bmatrix} 0 \\ -1 \\ 2 \end{bmatrix}$

3. Decide whether the following functions are linear or not.

   (a) $T : \mathbb{R}^2 \to \mathbb{R}^3$, $T(x_1, x_2) = \begin{bmatrix} x_1 x_2 \\ x_1 - x_2 \\ \frac{1}{2} x_2 \end{bmatrix}$.

   (b) $T : \mathbb{R}^3 \to \mathbb{R}^2$, $T(x_1, x_2, x_3) = \begin{bmatrix} x_1 + 3x_3 \\ x_2 + 1 \end{bmatrix}$.

   (c) $T : \mathbb{R}^3 \to \mathbb{R}^3$, $T(x_1, x_2, x_3) = \begin{bmatrix} x_1 + 3x_3 \\ x_2 - x_3 \\ x_2 \end{bmatrix}$.

(d) Fix $\mathbf{0} \neq \mathbf{v} \in \mathbb{R}^n$. Define $T : \mathbb{R}^n \to \mathbb{R}^n$ by $T(\mathbf{x}) = \mathbf{x} + \mathbf{v}$..

(e) $T : \mathbb{R}^2 \to \mathbb{R}^2$, $T(x, y) = \begin{bmatrix} y \\ x \end{bmatrix}$.

(f) $T : \mathbb{R}^2 \to \mathbb{R}^4$, $T(x, y) = \begin{bmatrix} x+y \\ x \\ y \\ x \end{bmatrix}$.

4. Let $T : \mathbb{R}^n \to \mathbb{R}^m$ be a linear map and V be a subspace of $\mathbb{R}^n$.

   (a) Prove that the null space $N(T)$ is a subspace of $\mathbb{R}^n$.

   (b) Prove that $T(V) = \{T(\mathbf{v}) \mid \mathbf{v} \in V\}$ is a subspace of $\mathbb{R}^m$.

5. If T is an isomorphism then T^{-1} is also an isomorphism i.e., T^{-1} which exists as a function by Theorem A.1.5 is also linear.

6. Prove the true statement(s) and find a counter example for the false statement(s).

   (a) Let $T : \mathbb{R}^n \to \mathbb{R}^m$ be a function such that $T(\mathbf{0}) \neq \mathbf{0}$ then T is not linear.

   (b) The map $T : \mathbb{R}^2 \to \mathbb{R}^2$ defined by $T(x, y) = \begin{bmatrix} x+1 \\ x-y \end{bmatrix}$ is a linear map.

   (c) The map $T : \mathbb{R}^2 \to \mathbb{R}^2$ given by a rotation by the angle 30° is a linear map.

   (d) The map $T : \mathbb{R}^n \to \mathbb{R}^n$ defined by $T(\mathbf{v}) = c\mathbf{v}$ $(c \in \mathbb{R})$ is a linear map.

   (e) The reflection with respect to the x-axis on $\mathbb{R}^2$ is a linear map.

   (f) Let $T : \mathbb{R}^n \to \mathbb{R}^m$ be linear map and S a finite subset of $\mathbb{R}^n$. If $T(S)$ is linearly independent then S is linearly independent.

7. Prove Theorem 2.1.4 (3)–(6).

8. Let $T : \mathbb{R}^n \to \mathbb{R}^m$ be a linear map. Prove that if $S = \{\mathbf{v}_1, \mathbf{v}_2, ..., \mathbf{v}_p\}$ is linearly dependent then so is $T(S) = \{T(\mathbf{v}_1), T(\mathbf{v}_2), ..., T(\mathbf{v}_p)\}$.

9. Let $A = \begin{bmatrix} 2 & -4 \\ 8 & 4 \end{bmatrix}$ and let $T_A : \mathbb{R}^2 \to \mathbb{R}^2$ be the corresponding matrix linear map. Show that T_A transforms the circle $x^2 + y^2 = r^2$ of radius r to an ellipse and find the equation of the ellipse.

10. Let $X(t) = [a_{ij}(t)], Y(t) = [b_{ij}(t)]$ be an $n \times n$ matrix where $a_{ij}(t), b_{ij}(t)$ are differentiable functions in the variable t. Let $\frac{d}{dt}X(t) = X'(t) = (a'_{ij}(t))$ be the matrix obtained by taking the derivative of each entry of $X(t)$. Prove that

$$\frac{d}{dt}(X(t)Y(t)) = X'(t)Y(t) + X(t)Y'(t)$$

2.2 Matrix Representation of Linear Maps

2.2.1 Standard Matrix of Linear Map

The definition of linear map is rather abstract. For practical use and computations we need to represent a linear map in a mathematical object, a matrix. We will show that every linear map $T : \mathbb{R}^n \to \mathbb{R}^m$ is a matrix transformation T_A for some matrix A. Consider an example.

Example 2.2.1. *Let $T : \mathbb{R}^2 \to \mathbb{R}^2$ be a linear map. By* Theorem 2.1.9, *T is determined by the values on the standard basis $\{\mathbf{e}^1, \mathbf{e}^2\}$. Let $T(\mathbf{e}^1) = \begin{bmatrix} 1 \\ 2 \end{bmatrix}, T(\mathbf{e}^2) = \begin{bmatrix} 3 \\ -1 \end{bmatrix}$. Then, for $\mathbf{x} = \begin{bmatrix} x_1 \\ x_2 \end{bmatrix}$, we have*

$$T(\mathbf{x}) = T(x_1\mathbf{e}^1 + x_2\mathbf{e}^2) = x_1T(\mathbf{e}^1) + x_2T(\mathbf{e}^2) = x_1 \begin{bmatrix} 1 \\ 2 \end{bmatrix} + x_2 \begin{bmatrix} 3 \\ -1 \end{bmatrix} = \begin{bmatrix} 1 & 3 \\ 2 & -1 \end{bmatrix} \begin{bmatrix} x_1 \\ x_2 \end{bmatrix}$$

Hence T is the matrix linear transformation corresponding to the matrix

$$A = [T(\mathbf{e}^1)\ T(\mathbf{e}^2)] = \begin{bmatrix} 1 & 3 \\ 2 & -1 \end{bmatrix}$$

Definition 2.2.2. *Let $T : \mathbb{R}^n \to \mathbb{R}^m$ be a linear map. Let $\{\mathbf{e}^1, \mathbf{e}^2, ..., \mathbf{e}^n\}$ be the standard basis of $\mathbb{R}^n$. Corresponding to T we associate an $m \times n$ matrix*

$$A_T := \begin{bmatrix} T(\mathbf{e}^1) & T(\mathbf{e}^2) & \cdots & T(\mathbf{e}^n) \end{bmatrix} \tag{2.13}$$

i.e., $[A_T]^{(i)} = T(\mathbf{e}^i)$. We call A_T the **standard matrix** *for T.*

Hence, if $T(\mathbf{e}^j) = \sum_{i=1}^m a_{ij}\mathbf{e}^i$ then

$$A_T^{(j)} = \begin{bmatrix} a_{1j} \\ a_{2j} \\ \vdots \\ a_{mj} \end{bmatrix} = T(\mathbf{e}^j) \tag{2.14}$$

Now we prove that every linear map is a matrix linear map of its standard matrix.

Theorem 2.2.3. *Let $T : \mathbb{R}^n \to \mathbb{R}^m$ be a linear map and A_T be the standard matrix for T. Then T is the matrix linear map associated to A_T.*

Proof. Let S be the matrix linear map associated to A_T. We need to show that $S = T$. By Theorem 2.1.9, it suffices to show they agree on the standard basis of $\mathbb{R}^n$. Now $S(\mathbf{e}^i) = A_T^{(i)}$. On the other hand, $T(\mathbf{e}^i) = A_T^{(i)}$ by definition of A_T. Hence $S = T$. □

Theorem 2.2.4. *Let $A \in M(m\times n)$. Then the standard matrix of the matrix linear map T_A associated with A is the same as A.*

Proof. The i-th column of the standard matrix of T_A is $T_A(\mathbf{e}^i) = A\mathbf{e}^i = A^{(i)}$. □

Definition 2.2.5. *Let $S, T : \mathbb{R}^n \to \mathbb{R}^m$ are linear maps and c be a scalar then we define the* **addition** *$S + T$ and the* **scalar multiplication** *cS by*

$$(S + T)(\mathbf{v}) = S(\mathbf{v}) + T(\mathbf{v}) \text{ and } (cS)(\mathbf{v}) = c(S(\mathbf{v}))\quad (\mathbf{v} \in \mathbb{R}^n) \tag{2.15}$$

Theorem 2.2.6. *Let $S, T : \mathbb{R}^n \to \mathbb{R}^m$ are linear maps and r be a scalar. Then the maps $S + T$ and rS are linear maps from $\mathbb{R}^n$ to $\mathbb{R}^m$ i.e., they belong to $L(n, m)$.*

Proof. Exercise 5. □

Theorem 2.2.7. *Let $S, T, U \in L(n,m)$ and r, s be scalars. Then we have*

(1) $S + T = T + S$	(2) $(S+T)+U = S+(T+U)$
(3) $S + 0 = S$	(4) $r(S+T) = rS + rT$
(5) $(r+s)S = rS + sS$	(6) $r(sS) = (rs)S$

Proof. To prove the identity (a), for example, we need to prove $(S+T)(\mathbf{v}) = (T+S)(\mathbf{v})$. By definition of addition of linear maps (2.15), we have $(S+T)(\mathbf{v}) = S(\mathbf{v}) + T(\mathbf{v})$. Since $S(\mathbf{v}), T(\mathbf{v})$ are elements of $\mathbb{R}^m$ we have $S(\mathbf{v}) + T(\mathbf{v}) = T(\mathbf{v}) + S(\mathbf{v})$ and the latter is the same as $(T+S)(\mathbf{v})$.

The proofs of other equalities are Exercise 6. □

Practice Problem 2.2.1. *Prove the equality $(S+T)+U = S+(T+U)$.*

Let $M(m\times n)$ be the set of all $m\times n$ matrices and let $L(n,m)$ be the set of all linear maps $T : \mathbb{R}^n \to \mathbb{R}^m$. Then we get maps

$$\Phi : M(m\times n) \to L(n,m), \quad A \mapsto T_A$$
$$\Psi : L(n,m) \to M(m\times n), \quad T \mapsto A_T$$

Theorem 2.2.8. *The maps Φ and Ψ are bijections between $M(m\times n)$ and $L(n,m)$ and they are inverses to each other i.e., $\Psi \circ \Phi(A) = A$ and $\Phi \circ \Psi(T) = T$. Further, Φ and Ψ satisfy*

$$\Phi(A+B) = \Phi(A) + \Phi(B), \quad \Phi(rA) = r\Phi(A) \quad \text{and}$$
$$\Psi(S+T) = \Psi(S) + \Psi(T), \quad \Psi(rS) = r\Psi(S)$$

That is

$$T_{(A+B)} = T_A + T_B, \qquad T_{rA} = rT_A \tag{2.16}$$
$$A_{(S+T)} = A_S + A_T, \qquad A_{rS} = rA_S \tag{2.17}$$

Proof. By Theorem 2.2.3 we have $\Phi \circ \Psi = I$. Now, $\Psi \circ \Phi = I$ by Theorem 2.2.3 and Theorem 2.2.4. Hence the maps Φ and Ψ are bijections by Theorem A.1.5.

To prove the first equality of (2.16) we need to show $T_{(A+B)}(\mathbf{v}) = T_A(\mathbf{v}) + T_B(\mathbf{v})$. Now $T_{(A+B)}(\mathbf{v}) = (A+B)(\mathbf{v})$ and $T_A(\mathbf{v}) + T_B(\mathbf{v}) = A\mathbf{v} + B\mathbf{v} = (A+B)\mathbf{v}$. As required.

To prove the first equality of (2.17) we need to show the i-th columns of the both sides are equal. The i-th column is obtained by evaluating at $\mathbf{e}^i$. Hence we need to show $(S+T)(\mathbf{e}^i) = S(\mathbf{e}^i) + T(\mathbf{e}^i)$ which is obvious.

Other equalities are similarly proved. □

Corollary 2.2.9. *Let $A, B \in M(m\times n)$. Then A and B are equal if and only if the corresponding linear maps T_A and T_B are the same.*

If $S, T \in L(n,m)$ then $S = T$ if and only if their standard matrices A_S and A_T are the same.

Proof. This follows from the fact that Φ and Ψ are bijections. □

The notion of linear map is an abstract concept and a matrix is an concrete mathematical object. Through the map Φ we express the abstract object, a linear map, in a concrete mathematical object, a matrix. And vice versa. Sometimes it is fruitful to use concrete object and sometimes it is convenient to use abstract gadget. This idea of concrete versus abstract will be prevailing throughout the text.

2.2.2 Matrix Multiplication and Composition of Linear Maps

Definition 2.2.10. *Let $A \in M(m\times n)$ and $B \in M(n\times p)$. The* **matrix multiplication** *AB is an $m\times p$ matrix defined by*

$$AB = A\begin{bmatrix}B^{(1)} & B^{(2)} & \cdots & B^{(p)}\end{bmatrix} = \begin{bmatrix}AB^{(1)} & AB^{(2)} & \cdots & AB^{(p)}\end{bmatrix} = \begin{bmatrix}A_{(1)}B \\ A_{(2)}B \\ \vdots \\ A_{(m)}B\end{bmatrix} \tag{2.18}$$

Entrywise the matrix multiplication can be defined by

$$(AB)_{ij} = \sum_{k=1}^{n} A_{ik}B_{kj} = A_{(i)}B^{(j)} \tag{2.19}$$

If k is a positive integer and A is an $n\times n$ matrix. Then we define the **k-th power** *A^k of A to be the k-times repeated product of A.*

The matrix multiplication AB is defined only when the number of rows is the same as the number of columns of B. In general, $AB \neq BA$ even if both AB and BA are defined (Exercise 7).

Example 2.2.11. *Let $A = \begin{bmatrix}1 & 2 & -1\\ 0 & 3 & 5\end{bmatrix}$ and $B = \begin{bmatrix}3 & -1\\ 1 & 2\\ 2 & 1\end{bmatrix}$. Then*

$$\begin{bmatrix}1 & 2 & -1\\ 0 & 3 & 5\end{bmatrix}\begin{bmatrix}3 & -1\\ 1 & 2\\ 2 & 1\end{bmatrix} = \left[A\begin{bmatrix}3\\1\\2\end{bmatrix} \quad A\begin{bmatrix}-1\\2\\1\end{bmatrix}\right] = \begin{bmatrix}\begin{bmatrix}1 & 2 & -1\end{bmatrix}B\\ \begin{bmatrix}0 & 3 & 5\end{bmatrix}B\end{bmatrix} = \begin{bmatrix}3 & 2\\ 13 & 11\end{bmatrix}$$

Practice Problem 2.2.2. *Find AB where $A = \begin{bmatrix}1 & -2\\ 3 & 1\\ 2 & 4\end{bmatrix}$ and $B = \begin{bmatrix}a & b\\ x & y\end{bmatrix}$.*

Example 2.2.12. *Let $A \in M(m\times n)$. Let $\mathbf{e}^i \in \mathbb{R}^n$ (resp. $\mathbf{e}_i \in \mathbb{R}_m$) be the column (resp. row) with 1 in the i-th position and 0 elsewhere* (Example 1.3.23)

$$\mathbf{e}^i = \begin{bmatrix}0\\ \vdots\\ 1\\ \vdots\\ 0\end{bmatrix}, \quad \mathbf{e}_i = [0, ..., 1, ..., 0]$$

Then

$$A\mathbf{e}^i = A^{(i)} \text{ and } \mathbf{e}_j A = A_{(j)} \tag{2.20}$$

If $A, B \in M(m\times n)$ then

$$A = B \text{ if and only if } A\mathbf{x} = B\mathbf{x} \text{ for all } \mathbf{x} \in \mathbb{R}^n \tag{2.21}$$

In fact, if we take $\mathbf{x} = \mathbf{e}^i$ then we have $A^{(i)} = B^{(i)}$ for all i.

Theorem 2.2.13. *Let A be an $m\times n$ matrix and B be an $n\times p$ matrix. If the columns of B are linearly dependent then so are the columns of AB.*

Proof. By Definition 2.2.10, the columns of AB are $\{T_A(B^{(1)}), T_A(B^{(2)}), ..., T_A(B^{(p)})\}$ which is linearly dependent by Theorem 2.1.8. □

Theorem 2.2.14. *Let $S : \mathbb{R}^n \to \mathbb{R}^m$ and $T : \mathbb{R}^m \to \mathbb{R}^p$ be linear maps. Then the composition $T \circ S : \mathbb{R}^n \to \mathbb{R}^p$ is a linear map.*

Proof. Let $\mathbf{u}_1, \mathbf{u}_2 \in R^n$. Then

$$\begin{aligned} T \circ S(\mathbf{u}_1 + \mathbf{u}_2) &= T(S(\mathbf{u}_1 + \mathbf{u}_2)) = T(S(\mathbf{u}_1) + S(\mathbf{u}_2)) \\ &= T(S(\mathbf{u}_1)) + T(S(\mathbf{u}_2)) = T \circ S(\mathbf{u}_1) + T \circ S(\mathbf{u}_2) \end{aligned}$$

Similarly, we have

$$T \circ S(c\mathbf{u}) = TS(cu) = T(cS(\mathbf{u})) = cT(S(\mathbf{u})) = cT \circ S(\mathbf{u})$$

for $c \in \mathbb{R}, \mathbf{u} \in \mathbb{R}^n$. □

Theorem 2.2.15. *Let $S : \mathbb{R}^n \to \mathbb{R}^m$ and $T : \mathbb{R}^m \to \mathbb{R}^p$ be linear maps. Let $A \in M(m\times n)$ and $B \in M(p\times m)$ be such that $S = T_A$ and $T = T_B$. Then $T \circ S$ is the matrix linear map associated with BA:*

$$T_{BA} = T_B \circ T_A$$

Proof. We have maps $\mathbb{R}^n \xrightarrow{T_A} \mathbb{R}^m \xrightarrow{T_B} \mathbb{R}^p$ and $T_{BA} : \mathbb{R}^n \to \mathbb{R}^p$ since BA is an $p\times n$ matrix. To prove the equality we evaluate both sides at $\mathbf{v} \in \mathbb{R}^n$: $T_{BA}(\mathbf{v}) = (BA)\mathbf{v}$ and $T_B \circ T_A(\mathbf{v}) = T_B(A\mathbf{v}) = B(A\mathbf{v}) = (BA)\mathbf{v}$. Whence the equality. □

Theorem 2.2.15 says that the matrix multiplication AB corresponds to the composition of the linear maps T_A, T_B associated with the matrices A, B.

Theorem 2.2.16. *Whenever the following matrix operations are defined the following identities hold:*

(1) $A(BC) = (AB)C$ (associative law of multiplication)

(2) $A(B + C) = AB + AC$ (left distributive law)

(3) $(B + C)A = BA + CA$ (rightdistributive law)

(4) $r(AB) = (rA)B = A(rB)$ for any scalar r

(5) $I_m A = A = AI_n$ $(A \in M(m\times n))$ where I_m is the $m\times m$ identity matrix

Proof. To prove the matrix identity (1) we need to prove both sides have the same (i, j) entry:

$$\begin{aligned} [(AB)C]_{ij} &= \sum_k (AB)_{ik} C_{kj} \\ &= \sum_k \left(\sum_l A_{il} B_{lk} \right) C_{kj} \\ &= \sum_{l,k} A_{il} B_{lk} C_{kj} \\ &= \sum_l A_{il} \sum_k (B_{lk} C_{kj}) \\ &= \sum A_{il} (BC)_{lj} = [A(BC)]_{ij} \end{aligned}$$

Now we give an easier proof of (1) using Corollary 2.2.9 and associativity of composition of functions (Theorem A.1.3):

$$T_{(AB)C} = T_{(AB)} \circ T_C = (T_A \circ T_B) \circ T_C = T_A \circ (T_B \circ T_C) = T_{A(BC)}$$

Hence $(AB)C = A(BC)$.

The other identities are easy to prove and we leave the proofs as Exercise 8. □

Practice Problem 2.2.3. *Prove* $A(B+C) = AB + AC$ *for the matrices* A, B, C.

Corresponding theorem of Theorem 2.2.16 for linear maps will be:

Theorem 2.2.17. *If* S, T *and* U *are linear maps then the following identity, whenever they are defined, hold*

(1) $S(TU) = (ST)U$	(associativity of composition)
(2) $S(T+U) = ST + SU$	(left distributive law)
(3) $(T+U)S$	(right distributive law)
(4) $r(ST) = (rS)T = S(rT)$	for any scalar r
(5) $IS = S = SI$	where I is the identity map

Proof. Exercise 9. □

If $T : \mathbb{R}^n \to \mathbb{R}^n$ is a linear map then we define the **k-th power** T^k of T to be the k-times composition of T. If $T = T_A$ for some $A \in M(n)$ then, by Theorem 2.2.15, we have

$$(T_A)^k = T_{A^k}$$

Definition 2.2.18. *Let* A *be an* $m\times n$ *matrix. We define the* **transpose** A^t *of* A *to be the* $n\times m$ *matrix*

$$(A^t)_{ij} = A_{ji}$$

Hence the rows of A^t are the columns of A and the corresponding matrix linear map is from $\mathbb{R}^m$ to $\mathbb{R}^n$ say, $T_{A^t} : \mathbb{R}^m \to \mathbb{R}^n$ whereas $T_A : \mathbb{R}^n \to \mathbb{R}^m$.

Theorem 2.2.19. *Let* A, B *be* $m\times n$ *matrices and* r *be a scalar. Then*

(1) $(A^t)^t = A$

(2) $(A+B)^t = A^t + B^t$

(3) $(rA)^t = r(A^t)$

(4) $(AB)^t = B^t A^t$

Proof. Exercise 10. □

Practice Problem 2.2.4. *Prove* $(A^t)^t = A$.

Answers for Practice Problems 2.2

(2.2.2) $A = \begin{bmatrix} a-2x & b-2y \\ 3a+x & 3b+y \\ 2a+4x & 2b+4y \end{bmatrix}$

(2.2.3) We have to show the i, j-th place of the both sides are the same :

$$[A(B+C)]_{ij} = \sum_k A_{ik}(B+C)_{kj} = \sum_k A_{ik}(B_{kj}+C_{kj}) = \sum_k (A_{ik}B_{kj} + A_{ik}C_{kj})$$
$$= \sum_k A_{ik}B_{kj} + \sum_k A_{ik}C_{kj} = (AB)_{ij} + (AC)_{ij} = (AB+AC)_{ij}$$

(2.2.4) $((A^t)^t)_{ij} = (A^t)_{ji} = A_{ij}$. The others are also easy to show.

2.2 Exercises

1. Prove that the map T is linear and find the standard matrix for each T.

   (a) $T : \mathbb{R}^3 \to \mathbb{R}^3$ given by $T(x_1, x_2, x_3) = \begin{bmatrix} -x_1 & & +x_3 \\ x_1 & +x_2 & \\ & x_2 & -x_3 \end{bmatrix}$

   (b) $T : \mathbb{R}^2 \to \mathbb{R}^2$ given by $T(x, y) = \begin{bmatrix} y \\ x \end{bmatrix}$

   (c) $T : \mathbb{R}^3 \to \mathbb{R}^3$ is given by $T(x, y, z) = \begin{bmatrix} y \\ z \\ x \end{bmatrix}$

   (d) $T : \mathbb{R}^4 \to \mathbb{R}^3$ is given by $T(x, y, z, w) = \begin{bmatrix} x + w \\ y \ - z \\ z - w \end{bmatrix}$

   (e) $T : \mathbb{R}^2 \to \mathbb{R}^4$ is given by $T(x, y) = \begin{bmatrix} x - y \\ x + y \\ -x + y \\ -x - y \end{bmatrix}$

   (f) $T : \mathbb{R}^4 \to \mathbb{R}^4$ is given by $T(x_1, x_2, x_3, x_4) = \begin{bmatrix} x_3 \\ x_1 \\ x_2 \\ x_4 \end{bmatrix}$

2. Let A be an $n \times n$ and $D = \text{diag}[\lambda_1, \lambda_2, ..., \lambda_n]$ be the diagonal matrix. Then
$$AD = [\lambda_1 A^{(1)} \ \lambda_2 A^{(2)} \ \cdots \lambda_n A^{(n)}]$$

3. Let A be an $m \times n$ matrix and B be an $n \times p$ matrix. Prove that
$$(AB)^{(j)} = AB^{(j)} \text{ and } (AB)_{(i)} = A_{(i)}B$$

4. Let A, A' be $r \times r$ matrices and C, C' be $s \times s$ matrices and B, B' be $r \times s$ matrices. Prove that
$$\begin{bmatrix} A & B \\ 0 & C \end{bmatrix} \begin{bmatrix} A' & B' \\ 0 & C' \end{bmatrix} = \begin{bmatrix} AA' & AB' + BC' \\ 0 & CC' \end{bmatrix}$$

5. Prove Theorem 2.2.6.

6. Prove Theorem 2.2.7 (3)–(6).
7. Find two square matrices A, B such that $AB = \mathbf{0}$ but $BA \neq \mathbf{0}$.
8. Prove Theorem 2.2.16 (3)–(5).
9. Prove Theorem 2.2.17 (2)–(5).
10. Prove Theorem 2.2.19 (2)–(4).

2.3 Rank and Nullity of Matrices

In this section, we will show that the rank of A is the same as the number of free variables of the homogeneous equation $A\mathbf{x} = \mathbf{0}$ namely the number of pivot columns (Corollary 2.3.6). We saw in §1.3, that the nullity of A is the same as the number of nonpivot columns (Theorem 1.3.33). Hence the nullity and rank add up to the number of columns of A which is an important fact.

We first give a different characterization of bases.

Definition 2.3.1. *Let V be a subspace of $\mathbb{R}^n$ and $S \subseteq V$. Let $S_0 = \{\mathbf{v}_1, \mathbf{v}_2, ..., \mathbf{v}_p\}$ be a subset of S. We call S_0 a* **maximal linearly independent subset** *of S if S_0 is linearly independent and $S_0 \cup \{\mathbf{v}\}$ is linearly dependent for any $\mathbf{v} \in S \setminus S_0$.*

The subset S could be V itself. A maximal linearly independent subset of V will be called simply 'a maximal linearly independent subset'. Also note that if S is linearly independent then S itself is a maximal linearly independent subset of S.

Theorem 2.3.2. *Let S be a subset of V and S_0 be a subset of S.*

(1) *Suppose S_0 is a maximal linearly independent subset of S. Then* $\operatorname{Span}(S_0) = \operatorname{Span}(S)$. *On the other hand, if $S_0' \subsetneq S_0$ then* $\operatorname{Span}(S_0') \subsetneq \operatorname{Span}(S)$.

(2) *Suppose S span V. Then S_0 is a basis of V if and only if S_0 is a maximal linearly independent subset of S.*

Proof. (1) Obviously $\operatorname{Span}(S_0) \subseteq \operatorname{Span}(S)$ since $S_0 \subseteq S$. To prove $\operatorname{Span}(S_0) = \operatorname{Span}(S)$ we need to show $S \subseteq \operatorname{Span}(S_0)$ by Theorem 1.3.2. If $\mathbf{v} \in S$ then $S_0 \cup \{\mathbf{v}\}$ is linearly dependent. By Theorem 1.3.19 we see that $\mathbf{v} \in \operatorname{Span}(S_0)$ as contended.

For the second statement, if $\mathbf{v} \in S_0 \setminus S_0'$ then $S_0' \cup \{\mathbf{v}\}$ being a subset of S_0 is linearly independent. By Theorem 1.3.19 again, $\mathbf{v} \notin \operatorname{Span}(S_0')$ i.e., $\operatorname{Span}(S_0) \neq \operatorname{Span}(S)$.

(2) If S_0 is a maximal linearly independent subset of S then $\operatorname{Span}(S_0) = \operatorname{Span}(S) = V$ as we already observed. Hence S_0 span V. Now S_0 is linearly independent by definition. Hence S_0 is a basis of V.

Now if S_0 is a basis then S_0 is linearly independent by definition. To see maximality let $\mathbf{v} \in S \setminus S_0$. Since S_0 is a basis, $\mathbf{v}$ is a linear combination of a basis S_0; $\mathbf{v} \in \operatorname{Span}(S_0)$. By Theorem 1.3.19 we see that $S_0 \cup \{\mathbf{v}\}$ is linearly dependent. □

Hence we can say that a maximal linearly independent subset S_0 of S is an 'optimal' subset of S which spans the same subspace as S does. Any proper subset of S_0 cannot span the same subspace as S does. And any subset of S properly containing S_0 contains a redundant vector to span the subspace $\operatorname{Span}(S)$.

We will apply the theorem to the column space of a matrix. Since the number of a basis, which is uniquely determined (Corollary 4.1.4 cited in Chapter 1), of a subspace V is defined to be the dimension of V, by Theorem 2.3.2 we have:

Corollary 2.3.3. *Let $A \in M(m\times n)$. Then $\rho(A)$ is the number of a maximal linearly independent columns of A.*

Proof. Since $\mathrm{Col}(A)$ is spanned by the columns of A we see that the number of maximal linearly independent columns is the dimension of $\mathrm{Col}(A)$. □

We need to find an effective method to find a maximal linearly independent columns of a matrix. First we consider an example.

Example 2.3.4. *Let* $S = \left\{ \mathbf{v}_1 = \begin{bmatrix} 1 \\ 2 \\ -1 \end{bmatrix}, \mathbf{v}_2 = \begin{bmatrix} -2 \\ -4 \\ 2 \end{bmatrix}, \mathbf{v}_3 = \begin{bmatrix} 3 \\ 2 \\ 1 \end{bmatrix}, \mathbf{v}_4 = \begin{bmatrix} 1 \\ -2 \\ -3 \end{bmatrix}, \mathbf{v}_5 = \begin{bmatrix} 3 \\ 1 \\ 2 \end{bmatrix} \right\}.$
To find $V = \mathrm{Span}\,(S)$ we compute the reduced echelon form B of

$$A = [\mathbf{v}_1\ \mathbf{v}_2\ \mathbf{v}_3\ \mathbf{v}_4\ \mathbf{v}_5] = \begin{bmatrix} 1 & -2 & 3 & 1 & 3 \\ 2 & -4 & 2 & -2 & 1 \\ -1 & 2 & 1 & -3 & 2 \end{bmatrix}$$

to get

$$B = \begin{bmatrix} 1 & -2 & 0 & 0 & -\frac{3}{4} \\ 0 & 0 & 1 & 0 & \frac{5}{4} \\ 0 & 0 & 0 & 1 & 0 \end{bmatrix} \tag{2.22}$$

By Theorem 1.3.36 (2), *S spans $\mathbb{R}^3$. Let's find a maximal linearly independent subset S.*

A set consisting of a single nonzero vector is linearly independent. Hence $S_1 = \{\mathbf{v}_1\}$ is linearly independent. But it is not maximal; $\{\mathbf{v}_1, \mathbf{v}_2\}$ is linearly dependent but $\{\mathbf{v}_1, \mathbf{v}_3\}$ is linearly independent since $\mathbf{v}_3$ is not a constant multiple of $\mathbf{v}_1$. Is $S_2 = \{\mathbf{v}_1, \mathbf{v}_3\}$ a maximal linearly independent set ? Let's try $\mathcal{B} = \{\mathbf{v}_1, \mathbf{v}_3, \mathbf{v}_4\}$. Consider the matrix $B' = [\mathbf{v}_1\ \mathbf{v}_3\ \mathbf{v}_4]$. Then by (2.3.9), *we see that the reeduced echelon form of B' is the identity matrix. Hence $\mathcal{B}$ is linearly independent (a basis of course) by* Theorem 1.3.36.

The number of linearly independent vectors of $\mathbb{R}^3$ does not exceed 3 (Corollary 1.3.9). *Hence $\mathcal{B}$ is a maximal linearly independent subset of S and hence $\mathcal{B}$ is a basis of $\mathbb{R}^3$ by* Theorem 2.3.2.

We can explain the fact that $\mathcal{B}$ forms a basis more succinctly by using the linear equation $A\mathbf{x} = \mathbf{0}$. Let $I \subseteq \{1, 2, 3, 4, 5\}$. Then $A^{(I)}$ and $B^{(I)}$ (see Definition 1.2.2 *for the notation) are row equivalent. Hence the solutions to the equation $A^{(I)}\mathbf{x} = \mathbf{0}$ is the same as the solutions of $B^{(I)}\mathbf{x} = \mathbf{0}$* (Theorem 1.1.7). *For example, if $I = \{1, 3, 4\}$ then $B^{(I)}\mathbf{x} = \mathbf{0}$ has only the trivial solution and hence $A^{(I)}\mathbf{x} = \mathbf{0}$ has only the trivial solution. Therefore, the columns $C(A^{(I)}) = \{A^{(1)}, A^{(3)}, A^{(4)}\}$ are linearly independent. On the other hand, if $I = \{1, 3, 5\}$ then $B^{(I)}\mathbf{x} = \mathbf{0}$ has a solution $\mathbf{x} = \begin{bmatrix} -\frac{3}{4} \\ \frac{5}{4} \\ -1 \end{bmatrix}$; $-\frac{3}{4}B^{(1)} + \frac{5}{4}B^{(3)} - B^{(5)} = \mathbf{0}$ and hence $-3A^{(1)} + 5A^{(3)} - 4A^{(5)} = \mathbf{0}$.*

Therefore we conclude that for $I \subseteq \{1, 2, 3, 4, 5\}$, the linear relations among $C(A^{(I)})$ is the same as the linear relations among $C(B^{(I)})$. In particular, the set of columns $C(A^{(I)})$ is linearly independent if and only if the set of columns $C(B^{(I)})$ are linearly independent.

Practice Problem 2.3.1. *Find a maximal linearly independent subset of*

$$S = \left\{ \begin{bmatrix} 1 \\ 0 \\ -1 \end{bmatrix}, \begin{bmatrix} -1 \\ 0 \\ 1 \end{bmatrix}, \begin{bmatrix} 1 \\ 1 \\ 0 \end{bmatrix}, \begin{bmatrix} 0 \\ 1 \\ 1 \end{bmatrix} \right\}$$

Let $A \in M(m\times n)$. Recall (Definition 1.3.3) that the column space $\text{Col}(A)$ of A is the subspace spanned by the columns $C(A) = \{A^{(1)}, A^{(2)}, \cdots, A^{(n)}\} \subseteq \mathbb{R}^m$ of A. And the row space $\text{Row}(A)$ of A is the subspace spanned by the rows of A. Since the rows of A are the same as the columns of A^t we have

$$\begin{aligned}\text{Col}(A) &= \text{Span}\,\{A^{(1)}, A^{(2)}, \cdots, A^{(n)}\} \subseteq \mathbb{R}^m \\ \text{Row}(A) &= \text{Span}\,\{A_{(1)}, A_{(2)}, \cdots, A_{(m)}\} = \text{Col}(A^t) \subseteq \mathbb{R}_n\end{aligned}$$

The **rank** $\rho(A)$ of A is defined to be (Definition 1.3.30)

$$\rho(A) = \dim \text{Col}(A)$$

Hence $\rho(A)$ is the number of maximal linearly independent columns of the set of columns of A by Corollary 2.3.3.

Recall that the matrices $A, B \in M(m\times n)$ are row equivalent if B is obtained from A by performing elementary operations (Definition 1.1.4). If $I = \{i_1, i_2, ..., i_k\} \subseteq \{1, 2, ..., n\}$ then we write $A^{(I)}$ for the $m\times k$ matrix with the columns $A^{(i_1)}, A^{(i_2)}, ..., A^{(i_k)}$ (Definition 1.2.2).

Theorem 2.3.5. *Let A, B are row equivalent $m\times n$ matrices and $I = \{i_1, i_2, ..., i_k\}$. Then the linear relations among the columns of A^I are the same as the linear relations among the columns of B^I. That is the vector equations*

$$x_1 A^{(i_1)} + \cdots + x_k A^{(i_k)} = \mathbf{0} \tag{2.23}$$

$$x_1 B^{(i_1)} + \cdots + x_k B^{(i_k)} = \mathbf{0} \tag{2.24}$$

have the same solution.

In particular, $\{A^{(i_1)}, ..., A^{(i_k)}\}$ is linearly independent if and only if $\{B^{(i_1)}, ..., B^{(i_k)}\}$ is linearly independent.

Proof. Let $I = \{i_1, i_2, ..., i_k\}$. Then $A^{(I)}$ is row equivalent to $B^{(I)}$. Hence the equations $A^{(I)}\mathbf{x} = \mathbf{0}$ and $B^{(I)}\mathbf{x} = \mathbf{0}$ have the same solutions (Theorem 1.1.7). That is if $\mathbf{c} = \begin{bmatrix} c_1 \\ \vdots \\ c_k \end{bmatrix}$ is a solution for $A^{(I)}\mathbf{x} = \mathbf{0}$ then we have a linear relation $c_1 A^{(i_1)} + \cdots + c_k A^{(i_k)} = \mathbf{0}$. And since $\mathbf{c}$ is also a solution of (2.24), we see that the linear relation $c_1 B^{(i_1)} + \cdots + c_k B^{(i_k)} = \mathbf{0}$ holds.

Also the columns of $A^{(I)}$ are linearly independent if and only if $A^{(I)}\mathbf{x} = \mathbf{0}$ has only the trivial solution. Since $A^{(I)}\mathbf{x} = \mathbf{0}$ and $B^{(I)}\mathbf{x} = \mathbf{0}$ have the same solution we see that the columns of $A^{(I)}$ are linearly independent if and only if the columns of $B^{(I)}$ are linearly independent.

Hence $C(A^{(I)})$ is linearly independent if and only if $C(B^{(I)})$ is linearly independent. And $C(A^{(I)})$ is maximal linearly independent subset of the columns $C(A)$ of A if and only if $C(B^{(I)})$ is maximal linearly independent subset of the columns $C(B)$ of B. □

We extract important facts from the theorem:

Corollary 2.3.6. *Let A and B be row equivalent $m\times n$ matrices. Let $I \subseteq \{1, 2, ..., n\}$.*

(1) *Then $C(A^{(I)})$ is a basis of* $\text{Col}(A)$ *if and only if $C(B^{(I)})$ is a basis of* $\text{Col}(B)$. *In particular, $\rho(A) = \rho(B)$.*

(2) *Let B be a (reduced) echelon form for A and let $C(B^{(I)})$ be the pivot columns of B. Then the columns $C(A^{(I)})$ form a basis for* $\text{Col}(A)$.

Proof. (1) By Theorem 2.3.5, $C(A^{(I)})$ is a basis of $C(A)$ if and only if it is a maximal linearly independent subset. But a linear relation among $C(A^{(I)})$ is the same as the linear relation among $C(B^{(I)})$. Hence $C(A^{(I)})$ is a maximal linearly independent columns of $C(A)$ if and only if $C(B^{(I)})$ is a maximal linearly independent columns of $C(B)$. Whence our assertion.

(2) If B is the reduced echelon form of A then it is obvious that the pivot columns form a basis for Col(B). By (1), the pivot columns of A form a basis for Col(A). □

Corollary 2.3.7. *Let $A \in M(m{\times}n)$. Then*

$$\rho(A) \leq \min\{m, n\} \tag{2.25}$$

Proof. By Corollary 2.3.6, $\rho(A)$ is the same as the number of the pivot columns of A. Now the number of pivot columns of A is the same as the number of leading 1's of the reduced echelon form which does not exceed n and m. □

For the proof of the following corollary bear in mind Example 2.3.4.

Corollary 2.3.8. *The reduced echelon form of an $m{\times}n$ matrix A is unique.*

Proof. Suppose B_1, B_2 are reduced echelon form of A. Then $A \smile B_1, A \smile B_2$ and B_1, B_2 are row equivalent since row equivalence is an equivalence relation (Exercise 1.1.5). We claim that B_1, B_2 have the same pivot columns say, $I = \{i_1, i_2, ..., i_k\}$-th columns. In fact, suppose $B_1^{(i)}$ is the first pivot column of B_1 for which $B_2^{(i)}$ is a nonpivot column of B_2. Then $B_2^{(i)}$ is a linear combination of $B_2^{(j)}$ with $j < i$. But $B_1^{(i)}$, being a pivot column, is not a linear combination of preceding columns of B_1. This contradict to the fact that the linear relations among the columns of B_1 are the same as the linear relations among the columns of B_2 (Theorem 2.3.5).

The pivot columns of B_1, B_2 are of the form $\mathbf{e}^i$ and since they have the same pivot positions, the pivot columns of B_1, B_2 are the same i.e., $B_1^{(i)} = B_2^{(i)}$ for $i \in I$. If the j-th column is a nonpivot column then $B_1^{(j)}$ is a linear combination of pivot columns say $B_1^{(j)} = \sum_{i \in I} a_i B_1^{(i)}$. By Theorem 2.3.5, we have $B_2^{(j)} = \sum_{i \in I} a_i B_2^{(i)}$ which is equal to $\sum_{i \in I} a_i B_1^{(i)}$ since $B_1^{(i)} = B_2^{(i)}$ for $i \in I$. Hence $B_1^{(j)} = B_2^{(j)}$. □

Example 2.3.9. *In* Example 2.3.4, *the reduced echelon form of $A = [\mathbf{v}_1, \mathbf{v}_2, \mathbf{v}_3, \mathbf{v}_4, \mathbf{v}_5]$ was*

$$B = \begin{bmatrix} 1 & -2 & 0 & 0 & -\frac{3}{4} \\ 0 & 0 & 1 & 0 & \frac{5}{4} \\ 0 & 0 & 0 & 1 & 0 \end{bmatrix}$$

Since $\{B^{(1)}, B^{(3)}, B^{(4)}\}$ are pivot columns of B we see that the corresponding columns $\{\mathbf{v}_1, \mathbf{v}_3, \mathbf{v}_4\}$ of A is a basis for Col(A) *as we expected.*

Also observe that $\{B^{(2)}, B^{(4)}, B^{(5)}\}$ is a basis for Col(B). *Therefore, the corresponding columns $\{\mathbf{v}_2, \mathbf{v}_4, \mathbf{v}_5\}$ of A is also a basis for* Col(A).

Example 2.3.10. *We will use our computations of* Example 1.1.13. *We saw that the following matrices are row equivalent:*

$$A_1 = \begin{bmatrix} 1 & -1 & 3 & -1 & 1 & 0 \\ 0 & 1 & -1 & 2 & -2 & -1 \\ 3 & -1 & 1 & -1 & 1 & 1 \end{bmatrix} \qquad A_2 = \begin{bmatrix} 1 & -1 & 3 & -1 & 1 & 0 \\ 0 & 1 & -1 & 2 & -2 & -1 \\ 0 & 0 & -6 & -2 & 2 & 3 \end{bmatrix}$$

$$A_3 = \begin{bmatrix} 1 & -1 & 3 & -1 & 1 & 0 \\ 0 & 1 & 0 & \frac{7}{3} & -\frac{7}{3} & -\frac{3}{2} \\ 0 & 0 & 1 & \frac{1}{3} & -\frac{1}{3} & 3 \end{bmatrix} \qquad A_4 = \begin{bmatrix} 1 & 0 & 0 & \frac{1}{3} & -\frac{1}{3} & 0 \\ 0 & 1 & 0 & \frac{7}{3} & -\frac{7}{3} & -\frac{3}{2} \\ 0 & 0 & 1 & \frac{1}{3} & -\frac{1}{3} & -\frac{1}{2} \end{bmatrix}$$

By Theorem 2.3.5, *any linear relations among the columns of any one of* A_i *gives the same linear relation among the columns of the other matrices. For example, we have the relation* $\frac{1}{3}A_4^{(1)} + \frac{7}{3}A_4^{(2)} + \frac{1}{3}A_4^{(3)} + A_4^{(5)} = \mathbf{0}$;

$$A_4^{(1)} + 7A_4^{(2)} + A_4^{(3)} + 3A_4^{(5)} = \mathbf{0} \tag{2.26}$$

Hence

$$A_1^{(1)} + 7A_1^{(2)} + A_1^{(3)} + 3A_1^{(5)} = \mathbf{0} \tag{2.27}$$

which can be checked by direct computation also. In this way, we can find all linear relations among the columns of A_1. *Observe that the relation of* (2.27) *is not easy to find whereas the relation* (2.26) *is easy to get.*

Example 2.3.11. *Let* $A = \begin{bmatrix} 1 & 3 & 3 & 2 & -9 \\ -1 & -1 & 1 & -4 & 1 \\ 0 & 1 & 2 & -1 & 3 \\ 1 & 2 & 1 & 3 & 6 \end{bmatrix}$. *To find a basis for* Col(A) *we compute an echelon form for* A: $\begin{bmatrix} 1 & 3 & 3 & 2 & -9 \\ 0 & 1 & 2 & -1 & -4 \\ 0 & 0 & 0 & 0 & 1 \\ 0 & 0 & 0 & 0 & 0 \end{bmatrix}$. *Hence we can choose* $\{A^{(1)}, A^{(2)}, A^{(5)}\}$ *as a basis for* Col(A).

Now we can check that $\mathcal{B} = \{A^{(1)}, A^{(2)}, A^{(5)}\}$ *is in fact a basis using the definition of basis: The reduced echelon form for* A *is*

$$B = \begin{bmatrix} 1 & 0 & -3 & 5 & 0 \\ 0 & 1 & 2 & -1 & 0 \\ 0 & 0 & 0 & 0 & 1 \\ 0 & 0 & 0 & 0 & 0 \end{bmatrix}$$

It is not easy to find linear relations among the columns of A. *But, by* Theorem 2.3.5, *the linear relations of the columns of* A *is the same as the linear relations of the columns of* B. *We can easily find that* $B^{(3)} = -3B^{(1)} + 2B^{(2)}$. *Hence* $A^{(3)} = -3A^{(1)} + 2A^{(2)}$. *Also* $B^{(4)} = 5B^{(1)} - B^{(2)}$. *Hence* $A^{(4)} = 5A^{(1)} - A^{(2)}$ *and hence* $\mathcal{B}$ *span* Col(A). *For linear independence, suppose* $c_1A^{(1)} + c_2A^{(2)} + c_5A^{(5)} = \mathbf{0}$. *Then* $c_1B^{(1)} + c_2B^{(2)} + c_5B^{(5)} = c_1\mathbf{e}^1 + c_2\mathbf{e}^2 + c_5\mathbf{e}^3 = \mathbf{0}$ *also. Hence* $c_i = 0$. *Thus* $\mathcal{B}$ *is linearly independent.*

By using the solutions of the equation $A\mathbf{x} = \mathbf{0}$ *we can find more relations. From* B, *the solutions of the equation* $A\mathbf{x} = \mathbf{0}$ *is* $x_3 \begin{bmatrix} 3 \\ -2 \\ 1 \\ 0 \\ 0 \end{bmatrix} + x_4 \begin{bmatrix} -5 \\ 1 \\ 0 \\ 1 \\ 0 \end{bmatrix}$. *If we take* $x_3 = \frac{1}{2}, x_4 = 1$ *then we have the solution* $\begin{bmatrix} -\frac{7}{2} \\ 0 \\ \frac{1}{2} \\ 1 \\ 0 \end{bmatrix}$. *From this we have a relation* $B^{(4)} = \frac{7}{2}B^{(1)} - \frac{1}{2}B^{(3)}$ *and hence we also have the relation* $A^{(4)} = \frac{7}{2}A^{(1)} - \frac{1}{2}A^{(3)}$ *and so on.*

Practice Problem 2.3.2. *Referring to* Example 1.1.16, *find a basis for* $\mathrm{Col}(A_1), \mathrm{Col}(A_2)$ *and write the other columns as linear combinations of the bases*

$$A_1 = \begin{bmatrix} 1 & 2 & 0 & 1 & 2 & 1 \\ 2 & -1 & 1 & -1 & 0 & 1 \\ -1 & 3 & -1 & 2 & 2 & 0 \\ -3 & 1 & -1 & 0 & -2 & -2 \end{bmatrix}, \quad A_2 = \begin{bmatrix} 1 & 2 & 0 & 1 & 2 & 1 \\ 0 & -5 & 1 & -3 & -4 & -1 \\ 0 & 5 & -1 & 3 & 4 & 1 \\ -3 & 1 & -1 & 0 & -2 & -2 \end{bmatrix}$$

using the fact that the reduced echelon form of A_1 and A_2 is

$$B = \begin{bmatrix} 1 & 0 & 0 & 1 & 2 & 1 \\ 0 & 1 & 0 & 0 & 0 & 0 \\ 0 & 0 & 1 & -3 & -4 & -1 \\ 0 & 0 & 0 & 0 & 0 & 0 \end{bmatrix}$$

Remark. *Let B be the reduced form of a matrix A.* Corollary 2.3.6 **does not say that the pivot columns $\{B^{(i_1)}, ..., B^{(i_r)}\}$ of B is a basis for** $\mathrm{Col}(A)$. *They may not belong to* $\mathrm{Col}(A)$. *It says the corresponding columns $\{A^{(i_1)}, ..., A^{(i_r)}\}$ of A is a basis for* $\mathrm{Col}(A)$. *For example, the reduced echelon form of the matrix*

$$A = \begin{bmatrix} 1 & 2 & 0 & 1 & 0 & 2 \\ 0 & -1 & 1 & 0 & 1 & -1 \\ 1 & 1 & 1 & 1 & 1 & 1 \end{bmatrix} \tag{2.28}$$

is

$$B = \begin{bmatrix} \mathbf{1} & 0 & 2 & 1 & 2 & 0 \\ 0 & \mathbf{1} & -1 & 0 & -1 & 1 \\ 0 & 0 & 0 & 0 & 0 & 0 \end{bmatrix} \tag{2.29}$$

Then $\{B^{(1)}, B^{(2)}\}$ is a basis for $\mathrm{Col}(B)$. *Hence $\{A^{(1)}, A^{(2)}\}$ is a basis for* $\mathrm{Col}(A)$. *And* $2A^{(1)} - A^{(2)} = A^{(5)}$ *by* Theorem 2.3.5, *since the same relation holds for the columns of B. However* $\mathrm{Span}\{A^{(1)}, A^{(2)}\} \neq \mathrm{Span}\{B^{(1)}, B^{(2)}\}$. (Check this.)

Theorem 2.3.12. *Let $A \in M(m\times n)$. Then*

$$\rho(A) = \rho(A^t)$$

Proof. By Corollary 2.3.6, $\rho(A)$ is the number of pivot columns. On the other hand, the subspace spanned by the rows of A is the same as the subspace spanned by the rows of reduced echelon form B of A (Theorem 1.3.4). Hence

$$\begin{aligned} \rho(A^t) &= \dim \mathrm{Col}(A^t) = \dim \mathrm{Row}(A) && \text{(Theorem 1.3.4)} \\ &= \dim \begin{pmatrix} \text{subspace spanned by rows of} \\ \text{the reduced echelon form } B \text{ of } A \end{pmatrix} && \text{(Theorem 1.3.26)} \\ &= \#\ (\text{nonzero rows of } B) \\ &= \#\ (\text{pivot positions of } A) = \rho(A) && \text{(Theorem 1.3.26)} \end{aligned}$$ □

For a different proof of this fact see Exercise 6.5.19.

Corollary 2.3.13. *Let $A \in M(m\times n)$ and $\rho(A) = r$. Then there are r maximal linearly independent columns. Also there are r maximal linearly independent rows.*

Proof. Follows immediately form the fact $\rho(A) = \rho(A^t)$. □

Corollary 2.3.14. *Let $A \in M(n)$. Then the following conditions are equivalent.*

(1) $\rho(A) = n$

(2) *The columns of A are linearly independent.*

(3) *The rows of A are linearly independent.*

Proof. Follows immediately from the fact $\rho(A) = \rho(A^t)$. □

Theorem 2.2.13 implies that if $A, B \in M(n)$ and $\rho(B) < n$ then $\rho(AB) < n$. Next corollary strengthens this result.

Corollary 2.3.15. *Let $A, B \in M(n)$. If $\rho(A) < n$ or $\rho(B) < n$ then $\rho(AB) < n$. That is, if the columns (or rows) of A or B are linearly dependent then so are the columns (or rows) of AB.*

Proof. If $\rho(A) = \rho(A^t) < n$ then the rows of A are linearly dependent. Hence the set of columns of A^t is linearly dependent and hence the columns of B^tA^t are linearly dependent by Theorem 2.2.13. Therefore $\rho(AB) = \rho(B^tA^t) < n$.

If $\rho(B) < n$ then the columns of B are linearly dependent by Corollary 2.3.14. By Theorem 2.2.13, the columns of AB are linearly dependent. Hence $\rho(AB) < n$. □

Corollary 2.3.15 fortifies Theorem 2.2.13 and will be generalized in Theorem 4.2.26.

Example 2.3.16. *Let's compute the rank of the transpose A^t of A in* Example 2.3.11. *The transpose of A is* $A^t = \begin{bmatrix} 1 & -1 & 0 & 1 \\ 3 & -1 & 1 & 2 \\ 3 & 1 & 2 & 1 \\ 2 & -4 & -1 & 3 \\ -9 & 1 & 3 & 6 \end{bmatrix}$ *and the echelon form for A^t is* $\begin{bmatrix} 1 & -1 & 0 & 1 \\ 0 & 1 & \frac{1}{2} & -\frac{1}{2} \\ 0 & 0 & 1 & \frac{11}{7} \\ 0 & 0 & 0 & 0 \\ 0 & 0 & 0 & 0 \end{bmatrix}$.
As we expected from Theorem 2.3.12, *the rank of A^t is* 3.

Recall that the **nullity** $\nu(A)$ of $A \in M(m{\times}n)$ is defined by (Definition 1.3.3) $\nu(A) = \dim N(A)$ where $N(A) = \{\mathbf{x} \in \mathbb{R}^n | A\mathbf{x} = \mathbf{0}\}$. Now we prove an important theorem whose primitive form is Theorem 1.1.17.

Theorem 2.3.17 (Rank Theorem). *Let $A \in M(m{\times}n)$. Then*

$$\rho(A) + \nu(A) = n \tag{2.30}$$

Proof. Pivot columns of A will consists of a basis for $\mathrm{Col}(A)$ whose dimension is $\rho(A)$ (Corollary 2.3.6). On the other hand, nonpivot columns correspond to free variables and the number of which is the dimension of $N(A)$ (Theorem 1.3.33). They add up to the number of columns of A and there are n of them. □

For example, look at the reduced echelon form B (2.29) of Remark. The number of pivots (1's in boldface) is 2 which is the rank $\rho(A)$ of A and the number of nonpivot columns is 4 which is the nullity $\nu(A)$ of A. And they add up to 6 which is the same as the number of columns of B (or A).

Corollary 2.3.18. *Let $A \in M(n)$. Then $\nu(A) = \nu(A^t)$.*

Proof. Since $\nu(A) + \rho(A) = n = \nu(A^t) + \rho(A^t)$ and $\rho(A^t) = \rho(A)$ we have the desired result. □

Example 2.3.19. *To find the rank and the nullity of* $A = \begin{bmatrix} -1 & -1 & 0 & 1 & 1 & 0 \\ 0 & -1 & -1 & 1 & 0 & 1 \\ 0 & 1 & 1 & 1 & -1 & 1 \end{bmatrix}$ *we compute the reduced echelon form*

$$B = \begin{bmatrix} 1 & 0 & -1 & 0 & -1 & 1 \\ 0 & 1 & 1 & 0 & -\frac{1}{2} & 0 \\ 0 & 0 & 0 & 1 & -\frac{1}{2} & 1 \end{bmatrix}$$

From this we find that $\rho(A) = 3$ *and* $\nu(A) = 3$. *Hence* $\rho(A) + \nu(A) = 6 =$ *the number of columns of* A.

Answers for Practice Problems 2.3

(2.3.1) $\left\{ \begin{bmatrix} 1 \\ 0 \\ -1 \end{bmatrix}, \begin{bmatrix} 1 \\ 1 \\ 0 \end{bmatrix} \right\}$

(2.3.2) The set of columns $\{A_i^{(1)}, A_i^{(2)}, A_i^{(3)}\}$ is a basis for $\text{Col}(A_i)$ for $i = 1, 2$. And we can write other columns as a linear combinations of the bases: $A_i^{(4)} = A_i^{(1)} - 3A_i^{(3)}$, $A_i^{(5)} = 2A_i^{(1)} - 4A_i^{(3)}$, $A_i^{(6)} = A_i^{(1)} - A_i^{(3)}$ for $i = 1, 2$.

2.3 Exercises

1. Find a set of columns of A which form a basis of $\text{Col}(A)$. And express the other columns as a linear combination of the basis columns. Also find the nullity of A.

(a) $A = \begin{bmatrix} 2 & 2 & 1 \\ -1 & -1 & 0 \\ 1 & 1 & 1 \end{bmatrix}$ (b) $A = \begin{bmatrix} 2 & 1 & 2 & 0 \\ 1 & 1 & 2 & 1 \\ -1 & 0 & 0 & 1 \end{bmatrix}$

(c) $A = \begin{bmatrix} 1 & -1 & 1 \\ 0 & 1 & 1 \\ 2 & -1 & 3 \\ 1 & 0 & 2 \end{bmatrix}$ (d) $A = \begin{bmatrix} 1 & -1 & 0 & 2 \\ 3 & 1 & 1 & 1 \\ 2 & 2 & 1 & -1 \\ 0 & 4 & 1 & -5 \end{bmatrix}$

(e) $A = \begin{bmatrix} 1 & 2 & -1 & 0 & -1 \\ 1 & 0 & 1 & -1 & 0 \\ 0 & 1 & 1 & 1 & 0 \end{bmatrix}$ (f) $A = \begin{bmatrix} 1 & 0 & 1 & 1 & 0 \\ 0 & 1 & 1 & -1 & 1 \\ 0 & 1 & -1 & -1 & 1 \\ 0 & 0 & 1 & 1 & -1 \end{bmatrix}$

2. Prove the true statement(s) and find counter example(s) for the false statement(s).

(a) Let $A \in M(m \times n)$. Then $\rho(A) = 0$ if and only if A is the zero matrix.

(b) If A and B are row equivalent. If $B^{(1)}, B^{(2)}, B^{(4)}$ are the pivot columns then $\text{Span}\{A^{(1)}, A^{(2)}, A^{(4)}\} = \text{Span}\{B^{(1)}, B^{(2)}, B^{(4)}\}$.

(c) Let A be an $n \times n$ matrix. If $\rho(A) = n$ then the columns of A form a basis of $\mathbb{R}^n$.

(d) Let A be an $n \times n$ matrix. If $\rho(A) = n$ then the rows of A form a basis of $\mathbb{R}_n$.

(e) Let A be an $m \times n$ $(m, n \geq 3)$ matrix. If the set of columns $\{A^{(i_1)}, A^{(i_2)}, A^{(i_3)}\}$ of A is linearly independent then the corresponding rows $\{A_{(i_1)}, A_{(i_2)}, A_{(i_3)}\}$ are also linearly independent.

(f) For $A \in M(m \times n)$ we have $\dim \mathrm{Row}(A) = \dim \mathrm{Col}(A)$.

(g) For $A \in M(m \times n)$ we have $\nu(A) = \nu(A^t)$.

(h) There is $A \in M(3 \times 5)$ such that $\rho(A) = 4$.

3. Elementary column operations are defined by replacing 'row' by 'column' as in Definition 1.1.2. We also define two matrices A, B are column equivalent if B is obtained by performing elementary column operations to the matrix A. Prove that if A and B are column equivalent then $\rho(A) = \rho(B)$.

4. Let $A, B \in M(n)$. Prove that if the columns of B are linearly dependent then so are the columns of BA. (Cf. Theorem 2.2.13)

5. Let $A, B \in M(m \times n)$.

(a) Prove that $N(A) \cap N(B) \subseteq N(A + B)$.

(b) Prove that $\mathrm{Col}(A + B) \subseteq \mathrm{Col}(A) + \mathrm{Col}(B)$. (See Definition 1.2.18 for the sum of subspaces.)

2.4 Left and Right Inverses

In this section, we investigate relationship between the matrix $A \in M(m \times n)$ and the corresponding matrix linear map $T_A : \mathbb{R}^n \to \mathbb{R}^m$ associated to A (Theorem 2.1.10). The properties of the function T_A say, injectivity, surjectivity (Definition A.1.4), reflects the properties of the matrix A as we will investigate now.

First, we recall some basic facts about functions which we need in this chapter from Appendix A.

Theorem A.1.5. *Let $f : X \to Y$ be a function.*

(1) The function f is injective if and only if f has a left inverse i.e., there is a function $g : Y \to X$ such that $g \circ f = I_X$.

(2) The function f is surjective if and only if f has a right inverse i.e., there is a function $g : Y \to X$ such that $f \circ g = I_Y$.

(3) The function $f : X \to Y$ is bijective if and only if f is invertible. In this case, there is a unique function $g : Y \to X$ satisfying $g \circ f = I_X$ and $f \circ g = I_Y$.

Theorem 2.4.1. *Let $A \in M(m \times n)$. The following conditions are equivalent.*

(1) The matrix linear map $T_A : \mathbb{R}^n \to \mathbb{R}^m$ is surjective.

(2) For any $\mathbf{b} \in \mathbb{R}^m$ the equation $A\mathbf{x} = \mathbf{b}$ has a solution.

(3) $\rho(A) = m$

(4) The columns of A span $\mathbb{R}^m$.

Any of these conditions implies $n \geq m$.

Proof. The equivalence (1) $\Leftrightarrow$ (2) is clear.

(1) $\Rightarrow$ (3) Assume T_A is surjectve. By Theorem 2.1.10, $\mathrm{Col}(A) = R(T_A)$. Since we assumed T_A is surjective we have $R(T_A) = \mathbb{R}^m$. Hence $\rho(A) = \dim \mathrm{Col}(A) = \dim \mathbb{R}^m = m$.

(3) $\Rightarrow$ (1) Now assume $\rho(A) = m$. Then $\mathrm{Col}(A)$ is a subspace of $\mathbb{R}^m$ of dimension m. Hence by Theorem 4.1.12, cited in §1.3, we have $\mathrm{Col}(A) = \mathbb{R}^m$. Hence T_A is onto.

The equivalence of (3) and (4) is clear.

Now we show $n \geq m$. By Corollary 2.3.6, $\rho(A)$ is the same as the number of pivot columns. But the number of the columns of A is n. Hence the condition (3) implies $n \geq m$. Since the conditions (1), (2) and (4) are equivalent to (3) they all imply $n \geq m$. □

Theorem 2.4.2. *Let $A \in M(m\times n)$. The following conditions are equivalent.*

(1) *The matrix linear map $T_A : \mathbb{R}^n \to \mathbb{R}^m$ is injective.*

(2) $\nu(A) = 0$ *i.e.,* $N(A) = \mathbf{0}$

(3) *The columns of A are linearly independent.*

(4) $\rho(A) = n$

Any of these conditions implies $n \leq m$.

Proof. (1) $\Rightarrow$ (2) Assume T_A is injective. Since T_A is a linear map $T_A(\mathbf{0}) = \mathbf{0}$. If there is a nonzero $\mathbf{x} \in \mathbb{R}^n$ such that $T_A(\mathbf{x}) = \mathbf{0}$ then this contradicts to the fact that T_A is injective since, $\mathbf{0}$ and $\mathbf{x}$ map to the same value $\mathbf{0}$ under T_A.

(2) $\Rightarrow$ (1) Now suppose $N(A) = \mathbf{0}$. To show T_A is injective let $T_A(\mathbf{x}) = T_A(\mathbf{y})$. Then $\mathbf{0} = T_A(\mathbf{x}) - T_A(\mathbf{y}) = T_A(\mathbf{x} - \mathbf{y})$. Hence $\mathbf{x} - \mathbf{y} \in N(A) = \mathbf{0}$. Hence $\mathbf{x} = \mathbf{y}$.

(2) $\Leftrightarrow$ (3) Obvious since both conditions are equivalent to the fact that $A\mathbf{x} = \mathbf{0}$ has only the trivial solution.

(2) $\Leftrightarrow$ (4) Follows from Theorem 2.3.17.

Now, the condition (3) implies that every column of A is a pivot column. Thus $m \geq n$. □

Corollary 2.4.3. *Let $A \in M(m\times n)$. Then $T_A : \mathbb{R}^n \to \mathbb{R}^m$ is surjective if and only if $T_{A^t} : \mathbb{R}^m \to \mathbb{R}^n$ is injective. Or, $T_A : \mathbb{R}^n \to \mathbb{R}^m$ is injective if and only if $T_{A^t} : \mathbb{R}^m \to \mathbb{R}^n$ is surjective.*

Proof. By Theorem 2.4.1, T_A is surjective if and only if $\rho(A) = m$ which is the same as $\rho(A^t)$. By Theorem 2.4.2, this is equivalent to that T_{A^t} is injective. □

Corollary 2.4.4. *Let $A \in M(m\times n)$.*

(1) *If $n < m$ then $T_A : \mathbb{R}^n \to \mathbb{R}^m$ is not surjective.*

(2) *If $n > m$ then $T_A : \mathbb{R}^n \to \mathbb{R}^m$ is not injective.*

Proof. (1) If $n < m$ then $\rho(A) \leq n < m$ and hence T_A is not onto by Theorem 2.4.1.

(2) Our condition implies that $\nu(A) = n - \rho(A) \geq n - m > 0$ by Corollary 2.3.7. By Theorem 2.4.2, T_A is not injective. □

Theorem 2.4.5. *If $T_A : \mathbb{R}^n \to \mathbb{R}^m$ is an isomorphism then $m = n$. And for $A \in M(n)$ the following conditions are equivalent.*

(1) *The matrix linear map $T_A : \mathbb{R}^n \to \mathbb{R}^n$ is bijective i.e., T_A is an isomorphism.*

(2) *For any $\mathbf{b} \in \mathbb{R}^n$ the equation $A\mathbf{x} = \mathbf{b}$ has a unique solution.*

(3) $\rho(A) = n$ *i.e., T_A is surjective.*

(4) $\nu(A) = 0$ *i.e.,* T_A *is injective.*

(5) *The columns of* A *form a basis of* $\mathbb{R}^n$.

(6) *The rows of* A *form a basis of* $\mathbb{R}_n$.

Proof. If T_A is an isomorphism then $\rho(A) = m$ since T_A is surjective. Now $\nu(A) = 0$ since T_A is injective. Now by Theorem 2.3.17, $\nu(A) + \rho(A) = n$. Hence $\rho(A) = n = m$.

(1) $\Leftrightarrow$ (2) is clear.

(1) $\Rightarrow$ (3) This follows from Theorem 2.4.1.

(3) $\Rightarrow$ (4) By Theorem 2.3.17, $\nu(A) + \rho(A) = n$. Since $\rho(A) = n$ we have $\nu(A) = 0$.

(4) $\Rightarrow$ (1) The condition $\nu(A) = 0$ implies T_A is injective by Theorem 2.4.2. By Theorem 2.3.17, we have $\rho(A) = n$. By Theorem 2.4.1, T_A is surjective.

The proof of the equivalences of the conditions (5) and (6) to the other conditions is Exercise 3. □

Practice Problem 2.4.1. *Decide whether* T_{A_i} *is injective or surjective or bijective or neither of these.*

$$A_1 = \begin{bmatrix} 2 & 2 & 1 \\ -1 & -1 & 0 \\ 1 & -1 & 1 \end{bmatrix}, \qquad A_2 = \begin{bmatrix} 2 & 1 & 2 & 0 \\ 1 & 1 & 2 & 1 \\ -1 & 1 & 0 & 1 \end{bmatrix}, \qquad A_3 = \begin{bmatrix} 1 & -1 & 1 \\ 0 & 1 & 1 \\ 2 & -1 & 3 \\ 1 & 0 & 2 \end{bmatrix}$$

We can summarize the properties of $A \in M(m\times n)$ and the corresponding properties of $T_A : \mathbb{R}^n \to \mathbb{R}^m$ in the following table:

property of $A \in M(m\times n)$	property of $T_A : \mathbb{R}^n \to \mathbb{R}^m$
$\nu(A) = 0$	T_A is injective
$\rho(A) = m$	T_A is surjective
$m = n = \rho(A)$	T_A is an isomorphism

Theorem 2.4.6. *Let* $S = \{\mathbf{v}_1, \mathbf{v}_2, ..., \mathbf{v}_n\} \subseteq \mathbb{R}^n$. *The following conditions are equivalent.*

(1) *The set* S *is a basis of* $\mathbb{R}^n$.

(2) *The set* S *is linearly independent.*

(3) *The set* S *spans* $\mathbb{R}^n$.

Proof. Let $A = [\mathbf{v}_1, \mathbf{v}_2, ..., \mathbf{v}_n] \in M(n)$.

(1) $\Leftrightarrow$ (2) By Theorem 1.3.16, a linearly independent subset of vectors in $\mathbb{R}^n$ contains at most n vectors. Since S contains n elements S is a maximal linearly independent subset of $\mathbb{R}^n$. Hence S is a basis (Theorem 2.3.2). The implication (1) $\Rightarrow$ (2) is obvious.

(1) $\Leftrightarrow$ (3) Since the columns of A span $\mathbb{R}^n$ we see $\rho(A) = n$. By Theorem 2.4.2 the columns of A are linearly independent. The converse is obvious. □

Definition 2.4.7. *Let* $A \in M(m\times n)$. *A* **left inverse** *of* A *is* $B \in M(n\times m)$ *such that* $BA = I_n$. *A* **right inverse** *of* A *is* $C \in M(n\times m)$ *such that* $AC = I_m$.

If A is an $m\times n$ matrix and $T_A : \mathbb{R}^n \to \mathbb{R}^m$ is surjective then T_A has a right inverse as a function (Theorem A.1.5). Our question is then whether there is a right inverse which is also linear. The answer is affirmative.

Theorem 2.4.8. *Let $A \in M(m\times n)$. The following conditions are equivalent.*

(1) *The matrix linear map T_A is surjective.*

(2) *There is $B \in M(n\times m)$ such that $AB = I_m$ i.e., $T_A \circ T_B = I_{\mathbb{R}^m}$.*

Proof. First suppose T_A is surjective. Then $A\mathbf{x} = \mathbf{e}^i$ has a solution $\mathbf{c}_i$ for $i = 1, 2, ..., m$. Let $B = [\mathbf{c}_1 \cdots \mathbf{c}_m]$. Then $AB = A[\mathbf{c}_1\ \mathbf{c}_2\ \cdots\ \mathbf{c}_m] = [A\mathbf{c}_1\ A\mathbf{c}_2\ \cdots\ A\mathbf{c}_m] = I_m$.

Conversely, if there is $B \in M(n\times m)$ such that $AB = I_m$ then $T_A \circ T_B = T_{I_m} = I_{\mathbb{R}^m}$ is the identity function on $\mathbb{R}^m$. Since T_A has a right inverse, by Theorem A.1.5, T_A is surjective. □

Example 2.4.9. *We want to find a right inverse of*

$$A = \begin{bmatrix} 1 & 2 & 0 & 3 & -1 \\ 2 & 0 & 1 & 1 & 0 \\ 0 & 1 & 0 & 2 & -2 \end{bmatrix}$$

Let $\{\mathbf{e}^1, \mathbf{e}^2, \mathbf{e}^3\}$ be the standard basis for $\mathbb{R}^3$. First solve $A\mathbf{x} = \mathbf{e}^i$ for $i = 1, 2, 3$. If the solutions for the equations $A\mathbf{x} = \mathbf{e}^i$ $(i = 1, 2, 3)$ are $\mathbf{c}_1, \mathbf{c}_2, \mathbf{c}_3 \in \mathbb{R}^5$ then $B = [\mathbf{c}_1\, \mathbf{c}_2\, \mathbf{c}_3]$ is a 5×3 matrix such that $AB = [A\mathbf{c}_1\, A\mathbf{c}_2\, A\mathbf{c}_3] = I_3$.

For this we form the augmented matrix $[A|\mathbf{e}^i]$ and reduce to an echelon form. Let's take $i = 1$:

$$\left[\begin{array}{ccccc|c} 1 & 2 & 0 & 3 & -1 & 1 \\ 2 & 0 & 1 & 1 & 0 & 0 \\ 0 & 1 & 0 & 2 & -2 & 0 \end{array}\right] \xrightarrow[\text{row operation}]{\text{elementary}} \left[\begin{array}{ccccc|c} 1 & 0 & 0 & -1 & 3 & 1 \\ 0 & 1 & 0 & 2 & -2 & 0 \\ 0 & 0 & 1 & 3 & -6 & -2 \end{array}\right]$$

Substituting back we get $\begin{cases} x_1 = 1 + x_4 - 3x_5 \\ x_2 = -2x_4 + 2x_5 \\ x_3 = -2 - 3x_3 + 6x_5 \end{cases}$ *Setting $x_4 = x_5 = 0$ we get* $\mathbf{c}_1 = \begin{bmatrix} 1 \\ 0 \\ -2 \\ 0 \\ 0 \end{bmatrix}$.

We do this at once: Form $[A|I_3]$ and reduce it to the reduced echelon form

$$\left[\begin{array}{ccccc|ccc} 1 & 2 & 0 & 3 & -1 & 1 & 0 & 0 \\ 2 & 0 & 1 & 1 & 0 & 0 & 1 & 0 \\ 0 & 1 & 0 & 2 & -2 & 0 & 0 & 1 \end{array}\right] \xrightarrow[\text{row operations}]{\text{elementary}} \left[\begin{array}{ccccc|ccc} 1 & 0 & 0 & -1 & 3 & 1 & 0 & -2 \\ 0 & 1 & 0 & 2 & -2 & 0 & 0 & 1 \\ 0 & 0 & 1 & 3 & -6 & -2 & 1 & 4 \end{array}\right]$$

Hence we get the other solutions $\mathbf{c}_2 = \begin{bmatrix} 0 \\ 0 \\ 1 \\ 0 \\ 0 \end{bmatrix}, \mathbf{c}_3 = \begin{bmatrix} -2 \\ 1 \\ 4 \\ 0 \\ 0 \end{bmatrix}$. *Hence we can choose a right inverse B*

$$B = \begin{bmatrix} 1 & 0 & -2 \\ 0 & 0 & 1 \\ -2 & 1 & 4 \\ 0 & 0 & 0 \\ 0 & 0 & 0 \end{bmatrix}$$

of the matrix A.

As the computation shows if we choose a different solution then we get a different right inverse. Hence the right inverse of a matrix is not unique. We could also take $\begin{bmatrix} -1 & 2 & -2 \\ 0 & 0 & 1 \\ 1 & -2 & 4 \\ 1 & -1 & 0 \\ 1 & -1 & 0 \end{bmatrix}$ *for a right inverse of A.*

Practice Problem 2.4.2. *Find a right inverse B of* $A = \begin{bmatrix} 2 & 1 & 2 & 0 \\ 1 & 1 & 2 & 0 \\ -1 & 1 & 0 & 1 \end{bmatrix}$

Theorem 2.4.10. *Let $A \in M(m\times n)$. The following conditions are equivalent.*

(1) *The matrix linear map $T_A : \mathbb{R}^n \to \mathbb{R}^m$ is injective.*

(2) *There is $C \in M(n\times m)$ such that $CA = I_n$.*

Proof. (1) $\Rightarrow$ (2) Assume T_A is injective. Then $0 = \nu(A) = n - \rho(A)$. Hence $\rho(A) = n$. Since $\rho(A) = \rho(A^t)$ we see that $T_{A^t} : \mathbb{R}^m \to \mathbb{R}^n$ is surjective by Corollary 2.4.3. By Theorem 2.4.8, there is $B \in M(n\times m)$ such that $A^tB = I_n$. Taking the transpose we get $B^tA = I_n$ and let $C = B^t$.

(2) $\Rightarrow$ (1) Since $T_C \circ T_A = I_{\mathbb{R}^n}$ we see that T_A has a left inverse. By Theorem A.1.5, T_A is injective. □

Now let us find the left inverse under the condition (1) of Theorem 2.4.10 namely, $\rho(A) = n$. A left inverse B of A satisfy $BA = I_n$. In order to use the previous method, B should be on the right side of A. For this we take the transpose of $BA = I_n$ to get $A^tB^t = I_n$. Since T_A is injective we see that $T_{A^t} : \mathbb{R}^m \to \mathbb{R}^n$ is surjective by Corollary 2.4.3. Hence the equation $A^t\mathbf{x} = \mathbf{b}$ has a solution for any $\mathbf{b} \in \mathbb{R}^n$. Now we can apply previous method with A^t in place of A: Choose $B^t = [\mathbf{c}_1\ \mathbf{c}_2 \cdots\ \mathbf{c}_n]$ where $\mathbf{c}_i$ is a solution for $A^t\mathbf{x} = \mathbf{e}^i$. To get a left inverse B of A we take transpose $B = [\mathbf{c}_1\ \mathbf{c}_2 \cdots\ \mathbf{c}_n]^t$.

Example 2.4.11. *Let* $A = \begin{bmatrix} 1 & -2 & 0 \\ 2 & 0 & 1 \\ 0 & 1 & 0 \\ 3 & 1 & 2 \\ -1 & 0 & -2 \end{bmatrix}$. *To find a left inverse of A let $BA = I_3$ where B is a 3×5 matrix. Taking the transpose we get $A^tB^t = I_3$. We first find B^t: For this we form, as in* Example 2.4.9.

$$[A^t|I_3] = \left[\begin{array}{ccccc|ccc} 1 & 2 & 0 & 3 & -1 & 1 & 0 & 0 \\ -2 & 0 & 1 & 1 & 0 & 0 & 1 & 0 \\ 0 & 1 & 0 & 2 & -2 & 0 & 0 & 1 \end{array}\right] \xrightarrow[\text{row operations}]{\text{elementary}} \left[\begin{array}{ccccc|ccc} 1 & 0 & 0 & -1 & 3 & 1 & 0 & -2 \\ 0 & 1 & 0 & 2 & -2 & 0 & 0 & 1 \\ 0 & 0 & 1 & -1 & 6 & 2 & 1 & -4 \end{array}\right]$$

We have $(B^t)^{(1)} = \begin{bmatrix} 1 \\ 0 \\ 2 \\ 0 \\ 0 \end{bmatrix} + x_4 \begin{bmatrix} 1 \\ -2 \\ 1 \\ 1 \\ 0 \end{bmatrix} + x_5 \begin{bmatrix} -3 \\ 2 \\ -6 \\ 0 \\ 1 \end{bmatrix}$, $(B^t)^{(2)} = \begin{bmatrix} 0 \\ 0 \\ 1 \\ 0 \\ 0 \end{bmatrix} + x_4 \begin{bmatrix} 1 \\ -2 \\ 1 \\ 1 \\ 0 \end{bmatrix} + x_5 \begin{bmatrix} -3 \\ 2 \\ -6 \\ 0 \\ 1 \end{bmatrix}$ *and*

$(B^t)^{(3)} = \begin{bmatrix} -2 \\ 1 \\ -4 \\ 0 \\ 0 \end{bmatrix} + x_4 \begin{bmatrix} 1 \\ -2 \\ 1 \\ 1 \\ 0 \end{bmatrix} + x_5 \begin{bmatrix} -3 \\ 2 \\ -6 \\ 0 \\ 1 \end{bmatrix}$. *Hence we can choose*

$$B = \begin{bmatrix} 1 & 0 & 2 & 0 & 0 \\ 0 & 0 & 1 & 0 & 0 \\ -2 & 1 & -4 & 0 & 0 \end{bmatrix}$$

Practice Problem 2.4.3. *Find a left inverse B of* $A = \begin{bmatrix} 2 & 1 & -1 \\ 1 & 1 & 1 \\ 2 & 2 & 0 \\ 0 & 1 & 1 \end{bmatrix}$.

Answers for Practice Problems 2.4

(2.4.1) The reduced echelon forms of A_i are

$$B_1 = \begin{bmatrix} 1 & 0 & 0 \\ 0 & 1 & 0 \\ 0 & 0 & 1 \end{bmatrix}, \quad B_2 = \begin{bmatrix} 1 & 0 & 0 & -1 \\ 0 & 1 & 0 & 0 \\ 0 & 0 & 1 & 1 \end{bmatrix}, \quad B_3 = \begin{bmatrix} 1 & 0 & 2 \\ 0 & 1 & 1 \\ 0 & 0 & 0 \\ 0 & 0 & 0 \end{bmatrix}$$

Hence we get the following table

	T_{A_1}	T_{A_2}	T_{A_3}
injective	○	○	×
surjective	○	×	×
bijective	○	×	×

(2.4.2) $B = \begin{bmatrix} 1 & -1 & 0 \\ 1 & -1 & 1 \\ -1 & \frac{3}{2} & -\frac{1}{2} \\ 0 & 0 & 0 \end{bmatrix}$

(2.4.3) $B = \begin{bmatrix} 1 & 1 & -1 & 0 \\ -1 & -1 & \frac{3}{2} & 0 \\ 0 & 1 & -\frac{1}{2} & 0 \end{bmatrix}$

2.4 Exercises

1. Find a right inverse of

(a) $A = [1 \ \ 1 \ \ 1]$, (b) $A = \begin{bmatrix} 1 & -2 & 1 \\ -2 & 4 & 1 \end{bmatrix}$, (c) $A = \begin{bmatrix} 1 & 1 & 0 & 2 \\ 0 & 1 & 1 & 1 \\ 1 & 0 & 1 & 0 \end{bmatrix}$

(d) $A = \begin{bmatrix} 0 & 0 & 0 & 1 \\ 0 & 0 & 1 & 1 \\ 0 & 1 & 1 & 1 \end{bmatrix}$, (e) $A = \begin{bmatrix} 1 & 0 & 0 & 1 \\ 0 & 1 & 0 & 1 \\ 0 & 1 & 1 & 1 \end{bmatrix}$

2. Find a left inverse of

$$\text{(a) } A = \begin{bmatrix} 1 \\ 1 \\ 1 \end{bmatrix} \quad \text{(b) } A = \begin{bmatrix} 1 & -2 \\ -2 & 4 \\ 1 & 1 \end{bmatrix} \quad \text{(c) } A = \begin{bmatrix} 1 & -1 & -2 \\ 1 & 1 & 0 \\ 0 & 1 & 1 \\ 1 & 0 & 1 \end{bmatrix} \quad \text{(d) } A = \begin{bmatrix} 1 & -1 & 0 \\ 1 & 1 & 0 \\ 0 & 1 & 1 \\ 1 & 0 & 1 \\ 0 & 1 & 0 \end{bmatrix}$$

3. Let $A \in M(n)$. Prove that the following conditions are equivalent thereby completing the proof of Theorem 2.4.5.

   (a) The matrix linear map T_A is an isomorphism.

   (b) The columns of A form a basis of $\mathbb{R}^n$.

   (c) The rows of A form a basis of $\mathbb{R}_n$.

4. Prove the true statement(s) and find a counter example for the false statement(s).

   (a) A matrix A has a left inverse if and only if T_A has a left inverse as a function.

   (b) A matrix A has a right inverse if and only if T_A has a right inverse as a function.

   (c) The map T_A is surjective if and only if the rows of A are linearly independent.

   (d) The map T_A is injective if and only if the rows of A are linearly independent.

   (e) The map T_A is surjective if the columns $C(A)$ of A are linearly independent.

   (f) The map $T_A : \mathbb{R}^n \to \mathbb{R}^m$ is surjective if the columns of A span $\mathbb{R}^m$.

   (g) If A is a 3×4 matrix then A has no left inverse.

   (h) If A is a 4×3 matrix then A has no right inverse.

   (i) If $A \in M(m\times n)$ has a left inverse as well as a right inverse then $m = n$.

   (j) If $A \in M(n)$ has a right inverse then A also has a left inverse.

2.5 Invertible Matrices and Elementary Matrices

In this section, we will study square matrices having a left and a right inverse.

2.5.1 Invertible Matrices and Corresponding Linear Maps

Definition 2.5.1. *We say $A \in M(n)$ is* **invertible** (*or* **nonsingular**) *if there is $B \in M(n)$ with $AB = BA = I$. In this case B is called an* (*two sided*) **inverse** *of A. A matrix which is not invertible is said to be* **singular**.

A linear map $T : \mathbb{R}^n \to \mathbb{R}^n$ is **invertible** *if there is a linear map $S : \mathbb{R}^n \to \mathbb{R}^n$ such that $S \circ T = I_{\mathbb{R}^n} = T \circ S$ and S is called an* (*two side*) **inverse** *of T.*

Let A be an $n\times n$ matrix. Suppose $BA = AB = I_n$ and $CA = AC = I_n$ also for some $C \in M(n)$. Then $C = CI_n = C(AB) = (CA)B = I_nB = B$. Hence the inverse of an invertible matrix is unique and we will denote it by A^{-1}. Hence it is legitimate to write

$$AA^{-1} = I_n = A^{-1}A$$

Note that invertibility is defined only for the square matrices.

Example 2.5.2. *It is easy to check* $\begin{bmatrix} a & b \\ c & d \end{bmatrix}^{-1} = \dfrac{1}{ad-bc}\begin{bmatrix} d & -b \\ -c & a \end{bmatrix}$ *provided* $ad - bc \neq 0$.

Theorem 2.5.3. *For $A \in M(n)$ the following conditions are equivalent.*

(1) *The matrix linear map T_A is an isomorphism.*

(2) *The matrix linear map T_A is invertible.*

(3) *There is $B \in M(n)$ such that $AB = I_n$.*

(4) *There is $C \in M(n)$ such that $CA = I_n$.*

Further, B and C of (3), (4) *are equal i.e., $AB = I_n$ if and only if $BA = I_n$.*

Proof. (1) $\Rightarrow$ (2) If T_A is an isomorphism then it is a bijection. Hence T_A has an inverse by Theorem A.1.5 (3). By Exercise 2.2.7, the inverse of T_A is also linear. Hence T_A is invertible.

(2) $\Rightarrow$ (1) Immediate by Theorem A.1.5 (3).

(1) $\Rightarrow$ (3) Since T_A is bijective it is, in particular, surjective. Now (3) follows from Theorem 2.4.8.

(3) $\Rightarrow$ (4) Condition (3) implies T_A is surjective. Hence $\rho(A) = n$. By Theorem 2.3.17, $\nu(A) = 0$. Hence T_A is injective. Now (4) follows from Theorem 2.4.10.

(4) $\Rightarrow$ (1) Condition (4) implies T_A is surjective and hence $\rho(A) = n$. Since $\nu(A)+\rho(A) = n$ we have $\nu(A) = 0$. Thus T_A is injective also by Theorem 2.4.2.

For the last statement, suppose $AB = I_n$ and $CA = I_n$. Then $C = CI_n = C(AB) = (CA)B = I_nB = B$. □

For $A \in M(n)$ to be invertible it is necessary and sufficient that $\rho(A) = n$ (Theorem 2.4.5); the reduced echelon form of A is I_n. Now, how do we find A^{-1} when it is invertible ? We already know how to find a right inverse of a matrix when $\rho(A) = n$. The right inverse of $A \in M(n)$, if exists, is the same as the left inverse by Theorem 2.5.3. Hence, to find A^{-1} we form an $n\times 2n$ matrix $[A|I_n]$ and reduce it, by elementary row operations, to $[I_n|B]$ which amount to solving systems of linear equations $A\mathbf{x} = \mathbf{e}^i$ simultaneously. Then $B^{(i)}$ is the solution of $A\mathbf{x} = \mathbf{e}^i$ i.e., $AB^{(i)} = \mathbf{e}^i$. Hence the matrix $B \in M(n)$ obtained in this way is the inverse of A.

From the discussion and the examples above the following fact is obvious.

Theorem 2.5.4. *Let $A \in M(n)$ be invertible. If the reduced echelon form of $[A|I_n]$ is $[I_n|B]$ then $B = A^{-1}$.*

$$\left[\begin{array}{c|c} A & I_m \end{array}\right] \xrightarrow[\text{row operations}]{\text{elementary}} \left[\begin{array}{c|c} I_m & B \end{array}\right]$$

Example 2.5.5. *Let* $A = \begin{bmatrix} 1 & 0 & 1 \\ 0 & 1 & 1 \\ 1 & 1 & 1 \end{bmatrix}$. *To find A^{-1} we need to solve $A\mathbf{x} = \mathbf{e}^i$ $(i = 1, 2, 3)$ as in* Example 2.4.9. *As before, we solve this at once: we form the augmented matrix $[A|I]$ and perform elementary row operations:*

$$\left[\begin{array}{ccc|ccc} 1 & 0 & 1 & 1 & 0 & 0 \\ 0 & 1 & 1 & 0 & 1 & 0 \\ 1 & 1 & 1 & 0 & 0 & 1 \end{array}\right] \xrightarrow[\text{row operations}]{\text{elementary}} \left[\begin{array}{ccc|ccc} 1 & 0 & 0 & 0 & -1 & 1 \\ 0 & 1 & 0 & -1 & 0 & 1 \\ 0 & 0 & 1 & 1 & 1 & -1 \end{array}\right]$$

Hence we get

$$A^{-1} = \begin{bmatrix} 0 & -1 & 1 \\ -1 & 0 & 1 \\ 1 & 1 & -1 \end{bmatrix}$$

Practice Problem 2.5.1. *Find the inverse of* $A = \begin{bmatrix} 2 & 2 & 1 \\ -1 & -1 & 0 \\ 1 & -1 & 1 \end{bmatrix}$.

Theorem 2.5.6. *Let $A \in M(n)$. Then A is invertible if and only if $T_A : \mathbb{R}^n \to \mathbb{R}^n$ is an isomorphism. If A is invertible then*

$$(T_A)^{-1} = T_{A^{-1}}$$

Proof. If A is invertible then there is $B \in M(n)$ such that $AB = I_n = BA$. By Theorem 2.2.15 we have $T_{AB} = T_A \circ T_B = T_{I_n}$ and $T_{BA} = T_B \circ T_A = T_{I_n}$. Since $T_{I_n} = I_{\mathbb{R}^n}$ (Corollary 2.2.9) we have $T_B \circ T_A = I_{\mathbb{R}^n}$. Similarly, $T_A \circ T_C = I_{\mathbb{R}^n}$. Hence T_A is invertible. The converse follows from Theorem 2.5.3.

Now suppose A is invertible then $AA^{-1} = I_n = A^{-1}A$. Hence we have $I_{\mathbb{R}^n} = T_{I_n} = T_A \circ T_{A^{-1}}$. Similarly, $I_{\mathbb{R}^n} = T_{I_n} = T_{A^{-1}} \circ T_A$. Hence $(T_A)^{-1} = T_{A^{-1}}$. □

Next theorem characterize the invertible matrices in terms of systems linear equations.

Theorem 2.5.7. *Let $A \in M(n)$. The following conditions are equivalent.*

(1) *The matrix A is invertible.*

(2) *For any $\mathbf{b} \in \mathbb{R}^n$ the equation $A\mathbf{x} = \mathbf{b}$ has a unique solution.*

(3) *The reduced echelon form of A is the identity matrix; or an echelon form of A is upper triangular with nonzero diagonal entries.*

In this case, the solution to $A\mathbf{x} = \mathbf{b}$ is given by

$$\mathbf{c} = A^{-1}\mathbf{b}$$

Proof. (1) ⇔ (2) The equation $A\mathbf{x} = \mathbf{b}$ has a unique solution if and only if T_A is a bijection. And T_A is a bijection if and only if A is invertible (Theorem 2.5.6).

(2) ⇔ (3) Theorem 1.3.36.

If $\mathbf{c}$ is a solution to the equation $A\mathbf{x} = \mathbf{b}$ then $A\mathbf{c} = \mathbf{b}$. Multiplying A^{-1} both sides we get $\mathbf{c} = A^{-1}A\mathbf{c} = A^{-1}\mathbf{b}$. □

By (2) of Theorem 2.5.7 and Example 1.1.11, we see that a 2×2 matrix $\begin{bmatrix} a & b \\ c & d \end{bmatrix}$ is invertible if and only if $ad - cd \neq 0$. We will generalize this to the $n\times n$ matrices in Theorem 3.3.7.

Solution by the inverse matrix will be useful when we need to solve several equations

$$A\mathbf{x} = \mathbf{b}_1,\ A\mathbf{x} = \mathbf{b}_2,\ ...,\ A\mathbf{x} = \mathbf{b}_k$$

at once.

Example 2.5.8. *Let us solve the equation*

$$\begin{cases} x - 2y + z = 3 \\ 3x + 2z = -1 \\ 2x + y = 5 \end{cases}$$

by using

$$\begin{bmatrix} 1 & -2 & 1 \\ 3 & 0 & 2 \\ 2 & 1 & 0 \end{bmatrix}^{-1} = \frac{1}{7}\begin{bmatrix} 2 & -1 & 4 \\ -4 & 2 & -1 \\ -3 & 5 & -6 \end{bmatrix}$$

The solution will be given by $A^{-1}\mathbf{b} = \frac{1}{7}\begin{bmatrix} 27 \\ -19 \\ -44 \end{bmatrix}$ *where* $\mathbf{b} = \begin{bmatrix} 3 \\ -1 \\ 5 \end{bmatrix}$. *Therefore, we get the solution* $x = \frac{27}{7}, y = -\frac{19}{7}, z = -\frac{44}{7}$.

Practice Problem 2.5.2. *Find the solution of* $\begin{cases} ax + by = c \\ a'x + b'y = c' \end{cases}$ *using* Example 2.5.2 *where* $ab' - a'b \neq 0$.

We list some properties of the inverse matrices.

Theorem 2.5.9. *Let* $A, B \in M(n)$ *be invertible matrices. Then we have*

(1) *The inverse* A^{-1} *is also invertible and* $(A^{-1})^{-1} = A$.

(2) *The product* AB *is invertible and* $(AB)^{-1} = B^{-1}A^{-1}$.

(3) *The transpose* A^t *is invertible and* $(A^t)^{-1} = (A^{-1})^t$

Proof. Proofs for the statements are straightforward. For example, for (2) we simply check: $AB(B^{-1}A^{-1}) = A(BB^{-1})A^{-1} = AA^{-1} = I$. Hence $(AB)^{-1} = B^{-1}A^{-1}$. □

2.5.2 Elementary Matrices and Matrix Inverse

To find the inverse of A we formed an $n \times 2n$ matrix $[A|I_n]$ and reduce it, by elementary row operations, to $[I_n|B]$ which amount to solving systems of linear equations $A\mathbf{x} = \mathbf{e}^i$ simultaneously. Then the matrix $B \in M(n)$ obtained in this way is the inverse of A.

$$\left[\begin{array}{c|c} A & I_n \end{array}\right] \xrightarrow[\text{row operations}]{\text{elementary}} \left[\begin{array}{c|c} I_n & B \end{array}\right]$$

We will take another point of view for finding the inverse of a matrix.

Definition 2.5.10. *An* **elementary matrix** *is a square matrix obtained by applying one of the elementary row operations to the identity matrix* I_n. *So it is one of the following* $n \times n$ *matrices:*

$$E_{ij}(c) := \begin{pmatrix} \text{Add } c\text{-times the } i\text{-th row} \\ \text{of } I_n \text{ to the } j\text{-th row} (i \neq j) \end{pmatrix} = \begin{cases} a_{ii} = 1 \\ a_{ji} = c \\ 0 \text{ elsewhere} \end{cases}$$

$$I_{ij} := \begin{pmatrix} \text{interchange } i\text{-th and} \\ j\text{-th rows of } I_n \ (i \neq j) \end{pmatrix} = \begin{cases} a_{ij} = a_{ji} = 1 \\ a_{kk} = 1 (k \neq i, j) \\ 0 \text{ elsewhere} \end{cases}$$

$$M_i(c) := \begin{pmatrix} \text{multiply } c \text{ to the} \\ i\text{-th row of } I_n \end{pmatrix} = \begin{cases} a_{ii} = c \\ a_{kk} = 1 (k \neq i) \\ 0 \text{ elsewhere} \end{cases}$$

The transpose of the elementary matrices are also elementary and obtained by performing the corresponding elementary column operations to the identity matrix. Hence

$$E_{ij}(c)^t = E_{ji}(c),\ I_{ij}^t = I_{ij} \text{ and } M_i(c)^t = M_i(c) \tag{2.31}$$

For the multiple AB of two matrices we say A is **pre-multiplied** to B or B is **post-multiplied** to A.

Theorem 2.5.11. *Performing an elementary row (resp. column) operation to a matrix is the same as pre–(resp. post–) multiplying the (resp. transpose of) corresponding elementary matrices of the right size.*

Proof. The following computation would be good enough to convince you the truth of the theorem: Choose $A = \begin{bmatrix} a & b & c \\ 1 & 2 & 3 \\ x & y & z \end{bmatrix}$ and pre–multiply $I_{12}, M_2(5), E_{31}(2)$ and post–multiply their transposes to A. (Here is the reason why the column operation works: Consider $(AE^t)^t = EA^t$ where E is an elementary matrix. This means that post-multiplying E^t to A is the same as performing the corresponding row operation to A^t which amounts to perform the corresponding column operation to A.) □

As the elementary row operations are reversible (Theorem 1.1.5) the elementary matrices are invertible and their inverses are again elementary:

$$E_{ij}(c)^{-1} = E_{ij}(-c), \quad I_{ij}^{-1} = I_{ij}, \quad M_i(c)^{-1} = M_i(c^{-1}) \tag{2.32}$$

Theorem 2.5.12. *Let $A \in M(n)$. Then A is invertible if and only if A is a product of elementary matrices say*

$$A = E_k E_{k-1} \cdots E_1$$

where E_i's are elementary matrices. In this case,

$$A^{-1} = E_1^{-1} E_2^{-1} \cdots E_k^{-1}$$

and

$$E_1^{-1} E_2^{-1} \cdots E_k^{-1} [A|I_n] = [I_n|A^{-1}]$$

Proof. If A is invertible then, the augmented matrix $[A|I_n]$ reduces to, by elementary row operations, $[I_n|B]$. Performing elementary row operations amounts to pre-multiplying elementary matrices F_i:

$$F_k \cdots F_2 F_1 [A|I_n] = [F_k \cdots F_2 F_1 A | F_k \cdots F_2 F_1] = [I_n|B] \tag{2.33}$$

Hence $F_k \cdots F_2 F_1 = A^{-1}$. Thus $A = F_1^{-1} F_2^{-1} \cdots F_k^{-1}$. Since the inverse of an elementary matrix is also elementary we get the desired result by renaming the elementary matrices say, $E_i = F_{k-i}^{-1}$.

Conversely, suppose A is a product of elementary matrices $A = E_k E_{k-1} \cdots E_1$. Since the elementary matrices are invertible and a product of invertible matrices are also invertible (Theorem 2.5.9) we conclude that A is also invertible. □

As we observed in Chapter 1, there are many different ways to reduce a matrix into the reduced echelon form using elementary row operations. Hence the product expression in elementary matrices is not unique.

Example 2.5.13. *To find the inverses of $A = \begin{bmatrix} 1 & 1 & 0 \\ 0 & 1 & 1 \\ 0 & 1 & 0 \end{bmatrix}$ form an augmented matrix $[A|I]$ and perform the elementary row operations*

$$\left[\begin{array}{ccc|ccc} 1 & 1 & 0 & 1 & 0 & 0 \\ 0 & 1 & 1 & 0 & 1 & 0 \\ 0 & 1 & 0 & 0 & 0 & 1 \end{array}\right] \xrightarrow[-1R_3]{-1R_2+R_3} \left[\begin{array}{ccc|ccc} 1 & 1 & 0 & 1 & 0 & 0 \\ 0 & 1 & 1 & 0 & 1 & 0 \\ 0 & 0 & 1 & 0 & 1 & -1 \end{array}\right] \xrightarrow{-1R_3+R_2}$$

$$\left[\begin{array}{ccc|ccc} 1 & 1 & 0 & 1 & 0 & 0 \\ 0 & 1 & 0 & 0 & 0 & 1 \\ 0 & 0 & 1 & 0 & 1 & -1 \end{array}\right] \xrightarrow{-1R_2+R_1} \left[\begin{array}{ccc|ccc} 1 & 0 & 0 & 1 & 0 & -1 \\ 0 & 1 & 0 & 0 & 0 & 1 \\ 0 & 0 & 1 & 0 & 1 & -1 \end{array}\right]$$

Hence

$$A^{-1} = \begin{bmatrix} 1 & 0 & -1 \\ 0 & 0 & 1 \\ 0 & 1 & -1 \end{bmatrix}$$

To express A as a product of elementary matrices we keep track of the elementary row operations and interpret them in terms of elementary matrices using Theorem 2.5.11*:*

$$E_{21}(-1)E_{32}(-1)M_3(-1)E_{23}(-1)[A|I_3] = [I_3|E_{21}(-1)E_{32}(-1)M_3(-1)E_{23}(-1)] = [I_3|A^{-1}]$$

Hence $A^{-1} = E_{21}(-1)E_{32}(-1)M_3(-1)E_{23}(-1)$. Therefore

$$\begin{aligned} A &= [E_{21}(-1)E_{32}(-1)M_3(-1)E_{23}(-1)]^{-1} = E_{23}(-1)^{-1}M_3(-1)^{-1}E_{32}(-1)^{-1}E_{21}(-1)^{-1} \\ &= E_{23}(1)M_3(-1)E_{32}(1)E_{21}(1) \end{aligned}$$

Practice Problem 2.5.3. *Find the inverses of $A = \begin{bmatrix} 1 & 0 & 1 & 0 \\ 1 & 0 & 0 & 0 \\ 1 & 1 & 0 & 0 \\ 1 & 0 & 0 & 1 \end{bmatrix}$ and express A as a product of elementary matrices.*

2.5.3 Characterization of Invertible Matrices

We summarize the conditions for $A \in M(n)$ to be invertible. Another important characterization of the invertible matrices using determinant will be given in Theorem 3.3.7.

Theorem 2.5.14. *Let $A \in M(n)$. The following conditions are equivalent.*

(1) *The matrix A is invertible.*

(2) *The matrix A is row equivalent to I_n i.e., A has n pivot positions.*

(3) $\nu(A) = 0$

(4) $\rho(A) = n$

(5) *The columns of A form a basis of $\mathbb{R}^n$.*

(6) *The rows of A form a basis of $\mathbb{R}_n$.*

(7) *There is $B \in M(n)$ such that $AB = I_n$.*

(8) *There is $C \in M(n)$ such that $CA = I_n$.*

(9) *The matrix A^t is invertible.*

(10) *An echelon form of A is upper triangular with nonzero diagonal entries.*

Proof. (1) $\Leftrightarrow$ (2) This is Theorem 2.5.12.

(1) $\Leftrightarrow$ (3) $\Leftrightarrow$ (4) $\Leftrightarrow$ (5) $\Leftrightarrow$ (6) Theorem 2.4.5.

(4) $\Leftrightarrow$ (7) Follows from Theorem 2.4.8 since $\rho(A) = n$ is equivalent to the condition that T_A is surjective.

(4) $\Leftrightarrow$ (8) Follows from Theorem 2.4.10 since the injectivity of T_A is equivalent to the condition that $\nu(A) = 0$.

(1) $\Leftrightarrow$ (9) The matrix A is invertible if and only if there is B such that $AB = I = BA$. Taking the transpose this is equivalent to the condition that $B^tA^t = I = A^tB^t$ and *vice versa*.

(1) $\Leftrightarrow$ (10) Theorem 2.5.7. □

The heart of the theorem is: If $A \in M(n)$ then

$$T_A \text{ is invertible } \Leftrightarrow T_A \text{ is injective } \Leftrightarrow T_A \text{ is surjective}$$

Answers for Practice Problems 2.5

(2.5.1) $A^{-1} = \frac{1}{2}\begin{bmatrix} -1 & -3 & 1 \\ 1 & 1 & -1 \\ 2 & 4 & 0 \end{bmatrix}$

(2.5.2) $\begin{bmatrix} x \\ y \end{bmatrix} = A^{-1}\begin{bmatrix} c \\ c' \end{bmatrix} = \frac{1}{ab'-bc'}\begin{bmatrix} b' & -b \\ -a' & a \end{bmatrix}\begin{bmatrix} c \\ c' \end{bmatrix} = \begin{bmatrix} \frac{b'c-bc'}{ab'-bc'} \\ \frac{ac'-a'c}{ab'-bc'} \end{bmatrix}$. Hence $x = \dfrac{b'c-bc'}{ab'-bc'}$ and $y = \dfrac{ac'-a'c}{ab'-bc'}$.

(2.5.3) There are many different way to write A as a product of elementary matrices. Here is one: By computing the reduced echelon form of A we get

$$E_{31}(-1)E_{32}(1)E_{34}(1)M_3(-1)I_{23}E_{14}(-1)E_{13}(-1)E_{12}(-1)A = I_4$$

Hence

$$\begin{aligned} A &= E_{12}(-1)^{-1}E_{13}(-1)^{-1}E_{14}(-1)^{-1}I_{23}^{-1}M_3(-1)^{-1}E_{34}(1)^{-1}E_{32}(1)^{-1}E_{31}(-1)^{-1} \\ &= E_{12}(1)E_{13}(1)E_{14}(1)I_{23}M_3(-1)E_{34}(-1)E_{32}(-1)E_{31}(1) \end{aligned}$$

2.5 Exercises

1. Decide whether the following matrices are invertible or not. If invertible find its inverse and write the given matrix A as a product of elementary matrices.

   (a) $A = \begin{bmatrix} 1 & 3 \\ 1 & 2 \end{bmatrix}$ (b) $A = \begin{bmatrix} 0 & 0 & 1 \\ 0 & 1 & 0 \\ 1 & 0 & 0 \end{bmatrix}$ (c) $A = \begin{bmatrix} 1 & 0 & 1 \\ 0 & 1 & 1 \\ 2 & -1 & 1 \end{bmatrix}$

   (d) $A = \begin{bmatrix} 1 & 0 & 0 & 1 \\ 0 & 0 & 0 & -1 \\ -1 & 0 & 1 & 1 \\ 1 & 1 & 0 & 1 \end{bmatrix}$ (e) $A = \begin{bmatrix} 1 & 2 & 3 & 4 \\ 5 & 6 & 7 & 8 \\ 9 & 10 & 11 & 12 \\ 13 & 14 & 15 & 16 \end{bmatrix}$

2. Find the inverse of

   (a) $A = \begin{bmatrix} 1 & 2 \\ -1 & 1 \end{bmatrix}$ (b) $A = \begin{bmatrix} 0 & 0 & -1 \\ 0 & 1 & 0 \\ 1 & 0 & 1 \end{bmatrix}$ (c) $A = \begin{bmatrix} 1 & 0 & -1 \\ 1 & 1 & 0 \\ 1 & 1 & 1 \end{bmatrix}$

   (d) $A = \begin{bmatrix} 1 & 0 & 0 & 0 \\ 0 & 1 & 1 & 1 \\ 1 & 0 & 0 & 1 \\ 0 & 1 & 0 & 0 \end{bmatrix}$ (e) $A = \begin{bmatrix} 1 & 1 & 0 & 0 \\ 0 & 1 & 1 & 1 \\ 1 & 0 & 1 & 0 \\ 0 & 1 & 0 & 0 \end{bmatrix}$

3. Prove (2.31) and (2.32)

4. Let $A \in M(n)$ be an invertible matrix. Prove that $(A^t)^{-1} = (A^{-1})^t$ and $(A^{-1})^{-1} = A$ (Theorem 2.5.9 (1), (3)).

5. Let $A, B \in M(n)$. If $I_n - AB$ is invertible then so is $I_n - BA$. (Hint: Prove by contradiction: $I_n - BA$ is not invertible if and only if there is nonzero $\mathbf{x}$ such that $(I_n - BA)\mathbf{x} = \mathbf{0}$.)

6. Let $A, B \in M(n)$. Prove that if AB is invertible then so are A and B.

7. Prove the true statement(s) and find a counter example for the false statement(s).

   (a) If $A \in M(n)$ has a left inverse then A also has a right inverse.

   (b) If A is upper or lower triangular with nonzero diagonal entries then A is invertible.

   (c) The matrix $A_\theta = \begin{bmatrix} \cos\theta & -\sin\theta \\ \sin\theta & \cos\theta \end{bmatrix}$ is invertible for any θ.

   (d) If $A \in M(n)$ is invertible and c is a nonzero scalar then cA is also invertible.

   (e) If $A \in M(n)$ is invertible and B is row equivalent to A then B is also invertible.

   (f) The matrix $A \in M(n)$ is invertible if and only if the columns of A are linearly independent.

   (g) The matrix $A \in M(n)$ is invertible if and only if the rows of A are linearly independent.

   (h) For $A \in M(n)$, the matrix linear map T_A is invertible if and only if T_A is surjective.

   (i) For $A \in M(n)$, the matrix linear map T_A is invertible if and only if T_A is injective.

8. Let A be a $r \times r$ matrix and C be an $s \times s$ matrix and B be an $r \times s$ matrix. Prove that

$$\begin{bmatrix} A & B \\ 0 & C \end{bmatrix}$$

   is invertible if and only if A, C are invertible and find its inverse. (Hint: Use Exercise 2.2.4)

Chapter 3

Determinant

In this chapter, we study an important function which assigns a number to each square matrix so called the determinant. We will see that there are wide varieties of applications of determinant.

3.1 Determinant of Order 2

We start with the determinant of a 2×2 matrix. Let $A = \begin{bmatrix} a_{11} & a_{12} \\ a_{21} & a_{22} \end{bmatrix}$. We saw, in Example 1.1.11, that $A\mathbf{x} = \mathbf{b}$ has a unique solution if and only if $a_{11}a_{22} - a_{12}a_{21} \neq 0$. We define the quantity $a_{11}a_{22} - a_{12}a_{21}$ to be the determinant of A which we denote by $\det(A)$ or $|A|$.

Definition 3.1.1. *Let* $A = \begin{bmatrix} a_{11} & a_{12} \\ a_{21} & a_{22} \end{bmatrix}$. *We define the* **determinant** *of* A *by*

$$\det(A) = a_{11}a_{22} - a_{12}a_{21} \tag{3.1}$$

We can view the determinant as a function in the columns of A i.e., if $A = [\mathbf{v}_1\ \mathbf{v}_2]$ where $\mathbf{v}_1 = \begin{bmatrix} a_{11} \\ a_{21} \end{bmatrix}$, $\mathbf{v}_2 = \begin{bmatrix} a_{12} \\ a_{22} \end{bmatrix}$ then we can write

$$\det(A) = \det[\mathbf{v}_1\ \mathbf{v}_2]$$

We prove basic properties of determinant for 2×2 matrices which we will generalize soon.

Theorem 3.1.2. *Let* $\mathbf{v}_1, \mathbf{v}_1', \mathbf{v}_2, \mathbf{v}_2'$ *be column vectors in* $\mathbb{R}^2$ *and* c, d *be constants. Then*

(1) $\det[\mathbf{v}_1 + \mathbf{v}_1'\ \ \mathbf{v}_2] = \det[\mathbf{v}_1\ \mathbf{v}_2] + \det[\mathbf{v}_1'\ \mathbf{v}_2]$ *and* $\det[c\mathbf{v}_1\ \mathbf{v}_2] = c\det[\mathbf{v}_1\ \mathbf{v}_2]$

(2) $\det[\mathbf{v}_1\ \ \mathbf{v}_2 + \mathbf{v}_2'] = \det[\mathbf{v}_1\ \mathbf{v}_2] + \det[\mathbf{v}_1\ \mathbf{v}_2']$ *and* $\det[\mathbf{v}_1\ d\mathbf{v}_2] = d\det[\mathbf{v}_1\ \mathbf{v}_2]$

(3) $\det[\mathbf{v}_1\ \mathbf{v}_2] = -\det[\mathbf{v}_2\ \mathbf{v}_1]$

(4) $\det[\mathbf{e}^1\ \mathbf{e}^2] = 1$

Proof. (1) Let $\mathbf{v}_1 = \begin{bmatrix} a_{11} \\ a_{21} \end{bmatrix}$, $\mathbf{v}_1' = \begin{bmatrix} a_{11}' \\ a_{21}' \end{bmatrix}$ and $\mathbf{v}_2 = \begin{bmatrix} a_{12} \\ a_{22} \end{bmatrix}$, $\mathbf{v}_2' = \begin{bmatrix} a_{12}' \\ a_{22}' \end{bmatrix}$. For the first equality we compute:

$$\begin{aligned}\det[\mathbf{v}_1+\mathbf{v}_1'\ \ \mathbf{v}_2] &= \det\begin{bmatrix}a_{11}+a_{11}' & a_{12}\\ a_{21}+a_{21}' & a_{22}\end{bmatrix}\\ &= (a_{11}a_{22}-a_{21}a_{12})+(a_{11}'a_{22}-a_{21}'a_{12})\\ &= \det[\mathbf{v}_1\ \mathbf{v}_2]+\det[\mathbf{v}_1'\ \mathbf{v}_2]\end{aligned}$$

For the second equality we have $\det[c\mathbf{v}_1\ \mathbf{v}_2] = ca_{11}a_{22} - ca_{12}a_{21} = c\det[\mathbf{v}_1\ \mathbf{v}_2]$.

The equalities of (2),(3) and (4) are similarly proved and we leave the proofs as an exercise. □

Practice Problem 3.1.1. *Complete the proof of* Theorem 3.1.2.

Observe that the properties of (2) of Theorem 3.1.2 will follow from the properties (1) and (3). In fact, $\det[\mathbf{v}_1\ d\mathbf{v}_2] = -\det[d\mathbf{v}_2\ \mathbf{v}_1] = -d\det[\mathbf{v}_2\ \mathbf{v}_1] = d\det[\mathbf{v}_1\ \mathbf{v}_2]$. The first property of (2) can be similarly proved.

Properties (1) and (2) of Theorem 3.1.2 says that the determinant is linear in the first and the second column. Properties (1) and (2) say that the determinant (of order 2) is **bilinear** in the columns. Bilinearlity of determinant implies

$$\det[\mathbf{v}\ \ \mathbf{0}] = \det[\mathbf{0}\ \ \mathbf{w}] = 0 \text{ for any } \mathbf{v},\mathbf{w}\in\mathbb{R}^2 \tag{3.2}$$

In fact, $\det[\mathbf{v}\ \ \mathbf{0}] = \det[\mathbf{v}\ \ \mathbf{0}+\mathbf{0}] = \det[\mathbf{v}\ \ \mathbf{0}] + \det[\mathbf{v}\ \ \mathbf{0}]$. Hence $\det[\mathbf{v}\ \ \mathbf{0}] = 0$. Similarly, we can show $\det[\mathbf{0}\ \ \mathbf{w}] = 0$. Of course, we can prove (3.2) by direct computation.

The property (3) in Theorem 3.1.2 says that the determinant is **alternating**. From this we have

$$\det[\mathbf{v}\ \ \mathbf{v}] = 0 \text{ for any } \mathbf{v}\in\mathbb{R}^2 \tag{3.3}$$

since we have $\det[\mathbf{v}\ \mathbf{v}] = -\det[\mathbf{v}\ \mathbf{v}]$ by interchanging two columns.

Practice Problem 3.1.2. *Using the properties of* Theorem 3.1.2 *compute* $\det\begin{bmatrix}-2 & 1\\ 3 & -5\end{bmatrix}$.

Using the properties of Theorem 3.1.2 we can recover the definition of determinant.

Theorem 3.1.3. *Let $f : M(2)\to\mathbb{R}$ be a function satisfying the properties* (1),(2),(3) *and* (4) *in the variables $\mathbf{v}_1,\mathbf{v}_2\in\mathbb{R}^2$. Then $f(\mathbf{v}_1,\mathbf{v}_2) = \det[\mathbf{v}_1\ \mathbf{v}_2]$.*

Proof. Let $A = [\mathbf{v}_1\ \mathbf{v}_2]$ be a 2×2 matrix with the columns $\mathbf{v}_1 = \begin{bmatrix}a_{11}\\ a_{21}\end{bmatrix}, \mathbf{v}_2 = \begin{bmatrix}a_{12}\\ a_{22}\end{bmatrix}$ and let $\{\mathbf{e}^1,\mathbf{e}^2\}$ be the standard basis of $\mathbb{R}^2$. Then we can write $A = [a_{11}\mathbf{e}^1+a_{21}\mathbf{e}^2\ \ \ a_{12}\mathbf{e}^1+a_{22}\mathbf{e}^2]$. Hence by (1), (2) and (3) we have

$$\begin{aligned}f(A) &= a_{11}f[\mathbf{e}^1\ \ a_{12}\mathbf{e}^1+a_{22}\mathbf{e}^2] + a_{21}f[\mathbf{e}^2\ \ a_{12}\mathbf{e}^1+a_{22}\mathbf{e}^2]\\ &= a_{11}a_{12}f[\mathbf{e}^1\ \mathbf{e}^1] + a_{11}a_{22}f[\mathbf{e}^1\ \ \mathbf{e}^2] + a_{21}a_{12}f[\mathbf{e}^2\ \ \mathbf{e}^1] + a_{21}a_{22}f[\mathbf{e}^2\ \ \mathbf{e}^2]\\ &= a_{11}a_{22}f[\mathbf{e}^1\ \ \mathbf{e}^2] + a_{21}a_{12}f[\mathbf{e}^2\ \ \mathbf{e}^1]\end{aligned}$$

since $f[\mathbf{e}^i\ \ \mathbf{e}^i] = 0$ $(i = 1,2)$. Now $f[\mathbf{e}^1\ \ \mathbf{e}^2] = 1$ and $f[\mathbf{e}^2\ \ \mathbf{e}^1] = -1$ by (3) and (4), we have

$$f\,(A) = a_{11}a_{22} - a_{12}a_{21} = \det(A)$$ □

Next theorem characterizes the invertible matrices by determinant.

Theorem 3.1.4. *Let $A\in M(2)$. Then A is invertible if and only if* $\det(A)\neq 0$.

Proof. We saw in Example 1.1.11 that $A\mathbf{x} = \mathbf{b}$ has a unique solution if and only if $\det(A) \neq 0$. By Theorem 2.4.5, we conclude that A is invertible if and only if $\det(A) \neq 0$. $\square$

Answers for Practice Problems 3.1

(3.1.1) Let $\mathbf{v}_1 = \begin{bmatrix} a_1 \\ a_2 \end{bmatrix}$, $\mathbf{v}_2 = \begin{bmatrix} b_1 \\ b_2 \end{bmatrix}$. Then $\det[\mathbf{v}_1\ \mathbf{v}_2] = a_1b_2 - a_2b_1$. On the other hand, $\det[\mathbf{v}_2\ \mathbf{v}_1] = a_2b_1 - a_1b_2$. Hence $\det[\mathbf{v}_1\ \mathbf{v}_2] = -\det[\mathbf{v}_2\ \mathbf{v}_1]$. Proof of the other equalities are straightforward computations.

(3.1.2) We have $\mathbf{v}_1 = \begin{bmatrix} -2 \\ 3 \end{bmatrix} = -2\mathbf{e}^1 + 3\mathbf{e}^2$, $\mathbf{v}_2 = \begin{bmatrix} 1 \\ -5 \end{bmatrix} = \mathbf{e}^1 - 5\mathbf{e}^2$; $[\mathbf{v}_1\ \mathbf{v}_2] = \begin{bmatrix} -2 & 1 \\ 3 & -5 \end{bmatrix}$.

Hence we have

$$\begin{aligned}
&\det[\mathbf{v}_1\ \mathbf{v}_2] \\
=&\det[-2\mathbf{e}^1 + 3\mathbf{e}^2 \quad \mathbf{e}^1 - 5\mathbf{e}^2] \\
=&(-2)\det[\mathbf{e}^1 \quad \mathbf{e}^1 - 5\mathbf{e}^2] + (3)\det[\mathbf{e}^2 \quad \mathbf{e}^1 - 5\mathbf{e}^2] \\
=&(-2)\det[\mathbf{e}^1\ \mathbf{e}^1] + (-2)(-5)\det[\mathbf{e}^1\ \mathbf{e}^2] + (3)\det[\mathbf{e}^2\ \mathbf{e}^1] + (3)(-5)\det[\mathbf{e}^2\ \mathbf{e}^2] \\
=&(-2)0 + (-2)(-5) + (3)(-1) + (3)(-5)0 \\
=&7
\end{aligned}$$

3.1 Exercises

1. Find the condition on a for $A = \begin{bmatrix} 1 & 2 \\ -1 & a \end{bmatrix}$ to be invertible.

2. Let $A \in M(2)$. Prove that the following statements are equivalent.

   (a) $\det(A) \neq 0$

   (b) $\{A^{(1)}, A^{(2)}\}$ is linearly independent.

   (c) $\{A_{(1)}, A_{(2)}\}$ is linearly independent.

3. Let $A = \begin{bmatrix} a & b \\ c & d \end{bmatrix}$. Prove that if $ad - bd \neq 0$ then $A^{-1} = \frac{1}{\det(A)} \begin{bmatrix} d & -b \\ -c & a \end{bmatrix}$.

4. Let $A, B \in M(2)$. Prove the following formulas.

   (a) $\det(A^t) = \det(A)$

   (b) $\det(AB) = \det(A)\det(B)$

   (c) $\det(A^{-1}) = \frac{1}{\det(A)}$ if A is invertible.

5. Let $A = \begin{bmatrix} a_{11} & a_{12} \\ a_{21} & a_{22} \end{bmatrix}$. Let $I = \{1, 2\}$. For $i \in I$ let i' be the element of I which is distinct from i. Prove that

$$\begin{aligned}
\det(A) &= \sum_{i=1,2} (-1)^{i+1} a_{1i} a_{1i'} = \sum_{i=1,2} (-1)^{i+2} a_{2i} a_{2i'} \\
&= \sum_{i=1,2} (-1)^{i+1} a_{i1} a_{i'1} = \sum_{i=1,2} (-1)^{i+2} a_{i2} a_{i'2}
\end{aligned}$$

6. Let

$$A_\theta = \begin{bmatrix} \cos\theta & -\sin\theta \\ \sin\theta & \cos\theta \end{bmatrix} \tag{3.4}$$

   (a) Prove that $\det(A_\theta) = 1$ and $A_\theta A_\theta^t = I$.

   (b) Prove that if a 2×2 matrix satisfies the conditions of (a) then it is of the form (3.4). (Hint: If $A = \begin{bmatrix} a & b \\ c & d \end{bmatrix}$ satisfies (a) then by the conditions we have $a^2 + b^2 = c^2 + d^2 = 1$ which implies $(a, b), (c, d)$ are on the circle of radius 1 with the center $(0, 0)$ in $\mathbb{R}^2$.)

3.2 Determinant of Order n

Let $A = [a_{ij}] \in M(n)$ $(n \geq 1)$ and define $\tilde{A}_{ij} \in M(n-1)$ by

$$\tilde{A}_{ij} = \begin{pmatrix} (n-1)\times(n-1) \text{ matrix obtained by} \\ \text{deleting the } i\text{-th row and } j\text{-th column} \end{pmatrix} \tag{3.5}$$

More generally, if $I = \{1_1, i_2, ..., i_k\}$, $J = \{j_1, j_2, ..., j_l\}$ be subsets of $\{1, 2, ..., n\}$ then we define $\widetilde{A}_{IJ}$ by

$$\tilde{A}_{IJ} = \begin{pmatrix} (n-k)\times(n-l) \text{ matrix obtained by deleting the} \\ i\text{-th row for all } i \in I \text{ and } j\text{-th column for all } j \in J \end{pmatrix} \tag{3.6}$$

Definition 3.2.1. *Let $A = [a_{ij}] \in M(n)$ $(n \geq 1)$. If $n = 1$ then A is just a number and its determinant $|A|$ of A is the number itself. For $n = 2$, we define the determinant by (3.1). For $n > 2$, we define the* **determinant** *inductively by*

$$\begin{aligned} \det(A) &= (-1)^{1+1} a_{11} \det \tilde{A}_{11} + (-1)^{1+2} a_{12} \det \tilde{A}_{12} + \cdots + (-1)^{1+n} a_{1n} \det \tilde{A}_{1n} \\ &= \sum_{i=1}^{n} (-1)^{1+i} a_{1i} \det \tilde{A}_{1i} \end{aligned} \tag{3.7}$$

*We define the (i, j)-***cofactor** *C_{ij} or $C_{ij}(A)$ of A by*

$$C_{ij} = (-1)^{i+j} \det \tilde{A}_{ij} \tag{3.8}$$

Hence we can write

$$\det A = \sum_{i=1}^{n} a_{1i} C_{1i} = \sum_{i=1}^{n} (-1)^{1+i} a_{1i} |\tilde{A}_{1i}| \tag{3.9}$$

The determinant of 1×1 matrix, namely a real number is the number itself. The determinant $|A|$ of a 2×2 matrix $A = \begin{bmatrix} a_{11} & a_{12} \\ a_{21} & a_{22} \end{bmatrix}$ becomes, by (3.7)

$$|A| = (-1)^{1+1} a_{11} a_{22} + (-1)^{1+2} a_{12} a_{21} = a_{11} a_{22} - a_{12} a_{21}$$

which is the same as (3.1).

Bear in mind that the determinant is defined only for the square matrices.

Example 3.2.2. *The determinant of a* 3×3 *matrix is*

$$\det\begin{bmatrix} a_{11} & a_{12} & a_{13}\\ a_{21} & a_{22} & a_{23}\\ a_{31} & a_{32} & a_{33}\end{bmatrix} = a_{11}\begin{vmatrix} a_{22} & a_{23}\\ a_{32} & a_{33}\end{vmatrix} - a_{12}\begin{vmatrix} a_{21} & a_{23}\\ a_{31} & a_{33}\end{vmatrix} + a_{13}\begin{vmatrix} a_{21} & a_{22}\\ a_{31} & a_{33}\end{vmatrix}$$
$$= a_{11}a_{22}a_{33} + a_{12}a_{23}a_{31} + a_{13}a_{21}a_{32}$$
$$- a_{11}a_{23}a_{32} - a_{13}a_{22}a_{31} - a_{12}a_{21}a_{33}$$

Observe that the first indices of each term in the determinant expansion are $1, 2, 3$ *and the second indices are the permutations of* $1, 2, 3$.

Notice that the determinant of a 3×3 *matrix is the sum of product of three entries in the direction of diagonal and negative sum of product of three entries in the direction of skew diagonal entries. If the size of a matrix is bigger than* 3 *there is no such device.*

Practice Problem 3.2.1. *Find the determinant of* $A = \begin{bmatrix} 1 & -2 & -1\\ 1 & 1 & 2\\ 0 & -1 & 1\end{bmatrix}$.

As in Example 3.2.2, the number of terms in the determinant of a 3×3 matrix is 3 times the number of terms of the determinant of a 2×2 matrix which is 2. Hence the number of terms in determinant of a 3×3 matrix is $3\times2 = 3! = 6$. The number of terms in the determinant of a 4×4 matrix is then 4 times the number terms in the determinant of a 3×3 matrix. Hence the number of terms in the determinant of 4×4 matrix is $4\cdot3! = 4!$. In general, the number of terms in the determinant of an $n\times n$ matrix is $n!$ which grows very fast as n increases.

Hence for actual computation of determinant we need to simplify the matrix using the properties of determinant, which we will discuss, before we compute the determinant.

Example 3.2.3. *To find the determinant of* $A = \begin{bmatrix} -1 & 0 & 2 & 3\\ 0 & 0 & 1 & 1\\ 1 & 0 & 0 & 1\\ 1 & 1 & 1 & 1\end{bmatrix}$ *use formula (3.7).*

$$|A| = (-1)\det\begin{bmatrix} 0 & 1 & 1\\ 0 & 0 & 1\\ 1 & 1 & 1\end{bmatrix} + 0(-1)\det\begin{bmatrix} 0 & 1 & 1\\ 1 & 0 & 1\\ 1 & 1 & 1\end{bmatrix} + 2\det\begin{bmatrix} 0 & 0 & 1\\ 1 & 0 & 1\\ 1 & 1 & 1\end{bmatrix} + 3(-1)\begin{bmatrix} 0 & 0 & 1\\ 1 & 0 & 0\\ 1 & 1 & 1\end{bmatrix}$$
$$= -1(-1)^{1+1}(1) + 0\cdot(-1)^{1+2}(1) + (2)(-1)^{1+3}(1) + (3)(-1)^{1+4}(1) = -2$$

Practice Problem 3.2.2. *Find the determinants of*

$$A = \begin{bmatrix} 1 & 0 & -2 & 1\\ 0 & 1 & 1 & 0\\ 2 & -1 & 0 & -1\\ 1 & 1 & -1 & 1\end{bmatrix}, \qquad B = \begin{bmatrix} 1 & 1 & 0 & 1\\ 1 & 0 & 1 & 1\\ 0 & 1 & 1 & 1\\ 1 & 1 & 1 & 0\end{bmatrix}$$

Theorem 3.2.4. *Let* $A = [a_{ij}] \in M(n)$. *Then for any fixed* k, l $(1 \le k, l \le n)$

$$\det(A) = a_{k1}C_{k1} + a_{k2}C_{k2} + \cdots + a_{kn}C_{kn} \tag{3.10}$$
$$= a_{1l}C_{1l} + a_{2l}C_{2l} + \cdots + a_{nl}C_{nl} \tag{3.11}$$

The right hand side of (3.10) is called the **determinant expansion of A according to the k-th row** and the right hand side of (3.11) is called the **determinant expansion of A according to the l-th column**.

Before we prove the theorem we illustrate how we apply it and we will give a proof of (3.10) after Theorem 3.2.5 and the proof of (3.11) will be given after Theorem 3.3.10.

First we let $n = 2, 3$ and we apply (3.10) and (3.11): To compute $\det \begin{bmatrix} a & b \\ c & d \end{bmatrix}$ first use (3.10) with $k = 2$. Then, as we saw in Exercise 3.1.5,

$$\det \begin{bmatrix} a & b \\ c & d \end{bmatrix} = c(-1)^{2+1}b + d(-1)^{2+2}a = -bc + ad$$

Next if we use (3.11) with $l = 1$ then

$$\det \begin{bmatrix} a & b \\ c & d \end{bmatrix} = a(-1)^{1+1}d + c(-1)^{1+2}b = ad - bc$$

which proves the theorem for special cases.

Now we compute the determinant of $A = \begin{bmatrix} 1 & 5 & 0 \\ 2 & 4 & -1 \\ 0 & -2 & 0 \end{bmatrix}$ by using (3.10) with $k = 3$. Then

$$|A| = (-1)(-1)^{2+3} \begin{vmatrix} 1 & 5 \\ 0 & -2 \end{vmatrix} = -2$$

Now if we use (3.11) with $l = 3$ then

$$|A| = (-2)(-1)^{3+2} \begin{vmatrix} 1 & 0 \\ 2 & -1 \end{vmatrix} = 2 \cdot (-1) = -2$$

Also we can find the determinant of the matrix A in Example 3.2.3 by expanding the determinant according to the 2nd column:

$$|A| = 1 \cdot (-1)^{2+4} \det \begin{bmatrix} -1 & 2 & 3 \\ 0 & 1 & 1 \\ 1 & 0 & 1 \end{bmatrix} = -2$$

Practice Problem 3.2.3. *Let* $A = \begin{bmatrix} 3 & -7 & 8 & 9 & -6 \\ 0 & 2 & -5 & 7 & 3 \\ 0 & 0 & 1 & 5 & 0 \\ 0 & 0 & 2 & 4 & -1 \\ 0 & 0 & 0 & -2 & 0 \end{bmatrix}$. *Find* $\det A$ *using* Theorem 3.2.4.

A matrix $A = [a_{ij}] \in M(n)$ is called a **upper (resp. lower) triangular matrix** if $a_{ij} = 0$ for all $i > j$ (resp. $i < j$). Either of them we call a **triangular matrix** (Definition 2.1.1).

Theorem 3.2.5. *The determinant of a triangular matrix is the product of the diagonal entries. In particular,* $\det(I_n) = 1$.

Proof. Exercise 2. □

Example 3.2.6. $\det\begin{bmatrix}1&2&3&4\\0&2&3&4\\0&0&3&4\\0&0&0&4\end{bmatrix} = 1\cdot 2\cdot 3\cdot 4 = 24$

Before we give a general proof of Theorem 3.2.4, we prove it for $n = 3$. For the proof of arbitrary n we generalize the proof for $n = 3$.

Example 3.2.7. *Let* $A = \begin{bmatrix}a_{11}&a_{12}&a_{13}\\a_{21}&a_{22}&a_{23}\\a_{31}&a_{32}&a_{33}\end{bmatrix}$. *To prove* Theorem 3.2.4 *for the* 3×3 *matrix* A *we use induction on the size of the matrix. Hence we assume that* Theorem 3.2.4 *is true for the* 2×2 *matrices. To show the determinant expansion of* A *according to the third row is the same as the determinant expansion according to the first row we first expand* $|A|$ *according to the first row and then, by induction, expand the determinants of each vterm according to any row say, the second row which corresponds to the third row of* A. *Write, temporarily,* $|A|_k$ *for the determinant expansion according to the* k*-th row. Then*

$$\begin{vmatrix}a_{11}&a_{12}&a_{13}\\a_{21}&a_{22}&a_{23}\\a_{31}&a_{32}&a_{33}\end{vmatrix}_1 =$$

$$= (-1)^{1+1}a_{11}\begin{vmatrix}a_{11}&a_{12}&a_{13}\\a_{21}&\boxed{a_{22}}&\boxed{a_{23}}\\a_{31}&\boxed{a_{32}}&\boxed{a_{33}}\end{vmatrix}_2 + (-1)^{1+2}a_{12}\begin{vmatrix}a_{11}&a_{12}&a_{13}\\\boxed{a_{21}}&a_{22}&\boxed{a_{23}}\\\boxed{a_{31}}&a_{32}&\boxed{a_{33}}\end{vmatrix}_2 + (-1)^{1+3}a_{13}\begin{vmatrix}a_{11}&a_{12}&a_{13}\\\boxed{a_{21}}&\boxed{a_{22}}&a_{23}\\\boxed{a_{31}}&\boxed{a_{32}}&a_{33}\end{vmatrix}_2$$

$$= (-1)^{1+1}a_{11}\left((-1)^{2+1}a_{32}\begin{vmatrix}a_{11}&a_{12}&a_{13}\\a_{21}&a_{22}&\boxed{a_{23}}\\a_{31}&a_{32}&a_{33}\end{vmatrix} + (-1)^{2+2}a_{33}\begin{vmatrix}a_{11}&a_{12}&a_{13}\\a_{21}&\boxed{a_{22}}&a_{23}\\a_{31}&a_{32}&a_{33}\end{vmatrix}\right)$$

$$+ (-1)^{1+2}a_{12}\left((-1)^{2+1}a_{31}\begin{vmatrix}a_{11}&a_{12}&a_{13}\\a_{21}&a_{22}&\boxed{a_{23}}\\a_{31}&a_{32}&a_{33}\end{vmatrix} + (-1)^{2+2}a_{33}\begin{vmatrix}a_{11}&a_{12}&a_{13}\\\boxed{a_{21}}&a_{22}&a_{23}\\a_{31}&a_{32}&a_{33}\end{vmatrix}\right)$$

$$+ (-1)^{1+3}a_{13}\left((-1)^{2+1}a_{31}\begin{vmatrix}a_{11}&a_{12}&a_{13}\\a_{21}&\boxed{a_{22}}&a_{23}\\a_{31}&a_{32}&a_{33}\end{vmatrix} + (-1)^{2+2}a_{32}\begin{vmatrix}a_{11}&a_{12}&a_{13}\\\boxed{a_{21}}&a_{22}&a_{23}\\a_{31}&a_{32}&a_{33}\end{vmatrix}\right)$$

In our notation, the equalities above can be expressed as

$$\begin{aligned}
|A|_1 &= a_{11}(-1)^{1+1}\left|\tilde{A}_{11}\right|_2 + a_{12}(-1)^{1+2}\left|\tilde{A}_{12}\right|_2 + a_{13}(-1)^{1+3}\left|\tilde{A}_{13}\right|_2\\
&= a_{11}(-1)^{1+1}\left(a_{32}(-1)^{2+1}\left|\tilde{A}_{\{1,3\}\{1,2\}}\right| + a_{33}(-1)^{2+2}\left|\tilde{A}_{\{1,3\}\{1,3\}}\right|\right)\\
&\quad + a_{12}(-1)^{1+2}\left(a_{31}(-1)^{2+1}\left|\tilde{A}_{\{1,3\}\{1,2\}}\right| + a_{33}(-1)^{2+2}\left|\tilde{A}_{\{1,3\}\{2,3\}}\right|\right)\\
&\quad + a_{13}(-1)^{1+3}\left(a_{31}(-1)^{2+1}\left|\tilde{A}_{\{1,3\}\{1,3\}}\right| + a_{32}(-1)^{2+2}\left|\tilde{A}_{\{1,3\}\{2,3\}}\right|\right)\\
&= a_{11}(-1)^{1+1}\left(a_{32}(-1)^{2+1}a_{23} + a_{33}(-1)^{2+2}a_{22}\right)\\
&\quad + a_{12}(-1)^{1+2}\left(a_{31}(-1)^{2+1}a_{23} + a_{33}(-1)^{2+2}a_{21}\right)\\
&\quad + a_{13}(-1)^{1+3}\left(a_{31}(-1)^{2+1}a_{22} + a_{32}(-1)^{2+2}a_{12}\right)
\end{aligned}$$

Now expand $|A|$ according to the third row and then expand the determinant of each term according to the first row which, of course, corresponds to the first row of A:

$$
\left|\begin{matrix} a_{11} & a_{12} & a_{13} \\ a_{21} & a_{22} & a_{23} \\ a_{31} & a_{32} & a_{33} \end{matrix}\right|_3
$$

$$
= (-1)^{3+1}a_{31}\left|\begin{matrix} a_{11} & \boxed{a_{12}} & \boxed{a_{13}} \\ a_{21} & \boxed{a_{22}} & \boxed{a_{23}} \\ a_{31} & a_{32} & a_{33} \end{matrix}\right|_1 + (-1)^{3+2}a_{32}\left|\begin{matrix} \boxed{a_{11}} & a_{32} & \boxed{a_{13}} \\ \boxed{a_{21}} & a_{22} & \boxed{a_{23}} \\ a_{31} & a_{32} & a_{33} \end{matrix}\right|_1 + (-1)^{3+3}a_{33}\left|\begin{matrix} \boxed{a_{11}} & \boxed{a_{12}} & a_{13} \\ \boxed{a_{21}} & \boxed{a_{22}} & a_{23} \\ a_{31} & a_{32} & a_{33} \end{matrix}\right|_1
$$

$$
= (-1)^{3+1}a_{31}\left((-1)^{1+1}a_{12}\left|\begin{matrix} a_{11} & a_{12} & a_{13} \\ a_{21} & a_{22} & \boxed{a_{23}} \\ a_{31} & a_{32} & a_{33} \end{matrix}\right| + (-1)^{1+2}a_{13}\left|\begin{matrix} a_{11} & a_{12} & a_{13} \\ a_{21} & \boxed{a_{22}} & a_{23} \\ a_{31} & a_{32} & a_{33} \end{matrix}\right|\right)
$$

$$
+ (-1)^{3+2}a_{32}\left((-1)^{1+1}a_{11}\left|\begin{matrix} a_{11} & a_{32} & a_{13} \\ a_{21} & a_{22} & \boxed{a_{23}} \\ a_{31} & a_{32} & a_{33} \end{matrix}\right| + (-1)^{1+2}a_{13}\left|\begin{matrix} a_{11} & a_{32} & a_{13} \\ \boxed{a_{21}} & a_{22} & a_{23} \\ a_{31} & a_{32} & a_{33} \end{matrix}\right|\right)
$$

$$
+ (-1)^{3+3}a_{33}\left((-1)^{1+1}a_{11}\left|\begin{matrix} a_{11} & a_{12} & a_{13} \\ a_{21} & \boxed{a_{22}} & a_{23} \\ a_{31} & a_{32} & a_{33} \end{matrix}\right| + (-1)^{1+2}a_{12}\left|\begin{matrix} a_{11} & a_{12} & a_{13} \\ \boxed{a_{21}} & a_{22} & a_{23} \\ a_{31} & a_{32} & a_{33} \end{matrix}\right|\right)
$$

In our notations, the equalities above becomes:

$$
\begin{aligned}
|A|_3 &= a_{31}(-1)^{3+1}\left|\tilde{A}_{31}\right|_1 + a_{32}(-1)^{3+2}\left|\tilde{A}_{32}\right|_1 + a_{33}(-1)^{3+3}\left|\tilde{A}_{33}\right|_1 \\
&= a_{31}(-1)^{3+1}\left(a_{12}(-1)^{1+1}\left|\tilde{A}_{\{1,3\}\{1,2\}}\right| + a_{13}(-1)^{1+2}\left|\tilde{A}_{\{1,3\}\{1,3\}}\right|\right) \\
&\quad + a_{32}(-1)^{3+2}\left(a_{11}(-1)^{1+1}\left|\tilde{A}_{\{1,3\}\{1,2\}}\right| + a_{13}(-1)^{1+2}\left|\tilde{A}_{\{1,3\}\{2,3\}}\right|\right) \\
&\quad + a_{33}(-1)^{3+3}\left(a_{11}(-1)^{1+1}\left|\tilde{A}_{\{1,3\}\{1,3\}}\right| + a_{12}(-1)^{1+2}\left|\tilde{A}_{\{1,3\}\{2,3\}}\right|\right) \\
&= a_{31}(-1)^{3+1}\left(a_{12}(-1)^{1+1}\alpha_{23} + a_{13}(-1)^{1+2}a_{22}\right) \\
&\quad + a_{32}(-1)^{3+2}\left(a_{11}(-1)^{1+1}a_{23} + a_{13}(-1)^{1+2}a_{21}\right) \\
&\quad + a_{33}(-1)^{3+3}\left(a_{11}(-1)^{1+1}a_{22} + a_{12}(-1)^{1+2}a_{21}\right)
\end{aligned}
$$

Ignoring the sign both have the same terms $a_{1i}a_{3k}|\tilde{A}_{\{1,3\}\{i,k\}}|$. Now comparing the signs we conclude that $|A|_1 = |A|_3$.

The proof of Theorem 3.2.4 will be a generalization of the example above.

Proof of **Theorem 3.2.4**. We prove (3.10) and we need to show:

$$
\det(A) = a_{k1}C_{k1} + a_{k2}C_{k2} + \cdots + a_{kn}C_{kn}
$$

We already showed this for $n = 2$ in Exercise 3.1.5 and in the remark following Theorem 3.2.4, and for $n = 3$ in Example 3.2.7.

To prove this for all n assume our result is true for $n-1$ and we will prove for $A \in M(n)$. We want to prove $|A|_1 = |A|_i$ where $|A|_i$ denotes the determinant expansion according to the i-th row. First, consider the determinant expansion according to the first row:

$$
|A|_1 = a_{11}(-1)^{1+1}|\tilde{A}_{11}| + a_{12}(-1)^{1+2}|\tilde{A}_{12}| + \cdots
$$

Now by induction we can expand $|\tilde{A}_{1k}|$ according to the $(i-1)$st row which is $A_{(i)}$ with the k-th position deleted. Hence, using the notation of (3.6)

$$\begin{aligned}|A|_1 &= a_{11}(-1)^{1+1}\left(a_{i2}(-1)^{i-1+1}|\tilde{A}_{\{1i\}\{12\}}| + a_{i3}(-1)^{i-1+2}|\tilde{A}_{\{1i\}\{13\}}| + \cdots\right)\\ &\quad + a_{12}(-1)^{1+2}\left(a_{i1}(-1)^{i-1+1}|\tilde{A}_{\{1i\}\{12\}}| + a_{i3}(-1)^{i-1+2}|\tilde{A}_{\{1i\}\{13\}}| + \cdots\right) + \cdots\end{aligned}$$

Now

$$|A|_i = a_{i1}(-1)^{i+1}|\tilde{A}_{i1}| + a_{i2}(-1)^{i+2}|\tilde{A}_{i2}| + \cdots$$

and expand $|\tilde{A}_{il}|$ according to the first row:

$$\begin{aligned}|A|_i &= a_{i1}(-1)^{i+1}\left(a_{12}(-1)^{1+1}|\tilde{A}_{\{i1\}\{12\}}| + a_{13}(-1)^{1+2}|\tilde{A}_{\{1i\}\{13\}}| + \cdots\right)\\ &\quad + a_{i2}(-1)^{i+2}\left(a_{11}(-1)^{1+1}|\tilde{A}_{\{i1\}\{12\}}| + a_{13}(-1)^{1+2}|\tilde{A}_{\{1i\}\{13\}}| + \cdots\right) + \cdots\end{aligned}$$

Ignoring the signs, $|A|_1$ and $|A|_i$ contains the same terms $a_{1k}a_{il}|\tilde{A}_{\{1i\}\{kl\}}|$ $(1 \le l, k \le n)$. Now compare the sign of each term $a_{1k}a_{il}|\tilde{A}_{\{1i\}\{kl\}}|$ in $|A|_1$ and $|A|_i$: The exponent of -1 for the former and the latter are

$$(1+k)+(i-1)+\begin{cases} l & (l<k) \\ (l-1) & (l>k)\end{cases} = \begin{cases} i+k+l & (l<k) \\ i+k+l-1 & (l>k)\end{cases} \quad \text{and}$$

$$(i+l)+1+\begin{cases} k-1 & (l<k) \\ k & (l>k)\end{cases} = \begin{cases} i+k+l & (l<k) \\ i+k+l+1 & (l>k)\end{cases}$$

respectively. They have the same parity. Hence $|A|_1 = |A|_i$ for any i. □

Proof of (3.11) will be given after Theorem 3.3.10.

Answers for Practice Problems 3.2

(3.2.1) $|A| = 6$.
(3.2.2) $|A| = 0$ and $|B| = 3$.
(3.2.3) $|A| = 3\begin{vmatrix} 2 & -5 & 7 & 3 \\ 0 & 1 & 5 & 0 \\ 0 & 2 & 4 & -1 \\ 0 & 0 & -2 & 0 \end{vmatrix} = 3\cdot 2\begin{vmatrix} 1 & 5 & 0 \\ 2 & 4 & -1 \\ 0 & -2 & 0\end{vmatrix} = 3\cdot 2\cdot(-2)(-1)^{3+2}\begin{vmatrix} 1 & 0 \\ 2 & -1\end{vmatrix} = -12$

3.2 Exercises

1. Find the determinant of

(a) $\begin{bmatrix} 1 & 2 & 1 \\ 3 & 1 & 0 \\ 2 & 1 & 4 \end{bmatrix}$ (b) $\begin{bmatrix} 1 & -1 & 2 \\ 3 & 4 & -1 \\ 1 & 1 & 1 \end{bmatrix}$ (c) $\begin{bmatrix} 3 & -2 & 3 \\ 1 & 2 & 1 \\ 2 & 2 & 1 \end{bmatrix}$

(d) $\begin{bmatrix} 1 & 0 & 1 & 1 \\ 1 & 1 & 1 & 0 \\ 1 & 2 & 0 & 1 \\ 1 & 1 & 1 & 1 \end{bmatrix}$ (e) $\begin{bmatrix} 2 & 0 & -1 & 3 \\ 2 & 4 & 0 & 1 \\ 2 & 0 & 0 & 1 \\ 1 & 2 & 1 & 2 \end{bmatrix}$ (f) $\begin{bmatrix} 1 & 1 & 0 & 0 \\ 1 & 0 & 0 & 1 \\ 0 & 1 & 1 & 0 \\ 0 & 0 & 1 & 1 \end{bmatrix}$.

2. Prove Theorem 3.2.5. (Hint. Use induction and Theorem 3.2.4.)

3. Let A, C be square matrices size r, s and $B, B' \in M(r\times s)$. Prove that

$$\det\begin{bmatrix} A & B \\ 0 & I_s \end{bmatrix} = \det(A) \text{ and } \det\begin{bmatrix} I_r & B' \\ 0 & C \end{bmatrix} = \det(C)$$

4. Prove the true statement(s) and find a counter example for the false statement(s). In this question A denotes a square matrix.

   (a) If one of a row of A consists entirely of 0 then $|A| = 0$.

   (b) If one of a column of A consists entirely of 0 then $|A| = 0$.

   (c) If A is an upper triangular with $A_{ii} = 0$ for some $1 \le i \le n$ then $|A| = 0$.

   (d) If $A^{(i)} = \mathbf{e}^j$ then $\det(A) = (-1)^{i+j}|\tilde{A}_{ij}|$.

   (e) If $A_{(i)} = \mathbf{e}_j$ then $\det(A) = (-1)^{i+j}|\tilde{A}_{ij}|$.

   (f) If $A^{(i)} = \mathbf{e}^i, A^{(j)} = \mathbf{e}^j$ $(i < j)$ and if $I = \{i, j\}$ then $|A| = (-1)^{ij}|\tilde{A}_{II}|$ (See (3.6) for the notation.)

   (g) If $A^{(i)} = \mathbf{e}^j, A^{(j)} = \mathbf{e}^i$ $(i < j)$ and if $I = \{i, j\}$ then $|A| = |\tilde{A}_{II}|$.

   (h) If $A^{(i)} = \mathbf{e}^i, A_{(j)} = \mathbf{e}_j$ $(\imath < j)$ and if $I = \{i, j\}$ then $|A| = (-1)^{ij}|\tilde{A}_{II}|$.

   (i) If c is a constant then $|cA| = c|A|$.

   (j) If c is a constant and $A \in M(n)$ then $|cA| = c^n|A|$.

   (k) The determinant of a triangular matrix is the product of the diagonal entries.

3.3 Properties of Determinant

3.3.1 Basic Properties of Determinant

The following properties of determinant are useful to compute the determinant of a matrix.

Theorem 3.3.1. *Let $A \in M(n)$.*

(1) *If B is obtained from A by adding a constant times a row of A to another row then* $\det B = \det A$.

(2) *If B is obtained from A by interchanging two rows of A then* $\det B = -\det A$.

(3) *If B is obtained from A by multiplying a row of A by a constant c then* $\det B = c \det A$.

In terms of elementary matrices:

(1)′ $\det(E_{ij}(c)) = 1$ *and* $\det(E_{ij}(c)A) = \det(E_{ij}(c))\det(A)$

(2)′ $\det(I_{ij}) = -1$ *and* $\det(I_{ij}A) = \det(I_{ij})\det(A)$

(3)′ $\det(M_i(c) = c$ *and* $\det(M_i(c)A) = \det(M_i(c))\det(A)$

In (1), (2), (3) if we replace 'row' by 'column' then the same conclusions hold; in this case we have to post multiply the transpose of the elementary matrices in (1)′, (2)′, (3)′.

Practice Problem 3.3.1. *Prove the properties* (1), (2), (3) *for* $n = 2$.

Proof of **Theorem 3.3.1**. It is an exercise (Exercise 3) to prove that $|E_{ij}(c)| = 1, |I_{ij}| = -1$ and $|M_i(c)| = c$.

We use induction on the size n of A to prove (1), (2) and (3). The case $n = 2$ is proved in Practice Problem 3.3.1 above. Now we assume the results for the matrices of size $n - 1$ and assume $n \geq 3$. By Theorem 3.2.4, we have, for any k

$$\det B = B_{k1}C_{k1}(B) + B_{k2}C_{k2}(B) + \cdots + B_{kn}C_{kn}(B)$$

where $C_{kp}(B)$ denotes the (k, p)-cofactor of B. For (1), choose $k \notin \{i, j\}$. Then $\tilde{B}_{kp}$ is obtained from $\tilde{A}_{kp}$ by adding c times a row to another row $(p = 1, ..., n)$. Since $\tilde{A}_{kp}, \tilde{B}_{kp} \in M(n-1)$, by induction, we have $|\tilde{A}_{kp}| = |\tilde{B}_{kp}|$; $C_{kp}(B) = C_{kp}(A)$. Hence $\det B = \det A$.

For (2), choose k as above. The same argument proves (2).

For (3), choose $k = i$. Then $B_{kp} = cA_{kp}$ $(p = 1, 2, ..., n)$ and $C_{kp}(B) = C_{kp}(A)$ for $p = 1, 2, ..., n$. Therefore, we have

$$\begin{aligned}\det B &= B_{k1}C_{k1}(B) + B_{k2}C_{k2}(B) + \cdots + B_{kn}C_{kn}(B) \\ &= cA_{k1}C_{k1}(A) + cA_{k2}C_{k2}(A) + \cdots + cA_{kn}C_{kn}(A) \\ &= c\det(A)\end{aligned}$$

□

We record a consequence of Theorem 3.3.1 which will be generalized in Theorem 3.3.11.

Corollary 3.3.2. *If A, B are elementary matrices then* $\det(AB) = \det(A)\det(B)$.

Proof. In $(1)', (2)', (3)'$ of Theorem 3.3.1, simply let A to be an elementary matrix. □

Corollary 3.3.3. *If two rows (columns) of A are identical then* $\det(A) = 0$.

Proof. Let i, j-th rows are equal then $I_{ij}A = A$. Hence $\det(I_{ij}A) = \det(A)$. On the other hand, $\det(I_{ij}A) = -\det(A)$. Hence $\det(A) = -\det(A)$ and therefore $\det(A) = 0$.

For the statement about columns follows from Theorem 3.3.10 which will be proved shortly. □

Corollary 3.3.4. *Let $A, B \in M(n)$. Assume A and B are row equivalent. Then $|A| = 0$ if and only if $|B| = 0$.*

Proof. Exercise 4. □

Note that Corollary 3.3.4 does <u>not</u> say that $|A| = |B|$ if A and B are row equivalent.

Example 3.3.5. *To find the determinant of* $A = \begin{bmatrix} 1 & 0 & 1 & 1 \\ 0 & 0 & 1 & 1 \\ 1 & 0 & 0 & 1 \\ 1 & 1 & 1 & 1 \end{bmatrix}$ *add -1 times the 4th row to the first row without changing the determinant:* $|A| = \det \begin{bmatrix} 0 & -1 & 0 & 0 \\ 0 & 0 & 1 & 1 \\ 1 & 0 & 0 & 1 \\ 1 & 1 & 1 & 1 \end{bmatrix}$. *Now expand $|A|$ according to the 1st row:* $|A| = (-1)(-1)^{1+2} \det \begin{bmatrix} 0 & 1 & 1 \\ 1 & 0 & 1 \\ 1 & 1 & 1 \end{bmatrix}$. *At this point we can compute the 3×3 matrix or we can further reduce it, by adding (-1) times the first row to the third row to get* $\det \begin{bmatrix} 0 & 1 & 1 \\ 1 & 0 & 1 \\ 1 & 1 & 1 \end{bmatrix} = \det \begin{bmatrix} 0 & 1 & 1 \\ 1 & 0 & 1 \\ 1 & 0 & 0 \end{bmatrix} = 1(-1)^{1+3} \det \begin{bmatrix} 1 & 1 \\ 0 & 1 \end{bmatrix} = 1$. *Hence* $|A| = 1$.

Practice Problem 3.3.2. *Find the determinant of* $A = \begin{bmatrix} 2 & -1 & 1 & 0 \\ 2 & 0 & 1 & -1 \\ 1 & 2 & 3 & -1 \\ 2 & 4 & -2 & 1 \end{bmatrix}$

Example 3.3.6. *Let* $A = \begin{bmatrix} a & b & c \\ p & q & r \\ x & y & z \end{bmatrix}$. *If we add* (-1) *times the* 3rd *row to the second row and then add the third column to the first column then we get* $B = \begin{bmatrix} a+c & b & c \\ p-x+r-y & q-y & r-z \\ x+z & y & z \end{bmatrix}$. *Since the row and column operations of the first type* $cR_i + R_j$ *does not change the determinant by* Theorem 3.3.1, *we get* $|A| = |B|$.

3.3.2 Invertible Matrices and Determinant

Following theorem is a very useful criterion to test whether a given square matrix is invertible or not.

Theorem 3.3.7. *Let* $A \in M(n)$. *Then* A *is invertible if and only if* $\det A \neq 0$.

Proof. If A is invertible then A is a product of elementary matrices $E_1, E_2, ..., E_p$; $A = E_1 E_2 \cdots E_p$ (Theorem 2.5.12). And we know that $|E_i| \neq 0$. Hence, by Corollary 3.3.2, we see that $\det(A) = \prod \det(E_i) \neq 0$.

Now assume $\det(A) \neq 0$. Reduce, by elementary row operations, A to an echelon form U where U is upper triangular whose determinant is the product of the diagonals (Theorem 2.5.14). Thus no diagonal entry of U is zero. Hence A is invertible by Theorem 2.5.14. □

A list of equivalent conditions for $A \in M(n)$ to be invertible is given in Theorem 2.5.14. Here is another one which will be useful to check whether the columns or rows of a square matrix are linearly independent.

Corollary 3.3.8. *Let* $A \in M(n)$. *The following conditions are equivalent.*

(1) $|A| \neq 0$

(2) *The columns of* A *form a basis of* $\mathbb{R}^n$.

(3) *The rows of* A *form a basis of* $\mathbb{R}_n$.

Proof. (1) $\Leftrightarrow$ (2) By the theorem, $|A| \neq 0$ if and only if A is invertible. Now, by Theorem 2.5.14, this is equivalent to the fact that the columns of A form a basis. Equivalence of (2) and (3) follows from Theorem 2.5.14. □

Example 3.3.9. *To see if* $S = \left\{ \begin{bmatrix} 1 \\ 1 \\ 2 \end{bmatrix}, \begin{bmatrix} 2 \\ 1 \\ 0 \end{bmatrix}, \begin{bmatrix} 0 \\ 1 \\ 0 \end{bmatrix} \right\}$ *form a basis of* $\mathbb{R}^3$ *we simply compute the determinant of* $A = \begin{bmatrix} 1 & 2 & 0 \\ 1 & 1 & 1 \\ 2 & 0 & 0 \end{bmatrix}$; $|A| = 4$. *Hence* S *is a basis for* $\mathbb{R}^3$. *Also we see that the rows of* A *form a basis of* $\mathbb{R}_3$.

Practice Problem 3.3.3. *Decide whether the columns of the matrix* $A = \begin{bmatrix} 1 & 0 & 1 & 0 \\ 1 & 1 & 0 & 1 \\ 0 & 1 & -1 & 1 \\ 1 & 1 & 0 & 0 \end{bmatrix}$ *form a basis for* $\mathbb{R}^4$.

Theorem 3.3.10. *Let* $A \in M(n)$. *Then* $\det A = \det A^t$.

Proof. If $|A| = 0$ then A is not invertible and hence A^t is not invertible (Theorem 2.5.14). Therefore $|A^t| = 0$ by Theorem 3.3.7.

Now suppose A is invertible. Then we can write $A = E_1E_2 \cdots E_k$ where E_i are elementary matrices. Hence $A^t = E_k^tE_{k-1}^t \cdots E_1^t$ and hence $|A^t| = |E_k^t||E_{k-1}^t| \cdots |E_1^t| = |E_k||E_{k-1}| \cdots |E_1| = |A|$ since $|E_i^t| = |E_i|$ by (2.31). □

Now we proceed to prove (3.11) **of Theorem 3.2.4.** For $n = 2$, it is easy to check (3.11) by direct computation. In general, we have, by Theorem 3.3.10 and induction

$$\begin{aligned} |A| &= |A^t| \\ &= A_{l1}^t(-1)^{l+1}|\tilde{A}_{l1}^t| + A_{l2}^t(-1)^{l+2}|\tilde{A}_{l2}^t| + \cdots + A_{ln}^t(-1)^{l+n}|\tilde{A}_{1n}^t| \quad \text{(by induction)} \\ &= A_{1l}(-1)^{1+l}|\tilde{A}_{1l}| + A_{2l}(-1)^{2+l}|\tilde{A}_{2l}| + \cdots + A_{nl}(-1)^{n+l}|\tilde{A}_{nl}| \end{aligned}$$

which proves (3.11).

Now we prove one of the important properties which generalizes a provisional result Corollary 3.3.2.

Theorem 3.3.11. *For* $A, B \in M(n)$ *we have* $\det(AB) = \det(A)\det(B)$.

Proof. If A is not invertible then $\rho(A) < n$. Hence $\rho(AB) < n$ by Corollary 2.3.15. Thus, AB is not invertible (Theorem 2.4.5). In this case we have $\det(AB) = 0 = \det(A)\det(B)$ by Theorem 3.3.7.

If A is invertible then A is a product of elementary matrices $A = E_pE_{p-1} \cdots E_1$ (Theorem 2.5.12). Hence

$$\begin{aligned} |AB| &= |E_pE_{p-1} \cdots E_1B| = |E_p||E_{p-1} \cdots E_1B| \\ &= |E_p||E_{p-1}||E_{p-2} \cdots E_1B| = \cdots = |E_p||E_{p-1}| \cdots |E_1||B| \\ &= |E_pE_{p-1}| \cdots |E_1||B| = \cdots = |E_pE_{p-1} \cdots E_1||B| \\ &= |A||B| \end{aligned}$$

by Theorem 3.3.1. □

A different proof of Theorem 3.3.11 will be given in Corollary 6.6.20.

Corollary 3.3.12. *Let* $A, B \in M(n)$.

(1) *If* A *is invertible then*

$$\det(A^{-1}) = \frac{1}{\det A}$$

(2) *If* A, B *are* **similar** (*i.e., there is an invertible matrix* P *such that* $B = P^{-1}AP$) *then we have* $\det(A) = \det(B)$.

Proof. (1) Since $AA^{-1} = I$ we have $|A||A^{-1}| = 1$.

(2) Since $B = P^{-1}AP$ we have $\det B = \det P^{-1} \det A \det P = (\det P)^{-1} \det A \det P = \det A$ by Theorem 3.3.11 and (1). □

3.3.3 Characterizing Properties of Determinant

Observe that determinant is a function on the set of $n \times n$ matrices: $\det : M(n) \to \mathbb{R}$. One of the basic property is

Theorem 3.3.13. *Determinant is* **multilinear** *in the columns i.e., linear in each column:*

$$\det[\mathbf{a}_1 \ \cdots \ c\mathbf{a}_j \ \cdots \ \mathbf{a}_n] = c\det[\mathbf{a}_1 \ \cdots \ \mathbf{a}_j \ \cdots \ \mathbf{a}_n] \tag{3.12}$$

$$\det[\mathbf{a}_1 \ \cdots \mathbf{a}_j + \mathbf{a}_j' \cdots \ \mathbf{a}_n] = \det[\mathbf{a}_1 \ \cdots \ \mathbf{a}_j \ \cdots \ \mathbf{a}_n] + \det[\mathbf{a}_1 \ \cdots \ \mathbf{a}_j' \ \cdots \ \mathbf{a}_n] \tag{3.13}$$

Further, the determinant is multilinear in the rows when we view det *as a function of rows.*

Proof. To prove (3.12), expand the determinant with respect to the j-th column (Theorem 3.2.4):

$$\begin{aligned}\det[\mathbf{a}_1 \ \cdots \ c\mathbf{a}_j \ \cdots \ \mathbf{a}_n] &= \sum_k c a_{kj}(-1)^{k+j}|\tilde{A}_{kj}| \\ &= c\sum_k a_{kj}(-1)^{k+j}|\tilde{A}_{kj}| = c|A|\end{aligned}$$

where $A = [\mathbf{a}_1 \ \cdots \ \mathbf{a}_n]$.

The equality (3.13) can be proved similarly. Statement about rows follows from the determinant expansion according to the rows. □

The following theorem generalize Theorem 3.1.3.

Theorem 3.3.14. *The determinant* $\det : M(n) \to \mathbb{R}$ *is characterized by*

(1) *Multilinear in the columns i.e.,* (3.12) *and* (3.13) *hold for each* j.

(2) *Alternating i.e., interchanging any two columns changes the sign of the determinant.*

(3) $\det(I_n) = 1$

If we replace 'column' by 'row' then the same conclusion is true.

Proof. The property (1) is Theorem 3.3.13, the property (2) is Theorem 3.3.1 and the property (3) is Theorem 3.2.5. Hence det satisfies the properties (1), (2) and (3).

If $n = 2$ we showed that the properties (1) – (3) characterize the determinant in Theorem 3.1.3. For $n > 2$ suppose $f, g : M(n) \to \mathbb{R}$ be two functions satisfying the properties (1)–(3). We claim that $f = g$. First observe that if $A = [a_{ij}] \in M(n)$ then use multilinearity to expand $\det(A)$. Since

$$A = \left[A^{(1)} \ A^{(2)} \ \cdots \ A^{(n)}\right] = \left[\sum_{j=1}^n a_{j1}\mathbf{e}^j \ \ \sum_{j=1}^n a_{j2}\mathbf{e}^j \ \ \cdots \ \ \sum_{j=1}^n a_{jn}\mathbf{e}^j\right]$$

Since f satisfies the properties (1) – (3) we have

$$\begin{aligned}f(A) &= f\left(\left[\textstyle\sum_j a_{j1}\mathbf{e}^j \ \ \sum_j a_{j2}\mathbf{e}^j \ \ \cdots \ \ \sum_j a_{jn}\mathbf{e}^j\right]\right) \\ &= f\left(\textstyle\sum_j a_{j1}\mathbf{e}^j \ \ \sum_j a_{j2}\mathbf{e}^j \ \ \cdots \ \ \sum_j a_{jn}\mathbf{e}^j\right) \quad \text{(by linearity in each column)} \\ &= a_{11}f\left(\mathbf{e}^1 \ \textstyle\sum_j a_{j2}\mathbf{e}^j \ \cdots \ \sum_j a_{jn}\mathbf{e}^j\right) + a_{21}f\left(\mathbf{e}^2 \ \sum_j a_{j2}\mathbf{e}^j \ \cdots \ \sum_j a_{jn}\mathbf{e}^j\right) + \cdots \\ &= a_{11}a_{12}f\left(\mathbf{e}^1 \ \mathbf{e}^1 \ A^{(3)} \ \cdots\right) + a_{11}a_{22}f\left(\mathbf{e}^1 \ \mathbf{e}^2 \ A^{(3)} \ \cdots\right) + \cdots \\ &= \cdots \\ &= \sum_{(i_1,\ldots,i_n)} a_{1i_1}a_{2i_2}\cdots a_{ni_n}f\left(\mathbf{e}^{i_1} \ \mathbf{e}^{i_2} \ \cdots \ \mathbf{e}^{i_n}\right)\end{aligned}$$

where $\{i_1, i_2, ..., i_n\}$ are all distinct since it vanishes if any two of them are the same by (2), and it is a permutation of $\{1, 2, ..., n\}$. Similarly for g. Hence $f = g$ if

$$f\left(\mathbf{e}^{i_1} \quad \mathbf{e}^{i_2} \quad \cdots \quad \mathbf{e}^{i_n}\right) = g\left(\mathbf{e}^{i_1} \quad \mathbf{e}^{i_2} \quad \cdots \quad \mathbf{e}^{i_n}\right) \tag{3.14}$$

Now $\left[\mathbf{e}^{i_1} \quad \mathbf{e}^{i_2} \quad \cdots \quad \mathbf{e}^{i_n}\right]$ is obtained by interchanging the columns of the identity matrix I_n successively. Each time we interchange two columns the value of f, g changes the sign. Hence by property (3), $f\left(\mathbf{e}^{i_1} \quad \mathbf{e}^{i_2} \quad \cdots \quad \mathbf{e}^{i_n}\right)$ is the same as $(-1)^k$ where k is the number of times of interchange of the columns of $[\mathbf{e}^{i_1} \ \mathbf{e}^{i_2} \ \cdots \ \mathbf{e}^{i_n}]$ to get the identity matrix I_n. Since the same is true for g also, (3.14) is true. Hence $f = g$.

Now the determinant satisfies the properties, and hence any function satisfying (1)–(3) is the same as the determinant.

Proof for the statement about the rows is Exercise 5. □

3.3.4 Explicit Formula for Determinant

Now we give an explicit description of determinant: Let $A = [a_{ij}] \in M(n)$. Then

$$A = \left[A^{(1)} \cdots A^{(i)} \cdots A^{(n)}\right] = \left[\sum_{j=1}^{n} a_{j1}\mathbf{e}^j \quad \cdots \quad \sum_{j=1}^{n} a_{ji}\mathbf{e}^j \quad \cdots \quad \sum_{j=1}^{n} a_{jn}\mathbf{e}^j\right]$$

Hence, by linearity in each column

$$\begin{aligned}
|A| &= \left|\sum_j a_{j1}\mathbf{e}^j \quad \sum_j a_{j2}\mathbf{e}^j \quad \cdots \quad \sum_j a_{jn}\mathbf{e}^j\right| \\
&= a_{11}\left|\mathbf{e}^1 \quad \sum_j a_{j2}\mathbf{e}^j \quad \cdots \quad \sum_j a_{jn}\mathbf{e}^j\right| + a_{12}\left|\mathbf{e}^2 \quad \sum_j a_{j2}\mathbf{e}^j \quad \cdots \quad \sum_j a_{jn}\mathbf{e}^j\right| + \cdots \\
&= a_{11}a_{21}\left|\mathbf{e}^1 \quad \mathbf{e}^1 \quad A^{(3)} \cdots\right| + a_{11}a_{22}\left|\mathbf{e}^1 \quad \mathbf{e}^2 \quad A^{(3)} \cdots\right| + \cdots \\
&= \cdots \\
&= \sum_{\sigma=(i_1,\ldots,i_n)} a_{1i_1}a_{2i_2}\cdots a_{ni_n}\left|\mathbf{e}^{i_1} \quad \mathbf{e}^{i_2} \quad \cdots \quad \mathbf{e}^{i_n}\right|
\end{aligned}$$

where the sum runs over all permutations $\sigma = (i_1, i_2, \ldots, i_n)$ of $\{1, 2, \ldots, n\}$. (See the definition below.)

Definition 3.3.15. *A* **permutation** *of* $\mathbf{n} = \{1, 2, \ldots, n\}$ *is a bijection* (Definition A.1.4) *from* $\mathbf{n}$ *onto itself. And the permutation* $\sigma = (i_1, i_2, \ldots, i_n)$ *is defined to be a bijection given by* $\sigma(k) = i_k$. *(Our notation for the permutation is not standard.) A* **transposition** *is a permutation interchanging two numbers of* $\mathbf{n}$ *and fixing the rest.*

A matrix of the form $[\mathbf{e}^{i_1} \ \mathbf{e}^{i_2} \ \cdots \ \mathbf{e}^{i_n}]$ *is called a* **permutation matrix**. *Hence a permutation matrix is obtained by permuting the columns of the identity matrix* I_n. *We define the* **sign of the permutation** $\sigma = (i_1, i_2, \ldots, i_n)$, $\operatorname{sign}(\sigma)$ *or* $|\sigma|$ *to be*

$$\operatorname{sign}(\sigma) = \det[\mathbf{e}^{i_1} \ \mathbf{e}^{i_2} \ \cdots \ \mathbf{e}^{i_n}] \tag{3.15}$$

It is a well known fact that every permutation is obtained by successive applications of transpositions. (For a proof of this fact see any undergraduate algebra text.) Since each time we interchange two columns the sign of the determinant changes and since successive interchange of columns of $\left[\mathbf{e}^{i_1} \ \mathbf{e}^{i_2} \cdots \mathbf{e}^{i_n}\right]$ reduces it to the identity matrix we see that

$$\det\left[\mathbf{e}^{i_1} \ \mathbf{e}^{i_2} \cdots \mathbf{e}^{i_n}\right] = \pm 1$$

If we obtain the identity matrix from $\left[\mathbf{e}^{i_1}\ \mathbf{e}^{i_2} \cdots \mathbf{e}^{i_n}\right]$ by interchanging the columns k-times then (Exercise 14)

$$\operatorname{sign}(\sigma) = (-1)^k$$

For example, we have

$$\operatorname{sign}(4132) = \det[\mathbf{e}^4\ \mathbf{e}^1\ \mathbf{e}^3\ \mathbf{e}^2] = \det\begin{bmatrix} 0 & 1 & 0 & 0 \\ 0 & 0 & 0 & 1 \\ 0 & 0 & 1 & 0 \\ 1 & 0 & 0 & 0 \end{bmatrix} = 1$$

We summarize our discussion in

Theorem 3.3.16. *For $A = [a_{ij}] \in M(n)$ we have*

$$\det(A) = \sum_{\sigma} \operatorname{sign}(\sigma) a_{1\sigma(1)} a_{2\sigma(2)} \cdots a_{n\sigma(n)} \tag{3.16}$$

$$= \sum_{\sigma} \operatorname{sign}(\sigma) a_{\sigma(1)1} a_{\sigma(2)2} \cdots a_{\sigma(n)n} \tag{3.17}$$

where σ runs over all permutations of $\mathbf{n} = \{1, 2, \ldots, n\}$.

Proof. The equality (3.17) follows from the fact that $|A| = |A^t|$ since $(A^t)_{\sigma(i)i} = A_{i\sigma(i)}$. □

The number of terms in Theorem 3.3.16 is $n!$ which grows very rapidly as n increases. For actual computation the formulas of Theorem 3.3.16 is useful only for $n = 2, 3$ which was given in the examples of §3.1 and §3.2.

The following fact is useful in computations of determinant.

Theorem 3.3.17. *Let A, C be square matrices.*

$$\det\begin{bmatrix} A & B \\ 0 & C \end{bmatrix} = \det(A)\det(C) = \det\begin{bmatrix} A & 0 \\ B' & C \end{bmatrix}$$

Proof. Suppose $\det(A) = 0$ then the columns of A are linearly dependent. Hence the columns of $M := \begin{bmatrix} A & B \\ 0 & C \end{bmatrix}$ are also linearly dependent. Therefore $\det(M) = 0 = \det(A)\det(C)$ (Theorem 3.3.7 and Theorem 2.4.5). Now suppose $\det(A) \neq 0$. Then, since A is invertible we can write

$$\begin{bmatrix} A & B \\ 0 & C \end{bmatrix} = \begin{bmatrix} A & 0 \\ 0 & I \end{bmatrix} \begin{bmatrix} I & A^{-1}B \\ 0 & C \end{bmatrix}$$

By Exercise 3.2.3, we also have $\det\begin{bmatrix} I & * \\ 0 & C \end{bmatrix} = \det(C)$. Hence by Exercise 2.2.4 we have,

$$\det\begin{bmatrix} A & B \\ 0 & C \end{bmatrix} = \det\begin{bmatrix} A & 0 \\ 0 & I \end{bmatrix} \det\begin{bmatrix} I & A^{-1}B \\ 0 & C \end{bmatrix} = \det(A)\det(C)$$

The proof of the second equality is an exercise. □

Practice Problem 3.3.4. *Prove that* $\det\begin{bmatrix} A & 0 \\ B' & C \end{bmatrix} = \det(A)\det(C)$.

To apply Theorem 3.3.17, first reduce the matrix we consider, by elementary row operations, to the block matrix of the form in the theorem.

Example 3.3.18. *To find the determinant of* $A = \begin{bmatrix} 0 & 0 & 2 & 2 \\ 1 & 1 & 2 & 1 \\ -1 & 2 & 1 & 3 \\ 0 & 0 & -1 & 1 \end{bmatrix}$, *interchanging the first and the third rows the sign of the determinant changes. Hence*

$$\det(A) = -\det \begin{bmatrix} -1 & 2 & 1 & 3 \\ 1 & 1 & 2 & 1 \\ 0 & 0 & 2 & 2 \\ 0 & 0 & -1 & 1 \end{bmatrix} = -\det \begin{bmatrix} -1 & 2 \\ 1 & 1 \end{bmatrix} \det \begin{bmatrix} 2 & 2 \\ -1 & 1 \end{bmatrix} = (-1)(-3)(4) = 12$$

Practice Problem 3.3.5. *Find the determinant of* $A = \begin{bmatrix} 1 & 2 & 0 & 0 \\ 1 & -1 & 2 & 5 \\ 1 & 1 & 0 & 0 \\ 2 & 1 & -3 & -1 \end{bmatrix}$.

There is another important function on the square matrices called the trace.

Definition 3.3.19. *Let* $A = [a_{ij}] \in M(n)$. *We define the* **trace** *of* A *to be the sum of diagonal entries of* A *i.e.,*

$$\mathrm{Tr}\,(A) = \sum_{i=1}^{n} a_{ii}$$

Theorem 3.3.20. *If* $A, B \in M(n, \mathbb{R})$. *Then*

$$\mathrm{Tr}\,(AB) = \mathrm{Tr}\,(BA)$$

Proof. We compute:

$$\begin{aligned} \mathrm{Tr}\,(AB) &= \sum_i (AB)_{ii} = \sum_i \left(\sum_k A_{ik} B_{ki} \right) = \sum_{i,k} A_{ik} B_{ki} \\ &= \sum_{i,k} B_{ki} A_{ik} = \sum_k (BA)_{kk} = \mathrm{Tr}\,(BA) \end{aligned}$$ □

Corollary 3.3.21. *Let* $A, B \in M(n)$. *If* A, B *are similar then* $\mathrm{Tr}\, A = \mathrm{Tr}\, B$.

Proof. Let $B = P^{-1}AP$. Then $\mathrm{Tr}\, B = \mathrm{Tr}\,(P^{-1}AP) = \mathrm{Tr}\,((AP)P^{-1}) = \mathrm{Tr}\, A$. □

Remark. *We identify* $M(n)$ *with* $\mathbb{R}^{n^2}$ *by identifying* $A = (x_{ij}) \in M(n)$ *with the column vector* $\mathbf{x} = \begin{bmatrix} A^{(1)} \\ \vdots \\ A^{(n)} \end{bmatrix}$ *in* $\mathbb{R}^{n^2}$. *Then we can view the determinant as a function* $\det : \mathbb{R}^{n^2} \to \mathbb{R}$. *By* Theorem 3.3.16, *we can write* det *as a polynomial*

$$\det(\mathbf{x}) = \sum_\sigma \mathrm{sign}(\sigma)\, x_{\sigma(1)1} x_{\sigma(2)2} \cdots x_{\sigma(n)n}$$

in n^2*-variables* $\mathbf{x} = (x_{11}, x_{21}, ... x_{n1}, x_{12}, ..., x_{n2}, ..., x_{nn})$. *In particular, the determinant is a continuous and differentiable function. The set of non-invertible matrices* $\mathbf{x}$ *is the zeros of* *the polynomial* $f(\mathbf{x}) = \sum_{\sigma} \text{sign}(\sigma) x_{\sigma(1)1} x_{\sigma(2)2} \cdots x_{\sigma(n)n}$ *which is 'very thin closed set' of* $\mathbb{R}^{n^2}$. *Its complement is the set of invertible matrices which form a 'big' open set. Hence we can say that the 'general'* $n \times n$ *matrices are invertible. Also, if* A *is invertible then the matrices 'sufficiently close' to* A *are all invertible.* (See Exercise 19 for example.)

Answers for Practice Problems 3.3

(3.3.1) Let $A = \begin{bmatrix} a & b \\ c & d \end{bmatrix}$. If B is obtained by adding α times the first row to the second row then $B = \begin{bmatrix} a & b \\ c + \alpha a & d + \alpha b \end{bmatrix}$. Hence $|B| = ad + \alpha ab - (bc + \alpha ab) = ad - bc = |A|$. If $B = \begin{bmatrix} c & d \\ a & b \end{bmatrix}$ then $\det B = bc - ad$ which is the same as $-\det A$. If B is obtained by multiplying α to the second row of A then $B = \begin{bmatrix} a & b \\ \alpha c & \alpha d \end{bmatrix}$. Hence $|B| = \alpha ad - \alpha cd = \alpha |A|$. Similarly if B is obtained by adding α times the second to the first row then $|B| = \alpha |A|$.

(3.3.2) $\det(A) = 39$.

(3.3.3) The columns of A are linearly independent if and only if A is invertible (Theorem 2.5.14). Now A is invertible if and only if $\det(A) \neq 0$. But we have $\det(A) = 0$. Hence the columns of A does not form a basis of $\mathbb{R}^4$.

(3.3.4) Let $M = \begin{bmatrix} A & 0 \\ B' & C \end{bmatrix}$. Then, using the first equality of Theorem 3.3.17, we compute $\det M = \det M^t = \begin{bmatrix} A^t & B^{t} \\ 0 & C^t \end{bmatrix} = \det(A^t)\det(C^t) = \det(A)\det(C)$.

(3.3.5) $\det(A) = 13$.

3.3 Exercises

1. Find the determinant of the following matrices.

(a) $A_1 = \begin{bmatrix} 1 & 0 & 2 & 1 \\ 1 & 0 & 1 & 1 \\ 1 & 2 & 1 & 0 \\ 1 & 1 & 0 & 0 \end{bmatrix}$ (b) $A_2 = \begin{bmatrix} 0 & 0 & 1 & -1 & 1 \\ 0 & 0 & 1 & 1 & 1 \\ 0 & 0 & 0 & -1 & 0 \\ 1 & 1 & 1 & 1 & 1 \\ 1 & -1 & 0 & 1 & 0 \end{bmatrix}$ (c) $A_3 = \begin{bmatrix} 1 & 2 & 3 & 4 & 5 \\ 6 & 7 & 8 & 9 & 10 \\ 11 & 12 & 13 & 14 & 15 \\ 16 & 17 & 18 & 19 & 20 \\ 21 & 22 & 23 & 24 & 25 \end{bmatrix}$

(d) $A_4 = \begin{bmatrix} 3 & 4 & 2 & 1 & 0 \\ 2 & 3 & 5 & 6 & 7 \\ 1 & 2 & 3 & 1 & 2 \\ 1 & 1 & 1 & 1 & 1 \\ 2 & 1 & 2 & 1 & 1 \end{bmatrix}$ (e) $A_5 = \begin{bmatrix} 1 & 2 & -1 \\ a & b & c \\ 2 & -1 & 0 \end{bmatrix}$ (f) $A_6 = \begin{bmatrix} 1 & 2 & -1 & 1 \\ 3 & -1 & 2 & -3 \\ 2 & -1 & 0 & 3 \\ a & b & c & d \end{bmatrix}$

2. Find the conditions on a, b, c, d for the following matrices to be invertible.

(a) $\begin{bmatrix} 1 & 2 & -1 \\ 3 & -2 & 1 \\ a & b & c \end{bmatrix}$ (b) $\begin{bmatrix} 1 & 2 & -2 & -1 \\ 1 & -2 & 1 & 0 \\ 1 & 1 & -1 & 0 \\ a & b & c & d \end{bmatrix}$ (c) $\begin{bmatrix} a & 2 & -1 & d \\ 3 & b & c & -3 \\ 2 & b & c & 3 \\ a & -1 & 3 & d \end{bmatrix}$

3. Prove that $|E_{ij}(c)| = 1, |I_{ij}| = -1$ and $|M_i(c)| = c$.

4. Prove Corollary 3.3.4.

5. Prove the statements about the rows of Theorem 3.3.14.

6. Prove the true statement(s) and find a counter example for the false statement(s) for the square matrices.

   (a) $\det(AB) = \det(BA)$
   (b) If $A, B \in M(n)$ are row equivalent then $\det(A) = \det(B)$.
   (c) If $A, B \in M(n)$ are similar then $\det(A) = \det(B)$.
   (d) A triangular matrix is invertible if and only if the diagonal entries are all nonzero.
   (e) If $\mathrm{Tr}\,(A) = \mathrm{Tr}\,(B) = 0$ then $\mathrm{Tr}\,(AB) = 0$.
   (f) Let $A \in M(n)$. Then $\rho(A) < n$ if and only if $\det(A) = 0$.
   (g) The determinant map $\det : M(n) \to \mathbb{R}$ is a surjective map.
   (h) The trace map $\mathrm{Tr} : M(n) \to \mathbb{R}$ is a surjective map.

7. Let $A \in M(n)$ and $c \in \mathbb{R}$. Prove that $\det(cA) = c^n \det(A)$.

8. Let $A \in M(n)$ be skew symmetric i.e., $A^t = -A$. If n is odd then A is not invertible.

9. For $A, B \in M(n), c \in \mathbb{R}$ prove that

$$\mathrm{Tr}\,(A + B) = \mathrm{Tr}\,(A) + \mathrm{Tr}\,(B), \quad \mathrm{Tr}\,(cA) = c\mathrm{Tr}\,(A)$$

10. Let $A \in M(n)$ and let $B = [A^{(n)}\ A^{(n-1)}\ \cdots\ A^{(1)}]$ be the matrix obtained by reversing the order of columns of A. Find $\det(B)$ in terms of $\det(A)$.

11. Let A, C be square matrices size r, s and $B, B' \in M(r \times s)$. Find the determinant of

$$\det \begin{bmatrix} B & A \\ I_s & 0 \end{bmatrix} \text{ and } \det \begin{bmatrix} B' & I_r \\ C & 0 \end{bmatrix}$$

12. Let M_i be square matrices. Let

$$A = \begin{bmatrix} M_1 & & & \\ & M_2 & & \\ & * & \ddots & \\ & & & M_k \end{bmatrix}$$

be the square matrix with M_i along the diagonal and the entries above M_i are zeros. Prove that $\det(A) = \prod \det(M_i)$.

13. Let A, B, C and D be matrices of size $n \times n, n \times m, m \times n$ and $m \times m$ respectively.

   (a) When A is invertible use the identity $\begin{bmatrix} A & B \\ C & D \end{bmatrix} = \begin{bmatrix} A & 0 \\ C & I_m \end{bmatrix} \begin{bmatrix} I_n & A^{-1}B \\ 0 & D - CA^{-1}B \end{bmatrix}$ to prove that

$$\det \begin{bmatrix} A & B \\ C & D \end{bmatrix} = \det(A)\det(D - CA^{-1}B)$$

(b) When D is invertible use the identity $\begin{bmatrix} A & B \\ C & D \end{bmatrix}\begin{bmatrix} I_n & 0 \\ -D^{-1}C & I_n \end{bmatrix} = \begin{bmatrix} A - BD^{-1}C & B \\ 0 & D \end{bmatrix}$ to prove that

$$\det \begin{bmatrix} A & B \\ C & D \end{bmatrix} = \det(D)\det(A - BD^{-1}C)$$

14. Let $\sigma = (i_1, i_2, \ldots, i_n)$ be a permutation of $(1, 2, \ldots, n)$. Prove that if k is the number of transpositions applied to $\sigma = (i_1, i_2, \ldots, i_n)$ to get the identity permutation $\iota = (1, 2, ..., n)$ then $\operatorname{sign}(\sigma) = (-1)^k$.

15. We say $A \in M(n)$ is **nilpotent** if $A^k = 0$ for some k. Prove that if A is nilpotent then $|A| = 0$.

16. We say $A \in M(n)$ is **orthogonal** if $A^tA = I_n$. Prove that if A is orthogonal then $|A| = \pm 1$.

17. We say $A \in M(n)$ is **skew symmetric** if $A^t = A$. Prove that if A is skew symmetric with odd order then $|A| = 0$.

18. Prove the equality

$$\begin{vmatrix} 1 & 1 & 1 & 1 & \cdots & 1 & 1 \\ b_1 & a_1 & a_1 & a_1 & \cdots & a_1 & a_1 \\ b_1 & b_2 & a_2 & a_2 & \cdots & a_2 & a_2 \\ b_1 & b_2 & b_3 & a_3 & \cdots & a_3 & a_3 \\ & & \cdots & \cdots & \cdots & & \\ b_1 & b_2 & b_3 & \cdots & b_{n-1} & a_{n-1} & a_{n-1} \\ b_1 & b_2 & b_3 & \cdots & b_{n-1} & b_n & a_n \end{vmatrix} = (a_1 - b_1)(a_2 - b_2)\cdots(a_n - b_n)$$

19. Let $a_{ij}(t)$ be continuous functions and let $A(t) = (a_{ij}(t))$. (For the determinant with entries in a commutative ring see the remark preceding Theorem B.2.6.)

   (a) Prove that if $A(0)$ is invertible then there is $\epsilon > 0$ such that $A(t)$ is invertible for all t with $|t| < \epsilon$.

   (b) Suppose $a_{ij}(t)$ are differentiable. Let A_i' be the matrix obtained from A by replacing the i-th column by $(A^{(i)})' = \begin{bmatrix} a_{1i}'(t) \\ \vdots \\ a_{ni}'(t) \end{bmatrix}$ where $a_{ij}'(t)$ denotes the derivative with respect to t. Prove that $\det(A(t))$ is differentiable and

$$\det(A)' = \sum_{1 \le i \le n} \det(A_i')$$

3.4 Applications of Determinant

3.4.1 Determinant of Submatrix and Rank

Definition 3.4.1. *Let $A \in M(m \times n)$. For $k \le \min\{m, n\}$, a* **submatrix of order** k *is a matrix obtained by first choosing k columns and then choosing k rows. Or by removing $n-k$ columns and $m-k$ rows.*

For example, let $A = \begin{bmatrix} 1 & 2 & 4 & 2 & 3 \\ 2 & 1 & 1 & 0 & 0 \\ 0 & 0 & 9 & 5 & 2 \\ 1 & -1 & 6 & 3 & -1 \end{bmatrix}$. By removing the second, fourth and fifth columns and then removing the second and fourth rows we get, in the notation of (1.25)

$$\left(A^{(\{1,4\})}\right)_{(\{1,3\})} = \begin{bmatrix} 1 & 2 \\ 2 & 0 \\ 0 & 5 \\ 1 & 3 \end{bmatrix}_{(\{1,3\})} = \begin{bmatrix} 1 & 2 \\ & \\ 0 & 5 \\ & \end{bmatrix} = \begin{bmatrix} 1 & 2 \\ 0 & 5 \end{bmatrix}$$

a 2×2 submatrix of A.

We introduce another method to compute the rank of a matrix.

Theorem 3.4.2. *Let $A \in M(m\times n)$. Then*

$$\rho(A) = \max\left\{ k \;\middle|\; \begin{array}{c}\text{there is a submatrix of order} \\ k \text{ having nonzero determinant}\end{array} \right\}$$

Proof. Let $\rho(A) = r$. We have to show that (i) there is a submatrix A' of order r having nonzero determinant and (ii) if A'' is a submatrix of order $> r$ then $\det(A'') = 0$.

(i) Since $r = \text{rank}(A)$ there are r linearly independent columns (Corollary 2.3.3) and we denote the matrix consisting of those columns by A_1. Since $\rho(A_1^t) = \rho(A_1)$ there are r linearly independent rows of A_1. Let A' be the submatrix of A consisting of those r rows of A_1. Then $A' \in M(r)$ is invertible (Theorem 2.5.14). Hence by Theorem 3.3.7 we have $\det(A') \neq 0$.

(ii) Now suppose $s > r$ and $A'' \in M(s)$. We need to show that $|A''| = 0$. Suppose $A'' = A_{(I)}^{(J)}$ for some subsets I, J of $\{1, 2, ..., m\}, \{1, 2, ..., n\}$ consisting of s elements. Then the s columns of $A^{(J)}$ is linearly dependent since it contains more columns than $\rho(A) = r$ (Corollary 2.3.3). Now the columns of $A_{(I)}^{(J)}$ are still linearly dependent. Hence $|A''| = 0$. ☐

Corollary 3.4.3. *Let $A \in M(m\times n)$. Let I, J be subsets of $\{1, 2, ..., n\}, \{1, 2, ..., m\}$ consisting of the same number of elements. If* $\det\left(A_{(J)}^{(I)}\right) \neq 0$ *then the columns of $A^{(I)}$ and the rows of $A_{(J)}$ are linearly independent.*

Proof. Since the columns of $A_{(J)}^{(I)}$ are linearly independent the corresponding columns of $A^{(I)}$ is also linearly independent. In fact, if $\sum_{i_t \in I} a_t A^{(i_t)} = 0$ then $\sum_{i_t \in I} a_t A_{(J)}^{(i_t)} = 0$. Since the columns of $A_{(J)}^{(I)}$ are linearly independent we see that all $a_{i_t} = 0$. ☐

Example 3.4.4. *Let* $A = \begin{bmatrix} 1 & 0 & 2 & 1 \\ 1 & 1 & 1 & 0 \\ 0 & 1 & 0 & 1 \\ 0 & -1 & 1 & 1 \end{bmatrix}$. *One computes that* $\det(A) = 0$. *Let* $J = \{2, 3, 4\}$ *and* $I = \{1, 2, 3\}$ *then the* 3×3 *matrix* $A_{(J)}^{(I)} = \begin{bmatrix} 1 & 1 & 1 \\ 0 & 1 & 0 \\ 0 & -1 & 1 \end{bmatrix}$ *has determinant* 1. *Hence by* Theorem 3.4.2, $\rho(A) = 3$. *By* Corollary 3.4.3, *the columns* $\{A^{(1)}, A^{(2)}, A^{(3)}\}$ *and the rows* $\{A_{(2)}, A_{(3)}, A_{(4)}\}$ *are linearly independent.*

Practice Problem 3.4.1. *Find the rank of* $A = \begin{bmatrix} 0 & 1 & 0 & 1 \\ 1 & 0 & 2 & 1 \\ 1 & 2 & 2 & 3 \\ 1 & 1 & 2 & 2 \end{bmatrix}$ *by using* Theorm 3.4.2.

We will use the following theorem in §8.6. Let $A = [a_{ij}] \in M(n)$ and let A_k be the submatrix of A which is the upper left $k \times k$ submatrix. Namely, if $I_k = \{1, 2, ..., k\}$ then

$$A_k = A_{(I_k)}^{(I_k)} = \text{ upper left } k \times k \text{ submatrix of } A \tag{3.18}$$

Theorem 3.4.5. *Let* $A = [a_{ij}] \in M(n)$*. Then* $|A_k| \neq 0$ *for* $k = 1, 2, ..., n$ *if and only if there is an upper triangular matrix* P *with 1's on the diagonal and* $B = AP$ *is the lower triangular matrix with* $b_{ii} \neq 0$*. Further,* P *is uniquely determined and* $|A_k| = |B_k|$*.*

Proof. First assume $|A_k| \neq 0$ for $k = 1, 2, ..., n$. Then the condition for $k = 1$ implies that $a_{11} \neq 0$. The elementary column operation $-\frac{a_{12}}{a_{11}} C_1 + C_2$ applied to A, the 2×2 submatrix of the upper left corner of A becomes $'B = \begin{bmatrix} a_{11} & 0 & * \\ b_{21} & b_{22} & * \\ & * & * \end{bmatrix}$. Then $|A_2|$ is the same as the determinant $|'B_2|$ of the upper left 2×2 submatrix of $'B$ since $'B$ is obtained from A by post multiplying the transpose of $E_{12}(-\frac{a_{12}}{a_{11}})$ (Theorem 3.3.1). And $'B_{22} \neq 0$ since $|A_2| \neq 0$. Similarly we reduce the upper left 3×3 submatrix of $'B$ to $''B = \begin{bmatrix} a_{11} & 0 & 0 & * \\ b_{21} & b_{22} & 0 & * \\ * & * & * & * \end{bmatrix}$ by performing column operations of type $cCj + C_3$ $(j < 3)$ to $'B$ without changing the determinant. Since $|A_3| \neq 0$ we conclude that $''B_{33} \neq 0$.

Proceeding in this way, at the k-th stage we perform the column operations of the form $cC_j + C_k$ with $j < k$ to get say, ${}^kB$. Hence the determinant of the upper left $k \times k$ submatrix of ${}^kB$ remain unchanged i.e., $|A_k| = |{}^kB_k|$. Finally, we arrive at

$$B = \begin{bmatrix} b_{11} & 0 & \cdots & 0 \\ * & b_{22} & 0 & \\ * & * & \ddots & 0 \\ * & & * & b_{nn} \end{bmatrix}$$

We obtain B by performing the column operations which is the same as post multiplying the corresponding elementary matrices to A of the form $E_{ij}(c)$ with $i < j$ that is upper triangular. That is $AE_1E_2 \cdots E_m = B$ where E_i is of the type $E_{ij}(c)$ with $i < j$. Let $P = E_1E_2 \cdots E_m$. Then P is upper triangular with 1's on the diagonal.

For the converse, write P as a product of elementary matrices $E_{ij}(c)$ with $i < j$ (Check this.) Hence B is obtained from A by performing column operations. Hence, as before, the upper left $k \times k$ submatrix of B has the same determinant as $|A_k|$ (Exercise 5).

For uniqueness of P, let $P^{(1)} = \begin{bmatrix} 1 \\ 0 \\ \vdots \\ 0 \\ 0 \end{bmatrix}$, $P^{(2)} = \begin{bmatrix} x_{12} \\ 1 \\ 0 \\ 0 \\ \vdots \end{bmatrix}$, $P^{(3)} = \begin{bmatrix} x_{13} \\ x_{23} \\ 1 \\ 0 \\ \vdots \end{bmatrix}$ and let $P^{(i)} = \sum_{j=1}^{i-1} x_{ji}\mathbf{e}^j + \mathbf{e}^i$ be the columns of P. Since $AP^{(1)} = B^{(1)}$ we have $B^{(1)} = A^{(1)}$. And $AP^{(2)} = B^{(2)}$ implies, by comparing the second place, $a_{11}x_{12} + a_{12} = 0$ since B is lower triangular. Since $a_{11} \neq 0$ we have $x_{12} = -\frac{a_{12}}{a_{11}}$; $P^{(2)}$ is uniquely determined. Now $AP^{(3)} = B^{(3)}$ with lower triangular B implies that the equations $\begin{cases} a_{11}x_{13} + a_{12}x_{23} + a_{13} = 0 \\ a_{21}x_{13} + a_{22}x_{23} + a_{23} = 0 \end{cases}$ has a unique solution for x_{13}, x_{23} since $|A_2| \neq 0$; $P^{(3)}$ is uniquely determined. Proceeding in this way we see that $AP^{(i)} = B^{(i)}$ has a unique solution. Hence P is uniquely determined. □

Under the conditions of Theorem 3.4.5 the proof the theorem shows that we can write

$$AE_1E_2\cdots E_m = B$$

where B is lower triangular and E_i is an elementary matrix of the form $E_{ij}(c)$ $(i < j)$ which corresponds to the column operation $cC_i + C_j$. If we let $P = E_1E_2\cdots E_m$ then we can write

$$AP = B \tag{3.19}$$

where P is an upper triangular matrix with $P_{ii} = 1$ and B is lower triangular. If we write $U = P^{-1}, L = B$ then we can write

$$A = LU \tag{3.20}$$

where L is lower triangular and U is upper triangular with $U_{ii} = 1$.

Example 3.4.6. *Let* $A = \begin{bmatrix} 1 & 1 & 0 & 0 \\ 0 & 1 & 1 & 0 \\ 0 & 0 & 1 & 1 \\ 1 & 0 & 0 & 2 \end{bmatrix}$. *Then* $|A_1| = |A_2| = |A_3| = |A| = 1$. *Hence A satisfies the conditions of the theorem. The column operations* $-1C_1 + C_2, -1C_2 + C_3$ *and* $-1C_3 + C_4$ *reduces A to* $B = \begin{bmatrix} 1 & 0 & 0 & 0 \\ 0 & 1 & 0 & 0 \\ 0 & 0 & 1 & 0 \\ 0 & -1 & 1 & 1 \end{bmatrix}$. *Hence we have* $AP = B$ *where* $P = E_{12}(-1)E_{23}(-1)E_{34}(-1) = \begin{bmatrix} 1 & -1 & 1 & -1 \\ 0 & 1 & -1 & 1 \\ 0 & 0 & 1 & -1 \\ 0 & 0 & 0 & 1 \end{bmatrix}$ *which is the decomposition of the form* (3.19).

If we let $U = P^{-1} = \begin{bmatrix} 1 & 1 & 0 & 0 \\ 0 & 1 & 1 & 0 \\ 0 & 0 & 1 & 1 \\ 0 & 0 & 0 & 1 \end{bmatrix}$ *and* $L = B$ *then we have the decomposition* $A = LU$ *of* (3.20).

Practice Problem 3.4.2. *Let* $A = \begin{bmatrix} 1 & 1 & 1 & 0 \\ 0 & 1 & 1 & 1 \\ 1 & 0 & 1 & 1 \\ 1 & 1 & 0 & 1 \end{bmatrix}$. *Decompose A into the form* (3.20).

3.4.2 Cramer's Rule

In this section, we introduce a method to solve a system of linear equations known as Cramer's rule†. Let $A \in M(n), \mathbf{b} \in \mathbb{R}^n$. Consider the equation $A\mathbf{x} = \mathbf{b}$ where $\mathbf{x} = \begin{bmatrix} x_1 \\ \vdots \\ x_n \end{bmatrix}$.

†Gabriel Cramer (1704–1752) was a Genevan mathematician.

We write $A_i(\mathbf{b})$ for the matrix obtained by replacing the i-th column of A by $\mathbf{b}$ *viz*

$$A_i(\mathbf{b}) = [A^{(1)} \cdots A^{(i-1)} \;\; \mathbf{b} \;\; A^{(i+1)} \cdots A^{(n)}]$$

If we let $I_i(\mathbf{x})$ be the identity matrix with the i-th column replaced by $\mathbf{x}$ then

$$\begin{aligned} AI_i(\mathbf{x}) &= A[\mathbf{e}^1 \cdots \mathbf{x} \cdots \mathbf{e}^n] = [A\mathbf{e}^1 \cdots A\mathbf{x} \cdots A\mathbf{e}^n] = \\ &= [A^{(1)} \cdots A^{(i-1)} \;\; \mathbf{b} \;\; A^{(i+1)} \cdots A^{(n)}] = A_i(\mathbf{b}) \end{aligned}$$

By expanding the determinant of $I_i(\mathbf{x})$ according to the i-th row we have $|I_i(\mathbf{x})| = x_i$ and hence we have

$$|A_i(\mathbf{b})| = |AI_i(\mathbf{x})| = |A||I_i(\mathbf{x})| = x_i|A| \tag{3.21}$$

Theorem 3.4.7 (Cramer's Rule). *Let $A \in M(n)$ be an invertible matrix. For any $\mathbf{b} \in \mathbb{R}^n$ the equation $A\mathbf{x} = \mathbf{b}$ has a unique solution given by*

$$x_i = \frac{\det(A_i(\mathbf{b}))}{\det(A)}$$

Proof. Since $\det(A) \neq 0$ we simply divide (3.21) by $\det(A)$. □

Example 3.4.8. *Applying Cramer's rule to*

$$\begin{cases} a_1x + b_1y = c_1 \\ a_2x + b_2y = c_2 \end{cases}$$

we get

$$x = \frac{\begin{vmatrix} c_1 & b_1 \\ c_2 & b_2 \end{vmatrix}}{\begin{vmatrix} a_1 & b_1 \\ a_2 & b_2 \end{vmatrix}} = \frac{b_2c_1 - b_1c_2}{a_1b_2 - a_2b_1}, \quad y = \frac{\begin{vmatrix} a_1 & c_1 \\ a_2 & c_2 \end{vmatrix}}{\begin{vmatrix} a_1 & b_1 \\ a_2 & b_2 \end{vmatrix}} = \frac{a_1c_2 - a_2c_1}{a_1b_2 - a_2b_1}$$

which we saw in Example 1.1.11.

Example 3.4.9. *To solve the system*

$$\begin{cases} x \phantom{{}+y} + z = 3 \\ x + y \phantom{{}+z} = 2 \\ \phantom{x+{}} y + z = -1 \end{cases}$$

we apply Cramer's rule: Let

$$A = \begin{bmatrix} 1 & 0 & 1 \\ 1 & 1 & 0 \\ 0 & 1 & 1 \end{bmatrix}, \; A_1(\mathbf{b}) = \begin{bmatrix} 3 & 0 & 1 \\ 2 & 1 & 0 \\ -1 & 1 & 1 \end{bmatrix}, \; A_2(\mathbf{b}) = \begin{bmatrix} 1 & 3 & 1 \\ 1 & 2 & 0 \\ 0 & -1 & 1 \end{bmatrix}, \; A_3(\mathbf{b}) = \begin{bmatrix} 1 & 0 & 3 \\ 1 & 1 & 2 \\ 0 & 1 & -1 \end{bmatrix}$$

Then $\det(A) = 2$, $\det(A_1(\mathbf{b})) = 6$, $\det(A_2(\mathbf{b})) = -2$, $\det(A_3(\mathbf{b})) = 0$. *Hence we get* $x = 3, \quad y = -1, \quad z = 0$.

Example 3.4.10. *To solve the system*

$$\begin{cases} x - y \phantom{{}+z} - w = 1 \\ \phantom{x-{}} y + z \phantom{{}-w} = 1 \\ x \phantom{{}-y} - z + w = 1 \\ \phantom{x-{}} y + z - w = 1 \end{cases}$$

we apply Cramer's rule: Let

$$A = \begin{bmatrix} 1 & -1 & 0 & -1 \\ 0 & 1 & 1 & 0 \\ 1 & 0 & -1 & 1 \\ 0 & 1 & 1 & -1 \end{bmatrix}, \quad A_1(\mathbf{b}) = \begin{bmatrix} 1 & -1 & 0 & -1 \\ 1 & 1 & 1 & 0 \\ 1 & 0 & -1 & 1 \\ 1 & 1 & 1 & -1 \end{bmatrix}, \quad A_2(\mathbf{b}) = \begin{bmatrix} 1 & 1 & 0 & -1 \\ 0 & 1 & 1 & 0 \\ 1 & 1 & -1 & 1 \\ 0 & 1 & 1 & -1 \end{bmatrix},$$

$$A_3(\mathbf{b}) = \begin{bmatrix} 1 & -1 & 1 & -1 \\ 0 & 1 & 1 & 0 \\ 1 & 0 & 1 & 1 \\ 0 & 1 & 1 & -1 \end{bmatrix}, \quad A_4(\mathbf{b}) = \begin{bmatrix} 1 & -1 & 0 & 1 \\ 0 & 1 & 1 & 1 \\ 1 & 0 & -1 & 1 \\ 0 & 1 & 1 & 1 \end{bmatrix},$$

We have

$$|A| = 2,\ |A_1(\mathbf{b})| = 3,\ |A_2(\mathbf{b})| = 1,\ |A_3(\mathbf{b})| = 1,\ |A_4(\mathbf{b})| = 0$$

Hence we have the solution $x_1 = \frac{3}{2}, x_2 = \frac{1}{2}, x_3 = \frac{1}{2}, x_4 = 0$.

Practice Problem 3.4.3. *Solve the equation by Cramer's rule:*

$$\begin{cases} x + 3y + z = 3 \\ x + 2y - z = -2 \\ -x - y + z = 1 \end{cases}$$

3.4.3 A^{-1} by Adjoint Matrix

In §2.5, we saw how to get the inverse of an invertible matrix. Now we will introduce another method to find the inverse of an invertible matrix.

Suppose $A \in M(n)$ is an invertible matrix. Then we have

$$AA^{-1} = I_n = \begin{bmatrix} \mathbf{e}^1 & \mathbf{e}^2 & \cdots & \mathbf{e}^n \end{bmatrix}$$

On the other hand, by block multiplication (Exercise 2.2.3) we have

$$AA^{-1} = A\left[(A^{-1})^{(1)}\ (A^{-1})^{(2)} \cdots\ (A^{-1})^{(n)}\right] = \left[A(A^{-1})^{(1)}\ A(A^{-1})^{(2)} \cdots\ A(A^{-1})^{(n)}\right]$$

Comparing the last expressions of the two equalities above we have $A(A^{-1})^{(j)} = \mathbf{e}^j$ i.e., $(A^{-1})^{(j)}$ is the solution of the equation $A\mathbf{x} = \mathbf{e}^j$. The i-th place (i.e., solution for x_i) of $(A^{-1})^{(j)}$ is $\dfrac{\det(A_i(\mathbf{e}^j))}{\det(A)}$ (Theorem 3.4.7):

$$\left(\text{the } i\text{-th place of } (A^{-1})^{(j)}\right) = \frac{\det(A_i(\mathbf{e}^j))}{\det(A)}$$

Now expand $\det(A_i(\mathbf{e}^j))$ according to the i-th column to get

$$\det(A_i(\mathbf{e}^j)) = (-1)^{i+j}|\tilde{A}_{ji}| := C_{ji} \tag{3.22}$$

which is the (j, i) cofactor (Definition 3.2.1). Hence

$$(A^{-1})_{ij} = \frac{(-1)^{i+j}|\tilde{A}_{ji}|}{\det(A)} = \frac{C_{ji}}{\det(A)} \tag{3.23}$$

Definition 3.4.11. *The determinant $|\tilde{A}_{ij}|$ of $(n-1)\times(n-1)$ matrix is called the (i,j)-* **minor** *of A. The* **adjoint**† *or* **adjugate** adj(A) *of A is obtained by taking the transpose of the matrix $C = (C_{ij})$*

$$\operatorname{adj}(A) = (C_{ji}) \tag{3.24}$$

where $C_{ij} = (-1)^{i+j}|\tilde{A}_{ij}|$ is the (i,j)-cofactor of A.

We have then,

Theorem 3.4.12. *Let $A \in M(n)$ be invertible. Then*

$$A^{-1} = \frac{1}{|A|}\operatorname{adj}(A) = \frac{1}{|A|}\begin{bmatrix} C_{11} & C_{21} & \cdots & C_{n1} \\ C_{12} & C_{22} & \cdots & C_{n2} \\ \vdots & \vdots & \ddots & \vdots \\ C_{1n} & C_{2n} & \cdots & C_{nn} \end{bmatrix} \tag{3.25}$$

We can prove that $\operatorname{adj}(A)A = (\det A)I_n$ (Exercise 6). For more about adjoint matrix see §6.6.2.

Example 3.4.13. *To find the inverse of $A = \begin{bmatrix} 2 & 1 & 3 \\ 1 & -1 & 1 \\ 1 & 4 & -2 \end{bmatrix}$ we compute the adjoint matrix:*

$$\operatorname{adj}(A) = \begin{bmatrix} (-1)^{1+1}\begin{bmatrix} -1 & 1 \\ 4 & -2 \end{bmatrix} = -2 & (-1)^{2+1}\begin{bmatrix} 1 & 3 \\ 4 & -2 \end{bmatrix} = 14 & (-1)^{3+1}\begin{bmatrix} 1 & 3 \\ -1 & 1 \end{bmatrix} = 4 \\ (-1)^{1+2}\begin{bmatrix} 1 & 1 \\ 1 & -2 \end{bmatrix} = 3 & (-1)^{2+2}\begin{bmatrix} 2 & 3 \\ 1 & -2 \end{bmatrix} = -7 & (-1)^{3+2}\begin{bmatrix} 2 & 3 \\ 1 & 1 \end{bmatrix} = 1 \\ (-1)^{1+3}\begin{bmatrix} 1 & -1 \\ 1 & 4 \end{bmatrix} = 5 & (-1)^{2+3}\begin{bmatrix} 2 & 1 \\ 1 & 4 \end{bmatrix} = -7 & (-1)^{3+3}\begin{bmatrix} 2 & 1 \\ 1 & -1 \end{bmatrix} = -3 \end{bmatrix}$$

and $\det(A) = 14$. *Hence,*

$$A^{-1} = \frac{1}{14}\begin{bmatrix} -2 & 14 & 4 \\ 3 & -7 & 1 \\ 5 & -7 & -3 \end{bmatrix}$$

Practice Problem 3.4.4. *Find the inverse of $A = \begin{bmatrix} 1 & 0 & 2 \\ 1 & 1 & 0 \\ 0 & 1 & -1 \end{bmatrix}$ by using the adjoint of A.*

3.4.4 Lagrange Interpolation and Vandermonde Matrix

Problem: Find a polynomial having prescribed values at a set of given points? Can we have control on the degree of the polynomial?

We are going to prove that given $n+1$ distinct points $x_0, x_1, ..., x_n$ there is a unique polynomial of degree $\leq n$ having the prescribed values $y_0, y_1, ..., y_n$ at $x_0, x_1, ..., x_n$.

†It should not be confused with 'adjoint' for the complex matrices (Definition 6.2.24).

The **Vandermonde matrix**† is of the form

$$V_n = \begin{bmatrix} 1 & x_0 & x_0^2 & \cdots & x_0^n \\ 1 & x_1 & x_1^2 & \cdots & x_1^n \\ 1 & x_2 & x_2^2 & \cdots & x_2^n \\ \cdots & \cdots & \cdots & \cdots & \cdots \\ 1 & x_n & x_n^2 & \cdots & x_n^n \end{bmatrix} \tag{3.26}$$

Let $\mathbf{c}_p = \begin{bmatrix} c_0 \\ c_1 \\ \vdots \\ c_n \end{bmatrix}$ and $\mathbf{y} = \begin{bmatrix} y_0 \\ y_1 \\ \vdots \\ y_n \end{bmatrix}$ and let $p(x) = c_0 + c_1 x + \cdots + c_n x^n$ then

$$V_n \mathbf{c}_p = \begin{bmatrix} \sum_i c_i x_0^i \\ \sum_i c_i x_1^i \\ \vdots \\ \sum_i c_i x_n^i \end{bmatrix} = \begin{bmatrix} p(x_0) \\ p(x_1) \\ \vdots \\ p(x_n) \end{bmatrix} \tag{3.27}$$

Hence

$$V_n\, \mathbf{c}_p = \mathbf{y} \ \text{ if and only if } \ p(x_i) = y_i \text{ for } 0 \le i \le n \tag{3.28}$$

We get an affirmative answer to our problem if the equation $V_n \mathbf{c}_p = \mathbf{y}$ has a solution. To prove that the solution $\mathbf{c}_p$ exist we will show the matrix V_n is invertible which amounts to prove that $\det(V_n) \neq 0$.

Theorem 3.4.14. *The determinant of the Vandermonde matrix is given by*

$$\det(V_n) = \prod_{0 \le i < j \le n} (x_j - x_i) \tag{3.29}$$

In particular, if x_i are all distinct then V_n is invertible.

Proof. We prove this by induction on n. For $n = 2$, $\det \begin{bmatrix} 1 & x_0 \\ 1 & x_1 \end{bmatrix} = (x_1 - x_0)$. Now assume the result for $n - 1$ and we prove (3.29) for n. We apply the sequence of column operations $-x_0 C_{n-1} + C_n$, $-x_0 C_{n-2} + C_{n-3}$, $\cdots$, $x_0 C_0 + C_1$ to the Vandermonde matrix V_n to get

$$\begin{bmatrix} 1 & 0 & 0 & \cdots & 0 \\ 1 & x_1 - x_0 & x_1^2 - x_0 x_1 & \cdots & x_1^n - x_0 x_1^{n-1} \\ 1 & x_2 - x_0 & x_2^2 - x_0 x_2 & \cdots & x_2^n - x_0 x_2^{n-1} \\ \cdots & \cdots & \cdots & \cdots & \cdots \\ 1 & x_n - x_0 & x_n^2 - x_0 x_n & \cdots & x_n^n - x_0 x_n^{n-1} \end{bmatrix}$$

†Alexandre-Théophile Vandermonde (1735–1796) was a French mathematician, musician and chemist who worked with Bézout and Lavoisier.

By expanding the determinant according to the first row we get

$$\det(V_n) = \det \begin{bmatrix} x_1 - x_0 & x_1^2 - x_0x_1 & \cdots & x_1^n - x_0x_1^{n-1} \\ x_2 - x_0 & x_2^2 - x_0x_2 & \cdots & x_2^n - x_0x_2^{n-1} \\ \cdots & \cdots & \cdots & \cdots \\ x_n - x_0 & x_n^2 - x_0x_n & \cdots & x_n^n - x_0x_n^{n-1} \end{bmatrix}$$

Each i-th row contains the common factor $(x_i - x_0)$. Factoring out those factors we have

$$\det(V_n) = \prod_{i=1}^{n}(x_i - x_0) \det \begin{bmatrix} 1 & x_1 & \cdots & x_1^{n-1} \\ 1 & x_2 & \cdots & x_2^{n-1} \\ & \cdots & \cdots & \\ 1 & x_n & \cdots & x_n^{n-1} \end{bmatrix} \tag{3.30}$$

The matrix on the right side of (3.30) is V_{n-1} and hence, by induction, $\det(V_{n-1}) = \prod_{1\le i<j\le n}(x_j - x_i)$. Taking product with $\prod_{i=1}^{n}(x_i - x_0)$ we get the desired result. □

Example 3.4.15. $\begin{vmatrix} 1 & x_0 & x_0^2 & x_0^3 \\ 1 & x_1 & x_1^2 & x_1^3 \\ 1 & x_2 & x_2^2 & x_2^3 \\ 1 & x_3 & x_3^2 & x_3^3 \end{vmatrix} = (x_3 - x_2)(x_3 - x_1)(x_3 - x_0)(x_2 - x_1)(x_2 - x_0)(x_1 - x_0)$

Practice Problem 3.4.5. *Find the determinant of* $A = \begin{bmatrix} 1 & 2 & 4 \\ 1 & 3 & 9 \\ 1 & 5 & 25 \end{bmatrix}$.

Theorem 3.4.16. *Given $n+1$ distinct points $x_0, x_1, \ldots, x_n$ there is a unique polynomial $p(x) = c_0 + c_1x + \cdots + c_nx^n$ of degree $\le n$ having the prescribed values $y_0, y_1, \ldots, y_n$ at $x_0, x_1, \ldots, x_n$ namely the curve $y = p(x)$ pass through the points (x_i, y_i) for $0 \le i \le n$.*

Proof. Since x_i's are distinct $\det(V_n) \neq 0$. By Theorem 3.3.7, V_n is invertible. Hence $V_n\mathbf{c}_p = \mathbf{y}$ has a unique solution. □

Now we want to find the polynomial $p(x)$ explicitly. We could try to find V_n^{-1}. But it is not easy. So we take a different approach first and then we consider V_n^{-1} next.

We want to find the polynomials $p_i(x)$ satisfying the properties

$$p_i(x_j) = \delta_{ij} = \begin{cases} 1 & (i = j) \\ 0 & (i \neq j) \end{cases} \tag{3.31}$$

That is, by (3.27), we have

$$V_n\mathbf{c}_{p_i} = \mathbf{e}^i \tag{3.32}$$

where $\{\mathbf{e}^i \mid i = 0, 1, \ldots, n\}$ is the standard basis of $\mathbb{R}^n$. In fact, if we let

$$p_i(x) = \frac{(x - x_0)\cdots(x - x_{i-1})(x - x_{i+1})\cdots(x - x_n)}{(x_i - x_0)\cdots(x_i - x_{i-1})(x_i - x_{i+1})\cdots(x_i - x_n)} \tag{3.33}$$

then $p_i(x)$ satisfies the desired properties. We call $p_i(x)$ a **Lagrange polynomial**[†]. Then

$$p(x) = y_0p_0(x) + y_1p_1(x) + \cdots + y_np_n(x) \tag{3.34}$$

is of degree $\leq n$ and we have

$$p(x_i) = y_i \quad (i = 0, 1, ..., n)$$

The polynomial (3.34) is called the **Lagrange interpolation formula**, and we showed above that the polynomial $p(x)$ is uniquely determined (Theorem 3.4.16). In terms of Vandermonde matrix the column of coefficients $\mathbf{c}_p$ of $p(x)$ is, by (3.28), given by

$$\mathbf{c}_p = V_n^{-1}\mathbf{y} \tag{3.35}$$

Example 3.4.17. *The Lagrange polynomials for $n = 2$ are*

$$p_0 = \frac{(x - x_1)(x - x_2)}{(x_0 - x_1)(x_0 - x_2)}, \; p_1 = \frac{(x - x_0)(x - x_2)}{(x_1 - x_0)(x_1 - x_2)}, \; p_2 = \frac{(x - x_0)(x - x_1)}{(x_2 - x_0)(x_2 - x_1)}$$

Hence

$$p(x) = y_0p_0(x) + y_1p_1(x) + y_2p_2(x)$$

is the polynomial of degree ≤ 2 having values y_0, y_1, y_2 at x_0, x_1, x_2.

Example 3.4.18. *We want to find the polynomial $p(x) = c_0 + c_1x + c_2x^2 + c_3x^3$ of degree ≤ 3 such that $p(1) = -1, p(2) = 4, p(3) = 5, p(5) = 0$. In order to apply (3.34), we compute the Lagrange polynomials*

$$\begin{aligned}
p_0(x) &= \frac{15}{4} - \frac{31}{8}x + \frac{5}{4}x^2 - \frac{1}{8}x^3 \\
p_1(x) &= -5 + \frac{23}{3}x - 3x^2 + \frac{1}{3}x^3 \\
p_2(x) &= \frac{5}{2} - \frac{17}{4}x + 2x^2 - \frac{1}{4}x^3 \\
p_3(x) &= -\frac{1}{4} + \frac{11}{24}x - \frac{1}{4}x^2 + \frac{1}{24}x^3
\end{aligned} \tag{3.36}$$

Hence

$$\begin{aligned}
p(x) &= (-1)p_0(x) + 4 \cdot p_1(x) + 5 \cdot p_2(x) + 0 \cdot p_3(x) \\
&= \frac{1}{24}(5x^3 - 78x^2 + 319x - 270)
\end{aligned} \tag{3.37}$$

Practice Problem 3.4.6. *Find the Lagrange polynomial for $x_0 = 1, x_1 = 4, x_2 = 5$ and find the polynomial $p(x)$ of degree ≤ 2 satisfying $p(1) = 3, p(4) = 1, p(5) = -1$.*

Now we consider the inverse V_n^{-1} of Vandermonde matrix V_n. Since $V_n^{-1}(\mathbf{e}^j) = (V_n^{-1})^{(j)}$ comprised of coefficients of $p_j(x)$ by (3.32), we can write $V_n^{-1}(\mathbf{e}^j) = \begin{bmatrix} c_{0j} \\ c_{1j} \\ \vdots \\ c_{nj} \end{bmatrix}$ where c_{ij} is the coefficient of x^i in $p_j(x)$. Now for any polynomial $f(x) = \sum a_ix^i$ we have $a_i = \frac{1}{i!}f^{(i)}(0)$

[†]Joseph Louis Lagrange (1736–1813) was a French mathematician.

where $f^{(i)}(0)$ denote the i-th derivative of $f(x)$ at $x = 0$. (Here, we let $0! = 1$.) Therefore, we have

$$(V_n^{-1})_{ij} = \frac{1}{i!} p_j^{(i)}(0) \qquad (i = 0, 1, ..., n) \tag{3.38}$$

Hence we obtain:

Theorem 3.4.19. *Let p_i be the Lagrange polynomial defined in* (3.33) *then the inverse of the Vandermonde matrix V_n of* (3.26) *is given by*

$$V_n^{-1} = \begin{bmatrix} p_0(0) & p_1(0) & p_2(0) & \cdots & p_n(0) \\ p_0'(0) & p_1'(0) & p_2'(0) & \cdots & p_n'(0) \\ \frac{1}{2!}p_0^{(2)}(0) & \frac{1}{2!}p_1^{(2)}(0) & \frac{1}{2!}p_2^{(2)}(0) & \cdots & \frac{1}{2!}p_n^{(2)}(0) \\ \vdots & \vdots & \vdots & & \vdots \\ \frac{1}{n!}p_0^{(n)}(0) & \frac{1}{n!}p_1^{(n)}(0) & \frac{1}{n!}p_2^{(n)}(0) & \cdots & \frac{1}{n!}p_n^{(n)}(0) \end{bmatrix} \tag{3.39}$$

Example 3.4.20. *For $n = 2$, we use Lagrange polynomial in* Example 3.4.17 *to find the inverse of Vandermonde matrix V_2.*

$$V_2^{-1} = \begin{bmatrix} \frac{x_1x_2}{(x_0-x_1)(x_0-x_2)} & \frac{x_0x_2}{(x_1-x_0)(x_1-x_2)} & \frac{x_0x_1}{(x_2-x_0)(x_2-x_1)} \\ \frac{-x_1-x_2}{(x_0-x_1)(x_0-x_2)} & \frac{-x_0-x_2}{(x_1-x_0)(x_1-x_2)} & \frac{-x_0-x_1}{(x_2-x_0)(x_2-x_1)} \\ \frac{1}{(x_0-x_1)(x_0-x_2)} & \frac{1}{(x_1-x_0)(x_1-x_2)} & \frac{1}{(x_2-x_0)(x_2-x_1)} \end{bmatrix}$$

Example 3.4.21. *Continuing* Example 3.4.18, *we have,*

$$V_3 = \begin{bmatrix} 1 & 1 & 1 & 1 \\ 1 & 2 & 4 & 8 \\ 1 & 3 & 9 & 27 \\ 1 & 5 & 25 & 125 \end{bmatrix}, \ \mathbf{c} = \begin{bmatrix} c_0 \\ c_1 \\ c_2 \\ c_3 \end{bmatrix}, \ \mathbf{y} = \begin{bmatrix} -1 \\ 4 \\ 5 \\ 0 \end{bmatrix} \ \text{and } |V_3| = 48$$

Then we need to find $\mathbf{c}$ *for which* $V_3\mathbf{c} = \mathbf{y}$. *By* Theorem 3.4.19 *and* (3.36)

$$V_3^{-1} = \begin{bmatrix} \frac{15}{4} & -5 & \frac{5}{2} & -\frac{1}{4} \\ -\frac{31}{8} & \frac{23}{3} & -\frac{17}{4} & \frac{11}{24} \\ \frac{5}{4} & -3 & 2 & -\frac{1}{4} \\ -\frac{1}{8} & \frac{1}{3} & -\frac{1}{4} & \frac{1}{24} \end{bmatrix}$$

Hence

$$V_3^{-1} \begin{bmatrix} -1 \\ 4 \\ 5 \\ 0 \end{bmatrix} = \begin{bmatrix} -\frac{45}{4} \\ \frac{319}{24} \\ -\frac{13}{4} \\ \frac{5}{24} \end{bmatrix}$$

which gives the polynomial (3.37).

Finding the inverse of a Vandermonde will be useful when we need to find several polynomials with several different values on given points. For example, we want to find polynomials with values

x_i	$f_1(x_i)$	$f_2(x_i)$	$f_3(x_i)$
1	2	3	4
2	1	3	2
3	2	5	6
5	3	4	1

Multiplying the columns of the table to the right side of V_3^{-1} *we get*

$$\begin{aligned} f_1(x) &= \frac{27}{4} - \frac{173}{24}x + \frac{11}{4}x^2 - \frac{7}{24}x^3 \\ f_2(x) &= \frac{31}{4} - \frac{193}{24}x + \frac{15}{4}x^2 - \frac{11}{24}x^3 \\ f_3(x) &= \frac{79}{4} - \frac{605}{24}x + \frac{43}{4}x^2 - \frac{31}{24}x^3 \end{aligned}$$

Practice Problem 3.4.7. *Do* Practice Problem 3.4.6 *again by using* Example 3.4.20.

3.4.5 Companion Matrix

Next we will compute a determinant of a matrix which we will use later.

Definition 3.4.22. *Let* $f(t) = t^k + a_{k-1}t^{k-1} + \cdots + a_1 t + a_0$. *The* **companion matrix** $C(f)$ *for the polynomial* $f(t)$ *is defined to be*

$$C(f) = \begin{bmatrix} 0 & 0 & 0 & \cdots & 0 & -a_0 \\ 1 & 0 & 0 & & & -a_1 \\ 0 & 1 & 0 & & \vdots & -a_2 \\ \vdots & & \ddots & \ddots & & \vdots \\ 0 & \cdots & 0 & 1 & 0 & -a_{k-2} \\ 0 & \cdots & 0 & 0 & 1 & -a_{k-1} \end{bmatrix} \tag{3.40}$$

For a square matrix $A \in M(n)$, *the polynomial* $\det(A - tI)$ *will be called the* **characteristic polynomial** *of* A.

Example 3.4.23. *If* $f(t) = -t^3 + t^2 + 1$ *then*

$$C(f) = \begin{bmatrix} 0 & 0 & -1 \\ 1 & 0 & 0 \\ 0 & 1 & -1 \end{bmatrix}$$

Theorem 3.4.24. *Let* $f(t) = t^k + a_{k-1}t^{k-1} + \cdots + a_1 t + a_0$ *and let* A *be the companion matrix* $C(f)$ *of* f. *Then*

$$\det(A - tI_k) = (-1)^k f(t)^* \tag{3.41}$$

*We view $A - tI_k$ as a matrix with entries in the ring of polynomials $\mathbb{R}[t]$ whose determinant defined in the same way as we did in §3.2. See Appendix B2.

Proof. First we have

$$A - tI_k = \begin{bmatrix} -t & 0 & 0 & \cdots & 0 & -a_0 \\ 1 & -t & 0 & & & -a_1 \\ 0 & 1 & -t & & & -a_2 \\ & & \ddots & \ddots & & \\ & & & 1 & -t & -a_{k-2} \\ 0 & \cdots & 0 & 0 & 1 & -a_{k-1} - t \end{bmatrix}$$

Apply the sequence of row operations to $A - tI$: $tR_k + R_{k-1}, tR_{k-1} + R_{k-2}, \ldots, tR_2 + R_1$. Then we obtain the matrix

$$\begin{bmatrix} 0 & 0 & 0 & \cdots & 0 & f(t) \\ 1 & 0 & 0 & \cdots & 0 & * \\ 0 & 1 & 0 & \cdots & 0 & * \\ & \cdots & & \cdots & & \vdots \\ 0 & 0 & 0 & \cdots & 1 & -a_{k-1} \end{bmatrix}$$

where $f(t) = -(t^k + a_{k-1}t^{k-1} + \cdots + a_1 t + a_0)$ and $*$ denote something in which we are not interested. Hence, by expanding the determinant according to the first row we see that

$$\det(A - tI) = (-1)^k (t^k + a_{k-1}t^{k-1} + \cdots + a_1 t + a_0)$$

□

Hence $C(f)$ is a matrix having the characteristic polynomial $(-1)^k f(t)$ where $k = \deg(f)$. If $f(t) = -(t + a)$ then $C(f)$ is a 1×1 matrix $[-a]$.

Answers for Practice Problems 3.4

(3.4.1) First, $|A| = 0$. If $I = \{1, 2, 3\}$ then $\det\left(A_{(I)}^{(I)}\right) = 2$. Hence $\rho(A) = 3$.

(3.4.2) $L = \begin{bmatrix} 1 & 0 & 0 & 0 \\ 0 & 1 & 0 & 0 \\ 1 & -1 & 1 & 0 \\ 1 & 0 & -1 & 3 \end{bmatrix}$, $U = \begin{bmatrix} 1 & 1 & 1 & 0 \\ 0 & 1 & 1 & 1 \\ 0 & 0 & 1 & 2 \\ 0 & 0 & 0 & 1 \end{bmatrix}$

(3.4.3) $x = 3, y = -1, z = 3$

(3.4.4) $A^{-1} = \begin{bmatrix} 1 & 2 & -2 \\ 1 & -1 & 2 \\ 1 & -1 & 1 \end{bmatrix}$

(3.4.5) $\det(A) = 6$

(3.4.6) $\begin{cases} p_0(x) = \frac{1}{12}x^2 - \frac{3}{4}x + \frac{5}{3} \\ p_1(x) = -\frac{1}{3}x^2 + 2x - \frac{5}{3} \\ p_2(x) = \frac{1}{4}x^2 - \frac{5}{4}x + 1 \end{cases}$ and $p(x) = -\frac{1}{3}x^2 + x + \frac{7}{3}$.

(3.4.7) $V_3 = \begin{bmatrix} 1 & 1 & 1 \\ 1 & 4 & 16 \\ 1 & 5 & 25 \end{bmatrix}$, $V_3^{-1} = \dfrac{1}{12}\begin{bmatrix} 20 & -20 & 12 \\ -9 & 24 & -15 \\ 1 & -4 & 3 \end{bmatrix}$, $\mathbf{y} = \begin{bmatrix} 3 \\ 1 \\ -1 \end{bmatrix}$. Hence $\begin{bmatrix} c_0 \\ c_1 \\ c_2 \end{bmatrix} =$ $V_3^{-1}\mathbf{y} = \dfrac{1}{12}\begin{bmatrix} 28 \\ 12 \\ -4 \end{bmatrix}$. Therefore $p(x) = -\frac{1}{3}x^2 + x + \frac{7}{3}$.

3.4 Exercises

1. Find the rank of the following matrices by using Theorem 3.4.2.

(a) $A = \begin{bmatrix} 1 & 0 & 0 \\ 1 & -1 & 2 \\ -1 & -1 & 2 \\ 0 & -2 & 4 \end{bmatrix}$ (b) $A = \begin{bmatrix} 1 & 1 & 0 & 1 & 0 \\ 2 & 1 & -1 & 0 & 2 \\ 0 & -1 & -1 & -2 & 2 \\ 2 & 0 & -2 & -2 & 4 \end{bmatrix}$

(c) $A = \begin{bmatrix} -1 & 1 & 1 \\ 1 & 0 & 1 \\ -1 & 2 & 3 \\ 0 & 1 & 2 \end{bmatrix}$ (d) $A = \begin{bmatrix} 1 & -1 & 2 & 0 \\ 2 & 2 & -1 & 1 \\ 3 & 1 & 1 & 1 \\ 0 & 4 & -5 & 1 \end{bmatrix}$

2. Solve the following system of equations by using Cramer's rule.

(a) $\begin{cases} y + z = a \\ 3x + y = b \\ x + y + z = c \end{cases}$ (b) $\begin{cases} x + 2y + z = 1 \\ x - z = b \\ -x + y + z = 2b \end{cases}$

(c) $\begin{cases} x + y + w = 1 \\ x + z + w = 1 \\ y + z + w = 1 \\ x + y + z = 1 \end{cases}$ (d) $\begin{cases} x + z + w = 1 \\ -x + y + w = 0 \\ y + z - w = 1 \\ x + y + z = -1 \end{cases}$

3. Find the polynomial $p(x)$ satisfying the conditions.

   (a) $\deg(p(x)) \leq 2$ with $p(1) = 0, p(2) = 5, p(3) = 4$.

   (b) $\deg(p(x)) \leq 3$ with $p(-1) = 0, p(0) = 1, p(1) = 4, p(2) = -1$.

   (c) $\deg(p(x)) \leq 4$ with $p(1) = 0, p(2) = 1, p(3) = -1, p(4) = -5$.

   (d) $\deg(p(x)) \leq 4$ with $p(1) = 0, p(2) = 1, p(3) = -1, p(4) = -1$.

4. Find A^{-1} by using the adjoint of A.

(a) $A = \begin{bmatrix} 1 & 1 & 1 \\ 0 & 1 & 0 \\ 0 & -1 & 1 \end{bmatrix}$ (b) $A = \begin{bmatrix} 1 & -1 & 0 \\ 0 & 1 & 1 \\ a & b & c \end{bmatrix}$ with $a + b \neq c$

(c) $A = \begin{bmatrix} 1 & 1 & 1 & 0 \\ 1 & 0 & 1 & 1 \\ 1 & 1 & 0 & 1 \\ 0 & 1 & 1 & 1 \end{bmatrix}$ (d) $A = \begin{bmatrix} 1 & 1 & 1 & -1 \\ 1 & 1 & -1 & 1 \\ 1 & -1 & 1 & 1 \\ -1 & 1 & 1 & 1 \end{bmatrix}$ (e) $A = \begin{bmatrix} 0 & 1 & 1 & 0 \\ 0 & 0 & 1 & 1 \\ 1 & 1 & 0 & 1 \\ 1 & 1 & 1 & 0 \end{bmatrix}$

5. Let $A, B \in M(n)$ be the matrix such that $B^{(i)} = A^{(i)} + \sum_{j<i} c_j A^{(j)}$. Prove that $|B_k| = |A_k|$.

6. Let $A \in M(n)$ and $\det(A) = d$. Prove that

$$A \operatorname{adj}(A) = dI_n = \operatorname{diag}[\, d, d, ..., d\,]$$

(Hint: The equality $(A \operatorname{adj}(A))_{ii} = d$ is easy to show. If $i \neq j$ then let A' be the matrix replacing the j-th row by the i-th row $A_{(i)}$. Then $|A'| = 0$ since $A'_{(i)} = A'_{(j)}$. Now expand A' according to the j-th row of A' to conclude $\sum_k a_{ik} \operatorname{adj}(A)_{kj} = 0$.)

7. Prove $\det \operatorname{adj}(A) = \det(A)^{n-1}$.

8. Let $A \in M(n)$. Prove that $\operatorname{adj}(A) = 0$ if and only if $\rho(A) < n-1$.

9. Prove that $\operatorname{adj}(cA) = c^{n-1}\operatorname{adj}(A)$.

10. Let $A = [a_{ij}], B = [b_{ij}]$ be $n \times n$ matrices. Let t be a variable and $\left.\frac{d}{dt}\right|_{t=0}$ denote the derivative at $t = 0$. Prove the formulas

    (a) $\left.\frac{d}{dt}\right|_{t=0} \det(A + tA) = n \det A$ (Hint: $\det(A + tA) = (1+t)^n \det A$)

    (b) $\left.\frac{d}{dt}\right|_{t=0} \det(A + tB) = \operatorname{Tr}\,((\operatorname{adj}A)^t B)$

    (Hint: You will need $\operatorname{Tr}\,((\operatorname{adj}A)^t B) = \sum C_{ij}(A) B_{ij}$ and

$$\sum_{\sigma:\sigma(i)=j} a_{1\sigma(1)} \ldots \widehat{a_{i\sigma(i)}} \ldots a_{n\sigma(n)} b_{ij} \qquad (\hat{a} \text{ means deletion of } a)$$
$$= \left.\frac{d}{dt}\right|_{t=0} (a_{1\sigma(1)} + tb_{1\sigma(1)})(a_{2\sigma(2)} + tb_{2\sigma(2)}) \cdots (a_{n\sigma(n)} + tb_{n\sigma(n)}).)$$

11. Find the characteristic polynomials of:

$$\text{(a) } A_1 = \begin{bmatrix} 1 & 0 & 1 \\ 1 & 1 & 0 \\ 0 & 0 & 1 \end{bmatrix}, \quad \text{(b) } A_2 = \begin{bmatrix} 0 & 1 & 1 & 0 \\ 0 & 0 & 1 & 1 \\ 1 & 1 & 0 & 1 \\ 1 & 1 & 1 & 0 \end{bmatrix}, \quad \text{(c) } A_3 = \begin{bmatrix} 0 & -1 & 0 & 0 & 0 \\ 1 & 0 & 0 & 0 & 0 \\ 0 & 0 & 0 & 0 & -1 \\ 0 & 0 & 1 & 0 & 0 \\ 0 & 0 & 0 & 1 & -1 \end{bmatrix}$$

Chapter 4

Bases and Linear Maps on Subspaces

So far, we have concerned about $\mathbb{R}^n$'s and linear maps between them. In this chapter, we will study subspaces of $\mathbb{R}^n$ and linear maps between the subspaces. We define bases for the subspaces of $\mathbb{R}^n$ and look at various aspect of the bases. We then represent a linear map between subspaces as a matrix using the bases of the subspaces. We also investigate the properties of the linear maps using the matrix representation of the linear maps.

4.1 Bases of Subspaces of $\mathbb{R}^n$ and Dimension

In this section, we collect basic facts about the bases of subspaces of $\mathbb{R}^n$. We already used some of the result without proofs. We recall some results in Chapter 1.

Theorem 4.1.1. *Let* $S = \{\mathbf{w}_1, \mathbf{w}_2, ..., \mathbf{w}_m\}$ *be a finite subset of* $\mathbb{R}^n$.

(1) *If* S *is linearly independent subset of* $\mathbb{R}^n$ *then* $m \leq n$.

(2) *If* S *spans* $\mathbb{R}^n$ *then* $m \geq n$.

(3) *The vector space* $\mathbb{R}^n$ *has a basis and if* $S = \{\mathbf{w}_1, \mathbf{w}_2, ..., \mathbf{w}_m\}$ *is a basis then* $m = n$.

Proof. This is a part of Theorem 1.3.36 except for the existence of basis of $\mathbb{R}^n$ of (3) which is guaranteed by the standard basis $\mathcal{E} = \{\mathbf{e}^1, \mathbf{e}^2, ..., \mathbf{e}^n\}$. □

Theorem 1.2.17. *Let* S *be a subset of* $\mathbb{R}^n$. *If* V *be a subspace of* $\mathbb{R}^n$ *then* $S \subseteq V$ *if and only if* $\text{Span}(S) \subseteq V$. *Namely,* $\text{Span}(S)$ *is the smallest subspace containing* S.

Theorem 1.3.19. *Let* $S = \{\mathbf{v}_1, \mathbf{v}_2, ..., \mathbf{v}_p\}$ *be a linearly independent set of nonzero vectors in a subspace* V. *Let* $\mathbf{0} \neq \mathbf{v}_{p+1} \in V$. *Then* $S \cup \{\mathbf{v}_{p+1}\}$ *is linearly dependent if and only if* $\mathbf{v}_{p+1} \in \text{Span}(S)$. *In this case,*

$$\text{Span}(S) = \text{Span}(S \cup \{\mathbf{v}_{p+1}\})$$

We use the theorems to prove:

Theorem 4.1.2. *Let* V *be a subspace of* $\mathbb{R}^n$. *Then there is a basis* $\mathcal{B} = \{\mathbf{v}_1, \mathbf{v}_2, ..., \mathbf{v}_m\}$ *of* V *and* $m \leq n$.

Proof. Since $\mathcal{B}$ is a basis, $\mathcal{B}$ is a linearly independent subset of $\mathbb{R}^n$. Hence $m \leq n$ by Theorem 4.1.1 (1).

Next we prove the existence of a basis for V. If V is the trivial subspace then the empty set is a basis. Now suppose V is nontrivial; there is $\mathbf{0} \neq \mathbf{v}_1 \in V$. Then $\text{Span}\{\mathbf{v}_1\} \subseteq V$ obviously. And either $\text{Span}\{\mathbf{v}_1\} = V$ or $\text{Span}\{\mathbf{v}_1\} \neq V$. If $\text{Span}\{\mathbf{v}_1\} = V$ then $\{\mathbf{v}_1\}$ is a basis of V. If $\text{Span}\{\mathbf{v}_1\} \neq V$ then there is $\mathbf{v}_2 \in V \smallsetminus \text{Span}\{\mathbf{v}_1\}$. Since $\mathbf{v}_2 \notin \text{Span}\{\mathbf{v}_1\}$ we conclude, by Theorem 1.3.19, $\{\mathbf{v}_1, \mathbf{v}_2\}$ is linearly independent.

Since $\text{Span}\{\mathbf{v}_1, \mathbf{v}_2\} \subseteq V$ either $\text{Span}\{\mathbf{v}_1, \mathbf{v}_2\} = V$ or $\text{Span}\{\mathbf{v}_1, \mathbf{v}_2\} \neq V$. In the former case, $\{\mathbf{v}_1, \mathbf{v}_2\}$ is a basis of V. In the latter case, there is nonzero $\mathbf{v}_3 \in V \smallsetminus \text{Span}\{\mathbf{v}_1, \mathbf{v}_2\}$. As before $\{\mathbf{v}_1, \mathbf{v}_2, \mathbf{v}_3\}$ is linearly independent.

We repeat the process. This process must stop since $\{\mathbf{v}_1, \mathbf{v}_2, ..., \mathbf{v}_k\}$ is linearly dependent if $k > n$ (Theorem 4.1.1 (1)). Suppose $\{\mathbf{v}_1, \mathbf{v}_2, ..., \mathbf{v}_k\}$ is linearly independent and $\{\mathbf{v}_1, \mathbf{v}_2, ..., \mathbf{v}_k, \mathbf{v}_{k+1}\}$ is linearly dependent for any $\mathbf{v}_{k+1} \in V$. We claim that $\{\mathbf{v}_1, \mathbf{v}_2, ..., \mathbf{v}_k\}$ is a basis for V. All we need to show is that $\text{Span}\{\mathbf{v}_1, \mathbf{v}_2, ..., \mathbf{v}_k\} = V$. In fact, if $\mathbf{v} \in V$ then $\{\mathbf{v}_1, \mathbf{v}_2, ..., \mathbf{v}_k, \mathbf{v}\}$ is linearly dependent and hence $\mathbf{v} \in \text{Span}\{\mathbf{v}_1, \mathbf{v}_2, ..., \mathbf{v}_k\}$ (Theorem 1.3.19) as contended. □

Following result is Theorem 1.3.24 for the subspaces.

Theorem 4.1.3. *Let V be a subspace of $\mathbb{R}^n$. Let $\mathcal{B} = \{\mathbf{v}_1, \mathbf{v}_2, ..., \mathbf{v}_k\}$ be a basis of V and $S = \{\mathbf{w}_1, \mathbf{w}_2, ..., \mathbf{w}_m\}$ be a subset of V. If $m > k$ then S is linearly dependent. Equivalently, if S is linearly independent then $m \leq k$.*

Proof. Since $\mathcal{B}$ is a basis we can write $\mathbf{w}_j = \sum_{i=1}^k c_{ij}\mathbf{v}_i$ $(1 \leq j \leq m, k < m)$ (Theorem 1.3.27). Let $C = [c_{ij}] \in M(k{\times}m)$. Then we have

$$[\mathbf{w}_1\ \mathbf{w}_2 \cdots \mathbf{w}_m] = [\mathbf{v}_1\ \mathbf{v}_2 \cdots\ \mathbf{v}_k] \begin{bmatrix} c_{11} & c_{12} & \cdots & c_{1m} \\ c_{21} & c_{22} & \cdots & c_{2m} \\ \cdots & \cdots & & \\ c_{k1} & c_{k2} & \cdots & c_{km} \end{bmatrix} \tag{4.1}$$

Since $m > k$, the columns $\{C^{(1)}, C^{(2)}, ..., C^{(m)}\} \subseteq \mathbb{R}^k$ of C are linearly dependent (Corollary 1.3.18). Hence there is nonzero $\mathbf{x} = \begin{bmatrix} x_1 \\ x_2 \\ \vdots \\ x_m \end{bmatrix}$ such that $\sum x_i C^{(i)} = C\mathbf{x} = \mathbf{0}$. Multiplying (4.1) by $\mathbf{x}$ to we get

$$[\mathbf{w}_1\ \mathbf{w}_2 \cdots \mathbf{w}_m] \begin{bmatrix} x_1 \\ x_2 \\ \vdots \\ x_m \end{bmatrix} = [\mathbf{v}_1\ \mathbf{v}_2 \cdots\ \mathbf{v}_k] \begin{bmatrix} c_{11} & c_{12} & \cdots & c_{1m} \\ c_{21} & c_{22} & \cdots & c_{2m} \\ \cdots & \cdots & & \\ c_{k1} & c_{k2} & & c_{km} \end{bmatrix} \begin{bmatrix} x_1 \\ x_2 \\ \vdots \\ x_m \end{bmatrix} = \mathbf{0}$$

Therefore $\sum x_i \mathbf{w}_i = \mathbf{0}$; S is linearly dependent. □

Corollary 4.1.4. *Let $\mathcal{B}, \mathcal{B}'$ be bases of V consisting of finite vectors. Then $\mathcal{B}$ and $\mathcal{B}'$ have the same number of elements.*

Proof. Since $\mathcal{B}$ is linearly independent and $\mathcal{B}'$ is a basis of V we have $\#\mathcal{B} \leq \#\mathcal{B}'$ by Theorem 4.1.3. Reversing the roles of $\mathcal{B}$ and $\mathcal{B}'$ we get $\#\mathcal{B} \geq \#\mathcal{B}'$. Therefore, $\mathcal{B}, \mathcal{B}'$ have the same number of vectors. □

Definition 4.1.5. *Let V be a subspace of $\mathbb{R}^n$. The number of basis of V is called the* **dimension** *of V and it is denoted by* $\dim(V)$.

We recall the definition of maximal linearly independent subset (Definition 2.3.1): Let V be a subspace of $\mathbb{R}^n$ and let S be a subset of V. A linearly independent subset S_0 of S is called a **maximal linearly independent subset** if $\mathbf{v} \in S \setminus S_0$ then $S_0 \cup \{\mathbf{v}\}$ is linearly dependent.

We showed that a maximal linearly independent subset of a spanning subset of V is a basis in Theorem 2.3.2. Existence of maximal linearly independent subset is guaranteed in more general situation (Theorem D.1.2). However we can find a maximal linearly independent subset of a finite spanning subset of a subspace of $\mathbb{R}^n$, as in the following theorem, using Theorem 1.3.19 repeatedly as in the proof of Theorem 4.1.2.

Theorem 4.1.6. *Let V be a subspace of $\mathbb{R}^n$ and S be a subset of V which spans V. There is a maximal linearly independent subset S_0 of S. And S_0 is a maximal linearly independent subset of S if and only if S_0 is a basis of V.*

Proof. To find a maximal linearly independent subset choose a nonzero $\mathbf{v}_1 \in S$. Then $S_1 = \{\mathbf{v}_1\}$ is a linearly independent. If $\operatorname{Span} S_1 = V$ then by Theorem 1.3.19 we see that $S_1 \cup \{\mathbf{v}_i\}$ is linearly dependent for any $\mathbf{v}_i \neq \mathbf{v}_1$. Hence S_1 is a maximal linearly independent subset of S. Now if $\operatorname{Span} S_1 \neq V$ then there is $\mathbf{v}_2 \in S$ such that $\mathbf{v}_2 \notin \operatorname{Span} S_1$. Otherwise $S \subseteq \operatorname{Span} S_1$ and hence $V = \operatorname{Span} S \subseteq \operatorname{Span} S_1$ which contradicts to our assumption. Now by Theorem 1.3.19, we see that $S_2 = \{\mathbf{v}_1, \mathbf{v}_2\}$ is linearly independent. If $V = \operatorname{Span} S_2$ then as before S_2 is a maximal linearly independent subset. If $\operatorname{Span} S_2 \neq V$ then as before there is $\mathbf{v}_3 \in S$ such that $\mathbf{v}_3 \notin \operatorname{Span} S_2$. By Theorem 1.3.19 again, $S_3 = \{\mathbf{v}_1, \mathbf{v}_2, \mathbf{v}_3\}$ is linearly independent. If S_3 spans V then S_3 is a maximal linearly independent subset and so on.

This process must stop because the number of linearly independent vectors is $\leq n$. And we reach to a stage that for any $\mathbf{v} \in S$ the subset $S_k \cup \{\mathbf{v}\}$ of S is linearly dependent. Then $S_0 := S_k$ is a maximal linearly independent subset of S.

The second statement is Theorem 2.3.2. □

Practice Problem 4.1.1. *Let* $S = \left\{ \begin{bmatrix} 1 \\ -1 \\ 0 \\ 1 \end{bmatrix}, \begin{bmatrix} 0 \\ 1 \\ 1 \\ 1 \end{bmatrix}, \begin{bmatrix} 1 \\ 0 \\ 1 \\ 2 \end{bmatrix}, \begin{bmatrix} 1 \\ 1 \\ 2 \\ 3 \end{bmatrix} \right\}$. *Find a maximal linearly independent subset of S.*

We will prove a slightly improved version of Theorem 4.1.6.

Theorem 4.1.7. *Let V be a subspace of $\mathbb{R}^n$ and S be a finite subset of V which span V. Suppose $T = \{\mathbf{v}_1, \mathbf{v}_2, \ldots, \mathbf{v}_m\}$ is a linearly independent set of vectors. Then there is a subset $S_0 = \{\mathbf{w}_1, \mathbf{w}_2, \ldots, \mathbf{w}_k\}$ of S such that $T \cup S_0$ is a basis of V.*

Proof. First suppose $T \cup \{\mathbf{w}\}$ is linearly dependent for any $\mathbf{w} \in S$. Then, we claim that T is a basis. For this we need to show that T spans V. Since for any $\mathbf{w} \in S$ we have $T \cup \{\mathbf{w}\}$ is linearly dependent we have $\mathbf{w} \in \operatorname{Span} T$ by Theorem 1.3.19. Hence $S \subseteq \operatorname{Span} T$ and hence $V = \operatorname{Span} S \subseteq \operatorname{Span} T$ (Theorem 1.2.17).

Next suppose there is $\mathbf{w} \in S$ such that $T \cup \{\mathbf{w}\}$ is linearly independent but $T \cup \{\mathbf{w}, \mathbf{w}'\}$ is linearly dependent for any $\mathbf{w}' \in S$. Then, we claim that $T \cup \{\mathbf{w}\}$ is a basis of V. For this we repeat the argument above. We show $T \cup \{\mathbf{w}\}$ spans V. Since $T \cup \{\mathbf{w}, \mathbf{w}'\}$ is linearly dependent for any $\mathbf{w}' \in S$ we see that $S \subseteq \operatorname{Span}(T \cup \{\mathbf{w}\})$ by Theorem 1.3.19. Hence by Theorem 1.2.17 again $\operatorname{Span}(T \cup \{\mathbf{w}\}) = V$.

Repeat this process until we get a subset $S_0 = \{\mathbf{w}_1, \mathbf{w}_2, \ldots, \mathbf{w}_k\}$ of S such that $T \cup S_0$ is linearly independent but $T \cup S_0 \cup \{\mathbf{w}'\}$ is linearly dependent for any $\mathbf{w}' \in S$. Then $S \subseteq \mathrm{Span}\,(T \cup S_0)$ by Theorem 1.3.19 and hence $T \cup S_0$ spans V. And it is linearly independent by construction. Hence $T \cup S_0$ is a basis. □

Corollary 4.1.8. *If $T = \{\mathbf{v}_1, \mathbf{v}_2, ..., \mathbf{v}_k\}$ is a linearly independent subset of V then there is a basis $\mathcal{B}$ of V containing T. In particular, $k \le \dim V$.*

Proof. Choose a basis $\mathcal{B}_0$ of V for S in Theorem 4.1.2. Since $\mathcal{B}_0$ span V there is a subset S_0 of $\mathcal{B}_0$ such that $T \cup S_0$ is a basis of V. □

Following theorem generalizes Theorem 2.4.6 for the subspaces instead of $\mathbb{R}^n$ itself.

Theorem 4.1.9. *Let V be a subspace of $\mathbb{R}^m$ with $\dim(V) = n$. Let $S = \{\mathbf{v}_1, \mathbf{v}_2, ..., \mathbf{v}_n\}$ be a subset of V.*

(1) *If S span V then S is a basis of V.*

(2) *If S is linearly independent then S is a basis of V.*

Proof. (1) Since S span V, by Theorem 4.1.6, we can choose elements from S to get a basis. But we have to choose them all because a basis must contain n elements; S is a basis.

(2) Let $\mathcal{B}$ be a basis of V. Since $\mathcal{B}$ span V, by Theorem 4.1.7, we can choose a subset S_0 of $\mathcal{B}$ having $n - n$ elements, say $S_0 = \emptyset$, such that $S \cup S_0$ is a basis. Hence S is a basis. □

Example 4.1.10. *Consider*

$$S = \left\{ \mathbf{v}_1 = \begin{bmatrix} 1 \\ 1 \\ 0 \end{bmatrix}, \mathbf{v}_2 = \begin{bmatrix} 2 \\ -1 \\ 1 \end{bmatrix}, \mathbf{v}_3 = \begin{bmatrix} -1 \\ 2 \\ -1 \end{bmatrix}, \mathbf{v}_4 = \begin{bmatrix} 0 \\ 3 \\ -1 \end{bmatrix}, \mathbf{v}_5 = \begin{bmatrix} 1 \\ 1 \\ 1 \end{bmatrix}, \mathbf{v}_6 = \begin{bmatrix} 0 \\ 1 \\ 1 \end{bmatrix} \right\}$$

First, observe that S spans $\mathbb{R}^3$. In fact, the determinant of the matrix $A = [\mathbf{v}_4\ \mathbf{v}_5\ \mathbf{v}_6]$ is nonzero. Hence the columns of the matrix A is a basis of $\mathbb{R}^3$ (Corollary 3.3.8). *Since S contains $\{\mathbf{v}_4, \mathbf{v}_5, \mathbf{v}_6\}$ we see that* $\mathrm{Span}\,(S) = \mathbb{R}^3$.

The single nonzero vector $\{\mathbf{v}_1\}$ is linearly independent. If we add the second one the set of vectors $\{\mathbf{v}_1, \mathbf{v}_2\}$ is still linearly independent since $\mathbf{v}_2$ is not a scalar multiple of $\mathbf{v}_1$. Next, $\{\mathbf{v}_1, \mathbf{v}_2, \mathbf{v}_3\}$ is linearly dependent for $\mathbf{v}_1 - \mathbf{v}_2 = \mathbf{v}_3$. (Check this also by considering $\det[\mathbf{v}_1\ \mathbf{v}_2\ \mathbf{v}_3]$.) *Try the next one: $\{\mathbf{v}_1, \mathbf{v}_2, \mathbf{v}_4\}$ is again linearly dependent for $2\mathbf{v}_1 - \mathbf{v}_2 = \mathbf{v}_4$. Now $\{\mathbf{v}_1, \mathbf{v}_2, \mathbf{v}_5\}$ is linearly independent since* $\det[\mathbf{v}_1\ \mathbf{v}_2\ \mathbf{v}_5] \neq 0$ (Corollary 3.3.7). *Adjoining $\mathbf{v}_6$ we get $\{\mathbf{v}_1, \mathbf{v}_2, \mathbf{v}_5, \mathbf{v}_6\}$ which is linearly dependent by* Corollary 1.3.18. *Hence $\{\mathbf{v}_1, \mathbf{v}_2, \mathbf{v}_5\}$ is a maximal linearly independent subset of S. Similarly $\{\mathbf{v}_1, \mathbf{v}_5, \mathbf{v}_6\}$ is also a maximal linearly independent subset of S.*

Practice Problem 4.1.2. *Let* $S = \left\{ \begin{bmatrix} 1 \\ 0 \\ -1 \end{bmatrix}, \begin{bmatrix} -1 \\ 1 \\ 0 \end{bmatrix}, \begin{bmatrix} 0 \\ 1 \\ -1 \end{bmatrix}, \begin{bmatrix} -1 \\ 2 \\ -1 \end{bmatrix}, \begin{bmatrix} 1 \\ 1 \\ 1 \end{bmatrix} \right\}$. *Show that S spans $\mathbb{R}^3$ and find two maximal linearly independent subsets of S.*

The following theorem is Theorem 1.3.28 for a subspace in place of $\mathbb{R}^n$.

Theorem 4.1.11. *A subset $\mathcal{B} = \{\mathbf{v}_1, \mathbf{v}_2, ..., \mathbf{v}_m\}$ is a basis if and only if every $\mathbf{v} \in V$ can be written uniquely as a linear combination of $\mathcal{B}$.*

Proof. The proof is the same as the proof of Theorem 1.3.28. □

Theorem 4.1.12. *Let $V_1 \subseteq V_2$ be subspaces of $\mathbb{R}^n$. Then*

(1) $\dim(V_1) \le \dim(V_2)$

(2) *If* $\dim(V_1) = \dim(V_2)$ *then* $V_1 = V_2$.

Proof. Let $\mathcal{B}$ be a basis of V_1. (1) Then $\mathcal{B}$, as a subset of V_2, is linearly independent. Hence $\dim V_1 = \#(\mathcal{B}) \le \dim V_2$ by Theorem 4.1.3.

(2) Since $\mathcal{B}$ contains $\dim(V_1) = \dim(V_2)$ vectors, $\mathcal{B}$ is a basis of V_2 by Theorem 4.1.9. Hence $\operatorname{Span}(\mathcal{B}) = V_2$. □

The following theorem shows that every subspace can be realized as a column space or a null space of a matrix. This theorem is not essential and one may skip this without hindering further reading of the book. See also Theorem 5.2.8 for a different approach.

Theorem 4.1.13. *Let V be a subspace of $\mathbb{R}^n$ of dimension p.*

(1) *If $p \le m$ then there is $A \in M(n\times m)$ such that $V = \operatorname{Col}(A)$.*

(2) *If $n - p \le m$ then there is $B \in M(m\times n)$ such that $V = N(B)$.*

Proof. (1) Let $\{\mathbf{v}_1, \mathbf{v}_2, ..., \mathbf{v}_p\}$ be a basis of V. Choose $m-p$ vectors $\{\mathbf{v}_{p+1}, ..., \mathbf{v}_m\}$ from V. Let $A = [\mathbf{v}_1\ \mathbf{v}_2\ \cdots\ \mathbf{v}_m]$. Then A is an $n\times m$ matrix and it is obvious that $\operatorname{Col}(A) = V$.

(2) If $n = p$ (i.e., $V = \mathbb{R}^n$) then we can simply take B to be an $m\times n$ zero matrix. Hence we assume $p < n$ and choose a basis $\{\mathbf{v}_1, \mathbf{v}_2, ..., \mathbf{v}_p\}$ of V. Let $A = [\mathbf{v}_1\ \mathbf{v}_2\ \cdots\ \mathbf{v}_p]$. The rank of A^t is p and hence $\nu(A^t) = n - p > 0$. Let $\{\mathbf{w}_1, ..., \mathbf{w}_{n-p}\}$ be a basis of $N(A^t)$. As $n - p \le m$ we choose $\mathbf{w}_{n-p+1}, ..., \mathbf{w}_m$ from $N(A^t)$. Let $B_0 = [\mathbf{w}_1\ \mathbf{w}_2\ \cdots\ \mathbf{w}_m] \in M(n\times m)$. Then $\rho(B_0) = \nu(A^t) = n - p$ and A^tB_0 is the zero matrix and hence B_0^tA is also the zero matrix; $B_0^tA^{(i)} = \mathbf{0}$ $(i = 1, 2, ..., p)$. Hence $V = \operatorname{Col}(A) \subseteq N(B_0^t)$.

Now $\rho(B_0) + \nu(B_0^t) = \rho(B_0^t) + \nu(B_0^t) = (n-p) + \nu(B_0^t) = n$. Hence $\nu(B_0^t) = p$; $\dim V = \dim N(B_0^t)$. Therefore if we let $B = B_0^t$ then $N(B) = \operatorname{Col}(A) = V$. □

Practice Problem 4.1.3. *Let* $S = \left\{ \begin{bmatrix} 1 \\ -1 \\ 0 \\ 1 \end{bmatrix}, \begin{bmatrix} 0 \\ 1 \\ 1 \\ 1 \end{bmatrix} \right\}$. *Find a 2×4 matrix whose null space is the same as* $\operatorname{Span}(S)$.

Answers for Practice Problems 4.1

(4.1.1) $S_0 = \left\{ \begin{bmatrix} 1 \\ -1 \\ 0 \\ 1 \end{bmatrix}, \begin{bmatrix} 0 \\ 1 \\ 1 \\ 1 \end{bmatrix} \right\}$

(4.1.2) We can choose $\left\{ \begin{bmatrix} 1 \\ 0 \\ -1 \end{bmatrix}, \begin{bmatrix} -1 \\ 1 \\ 0 \end{bmatrix}, \begin{bmatrix} 1 \\ 1 \\ 1 \end{bmatrix} \right\}$ and $\left\{ \begin{bmatrix} 1 \\ 0 \\ -1 \end{bmatrix}, \begin{bmatrix} 0 \\ 1 \\ -1 \end{bmatrix}, \begin{bmatrix} 1 \\ 1 \\ 1 \end{bmatrix} \right\}$.

(4.1.3) Let $A = \begin{bmatrix} 1 & 0 \\ -1 & 1 \\ 0 & 1 \\ 1 & 1 \end{bmatrix}$. Then $\rho(A) = \rho(A^t) = 2$. Hence $\nu(A^t) = 2$. To find a basis for $N(A^t)$ we compute the reduced echelon form B for $A^t = \begin{bmatrix} 1 & -1 & 0 & 1 \\ 0 & 1 & 1 & 1 \end{bmatrix}$: $B = \begin{bmatrix} 1 & 0 & 1 & 2 \\ 0 & 1 & 1 & 1 \end{bmatrix}$.

Hence $N(B)$ has basis $\left\{\begin{bmatrix}-1\\-1\\1\\0\end{bmatrix},\begin{bmatrix}-2\\-1\\0\\1\end{bmatrix}\right\}$. If we let $B_0=\begin{bmatrix}-1&-2\\-1&-1\\1&0\\0&1\end{bmatrix}$ then A^tB_0 is the zero matrix. Therefore $B_0^tA=\begin{bmatrix}0&0\\0&0\end{bmatrix}$; if we let $C=B_0^t=\begin{bmatrix}-1&-1&1&0\\-2&-1&0&1\end{bmatrix}$ then $N(C)=\mathrm{Col}(A)$.

4.1 Exercises

1. Show that S spans $\mathbb{R}^3$ or $\mathbb{R}^4$ and find two maximal linearly independent subsets of S.

   (a) $S=\left\{\begin{bmatrix}3\\2\\1\end{bmatrix},\begin{bmatrix}1\\1\\0\end{bmatrix},\begin{bmatrix}2\\1\\1\end{bmatrix},\begin{bmatrix}1\\2\\1\end{bmatrix},\begin{bmatrix}1\\1\\1\end{bmatrix}\right\}$

   (b) $S=\left\{\begin{bmatrix}1\\2\\1\end{bmatrix},\begin{bmatrix}1\\-1\\0\end{bmatrix},\begin{bmatrix}2\\1\\1\end{bmatrix},\begin{bmatrix}3\\0\\1\end{bmatrix},\begin{bmatrix}1\\5\\2\end{bmatrix},\begin{bmatrix}1\\1\\1\end{bmatrix}\right\}$

   (c) $S=\left\{\begin{bmatrix}1\\1\\1\\-1\end{bmatrix},\begin{bmatrix}1\\1\\3\\3\end{bmatrix},\begin{bmatrix}0\\0\\4\\2\end{bmatrix},\begin{bmatrix}2\\2\\4\\2\end{bmatrix},\begin{bmatrix}1\\0\\0\\1\end{bmatrix}\right\}$

   (d) $S=\left\{\begin{bmatrix}1\\1\\1\\-1\end{bmatrix},\begin{bmatrix}1\\1\\-1\\1\end{bmatrix},\begin{bmatrix}1\\1\\-3\\3\end{bmatrix},\begin{bmatrix}0\\0\\-4\\2\end{bmatrix},\begin{bmatrix}1\\1\\1\\1\end{bmatrix},\begin{bmatrix}1\\0\\0\\1\end{bmatrix}\right\}$

2. Let

$$\mathbf{v}_1=\begin{bmatrix}1\\2\\1\\0\end{bmatrix},\mathbf{v}_2=\begin{bmatrix}2\\1\\-1\\1\end{bmatrix},\mathbf{v}_3=\begin{bmatrix}-1\\1\\2\\-1\end{bmatrix},\mathbf{v}_4=\begin{bmatrix}1\\2\\1\\0\end{bmatrix},\mathbf{v}_5=\begin{bmatrix}1\\0\\-1\\1\end{bmatrix},\mathbf{v}_6=\begin{bmatrix}1\\1\\0\\1\end{bmatrix}$$

   and let $S=\{\mathbf{v}_1,\mathbf{v}_2,\mathbf{v}_3,\mathbf{v}_4,\mathbf{v}_5,\mathbf{v}_6\}$. Show that S spans the subspace

$$V:=\left\{\begin{bmatrix}x\\y\\z\\w\end{bmatrix}\middle|\, x-y+z=0\right\}$$

   of $\mathbb{R}^4$ and find a maximal linearly independent subset of S.

3. Let $S=\{\mathbf{v}_1,\mathbf{v}_2,...,\mathbf{v}_k\}$ be a linearly dependent set of vectors. Prove that every $\mathbf{v}_i\in S$ is a linear combination of $S\setminus\{\mathbf{v}_i\}$.

4. Prove the true statement(s) and find a counter example for the false statement(s).

   (a) Let V be a subspace of $\mathbb{R}^n$. If S is a linearly independent subset of V which span V then S is a maximal linearly independent subset of V.

(b) Let V be a subspace of dimension d in $\mathbb{R}^n$. If S is a set of vectors in V containing less than d elements then S is linearly independent.

(c) Let V be a subspace of $\mathbb{R}^3$. If there is a basis of V consisting of three vectors then $V = \mathbb{R}^3$.

(d) If V is a subspace of dimension 3 in $\mathbb{R}^6$. If $S = \{\mathbf{v}_1, \mathbf{v}_2, \mathbf{v}_3, \mathbf{v}_4\} \subseteq V$ then S can be linearly independent.

(e) If V is a subspace of dimension 3 in $\mathbb{R}^6$. If $S = \{\mathbf{v}_1, \mathbf{v}_2, ..., \mathbf{v}_k\}$ is a subset of V which span V then $k \geq 3$.

(f) Let V be a subspace of $\mathbb{R}^n$ and $S \subseteq V$. Let S_0 be a finite maximal linearly independent subset of S. Let T be a subset of S having the same number of vectors as S_0. Then T is a maximal linearly independent subset of S.

5. Let $A = \begin{bmatrix} 1 & -1 & 0 & 1 \\ -1 & 2 & 1 & 0 \\ 0 & 1 & 1 & 1 \end{bmatrix}$

(a) Find $B \in M(4\times4)$ such that $\text{Col}(A) = N(B)$.

(b) Find $C \in M(2\times3)$ such that $N(A) = \text{Col}(C)$.

6. Let $S = \{\mathbf{u}_1, \mathbf{u}_2, ..., \mathbf{u}_p\} \subseteq \mathbb{R}^n$. Let $\mathbf{v}_j = \sum_{i=1}^{p} a_{ij}\mathbf{u}_i$, $S' = \{\mathbf{v}_1, \mathbf{v}_2, ..., \mathbf{v}_p\}$ and $A = [a_{ij}] \in M(p)$.

(a) Prove that $\text{Span}\,(S') \subseteq \text{Span}\,(S)$.

(b) Prove that if A is invertible then $\text{Span}\,(S) = \text{Span}\,(T)$.

(c) Prove that A is invertible if S' is linearly independent.

4.2 Linear Maps between Subspaces

We saw in Chapter 2, that a linear maps from $\mathbb{R}^n$ to $\mathbb{R}^m$ are in one-to-one correspondence with the set of all $m\times n$ matrices (Theorem 2.2.8). The reason for this was that there was a natural bases, namely the standard bases for $\mathbb{R}^n$ and $\mathbb{R}^m$. However, there is no such a natural basis for a subspace of $\mathbb{R}^n$. Hence we need to choose a basis to represent a linear map between subspaces as a matrix. In this section, we study how to represent a linear map by a matrix by choosing bases of the subspaces.

4.2.1 Linear Maps and Matrix Representation

The definition of a linear map between subspaces are exactly the same as the definition of a linear map between $\mathbb{R}^n$'s (Definition 2.1.5).

Definition 4.2.1. *Let V, W be subspaces of $\mathbb{R}^N, \mathbb{R}^M$ respectively. A map $T : V \to W$ is a* **linear map (transformation, function)** *if for all $\mathbf{u}, \mathbf{v} \in V$ and $c \in \mathbb{R}$ we have*

$$T(\mathbf{u} + \mathbf{v}) = T(\mathbf{u}) + T(\mathbf{v}) \text{ and } T(c\mathbf{u}) = cT(\mathbf{u}) \tag{4.2}$$

A linear map T from V into itself is called a **linear operator**. *A bijective linear map is called an* **isomorphism**. *If there is an isomorphism between V and W then we say V and W a* **isomorphic**. *The* **identity map** *I_V (or I) on V is obviously an isomorphism. The* **zero map** *transforming every vector $\mathbf{v} \in V$ to the zero vector $\mathbf{0} \in W$ is also a linear map.*

Observe that if $T : V \to W$ is a linear map between subspaces then

$$T(\mathbf{0}) = \mathbf{0} \tag{4.3}$$

In fact, $T(\mathbf{0}) = T(\mathbf{0}+\mathbf{0}) = T(\mathbf{0}) + T(\mathbf{0})$. Hence $T(\mathbf{0}) = \mathbf{0}$.

First we show that a linear map is completely determined by its values on a basis which generalizes Theorem 2.1.9 to the subspaces in place of $\mathbb{R}^n$'s.

Theorem 4.2.2. *Let $\mathcal{B} = \{\mathbf{v}_1, \mathbf{v}_2, ..., \mathbf{v}_n\}$ be a basis of V.*

(1) Let $S, T : V \to W$ be two linear maps. If $S(\mathbf{v}_i) = T(\mathbf{v}_i)$ $(i = 1, 2, ..., n)$ then $S = T$.

(2) Let $\{\mathbf{w}_1, \mathbf{w}_2, ..., \mathbf{w}_n\}$ be a subset of W. Then there is a unique linear map $T : V \to W$ with $T(\mathbf{v}_i) = \mathbf{w}_i$ $(i = 1, ..., n)$.

Proof. (1) Let $\mathbf{v} \in V$. We need to show that $T(\mathbf{v}) = S(\mathbf{v})$. We can write $\mathbf{v} = \sum c_i\mathbf{v}_i$ in a unique way (Theorem 1.3.27). Since S and T are linear we have $T(\mathbf{v}) = T(\sum c_i\mathbf{v}_i) = \sum c_iT(\mathbf{v}_i) = \sum c_iS(\mathbf{v}_i) = S(\sum c_i\mathbf{v}_i) = S(\mathbf{v})$. Therefore $S = T$.

(2) Let $\mathbf{v} \in V$ and let $\mathbf{v} = \sum c_i\mathbf{v}_i$ as above. Define $T(\mathbf{v}) = \sum c_i\mathbf{w}_i$. Then obviously $T(\mathbf{v}_i) = \mathbf{w}_i$ and T defines a linear map. (Check this.) Uniqueness follows from (1). □

Theorem 4.2.3. *Let $S : U \to V$ and $T : V \to W$ be linear maps of subspaces.*

(1) The composition $T \circ S : U \to W$ is a linear map.

(2) If $T : V \to W$ is an isomorphism then $T^{-1} : W \to V$ is also an isomorphism. If S is also an isomorphism then so is $T \circ S$.

Proof. (1) Exercise.

(2) Since T is a bijection T^{-1} exists and is also a bijection (Theorem A.1.5). We need to prove that T^{-1} is linear. In fact, let $\mathbf{w}_1, \mathbf{w}_2 \in W$ and $c \in \mathbb{R}$. Then there are $\mathbf{v}_1, \mathbf{v}_2 \in V$ such that $\mathbf{w}_i = T(\mathbf{v}_i)$ $(i = 1, 2)$. Hence $T^{-1}(\mathbf{w}_1+\mathbf{w}_2) = T^{-1}(T(\mathbf{v}_1)+T(\mathbf{v}_2)) = T^{-1}(T(\mathbf{v}_1+\mathbf{v}_2)) = \mathbf{v}_1 + \mathbf{v}_2 = T^{-1}(\mathbf{w}_1) + T^{-1}(\mathbf{w}_2)$ since T is linear. Similarly $T^{-1}(c\mathbf{w}_1) = T^{-1}(cT(\mathbf{v}_1)) = T^{-1}(T(c\mathbf{v}_1)) = c\mathbf{v}_1 = cT^{-1}(\mathbf{w}_1)$.

The second statement follows from the fact that a composition of a bijection is a bijection (Theorem A.1.6) and (1). □

Definition 4.2.4. *Let $\mathcal{B} = \{\mathbf{v}_1, \mathbf{v}_2, ..., \mathbf{v}_n\}$ be a basis of a subspace V of $\mathbb{R}^N$ and let $\mathbf{v} \in V$. By* Theorem 4.1.11, *there are unique scalars $c_1, ..., c_n$ such that $\mathbf{v} = c_1\mathbf{v}_1 + \cdots + c_n\mathbf{v}_n$. We call the column vector $[\mathbf{v}]_\mathcal{B}$ formed by the scalars $c_1, ..., c_n$ the* **coordinate of v relative to** $\mathcal{B}$ *(or simply $\mathcal{B}$-coordinate):*

$$[\mathbf{v}]_\mathcal{B} = \begin{bmatrix} c_1 \\ \vdots \\ c_n \end{bmatrix} \in \mathbb{R}^n$$

The map $\phi_\mathcal{B} : V \to \mathbb{R}^n$ defined by $\phi_\mathcal{B}(\mathbf{v}) = [\mathbf{v}]_\mathcal{B}$ is called the **coordinate map** *relative to the basis $\mathcal{B}$.*

Observe that $[\mathbf{v}_i]_\mathcal{B} = \mathbf{e}^i$. Also observe that the expression $\mathbf{v} = \begin{bmatrix} a_1 \\ \vdots \\ a_n \end{bmatrix} \in \mathbb{R}^n$ is really the coordinate of $\mathbf{v}$ relative to the standard basis $\mathcal{E} = \{\mathbf{e}^1, \mathbf{e}^2, ..., \mathbf{e}^n\}$:

$$[\mathbf{v}]_{\mathcal{E}} = \begin{bmatrix} a_1 \\ \vdots \\ a_n \end{bmatrix}$$

Practice Problem 4.2.1. *Prove that $\phi_{\mathcal{B}} : V \to \mathbb{R}^n$ is a linear map.*

Theorem 4.2.5. *Let V be a subspace of $\mathbb{R}^N$ with a basis $\mathcal{B} = \{\mathbf{v}_1, \mathbf{v}_2, ..., \mathbf{v}_n\}$. Then the coordinate map $\phi_{\mathcal{B}} : V \to \mathbb{R}^n$ relative to $\mathcal{B}$ is an isomorphism.*

Proof. Let $\mathcal{B}$ be a basis of V and $\phi_{\mathcal{B}} : V \to \mathbb{R}^n$ be the coordinate map with respect to the basis $\mathcal{B}$. It suffices to show that $\phi_{\mathcal{B}}$ is bijective by Practice Problem 4.2.1. Suppose $\phi(\mathbf{v}) = \phi(\mathbf{v}')$. Let $\mathbf{v} = \sum a_i\mathbf{v}_i, \mathbf{v}' = \sum b_i\mathbf{v}_i$. Then $\phi_{\mathcal{B}}(\mathbf{v}) = \begin{bmatrix} a_1 \\ \vdots \\ a_n \end{bmatrix}$ and $\phi_{\mathcal{B}}(\mathbf{v}') = \begin{bmatrix} b_1 \\ \vdots \\ b_n \end{bmatrix}$. Since $\phi_{\mathcal{B}}(\mathbf{v}) = \phi_{\mathcal{B}}(\mathbf{v}')$ we have $a_i = b_i$ for all i; and hence $\mathbf{v} = \mathbf{v}'$. Thus $\phi_{\mathcal{B}}$ is injective. On the other hand, let $\mathbf{w} = \begin{bmatrix} c_1 \\ \vdots \\ c_n \end{bmatrix} \in \mathbb{R}^n$ and $\mathbf{v} = \sum c_i\mathbf{v}_i$. Then $\phi_{\mathcal{B}}(\mathbf{v}) = \mathbf{w}$. Therefore $\phi_{\mathcal{B}}$ is a bijection. □

Corollary 4.2.6. *Let V, W be subspaces of the same dimension then they are isomorphic.*

Proof. Let $\mathcal{B}$ and $\mathcal{C}$ be bases of V and W. Then the number of elements of $\mathcal{B}$ and $\mathcal{C}$ are the same say, equal to n. Then $\phi_{\mathcal{B}} : V \to \mathbb{R}^n$ and $\phi_{\mathcal{C}} : W \to \mathbb{R}^n$ are isomorphisms. Hence $\phi_{\mathcal{C}}^{-1} \circ \phi_{\mathcal{B}} : V \to W$ is an isomorphism by Theorem 4.2.3. □

The converse is also true (Theorem 4.2.25).

Practice Problem 4.2.2. *Give another proof of* Corollary 4.2.6 *by using* Theorem 4.2.2.

Definition 4.2.7. *Let $T : V \to W$ a linear map of subspaces. Let $\mathcal{B} = \{\mathbf{v}_1, \mathbf{v}_2, \dots, \mathbf{v}_n\}$ and $\mathcal{C} = \{\mathbf{w}_1, \mathbf{w}_2, \dots, \mathbf{w}_m\}$ be bases of V and W respectively. Let*

$$T(\mathbf{v}_j) = \sum_{i=1}^{m} a_{ij}\mathbf{w}_i \quad (j = 1, 2, \dots, n) \tag{4.4}$$

We get an $m \times n$ matrix

$$[T]_{\mathcal{B}}^{\mathcal{C}} := [a_{ij}] \in M(m\times n) \tag{4.5}$$

i.e., the j-th column of $[T]_{\mathcal{B}}^{\mathcal{C}}$ is the $\mathcal{C}$-coordinate of $T(\mathbf{v}_j)$:

$$\left([T]_{\mathcal{B}}^{\mathcal{C}}\right)^{(j)} = [T(\mathbf{v}_j)]_{\mathcal{C}} = \begin{bmatrix} a_{1j} \\ a_{2j} \\ \vdots \\ a_{mj} \end{bmatrix} \tag{4.6}$$

The matrix $[T]_{\mathcal{B}}^{\mathcal{C}}$ is called the **matrix representation of** *T with respect to the bases $\mathcal{B}, \mathcal{C}$. If $V = W, \mathcal{B} = \mathcal{C}$ then we simply write $[T]_{\mathcal{B}}$ for $[T]_{\mathcal{B}}^{\mathcal{C}}$.*

Example 4.2.8. *Let $A \in M(m\times n)$ and $T_A : \mathbb{R}^n \to \mathbb{R}^m$ be the matrix linear map associated to A. Let $\mathcal{E}^n, \mathcal{E}^m$ be the standard bases for $\mathbb{R}^n, \mathbb{R}^m$ then the matrix representation of T_A with respect to $\mathcal{E}^n, \mathcal{E}^m$ is A itself:*

$$[T_A]_{\mathcal{E}^n}^{\mathcal{E}^m} = A \tag{4.7}$$

Example 4.2.9. *Let*

$$V = \left\{ \begin{bmatrix} x \\ y \\ z \end{bmatrix} \in \mathbb{R}^3 \,\middle|\, x+y+z=0 \right\}$$

Then V is the null space of 1×3 matrix $A = [1\ \ 1\ \ 1]$. Since $\rho(A) = 1$ we see that $\dim V = 2$ *by* Theorem 2.3.17. *Now, one finds that*

$$\mathcal{B} = \left\{ \mathbf{v}_1 = \begin{bmatrix} 1 \\ -1 \\ 0 \end{bmatrix}, \mathbf{v}_2 = \begin{bmatrix} 1 \\ 0 \\ -1 \end{bmatrix} \right\}$$

is a basis of V. And for $\mathbf{v} = \begin{bmatrix} a \\ b \\ c \end{bmatrix} \in V$, define $T : V \to \mathbb{R}^3$ by

$$T(\mathbf{v}) = \begin{bmatrix} a+b \\ b-c \\ a+c \end{bmatrix}$$

Let's find $[T]_{\mathcal{B}}^{\mathcal{E}}$ where $\mathcal{E} = \{\mathbf{e}^1, \mathbf{e}^2, \mathbf{e}^3\}$ is the standard basis of $\mathbb{R}^3$. We compute

$$T(\mathbf{v}_1) = \begin{bmatrix} 0 \\ -1 \\ 1 \end{bmatrix} = -\mathbf{e}^2 + \mathbf{e}^3 \text{ and } T(\mathbf{v}_2) = \begin{bmatrix} 1 \\ 1 \\ 0 \end{bmatrix} = \mathbf{e}^1 + \mathbf{e}^2$$

Thus

$$[T]_{\mathcal{B}}^{\mathcal{E}} = \begin{bmatrix} 0 & 1 \\ -1 & 1 \\ 1 & 0 \end{bmatrix}$$

Now let

$$W = \left\{ \begin{bmatrix} x \\ y \\ z \end{bmatrix} \,\middle|\, x-y-z=0 \right\}$$

Then the map T defined above maps V to W. (Check this.) *And we find that*

$$\mathcal{C} = \left\{ \mathbf{w}_1 = \begin{bmatrix} 1 \\ 1 \\ 0 \end{bmatrix}, \mathbf{w}_2 = \begin{bmatrix} 1 \\ 0 \\ 1 \end{bmatrix} \right\}$$

is a basis of W. As $T : V \to W$ is a linear map let's find $[T]_{\mathcal{B}}^{\mathcal{C}}$.

$$T(\mathbf{v}_1) = \begin{bmatrix} 0 \\ -1 \\ 1 \end{bmatrix} = -\mathbf{w}_1 + \mathbf{w}_2 \text{ and } T(\mathbf{v}_2) = \begin{bmatrix} 1 \\ 1 \\ 0 \end{bmatrix} = \mathbf{w}_1 + 0\mathbf{w}_2$$

Hence we have

$$[T]_{\mathcal{B}}^{\mathcal{C}} = \begin{bmatrix} -1 & 1 \\ 1 & 0 \end{bmatrix}$$

Example 4.2.10. *Let*

$$V = \left\{ \begin{bmatrix} x \\ y \\ z \\ w \end{bmatrix} \middle| \, x - y + z - w = 0 \right\}$$

Then V is the null space of 1×4 matrix $A = [1\ -1\quad 1\ -1]$. Since $\rho(A) = 1$, $\dim(V) = \nu(A) = 3$. Let

$$\mathcal{B} = \left\{ \mathbf{v}_1 = \begin{bmatrix} 1 \\ 1 \\ 0 \\ 0 \end{bmatrix}, \mathbf{v}_2 = \begin{bmatrix} 1 \\ 0 \\ -1 \\ 0 \end{bmatrix}, \mathbf{v}_3 = \begin{bmatrix} 1 \\ 0 \\ 0 \\ 1 \end{bmatrix} \right\}, \quad \mathcal{B}' = \left\{ \mathbf{v}_1' = \begin{bmatrix} 1 \\ 1 \\ 1 \\ 1 \end{bmatrix}, \mathbf{v}_2' = \begin{bmatrix} 0 \\ -1 \\ 0 \\ 1 \end{bmatrix}, \mathbf{v}_3' = \begin{bmatrix} 0 \\ 0 \\ 1 \\ 1 \end{bmatrix} \right\}$$

Then $\mathcal{B}, \mathcal{B}'$ are linearly independent subsets of V. Hence $\mathcal{B}, \mathcal{B}'$ are bases of V.

Let

$$W = \left\{ \begin{bmatrix} x \\ y \\ z \end{bmatrix} \middle| \, x - y + z = 0 \right\}$$

Then W is a subspace of $\mathbb{R}^3$ of dimension 2. Let

$$\mathcal{C} = \left\{ \mathbf{w}_1 = \begin{bmatrix} 1 \\ 1 \\ 0 \end{bmatrix}, \mathbf{w}_2 = \begin{bmatrix} 1 \\ 0 \\ -1 \end{bmatrix} \right\}, \quad \mathcal{C}' = \left\{ \mathbf{w}_1' = \begin{bmatrix} 1 \\ 2 \\ 1 \end{bmatrix}, \mathbf{w}_2' = \begin{bmatrix} 2 \\ 1 \\ -1 \end{bmatrix} \right\}$$

Then $\mathcal{C}, \mathcal{C}'$ are linearly independent subsets of W. Hence $\mathcal{C}, \mathcal{C}'$ are bases of W. For $\mathbf{v} = \begin{bmatrix} a \\ b \\ c \\ d \end{bmatrix}$ we define

$$T(\mathbf{v}) = \begin{bmatrix} a + b + c \\ b + c + d \\ -a + d \end{bmatrix}$$

Then $T(\mathbf{v}) \in W$. Hence T defines linear a map

$$T : V \to W$$

To find $[T]_{\mathcal{B}}^{\mathcal{C}}, [T]_{\mathcal{B}'}^{\mathcal{C}'}$ we compute

$$T(\mathbf{v}_1) = \begin{bmatrix} 2 \\ 1 \\ -1 \end{bmatrix} = \mathbf{w}_1 + \mathbf{w}_2, \ T(\mathbf{v}_2) = \begin{bmatrix} 0 \\ -1 \\ -1 \end{bmatrix} = -\mathbf{w}_1 + \mathbf{w}_2, \ T(\mathbf{v}_3) = \begin{bmatrix} 1 \\ 1 \\ 0 \end{bmatrix} = \mathbf{w}_1 + 0\mathbf{w}_2,$$

and

$$T(\mathbf{v}_1') = \begin{bmatrix} 3 \\ 3 \\ 0 \end{bmatrix} = 3\mathbf{w}_1' + 0\mathbf{w}_2', \ T(\mathbf{v}_2') = \begin{bmatrix} -1 \\ 0 \\ 1 \end{bmatrix} = 0\mathbf{w}_1' - \mathbf{w}_2', \ T(\mathbf{v}_3') = \begin{bmatrix} 1 \\ 2 \\ 1 \end{bmatrix} = 2\mathbf{w}_1' - \mathbf{w}_2',$$

Hence

$$[T]_{\mathcal{B}}^{\mathcal{C}} = \begin{bmatrix} 1 & -1 & 1 \\ 1 & 1 & 0 \end{bmatrix}, \quad [T]_{\mathcal{B}'}^{\mathcal{C}'} = \begin{bmatrix} 3 & 0 & 2 \\ 0 & -1 & -1 \end{bmatrix}$$

Practice Problem 4.2.3. *Let* $A = \begin{bmatrix} 0 & 1 & -1 \\ 1 & 1 & 0 \\ -1 & 0 & -1 \end{bmatrix}$ *and* $V = \mathrm{Col}(A)$. *Let* $A^{(i)} = \mathbf{v}_i$ *for* $i = 1, 2, 3$.

(1) *Show that* $\mathcal{B}_1 = \{\mathbf{v}_1, \mathbf{v}_2\}$ *and* $\mathcal{B}_2 = \{\mathbf{v}_2, \mathbf{v}_3\}$ *are bases for* V.

(2) *Define* $T : V \to \mathbb{R}^3$ *by*

$$T(\mathbf{v}_1) = \begin{bmatrix} 1 \\ -1 \\ 0 \end{bmatrix}, \quad T(\mathbf{v}_2) = \begin{bmatrix} 1 \\ -2 \\ 1 \end{bmatrix}$$

Find $T(\mathbf{v}_3)$ *and show that* $T(\mathbf{v}_1) - T(\mathbf{v}_2) - T(\mathbf{v}_3) = \mathbf{0}$.

(3) *Let* $\mathcal{E} = \{\mathbf{e}^1, \mathbf{e}^2, \mathbf{e}^3\}$ *be the standard basis for* $\mathbb{R}^3$ *and* $\mathcal{C} = \left\{ \begin{bmatrix} 1 \\ 1 \\ 0 \end{bmatrix}, \begin{bmatrix} 1 \\ 0 \\ 1 \end{bmatrix}, \begin{bmatrix} 0 \\ 1 \\ 1 \end{bmatrix} \right\}$ *be a basis of* $\mathbb{R}^3$. *Find* $[T]^{\mathcal{E}}_{\mathcal{B}_1}$ *and* $[T]^{\mathcal{C}}_{\mathcal{B}_2}$

Let $T : V \to W$ be a linear map and $\mathcal{B}, \mathcal{C}$ be bases of V, W. Let $\phi_{\mathcal{B}}(\mathbf{v}) = [\mathbf{v}]_{\mathcal{B}}$, $\phi_{\mathcal{C}}(\mathbf{w}) = [\mathbf{w}]_{\mathcal{C}}$ and $[T]^{\mathcal{C}}_{\mathcal{B}} = A$. Then $\phi_{\mathcal{C}} \circ T(\mathbf{v}) = [T(\mathbf{v})]_{\mathcal{C}}$ is the coordinate of $T(\mathbf{v})$ relative to $\mathcal{C}$. On the other hand, we have $T_A \circ \phi_{\mathcal{B}}(\mathbf{v}) = A[\mathbf{v}]_{\mathcal{B}}$. Our question is then, whether these two are equal. That is $\phi_{\mathcal{C}} \circ T = T_A \circ \phi_{\mathcal{B}}$? In this case, we say the diagram

$$\begin{array}{ccc} V & \xrightarrow{\ T\ } & W \\ {\scriptstyle \phi_{\mathcal{B}}}\big\downarrow & & \big\downarrow{\scriptstyle \phi_{\mathcal{C}}} \\ \mathbb{R}^n & \xrightarrow[T_A]{} & \mathbb{R}^m \end{array} \tag{4.8}$$

is **commutative**. The answer is affirmative:

Theorem 4.2.11. *Let* $T : V \to W$ *a linear map of subspaces. Let* $\mathcal{B} = \{\mathbf{v}_1, \mathbf{v}_2, \ldots, \mathbf{v}_n\}, \mathcal{C} = \{\mathbf{w}_1, \mathbf{w}_2, \ldots, \mathbf{w}_m\}$ *be bases of* V *and* W *respectively. Let* $\mathbf{v} \in V$ *then*

$$[T(\mathbf{v})]_{\mathcal{C}} = [T]^{\mathcal{C}}_{\mathcal{B}}[\mathbf{v}]_{\mathcal{B}} \tag{4.9}$$

Proof. Let $A = [T]^{\mathcal{C}}_{\mathcal{B}}$. The maps $\phi : \mathbf{v} \mapsto [T(\mathbf{v})]_{\mathcal{C}}$ and $\psi : \mathbf{v} \mapsto A[\mathbf{v}]_{\mathcal{B}}$ are both linear maps $V \to \mathbb{R}^m$. (Check!) To prove they are equal it suffices to show they coincide on $\mathcal{B}$ by Theorem 4.2.2. Now we have

$$\phi(\mathbf{v}_i) = [T(\mathbf{v}_i)]_{\mathcal{C}} = A^{(i)} \text{ and } \psi(\mathbf{v}_i) = A[\mathbf{v}_i]_{\mathcal{B}} = A\mathbf{e}^i = A^{(i)}$$

Hence they are equal. □

Example 4.2.12. *Let* A *be an* $m \times n$ *matrix. The map* $T_A : \mathbb{R}^n \to \mathbb{R}^m$ (Theorem 2.1.10 of (2.10)) *was defined by*

$$T_A(\mathbf{v}) = A\mathbf{v}$$

If $\mathcal{E}^m, \mathcal{E}^n$ *are the standard bases of* $\mathbb{R}^m, \mathbb{R}^n$ *then we can rewrite this as*

$$[T_A(\mathbf{v})]_{\mathcal{E}^m} = [T_A]^{\mathcal{E}^m}_{\mathcal{E}^n}[\mathbf{v}]_{\mathcal{E}^n}$$

Example 4.2.13. *In* Example 4.2.9, *let* $\mathbf{v} = \begin{bmatrix} 3 \\ -2 \\ -1 \end{bmatrix}$. *Then* $\mathbf{v} = 2\mathbf{v}_1 + \mathbf{v}_2$ *and hence* $[\mathbf{v}]_{\mathcal{B}} = \begin{bmatrix} 2 \\ 1 \end{bmatrix}$. *Hence*

$$[T]_{\mathcal{B}}^{\mathcal{C}}[\mathbf{v}]_{\mathcal{B}} = \begin{bmatrix} -1 & 1 \\ 1 & 0 \end{bmatrix} \begin{bmatrix} 2 \\ 1 \end{bmatrix} = \begin{bmatrix} -1 \\ 2 \end{bmatrix}$$

On the other hand,

$$[T(\mathbf{v})]_{\mathcal{C}} = \begin{bmatrix} 1 \\ -1 \\ 2 \end{bmatrix}_{\mathcal{C}} = \begin{bmatrix} -1 \\ 2 \end{bmatrix}$$

Definition 4.2.14. *Now let*

$$L(V, W) = \{T : V \to W \mid T \text{ linear map } \}$$

be the set of all linear maps from V *to* W. *If* $V = W$ *then we will simply write* $L(V)$ *for* $L(V, W)$.

We define two operations, addition and scalar multiplication on $L(V, W)$: *For* T, T_1, T_2 *in* $L(V, W)$ *and* $c \in \mathbb{R}$ *define*

$$(T_1 + T_2)(\mathbf{v}) = T_1(\mathbf{v}) + T_2(\mathbf{v}) \quad \text{and} \quad (cT)(\mathbf{v}) = c(T(\mathbf{v})) \tag{4.10}$$

Theorem 4.2.15. *Let* $T, T_1, T_2 \in L(V, W)$ *and* $c \in \mathbb{R}$. *Then*

$$T_1 + T_2 \in L(V, W) \quad \text{and} \quad cT \in L(V, W)$$

Proof. Exercise 2 □

For subspaces V, W with the bases $\mathcal{B}, \mathcal{C}$ and a linear map $T : V \to W$ we associate the matrix $[T]_{\mathcal{B}}^{\mathcal{C}}$. The correspondence $T \mapsto [T]_{\mathcal{B}}^{\mathcal{C}}$ defines a map from $L(V, W)$ to $M(m \times n)$.

$$\begin{aligned} \Phi : L(V, W) &\to M(m \times n) \\ T &\mapsto [T]_{\mathcal{B}}^{\mathcal{C}} \end{aligned}$$

On the other hand, for $A \in M(m \times n)$, there is a unique linear map $T \in L(V, W)$ with $T(\mathbf{v}_j) = \sum_i a_{ij}\mathbf{w}_i$ (Theorem 4.2.2).

$$\begin{aligned} \Psi : M(m \times n) &\to L(V, W) \\ A &\mapsto T \end{aligned}$$

As for the cases of $\mathbb{R}^n$ (Theorem 2.2.8), we will show that the map Φ is a bijection with its inverse Ψ. The only difference is that we do not need to choose bases in Theorem 2.2.8 since there are natural bases, the standard bases. However, we need choose particular bases for V, W this time.

Theorem 4.2.16. *Let* $\mathcal{B}, \mathcal{C}$ *be bases for* V, W *respectively. Then the map*

$$\Phi : L(V, W) \to M(m \times n)$$

defined by $\Phi(T) = [T]_{\mathcal{B}}^{\mathcal{C}}$ *is a bijection with its inverse* Ψ. *Further, we have* $\Phi(T_1 + T_2) = \Phi(T_1) + \Phi(T_2)$ *and* $\Phi(cT) = c\Phi(T)$ *for* $T, T_1, T_2 \in L(V, W)$ *and* $c \in \mathbb{R}$. *That is*

$$[T_1 + T_2]_{\mathcal{B}}^{\mathcal{C}} = [T_1]_{\mathcal{B}}^{\mathcal{C}} + [T_2]_{\mathcal{B}}^{\mathcal{C}} \quad \text{and} \quad [cT]_{\mathcal{B}}^{\mathcal{C}} = c[T]_{\mathcal{B}}^{\mathcal{C}} \tag{4.11}$$

Proof. To show Φ is injective suppose $\Phi(T) = \Phi(S)$. Then $[T]_{\mathcal{B}}^{\mathcal{C}} = [S]_{\mathcal{B}}^{\mathcal{C}}$ which means the $\mathcal{C}$-coordinate of $T(\mathbf{v}_i)$ and $S(\mathbf{v}_i)$ are the same; $T(\mathbf{v}_i) = S(\mathbf{v}_i)$. Hence by Theorem 4.2.2 we see that $T = S$.

To show Φ is surjective let $A = [a_{ij}] \in M(m \times n)$. Then $T(\mathbf{v}_j) = \sum_i a_{ij}\mathbf{w}_i$ defines a linear map $T \in L(V, W)$ (Theorem 4.2.2). Now it is obvious that $[T]_{\mathcal{B}}^{\mathcal{C}} = [a_{ij}]$.

To show (4.11), let $[T_1]_{\mathcal{B}}^{\mathcal{C}} = [a_{ij}]$, $[T_2]_{\mathcal{B}}^{\mathcal{C}} = (b_{ij})$. Then

$$(T_1 + T_2)(\mathbf{v}_j) = T_1(\mathbf{v}_j) + T_2(\mathbf{v}_j) = \sum_i a_{ij}\mathbf{w}_i + \sum_i b_{ij}\mathbf{w}_i = \sum_i (a_{ij} + b_{ij})\mathbf{w}_i$$

Hence $[T_1 + T_2]_{\mathcal{B}}^{\mathcal{C}} = [a_{ij} + b_{ij}]$. On the other hand,

$$[T_1]_{\mathcal{B}}^{\mathcal{C}} + [T_2]_{\mathcal{B}}^{\mathcal{C}} = [a_{ij}] + [b_{ij}] = [a_{ij} + b_{ij}]$$

Therefore $[T_1 + T_2]_{\mathcal{B}}^{\mathcal{C}} = [T_1]_{\mathcal{B}}^{\mathcal{C}} + [T_2]_{\mathcal{B}}^{\mathcal{C}}$.

We leave the verification of the equality $[cT]_{\mathcal{B}}^{\mathcal{C}} = c[T]_{\mathcal{B}}^{\mathcal{C}}$ and the fact that Ψ is an inverse of Φ as an exercise. □

Corollary 4.2.17. *Let $S, T : V \to W$ be linear maps. Let $\mathcal{B}, \mathcal{C}$ be bases of V, W respectively. Then $S = T$ if and only if $[S]_{\mathcal{B}}^{\mathcal{C}} = [T]_{\mathcal{B}}^{\mathcal{C}}$. In particular, if $W = V, \mathcal{C} = \mathcal{B}$ and $[T]_{\mathcal{B}} = I_n$ then $T = I_V$.*

Next we show, as for the $\mathbb{R}^n$'s (Theorem 2.2.15), that the composition of linear maps correspond, under Φ, to the matrix multiplications of matrices. The corresponding theorem for abstract vector spaces will be given in Theorem 6.4.16.

Theorem 4.2.18. *Let U, V, W be subspaces with bases $\mathcal{A}, \mathcal{B}, \mathcal{C}$ respectively. If $U \xrightarrow{S} V \xrightarrow{T} W$ are linear maps then we have*

$$[T \circ S]_{\mathcal{A}}^{\mathcal{C}} = [T]_{\mathcal{B}}^{\mathcal{C}}[S]_{\mathcal{A}}^{\mathcal{B}} \tag{4.12}$$

i.e., the composition of linear maps corresponds to the multiplication of matrices.

In particular, if $U = V = W$ and $\mathcal{A} = \mathcal{B} = \mathcal{C}$ then

$$[T \circ S]_{\mathcal{B}} = [T]_{\mathcal{B}}[S]_{\mathcal{B}} \tag{4.13}$$

Proof. Let $A = [S]_{\mathcal{A}}^{\mathcal{B}}$, $B = [T]_{\mathcal{B}}^{\mathcal{C}}$. We will use Theorem 4.2.11 repeatedly:

$$\begin{array}{ccccc} U & \xrightarrow{\ S\ } & V & \xrightarrow{\ T\ } & W \\ {\scriptstyle \phi_{\mathcal{A}}}\downarrow & & {\scriptstyle \phi_{\mathcal{B}}}\downarrow & & {\scriptstyle \phi_{\mathcal{C}}}\downarrow \\ \mathbb{R}^n & \xrightarrow[T_A]{} & \mathbb{R}^m & \xrightarrow[T_B]{} & \mathbb{R}^p \end{array} \tag{4.14}$$

It suffices to show that $[T \circ S]_{\mathcal{A}}^{\mathcal{C}}[\mathbf{u}]_{\mathcal{A}} = BA[\mathbf{u}]_{\mathcal{A}}$ for all $\mathbf{u} \in U$. By Theorem 4.2.11, we have $\phi_{\mathcal{B}} \circ S = T_A \circ \phi_{\mathcal{A}}$ and $\phi_{\mathcal{C}} \circ T = T_B \circ \phi_{\mathcal{B}}$ i.e., $[S(\mathbf{u})]_{\mathcal{B}} = A[\mathbf{u}]_{\mathcal{A}}$ and $[T(\mathbf{v})]_{\mathcal{C}} = B[\mathbf{v}]_{\mathcal{B}}$. Hence we have

$$[T \circ S]_{\mathcal{A}}^{\mathcal{C}}[\mathbf{u}]_{\mathcal{A}} = [T \circ S(\mathbf{u})]_{\mathcal{C}} = [T(S(\mathbf{u}))]_{\mathcal{C}} = B[S(\mathbf{u}))]_{\mathcal{B}} = BA[\mathbf{u}]_{\mathcal{A}}$$

Thus $[T \circ S]_{\mathcal{A}}^{\mathcal{C}} = BA$. □

4.2.2 Null Space, Range Space

We define the null space and the range space in the same way as we did for the linear maps between $\mathbb{R}^n$'s (Definition 2.1.5).

Definition 4.2.19. *Let $T : V \to W$ be a linear map of subspaces. The* **null space (kernel)** *$N(T)$ and the* **range space** *$R(T)$ of a linear map T is defined by*

$$N(T) = \{\mathbf{v} \in V \mid T(\mathbf{v}) = 0\} \tag{4.15}$$

$$R(T) = \{T(\mathbf{v}) \mid \mathbf{v} \in V\} \tag{4.16}$$

By Theorem 2.1.7, the range and the null space a linear map $T : \mathbb{R}^n \to \mathbb{R}^m$ are subspaces. The same is true for the range and null space of a linear map between subspaces of $\mathbb{R}^n$.

For any function $T : V \to W$ and a subset S of V we write

$$T(S) = \{T(\mathbf{v}) \mid \mathbf{v} \in S\} \subseteq W \tag{4.17}$$

Theorem 4.2.20. *Let $T : V \to W$ be a linear map. The null space $N(T)$ and the range $R(T)$ of T are subspaces of V and W respectively. If S is a subspace of V then $T(S)$ is a subspace of W. Further, if S span V then $R(T) = \mathrm{Span}\,(T(S))$.*

Proof. Exercise 3. □

Definition 4.2.21. *The dimension of $N(T)$ is called the* **nullity** *of T and the dimension of $R(T)$ is called the* **rank** *of T. And we write, as before*

$$\nu(T) = \dim N(T), \quad \rho(T) = \dim R(T)$$

Example 4.2.22. *Suppose $T = T_A$ for some $A \in M(m{\times}n)$ then we showed in* Theorem 2.1.10 *that*

$$R(T_A) = \mathrm{Col}(A) \text{ and } N(T_A) = N(A)$$

Hence $\rho(T_A) = \rho(A)$ and $\nu(T_A) = \nu(A)$.

The following theorem extends the results of Theorem 2.4.1 and Theorem 2.4.2 to the subspaces.

Theorem 4.2.23. *Let $T : V \to W$ be a linear map.*

(1) *The linear map T is injective if and only if $N(T) = \{\mathbf{0}\}$ i.e., $\nu(T) = 0$.*

(2) *The linear map T is surjective if and only if $R(T) = W$ i.e., $\rho(T) = \dim W$.*

Proof. (1) First assume T is injective. Since T is linear we have $T(\mathbf{0}) = \mathbf{0}$ and since T is injective there is no $\mathbf{v} \neq \mathbf{0}$ such that $T(\mathbf{v}) = \mathbf{0}$. Hence $N(T) = \{\mathbf{0}\}$.

Conversely, suppose $N(T) = \mathbf{0}$ and $T(\mathbf{v}_1) = T(\mathbf{v}_2)$. Then $T(\mathbf{v}_2 - \mathbf{v}_1) = \mathbf{0}$. Hence $\mathbf{v}_2 - \mathbf{v}_1 \in N(T) = \{\mathbf{0}\}$ and hence $\mathbf{v}_2 - \mathbf{v}_1 = \mathbf{0}$ i.e., $\mathbf{v}_1 = \mathbf{v}_2$

(2) Obvious. □

Theorem 4.2.24. *Let $T : V \to W$ be a linear map and let $S = \{\mathbf{v}_1, \mathbf{v}_2, ..., \mathbf{v}_n\} \subseteq V$.*

(1) *The linear map T is injective if and only if $T(S)$ is linearly independent whenever S is linearly independent.*

(2) *The linear map T is surjective if and only if $T(S)$ spans W whenever S spans V.*

(3) *The linear map T is an isomorphism if and only if $T(S)$ is a basis of W whenever S is a basis of V. In particular, $\dim V = \dim W$ if and only if there is an isomorphism between V and W.*

Proof. (1) First, suppose T is injective. Let $\{\mathbf{v}_1, \cdots, \mathbf{v}_n\}$ be a linearly independent set of vectors in V and let $a_1T(\mathbf{v}_1) + \cdots + a_nT(\mathbf{v}_n) = \mathbf{0}$. Then by linearity of T we see that $T(a_1\mathbf{v}_1+\cdots+a_n\mathbf{v}_n) = \mathbf{0}$. Since T is injective, we have $a_1\mathbf{v}_1+\cdots+a_n\mathbf{v}_n = \mathbf{0}$. As $\{\mathbf{v}_1, \cdots, \mathbf{v}_n\}$ is linearly independent we have $a_1 = a_2 = \cdots = a_n = 0$ which says $\{T(\mathbf{v}_1), \cdots, T(\mathbf{v}_n)\}$ is linearly independent.

Now assume T sends every linearly independent set to a linearly independent set. Suppose $T(\mathbf{v}) = \mathbf{0}$ and $\mathbf{v} \neq \mathbf{0}$. Then the set $\{\mathbf{v}\}$ consisting of a single vector is linearly independent and by assumption $\{T(\mathbf{v}) = \mathbf{0}\}$ must be linearly independent. But any set containing the zero vector is linearly dependent. This is a contradiction. Hence $T(\mathbf{v}) \neq \mathbf{0}$.

We leave the proofs of (2) and (3) as Exercise 4. □

Theorem 4.2.25. *Let $T : V \to W$ be a linear map and let V' be a subspace of V.*

(1) *If $S = \{\mathbf{v}_1, \mathbf{v}_2, ..., \mathbf{v}_n\}$ spans V' then $T(V') = \operatorname{Span} T(S)$ and*

$$\dim T(V') \leq \dim V' \tag{4.18}$$

(2) *If T is injective then $\dim V' = \dim T(V')$. If T is an isomorphism then $\dim V = \dim W$.*

Proof. (1) Let $\mathbf{w} \in T(V')$. Then there is $\mathbf{v} \in V'$ such that $T(\mathbf{v}) = \mathbf{w}$. Since S spans V' we can write $\mathbf{v} = \sum c_i\mathbf{v}_i$ for some constants c_i. Hence $\mathbf{w} = T(\mathbf{v}) = \sum c_iT(\mathbf{v}_i)$; $T(V') \subseteq \operatorname{Span}(T(S))$. Now if $\mathbf{w} \in \operatorname{Span}(T(S))$ then we can write $\mathbf{w} = \sum c_iT(\mathbf{v}_i)$. Since T is linear and S span V' we have $\mathbf{w} = T(\sum c_i\mathbf{v}_i) \in T(V')$. Therefore $T(V') = \operatorname{Span}(T(S))$.

To prove the inequality (4.18), suppose $S = \{\mathbf{v}_1, \mathbf{v}_2, ..., \mathbf{v}_n\}$ is a basis of V' so that $\dim V' = n$. Then since $T(S)$ spans $T(V')$ and $\dim T(V')$ is the number of maximal linearly independent subset of $T(S)$ we see that $\dim T(V') \leq n$ (Theorem 4.1.7).

(2) Let $\mathcal{B} = \{\mathbf{v}_1, \mathbf{v}_2, ..., \mathbf{v}_n\}$ be a basis of V'. Then by (1), $T(\mathcal{B})$ spans $T(V')$. By Theorem 4.2.24 (1), $T(\mathcal{B})$ is also linear independent and hence $T(\mathcal{B})$ is a basis of $T(V')$; $\dim V' = \dim T(V')$.

If T is an isomorphism then T sends a basis of V to a basis of W. Since T is a bijection those bases have the same number of vectors; $\dim V = \dim W$. □

The following theorem generalizes Corollary 2.3.15.

Theorem 4.2.26. *Let $S : U \to V$ and $T : V \to W$ be linear maps. Then*

(1) $\rho(T \circ S) \leq \min\{\rho(S), \rho(T)\}$.

(2) *If S is an isomorphism then $\rho(T \circ S) = \rho(T)$ and $\nu(T \circ S) = \nu(T)$.*

(3) *If T is an isomorphism then $\rho(T \circ S) = \rho(S)$ and $\nu(T \circ S) = \nu(S)$.*

Proof. (1) First $\rho(TS) = \dim TS(U) \leq \dim S(U) = \rho(S)$ since the linear map decrease the dimension of subspaces (Theorem 4.2.25). On the other hand, since $TS(U) \subseteq T(V)$ we have $\rho(TS) = \dim TS(U) \leq \dim T(V) = \rho(T)$.

(2) Since S is an isomorphism, $S(U) = V$. Hence $TS(U) = T(V)$ and hence $\rho(TS) = \dim TS(U) = \dim T(V) = \rho(T)$.

For the nullity, S restricted to $S^{-1}(N(T))$ induces an isomorphism $S^{-1}(N(T)) \overset{\cong}{\to} N(T)$. Now since $S^{-1}(N(T)) = N(T \circ S)$ (Exercise 5) we have $\nu(T \circ S) = \nu(S)$.

(3) is similarly proved using Theorem 4.2.25. □

The proof of Theorem 4.2.26 used the abstract concepts like vector space and linear maps. We will use the theorem to prove its matrix version in Corollary 4.2.27 which we consider to be more concrete. Proving this in opposite direction – first prove a matrix version and then deduce from it a version of vector spaces and linear maps – will be more complicated. Explicit computation in mathematics is, most of the time, more powerful than logical inference of abstract notions. This is an example showing that this is not always the case.

Corollary 4.2.27. *Let $A \in M(m\times n), B \in M(n\times p)$. Then*

(1) $\rho(AB) \leq \min\{\rho(A), \rho(B)\}$

(2) *If B is invertible then $\rho(AB) = \rho(A)$.*

(3) *If A is invertible then $\rho(AB) = \rho(B)$.*

Proof. Simply take $T = T_A$ and $S = T_B$ in Theorem 4.2.26. □

Corollary 4.2.28. *Let $A \in M(m\times n)$. Then $\rho(A) = r$ if and only if there are invertible matrices P, Q of sizes m, n such that $A = PJ_rQ$ where $J_r \in M(m\times n)$ is of the form*

$$J_r = \begin{bmatrix} I_r & 0 \\ 0 & 0 \end{bmatrix} \tag{4.19}$$

This can be achieved by performing elementary row and column operations

$$E_1E_2\cdots E_kAF_1F_2\cdots F_l = J_r \quad \text{or } A = E'_kE'_{k-1}\cdots E'_1J_rF'_lF'_{l-1}\cdots F'_1 \tag{4.20}$$

where $E_i, F_j, E'_i = E_i^{-1}, F'_j = F_j^{-1}$ are elementary matrices and we can choose the invertible matrices $P = E'_kE'_{k-1}\cdots E'_1$, $Q = F'_lF'_{l-1}\cdots F'_1$.

Proof. Writing a formal proof of this will be tedious. Instead an illustration of our result will be suffice.

If $\rho(A) = 3$ then, by elementary row operations, A can be reduced to an reduced echelon form which looks like

$$\begin{bmatrix} \mathbf{1} & 2 & 0 & 0 & 0 & 1 \\ 0 & 0 & \mathbf{1} & 0 & 1 & 2 \\ 0 & 0 & 0 & \mathbf{1} & 0 & 1 \\ 0 & 0 & 0 & 0 & 0 & 0 \end{bmatrix}$$

Now by elementary column operations we can bring the matrix into the form

$$\begin{bmatrix} \mathbf{1} & 0 & 0 & 0 & 0 & 0 \\ 0 & \mathbf{1} & 0 & 0 & 0 & 0 \\ 0 & 0 & \mathbf{1} & 0 & 0 & 0 \\ 0 & 0 & 0 & 0 & 0 & 0 \end{bmatrix}$$

Conversely, suppose A can be reduced to the form (4.19) by elementary row and columns operations. Elementary operations are achieved by pre- or post-multiplying elementary matrices (Theorem 2.5.11). Elementary matrices are invertible and hence by Corollary 4.2.27, we have $\rho(A) = r$. □

Corollary 4.2.29. *Let $A \in M(n)$, $B \in M(r{\times}n)$ and $C \in M(n{\times}s)$. Then*

$$\rho(BA) = \rho(BA^t) \text{ and } \rho(AC) = \rho(A^tC)$$

Proof. By Corollary 4.2.28, we can write $A = PJ_rQ$ for some invertible matrices P, Q of size n. Since $\rho(A) = \rho(A^t)$ we see that $A^t = P'J_rQ'$ for some invertible matrices P', Q' of size n. By Corollary 4.2.27, we have $\rho(BA) = \rho(BPJ_rQ) = \rho(BJ_r)$ and $\rho(BA^t) = \rho(BP'J_rQ') = \rho(BJ_r)$. Hence $\rho(BA) = \rho(BA^t)$. The second equality is similarly proved. □

The following theorem relates the information of a linear map and its matrix representation. And we can consider the theorem to be a linear map version of Theorem 2.3.17.

Theorem 4.2.30. *Let V, W be subspaces and let $T : V \to W$ be a linear map. Let $\mathcal{B}$ and $\mathcal{C}$ be bases of V and W respectively and let $A = [T]_{\mathcal{B}}^{\mathcal{C}}$.*

(1) *Then*

$$\rho(T) = \rho(A) \text{ and } \nu(T) = \nu(A) \tag{4.21}$$

(2) **(Rank Theorem)** *We have*

$$\rho(T) + \nu(T) = n \tag{4.22}$$

Proof. (1) Let $\phi_{\mathcal{B}}(\mathbf{v}) = [\mathbf{v}]_{\mathcal{B}}$, $\phi_{\mathcal{C}}(\mathbf{w}) = [\mathbf{w}]_{\mathcal{C}}$ and $A = [T]_{\mathcal{B}}^{\mathcal{C}}$. Then we have a commutative diagram (Theorem 4.2.11)

$$\begin{array}{ccc} V & \xrightarrow{\ T\ } & W \\ {\scriptstyle \phi_{\mathcal{B}}}\downarrow & & \downarrow{\scriptstyle \phi_{\mathcal{C}}} \\ \mathbb{R}^n & \xrightarrow[T_A]{} & \mathbb{R}^m \end{array}$$

i.e., $\phi_{\mathcal{C}} \circ T = T_A \circ \phi_{\mathcal{B}}$. Since $\phi_{\mathcal{B}}$, $\phi_{\mathcal{C}}$ are isomorphisms, by Theorem 4.2.26 (2),(3) we have

$$\rho(T) = \rho(\phi_{\mathcal{C}} \circ T) = \rho(T_A \circ \phi_{\mathcal{B}}) = \rho(T_A) = \rho(A)$$

Similarly, $\nu(T) = \nu(A)$.

(2) The equality (4.22) follows from (4.21) together with the corresponding theorem for the matrices (Theorem 2.3.17). □

Corollary 4.2.31. *Let $T : V \to W$ be a liner map and let $\mathcal{B}, \mathcal{C}$ be bases for V, W. The following conditions are equivalent.*

(1) *The linear map T is surjective (resp. injective, resp. isomorphism).*

(2) *The matrix $[T]_{\mathcal{B}}^{\mathcal{C}}$ has a right (resp. left, resp. two side) inverse.*

(3) *The linear map T has a right (resp. left, resp. two side) linear inverse.*

(4) $\rho(T) = \dim W$ *(resp.* $\rho(T) = \dim V$, *resp.* $\nu(T) = 0$ *and* $\rho(T) = \dim W$*)*

Proof. (1) ⇒ (2) If T is surjective then $\rho(T) = \dim W$ and $\rho(T) = \rho[T]_{\mathcal{B}}^{\mathcal{C}}$ by Theorem 4.2.30. By Theorem 2.4.8, there is a matrix B such that $[T]_{\mathcal{B}}^{\mathcal{C}}B = I$.

(2) ⇒ (3) By Theorem 4.2.16, there is a linear map $S : W \to V$ corresponding to a right inverse B of $[T]_{\mathcal{B}}^{\mathcal{C}}$ such that $T \circ S = I_W$.

(3) ⇒ (1) Obvious.

The equivalence of (1) and (4) follows from the rank theorem. Similar proofs work for the injective maps and the isomorphisms. □

Corollary 4.2.32. *Let V, W be subspaces of the same dimension. Let $T : V \to W$ be a linear map and $\mathcal{B}, \mathcal{C}$ be bases for V, W. Then the following conditions are equivalent.*

(1) *The linear map T is an isomorphism.*

(2) *The linear map T is injective.*

(3) *The linear map T is surjective.*

(4) *The matrix representation $[T]_{\mathcal{B}}^{\mathcal{C}}$ is invertible.*

In this case, we have

$$\left([T]_{\mathcal{B}}^{\mathcal{C}}\right)^{-1} = [T^{-1}]_{\mathcal{C}}^{\mathcal{B}}$$

In particular, if $V = W$, $\mathcal{B} = \mathcal{C}$ then

$$\left([T]_{\mathcal{B}}\right)^{-1} = [T^{-1}]_{\mathcal{B}}$$

Proof. The implication (1) $\Rightarrow$ (2) is obvious. For (2) $\Rightarrow$ (1) we need to show surjectivity of T. For this we use rank theorem: $\rho(T) + \nu(T) = \dim V = \dim W$. Since T is injective we have $\nu(T) = 0$. Hence $\rho(T) = \dim W$; T is surjective also.

The equivalence (1) $\Leftrightarrow$ (3) is similarly proved.

(1) $\Rightarrow$ (4) Now if T is an isomorphism then T is invertible. Since $I_V = T^{-1}T$ we have $I_n = [I_V]_{\mathcal{B}} = [T^{-1}T]_{\mathcal{B}} = [T^{-1}]_{\mathcal{C}}^{\mathcal{B}}[T]_{\mathcal{B}}^{\mathcal{C}}$ by Theorem 4.2.18. Hence $[T]_{\mathcal{B}}^{\mathcal{C}}$ is invertible and $[T^{-1}]_{\mathcal{C}}^{\mathcal{B}} = \left([T]_{\mathcal{B}}^{\mathcal{C}}\right)^{-1}$.

(4) $\Rightarrow$ (1) Now suppose $A = [T]_{\mathcal{B}}^{\mathcal{C}}$ is invertible. Let $\mathcal{B} = \{\mathbf{v}_1, ..., \mathbf{v}_n\}, \mathcal{C} = \{\mathbf{w}_1, ..., \mathbf{w}_n\}$. Let $A^{-1} = B = (b_{ij})$ and define $S : V \to W$ by $S(\mathbf{w}_j) = \sum b_{ij}\mathbf{v}_i$. Then $[S]_{\mathcal{C}}^{\mathcal{B}} = B$ and $[ST]_{\mathcal{B}} = [S]_{\mathcal{C}}^{\mathcal{B}}[T]_{\mathcal{B}}^{\mathcal{C}} = BA = I_n$. Hence $ST = I_V$ by Corollary 4.2.17. □

Example 4.2.33. *Consider the linear map T in* Example 4.2.10. *The rank of T is given by the rank of the matrix $[T]_{\mathcal{B}}^{\mathcal{C}} = \begin{bmatrix} 1 & -1 & 1 \\ 1 & 1 & 0 \end{bmatrix}$ which is 2. Also note that the rank of the matrix in different bases $[T]_{\mathcal{B}'}^{\mathcal{C}'} = \begin{bmatrix} 3 & 0 & 2 \\ 0 & -1 & -1 \end{bmatrix}$ is also 2 as it should be.*

Example 4.2.34. *Let's consider* Example 4.2.9 *again. We had*

$$[T]_{\mathcal{B}}^{\mathcal{C}} = \begin{bmatrix} -1 & 1 \\ 1 & 0 \end{bmatrix}$$

Since $\det[T]_{\mathcal{B}}^{\mathcal{C}} = -1$ *we see that $[T]_{\mathcal{B}}^{\mathcal{C}}$ is invertible and hence T is an isomorphism* (Corollary 4.2.32). *Further we have*

$$[T^{-1}]_{\mathcal{C}}^{\mathcal{B}} = \left([T]_{\mathcal{B}}^{\mathcal{C}}\right)^{-1} = \begin{bmatrix} 0 & 1 \\ 1 & 1 \end{bmatrix}$$

We can use this to find T^{-1} explicitly: The equality above says

$$T^{-1}(\mathbf{w}_1) = \mathbf{v}_2 = \begin{bmatrix} 1 \\ 0 \\ -1 \end{bmatrix} \text{ and } T^{-1}(\mathbf{w}_2) = \mathbf{v}_1 + \mathbf{v}_2 = \begin{bmatrix} 2 \\ -1 \\ -1 \end{bmatrix}$$

Every element of W is of the form $\mathbf{w} = \begin{bmatrix} a+b \\ a \\ b \end{bmatrix} = a\mathbf{w}_1 + b\mathbf{w}_2$. Hence

$$T^{-1}(\mathbf{w}) = aT^{-1}(\mathbf{w}_1) + bT^{-1}(\mathbf{w}_2) = a\begin{bmatrix} 1 \\ 0 \\ -1 \end{bmatrix} + b\begin{bmatrix} 2 \\ -1 \\ -1 \end{bmatrix} = \begin{bmatrix} a+2b \\ -b \\ -a-b \end{bmatrix}$$

Theorem 4.2.35. *Let $A \in M(m \times n)$ with $m \geq n$. The following conditions are equivalent.*

(1) *The matrix linear map $T_A : \mathbb{R}^n \to \mathbb{R}^m$ is injective i.e., $\rho(A) = n$.*

(2) *The matrix $A^t A \in M(n)$ is invertible.*

Proof. (1) $\Rightarrow$ (2) Since T_A is injective the reduced echelon form of A is of the form $\begin{bmatrix} I_n \\ 0 \end{bmatrix}$ (Theorem 1.3.36 and Theorem 2.4.2): Namely, pre-multiplying elementary matrices E_i's we get:

$$E_k E_{k-1} \cdots E_1 A = \begin{bmatrix} 1 & 0 & \cdots & 0 \\ 0 & 1 & 0 & \vdots \\ 0 & 0 & \ddots & 0 \\ 0 & 0 & 0 & 1 \\ 0 & 0 & 0 & 0 \\ \vdots & & & \vdots \\ 0 & 0 & 0 & 0 \end{bmatrix} = \begin{bmatrix} I_n \\ 0 \end{bmatrix} \tag{4.23}$$

Taking the transpose we get

$$A^t E_k^t E_{k-1}^t \cdots E_1^t = \begin{bmatrix} 1 & 0 & 0 & 0 & 0 & \cdots & 0 \\ 0 & 1 & \vdots & 0 & 0 & & 0 \\ \vdots & \vdots & 1 & 0 & 0 & & 0 \\ 0 & 0 & 0 & 1 & 0 & \cdots & 0 \end{bmatrix} = [I_n \ 0] \tag{4.24}$$

Multiplying (4.23) and (4.24) we get

$$A^t E_k^t E_{k-1}^t \cdots E_1^t E_k E_{k-1} \cdots E_1 A = [I_n \ 0] \begin{bmatrix} I_n \\ 0 \end{bmatrix} = \begin{bmatrix} 1 & 0 & 0 & 0 \\ 0 & 1 & 0 & 0 \\ 0 & 0 & \ddots & 0 \\ 0 & 0 & 0 & 1 \end{bmatrix} = I_n \tag{4.25}$$

Let $E = E_k^t E_{k-1}^t \cdots E_1^t E_k E_{k-1} \cdots E_1$. Then $T_{A^t} T_E T_A = I_n$. Since E is invertible T_E is an isomorphism. Hence $n = \rho(I_n) = \rho(T_{A^t} T_E T_A) = \rho(T_{A^t} T_A) = \rho(A^t A)$ by Corollary 4.2.27. Hence $A^t A \in M(n)$ is invertible by Theorem 2.5.14.

(2) $\Rightarrow$ (1) Now suppose $A^t A$ is invertible say, $B(A^t A) = I_n$ for some $B \in M(n)$. Then $T_{BA^t} T_A = I$ i.e., T_A has a left inverse; T_A is injective (Theorem 2.4.10). □

Theorem 4.2.36. *Let $A \in M(m \times n)$ with $m \leq n$. The following conditions are equivalent.*

(1) *The matrix linear map $T_A : \mathbb{R}^n \to \mathbb{R}^m$ is surjective i.e., $\rho(A) = m$.*

(2) *The matrix $AA^t \in M(m)$ is invertible.*

Proof. By Corollary 2.4.3, the condition (1) is equivalent to that T_{A^t} is injective. Now use Theorem 4.2.35. Checking the details is an exercise. □

For more general results than Theorem 4.2.35 and Theorem 4.2.36 see Exercise 8.2.10.

Example 4.2.37. *Let $A = \begin{bmatrix} 1 & 2 & 3 \\ 0 & 1 & 0 \end{bmatrix}$. Then*

$$A^t = \begin{bmatrix} 1 & 0 \\ 2 & 1 \\ 3 & 0 \end{bmatrix} \text{ and } AA^t = \begin{bmatrix} 14 & 2 \\ 2 & 1 \end{bmatrix}$$

Since $|AA^t| = 10$ we see that AA^t is invertible. Hence $T_A : \mathbb{R}^3 \to \mathbb{R}^2$ is surjective. This can be crosschecked by observing $\rho(A) = 2$.

Answers for Practice Problems 4.2

(4.2.1) Let $\mathbf{v}_1, \mathbf{v}_2 \in V$ and $c \in \mathbb{R}$. Let $\mathbf{v}_1 = \sum a_i \mathbf{v}_i$ and $\mathbf{v}_2 = \sum b_i \mathbf{v}_i$. Then $\mathbf{v}_1 + \mathbf{v}_2 = \sum (a_i + b_i)\mathbf{v}_i$. Hence $\phi_{\mathcal{B}}(\mathbf{v}_1 + \mathbf{v}_2) = \begin{bmatrix} a_1 + b_1 \\ \vdots \\ a_n + b_n \end{bmatrix}$. On the other hand, $\phi_{\mathcal{B}}(\mathbf{v}_1) + \phi_{\mathcal{B}}(\mathbf{v}_2) = \begin{bmatrix} a_1 \\ \vdots \\ a_n \end{bmatrix} + \begin{bmatrix} b_1 \\ \vdots \\ b_n \end{bmatrix} = \begin{bmatrix} a_1 + b_1 \\ \vdots \\ a_n + b_n \end{bmatrix}$. Also $\phi_{\mathcal{B}}(c\mathbf{v}_1) = \phi_{\mathcal{B}}(\sum ca_i \mathbf{v}_i) = \begin{bmatrix} ca_1 \\ \vdots \\ ca_n \end{bmatrix} = c \begin{bmatrix} a_1 \\ \vdots \\ a_n \end{bmatrix} = c\phi_{\mathcal{B}}(\mathbf{v}_1)$.

(4.2.2) Let $\mathcal{B} = \{\mathbf{v}_1, \mathbf{v}_2, ..., \mathbf{v}_n\}$ and $\mathcal{C} = \{\mathbf{w}_1, \mathbf{w}_2, ..., \mathbf{w}_n\}$ be bases for V and W. There are unique linear maps $S : V \to W$ and $T : W \to V$ with $S(\mathbf{v}_i) = \mathbf{w}_i$ and $T(\mathbf{w}_i) = \mathbf{v}_i$ by Theorem 4.2.2. Then $T \circ S$ is a linear map satisfying $T \circ S(\mathbf{v}_i) = \mathbf{v}_i$. Since the identity I_V also satisfy the same property $I_V(\mathbf{v}_i) = \mathbf{v}_i$ and there is a unique map satisfying the property we conclude that $T \circ S = I_V$. Similarly we have $S \circ T = I_W$. Hence S is an isomorphism.

(4.2.3) (1) It is easy to show that $\rho(A) = 2$. Now it is immediate that $\mathcal{B}_1, \mathcal{B}_2$ are linearly independent. Hence they are bases for V.

(2) We have $T(\mathbf{v}_3) = \begin{bmatrix} 0 \\ 1 \\ -1 \end{bmatrix}$. Since $\mathbf{v}_1 - \mathbf{v}_2 - \mathbf{v}_3 = \mathbf{0}$ we see that $T(\mathbf{v}_1) - T(\mathbf{v}_2) - T(\mathbf{v}_3) = \mathbf{0}$.

(3) $[T]_{\mathcal{B}_1}^{\mathcal{E}} = \begin{bmatrix} 1 & 0 \\ -1 & 1 \\ 0 & -1 \end{bmatrix}$ and $[T]_{\mathcal{B}_2}^{\mathcal{C}} = \begin{bmatrix} -1 & 1 \\ 2 & -1 \\ -1 & 0 \end{bmatrix}$.

4.2 Exercises

1. Let V, W be the subspaces defined by

$$V = \left\{ \begin{bmatrix} x \\ y \\ z \\ w \end{bmatrix} \in \mathbb{R}^4 \,\middle|\, x - y + w = 0 \right\}, \quad W = \left\{ \begin{bmatrix} x \\ y \\ z \end{bmatrix} \in \mathbb{R}^3 \,\middle|\, x + y + z = 0 \right\}$$

For $\mathbf{v} = \begin{bmatrix} a \\ b \\ c \\ d \end{bmatrix} \in \mathbb{R}^4$ define $T : \mathbb{R}^4 \to \mathbb{R}^3$ by $T(\mathbf{v}) = \begin{bmatrix} 2a - 2b + c \\ b - c \\ -a + d \end{bmatrix}$.

(a) Let $\mathbf{v}_1 = \begin{bmatrix} 1 \\ 1 \\ 0 \\ 0 \end{bmatrix}, \mathbf{v}_2 = \begin{bmatrix} -1 \\ 0 \\ 0 \\ 1 \end{bmatrix}, \mathbf{v}_3 = \begin{bmatrix} 0 \\ 0 \\ 1 \\ 0 \end{bmatrix}$ and $\mathcal{B} = \{\mathbf{v}_1, \mathbf{v}_2, \mathbf{v}_3\}$. Let $\mathbf{w}_1 = \begin{bmatrix} -1 \\ 1 \\ 0 \end{bmatrix}, \mathbf{w}_2 = \begin{bmatrix} -1 \\ 0 \\ 1 \end{bmatrix}$ and $\mathcal{C} = \{\mathbf{w}_1, \mathbf{w}_2\}$. Prove that $\mathcal{B}, \mathcal{C}$ are bases for V, W respectively.

(b) Prove that T maps V to W.

(c) Find $[T]_{\mathcal{B}}^{\mathcal{C}}$ and $\rho(T)$.

2. Prove Theorem 4.2.15.

3. Prove Theorem 4.2.20.

4. Prove (2) and (3) of Theorem 4.2.24.

5. Prove the equality $S^{-1}(N(T)) = N(T \circ S)$ in the proof of Theorem 4.2.26.

6. Deduce Theorem 2.2.13 from Theorem 4.2.26.

7. Let $V \overset{\phi}{\to} V \overset{T}{\to} W \overset{\psi}{\to} W$ be linear maps. Assume ϕ, ψ are isomorphisms.
   (a) Prove that T is injective if and only if $\psi \circ T \circ \phi$ is injective.
   (b) Prove that T is surjective if and only if $\psi \circ T \circ \phi$ is surjective.

8. Find $\rho(A), \rho(B)$ and $\rho(AB)$.
   (a) $A = \begin{bmatrix} 0 & 1 \\ 1 & 1 \\ 1 & 1 \\ 1 & 1 \end{bmatrix}, \quad B = \begin{bmatrix} 1 & 0 & 1 \\ 1 & 1 & -1 \end{bmatrix}$
   (b) $A = \begin{bmatrix} 0 & 1 & -3 \\ 1 & 1 & 2 \\ 1 & 1 & 1 \\ 1 & 1 & 2 \end{bmatrix}, \quad B = \begin{bmatrix} 1 & 0 & 1 \\ 1 & 1 & -1 \\ 3 & 1 & 1 \end{bmatrix}$
   (c) $A = \begin{bmatrix} 1 & -1 & 0 & 1 \\ 0 & 1 & 1 & 0 \\ 1 & 1 & 0 & 0 \end{bmatrix}, \quad B = \begin{bmatrix} 1 & 0 & 1 \\ 3 & 1 & 1 \\ 4 & 1 & 2 \\ 5 & 2 & 1 \end{bmatrix}$
   (d) $A = \begin{bmatrix} 1 & 2 & -1 & 0 & 0 & 1 \\ 0 & 0 & 1 & -1 & 1 & 0 \\ 1 & 1 & 1 & 1 & 0 & 0 \\ 1 & 1 & 2 & 0 & 1 & 0 \end{bmatrix}, \quad B = \begin{bmatrix} 1 & 0 & 1 \\ 2 & 1 & 0 \\ 3 & 1 & 1 \\ 4 & 1 & 2 \\ 5 & 2 & 1 \\ 6 & 2 & 2 \end{bmatrix}$

9. Prove the true statement(s) and find a counter example for the false statement(s).
   (a) For $A, B \in M(n)$ we have $\rho(AB) = \rho(BA)$.
   (b) If V, W are subspaces such that $\dim V \neq \dim W$ then there is no isomorphism between V and W.
   (c) If V, W are subspaces such that $\dim V \neq \dim W$ then there is no isomorphism $T : V \to W$.
   (d) If $A, B \in M(n)$ then $\rho(AB) \leq \rho(A)$.
   (e) If $A, B \in M(n)$ then $\rho(AB) \leq \min(\rho(A), \rho(B))$.
   (f) Let $A \in M(4\times2)$ and $B \in M(2\times4)$. Then $\det(AB) = 0$.
   (g) If $T : V \to W$ is an injective linear map then there is a linear map $S : W \to V$ such that $S \circ T = I_V$.
   (h) Let $A \in M(m\times n)$. If P, Q are invertible then $\rho(PAQ) = \rho(A)$.
   (i) For any $A \in M(m\times n)$ we have $\nu(A) = \nu(A^t)$.

10. Decide whether T_A is injective or surjective. Find the inverses of A^tA or AA^t if exists.

(a) $A = \begin{bmatrix} 1 & 1 \\ 2 & 0 \\ 3 & -1 \\ 4 & 4 \\ 5 & 1 \\ 6 & 0 \end{bmatrix}$ (b) $A = \begin{bmatrix} 1 & 0 & 1 \\ 0 & 2 & 0 \\ 1 & -1 & 1 \\ 0 & 1 & 1 \\ 0 & 1 & 0 \end{bmatrix}$ (c) $A = \begin{bmatrix} 1 & 2 & 3 \\ 2 & 1 & 1 \\ 1 & 1 & 1 \\ 1 & 0 & 1 \\ 2 & 1 & 0 \end{bmatrix}$

(d) $A = \begin{bmatrix} 1 & 2 & 3 & 4 \\ 1 & 0 & 0 & 2 \end{bmatrix}$ (e) $A = \begin{bmatrix} 1 & 1 & 0 & 1 & 1 \\ 1 & 0 & 0 & 1 & 0 \\ 0 & 0 & 1 & 1 & 1 \end{bmatrix}$

11. Deduce Corollary 2.3.15 from Theorem 4.2.26.

12. Let $S, T \in L(V, W)$.

    (a) Prove that $N(S) \cap N(T) \subseteq N(S+T)$.

    (b) Prove that $R(S+T) \subseteq R(S) + R(T)$. (See Definition 1.2.18 for the sum of subspaces.)

4.3 Matrix Representation under Change of Bases

Let $T : V \to W$ be a linear map. Let $\mathcal{B}, \mathcal{C}$ be bases of V, W respectively. We defined the matrix representation $[T]_{\mathcal{B}}^{\mathcal{C}}$ of T with respect to $\mathcal{B}$ and $\mathcal{C}$. If $\mathcal{B}', \mathcal{C}'$ are some other bases for V, W respectively then we get $[T]_{\mathcal{B}'}^{\mathcal{C}'}$. What's the relationship between $[T]_{\mathcal{B}}^{\mathcal{C}}$ and $[T]_{\mathcal{B}'}^{\mathcal{C}'}$?

We first consider a simple case: Let $A \in M(n)$ and $T_A : \mathbb{R}^n \to \mathbb{R}^n$ be the corresponding linear map. Let $\mathcal{B} = \{\mathbf{v}_1, \mathbf{v}_2, ..., \mathbf{v}_n\}$ be a basis of $\mathbb{R}^n$ and $\mathcal{E}$ be the standard basis of $\mathbb{R}^n$. We know that $[T_A]_{\mathcal{E}} = A$ (Example 4.2.8). We want to find $[T_A]_{\mathcal{B}}$. First we have

$$[T_A]_{\mathcal{B}} = \big[[T_A(\mathbf{v}_1)]_{\mathcal{B}}\ [T_A(\mathbf{v}_2)]_{\mathcal{B}}\ \cdots\ [T_A(\mathbf{v}_n)]_{\mathcal{B}}\big] = \big[[A\mathbf{v}_1]_{\mathcal{B}}\ [A\mathbf{v}_2]_{\mathcal{B}}\ \cdots\ [A\mathbf{v}_n]_{\mathcal{B}}\big] \tag{4.26}$$

by Definition 4.2.7. We need to find $[A\mathbf{v}_i]_{\mathcal{B}}$. For any $\mathbf{v}$, let $[A\mathbf{v}]_{\mathcal{B}} = \begin{bmatrix} c_1 \\ \vdots \\ c_n \end{bmatrix}$, say

$$A\mathbf{v} = c_1\mathbf{v}_1 + c_2\mathbf{v}_2 + \cdots + c_n\mathbf{v}_n \tag{4.27}$$

If we let $P = [\mathbf{v}_1\ \mathbf{v}_2\ \cdots \mathbf{v}_n]$ or $P = [I]_{\mathcal{B}}^{\mathcal{E}}$ then P is invertible (Theorem 2.4.5) and the equality (4.27) becomes

$$A\mathbf{v} = P\begin{bmatrix} c_1 \\ \vdots \\ c_n \end{bmatrix} = P[A\mathbf{v}]_{\mathcal{B}} \quad \text{or} \quad [A\mathbf{v}]_{\mathcal{B}} = P^{-1}A\mathbf{v} \tag{4.28}$$

In particular, if $A = I_n$, the identity matrix then

$$[\mathbf{v}]_{\mathcal{B}} = P^{-1}\mathbf{v} \tag{4.29}$$

If we replace $[A\mathbf{v}_i]_{\mathcal{B}} = P^{-1}A\mathbf{v}_i$ of (4.28) in (4.26) then we have

$$[T_A]_{\mathcal{B}} = [P^{-1}A\mathbf{v}_1\ P^{-1}A\mathbf{v}_2 \cdots P^{-1}A\mathbf{v}_n] = P^{-1}A[\mathbf{v}_1\ \mathbf{v}_2 \cdots \mathbf{v}_n] = P^{-1}AP = P^{-1}[T_A]_{\mathcal{E}}P$$

Hence we have shown

Theorem 4.3.1. *Let $A \in M(n)$ and $T_A : \mathbb{R}^n \to \mathbb{R}^n$ be the matrix linear map for A. Let $\mathcal{B} = \{\mathbf{v}_1, \mathbf{v}_2, ..., \mathbf{v}_n\}$ be a basis of $\mathbb{R}^n$ and $\mathcal{E}$ be the standard basis. If we let $P = [\mathbf{v}_1\ \mathbf{v}_2 \cdots \mathbf{v}_n]$ then we have*

$$[T_A]_{\mathcal{B}} = P^{-1}[T_A]_{\mathcal{E}}P = P^{-1}AP \tag{4.30}$$

Here is another way to prove (4.30): First, observe that $P = [I]^{\mathcal{E}}_{\mathcal{B}}$ where I is the identity map on $\mathbb{R}^n$. Since $IT_A = T_A = T_AI$ we have, by Theorem 4.2.18

$$[T_A]^{\mathcal{E}}_{\mathcal{B}} = [IT_A]_{\mathcal{B}} = [I]^{\mathcal{E}}_{\mathcal{B}}[T_A]^{\mathcal{B}}_{\mathcal{B}} = P[T_A]_{\mathcal{B}}$$

$$[T_A]^{\mathcal{E}}_{\mathcal{B}} = [T_AI]_{\mathcal{B}} = [T_A]^{\mathcal{E}}_{\mathcal{E}}[I]^{\mathcal{E}}_{\mathcal{B}} = AP$$

Hence we have the equality (4.30).

Example 4.3.2. *Let $V = \mathbb{R}^2$. Let $\mathbf{v}_1 = \begin{bmatrix} 2 \\ 1 \end{bmatrix}, \mathbf{v}_2 = \begin{bmatrix} -1 \\ 1 \end{bmatrix}$ and $\mathcal{B} = \{\mathbf{v}_1, \mathbf{v}_2\}$. Then $\mathcal{B}$ is a basis of $\mathbb{R}^2$ and let $\mathcal{E} = \{\mathbf{e}^1, \mathbf{e}^2\}$ be the standard basis of $\mathbb{R}^2$. Let $\mathbf{v} = \begin{bmatrix} 4 \\ 5 \end{bmatrix} = [\mathbf{v}]_{\mathcal{E}}$. To find the coordinate $[\mathbf{v}]_{\mathcal{B}}$ of $\mathbf{v}$ with respect to $\mathcal{B}$. Let $[\mathbf{v}]_{\mathcal{B}} = \begin{bmatrix} c_1 \\ c_2 \end{bmatrix}$. Hence*

$$c_1 \begin{bmatrix} 2 \\ 1 \end{bmatrix} + c_2 \begin{bmatrix} -1 \\ 1 \end{bmatrix} = \begin{bmatrix} 4 \\ 5 \end{bmatrix} \text{ or } P \begin{bmatrix} c_1 \\ c_2 \end{bmatrix} = \begin{bmatrix} 2 & -1 \\ 1 & 1 \end{bmatrix} \begin{bmatrix} c_1 \\ c_2 \end{bmatrix} = \begin{bmatrix} 4 \\ 5 \end{bmatrix}$$

where $P = [\mathbf{v}_1\ \mathbf{v}_2] = \begin{bmatrix} 2 & -1 \\ 1 & 1 \end{bmatrix}$. Then P is invertible and $P^{-1} = \frac{1}{3}\begin{bmatrix} 1 & 1 \\ -1 & 2 \end{bmatrix}$. Hence

$$[\mathbf{v}]_{\mathcal{B}} = P^{-1} \begin{bmatrix} 4 \\ 5 \end{bmatrix} = \begin{bmatrix} 3 \\ 2 \end{bmatrix}$$

Let $T : \mathbb{R}^2 \to \mathbb{R}^2$ be the linear map defined by $T(\mathbf{v}) = \begin{bmatrix} a - b \\ 2a + 3b \end{bmatrix}$ for $\mathbf{v} = \begin{bmatrix} a \\ b \end{bmatrix}$. Then $T = T_A$ with $A = \begin{bmatrix} 1 & -1 \\ 2 & 3 \end{bmatrix}$ then

$$[T_A]_{\mathcal{B}} = P^{-1}AP = \frac{1}{3}\begin{bmatrix} 1 & 1 \\ -1 & 2 \end{bmatrix}\begin{bmatrix} 1 & -1 \\ 2 & 3 \end{bmatrix}\begin{bmatrix} 2 & -1 \\ 1 & 1 \end{bmatrix} = \frac{1}{3}\begin{bmatrix} 8 & -1 \\ 13 & 4 \end{bmatrix}$$

We can rewrite (4.29) as

$$P[\mathbf{v}]_{\mathcal{B}} = [\mathbf{v}]_{\mathcal{E}} \text{ where } P = [I]^{\mathcal{E}}_{\mathcal{B}}$$

Hence P changes $\mathcal{B}$-coordinate of $\mathbf{v}$ into the standard coordinate of $\mathbf{v}$. And the columns of P comprised of the vectors of $\mathcal{B}$ namely $\mathbf{v}_1, \mathbf{v}_2$. For this reason we will call P the change of coordinate matrix from $\mathcal{B}$ to the standard basis $\mathcal{E} = \{\mathbf{e}^1, \mathbf{e}^2, ..., \mathbf{e}^n\}$ of $\mathbb{R}^n$ (Definition 4.3.3).

Practice Problem 4.3.1. *Let $\mathbf{v}_1 = \begin{bmatrix} 0 \\ 1 \\ 1 \end{bmatrix}, \mathbf{v}_2 = \begin{bmatrix} 1 \\ 0 \\ 1 \end{bmatrix}, \mathbf{v}_3 = \begin{bmatrix} 1 \\ 1 \\ 0 \end{bmatrix}$ and let $\mathcal{B} = \{\mathbf{v}_1, \mathbf{v}_2, \mathbf{v}_3\}$. Let $A = \begin{bmatrix} 1 & 0 & 1 \\ 0 & -1 & 1 \\ 0 & 2 & 1 \end{bmatrix}$. Prove that $\mathcal{B}$ is a basis for $\mathbb{R}^3$ and find the coordinate of $\mathbf{v} = \begin{bmatrix} 2 \\ 3 \\ 1 \end{bmatrix}$ relative to $\mathcal{B}$. And find $[T_A]_{\mathcal{B}}$.*

We generalize this to the subspaces in place of $\mathbb{R}^n$'s.

Definition 4.3.3. *Let* $\mathcal{B} = \{\mathbf{v}_1, \mathbf{v}_2, ..., \mathbf{v}_n\}$, $\mathcal{C} = \{\mathbf{w}_1, \mathbf{w}_2, ..., \mathbf{w}_n\}$ *be two bases of a subspace* V. *The* **change of coordinate matrix** $P_{\mathcal{B}}^{\mathcal{C}}$ **from** $\mathcal{B}$ **to** $\mathcal{C}$ *is given by* $[I_V]_{\mathcal{B}}^{\mathcal{C}}$.

Hence

$$P_{\mathcal{B}}^{\mathcal{C}} = [I_V]_{\mathcal{B}}^{\mathcal{C}} = \big[[\mathbf{v}_1]_{\mathcal{C}}\ [\mathbf{v}_2]_{\mathcal{C}}\ \cdots\ [\mathbf{v}_n]_{\mathcal{C}}\big] \tag{4.31}$$

and by Theorem 4.2.11 we have

$$[\mathbf{v}]_{\mathcal{C}} = [I_V]_{\mathcal{B}}^{\mathcal{C}}[\mathbf{v}]_{\mathcal{B}} = P_{\mathcal{B}}^{\mathcal{C}}[\mathbf{v}]_{\mathcal{B}} \tag{4.32}$$

The columns of $P_{\mathcal{B}}^{\mathcal{C}}$, being the coordinates of a basis, are linearly independent. Thus $P_{\mathcal{B}}^{\mathcal{C}}$ is invertible. Since

$$P_{\mathcal{C}}^{\mathcal{B}} = [I_V]_{\mathcal{C}}^{\mathcal{B}} = \big[[\mathbf{w}_1]_{\mathcal{B}}\ [\mathbf{w}_2]_{\mathcal{B}}\ \cdots\ [\mathbf{w}_n]_{\mathcal{B}}\big] \tag{4.33}$$

we have $P_{\mathcal{B}}^{\mathcal{C}}P_{\mathcal{C}}^{\mathcal{B}} = [I]_{\mathcal{B}}^{\mathcal{C}}[I]_{\mathcal{C}}^{\mathcal{B}} = [I]_{\mathcal{B}}^{\mathcal{B}} = I_n$ by Theorem 4.2.18. Therefore

$$\left(P_{\mathcal{B}}^{\mathcal{C}}\right)^{-1} = P_{\mathcal{C}}^{\mathcal{B}} \tag{4.34}$$

Example 4.3.4. *Let* $\mathbf{v}_1 = \begin{bmatrix}-9\\1\end{bmatrix}, \mathbf{v}_2 = \begin{bmatrix}-5\\-1\end{bmatrix}; \mathbf{w}_1 = \begin{bmatrix}1\\-4\end{bmatrix}, \mathbf{w}_2 = \begin{bmatrix}3\\-5\end{bmatrix}$. *If we let* $\mathcal{B} = \{\mathbf{v}_1, \mathbf{v}_2\}$, $\mathcal{C} = \{\mathbf{w}_1, \mathbf{w}_2\}$ *then* $\mathcal{B}, \mathcal{C}$ *are bases for* $\mathbb{R}^2$. *To get the change of coordinate matrix* $P_{\mathcal{B}}^{\mathcal{C}}$ *from* $\mathcal{B}$ *to* $\mathcal{C}$ *we need to find* $[\mathbf{v}_1]_{\mathcal{C}} = \begin{bmatrix}x_1\\x_2\end{bmatrix}, [\mathbf{v}_2]_{\mathcal{C}} = \begin{bmatrix}y_1\\y_2\end{bmatrix}$ *i.e., we need solve the equations:*

$$\begin{cases}\mathbf{v}_1 = x_1\mathbf{w}_1 + x_2\mathbf{w}_2\\ \mathbf{v}_2 = y_1\mathbf{w}_1 + y_2\mathbf{w}_2\end{cases}$$

or

$$[\mathbf{w}_1\ \mathbf{w}_2]\begin{bmatrix}x_1\\x_2\end{bmatrix} = \mathbf{v}_1 \text{ and } [\mathbf{w}_1\ \mathbf{w}_2]\begin{bmatrix}y_1\\y_2\end{bmatrix} = \mathbf{v}_2$$

We solve this at once

$$[\mathbf{w}_1\ \mathbf{w}_2|\mathbf{v}_1\ \mathbf{v}_2] = \left[\begin{array}{cc|cc}1 & 3 & -9 & -5\\ -4 & -5 & 1 & -1\end{array}\right] \xrightarrow[\text{operations}]{\text{elementary row}} \left[\begin{array}{cc|cc}1 & 0 & 6 & 4\\ 0 & 1 & -5 & -3\end{array}\right]$$

Therefore $[\mathbf{v}_1]_{\mathcal{C}} = \begin{bmatrix}6\\-5\end{bmatrix}$, $[\mathbf{v}_2]_{\mathcal{C}} = \begin{bmatrix}4\\-3\end{bmatrix}$. *Hence*

$$P_{\mathcal{B}}^{\mathcal{C}} = \begin{bmatrix}6 & 4\\ -5 & -3\end{bmatrix} \text{ and } P_{\mathcal{C}}^{\mathcal{B}} = (P_{\mathcal{B}}^{\mathcal{C}})^{-1} = \tfrac{1}{2}\begin{bmatrix}-3 & 4\\ 5 & 6\end{bmatrix}$$

Let $\mathbf{v} = \begin{bmatrix}1\\1\end{bmatrix}_{\mathcal{E}}$. *Then* $P_{\mathcal{B}}^{\mathcal{E}} = [\mathbf{v}_1\ \mathbf{v}_2] = \begin{bmatrix}-9 & -5\\ 1 & -1\end{bmatrix}$ *and* $P_{\mathcal{B}}^{\mathcal{E}}[\mathbf{v}]_{\mathcal{B}} = [\mathbf{v}]_{\mathcal{E}}$. *Hence* $[\mathbf{v}]_{\mathcal{B}} = (P_{\mathcal{B}}^{\mathcal{E}})^{-1}[\mathbf{v}]_{\mathcal{E}} = \frac{1}{14}\begin{bmatrix}-1 & 5\\ -1 & -9\end{bmatrix}\begin{bmatrix}1\\1\end{bmatrix} = \frac{1}{7}\begin{bmatrix}2\\5\end{bmatrix}$. *And*

$$[\mathbf{v}]_{\mathcal{C}} = P_{\mathcal{B}}^{\mathcal{C}}[\mathbf{v}]_{\mathcal{B}} = \begin{bmatrix}6 & 4\\ -5 & -3\end{bmatrix}\begin{bmatrix}\frac{2}{7}\\ \frac{5}{7}\end{bmatrix} = \frac{1}{7}\begin{bmatrix}32\\-25\end{bmatrix}$$

Practice Problem 4.3.2. *Let* $\mathbf{v}_1 = \begin{bmatrix}1\\1\end{bmatrix}, \mathbf{v}_2 = \begin{bmatrix}-1\\1\end{bmatrix}$ *and* $\mathbf{w}_1 = \begin{bmatrix}1\\3\end{bmatrix}, \mathbf{w}_2 = \begin{bmatrix}1\\4\end{bmatrix}$. *Let* $\mathcal{B} = \{\mathbf{v}_1, \mathbf{v}_2\}$, $\mathcal{C} = \{\mathbf{w}_1, \mathbf{w}_2\}$. *Find* $P_{\mathcal{B}}^{\mathcal{C}}$ *and* $P_{\mathcal{C}}^{\mathcal{B}}$.

Next theorem relates the matrix representations of a linear map in different bases in terms of change of coordinate matrix generalizing Theorem 4.3.1.

Theorem 4.3.5. *Let* $T : V \to W$ *be a linear map between subspaces. Let* $\mathcal{B}, \mathcal{B}'$ *be bases of* V *and* $\mathcal{C}, \mathcal{C}'$ *be bases of* W. *Then*

$$P_{\mathcal{C}'}^{\mathcal{C}}[T]_{\mathcal{B}'}^{\mathcal{C}'} = [T]_{\mathcal{B}}^{\mathcal{C}} P_{\mathcal{B}'}^{\mathcal{B}} \quad \text{or} \quad [T]_{\mathcal{B}'}^{\mathcal{C}'} = P_{\mathcal{C}}^{\mathcal{C}'}[T]_{\mathcal{B}}^{\mathcal{C}} P_{\mathcal{B}'}^{\mathcal{B}} \tag{4.35}$$

Proof. We have $T \circ I_V = I_W \circ T$; $[T \circ I_V]_{\mathcal{B}'}^{\mathcal{C}} = [I_W \circ T]_{\mathcal{B}'}^{\mathcal{C}}$. Therefore, by Theorem 4.2.18 we have

$$[T \circ I_V]_{\mathcal{B}'}^{\mathcal{C}} = [T]_{\mathcal{B}}^{\mathcal{C}}[I_V]_{\mathcal{B}'}^{\mathcal{B}} \quad \text{and} \quad [I_W \circ T]_{\mathcal{B}'}^{\mathcal{C}} = [I_W]_{\mathcal{C}'}^{\mathcal{C}}[T]_{\mathcal{B}'}^{\mathcal{C}'}$$

Hence we get

$$[I_W]_{\mathcal{C}'}^{\mathcal{C}}[T]_{\mathcal{B}'}^{\mathcal{C}'} = [T]_{\mathcal{B}}^{\mathcal{C}}[I_V]_{\mathcal{B}'}^{\mathcal{B}}$$

which gives (4.35) by (4.34). □

Example 4.3.6. *In* Example 4.2.10 *the matrix representations in two different bases are*

$$[T]_{\mathcal{B}}^{\mathcal{C}} = \begin{bmatrix}1 & -1 & 1\\1 & 1 & 0\end{bmatrix}, \quad [T]_{\mathcal{B}'}^{\mathcal{C}'} = \begin{bmatrix}3 & 0 & 2\\0 & -1 & -1\end{bmatrix}$$

We find that the coordinate change matrices are

$$P_{\mathcal{B}'}^{\mathcal{B}} = [[\mathbf{v}_1']_{\mathcal{B}}\ [\mathbf{v}_2']_{\mathcal{B}}\ [\mathbf{v}_3']_{\mathcal{B}}] = \left[\begin{bmatrix}1\\-1\\1\end{bmatrix}\begin{bmatrix}-1\\0\\1\end{bmatrix}\begin{bmatrix}0\\-1\\1\end{bmatrix}\right] = \begin{bmatrix}1 & -1 & 0\\-1 & 0 & -1\\1 & 1 & 1\end{bmatrix}$$

$$P_{\mathcal{C}'}^{\mathcal{C}} = [[\mathbf{w}_1']_{\mathcal{B}}\ [\mathbf{w}_2']_{\mathcal{B}}] = \left[\begin{bmatrix}2\\-1\end{bmatrix}\begin{bmatrix}1\\1\end{bmatrix}\right] = \begin{bmatrix}2 & 1\\-1 & 1\end{bmatrix}$$

Hence

$$P_{\mathcal{C}'}^{\mathcal{C}}[T]_{\mathcal{B}'}^{\mathcal{C}'} = \begin{bmatrix}2 & 1\\-1 & 1\end{bmatrix}\begin{bmatrix}1 & \frac{1}{3} & 1\\1 & -\frac{2}{3} & 0\end{bmatrix} = \begin{bmatrix}3 & 0 & 2\\0 & -1 & -1\end{bmatrix}$$

$$[T]_{\mathcal{B}}^{\mathcal{C}} P_{\mathcal{B}'}^{\mathcal{B}} = \begin{bmatrix}1 & -1 & 1\\1 & 1 & 0\end{bmatrix}\begin{bmatrix}1 & -1 & 0\\-1 & 0 & -1\\1 & 1 & 1\end{bmatrix} = \begin{bmatrix}3 & 0 & 2\\0 & -1 & -1\end{bmatrix}$$

Practice Problem 4.3.3. *Let*

$$V = \left\{\begin{bmatrix}x\\y\\z\end{bmatrix} \in \mathbb{R}^3 \,\middle|\, x - y + 2z = 0\right\}, \quad W = \left\{\begin{bmatrix}x\\y\\z\end{bmatrix} \,\middle|\, 2x - y + z = 0\right\}$$

For $\mathbf{v} = \begin{bmatrix}a\\b\\c\end{bmatrix} \in V$, *define* $T : V \to \mathbb{R}^3$ *by*

$$T(\mathbf{v}) = \begin{bmatrix}a - b\\a - c\\b + c\end{bmatrix}$$

(1) *Show that T maps V to W.*

(2) *Let*

$$\mathcal{B} = \left\{ \mathbf{v}_1 = \begin{bmatrix} 1 \\ 1 \\ 0 \end{bmatrix}, \mathbf{v}_2 = \begin{bmatrix} 2 \\ 0 \\ -1 \end{bmatrix} \right\}, \quad \mathcal{B}' = \left\{ \mathbf{v}_1' = \begin{bmatrix} 1 \\ -1 \\ -1 \end{bmatrix}, \mathbf{v}_2' = \begin{bmatrix} 0 \\ 2 \\ 1 \end{bmatrix} \right\}$$

$$\mathcal{C} = \left\{ \mathbf{w}_1 = \begin{bmatrix} 0 \\ 1 \\ 1 \end{bmatrix}, \mathbf{w}_2 = \begin{bmatrix} -1 \\ -1 \\ 1 \end{bmatrix} \right\}, \quad \mathcal{C}' = \left\{ \mathbf{w}_1' = \begin{bmatrix} 1 \\ 1 \\ -1 \end{bmatrix}, \mathbf{w}_2' = \begin{bmatrix} 1 \\ 2 \\ 0 \end{bmatrix} \right\}$$

Show that $\mathcal{B}, \mathcal{B}'$ and $\mathcal{C}, \mathcal{C}'$ are bases of V and W respectively. Find $[T]_{\mathcal{B}}^{\mathcal{C}}$ and $[T]_{\mathcal{B}'}^{\mathcal{C}'}$.

(3) *Show that $P_{\mathcal{C}'}^{\mathcal{C}}[T]_{\mathcal{B}'}^{\mathcal{C}'} = [T]_{\mathcal{B}}^{\mathcal{C}} P_{\mathcal{B}'}^{\mathcal{B}}$.*

Corollary 4.3.7. *Let $\mathcal{B}$ and $\mathcal{C}$ be two bases of a vector space V. If $T : V \to V$ is a linear map then there is an $n \times n$ invertible matrix $P_{\mathcal{B}}^{\mathcal{C}}$ given by (4.31) satisfying*

$$[\mathbf{v}]_{\mathcal{C}} = P_{\mathcal{B}}^{\mathcal{C}}[\mathbf{v}]_{\mathcal{B}} \text{ and } P_{\mathcal{B}}^{\mathcal{C}} P_{\mathcal{C}}^{\mathcal{B}} = I_n \tag{4.36}$$

$$[T]_{\mathcal{C}} P_{\mathcal{B}}^{\mathcal{C}} = [T]_{\mathcal{B}}^{\mathcal{C}} = P_{\mathcal{B}}^{\mathcal{C}}[T]_{\mathcal{B}} \text{ or } [T]_{\mathcal{C}} = \left(P_{\mathcal{C}}^{\mathcal{B}}\right)^{-1} [T]_{\mathcal{B}} P_{\mathcal{C}}^{\mathcal{B}} \tag{4.37}$$

Proof. Simply take $\mathcal{C}$ of Theorem 4.3.5 to be our $\mathcal{B}$ and $\mathcal{B}', \mathcal{C}'$ to be our $\mathcal{C}$. □

Definition 4.3.8. *Two matrices $A, B \in M(n)$ are* **similar** *if there is an invertible matrix P such that $B = P^{-1}AP$. In this case, we write $A \sim B$.*

Thus the matrix representations of a linear map in different bases are similar.

Corollary 4.3.9. *Let $T : V \to V$ be a linear map. If $\mathcal{B}, \mathcal{B}'$ are bases of V then*

$$\det [T]_{\mathcal{B}} = \det [T]_{\mathcal{B}'} \quad \text{and} \quad \mathrm{Tr}\, [T]_{\mathcal{B}} = \mathrm{Tr}\, [T]_{\mathcal{B}'}$$

Proof. By Theorem 4.3.5, we have $[T]_{\mathcal{B}'} = (P_{\mathcal{B}'}^{\mathcal{B}})^{-1}[T]_{\mathcal{B}} P_{\mathcal{B}'}^{\mathcal{B}}$. By Corollary 3.3.12, we have $\det[T]_{\mathcal{B}} = \det[T]_{\mathcal{B}'}$. They have the same trace by Corollary 3.3.21. □

Example 4.3.10. *In* Example 4.3.2, *$T = T_A$ with $A = \begin{bmatrix} 1 & -1 \\ 2 & 3 \end{bmatrix}$. Hence* $\det [T_A]_{\mathcal{E}} = \det A =$ 5 *and* $\mathrm{Tr}\, [T_A]_{\mathcal{E}} = \mathrm{Tr}\,(A) = 4$. *On the other hand, $[T_A]_{\mathcal{B}} = \frac{1}{3}\begin{bmatrix} 8 & -1 \\ 13 & 4 \end{bmatrix}$. Hence* $\det [T_A]_{\mathcal{B}} = 5$ *and* $\mathrm{Tr}\, [T_A]_{\mathcal{B}} = 4$.

Let $T : V \to V$ be a linear map and $\mathcal{B}, \mathcal{C}$ be bases for V. By Corollary 4.3.9, $\det[T]_{\mathcal{B}} = \det[T]_{\mathcal{C}}$ and $\mathrm{Tr}\,[T]_{\mathcal{B}} = \mathrm{Tr}\,[T]_{\mathcal{C}}$. Hence the following definition is legitimate.

Definition 4.3.11. *Let V be a subspace with a basis $\mathcal{B}$ and $T : V \to V$ be a linear map. We define the* **determinant** *and the* **trace** *of the linear map T by*

$$\det(T) = \det [T]_{\mathcal{B}} \quad \text{and} \quad \mathrm{Tr}\,(T) = \mathrm{Tr}\,[T]_{\mathcal{B}}$$

Practice Problem 4.3.4. *Let $A = \begin{bmatrix} 1 & 2 & 0 \\ -2 & 2 & 1 \\ -2 & 4 & 0 \end{bmatrix}$. Let $\mathcal{B} = \{\mathbf{v}_1, \mathbf{v}_2, \mathbf{v}_3\}$ where $\mathbf{v}_1 = \begin{bmatrix} 1 \\ 0 \\ 2 \end{bmatrix}, \mathbf{v}_2 = \begin{bmatrix} 2 \\ 1 \\ 4 \end{bmatrix}, \mathbf{v}_3 = \begin{bmatrix} -1 \\ 0 \\ 1 \end{bmatrix}$ and $P = [\mathbf{v}_1\ \mathbf{v}_2\ \mathbf{v}_3]$. Show that $[T_A]_{\mathcal{B}} = P^{-1}AP$. Also show that $[T_A]_{\mathcal{B}}$ and $[T_A]_{\mathcal{E}} = A$ have the same trace and determinant.*

Answers for Practice Problems 4.3

(4.3.1) Let $P = \begin{bmatrix} 0 & 1 & 1 \\ 1 & 0 & 1 \\ 1 & 1 & 0 \end{bmatrix}$. Then $[\mathbf{v}]_{\mathcal{B}} = P^{-1}[\mathbf{v}]_{\mathcal{E}}$. Now $P^{-1} = \frac{1}{2}\begin{bmatrix} -1 & 1 & 1 \\ 1 & -1 & 1 \\ 1 & 1 & -1 \end{bmatrix}$.

Hence $[\mathbf{v}]_{\mathcal{B}} = P^{-1}[\mathbf{v}]_{\mathcal{E}} = \begin{bmatrix} 1 \\ 0 \\ 2 \end{bmatrix}$. And $[T_A]_{\mathcal{B}} = P^{-1}AP = \begin{bmatrix} 1 & 0 & 0 \\ 2 & 1 & 2 \\ -1 & 1 & 2 \end{bmatrix}$

(4.3.2) $[\mathbf{w}_1\ \mathbf{w}_2|\mathbf{v}_1\ \mathbf{v}_2] = \begin{bmatrix} 1 & 1 & 1 & -1 \\ 3 & 4 & 1 & 1 \end{bmatrix}$ reduces to $\begin{bmatrix} 1 & 0 & 3 & -5 \\ 0 & 1 & -2 & 4 \end{bmatrix}$. Hence $P_{\mathcal{B}}^{\mathcal{C}} = \begin{bmatrix} 3 & -5 \\ -2 & 4 \end{bmatrix}$ and $P_{\mathcal{C}}^{\mathcal{B}} = (P_{\mathcal{B}}^{\mathcal{C}})^{-1} = \frac{1}{2}\begin{bmatrix} 4 & 5 \\ 2 & 3 \end{bmatrix}$.

(4.3.3) (1) We need to show $T(\mathbf{v}) \in W$ for $\begin{bmatrix} a \\ b \\ c \end{bmatrix} \in V$. In fact, $2(a-b)-(a-c)+(b+c) = a-b+2c = 0$ since $\mathbf{v} \in V$.

(2) To see these are bases first check that they belong to V and W respectively. And then we can easily check that they are linearly independent. Since $\dim V = \dim W = 2$ they are bases.

To find the matrix representations: $T(\mathbf{v}_1) = \begin{bmatrix} 0 \\ 1 \\ 1 \end{bmatrix} = \mathbf{w}_1; T(\mathbf{v}_2) = \begin{bmatrix} 2 \\ 3 \\ -1 \end{bmatrix} = \mathbf{w}_1 - 2\mathbf{w}_2$.

Hence $[T]_{\mathcal{B}}^{\mathcal{C}} = \begin{bmatrix} 1 & 1 \\ 0 & -2 \end{bmatrix}$. Now $T(\mathbf{v}_1') = \begin{bmatrix} 2 \\ 2 \\ -2 \end{bmatrix} = 2\mathbf{w}_1'; T(\mathbf{v}_2') = \begin{bmatrix} -2 \\ -1 \\ 3 \end{bmatrix} = -3\mathbf{w}_1' + \mathbf{w}_2'$. Hence $[T]_{\mathcal{B}'}^{\mathcal{C}'} = \begin{bmatrix} 2 & -3 \\ 0 & 1 \end{bmatrix}$.

(3) To find $P_{\mathcal{C}'}^{\mathcal{C}}$ we compute $\mathbf{v}_1' = \begin{bmatrix} 1 \\ -1 \\ -1 \end{bmatrix} = -\mathbf{v}_1 + \mathbf{v}_2; \mathbf{v}_2' = \begin{bmatrix} 0 \\ 2 \\ 1 \end{bmatrix} = 2\mathbf{v}_1 - \mathbf{v}_2$. Hence $P_{\mathcal{C}'}^{\mathcal{C}} = \begin{bmatrix} -1 & 2 \\ 1 & -1 \end{bmatrix}$. For $P_{\mathcal{B}'}^{\mathcal{B}}$, we compute $\mathbf{w}_1' = \begin{bmatrix} 1 \\ 1 \\ -1 \end{bmatrix} = \mathbf{w}_2; \mathbf{w}_2' = \begin{bmatrix} 1 \\ 2 \\ 0 \end{bmatrix} = \mathbf{w}_1 - \mathbf{w}_2$. Hence $P_{\mathcal{B}'}^{\mathcal{B}} = \begin{bmatrix} 0 & 1 \\ -1 & -1 \end{bmatrix}$ and we have

$$P_{\mathcal{C}'}^{\mathcal{C}}[T]_{\mathcal{B}'}^{\mathcal{C}'} = \begin{bmatrix} 0 & 1 \\ -1 & -1 \end{bmatrix}\begin{bmatrix} 2 & -3 \\ 0 & 1 \end{bmatrix} = \begin{bmatrix} 0 & 1 \\ -2 & 2 \end{bmatrix} \quad \text{and}$$
$$[T]_{\mathcal{B}}^{\mathcal{C}}P_{\mathcal{B}'}^{\mathcal{B}} = \begin{bmatrix} 1 & 1 \\ 0 & -2 \end{bmatrix}\begin{bmatrix} -1 & 2 \\ 1 & -1 \end{bmatrix} = \begin{bmatrix} 0 & 1 \\ -2 & 2 \end{bmatrix}$$

(4.3.4) We have

$$T_A(\mathbf{v}_1) = -\frac{1}{3}\mathbf{v}_1 - \frac{4}{3}\mathbf{v}_3,\ T_A(\mathbf{v}_2) = -\frac{8}{3}\mathbf{v}_1 + 2\mathbf{v}_2 - \frac{8}{3}\mathbf{v}_3,\ T(\mathbf{v}_3) = -\frac{17}{3}\mathbf{v}_1 + 3\mathbf{v}_2 + \frac{4}{3}\mathbf{v}_3$$

Hence $[T_A]_{\mathcal{B}} = \dfrac{1}{3}\begin{bmatrix} -1 & -8 & -17 \\ 0 & 6 & 9 \\ -4 & -8 & 4 \end{bmatrix}$. We have $P = \begin{bmatrix} 1 & 2 & -1 \\ 0 & 1 & 0 \\ 2 & 4 & 1 \end{bmatrix}$ and $P^{-1} = \dfrac{1}{8}\begin{bmatrix} 4 & 0 & -2 \\ 2 & 0 & 1 \\ 4 & 8 & -6 \end{bmatrix}$.

Hence $P^{-1}AP = \frac{1}{3}\begin{bmatrix} -1 & -8 & -17 \\ 0 & 6 & 9 \\ -4 & -8 & 4 \end{bmatrix}$. We see that $[T_A]_{\mathcal{B}} = P^{-1}AP$ and both have determinant -8 and trace 3.

4.3 Exercises

1. Let $V = \left\{ \begin{bmatrix} x \\ y \\ z \end{bmatrix} \in \mathbb{R}^3 \,\middle|\, x - y + 2z = 0 \right\}$. For $\mathbf{v} = \begin{bmatrix} a \\ b \\ c \end{bmatrix} \in V$ define $T : V \to \mathbb{R}^3$ by

$$T(\mathbf{v}) = \begin{bmatrix} a-b \\ a-c \\ b-c \end{bmatrix}$$

   (a) Show that

$$\mathcal{B} = \left\{ \mathbf{v}_1 = \begin{bmatrix} 1 \\ 1 \\ 0 \end{bmatrix}, \mathbf{v}_2 = \begin{bmatrix} 2 \\ 0 \\ -1 \end{bmatrix} \right\}, \quad \mathcal{B}' = \left\{ \mathbf{v}_1' = \begin{bmatrix} 1 \\ -1 \\ -1 \end{bmatrix}, \mathbf{v}_2' = \begin{bmatrix} 0 \\ 2 \\ 1 \end{bmatrix} \right\}$$

   are bases of V.

   (b) Let $\mathcal{E} = \{\mathbf{e}^1, \mathbf{e}^2, \mathbf{e}^3\}$ be the standard basis of $\mathbb{R}^3$ and let

$$\mathcal{E}' = \left\{ \bar{\mathbf{e}}^1 = \begin{bmatrix} 1 \\ 0 \\ 0 \end{bmatrix}, \bar{\mathbf{e}}^2 = \begin{bmatrix} 1 \\ 1 \\ 0 \end{bmatrix}, \bar{\mathbf{e}}^3 = \begin{bmatrix} 1 \\ 1 \\ 1 \end{bmatrix} \right\}$$

   be a basis of $\mathbb{R}^3$. Find $[T]^{\mathcal{E}}_{\mathcal{B}}$ and $[T]^{\mathcal{E}'}_{\mathcal{B}'}$.

   (c) Prove that $P^{\mathcal{E}}_{\mathcal{E}'}[T]^{\mathcal{E}'}_{\mathcal{B}'} = [T]^{\mathcal{E}}_{\mathcal{B}} P^{\mathcal{B}}_{\mathcal{B}'}$.

2. Let

$$W = \left\{ \begin{bmatrix} x \\ y \\ z \end{bmatrix} \,\middle|\, x - y + z = 0 \right\}$$

   (a) Show that the map T of Exercise 1 maps V to W.

   (b) Let

$$\mathcal{C} = \left\{ \mathbf{w}_1 = \begin{bmatrix} 0 \\ 1 \\ 1 \end{bmatrix}, \mathbf{w}_2 = \begin{bmatrix} 1 \\ 1 \\ 0 \end{bmatrix} \right\}, \quad \mathcal{C}' = \left\{ \mathbf{w}_1' = \begin{bmatrix} 1 \\ 2 \\ 1 \end{bmatrix}, \mathbf{w}_2' = \begin{bmatrix} 1 \\ -1 \\ -2 \end{bmatrix} \right\}$$

   Show that $\mathcal{C}, \mathcal{C}'$ are bases for W and find $[T]^{\mathcal{C}}_{\mathcal{B}}$ and $[T]^{\mathcal{C}'}_{\mathcal{B}'}$.

   (c) Show that $P^{\mathcal{C}}_{\mathcal{C}'}[T]^{\mathcal{C}'}_{\mathcal{B}'} = [T]^{\mathcal{C}}_{\mathcal{B}} P^{\mathcal{B}}_{\mathcal{B}'}$.

3. Let $A, B \in M(n)$ are similar matrices.

   (a) Prove that there is a basis $\mathcal{B}$ of $\mathbb{R}^n$ such that $[T_A]_{\mathcal{B}} = B$.

(b) Prove that for any subspace V of dimension n there are bases $\mathcal{B}, \mathcal{B}'$ of V and a linear map $T : V \to V$ such that $[T]_{\mathcal{B}} = A$ and $[T]_{\mathcal{B}'} = B$.

4. Prove that similarity is an equivalence relations on $M(n)$. That is $A \sim A$ and $B \sim A$ if $A \sim B$. Also if $A \sim B$ and $B \sim A$ then $A \sim C$.

5. Prove the true statement(s) and find a counter example for the false statement(s).

   (a) Let $T : V \to V$ be a linear map. If $\mathcal{B}, \mathcal{B}'$ are two distinct bases then the matrix representations $[T]_{\mathcal{B}}$ and $[T]_{\mathcal{B}'}$ are always distinct.

   (b) If $T : V \to V$ is an isomorphism then there are two distinct bases $\mathcal{B}, \mathcal{B}'$ such that $[T]_{\mathcal{B}}$ and $[T]_{\mathcal{B}'}$ are distinct.

   (c) Let $A, B \in M(n)$. If there is an invertible matrix P such that $P^{-1}AP = P^{-1}BP$ then $A = B$.

   (d) Let $A, B \in M(n)$. If A, B are similar then they are matrix representations of a linear operator of $\mathbb{R}^n$ in two different bases.

   (e) Let $T : V \to W$ be a linear map between subspaces. Let $\mathcal{B}, \mathcal{B}'$ be bases of V and $\mathcal{C}, \mathcal{C}'$ be bases of W. Then $[T]_{\mathcal{B}'}^{\mathcal{C}'}$ and $[T]_{\mathcal{B}}^{\mathcal{C}}$ are similar.

   (f) If $P_{\mathcal{B}}^{\mathcal{C}}$ is the change of coordinate matrix from $\mathcal{B}$ to $\mathcal{C}$ then $P_{\mathcal{B}}^{\mathcal{C}}[\mathbf{v}]_{\mathcal{B}} = [\mathbf{v}]_{\mathcal{C}}$.

Chapter 5

Diagonalization and Geometry of $\mathbb{R}^n$

5.1 Eigenvectors and Diagonalization

Let $A \in M(n)$ and let $T_A : \mathbb{R}^n \to \mathbb{R}^n$ be a matrix linear map for A. Suppose there is a basis $\mathcal{B} = \{\mathbf{v}_i\}$ of $\mathbb{R}^n$ such that $A\mathbf{v}_i = \lambda_i \mathbf{v}_i$ for some constant λ_i. Then the matrix representation $[T_A]_{\mathcal{B}}$ is the diagonal diag$[\lambda_1, \lambda_2, ..., \lambda_n]$. In this case, we say A is diagonalizable.

Consider $[A] = \{[T_A]_{\mathcal{B}} \mid \mathcal{B} \text{ is a basis of } \mathbb{R}^n\}$. It consists of the matrix representations of the linear map T_A in all possible bases of $\mathbb{R}^n$. If A is diagonalizable then both A and diag$[\lambda_1, \lambda_2, ...\lambda_n]$ belong to $[A]$ and the diagonal matrix is a something special one. Hence we can consider diag$[\lambda_1, \lambda_2, ...\lambda_n]$ as a representative of $[A]$. However, as we will see soon, not all square matrices are diagonalizable. We will find conditions for A to be diagonalizable. In Chapter 7, we will show how we choose a representative when A is not diagonalizable. In this section, we investigate the conditions on A for which A is diagonalizable.

5.1.1 Eigenvectors and Eigenvalues

Definition 5.1.1. *Let $A \in M(n)$. A* <u>*nonzero*</u> *vector $\mathbf{v} \in \mathbb{R}^n$ is an* **eigenvector for** *A (or for T_A) if $T_A(\mathbf{v}) = \lambda \mathbf{v}$ for some scalar $\lambda \in \mathbb{R}$. The scalar λ is called the* **eigenvalue** *for A (or* **for** *T_A) and $\mathbf{v}$ is called an eigenvector corresponding to λ.*

In terms of the linear map $T_A : \mathbb{R}^n \to \mathbb{R}^n$, an eigenvector $\mathbf{v}$ with eigenvalue λ is 'stretched' or 'shrank' λ-times by the matrix linear transformation T_A.

If $\mathbf{v}$ is an eigenvector with the eigenvalue λ then so is $c\mathbf{v}$ for a nonzero scalar c. Hence if λ is an eigenvalue then there are infinitely many eigenvectors for λ. But $\{\mathbf{v}, c\mathbf{v}\}$ is linearly dependent. We want to find 'basically different' – linearly independent eigenvectors for λ as many as we can.

Example 5.1.2. (1) *Let* $A = \begin{bmatrix} 1 & 7 \\ 4 & -2 \end{bmatrix}$, $\mathbf{u} = \begin{bmatrix} 7 \\ 4 \end{bmatrix}$, $\mathbf{v} = \begin{bmatrix} 1 \\ -1 \end{bmatrix}$. *Then* $T_A(\mathbf{u}) = 5\mathbf{u}$ *and* $T_A(\mathbf{v}) = -6\mathbf{v}$. *Hence* $\mathbf{u}$ *is an eigenvector with the eigenvalue* $\lambda = 5$ *and* $\mathbf{v}$ *is an eigenvector corresponding to the eigenvalue* $\lambda = -6$.

Now $\mathcal{B} = \{\mathbf{u}, \mathbf{v}\}$ *is a basis of* $V = \mathbb{R}^2$. *The matrix representation of the linear map* $T_A : V \to V$ *in the basis* $\mathcal{B}$ *is* $[T_A]_{\mathcal{B}} = \begin{bmatrix} 5 & 0 \\ 0 & -6 \end{bmatrix}$. *If we let* $P = [I_V]^{\mathcal{E}}_{\mathcal{B}} = \begin{bmatrix} 7 & 4 \\ 1 & -1 \end{bmatrix} = [\mathbf{u}\ \mathbf{v}]$ *then*

$[T_A]_{\mathcal{B}} = P^{-1}[T_A]_{\mathcal{E}}P$ *by* Theorem 4.3.7.

(2) *Let* $A = \begin{bmatrix} \cos\theta & -\sin\theta \\ \sin\theta & \cos\theta \end{bmatrix}$. *Then* T_A *represents the rotation of the vectors of* $\mathbb{R}^2$ *by the angle* θ. *Hence every vector changes its direction by* T_A. *Hence* T_A *has no eigenvectors unless* $\theta = 0$ *or* $\theta = \pi$.

Theorem 5.1.3. *Let* $A \in M(n)$. *The following conditions are equivalent.*

(1) *A real number* λ *is an eigenvalue of* $A \in M(n)$

(2) *There is a nonzero* $\mathbf{v}$ *such that* $A\mathbf{v} = \lambda\mathbf{v}$.

(3) *A linear equation* $(A - \lambda I)(\mathbf{x}) = 0$ *has a nontrivial solution i.e.,* $N(A - \lambda I) \neq \mathbf{0}$.

(4) $\det(A - \lambda I) = 0$

In this case, the eigenvectors for the eigenvalue λ *are the nontrivial solutions for the equation* $(A - \lambda I)\mathbf{x} = \mathbf{0}$.

Proof. Exercise 1. □

Corollary 5.1.4. *A matrix* $A \in M(n)$ *has an eigenvalue* $\lambda = 0$ *if and only if* A *is not invertible.*

Proof. In fact, $\lambda = 0$ is an eigenvalue of A if and only if $N(A) \neq \mathbf{0}$ by Theorem 5.1.3. This is equivalent to that A is not invertible (Theorem 2.5.14 (3)). □

Definition 5.1.5. *For an eigenvalue* λ *of* $A \in M(n)$ *let*

$$E^A_\lambda = \{\mathbf{v} \in \mathbb{R}^n | A\mathbf{v} = \lambda\mathbf{v}\} \tag{5.1}$$

which we call the **eigenspace** *of* A *corresponding to* λ. *Often, we will simply write* E_λ *for* E^A_λ *if the reference to* A *is clear.*

If λ is an eigenvalue of $A \in M(n)$ then $E_\lambda = N(A - \lambda I)$. Since E_λ is a null space of a linear map it is a subspace of $\mathbb{R}^n$. Note that

$$E_\lambda = \{\text{eigenvectors for } \lambda\} \cup \{\mathbf{0}\}$$

If λ and μ are distinct eigenvalues for $A \in M(n)$ then

$$E_\lambda \cap E_\mu = \{\mathbf{0}\} \quad (\lambda \neq \mu) \tag{5.2}$$

In fact, if $\mathbf{v} \in E_\lambda \cap E_\mu$ then $\lambda\mathbf{v} = A\mathbf{v} = \mu\mathbf{v}$. Hence $(\lambda - \mu)\mathbf{v} = \mathbf{0}, (\lambda - \mu) \neq 0$. Thus $\mathbf{v} = \mathbf{0}$.

Also notice that

$$\dim E_\lambda = \nu(A - \lambda I) \tag{5.3}$$

and $E_\lambda \neq \mathbf{0}$ if and only if λ is an eigenvalue of A.

Theorem 5.1.6. *Let* $A \in M(n)$ *and* $\lambda_1, \lambda_1, ..., \lambda_k$ *are distinct eigenvalues of* A *with corresponding eigenvectors* $\mathbf{v}_1, \mathbf{v}_2, ..., \mathbf{v}_k$. *Then* $\{\mathbf{v}_1, \mathbf{v}_2, ..., \mathbf{v}_k\}$ *is linearly independent.*

Proof. Induction on k: If $k = 1$ then $\{\mathbf{v}_1\}$ is a single nonzero vector which is linearly independent. Suppose $k > 1$ and assume our result for $k - 1$. We need to prove the set of eigenvectors $\{\mathbf{v}_1, \mathbf{v}_2, ..., \mathbf{v}_k\}$ is linearly independent. Assume the contrary: there is a nontrivial linear relation

$$c_1\mathbf{v}_1 + c_2\mathbf{v}_2 + \cdots + c_k\mathbf{v}_k = \mathbf{0} \tag{5.4}$$

Now apply $A - \lambda_k I$ to (5.4):

$$c_1(\lambda_1 - \lambda_k)\mathbf{v}_1 + c_2(\lambda_2 - \lambda_k)\mathbf{v}_2 + \cdots + c_{k-1}(\lambda_{k-1} - \lambda_k)\mathbf{v}_{k-1} = \mathbf{0}$$

Since λ_i's are distinct and $\{\mathbf{v}_1, \mathbf{v}_2, ..., \mathbf{v}_{k-1}\}$ is linearly independent by induction assumption, we conclude that $c_1 = c_2 = \cdots = c_{k-1} = 0$. Hence $c_k\mathbf{v}_k = 0$ also by (5.4). Therefore all c_i's are zero. □

Corollary 5.1.7. *If $A \in M(n)$ has n distinct eigenvalues $\lambda_1, \lambda_2, ..., \lambda_n$ then there is a basis $\mathcal{B}$ consisting of the eigenvectors.*

Proof. If $\mathcal{B} = \{\mathbf{v}_1, ..., \mathbf{v}_n\}$ are eigenvectors corresponding to the distinct eigenvalues $\lambda_1, ..., \lambda_n$ then $\mathcal{B}$ is linearly independent. By Theorem 1.3.36 (3) or Theorem 4.1.9, $\mathcal{B}$ is a basis of $\mathbb{R}^n$ consisting of eigenvectors. □

Recall the definition of the sum of subspaces (Definition 1.2.18): If $V_1, V_2, ..., V_k$ are subspaces of $\mathbb{R}^n$ then we define

$$\sum_{i=1}^{k} V_i = \{\mathbf{v}_1 + \mathbf{v}_2 + \cdots + \mathbf{v}_k \mid \mathbf{v}_i \in V_i\}$$

Corollary 5.1.8. *Let $A \in M(n)$ and $\lambda_1, \lambda_2, ..., \lambda_k$ be the distinct eigenvalues of A and $E_{\lambda_1}, E_{\lambda_2}, ..., E_{\lambda_k}$ be the corresponding eigenspaces. If $\mathbf{v}_i \in E_{\lambda_i}$ and $\mathbf{v}_1 + \mathbf{v}_2 + \cdots + \mathbf{v}_k = \mathbf{0}$ then $\mathbf{v}_i = \mathbf{0}$ for all i. Further we have*

$$E_{\lambda_i} \cap \sum_{j \neq i} E_{\lambda_j} = \{\mathbf{0}\} \tag{5.5}$$

Proof. Suppose $\{\mathbf{v}_1, \mathbf{v}_2, ..., \mathbf{v}_j\}$ are nonzero vectors in the sum $\sum \mathbf{v}_i = \mathbf{0}$. By Theorem 5.1.6, $\{\mathbf{v}_1, \mathbf{v}_2, ..., \mathbf{v}_j\}$ is linearly independent. But $\mathbf{v}_1 + \mathbf{v}_2 + \cdots + \mathbf{v}_j = \mathbf{0}$ is a linear relation among $\{\mathbf{v}_1, \mathbf{v}_2, ..., \mathbf{v}_j\}$ which contradicts to linear independence of $\{\mathbf{v}_i\}$.

For the last statement, let $\mathbf{v}_i \in E_{\lambda_i} \cap \sum_{j \neq i} E_{\lambda_j}$. Then we can write $\mathbf{v}_i = \sum_{j \neq i} \mathbf{v}_j$; $\mathbf{v}_i - \sum_{j \neq i} \mathbf{v}_j = \mathbf{0}$. Hence all $\mathbf{v}_i$'s are zero by the previous statement. □

Notice that the equality (5.5) implies (5.2).

5.1.2 Characteristic Polynomial

A real number λ is an eigenvalue of A if and only if $\det(A - \lambda I) = 0$ (Theorem 5.1.3). Therefore the characteristic values are the roots of the polynomial $f(t) = \det(A - tI) = 0$. This motivates the following definition which we already saw in Definition 3.4.22:

Definition 5.1.9. *Let $A \in M(n)$. The polynomial*

$$f_A(t) = \det(A - tI)$$

in t is called the **characteristic polynomial** *of A. If $f_A(t) = (t - \lambda)^e g(t)$ with $g(\lambda) \neq 0$ then we say λ is an eigenvalue of A with* **multiplicity** *e. We say that a polynomial* **splits** *if it is a product of polynomials of degree 1 i.e., it is of the form $f(t) = a\prod(t - \lambda_i)^{e_i}$.*

We first remark that the determinant $\det(A - tI)$ is a determinant of a matrix which contains not only the real numbers but also the variable t. In Appendix B.2, we remarked that the determinant of a matrix with the entries of polynomials in t can be defined as we did for the real matrices. The matrices with real entries are then a special case where the entries of the matrices are the constant polynomials.

The second remark we want to make is that *in this chapter we consider only the real polynomials*

$$a_0 + a_1 t + \cdots + a_k t^k \ (a_i \in \mathbb{R},\ k \geq 0)$$

If $a_k \neq 0$ then we say the **degree** of f is k. The set of all real polynomials will be denoted by $\mathbb{R}[t]$. Hence when we say $f(t)$ is the product of two polynomials $f(t) = g(t)h(t)$ we assume $g, h \in \mathbb{R}[t]$. In this case, we say f is **divisible** by g, h and we write $g|f,\ h|f$. And in the expression $f_A(t) = (t-\lambda)^e g(t)$ we assume $\lambda \in \mathbb{R}$ and $g(t) \in \mathbb{R}[t]$. (See Appendix C 1.)

Theorem 5.1.10. *Let $A \in M(n)$. Then the characteristic polynomial $f_A(t)$ of A is a polynomial of degree n with the leading coefficient $(-1)^n$. The constant term of $f_A(t)$ is* $\det(A)$.

Proof. The characteristic polynomial is given by

$$f_A(t) = \det \begin{bmatrix} a_{11} - t & a_{12} & \cdots & a_{1n} \\ a_{21} & a_{22} - t & \cdots & a_{2n} \\ & \cdots & \cdots & \\ a_{n1} & \cdots & a_{n\,n-1} & a_{nn} - t \end{bmatrix}$$

The highest degree term in t is obtained by multiplying the diagonal entries: $\prod(a_{ii} - t) = (-1)^n t^n +$(lower degree terms) (See §3.3.4.) Hence the leading coefficient of $f_A(t)$ is $(-1)^n$. The constant term is then $f_A(0) = \det(A - 0I) = \det(A)$. □

It will turn out that the coefficient of t^{n-1} is $(-1)^{n-1}\mathrm{Tr}(A)$ (Theorem 6.7.8). Also notice that the polynomial $f_A(t)$ may not have a (real) root as in $f_A(t) = t^2 + 1$ where $A = \begin{bmatrix} 0 & 1 \\ -1 & 0 \end{bmatrix}$.

Theorem 5.1.11. *Let $A \in M(n)$. The eigenvalues are precisely the (real) roots of the characteristic polynomial $f_A(t)$. In particular, A has at most n (real) eigenvalues.*

If A is a triangular matrix then the eigenvalues of A are the entries on the diagonal.

Proof. Eigenvalues are the roots of $f_A(t)$ which is a polynomial of degree n. Hence $f_A(t)$ has at most n roots by Theorem C.1.4.

If A is triangular then $\det(A - tI) = \prod(a_{ii} - t)$. Hence the roots of the characteristic polynomial are the diagonal entries a_{ii}. □

In Chapter 6, we will consider the eigenvalues which is not necessarily a real number.

Example 5.1.12. *Let* $A = \begin{bmatrix} 5 & 3 & 1 & -1 \\ 1 & 3 & -1 & 0 \\ 0 & 0 & 5 & 4 \\ 0 & 0 & 0 & 1 \end{bmatrix}$. *The characteristic polynomial is*

$$\det(A - tI) = \det \begin{bmatrix} 5-t & 3 & 1 & -1 \\ 1 & 3-t & -1 & 0 \\ 0 & 0 & 5-t & 4 \\ 0 & 0 & 0 & 1-t \end{bmatrix} = \det \begin{bmatrix} 5-t & 3 \\ 1 & 3-t \end{bmatrix} \det \begin{bmatrix} 5-t & 4 \\ 0 & 1-t \end{bmatrix}$$

$$= \{(5-t)(3-t) - 3\}(5-t)(1-t) = (t-6)(t-2)(t-5)(t-1)$$

by Theorem 3.3.17. *Hence the eigenvalues for* A *are* $\lambda = 1, 2, 5, 6$.

Now we find the eigenvectors corresponding to the eigenvalues. An eigenvector $\mathbf{v}$ *for* $\lambda = 6$ *will satisfy* $(A - 6I)(\mathbf{v}) = \mathbf{0}$. *Hence* $\mathbf{v}$ *is a solution for*

$$(A - 6I)(\mathbf{x}) = \begin{bmatrix} -1 & 1 & 1 & -1 \\ 1 & -3 & -1 & 0 \\ 0 & 0 & -1 & 4 \\ 0 & 0 & 0 & -5 \end{bmatrix} \begin{bmatrix} x_1 \\ x_2 \\ x_3 \\ x_4 \end{bmatrix} = \begin{bmatrix} 0 \\ 0 \\ 0 \\ 0 \end{bmatrix}$$

We solve this to get an eigenvector $\begin{bmatrix} 3 \\ 1 \\ 0 \\ 0 \end{bmatrix}$. *For the eigenvalues* $\lambda = 2, 1, 5$ *the corresponding eigenvectors are* $\begin{bmatrix} 1 \\ -1 \\ 0 \\ 0 \end{bmatrix}, \begin{bmatrix} 7 \\ -6 \\ -5 \\ 5 \end{bmatrix}, \begin{bmatrix} 1 \\ -1 \\ 3 \\ 0 \end{bmatrix}$ *respectively. The eigenvectors form a basis of* $\mathbb{R}^4$ *by* Corollary 5.1.7.

Practice Problem 5.1.1. *Let* $A = \begin{bmatrix} 1 & 1 & 0 \\ 0 & 2 & 0 \\ -1 & 1 & 2 \end{bmatrix}$. *Find the characteristic polynomial and the eigenvalues. Also find linearly independent eigenvectors as many as you can.*

Recall Definition 4.3.8: For $A, B \in M(n)$ we say A and B are **similar** if there is an invertible matrix P such that $B = P^{-1}AP$. Similar matrices are 'similar' – one aspect of it is that similar matrices have the same determinant and trace (Corollary 3.3.12, Corollary 3.3.21). More over we have the same characteristic polynomial.

Theorem 5.1.13. *Let* $A, B \in M(n)$. *If* A *and* B *are similar then they have the same characteristic polynomials. In particular, they have the same eigenvalues.*

Proof. Suppose $A, B \in M(n)$ are similar say $B = P^{-1}AP$. Now

$$\begin{aligned} \det(B - tI) &= \det(P^{-1}AP - tI) = \det(P^{-1}(A - tI)P) = \\ &= \det P^{-1} \det(A - tI) \det P = \det(A - tI) \end{aligned}$$

by Theorem 3.3.11 and its corollary. □

Corollary 5.1.14. *Let* $T : V \to V$ *be a linear map on a subspace* V. *Let* $\mathcal{B}, \mathcal{C}$ *be bases for* V. *Then the characteristic polynomial of* $[T]_{\mathcal{B}}$ *and* $[T]_{\mathcal{C}}$ *are the same.*

Proof. Since $[T]_{\mathcal{B}}$ and $[T]_{\mathcal{C}}$ are similar (Theorem 4.3.7) they have the same characteristic polynomial. □

Since the characteristic polynomials of $[T]_{\mathcal{B}}$ and $[T]_{\mathcal{C}}$ are the same the characteristic polynomial is determined by the linear map T and does not depend on our choice of the bases. Hence we define

Definition 5.1.15. *Let* $T : V \to V$ *be a linear map on a subspace* V. *The* **characteristic polynomial** $f_T(t)$ *of* T *is defined by*

$$\det([T]_{\mathcal{B}} - tI)$$

for any basis $\mathcal{B}$ *of* V.

5.1.3 Diagonalization

Definition 5.1.16. *A matrix $D \in M(n)$ is called a* **diagonal matrix** (Definition 2.1.1) *if the only possible nonzero entries are on the diagonal. We will write the diagonal matrix D with $D_{ii} = \lambda_i$ by*

$$D = \text{diag}[\lambda_1, \lambda_2, ..., \lambda_n]$$

We say $A \in M(n)$ is **diagonalizable** *if there is a basis $\mathcal{B} = \{\mathbf{v}_1, \mathbf{v}_2, ..., \mathbf{v}_n\}$ such that $[T_A]_{\mathcal{B}}$ is a diagonal matrix.*

The diagonalizability of A is equivalent to that there is a basis $\mathcal{B} = \{\mathbf{v}_1, \mathbf{v}_2, ..., \mathbf{v}_n\}$ such that $A\mathbf{v}_i = \lambda_i \mathbf{v}_i$. Since then we have

$$[T_A]_{\mathcal{B}} = \text{diag}[\lambda_1, \lambda_2, ..., \lambda_n]$$

Hence $A \in M(n)$ is diagonalizable if and only if there is a basis consisting of eigenvectors.

Theorem 5.1.17. *Let $A \in M(n)$. The following conditions are equivalent.*

(1) *The matrix A is diagonalizable.*

(2) *There is a basis $\mathcal{B} = \{\mathbf{v}_1, \mathbf{v}_2, ..., \mathbf{v}_n\}$ consisting of eigenvectors.*

(3) *There is an invertible matrix $P \in M(n)$ such that $P^{-1}AP$ is diagonal.*

In this case, let $A\mathbf{v}_i = \lambda_i \mathbf{v}_i$ and let $P = [\mathbf{v}_1 \ \mathbf{v}_2 \ \cdots \ \mathbf{v}_n]$. Then

$$P^{-1}AP = [T_A]_{\mathcal{B}} = \text{diag}[\lambda_1, \lambda_2, ..., \lambda_n]$$

and the characteristic polynomial splits.

Proof. The equivalence (1) and (2) is the definition of diagonalizabllity.

For (2) $\Rightarrow$ (3), we may assume A is diagonalizable. Let $\mathcal{B} = \{\mathbf{v}_1, \mathbf{v}_2, ..., \mathbf{v}_n\}$ be linearly independent eigenvectors with the eigenvalues $\lambda_1, \lambda_2, ..., \lambda_n$ and let $P = [\mathbf{v}_1 \ \cdots \ \mathbf{v}_n]$. Then P is invertible since the columns are linearly independent. Now $[T]_{\mathcal{B}} = \text{diag}[\lambda_1, \lambda_2, ..., \lambda_n]$ which we denote by D. By Theorem 4.3.1 we have $[T]_{\mathcal{B}} = P^{-1}AP$. Hence $P^{-1}AP = D$.

To prove (3) $\Rightarrow$ (2) let $P = [\mathbf{v}_1 \ \cdots \ \mathbf{v}_n]$ be such that $P^{-1}AP$ is the diagonal matrix $D = \text{diag}[\lambda_1, \lambda_2, ..., \lambda_n]$. Then since P is invertible, the columns $\{P^{(i)} = \mathbf{v}_i \mid i = 1, 2, ..., n\}$ form a basis of $\mathbb{R}^n$. Now since $AP = PD$ and

$$\begin{aligned} AP &= A[\mathbf{v}_1 \ \cdots \ \mathbf{v}_n] = [A\mathbf{v}_1 \ \cdots \ A\mathbf{v}_n] \\ PD &= [\lambda_1\mathbf{v}_1 \ \lambda_2\mathbf{v}_2 \ \cdots \ \lambda_n\mathbf{v}_n] \end{aligned}$$

we have $A\mathbf{v}_i = \lambda_i\mathbf{v}_i$ i.e., $\mathbf{v}_i$ is an eigenvector.

To show the characteristic polynomial of a diagonalizable matrix A split suppose $P^{-1}AP = \text{diag}[\lambda_1, ..., \lambda_n]$. Then the characteristic polynomial of A is the same as the characteristic polynomial of the diagonal matrix $\text{diag}[\lambda_1, ..., \lambda_n]$ (Theorem 5.1.13). The characteristic polynomial of $\text{diag}[\lambda_1, ..., \lambda_n]$ is $f(t) = \pm\prod(\lambda_i - t)$ which is a splitting polynomial. □

Corollary 5.1.7 in different terminology becomes:

Corollary 5.1.18. *If $A \in M(n)$ has n distinct eigenvalues then A is diagonalizable.*

Proof. Immediate from Corollary 5.1.7 and the definition of diagonalizablity. □

Not all matrices are diagonalizable i.e., a matrix may not have enough linearly independent eigenvectors to form a basis. For example,

Example 5.1.19. (1) *The matrix* $A = \begin{bmatrix} 0 & 1 \\ 0 & 0 \end{bmatrix}$ *has eigenvalue* 0. *But there is only one linearly independent eigenvector* $\begin{bmatrix} 1 \\ 0 \end{bmatrix}$. *Since the solution to the equation* $A\mathbf{x} = \mathbf{0}$ *where* $\mathbf{x} = \begin{bmatrix} x_1 \\ x_2 \end{bmatrix}$ *is a constant multiples of* $\begin{bmatrix} 1 \\ 0 \end{bmatrix}$.

(2) *Let* $T_\theta : \mathbb{R}^2 \to \mathbb{R}^2$ *be the matrix linear map for* $A_\theta = \begin{bmatrix} \cos\theta & -\sin\theta \\ \sin\theta & \cos\theta \end{bmatrix}$. *Then* T *is the map of rotation by the angle* θ. *The characteristic polynomial of* A_θ *is* $f(t) = t^2 - (2\cos\theta)t + 1$ *which has real root only when* $\cos\theta = \pm 1$ *i.e.,* $\theta = 0$ *or* π. *For example, if* $\theta = \pi/2$ *then* $A = \begin{bmatrix} 0 & -1 \\ 1 & 0 \end{bmatrix}$. *It is obvious that* $\mathbf{v} \in V$ *and* $T_{\pi/2}(\mathbf{v})$ *cannot be collinear for any nonzero* $\mathbf{v}$. *Hence there are no eigenvectors. Consequently* $T_{\pi/2}$ *is not diagonalizable.*

Example 5.1.20. *Let* $A = \begin{bmatrix} 1 & 3 & 0 \\ 0 & -2 & 0 \\ 3 & 3 & -2 \end{bmatrix}$. *The characteristic polynomial of the matrix* A *is* $f_A(t) = -(t^3 + 3t^2 - 4) = -(t+2)^2(t-1)$. *It has roots* $\lambda_1 = -2$ *with multiplicity* 2 *and* $\lambda_2 = 1$ *with multiplicity* 1. *Let's find the eigenvectors for* $\lambda_1 = -2$. *An eigenvector* $\mathbf{v}$ *for* $\lambda_1 = -2$ *satisfies* $(A - (-2I))\mathbf{v} = \mathbf{0}$. *Hence we need to solve the equation* $(A + 2I)\mathbf{x} = \mathbf{0}$:

$$(A + 2I)\mathbf{x} = \begin{bmatrix} 3 & 3 & 0 \\ 0 & 0 & 0 \\ 3 & 3 & 0 \end{bmatrix} \begin{bmatrix} x_1 \\ x_2 \\ x_3 \end{bmatrix} = \begin{bmatrix} 0 \\ 0 \\ 0 \end{bmatrix}$$

and the equation $(A + 2I)\mathbf{x} = \mathbf{0}$ *has two linear independent solutions*

$$\left\{ \mathbf{v}_1 = \begin{bmatrix} 1 \\ -1 \\ 0 \end{bmatrix}, \mathbf{v}_2 = \begin{bmatrix} 0 \\ 0 \\ 1 \end{bmatrix} \right\}$$

To find the eigenvector for $\lambda_2 = 1$ *we need to solve the equation* $(A - I)\mathbf{x} = \mathbf{0}$:

$$(A - I)\mathbf{x} = \begin{bmatrix} 0 & 3 & 0 \\ 0 & -3 & 0 \\ 3 & 3 & -3 \end{bmatrix} \begin{bmatrix} x_1 \\ x_2 \\ x_3 \end{bmatrix} = \begin{bmatrix} 0 \\ 0 \\ 0 \end{bmatrix}$$

and the equation $(A - I)\mathbf{x} = \mathbf{0}$ *has a solution*

$$\left\{ \mathbf{v}_3 = \begin{bmatrix} 1 \\ 0 \\ 1 \end{bmatrix} \right\}$$

It is easy to show that $\mathcal{B} = \{\mathbf{v}_1, \mathbf{v}_2, \mathbf{v}_3\}$ *is linearly independent. Hence* A *is diagonalizable and*

$$[T_A]_{\mathcal{B}} = P^{-1}AP = \begin{bmatrix} -2 & 0 & 0 \\ 0 & -2 & 0 \\ 0 & 0 & 1 \end{bmatrix} \text{ where } P = \begin{bmatrix} 1 & 0 & 1 \\ -1 & 0 & 0 \\ 0 & 1 & 1 \end{bmatrix}$$

by Theorem 4.3.1.

Practice Problem 5.1.2. *Let* $A = \begin{bmatrix} 2 & 1 & 0 \\ 0 & 1 & 0 \\ 0 & 0 & 1 \end{bmatrix}$. *Decide whether A is diagonalizable or not. If A is diagonalizable then diagonalize A and find a basis consisting of eigenvectors. Also, find P such that $P^{-1}AP$ is a diagonal matrix.*

Theorem 5.1.21. *Let $A \in M(n)$ and λ be an eigenvalue of A with the multiplicity e. Let E_λ be the eigenspace for λ. Then*

$$\dim E_\lambda \leq e$$

Proof. Let $f_A(t) = \pm(t-\lambda)^e g(t)$ with $g(\lambda) \neq 0$. Choose a basis $\{\mathbf{v}_1, \mathbf{v}_2, ..., \mathbf{v}_k\}$ of E_λ. Extend this to a basis $\mathcal{B} = \{\mathbf{v}_1, \mathbf{v}_2, ..., \mathbf{v}_k, \mathbf{v}_{k+1}, ..., \mathbf{v}_n\}$ of $\mathbb{R}^n$ (Corollary 4.1.8). The matrix representation B of T_A with respect to $\mathcal{B}$ will be

$$B = \left[\begin{array}{cccc|c} \lambda & & & & \\ & \lambda & & & * \\ & & \ddots & & * \\ & & & \lambda & \\ \hline 0 & 0 & \cdots & 0 & \\ & & \cdots & & C \\ 0 & 0 & \cdots & 0 & \end{array}\right] \tag{5.6}$$

where the upper left corner is λI_k and C is a $(n-k)\times(n-k)$ matrix. If we let P be the matrix with the i-th column $\mathbf{v}_i$ then $B = P^{-1}AP$ (Theorem 4.3.1). Since A and B are similar, they have the same characteristic polynomial i.e., $f_A(t) = f_B(t)$ (Theorem 5.1.13). The characteristic polynomial of B is $f_B(t) = \pm(t-\lambda)^k f_C(t)$ (Theorem 3.3.17). Hence $(t-\lambda)^k | f_A(t) = \pm(t-\lambda)^e g(t)$ with $g(\lambda) \neq 0$. Therefore $k \leq e$. □

Theorem 5.1.22. *Let $A \in M(n)$. Suppose the characteristic polynomial $f(t)$ of A splits say $f(t) = (-1)^n \prod_{i=1}^k (t-\lambda_i)^{e_i}$. Then the following conditions are equivalent.*

(1) *The matrix A is diagonalizable.*

(2) $\dim E_{\lambda_i} = e_i$ $(1 \leq i \leq k)$

(3) $\rho(A - \lambda_i) = n - e_i$ $(1 \leq i \leq k)$

In this case, if $\mathcal{B}_i$ is a basis for $E_{\lambda_i} (1 \leq i \leq k)$ then $\bigcup_{i=1}^k \mathcal{B}_i$ is a basis of $\mathbb{R}^n$ consisting of eigenvectors of A.

Proof. Since $f(t) = (-1)^n \prod_{i=1}^k (t-\lambda_i)^{e_i}$ is of degree n (Theorem 5.1.10) we have $\sum e_i = n$.

(1) ⇒ (2) First, assume A is diagonalizable. Then there is a basis $\mathcal{B} = \{\mathbf{v}_1, \mathbf{v}_2, ..., \mathbf{v}_n\}$ of $\mathbb{R}^n$ consisting of eigenvectors (Theorem 5.1.17). Let $\mathcal{B}_i$ be the subset of $\mathcal{B}$ with the eigenvalue λ_i i.e., $\mathcal{B}_i \subseteq E_{\lambda_i}$, and k_i be the number of elements of $\mathcal{B}_i$. Then $\sum k_i = n$ and $k_i \leq e_i$ by Theorem 5.1.21. Now $\bigcup \mathcal{B}_i = \mathcal{B}$ and $\{\mathcal{B}_1, \mathcal{B}_2, ..., \mathcal{B}_k\}$ are disjoint subsets of $\mathcal{B}$ since $E_{\lambda_i} \cap E_{\lambda_j} = \mathbf{0}$ (Corollary 5.1.8) and $\mathcal{B}$ consists of nonzero vectors. Hence

$$n = \sum k_i \leq \sum e_i = n$$

and therefore $k_i = e_i$ since $k_i \leq e_i$. Thus $\dim E_{\lambda_i} = k_i = e_i$.

(2) ⇒ (1) Now assume $\dim E_{\lambda_i} = e_i$ and $\mathcal{B}_i$ be a basis of E_{λ_i} and let $\mathcal{B} = \cup \mathcal{B}_i$. We want to show $\mathcal{B}$ is a basis of $\mathbb{R}^n$. As before $\mathcal{B}_i$ are disjoint and the number of elements of $\mathcal{B}$ is $\sum \#(\mathcal{B}_i) = \sum e_i = n$. Now we claim that $\mathcal{B}$ is linearly independent. In fact, let $\mathcal{B}_i = \{\mathbf{v}_{ij}\}$

and let $\sum c_{ij}\mathbf{v}_{ij} = \mathbf{0}$. Let $\mathbf{w}_i = \sum_j c_{ij}\mathbf{v}_{ij}$. Then $\mathbf{w}_i \in E_{\lambda_i}$ and $\sum \mathbf{w}_i = \mathbf{0}$. Hence $\mathbf{w}_i = \mathbf{0}$ for all i by Corollary 5.1.8. Since $\mathcal{B}_i = \{\mathbf{v}_{ij}\}_j$ is a basis of E_{λ_i} we conclude that $c_{ij} = 0$ for all i, j.

Since $\mathcal{B}$ is linearly independent and consists of n vectors in $\mathbb{R}^n$, we see that $\mathcal{B}$ is a basis consisting of eigenvectors. Thus A is diagonalizable.

(2) $\Leftrightarrow$ (3) This follows from Theorem 2.3.17 applied to $A-\lambda I$: $\nu(A-\lambda I)+\rho(A-\lambda I) = n$ and $\nu(A-\lambda I) = \dim E_{\lambda_i}$.

We proved the last statement in the proof of the implication (2) $\Rightarrow$ (1). □

Example 5.1.23. *Let* $A = \begin{bmatrix} 1 & 2 & 0 & 1 \\ 0 & 1 & 0 & 0 \\ 2 & 1 & 2 & 0 \\ 0 & 0 & 0 & 1 \end{bmatrix}$. *The characteristic polynomial of A is* $f_A(t) = t^4 - 5t^3 + 9t^2 - 7t + 2 = (t-1)^3(t-2)$. *Hence the eigenvalues are $\lambda_1 = 1$ with multiplicity $e_1 = 3$ and $\lambda_2 = 2$ with multiplicity $e_2 = 1$. To find the eigenvectors for $\lambda_1 = 1$ we need to solve the equation $(A-I)\mathbf{x} = \mathbf{0}$:*

$$(A-I)\mathbf{x} = \begin{bmatrix} 0 & 2 & 0 & 1 \\ 0 & 0 & 0 & 0 \\ 2 & 1 & 1 & 0 \\ 0 & 0 & 0 & 0 \end{bmatrix} \begin{bmatrix} x_1 \\ x_2 \\ x_3 \\ x_4 \end{bmatrix} = \begin{bmatrix} 0 \\ 0 \\ 0 \\ 0 \end{bmatrix}$$

Since $\nu(A-I) = 2$ the equation $(A-I)\mathbf{x} = \mathbf{0}$ has two linearly independent solutions

$$\mathbf{v}_1 = \begin{bmatrix} 1 \\ 0 \\ -2 \\ 0 \end{bmatrix}, \quad \mathbf{v}_2 = \begin{bmatrix} 0 \\ 1 \\ -1 \\ -2 \end{bmatrix}$$

and hence $\dim E_{\lambda_1} = 2$ whereas $e_1 = 3$. Hence A is not diagonalizable.

For the eigenvalue $\lambda_2 = 2$, a solution to the equation $(A-2I)\mathbf{x} = \mathbf{0}$ is

$$\mathbf{v}_3 = \begin{bmatrix} 0 \\ 0 \\ 1 \\ 0 \end{bmatrix}$$

Here, again we see that not every square matrix A is diagonalizable. In Chapter 7, we will investigate the best possible way to represent a linear operator.

Practice Problem 5.1.3. *Decide diagonalizability of the following matrices.*

$$A = \begin{bmatrix} 1 & 0 & 0 & 2 \\ 0 & 3 & 0 & 0 \\ 0 & 0 & 3 & 0 \\ 0 & 0 & 0 & 3 \end{bmatrix}, \qquad B = \begin{bmatrix} 1 & 0 & 2 & 2 \\ 0 & 3 & 0 & 0 \\ 0 & 0 & 3 & 0 \\ 0 & 0 & 0 & 3 \end{bmatrix}, \qquad C = \begin{bmatrix} 1 & 0 & 0 & 0 \\ 0 & 3 & 0 & 2 \\ 0 & 0 & 3 & 0 \\ 0 & 0 & 0 & 3 \end{bmatrix}$$

At this point we want to point out an important theorem called Cayley-Hamilton Theorem deferring its proof until Chapter 6.

Theorem 6.8.14 (Cayley-Hamilton Theorem). *Let $A \in M(n)$ with the characteristic polynomial $f(t)$. Then $f(A)$ is the zero matrix.*

For example, if $A = \begin{bmatrix} 1 & -1 \\ 2 & 1 \end{bmatrix}$ then $f_A(t) = t^2 - 2t + 3$. Now

$$f_A(A) = A^2 - 2A + 3I = \begin{bmatrix} -1 & -2 \\ 4 & -1 \end{bmatrix} - 2\begin{bmatrix} 1 & -1 \\ 2 & 1 \end{bmatrix} + 3\begin{bmatrix} 1 & 0 \\ 0 & 1 \end{bmatrix} = \begin{bmatrix} 0 & 0 \\ 0 & 0 \end{bmatrix}$$

Answers for Practice Problems 5.1

(5.1.1) The characteristic polynomial is $f_A(t) = -(t-1)(t-2)^2$. Eigenvalues are $\lambda_1 = 1, \lambda_2 = 2$. For the eigenvalue $\lambda_1 = 1$ the corresponding eigenvector is $\mathbf{v}_1 = \begin{bmatrix} 1 \\ 0 \\ 1 \end{bmatrix}$. And $\left\{ \begin{bmatrix} 1 \\ 1 \\ 0 \end{bmatrix}, \begin{bmatrix} 1 \\ 1 \\ 1 \end{bmatrix} \right\}$ are the linearly independent eigenvectors for the eigenvalue $\lambda_2 = 2$.

(5.1.2) The characteristic polynomial is

$$\det(A - tI) = \begin{bmatrix} 2-t & 1 & 0 \\ 0 & 1-t & 0 \\ 0 & 0 & 1-t \end{bmatrix} = -(t-1)^2(t-2)$$

Eigenvalues are $\lambda_1 = 1, \lambda_2 = 2$ and their respective multiplicities are $e_1 = 2, e_2 = 1$.

We look for a basis of E_{λ_1}: Since $A - I = \begin{bmatrix} 1 & 1 & 0 \\ 0 & 0 & 0 \\ 0 & 0 & 0 \end{bmatrix}$ the eigenvectors for the eigenvalue $\lambda_1 = 1$ of A are $\left\{ \begin{bmatrix} 1 \\ -1 \\ 0 \end{bmatrix}, \begin{bmatrix} 0 \\ 0 \\ 1 \end{bmatrix} \right\}$

For E_{λ_2} we have $A - 2I = \begin{bmatrix} 0 & 1 & 0 \\ 0 & -1 & 0 \\ 0 & 0 & -1 \end{bmatrix}$. Hence the eigenvectors for $\lambda_2 = 2$ is $\left\{ \begin{bmatrix} 1 \\ 0 \\ 0 \end{bmatrix} \right\}$.

Therefore

$$\left\{ \begin{bmatrix} 1 \\ -1 \\ 0 \end{bmatrix}, \begin{bmatrix} 0 \\ 0 \\ 1 \end{bmatrix} \right\} \cup \left\{ \begin{bmatrix} 1 \\ 0 \\ 0 \end{bmatrix} \right\}$$

is a basis for $\mathbb{R}^3$ consisting of eigenvectors.

If we let $P = \begin{bmatrix} 1 & 0 & 1 \\ -1 & 0 & 0 \\ 0 & 1 & 0 \end{bmatrix}$ then

$$P^{-1}AP = \begin{bmatrix} 1 & 0 & 0 \\ 0 & 1 & 0 \\ 0 & 0 & 2 \end{bmatrix}$$

(5.1.3) Eigenvalues for A, B, C are $\lambda_1 = 1, \lambda_2 = 3$ with the multiplicities $e_1 = 1, e_2 = 3$. Since $\rho(A - I) = 3,\ \rho(A - 3I) = 1$, A is diagonalizable with the resulting diagonal matrix diag$[1, 3, 3, 3]$.

Since $\rho(B - I) = 3,\ \rho(B - 3I) = 1$, B is diagonalizable with the resulting diagonal matrix diag$[1, 3, 3, 3]$.

Since $\rho(C - I) = 3,\ \rho(C - 3I) = 2$, C is not diagonalizable.

5.1 Exercises

1. Prove Theorem 5.1.3.

2. Find the characteristic polynomials, eigenvalues and eigenvectors for the following matrices. Decide whether they are diagonalizable or not.

(a) $\begin{bmatrix} 2 & -5 & 0 \\ 1 & -2 & 0 \\ 1 & -2 & 1 \end{bmatrix}$ (b) $\begin{bmatrix} 5 & 0 & 4 \\ 0 & 1 & -4 \\ 4 & -4 & 3 \end{bmatrix}$ (c) $\begin{bmatrix} 2 & 1 & 0 \\ 0 & 1 & 0 \\ 1 & 1 & 1 \end{bmatrix}$ (d) $\begin{bmatrix} 0 & 1 & 0 & 0 \\ 0 & 0 & 1 & 0 \\ 0 & 0 & 0 & 1 \\ 1 & 0 & 0 & 0 \end{bmatrix}$

3. Prove the true statement(s) and find a counter example for the false statement(s)

   (a) If the determinant of $A \in M(n)$ is zero then A has an eigenvector.

   (b) A sum of two eigenvectors is also an eigenvector.

   (c) A sum of two eigenvectors corresponding to the same eigenvalue is an eigenvector.

   (d) Two row equivalent matrices have the same characteristic polynomial.

   (e) If the characteristic polynomial of $A \in M(n)$ does not split then A is not diagonalizable.

   (f) For each eigenvalue λ of $A \in M(n)$ there is an eigenvector corresponding to λ.

   (g) If $A, B \in M(n)$ are similar then A, B have the same eigenvalues.

   (h) If $A, B \in M(n)$ are similar then A, B have the same eigenvectors.

   (i) Every square matrix has an eigenvector.

4. Let $A \in M(n)$ be an upper triangular matrix. Prove that the characteristic polynomial $f_A(t)$ splits and the eigenvalues are the diagonal entries.

5. Let $A \in M(n)$. Prove that the characteristic polynomial of A and A^t are the same.

6. Let $A \in M(n)$ and λ be an eigenvalue of A.

   (a) Prove that λ^k is an eigenvalue of A^k.

   (b) Prove that $\lambda \neq 0$ if A is invertible.

   (c) Prove that if A is invertible then λ^{-1} is an eigenvalue of A^{-1}.

7. Let $A, B \in M(n)$ be similar say, $B = P^{-1}AP$. Let λ be an eigenvalue for A and hence for B as well. Let E_λ^A and E_λ^B be the eigenspaces for A and B.

   (a) Prove that $\mathbf{v} \in E_\lambda^B$ if and only if $P\mathbf{v} \in E_\lambda^A$.

   (b) Prove that the matrix linear map $T_P : E_\lambda^B \to E_\lambda^A$ of P defines an isomorphism.

8. Let $A \in M(n)$ and $B \in M(r), C \in M(s)$ $(r + s = n)$ and

$$A = \begin{bmatrix} B & 0 \\ 0 & C \end{bmatrix}, \quad A' = \begin{bmatrix} B & D \\ 0 & C \end{bmatrix}$$

   for some D. Prove that $f_A(t) = f_{A'}(t)$ and express it in terms of $f_B(t)$ and $f_C(t)$.

9. Let $A \in M(n)$ and c be a nonzero constant. Let

$$f_A(t) = (-1)^n(t^n + a_{n-1}t^{n-1} + a_{n-2}t^{n-2} + \cdots + a_0)$$

be the characteristic polynomial of A. Prove that the characteristic polynomial of cA is

$$f_{cA} = (-1)^n(t^n + a_{n-1}ct^{n-1} + a_{n-2}c^2t^{n-2} + \cdots + a_0c^n)$$

10. Let $A, B \in M(n)$. Prove that the characteristic polynomial $f_{AB}(t)$ of AB is the same as the characteristic polynomial $f_{BA}(t)$ of BA. (Hint: Use (4.19) of Corollary 4.2.28 to express A as a product of elementary matrices and the matrix of the form J_r. First prove the result when A is an elementary matrix or J_r of A.)

5.2 Inner Product, Length and Orthogonality

In $\mathbb{R}^n$, there is an extra structure other than the addition of vectors and the scalar multiplication, so called the dot product. Using the dot product we can define the length of a vector and the angle between two vectors which are geometric in nature.

5.2.1 Inner Product and Orthogonality

Definition 5.2.1. *Let* $\mathbf{u} = \begin{bmatrix} x_1 \\ x_2 \\ \vdots \\ x_n \end{bmatrix}, \mathbf{v} = \begin{bmatrix} y_1 \\ y_2 \\ \vdots \\ y_n \end{bmatrix}$ *be vectors in* $\mathbb{R}^n$. *The* **inner product** *or* **dot product** $\mathbf{u} \cdot \mathbf{v}$ *of* $\mathbf{u}$ *and* $\mathbf{v}$ *is defined by*

$$\mathbf{u} \cdot \mathbf{v} = \mathbf{u}^t\mathbf{v} = [x_1, x_2, ..., x_n] \begin{bmatrix} y_1 \\ y_2 \\ \vdots \\ y_n \end{bmatrix} = x_1y_1 + x_2y_2 + \cdots + x_ny_n \tag{5.7}$$

If $\mathbf{u}, \mathbf{v} \in \mathbb{R}_n$ *we define the dot product by*

$$\mathbf{u} \cdot \mathbf{v} = \mathbf{u}\mathbf{v}^t = x_1y_1 + x_2y_2 + \cdots + x_ny_n$$

Note that $\mathbf{u} \cdot \mathbf{0} = 0$ for all $\mathbf{u}$. Most of the time, we take dot products of the column vectors instead of the rows. The basic properties of dot product are:

Theorem 5.2.2. *Let* $\mathbf{u}, \mathbf{v}, \mathbf{w} \in \mathbb{R}^n$ *and* $c \in \mathbb{R}$. *Then*

(1) $\mathbf{u} \cdot \mathbf{v} = \mathbf{v} \cdot \mathbf{u}$

(2) $\mathbf{u} \cdot (\mathbf{v} + \mathbf{w}) = \mathbf{u} \cdot \mathbf{w} + \mathbf{u} \cdot \mathbf{w}$ *and* $(\mathbf{u} + \mathbf{v}) \cdot \mathbf{w} = \mathbf{u} \cdot \mathbf{w} + \mathbf{v} \cdot \mathbf{w}$

(3) $(c\mathbf{u}) \cdot \mathbf{v} = c(\mathbf{u} \cdot \mathbf{v}) = \mathbf{u} \cdot (c\mathbf{v})$

(4) $\mathbf{u} \cdot \mathbf{u} \geq 0$. *Further,* $\mathbf{u} \cdot \mathbf{u} = 0$ *if and only if* $\mathbf{u} = 0$.

Proof. Exercise 2. □

Definition 5.2.3. *The* **length** *or* **norm** $\|\mathbf{u}\|$ *of* $\mathbf{u}$ *is defined by*

$$\|\mathbf{u}\| = \sqrt{\mathbf{u} \cdot \mathbf{u}} = \sqrt{x_1^2 + x_2^2 + \cdots + x_n^2}$$

We can recover the inner product from the norm by the **polar identity** (Exercise 6 (a)).

$$\mathbf{u} \cdot \mathbf{v} = \frac{1}{4}\|\mathbf{u} + \mathbf{v}\|^2 - \frac{1}{4}\|\mathbf{u} - \mathbf{v}\|^2 \tag{5.8}$$

Geometrically, in $\mathbb{R}^2$ the norm $\|\mathbf{u}\|$ of $\mathbf{u} = \begin{bmatrix} a \\ b \end{bmatrix}$ represents the length $\sqrt{a^2 + b^2}$ of $\mathbf{u}$. On the other hand, let $\mathbf{v} \in \mathbb{R}^2$ be another vector and let θ be the angle between $\mathbf{u}$ and $\mathbf{v}$. Let $\alpha = \|\mathbf{u}\|, \beta = \|\mathbf{v}\|$ and $\gamma = \|\mathbf{v} - \mathbf{u}\|$. Then the law of cosines says

$$\gamma^2 = \alpha^2 + \beta^2 - 2\alpha\beta\cos\theta$$

On the other hand, $\gamma^2 = \|\mathbf{v} - \mathbf{u}\|^2 = \|\mathbf{v}\|^2 - 2\mathbf{u} \cdot \mathbf{v} + \|\mathbf{u}\|^2 = \alpha^2 - 2\mathbf{u} \cdot \mathbf{v} + \beta^2$. Hence we have

$$\mathbf{u} \cdot \mathbf{v} = \|\mathbf{u}\| \, \|\mathbf{v}\| \cos\theta \tag{5.9}$$

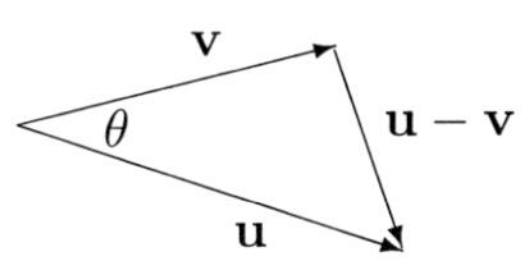

For example, if θ is the angle between $\mathbf{u} = \begin{bmatrix} 1 \\ 2 \end{bmatrix}$ and $\mathbf{v} = \begin{bmatrix} 2 \\ 5 \end{bmatrix}$. Then

$$\cos\theta = \frac{\mathbf{u} \cdot \mathbf{v}}{\|\mathbf{u}\| \, \|\mathbf{v}\|} = \frac{12}{\sqrt{5}\sqrt{29}}$$

If $\mathbf{u}$ and $\mathbf{v}$ are perpendicular, then $\mathbf{u} \cdot \mathbf{v} = 0$ by (5.9). More generally, we define,

Definition 5.2.4. *For nonzero* $\mathbf{u}, \mathbf{v} \in \mathbb{R}^n$*, we define the* **angle** θ *between* $\mathbf{u}$ *and* $\mathbf{v}$ *by*

$$\cos\theta = \frac{\mathbf{u} \cdot \mathbf{v}}{\|\mathbf{u}\| \, \|\mathbf{v}\|} \tag{5.10}$$

We say $\mathbf{u}, \mathbf{v}$ *are* **orthogonal** *or* **perpendicular** *if* $\mathbf{u} \cdot \mathbf{v} = 0$ *in which case we will write*

$$\mathbf{u} \perp \mathbf{v}$$

Two subsets S_1, S_2 *of* $\mathbb{R}^n$ *are said to be* **orthogonal** *if* $\mathbf{u}_1 \perp \mathbf{u}_2$ *for any* $\mathbf{u}_i \in S_i$, $(i = 1, 2)$. *In this case, we write* $S_1 \perp S_2$. *If* $S_2 = \{\mathbf{y}\}$ *then we will simply write* $S_1 \perp \mathbf{y}$.

It will follow from Cauchy Schwartz inequality (Theorem 5.2.28) that

$$|\cos\theta| \leq 1$$

The **distance** $d(\mathbf{u}, \mathbf{v})$ between $\mathbf{u}, \mathbf{v}$ is defined by

$$d(\mathbf{u}, \mathbf{v}) = \|\mathbf{u} - \mathbf{v}\| \tag{5.11}$$

Example 5.2.5. *Let* $\mathbf{u} = \begin{bmatrix} 1 \\ 2 \\ 1 \end{bmatrix}, \mathbf{v} = \begin{bmatrix} 2 \\ 1 \\ 0 \end{bmatrix}$. *Then* $\mathbf{u} \cdot \mathbf{v} = 4$, $\|\mathbf{u}\| = \sqrt{6}, \|\mathbf{v}\| = \sqrt{5}$. *Hence if* θ *is the angle between* $\mathbf{u}$ *and* $\mathbf{v}$ *then*

$$\cos\theta = \frac{4}{\sqrt{30}}$$

The distance between $\mathbf{u}$ *and* $\mathbf{v}$ *is*

$$d(\mathbf{u}, \mathbf{v}) = \|\mathbf{u} - \mathbf{v}\| = \sqrt{(-1)^2 + (1)^2 + (1)^2} = \sqrt{3}$$

For any scalar c, we have the formula

$$\|c\mathbf{u}\| = |c| \, \|\mathbf{u}\| \tag{5.12}$$

Since $\|c\mathbf{u}\| \geq 0$, (5.12) follows from $\|c\mathbf{u}\|^2 = c^2(x_1^2 + x_2^2 + \cdots + x_n^2) = c^2\|\mathbf{u}\|^2$.

Geometric justification of orthogonality is given by the following theorem.

Theorem 5.2.6 (Pythagorean Theorem). *Let* $\mathbf{u}, \mathbf{v} \in \mathbb{R}^n$. *Then*

$$\mathbf{u} \perp \mathbf{v} \;\; \textit{if and only if} \;\; \|\mathbf{u} + \mathbf{v}\|^2 = \|\mathbf{u}\|^2 + \|\mathbf{v}\|^2$$

Proof. This follow from the equality

$$\|\mathbf{u} + \mathbf{v}\|^2 = \mathbf{u} \cdot \mathbf{u} + 2\mathbf{u} \cdot \mathbf{v} + \mathbf{v} \cdot \mathbf{v} = \|\mathbf{u}\|^2 + 2\mathbf{u} \cdot \mathbf{v} + \|\mathbf{v}\|^2$$

□

Definition 5.2.7. *Let* S *be a subset of* $\mathbb{R}^n$. *The* **orthogonal complement** $S^\perp$ *of* S *is defined by*

$$S^\perp = \{\mathbf{u} \in \mathbb{R}^n \mid \mathbf{u} \perp \mathbf{x} \;\text{ for all } \mathbf{x} \in S\}$$

Let S_1, S_2 *be subsets of* $\mathbb{R}^n$.

First, observe that $S^\perp$ is a subspace even if S is just a subset. In fact, if $\mathbf{u}, \mathbf{v} \in S^\perp$ and $c \in \mathbb{R}$ then for any $\mathbf{x} \in S$ we have

$$(\mathbf{u} + \mathbf{v}) \cdot \mathbf{x} = \mathbf{u} \cdot \mathbf{x} + \mathbf{v} \cdot \mathbf{x} = 0 \text{ and } (c\mathbf{u}) \cdot \mathbf{x} = c(\mathbf{u} \cdot \mathbf{x}) = 0$$

For example, let $\mathbf{u} = \begin{bmatrix} 1 \\ 1 \\ 0 \end{bmatrix}$ and $S = \{\mathbf{u}\}$ then

$$S^\perp = \left\{ \begin{bmatrix} x \\ -x \\ y \end{bmatrix} \middle| \, x, y \in \mathbb{R} \right\} = \text{Span}\left\{ \begin{bmatrix} 1 \\ -1 \\ 0 \end{bmatrix}, \begin{bmatrix} 0 \\ 0 \\ 1 \end{bmatrix} \right\}$$

Next observations are (Exercise 3)

$$S^\perp = \text{Span}(S)^\perp \tag{5.13}$$

$$S \cap S^\perp = \begin{cases} \{\mathbf{0}\} & \text{if } \mathbf{0} \in S \\ \emptyset & \text{if } \mathbf{0} \notin S \end{cases} \tag{5.14}$$

If W is a subspace of $\mathbb{R}^n$ spanned by $\{\mathbf{u}_1, \mathbf{u}_2, ..., \mathbf{u}_k\}$ then $\mathbf{y} \perp W$ if and only if $\mathbf{y} \perp \mathbf{u}_i$ for all i. (Exercise 4)

Theorem 5.2.8. *Let $A \in M(m \times n)$. Then we have*

$$(\mathrm{Row}(A))^{\perp} = N(A)^t \subseteq \mathbb{R}_n \tag{5.15}$$

$$(\mathrm{Col}(A))^{\perp} = N(A^t) \subseteq \mathbb{R}^m \tag{5.16}$$

where $N(A)^t = \{\mathbf{u}^t \mid \mathbf{u} \in N(A)\} \subseteq \mathbb{R}_n$.

Proof. To prove (5.15) we notice that $\mathbf{x} \in N(A)^t \subseteq \mathbb{R}_n$ if and only if $A\mathbf{x}^t = 0$ which is equivalent to $A_{(i)} \cdot \mathbf{x} = \mathbf{0}$ for all i. That is $\mathbf{x} \in (\mathrm{Row}(A))^{\perp}$.

The proof of (5.16) is Exercise 5. □

Definition 5.2.9. *A set of (nonzero) vectors $S = \{\mathbf{u}_1, \mathbf{u}_2, ..., \mathbf{u}_p\} \subseteq \mathbb{R}^n$ is said to be an* **orthogonal** *if $\mathbf{u}_i \cdot \mathbf{u}_j = 0$ for all $i \neq j$. We say S is* **orthonormal** *if S is orthogonal and $\|\mathbf{u}_i\| = 1$ as well for all i.*

A basis $\mathcal{B} = \{\mathbf{u}_1, \mathbf{u}_2, ..., \mathbf{u}_p\}$ is called an **orthogonal** (**resp. orthonormal**) **basis** *if $\mathcal{B}$ is an orthogonal (resp.orthonormal) set of vectors.*

Orthogonal and orthonormal set of vectors are very convenient for various computations as illustrated in the following theorems.

Theorem 5.2.10. *If $S = \{\mathbf{u}_1, \mathbf{u}_2, ..., \mathbf{u}_p\}$ is an orthogonal set of nonzero vectors then S is linearly independent. If S is a subset of a subspace W dimension p then S is an orthogonal basis of W.*

Proof. Suppose $c_1\mathbf{u}_1 + \cdots + c_p\mathbf{u}_p = \mathbf{0}$. Taking the dot product $\cdot\mathbf{u}_i$ with this expression we have

$$0 = \mathbf{0} \cdot \mathbf{u}_i = c_1\mathbf{u}_1 \cdot \mathbf{u}_i + \cdots + c_p\mathbf{u}_p \cdot \mathbf{u}_i = c_i\mathbf{u}_i \cdot \mathbf{u}_i = c_i\|\mathbf{u}_i\|^2$$

Since $\|\mathbf{u}_i\|^2 \neq 0$ we have $c_i = 0$.

The second statement is immediate because S consists of p linearly independent vectors in p dimensional subspace. □

Theorem 5.2.11. *If $S = \{\mathbf{u}_1, \mathbf{u}_2, ..., \mathbf{u}_p\}$ is an orthogonal basis for a subspace W and if $\mathbf{y} \in W$ then*

$$\mathbf{y} = \frac{\mathbf{u}_1 \cdot \mathbf{y}}{\|\mathbf{u}_1\|^2}\mathbf{u}_1 + \frac{\mathbf{u}_2 \cdot \mathbf{y}}{\|\mathbf{u}_2\|^2}\mathbf{u}_2 + \cdots + \frac{\mathbf{u}_p \cdot \mathbf{y}}{\|\mathbf{u}_p\|^2}\mathbf{u}_p \tag{5.17}$$

In particular, if S is orthonormal then

$$\mathbf{y} = (\mathbf{u}_1 \cdot \mathbf{y})\mathbf{u}_1 + (\mathbf{u}_2 \cdot \mathbf{y})\mathbf{u}_2 + \cdots + (\mathbf{u}_p \cdot \mathbf{y})\mathbf{u}_p \tag{5.18}$$

Proof. Let $\mathbf{y} = c_1\mathbf{u}_1 + \cdots + c_p\mathbf{u}_p$. We need to determine c_i: Taking the dot product $\cdot\mathbf{u}_i$ with $\mathbf{y}$ we get

$$\mathbf{u}_i \cdot \mathbf{y} = \mathbf{u}_i \cdot \left(\sum_{j=1}^{p} c_j\mathbf{u}_j\right) = c_i\mathbf{u}_i \cdot \mathbf{u}_i$$

Hence $c_i = \dfrac{\mathbf{y} \cdot \mathbf{u}_i}{\mathbf{u}_i \cdot \mathbf{u}_i}$. □

Example 5.2.12. *Let $\mathbf{u}_1 = \begin{bmatrix} 1 \\ 3 \\ 1 \end{bmatrix}, \mathbf{u}_2 = \begin{bmatrix} 2 \\ -1 \\ 1 \end{bmatrix}, \mathbf{u}_3 = \begin{bmatrix} -4 \\ -1 \\ 7 \end{bmatrix}$. It is easy to show $S = \{\mathbf{u}_1, \mathbf{u}_2, \mathbf{u}_3\}$ is an orthogonal basis of $\mathbb{R}^3$.*

To express $\mathbf{y} = \begin{bmatrix} 1 \\ 1 \\ 1 \end{bmatrix}$ *as a linear combination of* S*, we compute:* $\mathbf{y} \cdot \mathbf{u}_1 = 5,\ \mathbf{y} \cdot \mathbf{u}_2 = 2,\ \mathbf{y} \cdot \mathbf{u}_3 = 2$ *and* $\|\mathbf{u}_1\|^2 = 11, \|\mathbf{u}_2\|^2 = 6, \|\mathbf{u}_3\|^2 = 66$*. Hence*

$$\begin{aligned} \mathbf{y} &= \frac{\mathbf{u}_1 \cdot \mathbf{y}}{\|\mathbf{u}_1\|^2}\mathbf{u}_1 + \frac{\mathbf{u}_2 \cdot \mathbf{y}}{\|\mathbf{u}_2\|^2}\mathbf{u}_2 + \frac{\mathbf{u}_3 \cdot \mathbf{y}}{\|\mathbf{u}_3\|^2}\mathbf{u}_3 \\ &= \frac{5}{11}\mathbf{u}_1 + \frac{1}{3}\mathbf{u}_2 + \frac{1}{33}\mathbf{u}_3 \end{aligned}$$

Practice Problem 5.2.1. *Let* $\mathbf{u}_1 = \begin{bmatrix} 0 \\ 1 \\ 1 \end{bmatrix}, \mathbf{u}_2 = \begin{bmatrix} 2 \\ -1 \\ 1 \end{bmatrix}, \mathbf{u}_3 = \begin{bmatrix} 1 \\ 1 \\ -1 \end{bmatrix}$ *and* $\mathcal{B} = \{\mathbf{u}_1, \mathbf{u}_2, \mathbf{u}_3\}$*. Show that* $\mathcal{B}$ *is an orthogonal basis for* $\mathbb{R}^3$ *and write* $\mathbf{v} = \begin{bmatrix} 1 \\ 1 \\ 0 \end{bmatrix}$ *as a linear combination of* $\mathcal{B}$*.*

Theorem 5.2.13. *Let* $A \in M(m{\times}n)$*. The following conditions are equivalent.*

(1) *The columns of* A *are orthonormal.*

(2) $A^tA = I_n$

(3) $(A\mathbf{x}) \cdot (A\mathbf{y}) = \mathbf{x} \cdot \mathbf{y}$ *for all* $\mathbf{x}, \mathbf{y} \in \mathbb{R}^n$

(4) $\|A\mathbf{x}\| = \|\mathbf{x}\|$ *for all* $\mathbf{x} \in \mathbb{R}^n$*.*

In this case, the matrix linear map T_A *is injective and* $n \leq m$*.*

Proof. (1) $\Rightarrow$ (2) is obvious.

(2) $\Rightarrow$ (3) $(A\mathbf{x}) \cdot (A\mathbf{y}) = (A\mathbf{x})^tA\mathbf{y} = \mathbf{x}^tA^tA\mathbf{y} \overset{(2)}{=} \mathbf{x}^tI_n\mathbf{y} = \mathbf{x}^t\mathbf{y} = \mathbf{x} \cdot \mathbf{y}$
(3) $\Rightarrow$ (1) $A^{(i)} \cdot A^{(j)} = (A\mathbf{e}^i)^t(A\mathbf{e}^j) = (A\mathbf{e}^i) \cdot (A\mathbf{e}^j) = \mathbf{e}^i \cdot \mathbf{e}^j = \delta_{ij}$
(3) $\Rightarrow$ (4) Obvious.
(4) $\Rightarrow$ (3) We use the polar identity (5.8).

$$\begin{aligned} A(\mathbf{u}) \cdot A(\mathbf{v}) &= \frac{1}{4}\|A(\mathbf{u}) + A(\mathbf{v})\|^2 - \frac{1}{4}\|A(\mathbf{u}) - A(\mathbf{v})\|^2 \\ &= \frac{1}{4}\|A(\mathbf{u} + \mathbf{v})\|^2 - \frac{1}{4}\|A(\mathbf{u} - \mathbf{v})\|^2 \\ &= \frac{1}{4}\|\mathbf{u} + \mathbf{v}\|^2 - \frac{1}{4}\|\mathbf{u} - \mathbf{v}\|^2 \qquad \text{(expanding the norm)} \\ &= \mathbf{u} \cdot \mathbf{v} \end{aligned}$$

Injectivity of T_A follows from Theorem 2.4.10 or Theorem 4.2.35 since A has a left inverse. Since T_A is injective the inequality $n \leq m$ follows from Theorem 2.4.2. □

If the columns of $A \in M(m{\times}n)$ are orthonormal then the corresponding matrix linear map $T_A : \mathbb{R}^n \to \mathbb{R}^m$ preserves dot product and hence preserves lengths, angles and orthogonality. We call such a map an **isometry** – a linear map preserving the dot product.

Theorem 5.2.14. *Let* $A \in M(n)$*. Then the following conditions are equivalent.*

(1) $A^tA = I_n$

(2) $AA^t = I_n$

(3) *The columns of* A *are orthonormal basis of* $\mathbb{R}^n$.

(4) *The rows of* A *are orthonormal basis of* $\mathbb{R}_n$.

(5) $(A\mathbf{x})\cdot(A\mathbf{y}) = \mathbf{x}\cdot\mathbf{y}$ *for all* $\mathbf{x},\mathbf{y}\in\mathbb{R}^n$

Proof. By Theorem 2.5.3, $A^tA = I_n$ if and only if $AA^t = I_n$. Hence (1) and (2) are equivalent. Now the condition (1) is equivalent to (3) by definition. The condition (2) is equivalent to (4) by definition. Also (2) is equivalent to (5) by Theorem 5.2.13. □

Definition 5.2.15. *A square matrix* $A\in M(n,\mathbb{R})$ *is* **orthogonal** *if* $AA^t = I_n = A^tA$ *i.e.,* $A^{-1} = A^t$. *If* A *is orthogonal we call the matrix linear map* T_A *an* **orthogonal matrix transformation**.

Example 5.2.16. *Let* $A_\theta = \begin{bmatrix}\cos\theta & -\sin\theta\\ \sin\theta & \cos\theta\end{bmatrix}$. *Then* A_θ *is an orthogonal matrix and the matrix transformation* T_θ *of* A_θ *represents the rotation by the angle* θ (Example 2.1.11). *Hence for any* $\mathbf{u},\mathbf{v}\in\mathbb{R}^2$ *we have* $(A_\theta\mathbf{u})\cdot(A_\theta\mathbf{v}) = \mathbf{u}\cdot\mathbf{v}$. (*Verify this also by direct computation.*)

Example 5.2.17. *Let* $A = \frac{1}{3}\begin{bmatrix}1 & 2 & 2\\ 2 & 1 & -2\\ 2 & -2 & 1\end{bmatrix}$ *and let* $\mathbf{x} = \begin{bmatrix}1\\1\\1\end{bmatrix}$, $\mathbf{y} = \begin{bmatrix}0\\1\\1\end{bmatrix}$. *Then* A *is an orthogonal matrix and* $\mathbf{x}\cdot\mathbf{y} = 2$. *And* $A\mathbf{x} = \frac{1}{3}\begin{bmatrix}5\\1\\1\end{bmatrix}$, $A\mathbf{y} = \frac{1}{3}\begin{bmatrix}4\\-1\\-1\end{bmatrix}$ *and hence* $A\mathbf{x}\cdot A\mathbf{y} = 2$.

5.2.2 Orthogonal Projections and Gram-Schmidt Process

Let $\mathbf{u},\mathbf{y}\in\mathbb{R}^n$ and $W = \mathrm{Span}\{\mathbf{u}\}$. The **orthogonal projection** $\hat{\mathbf{y}} = P_{\mathbf{u}}(\mathbf{y})$ of $\mathbf{y}$ onto $\mathbf{u}$ has to satisfies the property:

$$\hat{\mathbf{y}}\in W = \mathrm{Span}\{\mathbf{u}\} \text{ and } (\mathbf{y}-\hat{\mathbf{y}})\perp\mathbf{u} \tag{5.19}$$

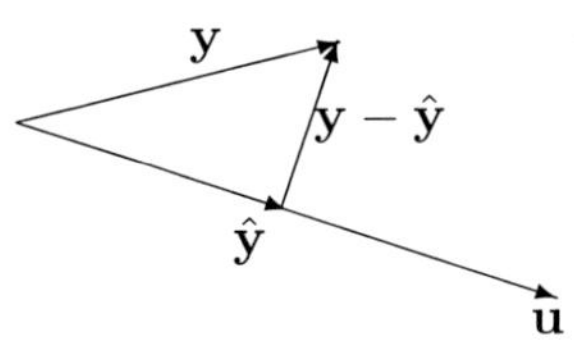

To find $\hat{\mathbf{y}}$ we let $\hat{\mathbf{y}} = \alpha\mathbf{u}$ be the (orthogonal) projection of $\mathbf{y}$ onto $\mathbf{u}$ i.e., $\mathbf{u}\cdot(\mathbf{y}-\alpha\mathbf{u}) = 0$; $\alpha\mathbf{u}\cdot\mathbf{u} = \mathbf{u}\cdot\mathbf{y}$. Hence $\alpha = \dfrac{\mathbf{u}\cdot\mathbf{y}}{\|\mathbf{u}\|^2}$ and hence we have

$$\hat{\mathbf{y}} = \frac{\mathbf{u}\cdot\mathbf{y}}{\|\mathbf{u}\|^2}\mathbf{u} \tag{5.20}$$

If we choose another basis for W say $\{c\mathbf{u}\}$ for some nonzero constant c. Then the corresponding $\hat{\mathbf{y}}$ is

$$\frac{c\mathbf{u}\cdot\mathbf{y}}{c\mathbf{u}\cdot c\mathbf{u}}c\mathbf{u} = \frac{\mathbf{u}\cdot\mathbf{y}}{\|\mathbf{u}\|^2}\mathbf{u}$$

Hence $\hat{\mathbf{y}}$ is independent of the choice of a basis of W. We will call $\hat{\mathbf{y}}$ the **orthogonal projection** of $\mathbf{y}$ onto W and we will write $\hat{\mathbf{y}} = P_W(\mathbf{y})$ or $P_{\mathbf{u}}(\mathbf{y})$ when we need to specify a basis of W:

$$\hat{\mathbf{y}} = P_W(\mathbf{y}) = P_{\mathbf{u}}(\mathbf{y}) = \frac{\mathbf{u}\cdot\mathbf{y}}{\|\mathbf{u}\|^2}\mathbf{u} \tag{5.21}$$

By (5.19), for any $\mathbf{w} \in W = \text{Span}\{\mathbf{u}\}$ we have

$$\mathbf{w} \perp (\mathbf{y} - \hat{\mathbf{y}}) \tag{5.22}$$

since $\mathbf{w}$ is a constant multiple of $\mathbf{u}$.

Another geometric interpretation of the orthogonal projection is that $\hat{\mathbf{y}}$ is the closest vector to $\mathbf{y}$ among the vectors in W;

$$\|\mathbf{y} - \hat{\mathbf{y}}\| \leq \|\mathbf{y} - \mathbf{w}\| \text{ for all } \mathbf{w} \in W \tag{5.23}$$

In fact, $\mathbf{y} - \mathbf{w} = (\mathbf{y} - \hat{\mathbf{y}}) + (\hat{\mathbf{y}} - \mathbf{w})$ and since $\hat{\mathbf{y}} - \mathbf{w} \in W$ we have $(\mathbf{y} - \hat{\mathbf{y}}) \perp (\hat{\mathbf{y}} - \mathbf{w})$. Therefore,

$$\|\mathbf{y} - \mathbf{w}\|^2 = \|\mathbf{y} - \hat{\mathbf{y}}\|^2 + \|\hat{\mathbf{y}} - \mathbf{w}\|^2$$

by Theorem 5.2.6. The inequality (5.23) follows since $\|\hat{\mathbf{y}} - \mathbf{w}\|^2 \geq 0$.

Example 5.2.18. *Let* $\mathbf{u} = \begin{bmatrix} 0 \\ 1 \\ 1 \end{bmatrix}$ *and* $W = \text{Span}\{\mathbf{u}\}$ *and let* $\mathbf{y} = \begin{bmatrix} 1 \\ 0 \\ 1 \end{bmatrix}$. *Then*

$$\hat{\mathbf{y}} = P_W(\mathbf{y}) = \frac{\mathbf{u}\cdot\mathbf{y}}{\|\mathbf{u}\|^2}\mathbf{u} = \frac{1}{2}\mathbf{u} = \frac{1}{2}\begin{bmatrix} 0 \\ 1 \\ 1 \end{bmatrix} \tag{5.24}$$

Hence $\mathbf{y} - \hat{\mathbf{y}} = \frac{1}{2}\begin{bmatrix} 2 \\ -1 \\ 1 \end{bmatrix}$ *and* $\mathbf{u}\cdot(\mathbf{y} - \hat{\mathbf{y}}) = 0$. *The distance between* $\mathbf{y}$ *and* $\hat{\mathbf{y}}$ *is*

$$d(\mathbf{y}, \hat{\mathbf{y}}) = \|\mathbf{y} - \hat{\mathbf{y}}\| = \frac{\sqrt{6}}{2}$$

Now, suppose W is a subspace of $\mathbb{R}^n$ with $\dim W \geq 1$. In terms of orthogonal projections, (5.17) of Theorem 5.2.11 says if $S = \{\mathbf{u}_1, \mathbf{u}_2, ..., \mathbf{u}_p\}$ is orthogonal basis for a subspace W and $\mathbf{y} \in W$ then we have

$$\mathbf{y} = \sum_{i=1}^{p} P_{\mathbf{u}_i}(\mathbf{y}) \tag{5.25}$$

Let $\mathbf{y} \in \mathbb{R}^n$ and W be a subspace of $\mathbb{R}^n$. To define the **orthogonal projection** $\hat{\mathbf{y}} = P_W(\mathbf{y})$ of $\mathbf{y}$ onto the subspace W with $\dim W \geq 1$ we also require the property (5.19):

$$\hat{\mathbf{y}} \in W \text{ and } (\mathbf{y} - \hat{\mathbf{y}}) \perp W \tag{5.26}$$

We will prove existence and uniqueness of $\hat{\mathbf{y}}$.

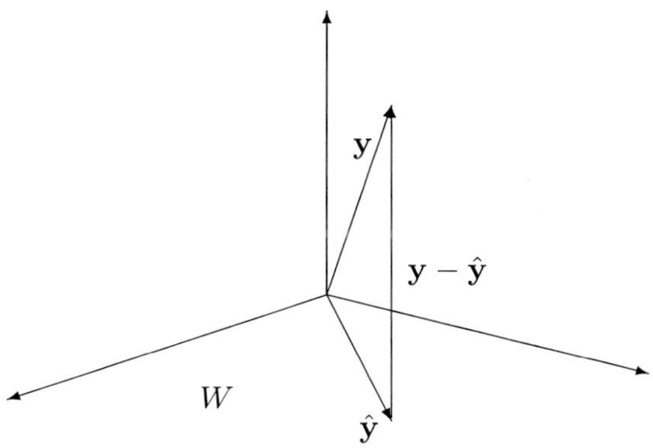

Suppose W has an orthogonal basis $S = \{\mathbf{u}_1, \mathbf{u}_2, ..., \mathbf{u}_p\}$ (Existence of such basis will be guaranteed by Theorem 5.2.26). Since $\hat{\mathbf{y}} \in W$ we let $\hat{\mathbf{y}} = \sum c_i \mathbf{u}_i$ and solve for c_i:

$$0 = (\mathbf{y} - \hat{\mathbf{y}}) \cdot \mathbf{u}_i = \mathbf{y} \cdot \mathbf{u}_i - c_i \mathbf{u}_i \cdot \mathbf{u}_i$$

Hence $c_i = \dfrac{\mathbf{u}_i \cdot \mathbf{y}}{\mathbf{u}_i \cdot \mathbf{u}_i}$. Thus the orthogonal projection $\hat{\mathbf{y}} = P_W(\mathbf{y})$ of $\mathbf{y}$ onto W is given by

$$P_W(\mathbf{y}) = \hat{\mathbf{y}} = \frac{\mathbf{u}_1 \cdot \mathbf{y}}{\|\mathbf{u}_1\|^2}\mathbf{u}_1 + \frac{\mathbf{u}_2 \cdot \mathbf{y}}{\|\mathbf{u}_2\|^2}\mathbf{u}_2 + \cdots + \frac{\mathbf{u}_p \cdot \mathbf{y}}{\|\mathbf{u}_p\|^2}\mathbf{u}_p \tag{5.27}$$

Shortly, we will show that (5.27) does not depend on the choice of the orthogonal basis S of W (Corollary 5.2.21).

If $S = \{\mathbf{u}_1, \mathbf{u}_2, ..., \mathbf{u}_p\}$ is orthonormal then (5.27) becomes

$$P_W(\mathbf{y}) = (\mathbf{u}_1 \cdot \mathbf{y})\mathbf{u}_1 + (\mathbf{u}_2 \cdot \mathbf{y})\mathbf{u}_2 + \cdots + (\mathbf{u}_p \cdot \mathbf{y})\mathbf{u}_p \tag{5.28}$$

Example 5.2.19. (1) *Let* $W = \text{Span}\{\mathbf{e}^1, \mathbf{e}^2\} \subseteq \mathbb{R}^3$ *and* $\mathbf{y} = \begin{bmatrix} 1 \\ 2 \\ 3 \end{bmatrix}$. *Then since* $\{\mathbf{e}^1, \mathbf{e}^2\}$ *is orthonormal we have*

$$P_W(\mathbf{y}) = (\mathbf{e}^1 \cdot \mathbf{y})\mathbf{e}^1 + (\mathbf{e}^2 \cdot \mathbf{y})\mathbf{e}^2 = \mathbf{e}^1 + 2\mathbf{e}^2 = \begin{bmatrix} 1 \\ 2 \\ 0 \end{bmatrix}$$

as we expected.

(2) *In* Example 5.2.12, *if we let* $W = \text{Span}\{\mathbf{u}_1, \mathbf{u}_2\}$ *then*

$$P_W(\mathbf{y}) = \frac{\mathbf{u}_1 \cdot \mathbf{y}}{\|\mathbf{u}_1\|^2}\mathbf{u}_1 + \frac{\mathbf{u}_2 \cdot \mathbf{y}}{\|\mathbf{u}_2\|^2}\mathbf{u}_2 = \frac{5}{11}\mathbf{u}_1 + \frac{1}{3}\mathbf{u}_2$$

(3) *In* Example 5.2.17, *let* $\mathbf{u}_1, \mathbf{u}_2$ *be the first two columns of* A *namely,* $\mathbf{u}_1 = \dfrac{1}{3}\begin{bmatrix} 1 \\ 2 \\ 2 \end{bmatrix}$ *and* $\mathbf{u}_2 = \dfrac{1}{3}\begin{bmatrix} 2 \\ 1 \\ -2 \end{bmatrix}$. *Let* $\mathbf{y} = \begin{bmatrix} 0 \\ 1 \\ 1 \end{bmatrix}$ *and* $W = \text{Span}\{\mathbf{u}_1, \mathbf{u}_2\}$. *Then*

$$P_W(\mathbf{y}) = (\mathbf{u}_1 \cdot \mathbf{y})\mathbf{u}_1 + (\mathbf{u}_2 \cdot \mathbf{y})\mathbf{u}_2 = \frac{4}{3}\mathbf{u}_1 - \frac{1}{3}\mathbf{u}_2 = \frac{1}{9}\begin{bmatrix} 2 \\ 7 \\ 10 \end{bmatrix}$$

Theorem 5.2.20. *Let W be a subspace of $\mathbb{R}^n$. Every $\mathbf{y} \in \mathbb{R}^n$ can be written uniquely as*

$$\mathbf{y} = \mathbf{y}' + \mathbf{z} \quad \text{where } \mathbf{y}' \in W \text{ and } \mathbf{z} \in W^{\perp} \tag{5.29}$$

Further, $\mathbf{y}'$ is given by $\mathbf{y}' = P_W(\mathbf{y})$ of (5.27).

The map

$$P_W : \mathbb{R}^n \to \mathbb{R}^n$$

is linear and

$$P_W(\mathbf{w}) = \mathbf{w} \text{ for all } \mathbf{w} \in W \tag{5.30}$$

$$P_W^2 = P_W \tag{5.31}$$

$$R(P_W) = W \text{ and } N(P_W) = W^{\perp} \tag{5.32}$$

Proof. For uniqueness, suppose $\mathbf{y}' + \mathbf{z} = \mathbf{y} = \mathbf{y}_1' + \mathbf{z}_1$. Then $\mathbf{y}' - \mathbf{y}_1' = \mathbf{z}_1 - \mathbf{z} \in W \cap W^{\perp} = \mathbf{0}$ by (5.14). Therefore $\mathbf{y}' = \mathbf{y}_1'$ and $\mathbf{z} = \mathbf{z}_1$.

For existence, we let $\mathbf{y}' = \hat{\mathbf{y}}$ of (5.27) and $\mathbf{z} = \mathbf{y} - \hat{\mathbf{y}}$. Then $\mathbf{y}', \mathbf{z}$ satisfies (5.29) by (5.26).

Linearity of the map P_W follows from the formula (5.27). The rest of the statements about the map P_W is Exercise 9. □

Corollary 5.2.21. *The expression* (5.27) *for $\hat{\mathbf{y}}$ does not depends on the choice of orthogonal basis $S = \{\mathbf{u}_1, \mathbf{u}_2, ..., \mathbf{u}_p\}$.*

Proof. A different expression using another orthogonal basis also satisfies (5.29). By uniqueness, both expressions are the same. □

Example 5.2.22. *Let* $\mathbf{u}_1 = \begin{bmatrix} 1 \\ -1 \\ 0 \\ 0 \end{bmatrix}$, $\mathbf{u}_2 = \begin{bmatrix} 0 \\ 1 \\ 0 \\ 0 \end{bmatrix}$, $\mathbf{u}_3 = \begin{bmatrix} 0 \\ 0 \\ 1 \\ -1 \end{bmatrix}$ *and* $\mathbf{y} = \begin{bmatrix} 1 \\ 1 \\ 1 \\ -1 \end{bmatrix}$. *Let* $W = \text{Span}\{\mathbf{u}_1, \mathbf{u}_2, \mathbf{u}_3\}$. *Then*

$$\mathcal{B} = \left\{ \mathbf{v}_1 = \begin{bmatrix} 1 \\ 0 \\ 0 \\ 0 \end{bmatrix}, \mathbf{v}_2 = \begin{bmatrix} 0 \\ 1 \\ 0 \\ 0 \end{bmatrix}, \mathbf{v}_3 = \frac{1}{\sqrt{2}} \begin{bmatrix} 0 \\ 0 \\ 1 \\ -1 \end{bmatrix} \right\}$$

$$\mathcal{C} = \left\{ \mathbf{w}_1 = \frac{1}{\sqrt{2}} \begin{bmatrix} 1 \\ -1 \\ 0 \\ 0 \end{bmatrix}, \mathbf{w}_2 = \frac{1}{\sqrt{2}} \begin{bmatrix} 1 \\ 1 \\ 0 \\ 0 \end{bmatrix}, \mathbf{w}_3 = \frac{1}{\sqrt{2}} \begin{bmatrix} 0 \\ 0 \\ 1 \\ -1 \end{bmatrix} \right\}$$

are orthonormal bases for W. The orthogonal projection of $\mathbf{y}$ onto W using $\mathcal{B}$ is

$$\sum (\mathbf{v}_i \cdot \mathbf{y}) \mathbf{v}_i = \begin{bmatrix} 1 \\ 0 \\ 0 \\ 0 \end{bmatrix} + \begin{bmatrix} 0 \\ 1 \\ 0 \\ 0 \end{bmatrix} + \begin{bmatrix} 0 \\ 0 \\ 1 \\ -1 \end{bmatrix} = \begin{bmatrix} 1 \\ 1 \\ 1 \\ -1 \end{bmatrix}$$

On the other hand, the orthogonal projection of $\mathbf{y}$ onto W using $\mathcal{C}$ gives us

$$\sum (\mathbf{w}_i \cdot \mathbf{y}) \mathbf{w}_i = 0 \cdot \begin{bmatrix} 1 \\ -1 \\ 0 \\ 0 \end{bmatrix} + \begin{bmatrix} 1 \\ 1 \\ 0 \\ 0 \end{bmatrix} + \begin{bmatrix} 0 \\ 0 \\ 1 \\ -1 \end{bmatrix} = \begin{bmatrix} 1 \\ 1 \\ 1 \\ -1 \end{bmatrix}$$

As in (5.23), where $\dim W = 1$, $P_W(\mathbf{y}) = \hat{\mathbf{y}}$ is the closest vector to $\mathbf{y}$ among the vectors in W. This will be true for any subspace W. And $\hat{\mathbf{y}}$ is characterized by this property.

Theorem 5.2.23. *Let W be a subspace of $\mathbb{R}^n$ and let $\mathbf{y} \in \mathbb{R}^n$. Then for all $\mathbf{w} \in W$ we have*

$$\|\mathbf{y} - \hat{\mathbf{y}}\| \leq \|\mathbf{y} - \mathbf{w}\| \tag{5.33}$$

Conversely, if $\mathbf{w}_0 \in W$ and $\|\mathbf{y} - \mathbf{w}_0\| \leq \|\mathbf{y} - \mathbf{w}\|$ for all $\mathbf{w} \in W$ then $\mathbf{w}_0 = \hat{\mathbf{y}}$.

Proof. The proof is the same as the proof of (5.23) in which W is of dimension 1. Since $\hat{\mathbf{y}}, \mathbf{w} \in W$ and $(\mathbf{y} - \hat{\mathbf{y}}) \perp W$ we have $(\mathbf{y} - \hat{\mathbf{y}}) \perp (\hat{\mathbf{y}} - \mathbf{v})$. By Pythagorean Theorem (Theorem 5.2.6) we have

$$\|\mathbf{y} - \mathbf{w}\|^2 = \|(\mathbf{y} - \hat{\mathbf{y}}) + (\hat{\mathbf{y}} - \mathbf{w})\|^2 = \|(\mathbf{y} - \hat{\mathbf{y}})\|^2 + \|(\hat{\mathbf{y}} - \mathbf{w})\|^2$$

Since $\|(\hat{\mathbf{y}} - \mathbf{w})\|^2 \geq 0$ we have the desired inequality.

For the converse, we have, as before

$$\|\mathbf{y} - \mathbf{w}_0\|^2 = \|(\mathbf{y} - \hat{\mathbf{y}}) + (\hat{\mathbf{y}} - \mathbf{w}_0)\|^2 = \|(\mathbf{y} - \hat{\mathbf{y}})\|^2 + \|(\hat{\mathbf{y}} - \mathbf{w}_0)\|^2$$

Since both $\|\mathbf{y} - \mathbf{w}_0\|$ and $\|\mathbf{y} - \hat{\mathbf{y}}\|$ are the smallest values of $\{\|\mathbf{y} - \mathbf{w}\| \mid \mathbf{w} \in W\}$ they are the same. Hence $\|(\hat{\mathbf{y}} - \mathbf{w}_0)\| = 0$ and hence $\hat{\mathbf{y}} = \mathbf{w}_0$. $\square$

Next we show how to compute the orthogonal projection when W is given by the column space of a matrix.

Theorem 5.2.24. *Let $A \in M(m \times n, \mathbb{R})$ and $W = \mathrm{Col}(A)$. If $\mathbf{y} \in \mathbb{R}^m$ then the orthogonal projection $P_W(\mathbf{y}) = \hat{\mathbf{y}} = A\tilde{\mathbf{x}} \in W$ of $\mathbf{y}$ onto W(i.e., $A\tilde{\mathbf{x}}$ is closest to $\mathbf{y}$ among all $A\mathbf{x}$ ($\mathbf{x} \in \mathbb{R}^n$)) is determined by*

$$A^t A\tilde{\mathbf{x}} = A^t\mathbf{y} \tag{5.34}$$

If the columns of A are linearly independent then A^tA is invertible and $\tilde{\mathbf{x}}$ is uniquely determined by

$$\tilde{\mathbf{x}} = (A^tA)^{-1}A^t\mathbf{y} \tag{5.35}$$

If the columns of A are orthonormal then $P_W(\mathbf{y}) = A\tilde{\mathbf{x}}$ is given by

$$P_W(\mathbf{y}) = AA^t\mathbf{y} \tag{5.36}$$

Proof. Since $W = \mathrm{Col}(A)$ the orthogonal projection $\hat{\mathbf{y}}$ of $\mathbf{y}$ onto W is of the form $\hat{\mathbf{y}} = A\tilde{\mathbf{x}}$. By (5.26), $\hat{\mathbf{y}}$ is characterized by $(\mathbf{y} - \hat{\mathbf{y}}) \perp W$. Since $W = \mathrm{Col}(A)$ this is equivalent to that $A^{(i)} \cdot (\mathbf{y} - \hat{\mathbf{y}}) = 0$ i.e., $A^t_{(i)}(\mathbf{y} - \hat{\mathbf{y}}) = 0$ for all i; $A^t\mathbf{y} = A^t\hat{\mathbf{y}}$. Since $\hat{\mathbf{y}} = A\tilde{\mathbf{x}}$ the latter condition is equivalent to $A^t\mathbf{y} = A^tA\tilde{\mathbf{x}}$ which is the condition (5.34).

Linear independence of the columns of A is equivalent to the condition (1) of Theorem 4.2.35. Hence A^tA is invertible and by multiplying $(A^tA)^{-1}$ to (5.34), $\tilde{\mathbf{x}}$ is given by (5.35).

If the columns of A are orthonormal then $A^tA = I_n$ (Theorem 5.2.13). Multiply A to the left sides of (5.34) to get $A\tilde{\mathbf{x}} = AA^t\mathbf{y}$. $\square$

Example 5.2.25. *Continuing* Example 5.2.17, *let* $\mathbf{u}_1 = \frac{1}{3}\begin{bmatrix} 1 \\ 2 \\ 2 \end{bmatrix}, \mathbf{u}_2 = \frac{1}{3}\begin{bmatrix} 2 \\ 1 \\ -2 \end{bmatrix}$. *Then* $U = [\mathbf{u}_1\ \mathbf{u}_2] = \frac{1}{3}\begin{bmatrix} 1 & 2 \\ 2 & 1 \\ 2 & -2 \end{bmatrix}$ *consists of orthonormal columns. If* $\mathbf{y} = \begin{bmatrix} 0 \\ 1 \\ 1 \end{bmatrix}$ *and* $W = \mathrm{Col}(U)$

then by (5.36) *we have*

$$P_W(\mathbf{y}) = UU^t\mathbf{y} = \frac{1}{9}\begin{bmatrix} 1 & 2 \\ 2 & 1 \\ 2 & -2 \end{bmatrix}\begin{bmatrix} 1 & 2 & 2 \\ 2 & 1 & -2 \end{bmatrix}\begin{bmatrix} 0 \\ 1 \\ 1 \end{bmatrix} = \frac{1}{9}\begin{bmatrix} 2 \\ 7 \\ 10 \end{bmatrix}$$

which is in accordance with Example 5.2.19 (3).

Practice Problem 5.2.2. *Let* $\mathbf{u}_1 = \frac{1}{\sqrt{2}}\begin{bmatrix} 1 \\ 1 \\ 0 \end{bmatrix}$, $\mathbf{u}_2 = \begin{bmatrix} 0 \\ 0 \\ 1 \end{bmatrix}$ *and* $U = [\mathbf{u}_1\ \mathbf{u}_2] = \frac{1}{\sqrt{2}}\begin{bmatrix} 1 & 0 \\ 1 & 0 \\ 0 & \sqrt{2} \end{bmatrix}$. *Show that columns of* U *are orthonormal and find* $P_W(\mathbf{y})$ *where* $\mathbf{y} = \begin{bmatrix} 1 \\ 0 \\ 1 \end{bmatrix}$ *and* $W = \mathrm{Col}(U)$.

Now we will describe a method to find a set of orthogonal vectors, so called Gram-Schmidt process. Let $\{\mathbf{w}_1, \mathbf{w}_2, ..., \mathbf{w}_p\} \subseteq \mathbb{R}^n$ be a linearly independent set of vectors and let $W_i = \mathrm{Span}\{\mathbf{w}_1, \mathbf{w}_2, ..., \mathbf{w}_i\}$. Let

$$\begin{aligned} \mathbf{v}_1 &= \mathbf{w}_1 \in W_1 \\ \mathbf{v}_2 &= \mathbf{w}_2 - P_{W_1}(\mathbf{w}_2) = \mathbf{w}_2 - \frac{\mathbf{v}_1 \cdot \mathbf{w}_2}{\mathbf{v}_1 \cdot \mathbf{v}_1}\mathbf{v}_1 \in W_2 \end{aligned}$$

By (5.29), we have $\mathbf{v}_2 \perp W_1$ and hence we have $\mathbf{v}_1 \perp \mathbf{v}_2$ and $\mathrm{Span}\{\mathbf{v}_1, \mathbf{v}_2\} = W_2$. Now let

$$\mathbf{v}_3 = \mathbf{w}_3 - P_{W_2}(\mathbf{w}_3) = \mathbf{w}_3 - \frac{\mathbf{v}_1 \cdot \mathbf{w}_3}{\mathbf{v}_1 \cdot \mathbf{v}_1}\mathbf{v}_1 - \frac{\mathbf{v}_2 \cdot \mathbf{w}_3}{\mathbf{v}_2 \cdot \mathbf{v}_2}\mathbf{v}_2 \in W_3$$

By (5.29) again, we have $\mathbf{v}_3 \perp W_2$ and hence $\{\mathbf{v}_1, \mathbf{v}_2, \mathbf{v}_3\}$ are mutually orthogonal. And also we have $\mathrm{Span}\{\mathbf{v}_1, \mathbf{v}_2, \mathbf{v}_3\} = W_3$. Repeating this we get Gram-Schmidt† orthogonalization.

Theorem 5.2.26 (Gram-Schmidt Orthogonalization Process). *Let* W *be a subspace of* $\mathbb{R}^n$ *and* $\{\mathbf{w}_1, \mathbf{w}_2, ..., \mathbf{w}_p\}$ *be a basis of* W. *Let* $W_i = \mathrm{Span}\{\mathbf{w}_1, \mathbf{w}_2, ..., \mathbf{w}_i\}$. *Then there is an orthogonal set of vectors* $\{\mathbf{v}_1, \mathbf{v}_2, ..., \mathbf{v}_p\}$ *in* W *such that*

$$\mathrm{Span}\{\mathbf{v}_1, \mathbf{v}_2, ..., \mathbf{v}_i\} = W_i \quad \text{and} \quad \mathbf{v}_{i+1} \perp W_i \tag{5.37}$$

The vector $\mathbf{v}_i$ *can be defined inductively as*

$$\begin{cases} \mathbf{v}_1 = \mathbf{w}_1 \text{ and} \\ \mathbf{v}_i = \mathbf{w}_i - P_{W_{i-1}}(\mathbf{w}_i) \\ \quad = \mathbf{w}_i - \dfrac{\mathbf{v}_1 \cdot \mathbf{w}_i}{\mathbf{v}_1 \cdot \mathbf{v}_1}\mathbf{v}_1 - \dfrac{\mathbf{v}_2 \cdot \mathbf{w}_i}{\mathbf{v}_2 \cdot \mathbf{v}_2}\mathbf{v}_2 - \cdots - \dfrac{\mathbf{v}_{i-1} \cdot \mathbf{w}_i}{\mathbf{v}_{i-1} \cdot \mathbf{v}_{i-1}}\mathbf{v}_{i-1} \end{cases} \tag{5.38}$$

Proof. From the discussion above the theorem is obvious except possibly the equality of (5.37) which is Exercise 12. □

Example 5.2.27. *Let* $\mathbf{y} = \begin{bmatrix} 1 \\ 1 \\ 1 \\ 1 \end{bmatrix}$ *and let* $\mathbf{w}_1 = \begin{bmatrix} 1 \\ 0 \\ -1 \\ 1 \end{bmatrix}$, $\mathbf{w}_2 = \begin{bmatrix} 1 \\ 1 \\ 0 \\ 1 \end{bmatrix}$, $\mathbf{w}_3 = \begin{bmatrix} 1 \\ 1 \\ 1 \\ 0 \end{bmatrix}$. *And let* $W = \mathrm{Span}\{\mathbf{w}_1, \mathbf{w}_2, \mathbf{w}_3\}$. *We want to find orthonormal basis for* W *and* $P_W(\mathbf{y})$.

†Jørgen Pedersen Gram (1850 – 1916) was a Danish actuary and mathematician.
Erhard Schmidt (1876 – 1959) was a German mathematician.

By Gram-Schmidt we compute:

$$\mathbf{v}_1 = \mathbf{w}_1;\ \mathbf{u}_1 = \frac{\mathbf{v}_1}{\|\mathbf{v}_1\|} = \frac{1}{\sqrt{3}}\begin{bmatrix}1\\0\\-1\\1\end{bmatrix}$$

$$\mathbf{v}_2 = \mathbf{w}_2 - P_{\mathbf{v}_1}(\mathbf{w}_2) = \mathbf{w}_2 - \frac{\mathbf{v}_1\cdot\mathbf{w}_2}{\mathbf{v}_1\cdot\mathbf{v}_1}\mathbf{v}_1 = \frac{1}{3}\begin{bmatrix}1\\3\\2\\1\end{bmatrix};\ \mathbf{u}_2 = \frac{1}{\sqrt{15}}\begin{bmatrix}1\\3\\2\\1\end{bmatrix}$$

$$\mathbf{v}_3 = \mathbf{w}_3 - P_{W_2}(\mathbf{w}_3) = \mathbf{w}_3 - \frac{\mathbf{v}_1\cdot\mathbf{w}_3}{\mathbf{v}_1\cdot\mathbf{v}_1}\mathbf{v}_1 - \frac{\mathbf{v}_2\cdot\mathbf{w}_3}{\mathbf{v}_2\cdot\mathbf{v}_2}\mathbf{v}_2 = \frac{1}{5}\begin{bmatrix}3\\-1\\1\\-2\end{bmatrix};\ \mathbf{u}_3 = \frac{1}{\sqrt{15}}\begin{bmatrix}3\\-1\\1\\-2\end{bmatrix}$$

Hence $\{\mathbf{u}_1, \mathbf{u}_2, \mathbf{u}_3\}$ *is an orthonormal basis for* W *and*

$$\begin{aligned}P_W(\mathbf{y}) &= (\mathbf{u}_1\cdot\mathbf{y})\mathbf{u}_1 + (\mathbf{u}_2\cdot\mathbf{y})\mathbf{u}_2 + (\mathbf{u}_3\cdot\mathbf{y})\mathbf{u}_3\\ &= \frac{1}{\sqrt{3}}\mathbf{u}_1 + \frac{7}{\sqrt{15}}\mathbf{u}_2 + \frac{1}{\sqrt{15}}\mathbf{u}_3 = \frac{1}{3}\begin{bmatrix}3\\4\\2\\2\end{bmatrix}\end{aligned}$$

As we see in this example, the computations in the applications of Gram–Schmidt process is rather complicated. Sometimes by altering a basis of W before we apply Gram–Schmidt process may ease the computations. Try that in the following Practice Problem.

Practice Problem 5.2.3. *Let* $\mathbf{w}_1 = \begin{bmatrix}1\\0\\1\\0\end{bmatrix}, \mathbf{w}_2 = \begin{bmatrix}1\\0\\0\\1\end{bmatrix}, \mathbf{w}_3 = \begin{bmatrix}1\\-1\\0\\0\end{bmatrix}$ *and* W *be the subspace spanned by* $\{\mathbf{w}_1, \mathbf{w}_2, \mathbf{w}_3\}$ *where and* $\mathbf{y} = \begin{bmatrix}1\\2\\0\\0\end{bmatrix}$. *Find an orthonormal basis for* W *and* $P_W(\mathbf{y})$.

5.2.3 Cauchy Schwartz Inequality and Triangle Inequality

We will prove Cauchy Schwartz† inequality for $\mathbb{R}^n$ whose proof depends on the orthogonal projections of Theorem 5.2.20. Cauchy Schwartz inequality for general inner product space will be proved in Chapter 8. Cauchy Schwartz inequality has many applications in an area of analysis.

Theorem 5.2.28. *Let* $\mathbf{u}, \mathbf{v} \in \mathbb{R}^n$. *Then*

(1) $\|P_{\mathbf{u}}(\mathbf{v})\| \le \|\mathbf{v}\|$

(2) **(Cauchy Schwartz Inequality)** $|\mathbf{u}\cdot\mathbf{v}| \le \|\mathbf{u}\|\,\|\mathbf{v}\|$

(3) **(Triangle Inequality)** $\|\mathbf{u}+\mathbf{v}\| \le \|\mathbf{u}\| + \|\mathbf{v}\|$

---

†Augustin-Louis Cauchy(1789 –1857) was a French mathematician, engineer and physicist.
Karl Hermann Amandus Schwarz (1843 –1921) was a German mathematician, known for his work in complex analysis.

Proof. (1) Taking the norm of $\mathbf{v} = P_{\mathbf{u}}(\mathbf{v}) + (\mathbf{v} - P_{\mathbf{u}}(\mathbf{v}))$ we have

$$\|\mathbf{v}\| = \|P_{\mathbf{u}}(\mathbf{v})\| + \|(\mathbf{v} - P_{\mathbf{u}}(\mathbf{v}))\|$$

since $P_{\mathbf{u}}(\mathbf{v}) \perp (\mathbf{v} - P_{\mathbf{u}}(\mathbf{v}))$ (Theorem 5.2.20). Since $\|(\mathbf{v} - P_{\mathbf{u}}(\mathbf{v}))\| \geq 0$ we get our inequality.

(2) By (5.21), we have $P_{\mathbf{u}}(\mathbf{v}) = \dfrac{\mathbf{v} \cdot \mathbf{u}}{\|\mathbf{u}\|^2}\mathbf{u}$. Taking the norm we get $\|P_{\mathbf{u}}(\mathbf{v})\| = \dfrac{|\mathbf{v} \cdot \mathbf{u}|}{\|\mathbf{u}\|^2}\|\mathbf{u}\| = \dfrac{|\mathbf{v} \cdot \mathbf{u}|}{\|\mathbf{u}\|}$. Hence by (1) we have

$$\|\mathbf{v}\| \geq \|P_{\mathbf{u}}(\mathbf{v})\| = \frac{|\mathbf{v} \cdot \mathbf{u}|}{\|\mathbf{u}\|}$$

which gives the desired inequality.

(3)
$$\begin{aligned} \|\mathbf{u} + \mathbf{v}\|^2 &= \mathbf{u} \cdot \mathbf{u} + 2\mathbf{u} \cdot \mathbf{v} + \mathbf{v} \cdot \mathbf{v} \\ &= \|\mathbf{u}\|^2 + 2\mathbf{u} \cdot \mathbf{v} + \|\mathbf{v}\|^2 \\ &\leq \|\mathbf{u}\|^2 + 2\|\mathbf{u}\|\,\|\mathbf{v}\| + \|\mathbf{v}\|^2 && \text{(Cauchy Schwartz)} \\ &= (\|\mathbf{u}\| + \|\mathbf{v}\|)^2 \end{aligned}$$

□

Example 5.2.29. *If* $\mathbf{x} = \begin{bmatrix} x_1 \\ x_2 \\ \vdots \\ x_n \end{bmatrix}$, $\mathbf{y} = \begin{bmatrix} y_1 \\ y_2 \\ \vdots \\ y_n \end{bmatrix}$ *then Cauchy Schwartz inequality becomes*

$$\left| \sum_{i=1}^{n} x_i y_i \right| \leq \sqrt{x_1^2 + x_2^2 + \cdots + x_n^2}\ \sqrt{y_1^2 + y_2^2 + \cdots + y_n^2}$$

and if $n = 2$ *this becomes*

$$|\, x_1 y_1 + x_2 y_2 \,| \leq \sqrt{x_1^2 + x_2^2}\ \sqrt{y_1^2 + y_2^2}$$

The triangle inequality becomes

$$\sqrt{\sum_{i=1}^{n}(x_i + y_i)^2} \leq \sqrt{\sum_{i=1}^{n} x_i^2} + \sqrt{\sum_{i=1}^{n} y_i^2}$$

and of $n = 2$ *this becomes*

$$\sqrt{(x_1 + y_1)^2 + (x_2 + y_2)^2} \leq \sqrt{x_1^2 + x_2^2} + \sqrt{y_1^2 + y_2^2}$$

For more applications of Cauchy Schwartz inequality to infinite dimensional vector spaces see §8.1.4.

5.2.4 Least Square Problem

Suppose we have an experimental data y_i at the times t_i $(i = 1, 2, \ldots, m)$. We want to find a line $y = ct + d$ that best fits the data collected.

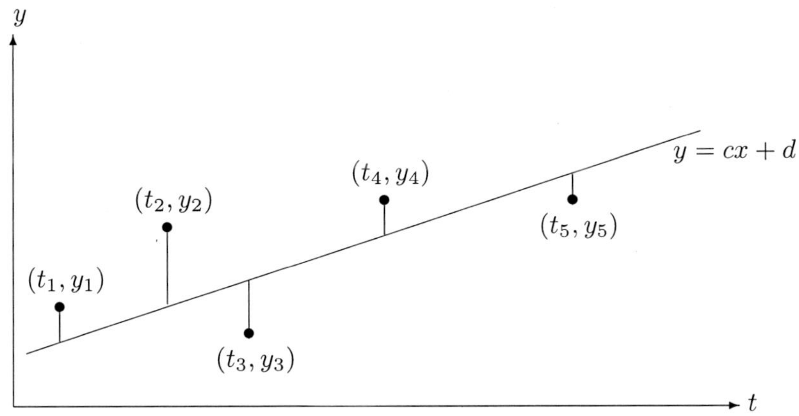

For this, we need to find a line $y = ct + d$ that minimize the sum of distance square of the vertical segment between the experimental data $\{(t_i, y_i)|\ i = 1, 2, ..., m\}$ and the point on the line. Hence we need to find c, d that minimize the quantity

$$E := \sum_{i=1}^{m} (y_i - ct_i - d)^2$$

Let

$$A = \begin{bmatrix} t_1 & 1 \\ t_2 & 1 \\ \vdots & \vdots \\ t_m & 1 \end{bmatrix}, \quad \mathbf{x} = \begin{bmatrix} c \\ d \end{bmatrix}, \quad \mathbf{y} = \begin{bmatrix} y_1 \\ y_2 \\ \vdots \\ y_m \end{bmatrix}$$

Then we can write

$$E = \|\mathbf{y} - A\mathbf{x}\|^2$$

and finding the line that best fits the data $\{(t_i, y_i)|\ i = 1, 2, ..., m\}$ amounts to find the vector $\tilde{\mathbf{x}} = \begin{bmatrix} c \\ d \end{bmatrix}$ that minimize E.

To minimize E we consider an $m \times n$ matrix A instead of a $m \times 2$ matrix for further applications. That will enable us to deal with the problems of finding a curve of higher degree that best fits the data. Hence our problem becomes: For a given matrix $A \in M(m \times n)$ and $\mathbf{y} \in \mathbb{R}^m$ find $\tilde{\mathbf{x}} \in \mathbb{R}^n$ that minimize $\|\mathbf{y} - A\mathbf{x}\|$ for all $\mathbf{x} \in \mathbb{R}^n$. That is

$$\|\mathbf{y} - A\tilde{\mathbf{x}}\| \leq \|\mathbf{y} - A\mathbf{x}\| \text{ for all } \mathbf{x} \in \mathbb{R}^n \tag{5.39}$$

The set of vectors $\{A\mathbf{x} \mid \mathbf{x} \in \mathbb{R}^n\}$ is the column space $W = \text{Col}(A)$ of A. Hence we need to find a vector $A\tilde{\mathbf{x}}$ in $\text{Col}(A)$ which is closest to $\mathbf{y}$. By Theorem 5.2.23, it is nothing but the orthogonal projection of $\mathbf{y}$ onto W;

$$A\tilde{\mathbf{x}} = P_W(\mathbf{y})$$

By Theorem 5.2.24 we obtain:

Theorem 5.2.30. *Let $A \in M(m\times n)$ and $\mathbf{y} \in \mathbb{R}^m$. Then there is $\tilde{\mathbf{x}} \in \mathbb{R}^n$ such that $A\tilde{\mathbf{x}}$ is closest to $\mathbf{y}$ among all $A\mathbf{x}$ ($\mathbf{x} \in \mathbb{R}^n$) i.e., $\|A\tilde{\mathbf{x}} - \mathbf{y}\| \le \|A\mathbf{x} - \mathbf{y}\|$ for all $\mathbf{x} \in \mathbb{R}^n$. We can choose $\tilde{\mathbf{x}}$ satisfying*

$$A^tA\tilde{\mathbf{x}} = A^t\mathbf{y} \tag{5.40}$$

Further, if $\operatorname{rank}(A) = n$ *then A^tA is invertible and*

$$\tilde{\mathbf{x}} = (A^tA)^{-1}A^t\mathbf{y}$$

Proof. This is simply a rephrasing of Theorem 5.2.24. □

Example 5.2.31. *Suppose we have data: $\{(1,2),(2,3),(3,5),(4,8)\}$. To find the best fitting line to these data let $A = \begin{bmatrix} 1 & 1 \\ 2 & 1 \\ 3 & 1 \\ 4 & 1 \end{bmatrix}$ and $\mathbf{y} = \begin{bmatrix} 2 \\ 3 \\ 5 \\ 8 \end{bmatrix}$. Since $\rho(A) = 2$ we see A^tA is invertible and $A^tA = \begin{bmatrix} 1 & 2 & 3 & 4 \\ 1 & 1 & 1 & 1 \end{bmatrix}\begin{bmatrix} 1 & 1 \\ 2 & 1 \\ 3 & 1 \\ 4 & 1 \end{bmatrix} = \begin{bmatrix} 30 & 10 \\ 10 & 4 \end{bmatrix}$. Hence*

$$\tilde{\mathbf{x}} = (A^tA)^{-1}A^t\mathbf{y} = \frac{1}{20}\begin{bmatrix} 4 & -10 \\ -10 & 30 \end{bmatrix}\begin{bmatrix} 1 & 2 & 3 & 4 \\ 1 & 1 & 1 & 1 \end{bmatrix}\begin{bmatrix} 2 \\ 3 \\ 5 \\ 8 \end{bmatrix} = \frac{1}{20}\begin{bmatrix} 40 \\ -10 \end{bmatrix} = \begin{bmatrix} 2 \\ -\frac{1}{2} \end{bmatrix}$$

Therefore, the best fitting line for the data is

$$y = 2x - \frac{1}{2}$$

Now in order to find the best fitting quadratic polynomial $y = cx^2 + dx + e$ to the collected data $\{(t_i, y_i)\}$ we need to minimize

$$E = \sum_{i=1}^{m}(y_i - ct_i^2 - dt_i - e)^2$$

If we let

$$A = \begin{bmatrix} t_1^2 & t_1 & 1 \\ t_2^2 & t_2 & 1 \\ \vdots & \vdots & \vdots \\ t_m^2 & t_m & 1 \end{bmatrix}, \quad \mathbf{x} = \begin{bmatrix} c \\ d \\ e \end{bmatrix}, \quad \mathbf{y} = \begin{bmatrix} y_1 \\ y_2 \\ \vdots \\ y_m \end{bmatrix}$$

then

$$E = \|\mathbf{y} - A\mathbf{x}\|^2$$

And we can apply Theorem 5.2.30 again.

Example 5.2.32. *Suppose we have data: $\{(1,2),(2,3),(3,5),(4,8)\}$. To find the best fitting polynomial of degree ≤ 2 to the data we let*

$$A = \begin{bmatrix} 1 & 1 & 1 \\ 4 & 2 & 1 \\ 9 & 3 & 1 \\ 16 & 4 & 1 \end{bmatrix}, \quad \tilde{\mathbf{x}} = \begin{bmatrix} a \\ b \\ c \end{bmatrix} \quad \text{and } \mathbf{y} = \begin{bmatrix} 2 \\ 3 \\ 5 \\ 8 \end{bmatrix}$$

We need to find $\tilde{\mathbf{x}} \in \mathbb{R}^3$ such that $A\tilde{\mathbf{x}}$ is closest to $\mathbf{y}$.

We use Theorem 5.2.30. *Now $\rho(A) = 3$ and hence A^tA is invertible*

$$A^tA = \begin{bmatrix} 354 & 100 & 30 \\ 100 & 30 & 10 \\ 30 & 10 & 4 \end{bmatrix}, \quad (A^tA)^{-1} = \frac{1}{20}\begin{bmatrix} 5 & -25 & 25 \\ -25 & 129 & -135 \\ 25 & -135 & 155 \end{bmatrix}$$

Hence

$$\tilde{\mathbf{x}} = (A^tA)^{-1}A^t\mathbf{y} = \frac{1}{2}\begin{bmatrix} 1 \\ -1 \\ 4 \end{bmatrix}$$

Therefore, the quadratic polynomial best fitting the data is

$$y = \frac{1}{2}x^2 - \frac{1}{2}x + 2$$

As can be seen from this example, the numbers involved get bigger as the degree of the polynomial we want get increases. We may need some aid from a machine.

Practice Problem 5.2.4. *Find the best fitting line and the quadratic polynomial for the data $\{(1,2),(3,4),\ (5,7),(6,9)\}$.*

Answers for Practice Problems 5.2

(5.2.1) We compute $\|\mathbf{u}_1\|^2 = 2, \|\mathbf{u}_2\|^2 = 6, \|\mathbf{u}_3\|^2 = 3$ and $\mathbf{v}\cdot\mathbf{u}_1 = 1, \mathbf{v}\cdot\mathbf{u}_2 = 1, \mathbf{v}\cdot\mathbf{u}_3 = 2$. Hence

$$\mathbf{v} = \frac{\mathbf{v}\cdot\mathbf{u}_1}{\|\mathbf{u}_1\|^2}\mathbf{u}_1 + \frac{\mathbf{v}\cdot\mathbf{u}_2}{\|\mathbf{u}_2\|^2}\mathbf{u}_2 + \frac{\mathbf{v}\cdot\mathbf{u}_3}{\|\mathbf{u}_3\|^2}\mathbf{u}_3 = \frac{1}{2}\mathbf{u}_1 + \frac{1}{6}\mathbf{u}_2 + \frac{2}{3}\mathbf{u}_3$$

(5.2.2) $P_W(\mathbf{y}) = UU^t\mathbf{y} = \frac{1}{2}\begin{bmatrix} 1 & 0 \\ 1 & 0 \\ 0 & \sqrt{2} \end{bmatrix}\begin{bmatrix} 1 & 1 & 0 \\ 0 & 0 & \sqrt{2} \end{bmatrix}\begin{bmatrix} 1 \\ 0 \\ 1 \end{bmatrix} = \frac{1}{2}\begin{bmatrix} 1 \\ 1 \\ 2 \end{bmatrix}$

(5.2.3) An orthonormal basis for W is $\{\mathbf{u}_1, \mathbf{u}_2, \mathbf{u}_3\}$ where $\mathbf{u}_1 = \frac{1}{\sqrt{2}}\begin{bmatrix} 1 \\ 0 \\ 1 \\ 0 \end{bmatrix}$, $\mathbf{u}_2 = \frac{1}{\sqrt{6}}\begin{bmatrix} 1 \\ 0 \\ -1 \\ 2 \end{bmatrix}$ and $\mathbf{u}_3 = \frac{1}{2\sqrt{3}}\begin{bmatrix} 1 \\ -3 \\ -1 \\ -1 \end{bmatrix}$. The orthogonal projection of $\mathbf{y}$ onto W is

$$P_W(\mathbf{y}) = (\mathbf{y}\cdot\mathbf{u}_1)\mathbf{u}_1 + (\mathbf{y}\cdot\mathbf{u}_2)\mathbf{u}_2 + (\mathbf{y}\cdot\mathbf{u}_3)\mathbf{u}_3 = \frac{1}{\sqrt{2}}\mathbf{u}_1 + \frac{1}{\sqrt{6}}\mathbf{u}_2 - \frac{1}{\sqrt{3}}\mathbf{u}_3 = \frac{1}{2}\begin{bmatrix} 1 \\ 1 \\ 1 \\ 1 \end{bmatrix}$$

If we replace $\{\mathbf{w}_1, \mathbf{w}_2, \mathbf{w}_3\}$ by $\left\{ \mathbf{w}_1' = \mathbf{w}_1 - \mathbf{w}_3 = \begin{bmatrix} 0 \\ 1 \\ 1 \\ 0 \end{bmatrix}, \mathbf{w}_2, \mathbf{w}_3 \right\}$ then $\mathbf{w}_1' \perp \mathbf{w}_2$ and we only need to find the third vector $\mathbf{v}_3$ to get an orthogonal basis: $\mathbf{v}_3 = \mathbf{w}_3 - P_{W_2}(\mathbf{w}_3) = \frac{1}{2}\begin{bmatrix} 1 \\ -1 \\ 1 \\ -1 \end{bmatrix}$.

Hence $\{\mathbf{w}_1', \mathbf{w}_2, \mathbf{v}_3\}$ is an orthogonal basis;

$$\left\{ \frac{1}{\sqrt{2}}\begin{bmatrix} 0 \\ 1 \\ 1 \\ 0 \end{bmatrix}, \frac{1}{\sqrt{2}}\begin{bmatrix} 1 \\ 0 \\ 0 \\ 1 \end{bmatrix}, \frac{1}{2}\begin{bmatrix} 1 \\ -1 \\ 1 \\ -1 \end{bmatrix} \right\}$$

is an orthonormal basis for W.

(5.2.4) Let $A = \begin{bmatrix} 1 & 1 \\ 3 & 1 \\ 5 & 1 \\ 6 & 1 \end{bmatrix}$, $\mathbf{y} = \begin{bmatrix} 2 \\ 4 \\ 7 \\ 9 \end{bmatrix}$. Then $A^tA = \begin{bmatrix} 71 & 15 \\ 15 & 4 \end{bmatrix}$ and $(A^tA)^{-1} = \frac{1}{59}\begin{bmatrix} 4 & -15 \\ -15 & 71 \end{bmatrix}$. Hence

$$(A^tA)^{-1}A^t\mathbf{y} = \frac{1}{59}\begin{bmatrix} 4 & -15 \\ -15 & 71 \end{bmatrix}\begin{bmatrix} 1 & 3 & 5 & 6 \\ 1 & 1 & 1 & 1 \end{bmatrix}\begin{bmatrix} 2 \\ 4 \\ 7 \\ 9 \end{bmatrix} = \frac{1}{59}\begin{bmatrix} 82 \\ 17 \end{bmatrix}$$

Hence the equation of the line best fitting the data is $y = \frac{82}{59}x + \frac{17}{59}$.

For the best fitting quadratic polynomial let $A = \begin{bmatrix} 1 & 1 & 1 \\ 9 & 3 & 1 \\ 25 & 5 & 1 \\ 36 & 6 & 1 \end{bmatrix}$, $\mathbf{x} = \begin{bmatrix} a \\ b \\ c \end{bmatrix}$, $\mathbf{y} = \begin{bmatrix} 2 \\ 4 \\ 7 \\ 9 \end{bmatrix}$.

Then

$$(A^tA)^{-1}A^t\mathbf{y} = \frac{1}{796}\begin{bmatrix} 109 \\ 347 \\ 1142 \end{bmatrix}$$

Hence the quadratic polynomial we look for is $y = \frac{1}{796}(109x^2 + 347x + 1142)$.

5.2 Exercises

1. Consider two vectors $\mathbf{u}, \mathbf{v}$ and let θ be the angle between $\mathbf{u}$ and $\mathbf{v}$. Find $\cos\theta$ and the distance between $\mathbf{u}$ and $\mathbf{v}$.

(a) $\mathbf{u} = \begin{bmatrix} 2 \\ 3 \end{bmatrix}, \mathbf{v} = \begin{bmatrix} -1 \\ 2 \end{bmatrix}$ (b) $\mathbf{u} = \begin{bmatrix} 2 \\ 1 \\ -1 \end{bmatrix}, \mathbf{v} = \begin{bmatrix} 1 \\ 1 \\ 0 \end{bmatrix}$ (c) $\mathbf{u} = \begin{bmatrix} 1 \\ 0 \\ 1 \\ 1 \end{bmatrix}, \mathbf{v} = \begin{bmatrix} -1 \\ 1 \\ 0 \\ 2 \end{bmatrix}$

2. Prove Theorem 5.2.2.

3. Prove the equalities (5.13) and (5.14).

4. Let W be a subspace of $\mathbb{R}^n$ spanned by $\{\mathbf{u}_1, \mathbf{u}_2, ..., \mathbf{u}_k\}$ and $\mathbf{y} \in \mathbb{R}^n$. Prove that $\mathbf{y} \perp W$ if and only if $\mathbf{y} \perp \mathbf{u}_i$ for all i.

5. Prove (5.16) of Theorem 5.2.8.

6. Prove the equalities for $\mathbf{u}, \mathbf{v} \in \mathbb{R}^n$.

   (a) $\mathbf{u} \cdot \mathbf{v} = \frac{1}{4}\|\mathbf{u}+\mathbf{v}\|^2 - \frac{1}{4}\|\mathbf{u}-\mathbf{v}\|^2$

   (b) $\|\mathbf{u}+\mathbf{v}\|^2 + \|\mathbf{u}-\mathbf{v}\|^2 = 2\|\mathbf{u}\|^2 + 2\|\mathbf{v}\|^2$

7. Let $\mathbf{u}, \mathbf{v} \in \mathbb{R}^n$. Prove that $\|\mathbf{u}+\mathbf{v}\| = 2\|\mathbf{u}\|$ and $\|\mathbf{u}\| = \|\mathbf{v}\|$ if and only if $\mathbf{u} = \mathbf{v}$. (Hint: Use Exercise 6 (b) above.)

8. In Example 5.2.18, use calculus to find c such that $\|\mathbf{y} - c\mathbf{u}\|^2$ is the smallest. Also find the minimum.

9. Complete the proof of Theorem 5.2.20.

10. For the set of vectors $S = \{\mathbf{w}_1, \mathbf{w}_2, ..., \mathbf{w}_p\}$ let $W_i = \text{Span}\{\mathbf{w}_1, \mathbf{w}_2, ..., \mathbf{w}_i\}$. Find an orthonormal basis of W_i $(i = 1, 2, ..., p)$ by using Gram Schmidt process.

    (a) $S = \left\{ \begin{bmatrix} 1 \\ 2 \end{bmatrix}, \begin{bmatrix} 3 \\ 4 \end{bmatrix} \right\}$ (b) $S = \left\{ \begin{bmatrix} 1 \\ 1 \\ 1 \end{bmatrix}, \begin{bmatrix} 1 \\ -1 \\ 2 \end{bmatrix}, \begin{bmatrix} 1 \\ 0 \\ 1 \end{bmatrix} \right\}$

    (c) $S = \left\{ \begin{bmatrix} 1 \\ -1 \\ 1 \end{bmatrix}, \begin{bmatrix} 1 \\ 0 \\ 1 \end{bmatrix}, \begin{bmatrix} 1 \\ 1 \\ -1 \end{bmatrix} \right\}$ (d) $S = \left\{ \begin{bmatrix} 1 \\ 1 \\ 1 \\ 1 \end{bmatrix}, \begin{bmatrix} 2 \\ 1 \\ 0 \\ 1 \end{bmatrix}, \begin{bmatrix} 1 \\ 1 \\ -1 \\ 0 \end{bmatrix} \right\}$

11. Find the orthogonal projection $P_W(\mathbf{y})$ of $\mathbf{y}$ to the subspace $W = \text{Span}(S)$.

    (a) $\mathbf{y} = \begin{bmatrix} -1 \\ 1 \end{bmatrix}$ and $S = \left\{ \begin{bmatrix} 1 \\ 2 \end{bmatrix} \right\}$

    (b) $\mathbf{y} = \begin{bmatrix} -1 \\ -1 \\ 1 \end{bmatrix}$ and $S = \left\{ \begin{bmatrix} 1 \\ 2 \\ 1 \end{bmatrix}, \begin{bmatrix} 1 \\ 1 \\ -1 \end{bmatrix} \right\}$

    (c) $\mathbf{y} = \begin{bmatrix} 1 \\ 2 \\ 3 \end{bmatrix}$ and $S = \left\{ \begin{bmatrix} 1 \\ 0 \\ 1 \end{bmatrix}, \begin{bmatrix} 1 \\ 1 \\ 1 \end{bmatrix} \right\}$

    (d) $\mathbf{y} = \begin{bmatrix} 1 \\ 1 \\ 1 \\ 1 \end{bmatrix}$ and $S = \left\{ \begin{bmatrix} 1 \\ 1 \\ 0 \\ 0 \end{bmatrix}, \begin{bmatrix} 1 \\ 1 \\ 1 \\ 0 \end{bmatrix} \right\}$

    (e) $\mathbf{y} = \begin{bmatrix} 1 \\ 2 \\ 3 \\ 4 \end{bmatrix}$ and $S = \left\{ \begin{bmatrix} 1 \\ 0 \\ 1 \\ 0 \end{bmatrix}, \begin{bmatrix} 0 \\ 1 \\ 0 \\ 1 \end{bmatrix}, \begin{bmatrix} 1 \\ 1 \\ 1 \\ -1 \end{bmatrix} \right\}$

12. Prove the equality $\text{Span}\{\mathbf{v}_1, \mathbf{v}_2, ..., \mathbf{v}_i\} = W_i$ of (5.37) in Theorem 5.2.26.

13. Find the best fitting polynomials of degree ≤ 1 and degree ≤ 2 for the data.

    (a) $\{(1,3),(3,5),(4,7)\}$
    (b) $\{(-1,2),(0,2),(1,3)\}$
    (c) $\{(-1,2),(0,2),(1,3),(2,4)\}$
    (d) $\{(-2,1),(-1,0),(0,2),(1,3),(2,3)\}$

14. Let W be a subspace of dimension p of $\mathbb{R}^n$.

    (a) Prove that there is a basis $\mathcal{B}$ of $\mathbb{R}^n$ such that $[P_W]_{\mathcal{B}} = \begin{bmatrix} I_p & 0 \\ 0 & 0 \end{bmatrix}$ where I_p is the $p{\times}p$ identity matrix. Deduce from this that a projection is an isomorphism if and only if it is the identity map.
    (b) Prove that $\|P_W(\mathbf{v})\| \leq \|\mathbf{v}\|$.

15. Let $S \subseteq T$ be a subsets of $\mathbb{R}^n$ and W be a subspace of $\mathbb{R}^n$. Prove the following.

    (a) $S^\perp \supseteq T^\perp$
    (b) $S \subseteq (S^\perp)^\perp$
    (c) $W = (W^\perp)^\perp$

16. Prove the true statement(s) and find a counter example for the false statement(s).

    (a) If A is an orthogonal matrix and if $S = \{\mathbf{v}_1, \mathbf{v}_2, ..., \mathbf{v}_k\}$ is orthogonal set of vectors then $T_A(S) = \{T_A(\mathbf{v}_1), T_A(\mathbf{v}_2), ..., T_A(\mathbf{v}_k)\}$ is also an orthogonal set of vectors.
    (b) Let $A \in M(m{\times}n)$. If the columns of A are orthonormal then $T_A : \mathbb{R}^m \to \mathbb{R}^n$ is an injective linear map.
    (c) Let $A \in M(m{\times}n)$. If the columns of A are orthogonal then $T_A : \mathbb{R}^m \to \mathbb{R}^n$ is an injective linear map.
    (d) If $S = \{\mathbf{v}_1, \mathbf{v}_2, ..., \mathbf{v}_k\}$ is an orthogonal set of vectors in $\mathbb{R}^n$ then $k \leq n$.
    (e) Let $A \in M(3{\times}2)$ and $A^tA = I_2$. Then we also have $AA^t = I_3$.
    (f) Let $A \in M(n)$. Then A is orthogonal if and only if $\|A\mathbf{x}\| = \|\mathbf{x}\|$.
    (g) If $A \in M(n)$ is an orthogonal matrix then $|A| = \pm 1$.
    (h) The determinant of an orthogonal projection P_W is 0 or 1.

Chapter 6

Vector Spaces and Linear Maps

So far we study the subspaces of $\mathbb{R}^n$ and linear maps between two subspaces. Our study of subspaces of $\mathbb{R}^n$ leads us to the definition of abstract vector space which is a formal statement that enables to distinguish the subspaces from the subsets of $\mathbb{R}^n$. As a consequence, a wide variety of objects can be realized as a vector space and we can analyze them mathematically. In this chapter, we state and prove the results we obtained in earlier chapters in terms of abstract vector spaces and linear maps. We also introduce tensor product and exterior product of vector spaces.

6.1 Definition of Vector Space and Basic Properties

6.1.1 Comments on Scalar Fields

Before we give a formal definition of vector space we make some remarks on the scalars. For the chapters 1–5, the scalars we considered were all real numbers. We widen our scope to extend the scalars to say, the complex numbers or restrict to the rational numbers. An abstract notion in mathematics including the scalars such as $\mathbb{Q}, \mathbb{R}$ and $\mathbb{C}$ is called a **field**. (See Appendix B.2 for the definition of fields.) In brief, a field is a set F with the operations, addition and multiplication which are associative, commutative and multiplication is distributive over addition. All of these properties are expected to hold for the fields $\mathbb{R}$, $\mathbb{C}$ and $\mathbb{Q}$. Probably one of the most distinguishing property of all is that every nonzero element $x \in F$ has a multiplicative inverse i.e., there is $y \in F$ such that $xy = 1$.

To impress the readers that there are abundance of examples of fields other than $\mathbb{Q}, \mathbb{R}$ and $\mathbb{C}$ we will give some examples. First consider

$$\mathbb{Q}[\sqrt{2}] = \{a + b\sqrt{2} \mid a, b \in \mathbb{Q}\}$$

which is a subset of $\mathbb{R}$. Sum and multiplication any two elements of $\mathbb{Q}[\sqrt{2}]$ belong to itself as can be easily checked:

$$(a + b\sqrt{2}) + (a' + b'\sqrt{2}) \in \mathbb{Q}[\sqrt{2}]$$
$$(a + b\sqrt{2})(a' + b'\sqrt{2}) \in \mathbb{Q}[\sqrt{2}]$$

Associativity, commutativity and distributivity hold because those properties are inherited from the field of real numbers. To check the existence of multiplicative inverse, let $a + b\sqrt{2}$

is a nonzero element of $\mathbb{Q}[\sqrt{2}]$. Then

$$\frac{1}{a+b\sqrt{2}} = \frac{a}{a^2-2b^2} - \frac{b}{a^2-2b^2}\sqrt{2}$$

which is again an element of $\mathbb{Q}[\sqrt{2}]$. Hence $\mathbb{Q}[\sqrt{2}]$ is a field. In the same way, if n is a square free integer[†] then $\mathbb{Q}[\sqrt{n}]$ becomes a field.

Next we will illustrate the existence of a finite field. Let $\mathbb{F}_2 := \{0,1\}$ and define the addition and multiplication by the table

$+$	0	1
0	0	1
1	1	0

$\times$	0	1
0	0	0
1	0	1

One can check that $\mathbb{F}_2$ is also a field, a field with 2 elements. Observe if $x \in \mathbb{F}_2$ then $x+x=0$. In particular, $1+1=2=0$ or $1=-1$ and we cannot divide by 2. In this case, we say the *characteristic of* $\mathbb{F}_2$ *is* 2.

It is known that for each prime p there exists a field $\mathbb{F}_{p^n}$ with p^n elements $(n=1,2,...)$ which contains $\mathbb{F}_p$, a field of p elements (Theorem B.2.5).

Any field contains, we know from abstract algebra, either $\mathbb{Q}$ or $\mathbb{F}_p$. If a field F contains $\mathbb{Q}$, as in the case of $\mathbb{R}$ or $\mathbb{C}$, then we say F is of **characteristic** 0 and we write $\text{char}(F)=0$. If F contains $\mathbb{F}_p$ then we say F is of **characteristic** p and we write $\text{char}(F)=p$. In this case, we have p times addition of any $x \in F$ is 0. In this case, we cannot divide by p. The only time when we need to worry about the characteristic of a field is when we need to divide by an integer. If the characteristic of a field is 0 then we do not need to worry about dividing by a nonzero integer. For more about fields see Appendix B.2.

Almost all the results of Chapter 1 through Chapter 5 are true if we replace the field $\mathbb{R}$ by any field F except the results of §5.2. First, we have difficulty in defining the inner product since we required the property $\mathbf{v}\cdot\mathbf{v} \geq 0$ which does not make sense for the general fields. In Chapter 8, we will define the inner product when the scalar is the field of complex numbers $\mathbb{C}$. If the field is of characteristic 2 then the polar identity (5.8) fails since we have to divide by 2. We cannot define the angle since $\cos\theta$ is not defined in the general fields. Also if the relation $\leq$ is not defined in the field then Cauchy Schwartz inequality (Theorem 5.2.28) does not make sense either.

Sometimes, we will repeat the results of the chapters 1–5 in the context of abstract vector space with some different proofs or we just quote the results of the earlier chapters without further proofs.

6.1.2 Definitions and Examples of Vector Spaces

First we give a formal definition of a vector space.

Definition 6.1.1. *Let F be a field. A set V, is a* **vector space over** F (*or* F**-vector space**) *if there are two operations called* **addition**, $\mathbf{u}+\mathbf{v}$ *of* $\mathbf{u},\mathbf{v} \in V$ *and* **scalar multiplication**, $a\cdot\mathbf{u}$ *of* $a \in F$ *and* $\mathbf{u} \in V$. *For all* $\mathbf{u},\mathbf{v} \in V, a, b \in F$ *these operations satisfy the properties* **(V1)–(V10)** *below*

[†]An integer n is square free if n is not divisible by a square of a prime.

(V1) $\mathbf{u}+\mathbf{v} \in V$ (closedness under addition).

(V2) $a\mathbf{v} \in V$ (closedness under scalar multiplication)

(V3) $\mathbf{u}+\mathbf{v} = \mathbf{v}+\mathbf{u}$ (commutativity of +)

(V4) $(\mathbf{u}+\mathbf{v})+\mathbf{w} = \mathbf{u}+(\mathbf{v}+\mathbf{w})$ (associativity of +)

(V5) *There is* $\mathbf{0}$ *such that* $\mathbf{u}+\mathbf{0} = \mathbf{u}$ (existence of additive identity)

(V6) *There is* $\mathbf{u}' \in V$ *such that* $\mathbf{u}+\mathbf{u}' = \mathbf{0}$ (existence of additive inverse)

(V7) $a(\mathbf{u}+\mathbf{v}) = a\mathbf{u}+a\mathbf{v}$ (distributivity of scalar multiplication)

(V8) $(a+b)\mathbf{u} = a\mathbf{u}+b\mathbf{u}$

(V9) $(ab)\mathbf{u} = a(b\mathbf{u})$

(V10) $1\mathbf{u} = \mathbf{u}$

We call an element of V *a* **vector** *and an element of* F *a* **scalar**.

The condition (V5) says, in particular, a vector space is nonempty. Further, the element $\mathbf{0}$ called a **zero vector** satisfying (V5) is unique. In fact, if $\bar{\mathbf{0}}$ is another element satisfying (V5) then $\mathbf{0} = \mathbf{0}+\bar{\mathbf{0}} = \bar{\mathbf{0}}$.

We start with some familiar examples.

Example 6.1.2. *A standard example of a vector space is* F^n *which consists of the column vectors*

$$F^n = \left\{ \begin{bmatrix} x_1 \\ x_2 \\ \vdots \\ x_n \end{bmatrix} \middle| \; x_i \in F \right\} \tag{6.1}$$

with the usual addition and scalar multiplication:

$$\begin{bmatrix} x_1 \\ x_2 \\ \vdots \\ x_n \end{bmatrix} + \begin{bmatrix} y_1 \\ y_2 \\ \vdots \\ y_n \end{bmatrix} = \begin{bmatrix} x_1+y_1 \\ x_2+y_2 \\ \vdots \\ x_n+y_n \end{bmatrix} \quad \text{and} \quad c \begin{bmatrix} x_1 \\ x_2 \\ \vdots \\ x_n \end{bmatrix} = \begin{bmatrix} cx_1 \\ cx_2 \\ \vdots \\ cx_n \end{bmatrix}$$

Unless otherwise stated, we view the vector space F^n *as a vector space over* F. *Of course, if* $F = \mathbb{R}$ *then* $F^n = \mathbb{R}^n$ *which we treated in earlier chapters.*

Also, observe that if $F = \mathbb{F}_p$ *then* F^n *contains* p^n *elements, a vector space with finite elements.*

Example 6.1.3. *Sometimes we need to consider the row vectors instead of columns. We let*

$$F_n = \{[x_1, x_2, ..., x_n] \mid x_i \in F\} \tag{6.2}$$

with the usual addition and scalar multiplication:

$$[x_1, x_2, ..., x_n] + [y_1, y_2, ..., y_n] = [x_1+y_1, x_2+y_2, ..., x_n+y_n],$$
$$c[x_1, x_2, ..., x_n] = [cx_1, cx_2, ..., cx_n]$$

Definition-Example 6.1.4. *An* $m \times n$ **matrix** A *over* F *is an array of scalars*

$$A = \begin{bmatrix} a_{11} & a_{12} & \cdots a_{1n} \\ a_{21} & a_{22} & \cdots a_{2n} \\ \cdots & \cdots & \\ a_{m1} & a_{m2} & \cdots a_{mn} \end{bmatrix}$$

with $a_{ij} \in F$*. We write* A_{ij} *for the* i,j*-th place of* A*. The set of all* $m \times n$ *matrices denoted by* $M(m \times n, F)$*. If* $m = n$ *then we will write* $M(n, F)$ *for* $M(m \times n, F)$*. If the reference to* F *is clear then we will simply write* $M(n), M(m \times n)$*.*

Two matrices $A, B \in M(m \times n, F)$ *are equal if* $A_{ij} = B_{ij}$ *for all* i, j*. The set of all* $m \times n$ *matrices* $M(m \times n, F)$ *is a vector space over* F *under usual addition and scalar multiplication: For* $A, B \in M(m \times n, F)$ *and* $c \in F$ *we define the* $m \times n$ *matrices* $A + B, cA$ *by* (Definition 2.1.2)

$$(A + B)_{ij} = A_{ij} + B_{ij} \text{ and } (cA)_{ij} = cA_{ij} \tag{6.3}$$

The zero vector in $M(m \times n, F)$ *is the matrix comprised of* 0*'s and the negative of* A *is* $-A$ *with* $(-A)_{ij} = -A_{ij}$*.*

Definition-Example 6.1.5. *Let* F *be a field. The set of all* **polynomials**

$$f(x) = a_0 + a_1x + \cdots + a_nx^n \quad (a_i \in F,\ a_n \neq 0, n \geq 0) \tag{6.4}$$

with coefficients in F *is denoted by* $F[x]$*:*

$$F[x] = \{a_0 + a_1x + \cdots + a_nx^n \mid a_i \in F\}$$

The **degree** $\deg(f)$ *of* $f(x)$ *in* (6.4) *is defined by*

$$\deg(f) = \begin{cases} n & \text{if } a_n \neq 0, n \geq 0 \text{ and } n \text{ is the largest such} \\ -\infty & \text{if } a_n = 0 \text{ for all } n \geq 0;\ f \text{ is the zero polynomial} \end{cases} \tag{6.5}$$

The set of polynomials $F[x]$ *becomes a vector space over* F *under usual addition and scalar multiplication: If* $f(x) = \sum a_ix^i, g(x) = \sum b_ix^i$ *and* $c \in F$ *then we define*

$$f(x) + g(x) = \sum (a_i + b_i)x^i,\ cf(x) = \sum ca_ix^i \tag{6.6}$$

The zero vector in $F[x]$ *is the zero polynomial and the additive inverse of* $f(x) = \sum a_ix^i$ *is* $-f(x) = \sum -a_ix^i$*.*

If we accept the convention that $-\infty + x = -\infty$ and $-\infty \leq x$ for all $x \in \{-\infty, 0, 1, 2, ...\}$ then the equality

$$\deg(fg) = \deg(f) + \deg(g) \tag{6.7}$$

is valid in all cases. For example, if $\deg(f) = n$ then $\deg(0) + \deg(f) = -\infty + n = -\infty$ and $\deg(0 \cdot f) = \deg(0) = -\infty$.

Definition-Example 6.1.6. *The set of polynomials with the coefficients in* F *of degree* $\leq n$ *is denoted by* $P_n(F)$ *(or simply by* P_n *if the reference to* F *is clear) i.e.,*

$$P_n(F) = \left\{ f(x) \in F[x] \,\middle|\, \deg f \leq n \right\}$$

and we define the addition and scalar multiplication by (6.6)*. Then* $P_n(F)$ *becomes a vector space over* F*. We also have*

$$F[x] = \bigcup_{n \geq 0} P_n(F)$$

Example 6.1.7. *Let F be a field and S be a nonempty set. Let*

$$\mathcal{F}(S,F) = \begin{pmatrix} \text{all functions} \\ f : S \to F \end{pmatrix}$$

For $f_1, f_2, f \in \mathcal{F}(S,F), c \in F$ we define addition and scalar multiplication by

$$(f_1 + f_2)(s) = f_1(s) + f_2(s), \qquad (cf)(s) = cf(s)$$

which make $\mathcal{F}(S,F)$ into a vector space over F.

If S is the set of natural numbers $\mathbb{N}$ and $F = \mathbb{R}$ then $\mathcal{S} := \mathcal{F}(\mathbb{N},\mathbb{R})$ is identified with the real sequences $\{(a_n)\}$ where $(a_n) = (a_1, a_2, ...)$ with $a_i \in \mathbb{R}$ by associating $f \in \mathcal{F}(\mathbb{N},\mathbb{R})$ with the sequence $(f(n))$.

If $S = F = \mathbb{R}$ then $\mathcal{F}(\mathbb{R},\mathbb{R}) = \{f : \mathbb{R} \to \mathbb{R} \mid f \text{ is a function}\}$. It contains the set of all continuous functions

$$C(\mathbb{R}) = \{f : \mathbb{R} \to \mathbb{R} \mid f \text{ is continuous}\}$$

The sum of two continuous functions and a constant multiple of a continuous function are also continuous. Hence, $C(\mathbb{R})$ is a vector space over $\mathbb{R}$.

We can replace 'continuous' by 'differentiable' or 'integrable' etc.

Example 6.1.8. (1) *Let $\mathbb{Q}[\sqrt{2}] = \{a + b\sqrt{2} \mid a, b \in \mathbb{Q}\}$ is a field as well as a vector space over $\mathbb{Q}$. In general, if $F \subseteq K$ are fields then K is a vector space over F.*

(2) *The set $\{\mathbf{0}\}$ consisting only of $\mathbf{0}$ whose existence is guaranteed by* (V5) *is a vector space, called the* **trivial vector space**. *Sometimes we denote the trivial vector space simply by $\mathbf{0}$.*

(3) *The field of complex numbers $\mathbb{C}$ is an $\mathbb{R}$-vector space as well as a $\mathbb{Q}$-vector space. Write $\mathbb{C}_\mathbb{R}$ for the $\mathbb{R}$-vector space $\mathbb{C}$.*

(4) *The field of real numbers $\mathbb{R}$ is <u>not</u> a vector space over $\mathbb{C}$*

Example 6.1.9. *Let $V = \mathbb{R}^2, F = \mathbb{R}$ and we define the addition and multiplication by some 'artificial' way say,*

$$\begin{bmatrix} x \\ y \end{bmatrix} + \begin{bmatrix} x' \\ y' \end{bmatrix} = \begin{bmatrix} x + x' \\ yy' \end{bmatrix}, \quad a\begin{bmatrix} x \\ y \end{bmatrix} = \begin{bmatrix} ax \\ ay \end{bmatrix} \tag{6.8}$$

Is V a vector space over $\mathbb{R}$? To see this, let $\mathbf{0} = \begin{bmatrix} a \\ b \end{bmatrix}$ be the element satisfying (V5). *Then*

$$\begin{bmatrix} x \\ y \end{bmatrix} + \begin{bmatrix} a \\ b \end{bmatrix} = \begin{bmatrix} x + a \\ yb \end{bmatrix} = \begin{bmatrix} x \\ y \end{bmatrix}$$

for all $\mathbf{v} = \begin{bmatrix} x \\ y \end{bmatrix}$. Hence $\begin{bmatrix} a \\ b \end{bmatrix} = \begin{bmatrix} 0 \\ 1 \end{bmatrix}$.

Now, by (V6), *for any $\mathbf{u}$ there is $\mathbf{u}'$ with $\mathbf{u} + \mathbf{u}' = \mathbf{0}$. If we let $\mathbf{u} = \begin{bmatrix} 1 \\ 0 \end{bmatrix}$ then $\mathbf{u}' = \begin{bmatrix} x' \\ y' \end{bmatrix}$ satisfies*

$$\mathbf{u} + \mathbf{u}' = \begin{bmatrix} 1 \\ 0 \end{bmatrix} + \begin{bmatrix} x' \\ y' \end{bmatrix} = \begin{bmatrix} 1 + x' \\ 0y' \end{bmatrix} = \begin{bmatrix} 0 \\ 1 \end{bmatrix}$$

But there is no y' satisfying the last equality. Hence V is not a vector space under the operations (6.8).

Practice Problem 6.1.1. *Decide whether the following operations make $\mathbb{R}^2$ into a vector space over $\mathbb{R}$ or not.*

(i) $\begin{bmatrix} x \\ y \end{bmatrix} + \begin{bmatrix} x' \\ y' \end{bmatrix} = \begin{bmatrix} 2x + x' \\ y + 2y' \end{bmatrix}, \quad a\begin{bmatrix} x \\ y \end{bmatrix} = \begin{bmatrix} ax \\ ay \end{bmatrix}.$

(ii) $\begin{bmatrix} x \\ y \end{bmatrix} + \begin{bmatrix} x' \\ y' \end{bmatrix} = \begin{bmatrix} x + x' \\ y - y' \end{bmatrix}, \quad a\begin{bmatrix} x \\ y \end{bmatrix} = \begin{bmatrix} ax \\ ay \end{bmatrix}.$

(iii) $\begin{bmatrix} x \\ y \end{bmatrix} + \begin{bmatrix} x' \\ y' \end{bmatrix} = \begin{bmatrix} xx' \\ yy' \end{bmatrix}, \quad a\begin{bmatrix} x \\ y \end{bmatrix} = \begin{bmatrix} ax \\ ay \end{bmatrix}.$

6.1.3 Basic Properties of Vector Spaces

Now we prove basic properties of vector spaces. We need to prove carefully the statements we want to prove using the definition of a vector space.

Theorem 6.1.10 (Cancellation law). *Let V be a vector space over a filed F and let $\mathbf{u}, \mathbf{v}, \mathbf{w} \in V$. If $\mathbf{u} + \mathbf{w} = \mathbf{v} + \mathbf{w}$ then we have $\mathbf{u} = \mathbf{v}$.*

Proof. By (V6), there is $\mathbf{w}' \in V$ such that $\mathbf{w} + \mathbf{w}' = \mathbf{0}$. Hence

$$\begin{aligned} \mathbf{u} &= \mathbf{u} + \mathbf{0} && \text{(V5)} \\ &= \mathbf{u} + (\mathbf{w} + \mathbf{w}') && \text{assumption} \\ &= (\mathbf{u} + \mathbf{w}) + \mathbf{w}' && \text{(V4)} \\ &= (\mathbf{v} + \mathbf{w}) + \mathbf{w}' && \text{assumption} \\ &= \mathbf{v} + (\mathbf{w} + \mathbf{w}') && \text{(V4)} \\ &= \mathbf{v} + \mathbf{0} && \text{(V5)} \\ &= \mathbf{v} \end{aligned}$$

□

Corollary 6.1.11. *Let V be a vector space over F.*

(1) *In* (V5) *the vector* $\mathbf{0}$ *is unique.*

(2) *There is a unique* $\mathbf{u}'$ *such that* $\mathbf{u} + \mathbf{u}' = \mathbf{0}$.

Proof. (1) We already observed this. Now we will use cancellation to show this. Suppose $\mathbf{0}$, $\mathbf{0}'$ both satisfy (V5). Then $\mathbf{0}+\mathbf{0}=\mathbf{0} =\mathbf{0}+\mathbf{0}'$. Hence by cancellation we have $\mathbf{0} = \mathbf{0}'$.

(2) Suppose $\mathbf{u}''$ also satisfy the equality say, $\mathbf{u} + \mathbf{u}' =\mathbf{0}= \mathbf{u} + \mathbf{u}''$. Then, by Theorem 6.1.10, we have $\mathbf{u}' = \mathbf{u}''$. □

Definition 6.1.12. *The unique vector* $\mathbf{0}$ *is called the* **zero vector** *and the unique vector* $\mathbf{u}'$ *corresponding to* $\mathbf{u}$ *in* Corollary 6.1.11 *will be denoted by* $-\mathbf{u}$.

Theorem 6.1.13. *Let V be a vector space over a field F.*

(1) *For any* $\mathbf{u} \in V$ *and* $c \in F$ *we have*

$$0\mathbf{u} = \mathbf{0} \text{ and } c\mathbf{0} = \mathbf{0}$$

(2) *For* $\mathbf{u} \in V, a \in F$ *we have*

$$(-a)\mathbf{u} = -(a\mathbf{u}) = a(-\mathbf{u})$$
$$-1\mathbf{u} = -\mathbf{u}$$

Proof. (1) We have $0\mathbf{u} \equiv \mathbf{0} + 0\mathbf{u} \overset{(V5)}{=} 0\mathbf{u} = (0+0)\mathbf{u} \overset{(V8)}{=} 0\mathbf{u} + 0\mathbf{u}$ where the second equality follows from the property $0+0=0$ of the field F. Hence by Theorem 6.1.10 we have $0\mathbf{u} = \mathbf{0}$.

For the second equality is Exercise 2.

(2) For the first equality we have $a\mathbf{u} + (-a)\mathbf{u} \overset{(V8)}{=} (a + (-a))\mathbf{u} = 0\mathbf{u} \overset{(1)}{=} \mathbf{0}$. Hence $(-a)\mathbf{u} = -(a\mathbf{u})$. In particular, $(-1)\mathbf{u} = -(1\mathbf{u}) = -\mathbf{u}$ by **(V10)**.

For the second equality we have $a\mathbf{u} + a(-\mathbf{u}) = a(\mathbf{u} + (-\mathbf{u})) = a\,\mathbf{0} = \mathbf{0}$, as required. □

Note that if F is of characteristic p and V is a vector space over F then for any $\mathbf{v} \in V$ the p-times addition of $\mathbf{v}$ becomes $\mathbf{v} + \cdots + \mathbf{v} = p\mathbf{v} = 0\mathbf{v} = \mathbf{0}$.

Example 6.1.14. *On* $\mathbb{R}^2$, *define*

$$\begin{bmatrix} x \\ y \end{bmatrix} + \begin{bmatrix} x' \\ y' \end{bmatrix} = \begin{bmatrix} x + x' \\ y + y' \end{bmatrix}, \quad a\begin{bmatrix} x \\ y \end{bmatrix} = \begin{bmatrix} ax \\ y \end{bmatrix}$$

Under these operations the zero vector should be $\begin{bmatrix} 0 \\ 0 \end{bmatrix}$. *If we let* $\mathbf{u} = \begin{bmatrix} 1 \\ 1 \end{bmatrix}$ *then* $0\mathbf{u} = \begin{bmatrix} 0 \\ 1 \end{bmatrix} \neq \begin{bmatrix} 0 \\ 0 \end{bmatrix}$. *Hence* $\mathbb{R}^2$ *with these operations is not a vector space since it violates* Theorem 6.1.13.

6.1.4 Subspaces

We define the subspaces of a vector space in the same manner as we did for $\mathbb{R}^n$ (Definition 1.2.10).

Definition 6.1.15. *Let V be a vector space over a field F. A subset W of V is called a* **subspace** (*over F, or* F**-subspace**) *of V if W is again a vector space over F under the same operation of V. A* <u>*nontrivial*</u> *subspace W of V is* **proper** *if $W \neq V$.*

Note that a subspace W of V is necessarily nonempty since $\mathbf{0} \in W$. The **trivial subspace** $\{\mathbf{0}\}$ and V itself are subspaces.

For a nonempty subset W of V to be a subspace it is necessary that

(1) if $\mathbf{u}, \mathbf{v} \in W$ then $\mathbf{u} + \mathbf{v} \in W$ and

(2) if $c \in F, \mathbf{u} \in W$ then $c\,\mathbf{u} \in W$.

We will show that these conditions are also sufficient.

Theorem 6.1.16. *A nonempty subset W of V is a subspace if and only if the following two conditions holds.*

(1) *Whenever* $\mathbf{u}, \mathbf{v} \in W$ *we have* $\mathbf{u} + \mathbf{v} \in W$.

(2) *If* $c \in F$ *and* $\mathbf{u} \in W$ *then* $c\mathbf{u} \in W$.

Proof. Obviously the conditions are necessary. Conversely assume the conditions hold for W. Then by (1) and (2), W is closed under addition and scalar multiplication. Since W is nonempty, choose $\mathbf{u} \in W$. Hence $\mathbf{0} = \mathbf{u} + (-1)\mathbf{u}$ belongs to W. Also whenever $\mathbf{u} \in W$, $-\mathbf{u} = (-1)\mathbf{u} \in W$ by (2). Now the properties (V3), (V4), (V7)–(V10) are inherited from the properties of the operations on V. □

The condition $W \neq \emptyset$ of Theorem 6.1.16 can be replaced by $\mathbf{0} \in W$ (Exercise 1).

Example 6.1.17. (1) *We saw, in* Chapter 1, *that the subspaces of* $\mathbb{R}^2$ *are* $\{\mathbf{0}\}$, *the lines through the origin and* $\mathbb{R}^2$ *itself.* (Example 1.3.31)

(2) *If* $A \in M(m\times n, F)$ *then*

$$N(A) = \{\mathbf{v} \in F^n \mid A\mathbf{v} = \mathbf{0}\}$$
$$\mathrm{Col}(A) = \mathrm{Span}\left\{A^{(1)}, A^{(2)}, \cdots, A^{(n)}\right\}$$

are subspaces of F^n *and* F^m *as we saw in* Chapter 2 *when* $F = \mathbb{R}$.

(3) *Let* $V = M(n, F)$. *Then* $W = \{A \in V \mid A = A^t\}$ *is a subspace where* $(A^t)_{ij} = A_{ji}$.

(4) *Let* $V = M(n, F)$. *Then* $W = \{A \in V \mid \mathrm{Tr}\,(A) = 0\}$ *is a subspace of* V *where* $\mathrm{Tr}(A) = \sum A_{ii}$.

(5) *Let* $V = F[X]$. *Then* $P_n(F) = \{f \in V \mid \deg f \le n\}$ *is a subspace of* $F[x]$ *for* $n = 0, 1, 2....$

Example 6.1.18. *Let* $V = \mathbb{C}^n$. *Then* V *is a vector space over* $F = \mathbb{C}$ *as well as a vector space over* $F = \mathbb{R}$ *in which case we write* $V_{\mathbb{R}}$ (Exercise 6.3.15). *The subset* $W := \mathbb{R}^n \subseteq \mathbb{C}^n$ *is not a* $\mathbb{C}$*-subspace since, for instance when* $n = 2$, $i\begin{bmatrix}1\\1\end{bmatrix} = \begin{bmatrix}i\\i\end{bmatrix} \notin \mathbb{R}^2$. *However,* W *is an* $\mathbb{R}$*-subspace. Also* $i\mathbb{R}^n = \{i\mathbf{v} \mid \mathbf{v} \in \mathbb{R}^n\} \subseteq \mathbb{C}^n$ *is an* $\mathbb{R}$*-subspace.*

Example 6.1.19. (1) *The vector space* $\mathcal{S} := \mathcal{F}(\mathbb{N}, \mathbb{R})$ *contains the convergent sequences* $(a_n)_{n\in\mathbb{N}}$ *(or simply* (a_n)*) which becomes a vector space under*

$$(a_n) + (b_n) = (a_n + b_n) \text{ and } c(a_n) = (ca_n) \tag{6.9}$$

(2) *Let* F *be* $\mathbb{R}$ *or* $\mathbb{C}$. *The set* $\mathcal{C}$ *of all sequences* (a_n) *such that the corresponding series* $\sum a_n$ *converses*

$$\mathcal{C} := \left\{(a_n) \;\middle|\; \sum a_n < \infty\right\}$$

form a vector space over F *under the operations* (6.9).

And the space of all convergent series contains the absolutely convergent series

$$\ell^1 := \left\{(a_n) \in \mathcal{C} \;\middle|\; \sum |a_i| < \infty\right\}$$

which we call the **ℓ^1-space** *is a subspace of* $\mathcal{C}$ (Exercise 6.1.5).

Also observe that for each n, *the finite sequences* $\{(a_1, a_2, ..., a_n, 0, 0, ...) \mid a_i \in \mathbb{R}\} \subseteq \ell^1$ *can be identified with the vector space* $\mathbb{R}_n$ *of the row vectors.*

(3) *The vector space* $\mathcal{F}(\mathbb{R}, \mathbb{R}) = \{f : \mathbb{R} \to \mathbb{R} \mid f \text{ is a function}\}$ *contains the subset of all continuous functions*

$$C(\mathbb{R}) := \{f : \mathbb{R} \to \mathbb{R} \mid f \text{ is continuous}\}$$

The sum of two continuous functions and a constant multiple of a continuous function are also continuous. Hence $C(\mathbb{R})$ *is a subspace of* $\mathcal{F}(\mathbb{R}, \mathbb{R})$. *We can replace 'continuous' by 'differentiable' or 'integrable' etc to get subspaces of* $\mathcal{F}(\mathbb{R}, \mathbb{R})$.

We define the **sum** of subspace as we did for the subspaces of $\mathbb{R}^n$ (Definition 1.2.18). If W_1, W_2 are subspaces of a vector space V then we define

$$W_1 + W_2 = \{\mathbf{w}_1 + \mathbf{w}_2 \mid \mathbf{w}_i \in W_i \ (i = 1, 2)\}$$

The following is a vector space version of Theorem 1.2.19.

Theorem 6.1.20. *Let W_1, W_2 be subspaces of V. Then*

(1) *The intersection $W_1 \cap W_2$ is a subspace of V.*

(2) *The sum $W_1 + W_2$ is a subspace of V.*

Proof. Exercise 4. □

Example 6.1.21. (1) *Let*

$$W_1 = \left\{ \begin{bmatrix} x \\ y \\ 0 \end{bmatrix} \middle| \, x, y \in \mathbb{R} \right\}, \quad W_2 = \left\{ \begin{bmatrix} 0 \\ 0 \\ z \end{bmatrix} \middle| \, z \in \mathbb{R} \right\}$$

Then we have $W_1 + W_2 = \mathbb{R}^3$ and $W_1 \cap W_2 = \{\mathbf{0}\}$.

(2) *Let*

$$W_1 = \left\{ \begin{bmatrix} x \\ x \\ 0 \end{bmatrix} \middle| \, x \in \mathbb{R} \right\}, \quad W_2 = \left\{ \begin{bmatrix} x \\ 2z \\ z \end{bmatrix} \middle| \, x, z \in \mathbb{R} \right\}$$

Then $W_1 + W_2 = \mathbb{R}^3$ and

$$W_1 \cap W_2 = \left\{ \begin{bmatrix} x \\ 0 \\ 0 \end{bmatrix} \middle| \, x \in \mathbb{R} \right\}$$

Practice Problem 6.1.2. *Find an example of two subspaces of $\mathbb{R}^2$ whose union is not a subspace.*

Answers for Practice Problems 6.1

(6.1.1) (i) If we take $\begin{bmatrix} x \\ y \end{bmatrix} = \begin{bmatrix} 1 \\ 2 \end{bmatrix}$ and $\begin{bmatrix} x' \\ y' \end{bmatrix} = \begin{bmatrix} 1 \\ 1 \end{bmatrix}$ then

$$\begin{bmatrix} 1 \\ 2 \end{bmatrix} + \begin{bmatrix} 1 \\ 1 \end{bmatrix} = \begin{bmatrix} 3 \\ 4 \end{bmatrix} \text{ and } \begin{bmatrix} 1 \\ 1 \end{bmatrix} + \begin{bmatrix} 1 \\ 2 \end{bmatrix} = \begin{bmatrix} 3 \\ 5 \end{bmatrix}$$

Hence commutativity of addition fails; this is not a vector space.

(ii) We have $\begin{bmatrix} 1 \\ 2 \end{bmatrix} + \begin{bmatrix} 2 \\ 1 \end{bmatrix} = \begin{bmatrix} 3 \\ 1 \end{bmatrix}$ and $\begin{bmatrix} 2 \\ 1 \end{bmatrix} + \begin{bmatrix} 1 \\ 2 \end{bmatrix} = \begin{bmatrix} 3 \\ -1 \end{bmatrix}$. Hence commutativity of addition fails. Hence this is not a vector space.

(iii) If $\mathbf{0} = \begin{bmatrix} a \\ b \end{bmatrix}$ then $\begin{bmatrix} a \\ b \end{bmatrix} + \begin{bmatrix} x \\ y \end{bmatrix} = \begin{bmatrix} ax \\ by \end{bmatrix} = \begin{bmatrix} x \\ y \end{bmatrix}$ for all $\begin{bmatrix} x \\ y \end{bmatrix}$. Hence $\mathbf{0} = \begin{bmatrix} 1 \\ 1 \end{bmatrix}$. Now let $\mathbf{u} = \begin{bmatrix} 1 \\ 0 \end{bmatrix}$ and let $\mathbf{u}' = \begin{bmatrix} x \\ y \end{bmatrix}$ of (V6). Hence $\mathbf{u} + \mathbf{u}' = \begin{bmatrix} x \\ 0 \end{bmatrix} = \begin{bmatrix} 1 \\ 1 \end{bmatrix}$; there is no $\mathbf{u}'$ satisfying $\mathbf{u} + \mathbf{u}' = \mathbf{0}$. Therefore, this is not a vector space.

(6.1.2) Let $W_1 = \left\{ \begin{bmatrix} x \\ 0 \end{bmatrix} \middle| \, x \in \mathbb{R} \right\}, W_2 = \left\{ \begin{bmatrix} 0 \\ y \end{bmatrix} \middle| \, y \in \mathbb{R} \right\}$ then

$$W_1 \cup W_2 = \left\{ \begin{bmatrix} x \\ y \end{bmatrix} \middle| \, x = 0 \text{ or } y = 0 \right\}$$

is not a subspace of $\mathbb{R}^2$. In fact, $\mathbf{v}_1 = \begin{bmatrix} 1 \\ 0 \end{bmatrix} \in W_1, \mathbf{v}_2 = \begin{bmatrix} 0 \\ 1 \end{bmatrix} \in W_2$ but $\mathbf{v}_1 + \mathbf{v}_2 = \begin{bmatrix} 1 \\ 1 \end{bmatrix}$ does not belong to $W_1 \cup W_2$.

6.1 Exercises

1. Prove that the condition $W \neq \emptyset$ of Theorem 6.1.16 can be replaced by the condition that $\mathbf{0} \in W$.

2. Let V be a vector space over F. Prove that for any $c \in F$ we have $c\mathbf{0} = \mathbf{0}$.

3. Consider the 'additions' on $\mathbb{R}^2$ defined by

   (A1) $\begin{bmatrix} x \\ y \end{bmatrix} + \begin{bmatrix} x' \\ y' \end{bmatrix} = \begin{bmatrix} x + x' \\ y + y' \end{bmatrix}$ (A2) $\begin{bmatrix} x \\ y \end{bmatrix} + \begin{bmatrix} x' \\ y' \end{bmatrix} = \begin{bmatrix} x + x' \\ yy' \end{bmatrix}$

   (A3) $\begin{bmatrix} x \\ y \end{bmatrix} + \begin{bmatrix} x' \\ y' \end{bmatrix} = \begin{bmatrix} x + 2x' \\ y + y' \end{bmatrix}$ (A4) $\begin{bmatrix} x \\ y \end{bmatrix} + \begin{bmatrix} x' \\ y' \end{bmatrix} = \begin{bmatrix} xy + x' \\ y + y' \end{bmatrix}$

   and the 'scalar multiplications' on $\mathbb{R}^2$ defined by

   (S1) $c\begin{bmatrix} x \\ y \end{bmatrix} = \begin{bmatrix} cx \\ cy \end{bmatrix}$ (S2) $c\begin{bmatrix} x \\ y \end{bmatrix} = \begin{bmatrix} cx \\ 0 \end{bmatrix}$

   (S3) $c\begin{bmatrix} x \\ y \end{bmatrix} = \begin{bmatrix} cx \\ y \end{bmatrix}$ (S4) $c\begin{bmatrix} x \\ y \end{bmatrix} = \begin{bmatrix} cx \\ 2cy \end{bmatrix}$

   Which of the following combinations of 'addition' and 'scalar multiplication' makes $\mathbb{R}^2$ a vector space? If it is a vector space then prove the fact carefully and find the zero vector and the additive inverse $-\mathbf{u}$ of $\mathbf{u} = \begin{bmatrix} x \\ y \end{bmatrix}$. If it is not a vector space then show why it fails to be a vector space.

   (a) (A2) and (S1)
   (b) (A1) and (S2)
   (c) (A1) and (S3)
   (d) (A3) and (S1)
   (e) (A1) and (S4)
   (f) (A4) and (S1)

4. Prove Theorem 6.1.20.

5. Prove that the convergent series and ℓ^1 are vector spaces.

6. Prove that each of the following subset W of the vector space V is a subspace.

   (a) $W = \{A \in M(n, F) \mid \mathrm{Tr}\,(A) = 0\}$ and $V = M(n, F)$
   (b) $W = \{A \in M(n, F) \mid A = A^t\}$ and $V = M(n, F)$
   (c) $W = \{A \in M(n, F) \mid A^t = -A\}$ and $V = M(n, F)$
   (d) $W = \{A \in M(n, F) \mid A \text{ is upper triangular}\}$ and $V = M(n, F)$
   (e) $W = \{A \in M(n, F) \mid A \text{ is diagonal}\}$ and $V = M(n, F)$
   (f) $W = \{f \in V \mid f(0) = f(1) = 0\}$ and $V = C(\mathbb{R})$

7. Prove the true statement(s) and find a counter example for the false statement(s).

   (a) The real numbers $\mathbb{R}$ is a vector space over $\mathbb{Q}$.

(b) The real numbers $\mathbb{R}$ is a vector space over $\mathbb{C}$.

(c) The real numbers $\mathbb{R}$, as a subset of $\mathbb{C}$, is a subspace of $\mathbb{C}$ over $\mathbb{R}$.

(d) The real numbers $\mathbb{R}$, as a subset of $\mathbb{C}$, is a subspace of $\mathbb{C}$ over $\mathbb{C}$.

(e) The complex numbers $\mathbb{C}$ is a vector space over $\mathbb{Q}$.

(f) The rational numbers $\mathbb{Q}$ is a vector space over $\mathbb{R}$.

(g) The set of all real convergent sequences form a vector space under usual addition and constant multiple.

(h) Every vector space contains an infinite number of elements.

8. Let W_1, W_2 be subspaces of a vector space V. Prove that $W_1 \cup W_2$ is a subspace if and only if $W_1 \subseteq W_2$ or $W_2 \subseteq W_1$.

9. Let V be a vector space over F and U be a subspace.

(a) For $\mathbf{v}_1, \mathbf{v}_2 \in V$ we define $\mathbf{v}_1 \sim_U \mathbf{v}_2$ (or simply $\mathbf{v}_1 \sim \mathbf{v}_2$) if and only $\mathbf{v}_1 - \mathbf{v}_2 \in U$. Prove that $\sim$ is an equivalence relation on V. And prove that the equivalence class $\bar{\mathbf{v}}$ of $\mathbf{v}$ is given by $\bar{\mathbf{v}} = \mathbf{v} + U = \{\mathbf{v} + \mathbf{u} \mid \mathbf{u} \in U\}$ (Definition A.2.1).

(b) Prove that $\bar{\mathbf{v}}_1 = \bar{\mathbf{v}}_2$ if and only if $\mathbf{v}_1 \sim \mathbf{v}_2$.

(c) Let V/U be the set of all equivalence classes $V/U = \{\bar{\mathbf{v}} \mid \mathbf{v} \in V\}$. Define

$$\bar{\mathbf{v}}_1 + \bar{\mathbf{v}}_2 = \overline{\mathbf{v}_1 + \mathbf{v}_2} \text{ and } c\bar{\mathbf{v}}_1 = \overline{c\mathbf{v}_1}$$

Prove that these operations are well defined i.e., if $\bar{\mathbf{v}}_1 = \bar{\mathbf{v}}'_1$, $\bar{\mathbf{v}}_2 = \bar{\mathbf{v}}'_2$ then $\overline{\mathbf{v}_1 + \mathbf{v}_2} = \overline{\mathbf{v}'_1 + \mathbf{v}'_2}$ and $\overline{c\mathbf{v}_1} = \overline{c\mathbf{v}'_1}$.

(d) Prove that V/U is a vector space under the operations defined above. The vector space V/U is called the **quotient space** of V by U.

6.2 Bases and Direct Sum of Subspaces

We define a basis of a vector space as we did for the subspaces of $\mathbb{R}^n$ in Chapter 2. We introduce the concept of direct sum of subspaces which will be important later on.

6.2.1 Spanning Set and Linear Independence

We recall the definition of spanning set of a subspace (Definition 1.3.1) and linear independence (Definition 1.3.14) in which the coefficients were the real numbers whereas, in this chapter, we will allow the coefficients to be in the field on which the vector space is defined.

Definition 6.2.1. *Let V be a vector space over F and S be a subset of V. For $\mathbf{v} \in V$, we say $\mathbf{v}$ is a* **linear combination** *of S if*

$$\mathbf{v} = a_1\mathbf{u}_1 + a_2\mathbf{u}_2 + \cdots + a_n\mathbf{u}_n$$

where $\mathbf{u}_i \in S, a_i \in F$. The set of all linear combinations of S, called the **span** *of S, will be denoted by* $\mathrm{Span}_F(S)$ *or simply by* $\mathrm{Span}(S)$.

$$\mathrm{Span}_F(S) = \left\{\sum c_i\mathbf{v}_i \;\middle|\; c_i \in F, \mathbf{v}_i \in S\right\}$$

The span Span (S) *of* S *is a subspace of* V (Theorem 6.2.4). *We say a subspace* W *of* V *is* **spanned** (*or* **generated**) *by* S *over* F *if* Span $(S) = W$. *We call* S *a set of* **generators** *or* **spanning set** *of* W. *We define, as in* §1.2.3,

$$\text{Span}\,(\emptyset) = \{\mathbf{0}\} \tag{6.10}$$

Observe that if $S = \{\mathbf{u}_1, \mathbf{u}_2, \ldots, \mathbf{u}_n\}$ and $W_i = \text{Span}\,\{\mathbf{u}_i\}$ then

$$\text{Span}\,(S) = W_1 + W_2 + \cdots + W_n$$

Definition 6.2.2. *We say a subset* S *of* V *is* **linearly independent over** F *if for any subset* $\{\mathbf{v}_1, \mathbf{v}_2, ..., \mathbf{v}_k\}$ *of* S *satisfies the property:*

$$\text{If } c_1\mathbf{v}_1 + c_2\mathbf{v}_2 + \cdots + c_k\mathbf{v}_k = \mathbf{0} \text{ with } c_i \in F \text{ then } c_1 = c_2 = \cdots = c_k = 0.$$

We say S *is* **linearly dependent** *if* S *is not linearly independent i.e., there are* $\mathbf{v}_i \in S$ *and* $c_i \in F$ *not all zero such that* $c_1\mathbf{v}_1 + c_2\mathbf{v}_2 + \cdots + c_n\mathbf{v}_n = \mathbf{0}$.

The empty set $\emptyset$ is linearly independent because linearly dependent set is necessarily nonempty by definition. Also note that if $\mathbf{0} \in S$ then S is linearly dependent since $c\mathbf{0} = \mathbf{0}$ for any $c \in F$ (Exercise 6.1.2).

If $\mathbf{v} \neq \mathbf{0}$ then $\{\mathbf{v}\}$ is linearly independent. Add one more nonzero vector $\mathbf{w}$ and suppose $\{\mathbf{v}, \mathbf{w}\}$ is linearly dependent namely $c\mathbf{v} + d\mathbf{w} = \mathbf{0}$ for some $c, d \in F$. In this case, $c \neq 0$ and $d \neq 0$. For if one of c, d is zero say, $c = 0$ then $d\mathbf{w} = 0$ and hence $d = 0$ for $\mathbf{w} \neq \mathbf{0}$. Therefore, $\mathbf{v} = -\frac{d}{c}\mathbf{w}$ i.e., $\mathbf{v}$ is a constant multiple of $\mathbf{w}$. Conversely if one of $\mathbf{v}, \mathbf{w}$ is a constant multiple of the other say $\mathbf{v} = c\mathbf{w}$ then $\{\mathbf{v}, \mathbf{w}\}$ is linearly dependent since $\mathbf{v} - d\mathbf{w} = \mathbf{0}$. In conclusion, $\{\mathbf{v}, \mathbf{w}\}$ is linearly dependent if and only if one of them is a constant multiple of the other i.e., $\mathbf{v} \in \text{Span}\,\{\mathbf{w}\}$. This will be generalized in Theorem 6.2.6.

Example 6.2.3. (1) *Let* $\mathbf{v}_1 = 1 + x$, $\mathbf{v}_2 = x - x^2$, $S = \{\mathbf{v}_1, \mathbf{v}_2\} \subseteq P_2$. *Let* $\mathbf{v} = 1 + x + x^2$. *Is* S *linearly independent ? Is* $\mathbf{v} \in \text{Span}\,(S)$*?*

To see if S *is linearly independent, let* $a\mathbf{v}_1 + b\mathbf{v}_2 = \mathbf{0}$. *Then* $a + (a+b)x - bx^2 = 0$. *Hence* $a = b = 0$ *and hence* S *is linearly independent.*

To find Span (S) *consider the linear combination* $a\mathbf{v}_1 + b\mathbf{v}_2 = a + (a+b)x - bx^2$. *Hence* $\alpha + \beta x + \gamma x^2 \in \text{Span}\,(S)$ *if and only if* $\alpha = \beta + \gamma$. *Therefore*

$$\text{Span}\,(S) = \{\alpha + \beta x + \gamma x^2 \mid \alpha = \beta + \gamma\}$$

Since $\mathbf{v} = 1 + x + x^2$ *does not satisfies the condition we conclude that* $\mathbf{v} \notin \text{Span}\,(S)$.

(2) *Let* $\mathbf{v}_1 = \begin{bmatrix} 1 & 0 \\ 0 & -1 \end{bmatrix}, \mathbf{v}_2 = \begin{bmatrix} 0 & 1 \\ 1 & 0 \end{bmatrix}, \mathbf{v}_3 = \begin{bmatrix} 0 & 1 \\ -1 & 0 \end{bmatrix}$ *and* $S = \{\mathbf{v}_1, \mathbf{v}_2, \mathbf{v}_3\}$. *Is* S *linearly dependent ? What is* Span (S) *?*

To see if S *is linearly independent suppose* $a\mathbf{v}_1 + b\mathbf{v}_2 + c\mathbf{v}_3 = \mathbf{0}$. *Then since*

$$a\mathbf{v}_1 + b\mathbf{v}_2 + c\mathbf{v}_3 = \begin{bmatrix} a & b+c \\ b-c & -a \end{bmatrix} \tag{6.11}$$

we have $a = 0, b + c = 0, b - c = 0$. *Therefore* $a = b = c = 0$ *and hence* S *is linearly independent.*

To find Span (S) *look at* (6.11). *How do we characterize the* $2{\times}2$ *matrices of the form* (6.11)*? The first observation is that the sum of the diagonal entries is* 0. *Is there any other* *conditions? If, for any real numbers x, y, the equation $\begin{cases} b+c=x \\ b-c=y \end{cases}$ is solvable in b, c then there is no further restriction—we can choose any numbers for the skew diagonal entries. It is immediate that the equation has a solution. Hence*

$$\text{Span}(S) = \left\{ \begin{bmatrix} a & b \\ c & -a \end{bmatrix} \middle| a, b, c \in \mathbb{R} \right\} = \{ A \in M(2, F) \mid \text{Tr}(A) = 0 \}$$

(3) *If $V = F[x]$ then $\{1, x, x^2, \ldots, x^n\}$ is linearly independent since a polynomial is zero if and only if all the coefficients are 0, by definition. Since we can choose n arbitrarily, there are infinitely many linearly independent vectors.*

We proved the following theorem for $\mathbb{R}^n$ in Theorem 1.2.17.

Theorem 6.2.4. *Let S be a subset of V.*

(1) *The span* Span (S) *of S is a subspace of V.*

(2) *If W is a subspace and $S \subseteq W$ then* Span $(S) \subseteq W$ *i.e.,* Span (S) *is the smallest subspace containing S.*

Proof. Exercise 1. □

The following theorem is a vector space version of Theorem 1.3.15.

Theorem 6.2.5. *Let $S_1 \subseteq S_2 \subseteq V$ be subsets.*

(1) *If S_1 is linearly dependent then so is S_2.*

(2) *If S_2 is linearly independent then so is S_1.*

(3) *If S_1 spans V then so does S_2.*

Proof. The second statement is the contrapositive of the first one which is obvious. The third statement is also obvious. □

From this we see that the more vectors there are the more likely the vectors are linearly dependent, and the less vectors there are the more likely the vectors are linearly independent.

The following theorem is Theorem 1.3.19 in terms of vector space. We repeat its proof for convenience.

Theorem 6.2.6. *Let V be a vector space and S a linearly independent set and $\mathbf{v} \in V$. Then $S \cup \{\mathbf{v}\}$ is linearly dependent if and only if $\mathbf{v} \in$* Span (S).

Proof. Suppose $S \cup \{\mathbf{v}\}$ is linearly dependent. Then there are constants $a_0, a_1, \ldots, a_n$ not all zero such that

$$a_0\mathbf{v} + a_1\mathbf{v}_1 + \cdots + a_n\mathbf{v}_n = 0$$

Note $a_0 \neq 0$ for if $a_0 = 0$ then $a_1\mathbf{v}_1 + \cdots + a_n\mathbf{v}_n = 0$ which contradict to the fact that S is linearly independent. So $\mathbf{v} = -\frac{1}{a_0}(a_1\mathbf{v}_1 + \cdots + a_n\mathbf{v}_n)$ which implies $\mathbf{v} \in \text{Span}(S)$.

Conversely, if $\mathbf{v} \in \text{span}(S)$ then $\mathbf{v} = a_1\mathbf{v}_1 + \cdots + a_n\mathbf{v}_n$ for some $a_i \in F, \mathbf{v}_i \in S$. Hence $\mathbf{v} - a_1\mathbf{v}_1 - \cdots - a_n\mathbf{v}_n = 0$. Therefore $S \cup \{\mathbf{v}\}$ is linearly dependent. □

6.2.2 Bases and Dimension.

We define the bases of a vector space in exactly the same manner as we did for the subspaces of $\mathbb{R}^n$ (Definition 1.3.21). This subsection can be viewed as a vector space version of §4.1. One difference is that we could prove the existence of a basis for a subspace of $\mathbb{R}^n$ but we need to resort to a general theorem for the existence of a basis for an abstract vector space.

Definition 6.2.7. *Let V be a vector space over a field F. An ordered subset $\mathcal{B}$ of V is a* **basis** *of V if $\mathcal{B}$ is linearly independent and span V.*

Example 6.2.8. (1) *Let $\mathbf{e}^i$ be the column whose i-th place is 1 and zero elsewhere as usual. Then $\mathcal{E} = \{\mathbf{e}^1, \mathbf{e}^2, \dots, \mathbf{e}^n\}$ is a basis for F^n which we call the* **standard basis**. *Because a basis comes with a specific ordering the basis $\{\mathbf{e}^n, \mathbf{e}^{n-1}, \dots, \mathbf{e}^1\}$ is different from $\mathcal{E}$.*

(2) *Let $P_n(F) = \{f \in F[x] | \deg(f) \leq n\}$ be the vector space of polynomials of degree $\leq n$. Then $\mathcal{B} = \{1, x, x^2, \dots, x^n\}$ is a basis of $P_n(F)$ which we also call the* **standard basis** *of $P_n(F)$.*

(3) *If $V = F[x]$ then $\mathcal{B} = \{1, x, x^2, \dots\}$ is a basis containing infinitely many vectors.*

(4) *Let E_{ij} be the $m \times n$ matrix whose (i, j)-th position is 1 and 0 elsewhere. Then*

$$\{E_{ij} \mid 1 \leq i \leq m, 1 \leq j \leq n\}$$

is a basis for $M(m \times n, F)$ which we call the **standard basis** *of $M(m \times n)$.*

By Theorem 4.1.2 every subspace of $\mathbb{R}^n$ has a basis. In general, every vector space has a basis. For its proof we refer to Appendix D.

Theorem D.2.1. *Let V be a vector space over a field F. Then V has a basis.*

We characterize a basis of a vector space. in different terminology.

Definition 6.2.9. *Let V be a vector space and $T \subseteq S$ be subsets of V. We say T is a* **maximal linearly independent subset in** *S if T is linearly independent but $T \cup \{\mathbf{v}\}$ is linearly dependent for any vector $\mathbf{v} \in S \smallsetminus T$* (Definition 2.3.1). *We call T a* **minimal spanning (generating) set** *of S if* $\mathrm{Span}\,(T) = \mathrm{Span}\,(S)$ *but* $\mathrm{Span}\,(T \smallsetminus \{\mathbf{v}\}) \neq \mathrm{Span}\,(S)$ *for any $\mathbf{v} \in T$.*

We showed in Theorem 4.1.6 that a maximal linearly independent subset is the same as a basis. Now we show that we can also characterize a basis as a minimal spanning set.

Theorem 6.2.10. *Let V be a vector space and S spans V. For a subset T of S the following conditions are equivalent.*

(1) *The subset T is a basis of V.*

(2) *The subset T is a maximal linearly independent set in S.*

(3) *The subset T is a minimal spanning set of S.*

Proof. (1)⇒(2) If T is a basis then T is linearly independent by definition. If $\mathbf{v} \in S$ then $\mathbf{v}$ is a linear combination of elements of T say $\mathbf{v} = \sum_{\mathbf{v}_i \in T} a_i \mathbf{v}_i$ $(a_i \in F)$. Hence $T \cup \{\mathbf{v}\}$ is linearly dependent by Theorem 6.2.6. Hence T is a maximal linearly independent set.

(2) ⇒ (3) First, we need to show that T span $\mathrm{Span}\,(S)$. If $\mathbf{v} \in S \smallsetminus T$ then $T \cup \{\mathbf{v}\}$ is linearly dependent. Hence $\mathbf{v} \in \mathrm{Span}\,(T)$ (Theorem 1.3.19). Therefore $S \subseteq \mathrm{Span}\,(T)$ and

hence $\mathrm{Span}(S) = \mathrm{Span}(T)$. Next, we have to show that if $\mathbf{v} \in T$ then $\mathrm{Span}(T \setminus \{\mathbf{v}\}) \neq \mathrm{Span}(S)$. For this assume the contrary: There is $\mathbf{v} \in T$ such that $\mathrm{Span}(T \setminus \{\mathbf{v}\}) = \mathrm{Span}(S)$. Then since $\mathbf{v} \in S$ we see that $\mathbf{v} \in \mathrm{Span}(T \setminus \{\mathbf{v}\})$ which says that $\mathbf{v}$ is a linear combination of $T \setminus \{\mathbf{v}\}$ which contradict to the fact that T is linearly independent.

(3)⇒(1) If T is a minimal spanning subset of S then T spans $V = \mathrm{Span}(S)$ by definition. To see T is linearly independent suppose $\mathbf{0} = \sum a_i\mathbf{v}_i$ ($\mathbf{v}_i \in T, a_i \in F$). If there is i such that $a_i \neq 0$ then $\mathbf{v}_i = -\frac{1}{a_i}\sum_{j\neq i} a_j\mathbf{v}_j$. Hence $T \subseteq \mathrm{Span}(T \setminus \{\mathbf{v}_i\})$; $\mathrm{Span}(T \setminus \{\mathbf{v}_i\}) = \mathrm{Span}(T) = V$ which means that T is not minimal. □

Corollary 6.2.11. *Let S be a finite set of vectors which spans V. Then there is a subset $\mathcal{B}$ of S which is a basis.*

Proof. If $\mathrm{Span}(S \setminus \{\mathbf{v}\}) \neq V$ for any $\mathbf{v} \in S$ then S is a minimal spanning set. Hence S is a basis. Suppose there is $\mathbf{v}_1 \in S$ for which $\mathrm{Span}(S \setminus \{\mathbf{v}_1\}) = V$ but there is no $\mathbf{v}_2 \in S$ such that $\mathrm{Span}(S \setminus \{\mathbf{v}_1, \mathbf{v}_2\}) = V$ then $S \setminus \{\mathbf{v}_1\}$ is a minimal spanning set and hence it is a basis. We keeping removing the 'redundant' vectors to get a minimal spanning set $\mathcal{B} = S \setminus \{\mathbf{v}_1, \mathbf{v}_2, ..., \mathbf{v}_k\}$. □

In the proof of the corollary we could choose a subset of S which is maximal linearly independent as in Theorem 4.1.6.

The following theorem is an abstract vector space version of Theorem 1.3.28. The proof is exactly the same as before and we omit the proof.

Theorem 6.2.12. *Let V be a vector space and let $\mathcal{B} = \{\mathbf{v}_1, \mathbf{v}_2, \ldots, \mathbf{v}_n\}$ be a subset of V. Then $\mathcal{B}$ is a basis of V if and only if for each $\mathbf{v} \in V$ can be written uniquely as $\mathbf{v} = c_1\mathbf{v}_1 + c_2\mathbf{v}_2 + \cdots + c_n\mathbf{v}_n$ ($c_i \in F$).*

We could prove the existence of a basis for the subspaces of $\mathbb{R}^n$ (Theorem 4.1.2). However, as we remarked earlier, for the abstract vector space we resort to the general theorem of existence of a basis for a vector space (Theorem D.2.1). The following theorem replaces Theorem 4.1.3 for the vector spaces. Observe that the proof of Theorem 4.1.3 fails for the abstract vector spaces.

Theorem 6.2.13. *Let V be a vector space and $S = \{\mathbf{v}_1, \mathbf{v}_2, \ldots, \mathbf{v}_n\}$ be a finite set which span V and let $T = \{\mathbf{w}_1, \mathbf{w}_2, \ldots, \mathbf{w}_m\}$ be a linearly independent vectors. Then there is a subset S_0 of S consisting of $n - m$ elements such that $T \cup S_0$ span V. In particular, $n \geq m$.*

Proof. We prove by induction on m. Suppose $m = 1$ and hence $T = \{\mathbf{w}_1\}$. Since S span V, we write $\mathbf{w}_1 = \sum_{i=1}^n b_i\mathbf{v}_i$. One of b_i is nonzero say, $b_1 \neq 0$ (If all $b_i = 0$ then $\mathbf{w}_1 = \mathbf{0}$ which is absurd.) Hence we have

$$\mathbf{v}_1 = -b_1^{-1}\mathbf{w}_1 + \sum_{i=2}^n b_1^{-1}b_i\mathbf{v}_i$$

Therefore $\mathbf{v}_1 \in \mathrm{Span}\{\mathbf{w}_1, \mathbf{v}_2, \mathbf{v}_3, ..., \mathbf{v}_n\}$ and hence $\mathrm{Span}\{\mathbf{w}_1, \mathbf{v}_2, \mathbf{v}_3, ..., \mathbf{v}_n\} = V$.

Now suppose $m = 2$ and $T = \{\mathbf{w}_1, \mathbf{w}_2\}$. By the previous paragraph there are $\mathbf{v}_2, ..., \mathbf{v}_n$ in S such that $\{\mathbf{w}_1, \mathbf{v}_2, \mathbf{v}_3, ..., \mathbf{v}_n\}$ spans V. Hence we can write $\mathbf{w}_2 = a_1\mathbf{w}_1 + \sum_{i=2}^n b_i\mathbf{v}_i$. There is an i such that $b_i \neq 0$, for if all $b_i = 0$ then $a_1\mathbf{w}_1 - \mathbf{w}_2 = \mathbf{0}$ which contradicts to the fact that T is linearly independent. Assume $b_2 \neq 0$. Hence we can write

$$\mathbf{v}_2 = b_2^{-1}a_1\mathbf{w}_1 - b_2^{-1}\mathbf{w}_2 + \sum_{i=3}^n b_2^{-1}b_i\mathbf{v}_i$$

and hence $\mathbf{v}_2 \in \operatorname{Span}\{\mathbf{w}_1, \mathbf{w}_2, \mathbf{v}_3, \mathbf{v}_4, ..., \mathbf{v}_n\}$ and $V = \operatorname{Span}\{\mathbf{w}_1, \mathbf{w}_2, \mathbf{v}_3, \mathbf{v}_4, ..., \mathbf{v}_n\}$.

Proceeding in this way we get our result. It is an exercise to complete the proof of formal induction steps (Exercise 4). □

Corollary 6.2.14. *Let $\mathcal{B}, \mathcal{B}'$ be bases of V consisting of finite vectors. Then $\mathcal{B}$ and $\mathcal{B}'$ have the same number of elements.*

Proof. Since $\mathcal{B}$ is linearly independent and $\mathcal{B}'$ span V we have $\#\mathcal{B} \leq \#\mathcal{B}'$ by Theorem 6.2.13. Now reversing the roles of $\mathcal{B}$ and $\mathcal{B}'$ to get $\#\mathcal{B} \geq \#\mathcal{B}'$ □

More generally, if $\mathcal{B}, \mathcal{B}'$ are bases of a vector space V then there is a bijection between $\mathcal{B}$ and $\mathcal{B}'$ even if they are infinite (Theorem D.2.3).

Definition 6.2.15. *A vector space V over a field F is said to be* **finite dimensional** *if it has a basis consisting of finite vectors. If the number of a basis is n then we say that V is of* **dimension** *n and we write $dim_F(V) = n$(or simply $dim(V) = n$). A vector space not having a finite basis (i.e., no finite subset of V spans V) is called an* **infinite dimensional** *vector space.*

Most of the time we will deal with the vector spaces of finite dimension. Infinite dimensional vector spaces with extra structures are treated in the area of analysis.

Corollary 6.2.16. *Let $\dim V = n$ and let S be a set of vectors having n elements.*

(1) *If S span V then S is a basis of V.*

(2) *If S is linearly independent then S is a basis of V.*

Proof. (1) By Corollary 6.2.11, we can choose a subset $\mathcal{B}$ of S which is a basis. Since $\dim(V) = n$, $\mathcal{B}$ has n elements. Because the number of elements of S is also n, we conclude that $\mathcal{B} = S$.

(2) Let $\mathcal{B}$ be a basis of V. Then, by Theorem 6.2.13, there is a subset S_0 of $\mathcal{B}$ consisting of $n - n = 0$ vectors such that $S \cup S_0 = S$ spans V. Since S is linearly independent as well, S itself is a basis. □

Abstract vector space version of Theorem 4.1.3 and Corollary 4.1.8 is:

Corollary 6.2.17. *Let V be a vector space of dimension n. Let S be a finite subset of V.*

(1) *If S is linearly independent then the number of elements of S is at most n.*

(2) *If S spans V then the number of elements of S is at least n.*

Proof. Exercise 5. □

We have an abstract vector space version of Theorem 4.1.7.

Corollary 6.2.18. *Let V be a vector space of dimension n. If $T = \{\mathbf{v}_1, \mathbf{v}_2, \ldots, \mathbf{v}_k\}$ $(k \leq n)$ is a linearly independent then there is a set of vectors $S_0 = \{\mathbf{v}_{k+1}, \ldots, \mathbf{v}_n\}$ such that $T \cup S_0$ is a basis.*

Proof. Choose a basis $\mathcal{B}$ consisting of n vectors. By Theorem 6.2.13, we can choose a subset $S_0 = \{\mathbf{v}_{k+1}, \ldots, \mathbf{v}_n\}$ of $\mathcal{B}$ such that $T \cup S_0$ spans V. Since $T \cup S_0$ consists of n elements it is a basis by Corollary 6.2.16. □

Example 6.2.19. (1) *If* $V = \{\mathbf{0}\}$ *then the empty set is a basis. In fact, the empty set is linearly independent* (Remark after Definition 6.2.2) *and the empty set spans* V *by* (6.10). *Hence the dimension of the trivial vector space is of dimension* 0 *and conversely.*

(2) $\dim_F F^n = n, \ \dim_F M(m \times n, F) = nm.$

(3) $\dim_{\mathbb{C}}(\mathbb{C}) = 1, \ \dim_{\mathbb{R}}(\mathbb{C}) = 2$

(4) $\dim P_n(F) = n+1, \ \dim_F F[x] = \infty$

Example 6.2.20. *Consider the vector space* $V = \mathbb{R}$ *over the rational numbers* $F = \mathbb{Q}$. *There are infinitely many irrational numbers* $\sqrt{2}, \sqrt{3}, \pi, e, \ldots$. *We expect that there are infinitely many linearly independent real numbers over* $\mathbb{Q}$ *which would imply that* $\dim \mathbb{R}_{\mathbb{Q}} = \infty$. *Now* $\sqrt{2}, \sqrt{3}$ *are linearly independent since* $\sqrt{2}$ *is not a rational multiple of* $\sqrt{3}$. *(Proof?)*

Let's try to show $\{\sqrt{2}, \sqrt{3}, \sqrt{5}\}$ *is linearly independent. We need to show that if* $a_1\sqrt{2} + a_2\sqrt{3} + a_3\sqrt{5} = 0$ *where* $a_i \in \mathbb{Q}$ *then* $a_i = 0$ *for all* i. *It is not immediate that this is in fact true. We need a little bit of field theory to prove that* $\{\sqrt{n} \mid n \text{ is square free}\}$ *is linearly independent. However these are not enough to span* V *over* F (Exercise 6.3.18).

If one knows that the numbers like $e, \pi, \ldots$ *are not a root of any polynomial having rational coefficients then it is immediate that* $\{1, \pi, \pi^2, \ldots\}$ *and* $\{1, e, e^2, \ldots\}$ *are linearly independent. Hence* $\dim_{\mathbb{Q}} \mathbb{R} = \infty$.

We will show that $\dim_{\mathbb{Q}}\mathbb{R} = \infty$ *in* Example 6.3.8 *by using a different method.*

Example 6.2.21. *The vector space* $V = \mathbb{C}^n$ *is a vector space over* $\mathbb{C}$. *The vector space* V *is a vector space over* $\mathbb{R}$ *as well which we denote by* $V_{\mathbb{R}}$ *See* (Exercise 6.3.15 for formal definition). *We can choose* $\{\mathbf{e}^1, \mathbf{e}^2, \ldots, \mathbf{e}^n\}$ *as a basis of* $\mathbb{C}$*-vector space* V. *On the other hand, we can choose* $\{\mathbf{e}^1, \mathbf{e}^2, \ldots, \mathbf{e}^n\} \cup \{i\mathbf{e}^1, i\mathbf{e}^2, \ldots, i\mathbf{e}^n\}$ *as a basis for the* $\mathbb{R}$*-vector space* $V_{\mathbb{R}}$. *If* $S = \{\mathbf{u}_1, \mathbf{u}_2, \ldots, \mathbf{u}_k\}$ *is a finite subset of* V. *Then*

$$\operatorname{Span}_{\mathbb{R}}(S) = \left\{ \sum a_i \mathbf{u}_i \;\middle|\; a_i \in \mathbb{R} \right\}$$

is an $\mathbb{R}$*-subspace of* V *whereas* $\operatorname{Span}(S) = \operatorname{Span}_{\mathbb{C}}(S)$ *is a* $\mathbb{C}$*-subspace of* V. *Also note that* $\mathbb{R}$*-linear dependence of* S *implies* $\mathbb{C}$*-linear dependence.*

If $S = \{\mathbf{u}_1, \mathbf{u}_2, \ldots, \mathbf{u}_k\} \subseteq \mathbb{R}^n \subseteq \mathbb{C}^n$ *then* S *is linearly independent over* $\mathbb{R}$ *if and only if* S *is linearly independent over* $\mathbb{C}$. *In fact, if* S *is* $\mathbb{C}$*-linearly independent then obviously it is* $\mathbb{R}$*-linearly independent. Conversely, assume* S *is* $\mathbb{R}$*-linearly independent. Suppose* $\sum c_k \mathbf{u}_k = \mathbf{0}$ *and let* $c_k = a_k + ib_k$. *Then* $\sum a_k \mathbf{u}_k + i \sum b_k \mathbf{u}_k = \mathbf{0}$. *Hence* $\sum a_k \mathbf{u}_k = \mathbf{0}$ *and* $\sum b_k \mathbf{u}_k = \mathbf{0}$. *Therefore,* $a_k = b_k = 0$ *for all* k *i.e.,* $c_k = 0$ *for all* k. *If* S *is not a subset of* $\mathbb{R}^n$ *then the corresponding statement is false. For example,* $\left\{ \begin{bmatrix} 1 \\ 1 \end{bmatrix}, \begin{bmatrix} i \\ i \end{bmatrix} \right\}$ *is* $\mathbb{R}$*-linearly independent but* $\mathbb{C}$*-linearly dependent.*

Theorem 6.2.22. *Let* W *be a subspace of a finite dimensional vector space* V. *Then*

(1) $\dim W \leq \dim V$

(2) *If* $\dim W = \dim V$ *then* $W = V$.

Proof. (1) If $\mathcal{C} = \{\mathbf{v}_1, \mathbf{v}_2, \ldots, \mathbf{v}_k\}$ is a basis of W then $\mathcal{C}$ is linearly independent in V. Hence by Corollary 6.2.18, we can extend $\mathcal{C}$ to get a basis $\mathcal{C} \cup \{\mathbf{v}_{k+1}, \ldots, \mathbf{v}_n\}$ is a basis of V.

(2) We need to show $W \supseteq V$. Let $\mathcal{C}$ be a basis of W. Then $\#\mathcal{C} = \dim V$ and $\mathcal{C}$ is linearly independent in V. By Corollary 6.2.16 we see that $\mathcal{C}$ is a basis of V. Therefore, $W = \operatorname{Span}(\mathcal{C}) = V$. □

Example 6.2.23. *Let W be the subspace of $M(2,\mathbb{R})$ consisting of* **symmetric matrices** *i.e., $A = A^t$.*

$$W = \{A \in M(2,\mathbb{R}) \mid A = A^t\}$$

We look for a basis for W. Consider the symmetric matrices

$$\mathcal{B} = \left\{\mathbf{u}_1 = \begin{bmatrix} 1 & 0 \\ 0 & 0 \end{bmatrix}, \mathbf{u}_2 = \begin{bmatrix} 0 & 0 \\ 0 & 1 \end{bmatrix}, \mathbf{u}_3 = \begin{bmatrix} 0 & 1 \\ 1 & 0 \end{bmatrix}\right\} \tag{6.12}$$

We claim that $\mathcal{B}$ is a basis for W. Let $A = \begin{bmatrix} a & b \\ c & d \end{bmatrix}$ be a symmetric matrix. Then $b = c$. Hence $A = \begin{bmatrix} a & b \\ b & d \end{bmatrix}$. Therefore $A = a\mathbf{u}_1 + d\mathbf{u}_2 + b\mathbf{u}_3$. Now it is easy to show that $\mathcal{B}$ is linearly independent.

Practice Problem 6.2.1. *Let W be the subspace of $M(2,\mathbb{R})$ consisting of* **skew symmetric matrices** *i.e,. $A = -A^t$.*

$$W = \{A \in M(2,\mathbb{R}) \mid A = -A^t\} \tag{6.13}$$

Find a basis for W.

Probably, this is a good place to explain why the characteristic of the scalar field is necessary. If F is a field of characteristic 2 instead of $\mathbb{R}$ then $a + a = 0$ for all $a \in F$ and hence $A = -A$ for $A \in M(n, F)$. Therefore, if the characteristic of F is 2 then the symmetric matrices are the same as the skew symmetric matrices.

Definition 6.2.24. *Let $A = [a_{ij}] \in M(n,\mathbb{C})$. We define the* **conjugate** *$\bar{A}$ and* **transpose conjugate** *or* **adjoint** *A^* of A by*

$$(\bar{A})_{ij} = \bar{a}_{ij}, \ (A^*)_{ij} = \bar{a}_{ji}$$

We say A is **Hermitian** *(resp.* **skew Hermitian***) if $A^* = A$ (resp $A^* = -A$).*

Let $\mathbf{v} = \begin{bmatrix} z_1 \\ z_2 \\ \vdots \\ z_n \end{bmatrix} \in \mathbb{C}^n$ with $z_k = a_k + ib_k$. Define the **real** *and* **imaginary part** *and the* **conjugate** *of $\mathbf{v}$ by*

$$\mathbf{v}_{re} = \begin{bmatrix} a_1 \\ a_2 \\ \vdots \\ a_n \end{bmatrix} \in \mathbb{R}^n, \ \mathbf{v}_{im} = \begin{bmatrix} b_1 \\ b_2 \\ \vdots \\ b_n \end{bmatrix} \in \mathbb{R}^n, \ \bar{\mathbf{v}} = \begin{bmatrix} \bar{z}_1 \\ \bar{z}_2 \\ \vdots \\ \bar{z}_n \end{bmatrix} \in \mathbb{C}^n \tag{6.14}$$

Hence $\mathbf{v} = \mathbf{v}_{re} + i\mathbf{v}_{im}$.

Basic formulas for the conjugate and adjoint matrices are:

Theorem 6.2.25. *Let $A, B \in M(n,\mathbb{C})$. Then we have*

(1) $(A \pm B)^* = A^* \pm B^*$ (2) $(cA)^* = \bar{c}A^*$

(3) $\overline{AB} = \bar{A}\bar{B}, \ (AB)^* = B^*A^*$ (4) $A^{**} = A$

(5) $\det(\bar{A}) = \det(A^*) = \overline{\det(A)}$ (6) $\mathrm{Tr}\,(\bar{A}) = \mathrm{Tr}\,(A^*) = \overline{\mathrm{Tr}\,(A)}$

Proof. Proofs for (1)–(4) and (6) are Exercise 7.

To show the equality $\det(\bar{A}) = \overline{\det(A)}$ of (5) let $A = [a_{ij}]$. By Theorem 3.3.16

$$\det(A) = \sum_{\sigma} \text{sign}(\sigma) a_{1\sigma(1)} a_{2\sigma(2)} \cdots a_{n\sigma(n)}$$

where σ runs over all permutations of $\{1, 2, \ldots, n\}$. Hence

$$\begin{aligned}
\det(\bar{A}) &= \sum_{\sigma} \text{sign}(\sigma) \bar{a}_{1\sigma(1)} \bar{a}_{2\sigma(2)} \cdots \bar{a}_{n\sigma(n)} \\
&= \sum_{\sigma} \text{sign}(\sigma) \overline{a_{1\sigma(1)} a_{2\sigma(2)} \cdots a_{n\sigma(n)}} \\
&= \overline{\sum_{\sigma} \text{sign}(\sigma) a_{1\sigma(1)} a_{2\sigma(2)} \cdots a_{n\sigma(n)}} = \overline{\det(A)}
\end{aligned}$$

Using this the first equality of (5) follows as $\det(A^*) = \det(\overline{A^t}) = \overline{\det(A^t)} = \overline{\det(A)}$. □

Observe that if A is Hermitian then, by (5) of the theorem, we see that $\det(A)$ is real.

Example 6.2.26. *Consider the set W of all 2×2 Hermitian matrices. For example, the matrix $\begin{bmatrix} 2 & 1-i \\ 1+i & 3 \end{bmatrix}$ is Hermitian. The real symmetric matrices are also Hermitian. The set of 2×2 matrices $M(2, \mathbb{C})$ is a 4 dimensional vector space over $\mathbb{C}$ and 8 dimensional vector space over $\mathbb{R}$. (Check this.) Is W a subspace of $M(2, \mathbb{C})$?. By (2) of* Theorem 6.2.25, *W is not a $\mathbb{C}$-subspace. However, W is an $\mathbb{R}$-subspace. We want to find a basis for W. Since real symmetric matrices are Hermitian we may choose $\{\mathbf{u}_1, \mathbf{u}_2, \mathbf{u}_3\}$ of* Example 6.2.23 *as a part of a basis. We get another simple Hermitian matrix*

$$\mathbf{v} = \begin{bmatrix} 0 & i \\ -i & 0 \end{bmatrix} \tag{6.15}$$

Let's see if $S = \{\mathbf{u}_1, \mathbf{u}_2, \mathbf{u}_3, \mathbf{v}\}$ span W. Suppose $A = \begin{bmatrix} a & b \\ c & d \end{bmatrix}$ be a Hermitian matrix. Then $a, d \in \mathbb{R}$ and $c = \bar{b}$. Let $c = \alpha + \beta i, b = \alpha - \beta i$. Hence

$$A = \begin{bmatrix} a & \alpha + \beta i \\ a - \beta i & d \end{bmatrix} = a\mathbf{u}_1 + d\mathbf{u}_2 + \alpha\mathbf{u}_3 + \beta\mathbf{v}$$

It is easy to show S is linearly independent. (Check this.) In conclusion W is a real subspace of $M(2, \mathbb{C})$ of dimension 4.

Example 6.2.27. *Let $A = \begin{bmatrix} i & 1-i \\ 2+i & 1 \end{bmatrix}$. Then $\bar{A} = \begin{bmatrix} -i & 1+i \\ 2-i & 1 \end{bmatrix}$. Now* $\det(A) = -3+2i$ *and* $\det(\bar{A}) = -3 - 2i$. *Hence* $\det(\bar{A}) = \overline{\det(A)}$.

6.2.3 Direct Sum of Subspaces

Suppose W_1, W_2 are subspaces of V. We saw in Theorem 6.1.20 that the sum of two subspaces $W_1 + W_2$ is again a subspace. We first compute the dimension of the sum $W_1 + W_2$ in terms of $\dim W_1$, $\dim W_2$ and $\dim(W_1 \cap W_2)$.

Theorem 6.2.28. *Let V is a vector space and W_1, W_2 be subspaces of V. Then $W_1 + W_2$ is a subspace of dimension* $\dim W_1 + \dim W_2 - \dim(W_1 \cap W_2)$

$$\dim(W_1 + W_2) = \dim W_1 + \dim W_2 - \dim(W_1 \cap W_2) \tag{6.16}$$

Proof. First we recall $W_1 + W_2 = \{\mathbf{w}_1 + \mathbf{w}_2 \mid \mathbf{w}_i \in W_i\}$ is a subspace of V (Theorem 6.1.20). Choose a basis $\{\mathbf{u}_1, \ldots, \mathbf{u}_k\}$ of $W_1 \cap W_2$. Extend this to a basis $\{\mathbf{u}_1, \ldots, \mathbf{u}_k, \mathbf{v}_1, \ldots, \mathbf{v}_m\}$ of W_1 and to a basis $\{\mathbf{u}_1, \ldots, \mathbf{u}_k, \mathbf{w}_1, \ldots, \mathbf{w}_p\}$ of W_2.

To prove the equality (6.16) it suffices to show $\mathcal{B} = \{\mathbf{u}_1, \ldots, \mathbf{u}_k, \mathbf{v}_1, \ldots, \mathbf{v}_m, \mathbf{w}_1, \ldots, \mathbf{w}_p\}$ is a basis of $W_1 + W_2$. First we show that $\mathcal{B}$ span $W_1 + W_2$. In fact, Let $\mathbf{v} \in W_1 + W_2$. Then $\mathbf{v} = \mathbf{w}_1 + \mathbf{w}_2$ for some $\mathbf{w}_i \in W_i$ $(i = 1, 2)$. Now $\mathbf{w}_1 = a_1\mathbf{u}_1 + \cdots + a_k\mathbf{u}_k + b_1\mathbf{v}_1 + \cdots + b_m\mathbf{v}_m$. Similarly $\mathbf{w}_2 = a_1'\mathbf{u}_1 + \cdots + a_k'\mathbf{u}_k + c_1\mathbf{w}_1 + \cdots + c_m\mathbf{w}_m$. Hence

$$\mathbf{v} = (a_1 + a_1')\mathbf{u}_1 + \cdots + (a_k + a_k')\mathbf{u}_k + b_1\mathbf{v}_1 + \cdots + b_m\mathbf{v}_m + c_1\mathbf{w}_1 + \cdots + c_m\mathbf{w}_m$$

which shows that $\mathcal{B}$ span $W_1 + W_2$.

Next, we show that $\mathcal{B}$ is linearly independent. For this, let

$$a_1\mathbf{u}_1 + \cdots + a_k\mathbf{u}_k + b_1\mathbf{v}_1 + \cdots + b_m\mathbf{v}_m + c_1\mathbf{w}_1 + \cdots + c_m\mathbf{w}_m = \mathbf{0}$$

Then we have $a_1\mathbf{u}_1 + \cdots + a_k\mathbf{u}_k + b_1\mathbf{v}_1 + \cdots + b_m\mathbf{v}_m = -(c_1\mathbf{w}_1 + \cdots + c_m\mathbf{w}_m)$. The left hand side of the equality is in W_1 and the right hand side of the equality is in W_2. Hence $c_1\mathbf{w}_1 + \cdots + c_m\mathbf{w}_m \in W_1 \cap W_2$ and therefore it can be uniquely written as a sum $d_1\mathbf{u}_1 + \cdots + d_k\mathbf{u}_k$ since $\{\mathbf{u}_1, \ldots, \mathbf{u}_k\}$ is a basis of $W_1 \cap W_2$; $c_1\mathbf{w}_1 + \cdots + c_m\mathbf{w}_m = d_1\mathbf{u}_1 + \cdots + d_k\mathbf{u}_k$. Hence $-d_1\mathbf{u}_1 - \cdots - d_k\mathbf{u}_k + c_1\mathbf{w}_1 + \cdots + c_m\mathbf{w}_m = \mathbf{0}$ which implies that all the coefficients are zeros since $\{\mathbf{u}_1, \ldots, \mathbf{u}_k, \mathbf{w}_1, \ldots, \mathbf{w}_p\}$ is a basis of W_2. In particular, $c_1 = \cdots = c_m = 0$. Consequently $a_1\mathbf{u}_1 + \cdots + a_k\mathbf{u}_k + b_1\mathbf{v}_1 + \cdots + b_m\mathbf{v}_m = \mathbf{0}$. Since $\{\mathbf{u}_1, \ldots, \mathbf{u}_k, \mathbf{v}_1, \ldots, \mathbf{v}_m\}$ is a basis of W_1 we see $a_1 = \cdots = a_k = b_1 = \cdots = b_m = 0$. Hence $\mathcal{B}$ is linearly independent. □

A sum of two subspaces will be more useful when their intersection is trivial.

Definition 6.2.29. *Let W_1, W_2 be subspaces of V. We say that the sum $W_1 + W_2 := W$ is a* **direct sum** *if*

$$W_1 \cap W_2 = \{\mathbf{0}\} \tag{6.17}$$

In this case, we write

$$W = W_1 \oplus W_2$$

To understand a vector space V, one of the idea is to write V as a direct sum of smaller subspaces which are easier to understand. We give equivalent conditions for a vector space to be a direct sum of subspaces.

Theorem 6.2.30. *The following conditions are equivalent.*

(1) $V = W_1 \oplus W_2$

(2) $V = W_1 + W_2$ *and* $\dim W_1 + \dim W_2 = \dim V$

(3) *We have $V = W_1 + W_2$. And if $\mathbf{w}_1 + \mathbf{w}_2 = \mathbf{0}$ $(\mathbf{w}_i \in W_i)$ then $\mathbf{w}_1 = \mathbf{w}_2 = \mathbf{0}$.*

(4) *For any $\mathbf{v} \in V$ there are unique $\mathbf{w}_1 \in W_1$ and $\mathbf{w}_2 \in W_2$ such that $\mathbf{v} = \mathbf{w}_1 + \mathbf{w}_2$.*

(5) *If $\mathcal{B}_1, \mathcal{B}_2$ are bases of W_1, W_2 then $\mathcal{B}_1, \mathcal{B}_2$ are disjoint and $\mathcal{B}_1 \cup \mathcal{B}_2$ is a basis of V.*

Proof. (1) $\Leftrightarrow$ (2) Obvious by Theorem 6.2.28.

(1) $\Rightarrow$ (3) If $\mathbf{w}_1 + \mathbf{w}_2 = \mathbf{0}$ then $\mathbf{w}_1 = -\mathbf{w}_2 \in W_1 \cap W_2 = \{\mathbf{0}\}$. Hence $\mathbf{w}_1 = \mathbf{w}_2 = \mathbf{0}$.

(3) $\Rightarrow$ (4) We only need to show that the expression $\mathbf{v} = \mathbf{w}_1 + \mathbf{w}_2$ is unique. In fact, if $\mathbf{v} = \mathbf{w}_1 + \mathbf{w}_2 = \mathbf{w}_1' + \mathbf{w}_2'$. Then $(\mathbf{w}_1 - \mathbf{w}_1') + (\mathbf{w}_2 - \mathbf{w}_2') = \mathbf{0}$. Hence $\mathbf{w}_1 = \mathbf{w}_1', \quad \mathbf{w}_2 = \mathbf{w}_2'$.

(4) $\Rightarrow$ (5) Let $\mathbf{v} \in V$ and write $\mathbf{v} = \mathbf{w}_1 + \mathbf{w}_2$ ($\mathbf{w}_i \in W_i$) in a unique way. Let $\mathcal{B}_1 = \{\mathbf{w}_j^{(1)}\}, \mathcal{B}_2 = \{\mathbf{w}_j^{(2)}\}$ be bases of W_1, W_2. Then we can write $\mathbf{w}_i = \sum c_j^{(i)}\mathbf{w}_j^{(i)}$ uniquely for $i = 1, 2$. Hence we can write uniquely as $\mathbf{v} = \sum c_j^{(1)}\mathbf{w}_j^{(1)} + \sum c_j^{(2)}\mathbf{w}_j^{(2)}$. Hence $\mathcal{B}_1 \cup \mathcal{B}_2$ is a basis of V (Theorem 6.2.12). To show $\mathcal{B}_1$, $\mathcal{B}_2$ are disjoint let $\mathbf{w} \in \mathcal{B}_1 \cap \mathcal{B}_2$. Then we can write $\mathbf{w} = \mathbf{w} + \mathbf{0} = \mathbf{0} + \mathbf{w}$ as sums of elements of W_1, W_2. Since this expression is unique we have $\mathbf{w} = \mathbf{0}$. But a basis cannot contain the zero vector; there is no such $\mathbf{w}$.

(5) $\Rightarrow$ (2) We have $\dim V = \#(\mathcal{B}_1 \cup \mathcal{B}_2) = \#\mathcal{B}_1 + \#\mathcal{B}_2 = \dim W_1 + \dim W_2$ since $\mathcal{B}_1$ and $\mathcal{B}_2$ are disjoint. Since $\mathcal{B}_1 \cup \mathcal{B}_2$ is a basis of V, it is immediate that $V = W_1 + W_2$.

To show $W_1 \cap W_2 = \mathbf{0}$ let $\mathbf{w} \in W_1 \cap W_2$. Using the notation for $\mathcal{B}_1, \mathcal{B}_2$ above, write $\sum c_j^{(1)}\mathbf{w}_j^{(1)} = \mathbf{w} = \sum c_j^{(2)}\mathbf{w}_j^{(2)}$. Then we have $\sum c_j^{(1)}\mathbf{w}_j^{(1)} - \sum c_j^{(2)}\mathbf{w}_j^{(2)} = \mathbf{0}$. Since $\mathcal{B}_1 \cup \mathcal{B}_2$ is a basis for V we see that all the coefficients are zeros. Hence $\mathbf{w} = \mathbf{0}$. □

Example 6.2.31. *Let $V = \mathbb{R}^2$. Let*

$$W_1 = \left\{ \begin{bmatrix} x \\ 0 \end{bmatrix} \middle| \, x \in \mathbb{R} \right\}, \; W_2 = \left\{ \begin{bmatrix} 0 \\ y \end{bmatrix} \middle| \, y \in \mathbb{R} \right\}$$

$$W_3 = \left\{ \begin{bmatrix} x \\ x \end{bmatrix} \middle| \, x \in \mathbb{R} \right\}, \; W_4 = \left\{ \begin{bmatrix} x \\ 2x \end{bmatrix} \middle| \, x \in \mathbb{R} \right\}$$

Then $V = W_1 \oplus W_2 = W_2 \oplus W_3 = W_3 \oplus W_4 = \cdots$

Example 6.2.32. *Let W be a subspace of $\mathbb{R}^n$ and let $W^\perp$ be the orthogonal complement* (Definition 5.2.7)*. Then, by* Theorem 5.2.20, *every $\mathbf{y} \in \mathbb{R}^n$ can be written uniquely as $\mathbf{y} = \hat{\mathbf{y}} + \mathbf{z}$ with $\hat{\mathbf{y}} \in W, \; \mathbf{z} \in W^\perp$. Hence $\mathbb{R}^n = W \oplus W^\perp$.*

Example 6.2.33. *Let $V = \mathbb{R}^3$. Let*

$$W_1 = \left\{ \begin{bmatrix} x \\ y \\ z \end{bmatrix} \middle| \, x + y + z = 0 \right\}, \quad W_2 = \left\{ \begin{bmatrix} x \\ y \\ z \end{bmatrix} \middle| \, x + z = 0 \right\}$$

$$W_3 = \left\{ \begin{bmatrix} 0 \\ y \\ y \end{bmatrix} \middle| \, y \in \mathbb{R} \right\}, \qquad W_4 = \left\{ \begin{bmatrix} x \\ x \\ 0 \end{bmatrix} \middle| \, x \in \mathbb{R} \right\}$$

Then $W_1 \cap W_3 = W_1 \cap W_4 = \{\mathbf{0}\}$ but $W_1 \cap W_2 \neq \{\mathbf{0}\}$. We choose bases $\mathcal{B}_i$ for W_i ($i = 1, 2, 3, 4$):

$$\mathcal{B}_1 = \left\{ \mathbf{v}_1 = \begin{bmatrix} 1 \\ 0 \\ -1 \end{bmatrix}, \mathbf{v}_2 = \begin{bmatrix} 1 \\ -1 \\ 0 \end{bmatrix} \right\}, \quad \mathcal{B}_2 = \left\{ \mathbf{v}_1' = \begin{bmatrix} 1 \\ 0 \\ -1 \end{bmatrix}, \mathbf{v}_2' = \begin{bmatrix} 0 \\ 1 \\ 0 \end{bmatrix} \right\}$$

$$\mathcal{B}_3 = \left\{ \mathbf{v}_3 = \begin{bmatrix} 0 \\ 1 \\ 1 \end{bmatrix} \right\}, \qquad \mathcal{B}_4 = \left\{ \mathbf{v}_3' = \begin{bmatrix} 1 \\ 1 \\ 0 \end{bmatrix} \right\}$$

Then $\mathcal{B} := \{\mathbf{v}_1, \mathbf{v}_2, \mathbf{v}_3\}$ is linearly independent. Hence $\mathcal{B}$ is a basis of V; $W_1 + W_3 = V$. Thus $V = W_1 \oplus W_3$. Similarly, $V = W_1 \oplus W_4$.

Practice Problem 6.2.2. *Show that* $W_2 \oplus W_3 = V$ *in* Example 6.2.33.

Example 6.2.34. *Let* $V = M(2,\mathbb{R})$. *Let* W, W' *be the subspace of symmetric and skew symmetric matrices. Then* $\mathcal{B} = \left\{\mathbf{u}_1 = \begin{bmatrix}1 & 0\\ 0 & 0\end{bmatrix}, \mathbf{u}_2 = \begin{bmatrix}0 & 0\\ 0 & 1\end{bmatrix}, \mathbf{u}_3 = \begin{bmatrix}0 & 1\\ 1 & 0\end{bmatrix}\right\}$ *is a basis for* W *and* $\mathcal{B}' = \left\{\mathbf{v} = \begin{bmatrix}0 & 1\\ -1 & 0\end{bmatrix}\right\}$ *is a basis for* W' (Practice Problem 6.2.1).

If $A = \begin{bmatrix}a & b\\ c & d\end{bmatrix} \in V$ *then a simple computation shows*

$$\begin{aligned} A = \begin{bmatrix}a & b\\ c & d\end{bmatrix} &= \left(a\begin{bmatrix}1 & 0\\ 0 & 0\end{bmatrix} + d\begin{bmatrix}0 & 0\\ 0 & 1\end{bmatrix} + \frac{b+c}{2}\begin{bmatrix}0 & 1\\ 1 & 0\end{bmatrix}\right) + \frac{b-c}{2}\begin{bmatrix}0 & 1\\ -1 & 0\end{bmatrix} \\ &= \begin{bmatrix}a & \frac{b+c}{2}\\ \frac{b+c}{2} & d\end{bmatrix} + \begin{bmatrix}0 & \frac{b-c}{2}\\ \frac{-b+c}{2} & 0\end{bmatrix} \\ &= \frac{A+A^t}{2} + \frac{A-A^t}{2} \end{aligned} \tag{6.18}$$

with $\frac{A+A^t}{2}$ *symmetric and* $\frac{A-A^t}{2}$ *skew symmetric. Hence every* $A \in M(2,\mathbb{R})$ *is a linear combination of* $\mathcal{B}$ *and* $\mathcal{B}'$. *Since* $\dim V = 4$ *and the four vectors* $\mathcal{B} \cup \mathcal{B}'$ *spans* V *we conclude that* $\mathcal{B}_1 \cup \mathcal{B}_2$ *is a basis for* V (Corollary 6.2.16). *Therefore,* $V = W \oplus W'$.

As we mention before, if the characteristic of F is 2 then the symmetric matrices are the same as the skew symmetric matrices and $\mathbf{v} = \mathbf{u}_3$. And therefore, if the characteristic of F is 2 then $M(2,F)$, unlike $M(2,\mathbb{R})$ as in Example 6.2.34, is not a direct sum of symmetric and skew symmetric matrices.

To define the direct sum $V = W_1 \oplus W_2 \oplus W_3$ of three subspaces W_1, W_2, W_3 we define inductively: Let $W = W_1 \oplus W_2$ and we define $W_1 \oplus W_2 \oplus W_3 = W \oplus W_3$. For $W = W_1 \oplus W_2$ we require $W_1 \cap W_2 = \{\mathbf{0}\}$ and we require $W \cap W_3 = \{\mathbf{0}\}$ for $V = W \oplus W_3$. Hence we define the direct sum $V = W_1 \oplus W_2 \oplus W_3$ by

$$\begin{cases} V = W_1 + W_2 + W_3 \text{ and} \\ W_1 \cap W_2 = \{\mathbf{0}\}, \ (W_1 + W_2) \cap W_3 = \{\mathbf{0}\} \end{cases}$$

More generally we define the direct sum of finite subspaces as

Definition 6.2.35. *Let* $W_1, W_2, \ldots, W_k$ *be subspaces of* V. *If*

(1) $\left(\sum_{j<i} W_j\right) \cap W_i = \{\mathbf{0}\}$ $(i = 2, 3, \ldots, k)$

then we say the sum $\sum W_i$ *is a* **direct sum** *and we write* $\oplus_i W_i$. *Further, if*

(2) $V = W_1 + W_2 + \cdots + W_k$

then we say V *is a* **direct sum** *of* $W_1, W_2, \ldots, W_k$. *In this case, we will write*

$$V = W_1 \oplus W_2 \oplus \cdots \oplus W_k = \bigoplus_{i=1}^{k} W_i$$

Now we state generalization of Theorem 6.2.30 for any finite direct sum of subspaces.

Theorem 6.2.36. *Let* $W_1, W_2, \ldots, W_k$ *be subspaces of* V. *Then the following conditions are equivalent:*

(1) $V = W_1 \oplus W_2 \oplus \cdots \oplus W_k$

(2) $V = \sum_{i=1}^k W_i$ *and* $\dim V = \sum_{i=1}^k \dim W_i$.

(3) $V = W_1 + W_2 + \cdots + W_k$ *and if* $\sum_{i=1}^k \mathbf{w}_i = \mathbf{0}$ *with* $\mathbf{w}_i \in W_i$ *then* $\mathbf{w}_i = \mathbf{0}$ *for all* i.

(4) *Each* $\mathbf{v} \in V$ *can be written uniquely as* $\mathbf{v} = \sum_{i=1}^k \mathbf{w}_i$ *with* $\mathbf{w}_i \in W_i$.

(5) *If* $\mathcal{B}_i$ *is a basis of* W_i $(1 \le i \le k)$ *then* $\{\mathcal{B}_i\}$ *is disjoint and* $\bigcup_i \mathcal{B}_i$ *is a basis of* V.

Proof. (1) $\Rightarrow$ (2) If $V = W_1 \oplus \cdots \oplus W_k$ then $\dim V = \sum_{i=1}^k \dim W_i$ by Theorem 6.2.28 and induction.

(2) $\Rightarrow$ (1) We only need to show that $(\sum_{j<l} W_j) \cap W_l = \mathbf{0}$ for all l. Assume the contrary: $(\sum_{j<l} W_j) \cap W_l \neq \mathbf{0}$. Let $W = \sum_{j<l} W_j$ and $W' = \sum_{j \ge l} W_j$. Then our assumption says $W \cap W' \neq \mathbf{0}$. And by Theorem 6.2.28,

$$\begin{aligned}\dim V = \dim(W + W') &= \dim W + \dim W' - \dim(W \cap W') \\ &\le \sum_{j<l} \dim W_i + \sum_{j \ge l} \dim W_j - \dim(W \cap W') < \dim V\end{aligned}$$

since $\dim(W \cap W') \neq 0$. A contradiction. Hence $(\sum_{j<l} W_j) \cap W_l = \mathbf{0}$ for all l.

The proofs of the other equivalences are Exercise 10. □

Corollary 6.2.37. *Let* $V = W_1 \oplus \cdots \oplus W_k$. *If* S_i *is a linearly independent subset of* W_i *then* $\cup S_i$ *is linearly independent.*

Proof. Extend S_i to a basis $\mathcal{B}_i$ of W_i. Then $\cup\mathcal{B}_i$ is a basis of V and $\cup S_i \subseteq \cup\mathcal{B}_i$. Since $\cup\mathcal{B}_i$ is linearly independent and hence $\cup S_i$, being a subset of a linearly independent set, is linearly independent (Theorem 6.2.5). □

Answers for Practice Problems 6.2

(6.2.1) Let $A = \begin{bmatrix} a & b \\ c & d \end{bmatrix}$ be a skew symmetric matrix. Then $A + A^t = \begin{bmatrix} 2a & b+c \\ b+c & 2d \end{bmatrix}$ is the zero matrix. Hence $a = d = 0$ and $b = -c$; $A = b\begin{bmatrix} 0 & 1 \\ -1 & 0 \end{bmatrix}$. Therefore $\left\{\begin{bmatrix} 0 & 1 \\ -1 & 0 \end{bmatrix}\right\}$ is a basis for W.

(6.2.2) Just check that $W_2 + W_3 = V$ and $W_2 \cap W_3 = \{\mathbf{0}\}$.

6.2 Exercises

1. Prove Theorem 6.2.4.

2. Let $S = \{\mathbf{v}_1, \mathbf{v}_2, ..., \mathbf{v}_n\}$ be a linearly independent set of vectors. Let $S_1, S_2, ..., S_k$ be a partition of S and $W_i = \mathrm{Span}\,(S_i)$ $(i = 1, 2, ..., k)$. Prove that if $\mathbf{w}_i$ is a nonzero vector in W_i then $\{\mathbf{w}_1, \mathbf{w}_2, ..., \mathbf{w}_k\}$ is linearly independent. (A **partition** $\{S_i \mid i = 1, 2, ..., k\}$ of S is a disjoint subsets S_i of S such that $\cup_i S_i = S$.)

3. Show that the following sets form a vector space and find their bases.

   (a) $\{A \in M(n, F) \mid \mathrm{Tr}\,(A) = 0\}$ where F is a field.

(b) $\{A \in M(n, \mathbb{R}) \mid A = A^t\}$

(c) $\{A \in M(n, \mathbb{R}) \mid A = -A^t\}$

(d) $\{A \in M(2, F) \mid AB = 0\}$ where $B = \begin{bmatrix} 1 & 2 \\ 0 & 0 \end{bmatrix}$ and F is a field with $\text{char}(F) \neq 2$.

(e) $M(n, \mathbb{C})$ as vector space over $\mathbb{R}$.

(f) $\mathbb{Q}[\sqrt{2}]$ over $F = \mathbb{Q}$.

4. Complete the formal induction steps of Theorem 6.2.13.

5. Prove Corollary 6.2.17.

6. Let $\{\mathbf{u}_1, \mathbf{u}_2, ..., \mathbf{u}_k\}$ be a subset of $\mathbb{C}^n$. Prove that $\{\mathbf{u}_1, \mathbf{u}_2, ..., \mathbf{u}_k\}$ is $\mathbb{C}$-linearly independent if and only if $\{\bar{\mathbf{u}}_1, \bar{\mathbf{u}}_2, ..., \bar{\mathbf{u}}_k\}$ is $\mathbb{C}$-linearly independent.

7. Complete the proof of Theorem 6.2.25.

8. Let $V = M(n, \mathbb{R})$ and W be the subspace of V of symmetric matrices and W' be the subspace of skew symmetric matrices. Prove that $V = W \oplus W'$.

9. In Definition 6.2.36 the condition $W_i \cap \sum_{j<i} W_j = \{\mathbf{0}\}$ for $i = 2, 3, \ldots, k$ is equivalent to the condition $W_i \cap \sum_{j \neq i} W_j = \{\mathbf{0}\}$ for $i = 1, 2, \ldots, k$. (Hint: We only need to prove the former condition imply the latter. Try small k's. If $k = 2$ our result is obvious. Hence try $k = 3, 4$ and find out a way to prove this formally.)

10. Complete the proof of Theorem 6.2.36.

11. Prove the true statement(s) and find a counterexample for the false statement(s).

(a) A basis cannot contain the zero vector.

(b) If $S := \{\mathbf{u}_1, \mathbf{u}_2, ..., \mathbf{u}_n\}$ is a $\mathbb{C}$-linearly independent set of vectors in $\mathbb{C}^n$ then, with the notation of (6.14), both the real part $\{(\mathbf{u}_1)_{re}, (\mathbf{u}_2)_{re}, ..., (\mathbf{u}_n)_{re}\}$ and the imaginary part $\{(\mathbf{u}_1)_{im}, (\mathbf{u}_2)_{im}, ..., (\mathbf{u}_n)_{im}\}$ of S are $\mathbb{R}$-linearly independent.

(c) Let $\mathbf{u}, \mathbf{v} \in \mathbb{C}^2$ be $\mathbb{C}$-linearly independent vectors. Then at least one of $\{\mathbf{u}_{re}, \mathbf{v}_{re}\}$ or $\{\mathbf{u}_{im}, \mathbf{v}_{im}\}$ is $\mathbb{R}$-linearly independent.

(d) If W is a $\mathbb{C}$-subspace of $V = \mathbb{C}^n$. If $\mathbf{u} \in W$ then $\bar{\mathbf{u}} \in W$.

(e) If $V = \mathbb{C}^n$ and W is a $\mathbb{C}$-subspace of V of dimension k then $W \cap \mathbb{R}^n$ is a real vector space of dimension of k.

(f) For subspaces W, W_1, W_2 of a vector space V with $W_1 \cap W_2 = \{\mathbf{0}\}$ we have the equality $W \cap (W_1 \oplus W_2) = (W \cap W_1) \oplus (W \cap W_2)$.

(g) The set of all Hermitian matrices of $M(n, \mathbb{C})$ form a $\mathbb{C}$-subspace.

(h) The set $\{A \in M(n, F) | \det(A) = 0\}$ is a subspace of $M(n, F)$.

(i) The determinant of a Hermitian matrix is real.

(j) Let V be a vector space of dimension n and W_i be subspaces of V. If $\sum \dim W_i = n$ then $V = W_1 \oplus W_2 \oplus \cdots \oplus W_k$.

(k) Let V be a vector space and W_i be subspaces of V with bases $\mathcal{B}_i$ $(i = 1, 2, ..., k)$. If $V = W_1 \oplus W_2 \oplus \cdots \oplus W_k$ then $\bigcup \mathcal{B}_i$ is a basis of V.

(l) If $W_1, W_2, ..., W_k$ are subspaces of V with $V = W_1+W_2+\cdots+W_k$ and $W_i \cap W_j = \mathbf{0}$ for $i \neq j$ then $V = W_1 \oplus W_2 \oplus \cdots \oplus W_k$.

(m) If $V = W \oplus W' = W \oplus W''$ then $W' = W''$.

(n) If $\mathcal{B} = \{\mathbf{v}_1, \mathbf{v}_2, ..., \mathbf{v}_n\}$ is a basis for a vector space V and $W_1 = \text{Span}\{\mathbf{v}_1, \mathbf{v}_2, ..., \mathbf{v}_k\}$, $W_2 = \text{Span}\{\mathbf{v}_{k+1}, \mathbf{v}_{k+2}, ..., \mathbf{v}_n\}$ then $V = W_1 \oplus W_2$.

(o) If W_1, W_2 are subspaces of a vector space V and $\dim W_1 + \dim W_2 = \dim V$ then $V = W_1 \oplus W_2$.

(p) Let $W_1, W_2, ..., W_k$ be subspaces of V. If $\sum_{i=1}^k \dim W_i = \dim V$ and $W_i \cap W_j = \mathbf{0}$ for $i \neq j$. Then $V = \bigoplus_{i=1}^k W_i$.

12. Prove that the vector space of all absolutely convergent real sequences ℓ^1 (Example 6.1.19) is a vector space of infinite dimension.

13. Let W_i $(1 \leq i \leq k)$ be a subspace of dimension 1 in V and let $W = \sum_i W_i$. Choose a nonzero $\mathbf{w}_i \in W_i$. Prove that $W = \bigoplus_i W_i$ if and only if $\{\mathbf{w}_1, \mathbf{w}_2, ..., \mathbf{w}_k\}$ is linearly independent.

14. Let V be a vector space and W be a subspace of V. Prove that there is a subspace W' such that $V = W \oplus W'$.

15. Prove that $M(n, \mathbb{C}) = M(n, \mathbb{R}) \oplus iM(n, \mathbb{R})$ when we view $M(n, \mathbb{C})$ as a vector space over $\mathbb{R}$.

16. Let $V = M(n, \mathbb{C})$. Let W, W' be the set of all Hermitian and skew Hermitian matrices (Definition 6.2.24).

$$W = \{A \in V \mid A^* = A\}$$
$$W' = \{A \in V \mid A^* = -A\}$$

(a) Prove that $W,\ W'$ are $\mathbb{R}$ subspace of V and $\dim_{\mathbb{R}} W = n^2$.

(b) Prove that $A \in W$ if and only if iA is skew Hermitian. And prove that $A \mapsto iA$ defines an isomorphism from W to W' and deduce that $\dim_{\mathbb{R}} W' = n^2$.

(c) Prove that if $A \in V$ then $A^+ := \dfrac{A + A^*}{2} \in W$ and $A^- := \dfrac{A - A^*}{2} \in W'$ and $A = A^+ + A^-$.

(d) Prove that $V = W \oplus W'$. (Compare this with Exercise 6.2.15.)

17. Let $V = P_3(\mathbb{R})$. Prove that W is a subspace of V and find a basis for W.

(a) $W = \{p(t) \mid p(1) = 0\}$

(b) $W = \{p(t) \mid p(-1) + p(1) = 0\}$

(c) $W = \{p(t) \mid p(-1) + p(0) + p(1) = 0\}$

18. Let W_1, W_2 be subspaces of V. Prove that, using the definition of direct sum, if $W_1 \cap W_2 = \{\mathbf{0}\}$ and $\dim W_1 + \dim W_2 = \dim V$ then $V = W_1 \oplus W_2$ without invoking to Theorem 6.2.36.

19. Let V be a vector space and U be a subspace. Let $\{\mathbf{v}_1, \mathbf{v}_2, ..., \mathbf{v}_r\}$ be a basis for U. Extend this to a basis of $\{\mathbf{v}_1, \mathbf{v}_2, ..., \mathbf{v}_n\}$ of V. Prove that $\{\bar{\mathbf{v}}_{r+1}, \bar{\mathbf{v}}_{r+2}, ..., \bar{\mathbf{v}}_n\}$ is a basis for V/U. Hence

$$\dim V/U = \dim V - \dim U$$

6.3 Linear Maps on Vector Spaces

We defined the linear maps between subspaces of $\mathbb{R}^n$ (Definition 4.2.1). In this section, we define the linear maps between abstract vector spaces. Always bear in mind the examples of the linear maps between subspaces of $\mathbb{R}^n$.

6.3.1 Linear Maps on Vector Spaces

Definition 6.3.1. *A map $T : V \to W$ of vector spaces over a field F is a (F)-***linear map** (**transformation, function**) *if*

$$T(\mathbf{u}+\mathbf{v}) = T(\mathbf{u}) + T(\mathbf{v}) \text{ and } T(c\mathbf{u}) = cT(\mathbf{u}) \tag{6.19}$$

for all $\mathbf{u}, \mathbf{v} \in V$ and $c \in F$. If $W = V$ then T is called a **linear operator** *on V. If $W = F$ then T is called a* **linear functional**. *A bijective linear map T is called an* **isomorphism**. *We indicate an isomorphism T by*

$$T : V \xrightarrow{\cong} W$$

If there is an isomorphism between V and W then we write $V \cong W$ and we say V and W are **isomorphic**.

Let $T : V \to W$ be a linear map. Then since $T(\mathbf{0}) = T(\mathbf{0}+\mathbf{0}) = T(\mathbf{0}) + T(\mathbf{0})$ we have, by cancellation

$$T(\mathbf{0}) = \mathbf{0} \tag{6.20}$$

If $T : V \to W$ is a linear map then by induction we have

$$T(a_1\mathbf{v}_1 + \cdots + a_n\mathbf{v}_n) = a_1T(\mathbf{v}_1) + \cdots + a_nT(\mathbf{v}_n) \tag{6.21}$$

Example 6.3.2. (1) *If $T : V \to W$ is defined by $T(\mathbf{v}) = \mathbf{0}$ for all $\mathbf{v} \in V$ then T is obviously a linear map and we call T the* **zero (linear) map** *and we denote it by* 0. *If T is a linear operator on V defined by $T(\mathbf{v}) = \mathbf{v}$ then T is also a linear map which we call the* **identity map** *and we will denote it by I_V.*

(2) *Let $V = \mathbb{C}^2$. We can view V as a vector space over $\mathbb{C}$ as well as a vector space over $\mathbb{R}$. With the notations of* (6.14), *we define $T : V \to V$ by $T(\mathbf{v}) = \bar{\mathbf{v}}$. Then T is not a linear map as $\mathbb{C}$-vector space but T is an $\mathbb{R}$-linear map on the $\mathbb{R}$-vector space V.*

Example 6.3.3. *Let $V_1 = M(2,\mathbb{R}), V_2 = P_3(\mathbb{R}), V_3 = \mathbb{R}^4$. Define $T_1 : V_1 \to V_2$ and $T_2 : V_1 \to V_3$ be the maps defined by $T_1(A) = a + bx + cx^2 + dx^3$ and $T_2(A) = \begin{bmatrix} a \\ b \\ c \\ d \end{bmatrix}$ where $A = \begin{bmatrix} a & b \\ c & d \end{bmatrix}$. Then T_1, T_2 are isomorphisms.*

Example 6.3.4. *Let $A \in M(m \times n, F)$. Define the linear map*

$$T_A : F^n \to F^m$$

by $T_A(\mathbf{v}) = A\mathbf{v}$ which we call the **matrix linear map** *of A. Then T_A is a linear map as we saw before* (Theorem 2.1.10) *when $F = \mathbb{R}$.*

Example 6.3.5. *Let $V = F[x]$ The maps $T : V \to V$ defined by*

$$T(f(x)) = f'(x) \text{ and } T(f(x)) = \int_0^x f(t)\,dt$$

are linear maps.

We recall the definition of coordinate maps on subspaces of $\mathbb{R}^n$ (Definition 4.2.4).

Definition 6.3.6. *Let V be a vector space and $\mathcal{B} = \{\mathbf{v}_1, \mathbf{v}_2, ..., \mathbf{v}_n\}$ be a basis of V. If $\mathbf{v} \in V$ then we can write uniquely $\mathbf{v}$ as a linear combination of $\mathcal{B}$ say $\mathbf{v} = a_1\mathbf{v}_1 + a_2\mathbf{v}_2 + \cdots + a_n\mathbf{v}_n$. We define the* **coordinate** *$[\mathbf{v}]_{\mathcal{B}}$ of $\mathbf{v}$ with respect to the basis $\mathcal{B}$ to be the column vector*

$$[\mathbf{v}]_{\mathcal{B}} = \begin{bmatrix} a_1 \\ a_2 \\ \vdots \\ a_n \end{bmatrix} \tag{6.22}$$

We define the **coordinate map** *$\phi : V \to F^n$ by*

$$\phi(\mathbf{v}) = [\mathbf{v}]_{\mathcal{B}} \tag{6.23}$$

For example, if $V = P_2(F)$ and $\mathcal{B} = \{1, x, x^2\}$. If $f(x) = 2x + x^2$ then $[f(x)]_{\mathcal{B}} = \begin{bmatrix} 0 \\ 2 \\ 1 \end{bmatrix}$.

And if $\mathbf{v} = \begin{bmatrix} -1 \\ 0 \\ 2 \end{bmatrix}$ and $\mathcal{E}$ is the standard basis of $\mathbb{R}^3$ then $[\mathbf{v}]_{\mathcal{E}} = \begin{bmatrix} -1 \\ 0 \\ 2 \end{bmatrix}$.

The following is a vector space version of Theorem 4.2.1.

Theorem 6.3.7. *Let V be a vector space over F of dimension n. Then the coordinate map $\phi : V \to F^n$ is an isomorphism.*

Proof. Exercise 5. □

For the following example one needs some knowledge of cardinality in the theory of sets (Appendix D).

Example 6.3.8. *In* Example 6.2.20, *we saw that* $\dim_{\mathbb{Q}} \mathbb{R} = \infty$. *We will prove this by cardinality computations. Suppose* $\dim_{\mathbb{Q}} \mathbb{R}$ *is finite say n. Then $\mathbb{R} \cong \mathbb{Q}^n$. On the other hand, the cardinality of $\mathbb{Q}$ is $\aleph_0$ and the cardinality of $\mathbb{Q}^n$ is also $\aleph_0$ by* Theorem D.1.9. *But the cardinality of $\mathbb{R}$ is $\aleph_1$ which is a contradiction* (Theorem D.1.5). *It turns out that the cardinality of a basis of $\mathbb{R}$ over $\mathbb{Q}$ is $\aleph_1$* (Exercise 18).

The following theorem is a vector space version of Theorem 4.2.2. The same proof works for this case also and we omit the proof.

Theorem 6.3.9. *Let V, W be vector spaces and $\mathcal{B} = \{\mathbf{v}_1, \mathbf{v}_2, \ldots, \mathbf{v}_n\}$ be a basis of V.*

(1) *Let $\{\mathbf{w}_1, \mathbf{w}_2, ..., \mathbf{w}_n\}$ be n vectors in W. Then there is a unique linear map $T : V \to W$ satisfying $T(\mathbf{v}_i) = \mathbf{w}_i$ $(i = 1, 2, ..., n)$.*

(2) *If $S, T : V \to W$ are linear maps and $T(\mathbf{v}_i) = S(\mathbf{v}_i)$ for all i then $S = T$ i.e., a linear map is completely determined by its values on a basis $\mathcal{B}$ of V.*

In particular, if $T(\mathbf{v}_i) = \mathbf{0}$ for all $\mathbf{v}_i$ then T is the zero map.

In Theorem 6.3.9 (2) we can replace "a basis $\mathcal{B}$" by a "a spanning set $\mathcal{B}$". Hence we can say two linear maps are equal if they agree on a spanning set.

Now we define the null space and the range space of a linear map on vector spaces in the same way as we did for $\mathbb{R}^n$'s (Definition 2.1.5).

Let $T : V \to W$ be a linear map. We define the **null space** $N(T)$ and the **range space** $R(T)$ by

$$N(T) = \{\mathbf{v} \in V \mid T(\mathbf{v}) = 0\}$$
$$R(T) = \{T(\mathbf{v}) \in W \mid \mathbf{v} \in V\}$$

These are subspaces of V, W as for the case of $\mathbb{R}^n, \mathbb{R}^m$ whose proof will be Exercise 6.

Theorem 6.3.10. *If $T : V \to W$ is linear then $N(T)$ and $R(T)$ are subspaces of V and W respectively.*

We call the dimension of $N(T)$ the **nullity of** T and the dimension of $R(T)$ the **rank** of T. We write

$$\nu(T) = \dim N(T) \quad \text{and} \quad \rho(T) = \dim R(T)$$

Let $A \in M(m \times n)$ and $T_A : F^n \to F^m$ be the associated linear map then we have

$$R(T_A) = \text{Col}(A) \text{ and } \rho(A) = \dim R(T_A)$$
$$N(T_A) = N(A) \text{ and } \nu(A) = \dim N(A)$$

whose proof is the same as the proof of Theorem 2.1.10 for $F = \mathbb{R}$.

Definition 6.3.11. *Let $W_1, W_2, \ldots, W_k$ be vector spaces over F. The* **direct product** *$W_1 \times W_2 \times \cdots \times W_k$ of $W_1, W_2, \ldots, W_k$ is the Cartesian product of W_i's with the operations defined by*

$$(\mathbf{w}_1, ..., \mathbf{w}_k) + (\mathbf{w}'_1, ..., \mathbf{w}'_k) = (\mathbf{w}_1 + \mathbf{w}'_1, ..., \mathbf{w}_k + \mathbf{w}'_k)$$
$$c(\mathbf{w}_1, ..., \mathbf{w}_k) = (c\mathbf{w}_1, ..., c\mathbf{w}_k)$$

Often, the direct product $W_1 \times W_2 \times \cdots \times W_k$ will be denoted by $\prod_{i=1}^k W_i$.

There are linear maps

$$p_i : \prod_{j=1}^k W_j \to W_i \quad \text{and} \quad e_i : W_i \to \prod_{j=1}^k W_j \tag{6.24}$$

defined by $p_i(\mathbf{w}_1, ..., \mathbf{w}_k) = \mathbf{w}_i$ and $e_i(\mathbf{w}_i) = (\mathbf{0}, ..., \mathbf{w}_i, ..., \mathbf{0})$ where $\mathbf{w}_i$ is in the i-th place and zero elsewhere. Then p_i is surjective and e_i is injective. Also $R(p_i) = W_i$ and e_i is an isomorphism from W_i to $R(e_i)$.

Suppose $W_1, W_2, ..., W_k$ are subspaces of a vector space V with $(\sum_{i<j} W_i) \cap W_j = \mathbf{0}$ ($j = 2, 3, ..., k$). Then $\oplus W_i$ is a subspace of V and there are maps

$$\pi_i : \bigoplus_{j=1}^k W_j \to W_i \quad \text{and} \quad \iota_i : W_i \to \bigoplus_{j=1}^k W_j \tag{6.25}$$

defined by $\pi_i(\mathbf{w}_1 + \cdots + \mathbf{w}_k) = \mathbf{w}_i$ and $\iota_i(\mathbf{w}_i) = \mathbf{w}_i$ which is the inclusion map. It is immediate that the maps p_i, e_i, π_i and ι_i are linear. If $V = \oplus W_i$ then we can view π_i as linear operators on V. Further, we have

$$W_i = R(\pi_i) = R(\iota_i)$$

To see $W_1 = R(\pi_1)$ we have obviously, $R(\pi_1) \subseteq W_1$. On the other hand, if $\mathbf{w}_1 \in W_1$ then $\mathbf{w}_1 = \pi_1(\mathbf{w}_1)$ and hence $W_1 \subseteq R(\pi_1)$. The equality $W_i = R(\iota_i)$ is similarly proved.

Notice that the direct sum $\oplus W_i$ is defined for the subspaces W_i of a given vector space V. On the other hand, the product $\prod W_i$ is defined for any vector spaces W_i.

We introduce **Kronecker delta**† which is defined by

$$\delta_{ij} = \begin{cases} 1 & \text{if } i = j \\ 0 & \text{if } i \neq j \end{cases}$$

Here is another characterization of a direct sum complementing Theorem 6.2.36 .

Theorem 6.3.12. *Let V be a vector space and W_i $(1 \leq i \leq k)$ be subspaces of V. The following conditions are equivalent.*

(1) $V = W_1 \oplus W_2 \oplus \cdots \oplus W_k$

(2) *There are linear maps $\pi_i : V \to V$ such that $W_i = R(\pi_i)$ and*

$$\begin{aligned} \pi_1 + \pi_2 + \cdots + \pi_k &= I_V \\ \pi_i \pi_j &= \delta_{ij} \pi_i \end{aligned} \tag{6.26}$$

where δ_{ij} is the Kronecker delta.

Proof. We show this for $k = 2$ leaving the proof for arbitrary k as an exercise.

(1) $\Rightarrow$ (2) We define π_i by (6.25). By Theorem 6.2.30 (4), for any $\mathbf{v} \in V$ can be written uniquely as $\mathbf{v} = \mathbf{w}_1 + \mathbf{w}_2$ $(\mathbf{w}_i \in W_i)$. We define $\pi_i(\mathbf{v}) = \mathbf{w}_i$. One checks easily that π_i $(i = 1, 2)$ are linear maps. Now, since $\mathbf{v} = \mathbf{w}_1 + \mathbf{w}_2 = \pi_1(\mathbf{v}) + \pi_2(\mathbf{v}) = (\pi_1 + \pi_2)(\mathbf{v})$ we have $\pi_1 + \pi_2 = I_V$. To see $\pi_1^2 = \pi_1$ we compute, $\pi_1^2(\mathbf{v}) = \pi_1(\mathbf{w}_1) = \pi_1(\mathbf{w}_1 + \mathbf{0}) = \mathbf{w}_1 = \pi_1(\mathbf{v})$. Similarly, $\pi_2^2 = \pi_2$.

To see $\pi_1 \circ \pi_2 = 0$ we compute $\pi_1\pi_2(\mathbf{v}) = \pi_1\pi_2(\mathbf{w}_1 + \mathbf{w}_2) = \pi_1(\mathbf{w}_2) = \pi_1(\mathbf{0} + \mathbf{w}_2) = \mathbf{0}$. Similarly, $\pi_2 \circ \pi_1 = 0$.

(2) $\Rightarrow$ (1) By Theorem 6.2.30 (3), first we need to show $V = W_1 + W_2$. This follows form $\pi_1 + \pi_2 = I_V$; $\mathbf{v} = \pi_1(\mathbf{v}) + \pi_2(\mathbf{v})$ and $\pi_i(\mathbf{v}) \in W_i$. To prove $W_1 \cap W_2 = \mathbf{0}$ suppose $\mathbf{v} \in W_1 \cap W_2$. Then $\mathbf{v} = \pi_1(\mathbf{v}) + \pi_2(\mathbf{v})$ since $\pi_1 + \pi_2 = I_V$. On the other hand, since $\mathbf{v} \in W_1$ we can write $\mathbf{v} = \pi_1(\mathbf{v}')$ for some $\mathbf{v}' \in V$. Hence $\pi_2(\mathbf{v}) = \pi_2\pi_1(\mathbf{v}') = \mathbf{0}$ for $\pi_2\pi_1 = 0$. Similarly, $\pi_1(\mathbf{v}) = \mathbf{0}$. Therefore, $\mathbf{v} = \pi_1(\mathbf{v}) + \pi_2(\mathbf{v}) = \mathbf{0}$. □

The maps p_i, π_i are called '**projections**'. We will call the linear map π_i of Theorem 6.3.12 the i-th **projection** according to the direct sum decomposition $V = \oplus_i W_i$. For example, the orthogonal projection (5.27) onto a subspace W of $\mathbb{R}^n$ is a projection onto the subspace W according to the decomposition $\mathbb{R}^n = W \oplus W^\perp$ (Example 6.2.32). We need to exercise some caution since the term 'projection' is used in various contexts. Cf. Definition 6.3.15.

Example 6.3.13. *We continue* Example 6.2.33 *where we let $V = \mathbb{R}^3$ and*

$$W_1 = \left\{ \begin{bmatrix} x \\ y \\ z \end{bmatrix} \middle| \; x + y + z = 0 \right\}, W_3 = \left\{ \begin{bmatrix} 0 \\ y \\ y \end{bmatrix} \middle| \; y \in \mathbb{R} \right\}, W_4 = \left\{ \begin{bmatrix} x \\ x \\ 0 \end{bmatrix} \middle| \; x \in \mathbb{R} \right\}$$

†Leopold Kronecker (1823 – 1891) was a German mathematician who worked on number theory, algebra and logic.

We saw $V = W_1 \oplus W_3 = W_1 \oplus W_4$. *Let* $\mathbf{v} = \begin{bmatrix} 1 \\ 1 \\ 1 \end{bmatrix}$. *We want to find the projections* $\pi_i(\mathbf{v})$ $(i = 1, 2)$ *onto* W_1 *and* W_3 *according to the decomposition* $V = W_1 \oplus W_3$. *Choose bases* $\left\{ \begin{bmatrix} 1 \\ 0 \\ -1 \end{bmatrix}, \begin{bmatrix} 1 \\ -1 \\ 0 \end{bmatrix} \right\}$ *for* W_1 *and* $\left\{ \begin{bmatrix} 0 \\ 1 \\ 1 \end{bmatrix} \right\}$ *for* W_3. *Now* $\mathbf{v} = \frac{1}{2}\begin{bmatrix} 1 \\ 0 \\ -1 \end{bmatrix} + \frac{1}{2}\begin{bmatrix} 1 \\ -1 \\ 0 \end{bmatrix} + \frac{3}{2}\begin{bmatrix} 0 \\ 1 \\ 1 \end{bmatrix}$.
Hence

$$\pi_1(\mathbf{v}) = \frac{1}{2}\begin{bmatrix} 1 \\ 0 \\ -1 \end{bmatrix} + \frac{1}{2}\begin{bmatrix} 1 \\ -1 \\ 0 \end{bmatrix} = \frac{1}{2}\begin{bmatrix} 2 \\ -1 \\ -1 \end{bmatrix}, \quad \pi_2(\mathbf{v}) = \frac{3}{2}\begin{bmatrix} 0 \\ 1 \\ 1 \end{bmatrix}$$

Now we want to find the projections $\pi_i(\mathbf{v})$ $(i = 1, 2)$ *onto* W_1 *and* W_4 *according to the decomposition* $V = W_1 \oplus W_4$. *We write* $\mathbf{v} = -\mathbf{v}_1 + \frac{1}{2}\mathbf{v}_2 + \frac{3}{2}\mathbf{v}_3'$ *Hence*

$$\pi_1(\mathbf{v}) = -\begin{bmatrix} 1 \\ 0 \\ -1 \end{bmatrix} + \frac{1}{2}\begin{bmatrix} 1 \\ -1 \\ 0 \end{bmatrix} = \frac{1}{2}\begin{bmatrix} -1 \\ -1 \\ 2 \end{bmatrix}, \quad \pi_2(\mathbf{v}) = \frac{3}{2}\begin{bmatrix} 1 \\ 1 \\ 0 \end{bmatrix}$$

Practice Problem 6.3.1. *In* Example 6.3.13, *find* $\pi_i(\mathbf{v})$ *for* $\mathbf{v} = \begin{bmatrix} a \\ b \\ c \end{bmatrix}$ *according to the decomposition* $V = W_1 \oplus W_3$.

Practice Problem 6.3.2. *Referring to* Example 6.2.34, *find the projections* π, π' *according to the direct sum decomposition* $V = W \oplus W'$.

Let W_i $(1 \le i \le k)$ be subspaces of a vector space V. Then we can form the Cartesian product $\prod_{i=1}^k W_i$. Also we can form a subspace $\bigoplus_{i=1}^k W_i$. It will turns out that they are almost the same.

Now suppose W_i $(1 \le i \le k)$ are vector spaces. We will show that the product $\prod_{i=1}^k W_i$ can be viewed as a direct sum. Let W_i $(1 \le i \le k)$ be vector spaces and let $V = \prod_{i=1}^k W_i$. let $W_i' = R(e_i)$ be the range of e_i of (6.24). Then $e_i : W_i \to W_i'$ is an isomorphism and $W_i' \subseteq V$. Also it is obvious that $(\sum_{i<j} W_i') \cap W_j' = \mathbf{0}$ $(j = 2, 3, ..., k)$. Hence we can form a direct sum $\bigoplus_{i=1}^k W_i'$ in V. We will show $\prod_{i=1}^k W_i$ and $\bigoplus_{i=1}^k W_i'$ are isomorphic.

Theorem 6.3.14. (1) *Let* $W_1, W_2, \ldots, W_k$ *be subspaces of a vector space* V *over* F *such that* $(\sum_{i<j} W_i) \cap W_j = \mathbf{0}$ $(j = 2, 3, ..., k)$. *Then there is an isomorphism*

$$\phi : W_1 \times \cdots \times W_k \to W_1 \oplus \cdots \oplus W_k \tag{6.27}$$

defined by $\phi(\mathbf{w}_1, \ldots, \mathbf{w}_k) = (\mathbf{w}_1 + \cdots + \mathbf{w}_k)$.

(2) *Let* $W_1, W_2, \ldots, W_k$ *be vector spaces over* F. *Let* $e_i : W_i \to \prod W_i$ *be the map defined in* (6.24) *and let* $W_i' = R(e_i)$, *the range of* e_i. *Then* W_i' *is a subspace of* $\prod W_i$ *which is isomorphic to* W_i *and there is an isomorphism*

$$\psi : W_1 \times \cdots \times W_k \to W_1' \oplus \cdots \oplus W_k' \tag{6.28}$$

defined by $\psi(\mathbf{w}_1, ..., \mathbf{w}_k) = \sum e_i(\mathbf{w}_i)$.

Proof. (1) First, ϕ is surjective since $\phi(\mathbf{w}_1, ..., \mathbf{w}_k) = \sum_i \mathbf{w}_i$ and every element of $\bigoplus_i W_i$ is of this form. And if $\phi(\mathbf{w}_1, \ldots, \mathbf{w}_k) = \mathbf{0}$ then $\mathbf{w}_1 + \cdots + \mathbf{w}_k = \mathbf{0}$ and hence $\mathbf{w}_i = \mathbf{0}$ for all i by Theorem 6.2.36. Hence ϕ is an isomorphism.

(2) Since e_i is injective we see that W_i is isomorphic to its range W'_i. To show $(\sum_{i<j} W'_i) \cap W'_j = \mathbf{0}$ $(j = 2, 3, ..., k)$ let $\mathbf{w} \in (\sum_{i<j} W'_i) \cap W'_j$. We can write $\mathbf{w} = \sum_{i<j} e_i(\mathbf{w}_i) = e_j(\mathbf{w}_j)$; $\sum_{i \le j} e_i(\mathbf{w}_i) = \mathbf{0}$ in $\oplus_{i \le j} W'_i$. Hence $e_i(\mathbf{w}_i) = \mathbf{0}$ by Theorem 6.2.36. Hence $\mathbf{w} = \mathbf{0}$.

To see ψ is injective suppose $\sum e_i(\mathbf{w}_i) = \mathbf{0}$. Then $e_i(\mathbf{w}_i) = \mathbf{0}$. Since e_i is injective we conclude that $\mathbf{w}_i = \mathbf{0}$. Surjectivity of ψ is obvious. □

When only finitely many vector spaces are involved we often identify the direct sum with direct product using the isomorphisms ϕ and ψ. If infinitely many vector spaces W_i are involved then the direct product $\prod W_i$ is not the same as the direct sum $\oplus W_i$. For the infinite direct products and sums of vector spaces see Exercise 12.

Let $V = W_1 \oplus W_2$ and let $\mathbf{v} = \mathbf{w}_1 + \mathbf{w}_2$ $(\mathbf{w}_i \in W_i,\ i = 1, 2)$ then the map $T_i : V \to V$ defined by $T_i(\mathbf{v}) = \mathbf{w}_i$ is called the projection onto W_i (Theorem 6.3.12) according to the decomposition $V = W_1 \oplus W_2$. By (6.26), we have

$$T_i^2 = T_i \ (i = 1, 2)$$

Also observe that $R(T_i) = W_i$ and hence T_i is the projection onto its range $R(T_i)$.

Definition 6.3.15. *A linear operator $T : V \to V$ satisfying $T^2 = T$ is called a* **projection***.*

If W is a subspace of $\mathbb{R}^n$ then, by Theorem 5.2.20, the orthogonal projection onto W is a projection. By the remark above, a projection onto a subspace according to a direct sum decomposition is a projection. The converse is also true, namely a projection is a projection according to some direct sum decomposition as the folowing theorem shows.

Theorem 6.3.16. *Let T be a linear operator on V. Then T is a projection according to a direct sum decomposition if and only if $T^2 = T$. If T is a projection then $V = R(T) \oplus N(T)$ and T is the projection onto $R(T)$ according to the decomposition $V = R(T) \oplus N(T)$.*

Proof. If T is a projection according to a direct sum decomposition then we already saw in (6.26), that $T^2 = T$.

Conversely suppose $T^2 = T$. First we claim that $V = R(T) \oplus N(T)$. In fact, let $\mathbf{v} \in V$ and write $\mathbf{v} = T(\mathbf{v}) + (\mathbf{v} - T(\mathbf{v}))$. Then $T(\mathbf{v}) \in R(T)$ and $\mathbf{v} - T(\mathbf{v}) \in N(T)$ since $T(\mathbf{v} - T(\mathbf{v})) = T(\mathbf{v}) - T^2(\mathbf{v}) = \mathbf{0}$. Hence $V = R(T) + N(T)$. To see $R(T) \cap N(T) = \mathbf{0}$ let $\mathbf{v} \in R(T) \cap N(T)$. Then, as $\mathbf{v} \in R(T)$, we have $\mathbf{v} = T(\mathbf{v}')$ for some $\mathbf{v}' \in V$. Thus $\mathbf{v} = T(\mathbf{v}') = T^2(\mathbf{v}') = T(T(\mathbf{v}')) = T(\mathbf{v}) = \mathbf{0}$ as $\mathbf{v} \in N(T)$. Hence $R(T) \cap N(T) = \mathbf{0}$ and hence $V = N(T) \oplus R(T)$. And T maps $\mathbf{v}$ to $T(\mathbf{v})$. Therefore, T is a projection onto $R(T)$. □

Example 6.3.17. *Consider again* Example 6.2.34*: $V = M(2, \mathbb{R}) = W \oplus W'$ where W is the subspace of symmetric matrices and W' is the subspace of skew symmetric matrices. In* Practice Problem 6.3.2*, we saw $\pi(A) = \frac{A+A^t}{2}$ and $\pi'(A) = \frac{A-A^t}{2}$. Now $\pi^2(A) = \pi(\frac{A+A^t}{2}) = \frac{A+A^t}{2} = \pi(A)$ (Check this). Hence $\pi^2 = \pi$. Similarly $(\pi')^2 = \pi'$.*

Practice Problem 6.3.3. *For π_i of* Practice Problem 6.3.1 *prove that $\pi_i^2 = \pi_i$ $(i = 1, 2)$.*

6.3.2 Rank and Nullity

We recall Theorem 4.2.25 whose proof works for the general vector spaces also.

Theorem 6.3.18. *Let $T : V \to W$ be a linear map and let V' be a subspace of V.*

(1) *If $S = \{\mathbf{v}_1, \mathbf{v}_2, ..., \mathbf{v}_n\}$ spans V' then $T(V') = \operatorname{Span} T(S)$ and*

$$\dim T(V') \leq \dim V' \tag{6.29}$$

(2) *If T is injective then $\dim V' = \dim T(V')$. If T is an isomorphism then $\dim V = \dim W$.*

Example 6.3.19. *Let $V = P_2(\mathbb{R})$, the vector space of real polynomials degree ≤ 2. Let $T : V \to V$ be the linear map defined by $T(f) = f'$. Then $N(T) = \mathbb{R}$, the constant polynomials and $R(T)$ is the polynomials of degree ≤ 1; $R(T) = P_1(\mathbb{R})$. And $\dim R(T) = 2 \leq \dim V = 3$.*

The following theorem is a vector space version of Theorem 4.2.26 which follows from Theorem 6.3.18. The proof of Theorem 4.2.26 also works for the abstract vector spaces too.

Theorem 6.3.20. *Let $S : U \to V$ and $T : V \to W$ be linear maps. Then*

(1) $\rho(T \circ S) \leq \min\{\rho(S), \rho(T)\}$.

(2) *If S is an isomorphism then $\rho(T \circ S) = \rho(T)$ and $\nu(T \circ S) = \nu(T)$.*

(3) *If T is an isomorphism then $\rho(T \circ S) = \rho(S)$ and $\nu(T \circ S) = \nu(S)$.*

The following theorem is a vector space version of Theorem 4.2.30. We proved the theorem by first representing the linear map as a matrix and then use Theorem 2.3.17. But this time we will show how the proof can be carried on without representing the linear map as a matrix.

Theorem 6.3.21 (Rank Theorem). *If $T : V \to W$ is a linear map then*

$$\nu(T) + \rho(T) = \dim(V)$$

Proof. Let $\dim V = n \geq \dim N(T) = k$. Let $\mathcal{B}' = \{\mathbf{v}_1, \mathbf{v}_2, \ldots, \mathbf{v}_k\}$ be a basis of $N(T)$ and extend this to a basis $\mathcal{B} = \{\mathbf{v}_1, \mathbf{v}_2, \ldots, \mathbf{v}_n\}$ of V (Corollary 6.2.18). We claim that $\mathcal{B}'' = \{T(\mathbf{v}_{k+1}), T(\mathbf{v}_{k+2}), \ldots, T(\mathbf{v}_n)\}$ is a basis for $R(T)$. (Then we are done!) In fact, first we show that $\mathcal{B}''$ span $R(T)$. Let $T(\mathbf{v}) \in R(T)$. Write $\mathbf{v} = a_1\mathbf{v}_1 + \cdots + a_n\mathbf{v}_n$. Hence

$$T(\mathbf{v}) = a_{k+1}T(\mathbf{v}_{k+1}) + a_{k+2}T(\mathbf{v}_{k+2}) + \cdots + a_nT(\mathbf{v}_n)$$

since $\mathbf{v}_1, \ldots, \mathbf{v}_k \in N(T)$ and hence $\mathcal{B}''$ span $R(T)$.

Now we show that $\mathcal{B}''$ is linearly independent. Let $a_{k+1}T(\mathbf{v}_{k+1}) + \cdots + a_nT(\mathbf{v}_n) = \mathbf{0}$. Then, since $T(a_{k+1}\mathbf{v}_{k+1} + \cdots + a_n\mathbf{v}_n) = \mathbf{0}$, we have

$$a_{k+1}\mathbf{v}_{k+1} + \cdots + a_n\mathbf{v}_n \in N(T)$$

Since $\mathcal{B}'$ is a basis of $N(T)$ we can write this as a linear combination of $\mathcal{B}'$:

$$a_{k+1}\mathbf{v}_{k+1} + \cdots + a_n\mathbf{v}_n = a_1\mathbf{v}_1 + \cdots + a_k\mathbf{v}_k$$

for some $a_1, a_2, \ldots, a_k \in F$. Therefore

$$a_1\mathbf{v}_1 + \cdots + a_k\mathbf{v}_k - a_{k+1}\mathbf{v}_{k+1} - \cdots - a_n\mathbf{v}_n = \mathbf{0}$$

Now since $\mathcal{B}$ is a basis we have $a_i = 0$ for all i. □

For a proof of Theorem 6.3.21 using the matrix representation of the linear map T see Theorem 6.4.4.

The following theorem stated in terms of vector spaces and linear maps is a restatement of Theorem 2.4.2 which is stated in terms of matrix linear maps.

Theorem 6.3.22. *Let $T : V \to W$ be a linear map. Then the following conditions are equivalent:*

(1) *The linear map T is injective.*

(2) $\nu(T) = 0$ *i.e.,* $N(T) = \{\mathbf{0}\}$

(3) $\rho(T) = \dim(V)$

Proof. (1) $\Rightarrow$ (2) Assume T is injective and let $\mathbf{v} \in N(T)$. Then $T(\mathbf{v}) = \mathbf{0}$. Since $T(\mathbf{0}) = \mathbf{0}$ the injectivity of T implies $\mathbf{v} = \mathbf{0}$.

(2) $\Rightarrow$ (1) Assume $N(T) = \{\mathbf{0}\}$ and let $T(\mathbf{v}) = T(\mathbf{w})$. Then $\mathbf{0} = T(\mathbf{v}) - T(\mathbf{w}) = T(\mathbf{v}-\mathbf{w})$. Hence $\mathbf{v} - \mathbf{w} \in N(T) = \{\mathbf{0}\}$. Hence $\mathbf{v} = \mathbf{w}$.

(2) $\Leftrightarrow$ (3) Theorem 6.3.21 □

Corollary 6.3.23. *If $T : V \to W$ be an injective linear map then* $\dim(V) \leq \dim(W)$.

Proof. Since T is injective $\nu(T) = 0$ and hence $\dim(V) = \rho(T) \leq \dim(W)$. □

The following theorem is a variation of Theorem 2.4.1 stated in terms of vector space and linear maps.

Theorem 6.3.24. *Let $T : V \to W$ be a linear map. Then the following conditions are equivalent.*

(1) *The linear map T is surjective.*

(2) $\rho(T) = \dim(W)$

(3) $\nu(T) = \dim(V) - \dim(W)$

Proof. (1) $\Leftrightarrow$ (2) Since $R(T) \subseteq W$ and they are of the same dimension they are equal by Theorem 6.2.22.

(2) $\Leftrightarrow$ (3) By Theorem 6.3.21 we have $\nu(T) = \dim(V) - \rho(T)$. Hence $\rho(T) = \dim(W)$ if and only if $\nu(T) = \dim(V) - \dim(W)$. □

Corollary 6.3.25. *If $T : V \to W$ is a surjective linear map then* $\dim(V) \geq \dim(W)$.

Proof. This follows from Theorem 6.3.21 since $\nu(T) \geq 0$ always and $\rho(T) = \dim W$ as T is surjective. □

The following theorem is Theorem 2.4.5 in abstract terminology.

Theorem 6.3.26. *If $T : V \to W$ is a linear map with* $\dim V = \dim W < \infty$. *Then the following conditions are equivalent.*

(1) *The linear map T is an isomorphism.*

(2) *The linear map T is surjective.*

(3) *The linear map T is injective.*

Proof. (1) $\Leftrightarrow$ (2) If T is onto then $\rho(T) = \dim W = \dim V$. By Theorem 6.3.21 we have $\nu(T) = 0$. Hence T is injective and hence T is an isomorphism. The converse is obvious.

(2) $\Leftrightarrow$ (3) By Theorem 6.3.22, T is injective if and only if $\nu(T) = 0$. By Theorem 6.3.21 this is equivalent to $\rho(T) = \dim W$ since $\dim V = \dim W$. By Theorem 6.3.24 this is equivalent to that T is surjective. □

Finite dimensionality is essential in Theorem 6.3.26. See Exercise 13.

Example 6.3.27. *Let* $A_\theta = \begin{bmatrix} \cos\theta & -\sin\theta \\ \sin\theta & \cos\theta \end{bmatrix}$ *and let* $T_\theta : \mathbb{R}^2 \to \mathbb{R}^2$ *be the matrix linear map corresponding to* A_θ*. Hence* $T_\theta(\mathbf{v}) = A_\theta \mathbf{v}$ *for* $\mathbf{v} \in \mathbb{R}^2$*. Then* T_θ *is the rotation by the angle* θ*. Hence it is injective. Therefore* T_θ *is an isomorphism.*

Next we show that the rank, nullity and the determinant of the conjugate and the adjoint of a complex matrix is unchanged.

Theorem 6.3.28. *Let* $A \in M(m \times n, \mathbb{C})$*. Then we have*

$$\rho(A) = \rho(\bar{A}) = \rho(A^*)$$

Further, if $m = n$ *then also we have*

$$\nu(A) = \nu(\bar{A}) = \nu(A^*)$$

Proof. By rank theorem (Theorem 6.3.21) it suffices to prove only for the ranks. Since $\rho(B) = \rho(B^t)$ for any matrix B by Theorem 2.3.12 (with arbitrary scalar field instead of $\mathbb{R}$), we have $\rho(A^*) = \rho\big((\bar{A})^t \big) = \rho(\bar{A})$. Hence we only need to prove $\rho(A) = \rho(\bar{A})$. Now $\{\mathbf{v}_1, \mathbf{v}_2, ..., \mathbf{v}_n\}$ is linearly dependent if and only if $\{\bar{\mathbf{v}}_1, \bar{\mathbf{v}}_2, ..., \bar{\mathbf{v}}_n\}$ is linearly dependent (Exercise 6.2.6). Therefore the number of maximal linearly independent columns of A is the same as the number of maximal linearly independent columns of $\bar{A}$. Therefore $\rho(A) = \rho(\bar{A})$.

Proof for the last statement is an exercise. □

The following theorem is Theorem 4.2.24 for the abstract vector spaces. The same proof is valid without any change.

Theorem 6.3.29. *Let* $S = \{\mathbf{v}_1, \mathbf{v}_2, \ldots, \mathbf{v}_n\}$ *be a subset of* V *and* $T : V \to W$ *be a linear map. Let* $T(S) = \{T(\mathbf{v}_1), T(\mathbf{v}_2), \ldots, T(\mathbf{v}_n)\}$*. Then*

(1) *The linear map* T *is injective if and only if* $T(S)$ *is linearly independent whenever* S *is linearly independent.*

(2) *The linear map* T *is surjective if and only if* $T(S)$ *spans* W *whenever* S *span* V*.*

(3) *The linear map* T *is an isomorphism if and only if* $T(S)$ *is a basis of* W *whenever* S *is a basis.*

Proof. The proof is the same as the proof of Theorem 4.2.24. □

The following result is a vector space version of Corollary 4.2.6.

Corollary 6.3.30. *The vector spaces* V *and* W *are isomorphic if and only if* $\dim V = \dim W$*.*

Proof. Suppose V, W are isomorphic say $T : V \to W$ is an isomorphism. If $\mathcal{B}$ a basis of V then $T(\mathcal{B})$ is a basis of W and $\#\mathcal{B} = \#T(\mathcal{B})$. Hence $\dim V = \dim W$.

Conversely, suppose $\dim V = \dim W$. If $\mathcal{B} = \{\mathbf{v}_1, \mathbf{v}_2, \ldots, \mathbf{v}_n\}$, $\mathcal{C} = \{\mathbf{w}_1, \mathbf{w}_2, \ldots, \mathbf{w}_n\}$ are bases of V, W then $T(\mathbf{v}_i) = \mathbf{w}_i$ defines an isomorphism $T : V \to W$ by Theorem 6.3.29. □

Answers for Practice Problems 6.3

(6.3.1) We need to write $\mathbf{v}$ as a linear combination of

$$\left\{\mathbf{v}_1 = \begin{bmatrix} 1 \\ 0 \\ -1 \end{bmatrix}, \mathbf{v}_2 = \begin{bmatrix} 1 \\ -1 \\ 0 \end{bmatrix}\right\} \text{ and } \left\{\mathbf{v}_3 = \begin{bmatrix} 0 \\ 1 \\ 1 \end{bmatrix}\right\}$$

Hence we need to solve $x\mathbf{v}_1 + y\mathbf{v}_2 + z\mathbf{v}_3 = \mathbf{v}$:

$$\begin{bmatrix} 1 & 1 & 0 \\ 0 & -1 & 1 \\ -1 & 0 & 1 \end{bmatrix} \begin{bmatrix} x \\ y \\ z \end{bmatrix} = \begin{bmatrix} a \\ b \\ c \end{bmatrix}$$

Solving this we have $x = \frac{1}{2}(a+b-c), y = \frac{1}{2}(a-b+c), z = \frac{1}{2}(a+b+c)$. Hence

$$\pi_1(\mathbf{v}) = x\mathbf{v}_1 + y\mathbf{v}_2 = \frac{1}{2}\begin{bmatrix} 2a \\ -a+b-c \\ -a-b+c \end{bmatrix}, \quad \pi_2(\mathbf{v}) = \frac{1}{2}\begin{bmatrix} 0 \\ a+b+c \\ a+b+c \end{bmatrix}$$

(6.3.2) For $A \in V$ we have $\pi(A) = \dfrac{A + A^t}{2}$ and $\pi'(A) = \dfrac{A - A^t}{2}$.

(6.3.3) Let $\mathbf{v} = \begin{bmatrix} a \\ b \\ c \end{bmatrix}$. By Practice Problem 6.3.1, we have

$$\pi_1(\mathbf{v}) = \frac{1}{2}\begin{bmatrix} 2a \\ -a+b-c \\ -a-b+c \end{bmatrix}, \quad \pi_2(\mathbf{v}) = \frac{1}{2}\begin{bmatrix} 0 \\ a+b+c \\ a+b+c \end{bmatrix}$$

Now one checks easily that $\pi_1^2(\mathbf{v}) = \pi_1(\mathbf{v})$ and also $\pi_2^2(\mathbf{v}) = \pi_2(\mathbf{v})$.

6.3 Exercises

1. Let $V = P_2(F)$. Determine whether the following maps $T : V \to V$ are linear or not.

   (a) $T(a_0 + a_1x + a_2x^2) = a_0^2 + a_1^2x + a_2^2x^2$.

   (b) $T(a_0 + a_1x + a_2x^2) = a_0 + (a_1 + 1)x + (a_2 + 2)x^2$.

   (c) $T(a_0 + a_1x + a_2x^2) = a_1x + 2a_2x^2$.

   (d) $T(a_0 + a_1x + a_2x^2) = (a_0 + a_1) + (a_1 + a_2)x + a_2x^2$.

2. Determine whether the following maps $T : V \to W$ are linear or not.

   (a) $V = M(2, F), T : V \to V$ where $T\left(\begin{bmatrix} a & b \\ c & d \end{bmatrix}\right) = \begin{bmatrix} a & 2c \\ -b & d \end{bmatrix}$

(b) $V = M(2,F), W = \mathbb{R}^3, T : V \to W$ where $T\left(\begin{bmatrix} a & b \\ c & d \end{bmatrix}\right) = \begin{bmatrix} a \\ b-c \\ d \end{bmatrix}$

(c) $V = M(2,F), W = \mathbb{R}^2, T : V \to W$ where $T\left(\begin{bmatrix} a & b \\ c & d \end{bmatrix}\right) = \begin{bmatrix} ab \\ cd \end{bmatrix}$

(d) $V = M(2,F), T : V \to V$ where $T(A) = A + A^t$

(e) $V = M(2,F), T : V \to V$ where $T(A) = A - A^t$

(f) $V = M(2,F), W = P_2(F), T : V \to W$ where $T\left(\begin{bmatrix} a & b \\ c & d \end{bmatrix}\right) = a + bcx + dx^2$

(g) $V = M(2,F), W = F^4,\ T : V \to W$ where $T\left(\begin{bmatrix} a & b \\ c & d \end{bmatrix}\right) = \begin{bmatrix} a \\ b \\ c \\ d \end{bmatrix}$

(h) $V = W = \mathbb{C}^n,\ T : V \to V$ where $T(\mathbf{v}) = \bar{\mathbf{v}}$

3. Prove the true statement(s) and find a counter example for the false statement(s).

   (a) If T_1, T_2 are linear operators on V then $\rho(T_1T_2) = \rho(T_2T_1)$.

   (b) If $S = \{\mathbf{v}_1, \mathbf{v}_2, ..., \mathbf{v}_n\} \subseteq \mathbb{C}^n$ is $\mathbb{C}$-linearly independent then S is $\mathbb{R}$-linearly independent.

   (c) If $\mathcal{B} = \{\mathbf{v}_1, \mathbf{v}_2, ..., \mathbf{v}_n\}$ is a basis for V over $\mathbb{Q}$ then $\mathcal{B}' = \{\mathbf{v}_1, 2\mathbf{v}_2, 3\mathbf{v}_3 ..., n\mathbf{v}_n\}$ is also a basis for V.

   (d) If $T : V \to W$ is a linear map and $\dim V = \dim R(T)$ then T is injective.

   (e) If $\dim V < \dim W$ then there is no surjective linear map $T : V \to W$.

   (f) Let $T : V \to V$ be a linear map. If $R(T) \cap N(T) = \{\mathbf{0}\}$ then $T^2 = T$.

   (g) If $T : V \to W$ is nonzero linear and $\dim W = 1$ then T is surjective.

   (h) If $\dim V \neq \dim W$ then there is no isomorphism $T : V \to W$.

   (i) If $\dim V = \dim W$ then there is always an isomorphism $T : V \to W$.

   (j) Let $f(x) \in \mathbb{R}[x]$. A polynomial map $T_f : \mathbb{R} \to \mathbb{R}$ defined by $T_f(x) = f(x)$ is linear if $\deg(f) = 1$.

   (k) There is a linear map $T : V \to W$ such that $\rho(T) > \dim V$.

   (l) If $S : U \to V,\ T : V \to W$ are linear maps such that $\rho(T \circ S) = \rho(S)$ then S is an isomorphism.

   (m) The identity linear operator and the zero linear operator on a vector space V are also projections.

4. Find the projections of $\mathbf{v}$ according to the decomposition $V = W_1 \oplus W_2$ in the following cases. You need to check that $V = W_1 \oplus W_2$ in each case.

   (a) $V = \mathbb{R}^3, W_1 = \left\{ \begin{bmatrix} x \\ y \\ z \end{bmatrix} \middle| \begin{matrix} 2x + y - z = 0 \\ -x + y = 0 \end{matrix} \right\},\ W_2 = \left\{ \begin{bmatrix} x \\ y \\ z \end{bmatrix} \middle| \ y - z = 0 \right\};\ \mathbf{v} = \begin{bmatrix} 1 \\ 2 \\ 1 \end{bmatrix}$

   (b) $V = \mathbb{R}^3,\ W_1 = \left\{ \begin{bmatrix} x \\ y \\ z \end{bmatrix} \middle| \begin{matrix} 2x + y - z = 0 \\ -x + y = 0 \end{matrix} \right\},\ W_2 = \left\{ \begin{bmatrix} x \\ y \\ z \end{bmatrix} \middle| \ y + z = 0 \right\};\mathbf{v} = \begin{bmatrix} 1 \\ 2 \\ 1 \end{bmatrix}$

(c) $V = M(2, F)$ where W_1 is the subspace of symmetric matrices and W_2 is the subspace of skew symmetric matrices as in Example 6.2.34. And $\mathbf{v} = \begin{bmatrix} a & b \\ c & d \end{bmatrix}$.

(d) $V = M(2, F)$ with the $\text{char}(F) \neq 2$.

$$W_1 = \left\{ \begin{bmatrix} a & 0 \\ c & d \end{bmatrix} \middle| a + d = 0 \right\}, \ W_2 = \left\{ \begin{bmatrix} a & b \\ 0 & d \end{bmatrix} \middle| a - d = 0 \right\}; \ \mathbf{v} = \begin{bmatrix} a & b \\ c & d \end{bmatrix}$$

5. Prove Theorem 6.3.7.

6. Prove Theorem 6.3.10.

7. Prove that a projection T (Definition 6.3.15) on a vector space V is an isomorphism if and only if $T = I_V$.

8. Let $V = P_4(\mathbb{R})$. For $f(x) = a_0 + a_1x + a_2x^2 + a_3x^3 + a_4x^4$ define $T : V \to V$ by

$$T(f)(x) = a_0 + \frac{a_1 + a_2}{2}x^2 + \frac{a_3 + a_4}{2}x^4$$

Prove that T is a projection i.e., $T^2 = T$. And write V as a direct sum of $W_1 := R(T)$ and $W_2 := N(T)$ and describe the projections onto W_1 and W_2 according to the direct sum $V = W_1 \oplus W_2$.

9. Let V be a vector space over F and U be a subspace of V. Let $T : U \to W$ be a linear map. Prove that there is a linear map $T' : V \to W$ such that T and T' agree on U.

10. Provide a proof of Theorem 6.3.12 for the direct sum of k subspaces. Also show that $N(\pi_j) = \oplus_{j \neq i} W_j$.

11. Let $T : V \to W$ be a linear map. Let V', W' be subspaces of V, W. Prove that $T(V'), T^{-1}(W')$ are subspaces of W, V. Here, $T^{-1}(W')$ denotes the inverse image $T^{-1}(W') = \{\mathbf{v} \in V \mid T(\mathbf{v}) \in W'\}$.

12. Let $\{W_i\}_{i \in I}$ be an infinite family of vector spaces. We define the direct product $\prod_{i \in I} W_i$ to be the Cartesian product and the direct sum $\bigoplus_{i \in I} W_i$ by

$$\bigoplus_{i \in I} W_i = \{(\mathbf{w}_i)_{i \in I} \mid \text{ only finitely many } \mathbf{w}_i \text{ are nonzero}\}$$

Show that $\bigoplus_{i \in I} W_i \neq \prod_{i \in I} W_i$.

13. Let $V = \mathbb{R}[x]$ be the vector space of the real polynomials. Find an example of an injective (resp. surjective) linear map $\phi : V \to V$ which is not surjective (resp. injective).

14. Prove (1) and (2) of Theorem 6.3.29.

15. Let W be a vector space over $\mathbb{C}$ with $\dim_{\mathbb{C}} W = m$. Then we can view W as an $\mathbb{R}$-vector space which we denote by $W_{\mathbb{R}}$ i.e., use only the real numbers for the scalar multiplication. We call the vector space $W_{\mathbb{R}}$ the vector space obtained from W by **restricting the scalars to $\mathbb{R}$.**

(a) Let $\mathcal{B} = \{\mathbf{w}_1, \mathbf{w}_2, ..., \mathbf{w}_m\}$ be a basis of $\mathbb{C}$-vector space W. Prove that the set of vectors

$$\mathcal{B}_{\mathbb{R}} = \{\mathbf{w}_1, \mathbf{w}_2, ..., \mathbf{w}_m, i\mathbf{w}_1, i\mathbf{w}_2, ..., i\mathbf{w}_m\}$$

is a basis of $W_{\mathbb{R}}$. (Hence $W_{\mathbb{R}}$ is an $\mathbb{R}$-vector space of dimension $2m$.)

(b) Let $T : W \to W'$ be a $\mathbb{C}$-linear map of $\mathbb{C}$-vector spaces. Show that the map T is also an $\mathbb{R}$-linear map from $W_{\mathbb{R}}$ to $W'_{\mathbb{R}}$ which we denote by $T_{\mathbb{R}}$.

$$T_{\mathbb{R}} : W_{\mathbb{R}} \to W'_{\mathbb{R}}$$

Show that if $S : W' \to W''$ then $(S \circ T)_{\mathbb{R}} = S_{\mathbb{R}} \circ T_{\mathbb{R}}$.

(c) Let $A \in M(n, \mathbb{C})$. Let $V = \mathbb{C}^n$ and $T = T_A : V \to V$. By (a), we see that $\mathcal{E} = \{\mathbf{e}^1, \mathbf{e}^2, ..., \mathbf{e}^n, i\mathbf{e}^1, i\mathbf{e}^2, ..., i\mathbf{e}^n\}$ is a basis for $V_{\mathbb{R}}$. Find the matrix representation $[T_{\mathbb{R}}]_{\mathcal{E}}$ of $T_{\mathbb{R}}$ with respect to $\mathcal{E}$. (For $\det T_{\mathbb{R}}$, see Theorem 7.3.21.)

(d) Prove that $N(T)_{\mathbb{R}} = N(T_{\mathbb{R}})$ and $R(T)_{\mathbb{R}} = R(T_{\mathbb{R}})$.

(e) Prove that T is injective (resp. surjective, isomorphism) if and only if the same is true for $T_{\mathbb{R}}$.

16. Let V be a vector space over $\mathbb{R}$. Let

$$V^{\mathbb{C}} = \{\mathbf{u} + i\mathbf{v} \mid \mathbf{u}, \mathbf{v} \in V,\ i^2 = -1\}$$

Define the addition and the multiplication by a complex number $x + iy$ on $V^{\mathbb{C}}$ by

$$(\mathbf{u} + i\mathbf{v}) + (\mathbf{u}' + i\mathbf{v}') = (\mathbf{u} + \mathbf{u}') + i(\mathbf{v} + \mathbf{v}')$$
$$(x + yi)(\mathbf{u} + i\mathbf{v}) = (x\mathbf{u} - y\mathbf{v}) + i(y\mathbf{u} + x\mathbf{v})$$

(a) Prove that $V^{\mathbb{C}}$ is a vector space over $\mathbb{C}$ and every element of $V^{\mathbb{C}}$ can be written as $\sum_k (a_k\mathbf{u}_k + ib_k\mathbf{u}_k)$ with $a_k, b_k \in \mathbb{R}$. We call the vector space $V^{\mathbb{C}}$ the vector space obtained from V by **extending the scalars to** $\mathbb{C}$ or **complexification** of V. (See Exercise 6.6.8 for more natural way to represent the complexification.)

(b) For $\mathbf{v} \in V$ write $\mathbf{v}'$ for $\mathbf{v} + i\mathbf{0} \in V^{\mathbb{C}}$. Suppose $\mathcal{B} = \{\mathbf{v}_1, \mathbf{v}_2, ..., \mathbf{v}_n\}$ is a basis for V. Prove that

$$\mathcal{B}^{\mathbb{C}} = \{\mathbf{v}'_1, \mathbf{v}'_2, ..., \mathbf{v}'_n\}$$

is a basis of $V^{\mathbb{C}}$ and deduce that $\dim_{\mathbb{C}} V^{\mathbb{C}} = n$.

(c) If $T : V_1 \to V_2$ is a linear map of real vector spaces. Then we define $T^{\mathbb{C}} : V_1^{\mathbb{C}} \to V_2^{\mathbb{C}}$ by

$$T^{\mathbb{C}}(\mathbf{u} + i\mathbf{v}) = T(\mathbf{u}) + iT(\mathbf{v})$$

Prove then that $T^{\mathbb{C}}$ is a $\mathbb{C}$-linear map of $\mathbb{C}$-vector spaces. Prove that if $S : V_2 \to V_3$ then $(S \circ T)^{\mathbb{C}} = S^{\mathbb{C}} \circ T^{\mathbb{C}}$.

(d) Let $\mathcal{B} = \{\mathbf{v}_1, \mathbf{v}_2, ..., \mathbf{v}_n\}$, $\mathcal{C} = \{\mathbf{w}_1, \mathbf{w}_2, ..., \mathbf{w}_m\}$ be bases of V_1, V_2 and $\mathcal{B}', \mathcal{C}'$ be the corresponding bases for $V_1^{\mathbb{C}}, V_2^{\mathbb{C}}$. Let $T : V_1 \to V_2$ be a linear map of $\mathbb{R}$-vector spaces. Prove that $[T^{\mathbb{C}}]_{\mathcal{B}'}^{\mathcal{C}'} = [T]_{\mathcal{B}}^{\mathcal{C}}$.

(e) When $T : V \to V$ be a linear operator on a finite dimensional real vector space. Prove that $\det(T) = \det(T^{\mathbb{C}})$. Also prove that the characteristic polynomial of T is the same as the characteristic polynomial of $T^{\mathbb{C}}$.

(f) Prove that
$$N(T)^{\mathbb{C}} = N(T^{\mathbb{C}}) \text{ and } R(T)^{\mathbb{C}} = R(T^{\mathbb{C}})$$

(g) Prove that T is injective (resp. surjective, isomorphism) if and only if the same is true for $T^{\mathbb{C}}$.

17. (a) Let V be a vector space over $\mathbb{R}$ of finite dimension. Prove that the natural map
$$\phi : V \to (V^{\mathbb{C}})_{\mathbb{R}}$$
defined by $\phi(\mathbf{v}) = \mathbf{v}+i\mathbf{0}$ is a linear map of real vector spaces. Is ϕ an isomorphism? Injective?

(b) Let W be a vector space over $\mathbb{C}$ of finite dimension. Prove that the natural map
$$\psi : (W_{\mathbb{R}})^{\mathbb{C}} \to W$$
defined by $\psi(\mathbf{w} + i\mathbf{w}') = \mathbf{w} + i\mathbf{w}'$ is a linear map of complex vector spaces. Is ψ an isomorphism?

18. The cardinality of a basis of a vector space is uniquely determined by Theorem D.2.3. We saw in Example 6.3.8 that $\dim_{\mathbb{Q}} \mathbb{R} = \infty$. Prove that the cardinality of a basis of $\mathbb{R}$ as a vector space over $\mathbb{Q}$ is $\aleph_1$.

19. Let $U, U_1 \subseteq U_2$ be subspaces of V. (See Exercise 6.1.9)

(a) Prove that there is a natural linear map
$$\Pi : V \to V/U$$
given by $\mathbf{v} \mapsto \bar{\mathbf{v}}$ with the null space $N(\Pi) = U$.

(b) Prove that $\pi(\mathbf{v} + U_1) = \mathbf{v} + U_2$ determines a well defined linear map
$$\pi : V/U_1 \to V/U_2$$

20. Let $T : V \to W$ be a linear map.

(a) Prove that $\bar{T} : V/N(T) \to W$ defined by $\bar{T}(\mathbf{v}) = T(\mathbf{v})$ is a well defined injective linear map. If T is surjective then $\bar{T}$ is an isomorphism.

(b) Let U be a subspace of V. Prove that $\bar{T} : V/U \to W$ given by $\bar{T}(\bar{\mathbf{v}}) = T(\mathbf{v})$ is a well defined linear map if and only if $U \subseteq N(T)$.

(c) Let U be a subspace of V. Deduce from (b) that there is a bijection between the set of linear maps $T : V \to W$ such that $U \subseteq N(T)$ and the set of all linear maps $S : V/U \to W$:
$$\left\{ \begin{array}{c} \text{linear map } V \overset{T}{\to} W \\ \text{with } U \subseteq N(T) \end{array} \right\} \xrightarrow{\text{bijection}} \left\{ \begin{array}{c} \text{linear map} \\ V/U \overset{S}{\to} W \end{array} \right\}$$

(d) Let X be a subspace of W. Prove that $\bar{T} : V/T^{-1}(X) \to W/X$ is an injective linear map.

21. Let U, V be subspaces of a vector space and $U_1,\ V_1$ be subspaces of $U,\ V$ respectively. Prove that there is a natural isomorphism
$$(U \oplus V)/(U_1 \oplus V_1) \to U/U_1 \oplus V/V_1$$
In particular, $(U \oplus V)/U$ is isomorphic to V.

6.4 Matrices and Linear Maps

The notion of linear map is an abstract mathematical gadget and often we need to represent an abstract object in a concrete mathematical object namely a matrix. We will represent a linear map of vector spaces as we did for the linear maps between subspaces of $\mathbb{R}^n$ in §4.2.

6.4.1 Linear Maps and Matrices

Let V be a vector space over F of dimension n with a basis $\mathcal{B} = \{\mathbf{v}_1, \mathbf{v}_2, \ldots, \mathbf{v}_n\}$. If $\mathbf{v} = a_1\mathbf{v}_1 + \cdots + a_n\mathbf{v}_n$ then the coordinate of $\mathbf{v}$ with respect to $\mathcal{B}$ is defined (Definition 6.3.6) by

$$[\mathbf{v}]_{\mathcal{B}} = \begin{bmatrix} a_1 \\ a_2 \\ \vdots \\ a_n \end{bmatrix} \in F^n \tag{6.30}$$

Let $T : V \to W$ be a linear map. Let $\mathcal{B} = \{\mathbf{v}_1, \ldots, \mathbf{v}_n\}$, $\mathcal{C} = \{\mathbf{w}_1, \ldots, \mathbf{w}_m\}$ be bases of V and W respectively. We define the matrix representation of a linear map as we did in Definition 4.2.7: Writing

$$T(\mathbf{v}_j) = \sum_{i=1}^{m} a_{ij}\mathbf{w}_i \quad (j = 1, 2, \ldots, n) \tag{6.31}$$

we get an $m \times n$ matrix over F

$$[T]_{\mathcal{B}}^{\mathcal{C}} := [a_{ij}] \in M(m{\times}n, F) \tag{6.32}$$

i.e., $[T(\mathbf{v}_j)]_{\mathcal{C}}$ is the j-th column of $[T]_{\mathcal{B}}^{\mathcal{C}}$. Namely,

$$\left([T]_{\mathcal{B}}^{\mathcal{C}}\right)^{(j)} = [T(\mathbf{v}_j)]_{\mathcal{C}} \tag{6.33}$$

The matrix $[T]_{\mathcal{B}}^{\mathcal{C}}$ is called the **matrix representation of** T with respect to $\mathcal{B}, \mathcal{C}$. If $V = W$ and $\mathcal{B} = \mathcal{C}$ then we simply write $[T]_{\mathcal{B}}$ for $[T]_{\mathcal{B}}^{\mathcal{C}}$.

Example 6.4.1. (1) *If $I_V : V \to V$ is the identity map then for any basis $\mathcal{B}$ we have $[I_V]_{\mathcal{B}} = I_n$, the $n{\times}n$ identity matrix where $n = \dim V$.*

(2) *Consider $\mathbb{C}$ as a vector space over $\mathbb{R}$ with the basis $\mathcal{B} = \{1, i\}$. Then the complex conjugation $T : \mathbb{C} \to \mathbb{C}$ defined by $T(z) = \bar{z}$ is $\mathbb{R}$-linear and $[T]_{\mathcal{B}} = \begin{bmatrix} 1 & 0 \\ 0 & -1 \end{bmatrix}$.*

Let $w = a + bi \in \mathbb{C}$. The complex multiplication $T_w : \mathbb{C} \to \mathbb{C}$ by w defined by $T(z) = zw$ is $\mathbb{R}$-linear and $[T]_{\mathcal{B}} = \begin{bmatrix} a & -b \\ b & a \end{bmatrix}$.

(3) *Let $V = P_2(\mathbb{R})$ and $T(f) = f'$. Let $\mathcal{B} = \{1, x, x^2\}$. To find $[T]_{\mathcal{B}}$ we compute $T(1) = 0, T(x) = 1, T(x^2) = 2x$. Hence*

$$[T(1)]_{\mathcal{B}} = \begin{bmatrix} 0 \\ 0 \\ 0 \end{bmatrix}, [T(x)]_{\mathcal{B}} = \begin{bmatrix} 1 \\ 0 \\ 0 \end{bmatrix}, [T(x^2)]_{\mathcal{B}} = \begin{bmatrix} 0 \\ 2 \\ 0 \end{bmatrix}$$

Hence

$$[T]_{\mathcal{B}} = \begin{bmatrix} 0 & 1 & 0 \\ 0 & 0 & 2 \\ 0 & 0 & 0 \end{bmatrix}$$

(4) *We recall the matrix representation of a linear map we treated earlier in* §4.2. *Let* $T = T_A : \mathbb{R}^3 \to \mathbb{R}^2$ *where* $A = \begin{bmatrix} 1 & 2 & 0 \\ -1 & 0 & 1 \end{bmatrix}$. *Let* $\mathcal{E}^3, \mathcal{E}^2$ *be the standard bases of* $\mathbb{R}^3, \mathbb{R}^2$. *We know that* $[T]_{\mathcal{E}^3}^{\mathcal{E}^2} = A$.

Now let $\mathcal{B} = \left\{ \mathbf{v}_1 = \begin{bmatrix} 1 \\ 0 \\ 0 \end{bmatrix}, \mathbf{v}_2 = \begin{bmatrix} 1 \\ 1 \\ 0 \end{bmatrix}, \mathbf{v}_3 \begin{bmatrix} 1 \\ 1 \\ 1 \end{bmatrix} \right\}, \mathcal{C} = \left\{ \mathbf{w}_1 = \begin{bmatrix} 1 \\ 1 \end{bmatrix}, \mathbf{w}_2 = \begin{bmatrix} 1 \\ -1 \end{bmatrix} \right\}$. *Let's find* $[T]_{\mathcal{B}}^{\mathcal{C}}$. *For this we compute*

$$T(\mathbf{v}_1) = A\mathbf{v}_1 = \begin{bmatrix} 1 & 2 & 0 \\ -1 & 0 & 1 \end{bmatrix} \begin{bmatrix} 1 \\ 0 \\ 0 \end{bmatrix} = \begin{bmatrix} 1 \\ -1 \end{bmatrix} = 0\mathbf{w}_1 + \mathbf{w}_2; \ [T(\mathbf{v}_1)]_{\mathcal{C}} = \begin{bmatrix} 0 \\ 1 \end{bmatrix}$$

Similarly,

$$T(\mathbf{v}_2) = \begin{bmatrix} 3 \\ -1 \end{bmatrix} = \mathbf{w}_1 + 2\mathbf{w}_2 \text{ and } T(\mathbf{v}_3) = \begin{bmatrix} 3 \\ 0 \end{bmatrix} = \frac{3}{2}\mathbf{w}_1 + \frac{3}{2}\mathbf{w}_2$$

Hence

$$[T(\mathbf{v}_2)]_{\mathcal{C}} = \begin{bmatrix} 1 \\ 2 \end{bmatrix}, \ [T(\mathbf{v}_3)]_{\mathcal{C}} = \begin{bmatrix} \frac{3}{2} \\ \frac{3}{2} \end{bmatrix}$$

and hence

$$[T]_{\mathcal{B}}^{\mathcal{C}} = \begin{bmatrix} 0 & 1 & \frac{3}{2} \\ 1 & 2 & \frac{3}{2} \end{bmatrix}$$

Practice Problem 6.4.1. *Let* $T : P_2(F) \to \mathbb{R}^2$ *be defined by* $T(f) = \begin{bmatrix} f(1) \\ f(2) \end{bmatrix}$. *Let* $\mathcal{B} = \{1, x, x^2\}$ *and* $\mathcal{E} = \{\mathbf{e}^1, \mathbf{e}^2\}$. *Show that* T *is linear and find* $[T]_{\mathcal{B}}^{\mathcal{E}}$.

The following theorem is a vector space version of Theorem 4.2.11. The proof of Theorem 4.2.11 works for this case. This time we will prove this by direct computation.

Theorem 6.4.2. *Let* $T : V \to W$ *be a linear map with bases* $\mathcal{B} = \{\mathbf{v}_1, \mathbf{v}_2, \ldots, \mathbf{v}_n\}$ *and* $\mathcal{C} = \{\mathbf{w}_1, \mathbf{w}_2, \ldots, \mathbf{w}_m\}$ *of* V *and* W *respectively. Then*

$$[T(\mathbf{v})]_{\mathcal{C}} = [T]_{\mathcal{B}}^{\mathcal{C}}[\mathbf{v}]_{\mathcal{B}} \tag{6.34}$$

If we let $A = [T]_{\mathcal{B}}^{\mathcal{C}}$ *and* $\phi_{\mathcal{B}}, \phi_{\mathcal{C}}$ *be the coordinate maps relative to* $\mathcal{B}, \mathcal{C}$ *then* (6.34) *says the diagram*

$$\begin{array}{ccc} V & \xrightarrow{\ T\ } & W \\ \phi_{\mathcal{B}} \big\downarrow & & \big\downarrow \phi_{\mathcal{C}} \\ F^n & \xrightarrow[T_A]{} & F^m \end{array} \tag{6.35}$$

is commutative i.e., $\phi_{\mathcal{C}} \circ T = T_A \circ \phi_{\mathcal{B}}$.

Proof. Let $\mathbf{v} \in V$ and let $\mathbf{v} = c_1\mathbf{v}_1 + c_2\mathbf{v}_2 + \cdots + c_n\mathbf{v}_n$. Let $T(\mathbf{v}_j) = \sum_{i=1}^m a_{ij}\mathbf{w}_i$ so that $[T]_{\mathcal{B}}^{\mathcal{C}} = [a_{ij}] = A$. Then

$$\begin{aligned} T(\mathbf{v}) &= T\Big(\sum_{j=1}^n c_j\mathbf{v}_j\Big) = \sum_{j=1}^n c_jT(\mathbf{v}_j) \\ &= \sum_{j=1}^n c_j\Big(\sum_{i=1}^m a_{ij}\mathbf{w}_i\Big) = \sum_{i=1}^m \Big(\sum_{j=1}^n a_{ij}c_j\Big)\mathbf{w}_i \\ &= \sum_{i}^m A_{(i)}[\mathbf{v}_{\mathcal{B}}]\mathbf{w}_i \end{aligned}$$

Now $A_{(i)}[\mathbf{v}]_{\mathcal{B}}$ is the i-th component of $A[\mathbf{v}]_{\mathcal{B}}$ which is the coefficient of $\mathbf{w}_i$ in the last equality. Hence $[T(\mathbf{v})]_{\mathcal{C}} = A[\mathbf{v}]_{\mathcal{B}} = [T]_{\mathcal{B}}^{\mathcal{C}}[\mathbf{v}]_{\mathcal{B}}$. □

Example 6.4.3. *Let $V = P_3(\mathbb{R})$ and $T : V \to V$ be given by $T(f(x)) = (x+1)f'(x)$. Let $\mathcal{B} = \{1, x, x^2, x^3\}$ be the standard basis of V. Let $\mathbf{v} = 3x^2 + x^3$. Let's find $[T(\mathbf{v})]_{\mathcal{B}}$. To find $[T]_{\mathcal{B}}$ we compute: $T(1) = 0, T(x) = 1 + x, T(x^2) = 2x + 2x^2, T(x^3) = 3x^2 + 3x^3$. Hence*

$$[T]_{\mathcal{B}} = \begin{bmatrix} 0 & 1 & 0 & 0 \\ 0 & 1 & 2 & 0 \\ 0 & 0 & 2 & 3 \\ 0 & 0 & 0 & 3 \end{bmatrix} \text{ and } [\mathbf{v}]_{\mathcal{B}} = \begin{bmatrix} 0 \\ 0 \\ 3 \\ 1 \end{bmatrix}$$

Therefore

$$[T(\mathbf{v})]_{\mathcal{B}} = [T]_{\mathcal{B}}[\mathbf{v}]_{\mathcal{B}} = \begin{bmatrix} 0 \\ 6 \\ 9 \\ 3 \end{bmatrix}$$

On the other hand, $T(\mathbf{v}) = 6x + 9x^2 + 3x^3$. From this we see that $[T(\mathbf{v})]_{\mathcal{B}} = [T]_B[\mathbf{v}]_{\mathcal{B}}$.

The following theorem is an abstract vector space version of Theorem 4.2.30. We reproduce its proof. From this theorem we can deduce Rank Theorem as we did before.

Theorem 6.4.4. *Let $T : V \to W$ be a linear map and $\mathcal{B}, \mathcal{C}$ be bases for V, W. Let $A = [T]_{\mathcal{B}}^{\mathcal{C}}$ be the matrix representation of T relative to $\mathcal{B}, \mathcal{C}$. Then*

$$\rho(T) = \rho(A) \quad \text{and} \quad \nu(T) = \nu(A)$$

Proof. If we let $A = [T]_{\mathcal{B}}^{\mathcal{C}}$ then we have a commutative diagram (6.35)

$$\begin{array}{ccc} V & \xrightarrow{\ T\ } & W \\ {\scriptstyle \phi_{\mathcal{B}}}\downarrow & & \downarrow{\scriptstyle \phi_{\mathcal{C}}} \\ F^n & \xrightarrow[T_A]{} & F^m \end{array}$$

namely, $\phi_{\mathcal{C}} \circ T = T_A \circ \phi_{\mathcal{B}}$. Since $\phi_{\mathcal{B}}, \phi_{\mathcal{C}}$ are isomorphisms, by Theorem 6.3.20 (2),(3) we have

$$\rho(T) = \rho(\phi_{\mathcal{C}} \circ T) = \rho(T_A \circ \phi_{\mathcal{B}}) = \rho(T_A) = \rho(A)$$

Similarly, $\nu(T) = \nu(A)$. □

The following corollary is a immediate consequence of Theorem 6.4.4.

Corollary 6.4.5. *With the same notation of* Theorem 6.4.4 *we have,*

(1) *The linear map T is injective if and only if T_A is injective.*

(2) *The linear map T is surjective if and only if T_A is surjective.*

Example 6.4.6. *Consider the linear map $T : P_3 \to P_3$ of* Example 6.4.3. *Since* $\rho([T]_{\mathcal{B}}) = 3$ *we conclude that $\rho(T) = 3$. Hence T is neither injective nor surjective.*

Example 6.4.7. *Let $A \in M(n, \mathbb{R})$. Then $T = T_A$ is a linear operator $T : \mathbb{R}^n \to \mathbb{R}^n$. We can view T as a linear operator on $\mathbb{C}^n$ in which case we write* (Exercise 6.3.16)

$$T^{\mathbb{C}} : \mathbb{C}^n \to \mathbb{C}^n$$

Also we can view A as an element of $M(n, \mathbb{C})$ and we denote it by $A^{\mathbb{C}} \in M(n, \mathbb{C})$; $T^{\mathbb{C}} = T_{A^{\mathbb{C}}}$. Now we want to show that

$$\rho(T) = \rho(T^{\mathbb{C}}), \ \ \nu(T) = \nu(T^{\mathbb{C}})$$

In fact, by the rank theorem we have $\nu(T^{\mathbb{C}}) + \rho(T^{\mathbb{C}}) = n = \nu(T) + \rho(T)$. On the other hand, $\rho(T)$ is the number of maximal $\mathbb{R}$-linearly independent columns of $A \in M(n, \mathbb{R})$. By Example 6.2.21, *the columns of A, as a subset of $\mathbb{C}^n$ is $\mathbb{R}$-linearly independent if and only if they are $\mathbb{C}$-linearly independent. Hence $\rho(T^{\mathbb{C}}) = \rho(T)$ and therefore $\nu(T) = \nu(T^{\mathbb{C}})$.*

Definition 6.4.8. *Let V, W be a vector spaces over a field F. Let*

$$L_F(V, W) := \{ \text{ all } F\text{-linear maps } T : V \to W\}$$

be the set of all F-linear maps from V to W. If the reference to the field F is clear then we simply write $L(V, W)$ for $L_F(V, W)$. If $V = W$ then we will write $L(V)$ for $L(V, W)$.

Define addition and scalar multiplication on $L(V, W)$: For $T, T_1, T_2 \in L(V, W)$ and $c \in F$

$$(T_1 + T_2)(\mathbf{v}) = T_1(\mathbf{v}) + T_2(\mathbf{v}), \quad (cT)(\mathbf{v}) = c(T(\mathbf{v})) \tag{6.36}$$

for all $\mathbf{v} \in V$.

Theorem 6.4.9. *The set of linear maps $L(V, W)$ from V to W is a vector space over F under the operations defined by* (6.36).

Proof. Exercise 2. □

The matrix representation of linear maps (6.31) with respect to the bases $\mathcal{B}, \mathcal{C}$ defines a map

$$\begin{aligned} \Phi : L(V, W) &\to M(m \times n, F) \\ T &\mapsto [T]_{\mathcal{B}}^{\mathcal{C}} \end{aligned} \tag{6.37}$$

We saw in Theorem 4.2.16 that Φ is a bijection when $F = \mathbb{R}$ and V, W are subspaces of $\mathbb{R}^n$. We know that $M(m \times n, F)$ is a vector space (Example 6.1.4) over F. And $L(V, W)$ is a vector space over F as well. We will prove that Φ is actually an isomorphism of vector spaces. The proof is almost same as the proof of Theorem 4.2.16. One difference is that we did not recognize $L(V, W)$ and $M(m \times n, F)$ as vector spaces.

Theorem 6.4.10. *The map*

$$\Phi : L(V, W) \xrightarrow{\cong} M(m \times n, F) \ ; \quad T \mapsto [T]_{\mathcal{B}}^{\mathcal{C}}$$

is an isomorphism of vector spaces over F. In particular,

$$[T_1 + T_2]_{\mathcal{B}}^{\mathcal{C}} = [T_1]_{\mathcal{B}}^{\mathcal{C}} + [T_2]_{\mathcal{B}}^{\mathcal{C}}, \ [cT]_{\mathcal{B}}^{\mathcal{C}} = c[T]_{\mathcal{B}}^{\mathcal{C}}$$

$$T = S \text{ if and only if } [T]_{\mathcal{B}}^{\mathcal{C}} = [S]_{\mathcal{B}}^{\mathcal{C}}$$

Proof. We need to check: (1) Φ is linear, (2) Φ is injective and (3) Φ is surjective.

(1) To show Φ is linear let $[T_1]_{\mathcal{B}}^{\mathcal{C}} = [a_{ij}]$, $[T_2]_{\mathcal{B}}^{\mathcal{C}} = [b_{ij}]$. Then

$$(T_1 + T_2)(\mathbf{v}_j) = T_1(\mathbf{v}_j) + T_2(\mathbf{v}_j) = \sum_i a_{ij}\mathbf{w}_i + \sum_i b_{ij}\mathbf{w}_i = \sum_i (a_{ij} + b_{ij})\mathbf{w}_i$$

Hence $[T_1 + T_2]_{\mathcal{B}}^{\mathcal{C}} = (a_{ij} + b_{ij})$. On the other hand,

$$[T_1]_{\mathcal{B}}^{\mathcal{C}} + [T_2]_{\mathcal{B}}^{\mathcal{C}} = [a_{ij}] + [b_{ij}] = [a_{ij} + b_{ij}]$$

Therefore $[T_1 + T_2]_{\mathcal{B}}^{\mathcal{C}} = [T_1]_{\mathcal{B}}^{\mathcal{C}} + [T_2]_{\mathcal{B}}^{\mathcal{C}}$. We leave the verification of $[cT]_{\mathcal{B}}^{\mathcal{C}} = c[T]_{\mathcal{B}}^{\mathcal{C}}$ as an exercise.

(2) To show Φ is injective suppose $\Phi(T) = [T]_{\mathcal{B}}^{\mathcal{C}} = 0$. Then $T(\mathbf{v}_i) = \mathbf{0}$ for all $\mathbf{v}_i \in \mathcal{B}$. Hence $T = 0$ by Theorem 6.3.9. Hence by Theorem 6.3.22, Φ is injective.

(3) To show Φ is surjective let $A = [a_{ij}] \in M(m \times n)$. Then $T(\mathbf{v}_j) = \sum_i a_{ij}\mathbf{w}_i$ defines a linear map (Theorem 6.3.9). Now it is obvious that $[T]_{\mathcal{B}}^{\mathcal{C}} = [a_{ij}]$. □

Corollary 6.4.11. *If V, W are vector spaces of dimension n, m respectively over F then $\dim_F L_F(V, W) = mn$.*

Proof. We saw before that $\dim M(m \times n, F) = nm$ (Example 6.2.8) and isomorphic vector spaces are of the same dimension (Corollary 6.3.30). □

Next we consider the compositions of linear maps.

Theorem 6.4.12. *Let U, V, W be vector spaces. If $U \xrightarrow{S} V \xrightarrow{T} W$ are linear maps (resp. isomorphisms) then the composition $T \circ S : U \to W$ is linear (resp. isomorphism). (We will often write simply TS for $T \circ S$.)*

Proof. To show, for example, $TS(\mathbf{v}_1 + \mathbf{v}_2) = TS(\mathbf{v}_1) + TS(\mathbf{v}_2)$ we compute

$$\begin{aligned} TS(\mathbf{v}_1 + \mathbf{v}_2) &= T(S(\mathbf{v}_1 + \mathbf{v}_2)) \\ &= T(S(\mathbf{v}_1) + S(\mathbf{v}_2)) \\ &= TS(\mathbf{v}_1) + TS(\mathbf{v}_2) \end{aligned}$$

The equality $TS(c\mathbf{v}) = cTS(\mathbf{v})$ is similarly proved.

The proofs of the rest of the statements are the same as the proof of Theorem 4.2.3 and we leave the details for the readers (Exercise 3). □

The following theorem is the replica of Theorem 2.2.17 in which the linear maps are between $\mathbb{R}^n$'s whereas the linear maps in the following theorem are between vector spaces.

Theorem 6.4.13. *Let T, T_1, T_2 be linear maps on vector spaces over a field F and let $c \in F$. Then the following equality hold whenever the operations make sense.*

(1) $T(T_1T_2) = (TT_1)T_2$

(2) $T(T_1 + T_2) = TT_1 + TT_2$ *and* $(T_1 + T_2)T = T_1T + T_2T$

(3) $c(T_1T_2) = (cT_1)T_2 = T_1(cT_2)$

(4) $TI = IT = T$

Proof. Exercise. □

We recall the definition of matrix multiplication (Definition 2.2.10). Let $A \in M(m{\times}n, F)$ and $B \in M(n{\times}p, F)$. The **multiplication (product)** $AB \in M(m{\times}p, F)$ is defined by

$$(AB)_{ij} = \sum_{k=1}^{n} A_{ik}B_{kj} \quad (1 \le i \le m, 1 \le j \le p)$$

Recall our notation: For $A \in M(m{\times}n)$ we write

$$A_{(i)} = \text{the } i\text{-th row of } A, \qquad A^{(j)} = \text{the } j\text{-th column of } A$$

With this notation we have

$$(AB)_{ij} = A_{(i)}B^{(j)}$$

where the right side of the equality is the product of $1{\times}n$ and $n{\times}1$ matrices.

The properties of matrix multiplications of Chapter 2 (Theorem 2.2.16) remain true without any change.

Theorem 6.4.14. *The following equalities hold whenever the matrix operations make sense.*

(1) $A(BC) = (AB)C$

(2) $A(B + C) = AB + AC$ *and* $(B + C)A = BA + CA$

(3) $c(AB) = (cA)B = A(cB)$ *for any scalar* c

(4) $IA = A = AI$ *where* I *is the identity matrix of the right size*

Next we consider the matrix linear maps (Example 6.3.4) corresponding to the matrices.

Theorem 6.4.15. *Let $A \in M(m{\times}n)$ and $T_A : F^n \to F^m$ be the corresponding matrix linear map. Let $\mathcal{E}^n, \mathcal{E}^m$ be the standard bases of F^n, F^m. Then*

(1) $[T_A]_{\mathcal{E}^n}^{\mathcal{E}^m} = A$

(2) $T_A = T_B$ *if and ony if* $A = B$

(3) $T_{A+B} = T_A + T_B$

(4) *If* $T : F^n \to F^m$ *is linear then there is a unique* $A \in M(m{\times}n)$ *such that* $T = T_A$.

(5) $T_{AB} = T_A \circ T_B$ *if* AB *is defined*

(6) $T_{I_n} = I_{F^n}$

Proof. If $F = \mathbb{R}$ then the equality (3) is Theorem 2.2.8 and (5) is Theorem 2.2.15. We leave the detailed verifications of these properties as an exercise. □

Now we show that the composition of linear maps corresponds to the multiplication of matrices. The following theorem is a vector space version of Theorem 4.2.18 and the same proof works for this case. This time we give a proof by direct computation.

Theorem 6.4.16. *Let U, V, W be vector spaces over F with bases $\mathcal{A}, \mathcal{B}, \mathcal{C}$ respectively. If $U \xrightarrow{S} V \xrightarrow{T} W$ are linear maps then*

$$[T \circ S]_{\mathcal{A}}^{\mathcal{C}} = [T]_{\mathcal{B}}^{\mathcal{C}}[S]_{\mathcal{A}}^{\mathcal{B}}$$

i.e., the composition of linear maps corresponds to the multiplication of matrices.

In particular, if $U = V = W$ and $\mathcal{A} = \mathcal{B} = \mathcal{C}$ then

$$[T \circ S]_{\mathcal{B}} = [T]_{\mathcal{B}}[S]_{\mathcal{B}}$$

Proof. Let $\mathcal{A} = \{\mathbf{u}_1, \mathbf{u}_2, \dots, \mathbf{u}_n\}$, $\mathcal{B} = \{\mathbf{v}_1, \mathbf{v}_2, \dots, \mathbf{v}_m\}$ and $\mathcal{C} = \{\mathbf{w}_1, \mathbf{w}_2, \dots, \mathbf{w}_p\}$ be bases for U, V, W. Let $A = [S]_{\mathcal{A}}^{\mathcal{B}} \in M(m\times n)$, $B = [T]_{\mathcal{B}}^{\mathcal{C}} \in M(p\times m)$. Then $BA \in M(p\times n)$ and we have

$$S(\mathbf{u}_j) = \sum_{i=1}^{m} A_{ij}\mathbf{v}_i \quad \text{and} \quad T(\mathbf{v}_j) = \sum_{i=1}^{p} B_{ij}\mathbf{w}_i$$

Hence

$$(T \circ S)(\mathbf{u}_j) = T(S(\mathbf{u}_j)) = T(\sum_{i=1}^{m} A_{ij}\mathbf{v}_i) = \sum_{i=1}^{m} A_{ij}T(\mathbf{v}_i) = \sum_{i=1}^{m} A_{ij}\Big(\sum_{k=1}^{p} B_{ki}\mathbf{w}_k\Big)$$

$$= \sum_{i,k} A_{ij}B_{ki}\mathbf{w}_k = \sum_{k=1}^{p}\Big(\sum_{i=1}^{m} B_{ki}A_{ij}\Big)\mathbf{w}_k = \sum_{k=1}^{p}(BA)_{kj}\mathbf{w}_k$$

Therefore $[T \circ S]_{\mathcal{A}}^{\mathcal{C}} = BA = [T]_{\mathcal{B}}^{\mathcal{C}}[S]_{\mathcal{A}}^{\mathcal{B}}$. □

Example 6.4.17. *Define a map $S : P_3 \to P_3$ by $S(f(x)) = f(x+1)$ and define $T : P_3 \to \mathbb{R}^3$ by $T(f(x)) = \begin{bmatrix} f(-1) \\ f(0) \\ f(1) \end{bmatrix}$. Let $\mathcal{B} = \{1, x, x^2, x^3\}$ be the standard basis for P_3 and $\mathcal{E}$ be the standard basis for $\mathbb{R}^3$. Let $A = [S]_{\mathcal{B}}, B = [T]_{\mathcal{B}}^{\mathcal{E}}$. Then we have*

$$A = \begin{bmatrix} 1 & 1 & 1 & 1 \\ 0 & 1 & 2 & 3 \\ 0 & 0 & 1 & 3 \\ 0 & 0 & 0 & 1 \end{bmatrix}, \quad B = \begin{bmatrix} 1 & -1 & 1 & -1 \\ 1 & 0 & 0 & 0 \\ 1 & 1 & 1 & 1 \end{bmatrix}, \quad BA = \begin{bmatrix} 1 & 0 & 0 & 0 \\ 1 & 1 & 1 & 1 \\ 1 & 2 & 4 & 8 \end{bmatrix}$$

On the other hand,

$$[TS]_{\mathcal{B}}^{\mathcal{E}} = \begin{bmatrix} 1 & 0 & 0 & 0 \\ 1 & 1 & 1 & 1 \\ 1 & 2 & 4 & 8 \end{bmatrix}$$

Recall the definition of invertible matrix (Definition 2.5.1).

Definition 6.4.18. *A matrix $A \in M(n)$ is* **invertible** *if there is $B \in M(n)$ such that $AB = BA = I_n$ where I_n is the $n\times n$ identity matrix. A linear map $T : V \to W$ is* **invertible** *if T is invertible as a function i.e., there is a function $S : W \to V$ with $ST = I_V$ and $TS = I_W$.*

A matrix $A \in M(n)$ is **nilpotent** *if $A^k = 0$ for some k. And a linear operator $T \in L(V)$ is* **nilpotent** *if $T^k = 0$ for some positive integer k.*

If $A \in M(n, F)$ and $AB = I_n$ for some $B \in M(n, F)$ then $BA = I_n$ and B is uniquely determined, as we saw in Theorem 2.5.3. And we denote the matrix B by A^{-1} which we call the **inverse** of A. And note that if A, B are invertible then so is AB as in Theorem 2.5.9 and

$$(AB)^{-1} = B^{-1}A^{-1} \text{ and } (A^{-1})^{-1} = A \tag{6.38}$$

$$(A^t)^{-1} = (A^{-1})^t \tag{6.39}$$

An upper or lower triangular matrix with 0's on the diagonal is nilpotent. For example, if $A = \begin{bmatrix} 0 & 1 & 1 \\ 0 & 0 & 1 \\ 0 & 0 & 0 \end{bmatrix}$ then $A^3 = 0$. Hence A is nilpotent.

A nilpotent matrix is never invertible. In fact, suppose $A^p = 0$ but $A^{p-1} \neq 0$. If A were invertible, say $AB = I_n = BA$ then, multiplying A^{p-1} to the left, we get $A^{p-1}AB = 0B = 0$. On the other hand, $A^{p-1}I_n = A^{p-1} \neq 0$ which is a contradiction.

Can an invertible matrix nilpotent? (Exercise.)

By Theorem A.1.5, a linear map T is invertible if and only if T is a bijection i.e., an isomorphism. Also the inverse of T is uniquely determined and we denote it by T^{-1}.

Theorem 6.4.19. *If $T : V \to W$ is an invertible linear map then T^{-1} is an isomorphism.*

Proof. We need to show that T^{-1} is linear. Since T is a bijection $T^{-1}(\mathbf{w}_1+\mathbf{w}_2) = T^{-1}(\mathbf{w}_1)+T^{-1}(\mathbf{w}_2)$ if and only if $T(T^{-1}(\mathbf{w}_1+\mathbf{w}_2)) = T(T^{-1}(\mathbf{w}_1)+T^{-1}(\mathbf{w}_2))$. Now $TT^{-1}(\mathbf{w}_1+\mathbf{w}_2) = \mathbf{w}_1 + \mathbf{w}_2$ and

$$\begin{aligned} T(T^{-1}(\mathbf{w}_1) + T^{-1}(\mathbf{w}_2)) &= T(T^{-1}(\mathbf{w}_1) + T^{-1}(\mathbf{w}_2)) \\ &= TT^{-1}(\mathbf{w}_1) + TT^{-1}(\mathbf{w}_2) = \mathbf{w}_1 + \mathbf{w}_2 \end{aligned}$$

Therefore $T^{-1}(\mathbf{w}_1+\mathbf{w}_2) = T^{-1}(\mathbf{w}_1)+T^{-1}(\mathbf{w}_2)$. Similarly $T^{-1}(c\mathbf{w}) = cT^{-1}(\mathbf{w})$. Hence T^{-1} is linear.

Further since T is a bijection so is T^{-1} and since T^{-1} is linear we see that T^{-1} is an isomorphism. □

If S, T are invertible then so is TS and

$$(TS)^{-1} = S^{-1}T^{-1} \quad \text{and} \quad (T^{-1})^{-1} = T \tag{6.40}$$

Since $(TS)S^{-1}T^{-1} = I$ and $T(T^{-1}) = I$.

So far there is no definition for linear operators corresponding to the transpose of a square matrix. Hence there is no formula for linear maps corresponding to (6.39). In §6.5, we will define a linear map corresponding to the transpose of a matrices (Theorem 6.5.15).

Now we investigate the relationship between the invertible linear maps and their matrix representations.

Theorem 6.4.20. *Let V, W be vector spaces of the same dimension with the bases $\mathcal{B}, \mathcal{C}$. Let $T : V \to W$ be a linear map and let $A = [T]_{\mathcal{B}}^{\mathcal{C}}$ be the matrix representation of T with respect to $\mathcal{B}, \mathcal{C}$. Then there is a linear map $S : W \to V$ such that $ST = I_V$ if and only if A is invertible. In this case, we have $TS = I_W$ (i.e., T is invertible) and*

$$[S]_{\mathcal{C}}^{\mathcal{B}} = A^{-1}$$

Proof. Let $B = [S]_{\mathcal{C}}^{\mathcal{B}}$. Then, by Theorem 6.4.16, we have

$$I_n = [I_V]_{\mathcal{B}} = [ST]_{\mathcal{B}} = [S]_{\mathcal{C}}^{\mathcal{B}}[T]_{\mathcal{B}}^{\mathcal{C}}$$

Hence $I_n = BA$ and hence A is invertible by Theorem 2.5.3 and $A^{-1} = B$.

For the converse, let $AB = BA = I_n$ for some $B \in M(n)$. To define S let $\mathcal{B} = \{\mathbf{v}_1, \mathbf{v}_2, \dots, \mathbf{v}_n\}$, $\mathcal{C} = \{\mathbf{w}_1, \mathbf{w}_2, \dots, \mathbf{w}_n\}$. Then by Theorem 6.3.9, there is a unique linear map $S \in L(W, V)$ such that $S(\mathbf{w}_j) = \sum B_{ij}\mathbf{v}_i$ i.e., $[S]_{\mathcal{C}}^{\mathcal{B}} = B$. Now

$$[ST]_{\mathcal{B}} = [S]_{\mathcal{C}}^{\mathcal{B}}[T]_{\mathcal{B}}^{\mathcal{C}} = BA = I_n = [I_V]_{\mathcal{B}}$$

Hence $ST = I_V$ by Theorem 6.4.10. Since $AB = I_n$ the same argument shows that $TS = I_V$ as well. □

Definition 6.4.21. *Let $T : V \to V$ (resp. $A \in M(n, F)$) be a linear operator (resp. square matrix) on a vector space V over F. We write T^m (resp. A^m) for the m-times composition (resp. m-th power) of T (resp. A) which is again a linear operator on V (resp. square matrix). Let $f(t) = \sum_{i=0}^{k} a_i t^i \in F[t]$ be a polynomial. We define*

$$\begin{aligned} f(T) &= a_0 I_V + a_1 T + a_2 T^2 + \cdots + a_k T^k \\ f(A) &= a_0 I_n + a_1 A + a_2 A^2 + \cdots + a_k A^k \end{aligned} \tag{6.41}$$

which is again a linear operator on V (resp. square matrix).

For $f, g \in F[t]$ we have (Exercise 4)

$$(f+g)(T) = f(T) + g(T), \quad (fg)(T) = f(T)g(T) \tag{6.42}$$

Here, $(fg)(T)$ is obtained by first taking the product $f(t)g(t)$ of two polynomials and then replace t by T. On the other hand, $f(T)g(T)$ is the composition of two linear operators $f(T)$ and $g(T)$.

Corollary 6.4.22. *Let V be a vector space with a basis $\mathcal{B}$. Then the map* (6.37)

$$\Phi : L(V) \to M(n, F)$$

defined by $\Phi(T) = [T]_{\mathcal{B}}$ is an isomorphism of vector spaces satisfying

$$\Phi(S \circ T) = \Phi(S)\,\Phi(T) \text{ i.e., } [S \circ T]_{\mathcal{B}} = [S]_{\mathcal{B}}[T]_{\mathcal{B}} \tag{6.43}$$

Further, T is invertible (resp. nilpotent) if and only if $[T]_{\mathcal{B}}$ is invertible (resp. nilpotent). In this case, as in Theorem 4.2.32 *we have*

$$[T^{-1}]_{\mathcal{B}} = ([T]_{\mathcal{B}})^{-1} \tag{6.44}$$

If $f(t) \in F[t]$ then

$$[f(T)]_{\mathcal{B}} = f([T]_{\mathcal{B}}) \tag{6.45}$$

Let $\phi_{\mathcal{B}} : V \to F^n$ be the coordinate map with respect to the basis $\mathcal{B}$. If X is a subset of V then we write $\bar{X}$ for the image $\phi_{\mathcal{B}}(X) \subseteq F^n$ of $X \subseteq V$ under $\phi_{\mathcal{B}}$. Then $T(X) = \mathbf{0}$ if and only if $[T]_{\mathcal{B}}(\bar{X}) = \mathbf{0}$ for any linear operator T on V.

Proof. Exercise 5. □

Corollary 6.4.23. *Let $A \in M(n)$. Then A is invertible if and only if T_A is invertible. Further*

$$T_{A^{-1}} = (T_A)^{-1}$$

Proof. See the proof of Theorem 2.5.6. □

Example 6.4.24. *Let $V = P_2(\mathbb{R})$ and $W = \mathbb{R}^3$ and $\mathcal{B} = \{1, x, x^2\}$ and $\mathcal{C} = \{\mathbf{e}^1, \mathbf{e}^2, \mathbf{e}^3\}$. And $T : V \to W$ be the map defined by $T(f) = \begin{bmatrix} f(0) \\ f(1) \\ f(2) \end{bmatrix}$. We want to show T is invertible and find $T^{-1}(\mathbf{v})$ for $\mathbf{v} = \begin{bmatrix} a \\ b \\ c \end{bmatrix}$.*

By the definition of T we have

$$T(1) = \begin{bmatrix} 1 \\ 1 \\ 1 \end{bmatrix}, \quad T(x) = \begin{bmatrix} 0 \\ 1 \\ 2 \end{bmatrix}, \quad T(x^2) = \begin{bmatrix} 0 \\ 1 \\ 4 \end{bmatrix}$$

Therefore

$$[T]_{\mathcal{B}}^{\mathcal{C}} = \begin{bmatrix} 1 & 0 & 0 \\ 1 & 1 & 1 \\ 1 & 2 & 4 \end{bmatrix}$$

which is invertible. And hence T is invertible.

Now by Theorem 6.4.20 *we have*

$$[T^{-1}]_{\mathcal{C}}^{\mathcal{B}} = ([T]_{\mathcal{B}}^{\mathcal{C}})^{-1} = \frac{1}{2} \begin{bmatrix} 2 & 0 & 0 \\ -3 & 4 & -1 \\ 1 & -2 & 1 \end{bmatrix}$$

Hence $[T^{-1}(\mathbf{e}^1)]_{\mathcal{B}} = \frac{1}{2} \begin{bmatrix} 2 \\ -3 \\ 1 \end{bmatrix}$ and hence $T^{-1}(\mathbf{e}^1) = 1 - \frac{3}{2}x + \frac{1}{2}x^2$. Similarly,

$$T^{-1}(\mathbf{e}^2) = 2x - x^2 \text{ and } T^{-1}(\mathbf{e}^3) = -\frac{1}{2}x + \frac{1}{2}x^2$$

Thus,

$$\begin{aligned} T^{-1}(\mathbf{v}) &= aT^{-1}(\mathbf{e}^1) + bT^{-1}(\mathbf{e}^2) + cT^{-1}(\mathbf{e}^3) \\ &= a\left(1 - \frac{3}{2}x + \frac{1}{2}x^2\right) + b(2x - x^2) + c\left(-\frac{1}{2}x + \frac{1}{2}x^2\right) \\ &= a + \left(-\frac{3}{2}a + 2b - \frac{1}{2}c\right)x + \left(\frac{1}{2}a - b + \frac{1}{2}c\right)x^2 \end{aligned}$$

Practice Problem 6.4.2. *Let $V = M(2, F)$ and $T : V \to V$ be the linear map defined by*

$$T\left(\begin{bmatrix} a & b \\ c & d \end{bmatrix}\right) = \begin{bmatrix} a & b+c \\ c+d & d \end{bmatrix}$$

Show that T is an isomorphism and find $T^{-1}(\mathbf{v})$ where $\mathbf{v} = \begin{bmatrix} 1 & 1 \\ 0 & 1 \end{bmatrix}$.

6.4.2 Change of Coordinate Matrix.

This section is almost the same as §4.3 which is about subspaces of $\mathbb{R}^n$ whereas this section concerns about abstract vector space over a field. We recall Definition 4.3.3:

Definition 6.4.25. *Let* $\mathcal{B} = \{\mathbf{v}_1, \mathbf{v}_2, \ldots, \mathbf{v}_n\}$, $\mathcal{B}' = \{\mathbf{v}'_1, \mathbf{v}'_2 \ldots, \mathbf{v}'_n\}$ *be bases of* V. *Let*

$$\mathbf{v}'_j = I(\mathbf{v}'_j) = \sum_{i=1}^{n} p_{ij}\mathbf{v}_i \text{ i.e., } [I_V]^{\mathcal{B}}_{\mathcal{B}'} = (p_{ij}) := P$$

The matrix P *is called the* **change of coordinate matrix** *from* $\mathcal{B}'$ *to* $\mathcal{B}$.

The matrix $P = [I]^{\mathcal{B}}_{\mathcal{B}'}$ is invertible since I is invertible and

$$\left([I]^{\mathcal{B}}_{\mathcal{B}'}\right)^{-1} = [I]^{\mathcal{B}'}_{\mathcal{B}} \tag{6.46}$$

In fact, $[I]^{\mathcal{B}}_{\mathcal{B}'}[I]^{\mathcal{B}'}_{\mathcal{B}} = [I]^{\mathcal{B}}_{\mathcal{B}} = I_n$ by Theorem 6.4.16.

Since $[T(\mathbf{v})]_{\mathcal{C}} = [T]^{\mathcal{C}}_{\mathcal{B}}[\mathbf{v}]_{\mathcal{B}}$ (Theorem 6.4.2) for $\mathbf{v} \in V$ we have

$$[\mathbf{v}]_{\mathcal{B}} = P[\mathbf{v}]_{\mathcal{B}'} \text{ or } [\mathbf{v}]_{\mathcal{B}'} = P^{-1}[\mathbf{v}]_{\mathcal{B}}$$

Theorem 6.4.26. (1) *Let* $T : V \to W$ *be a linear map. Let* $\mathcal{B}, \mathcal{B}'$ *be bases of* V *and* $\mathcal{C}, \mathcal{C}'$ *be bases of* W. *Then*

$$[T]^{\mathcal{C}'}_{\mathcal{B}'} = Q[T]^{\mathcal{C}}_{\mathcal{B}}P^{-1} \text{ or } [T]^{\mathcal{C}}_{\mathcal{B}} = Q^{-1}[T]^{\mathcal{C}'}_{\mathcal{B}'}P \tag{6.47}$$

where $P = [I_V]^{\mathcal{B}'}_{\mathcal{B}}$ *and* $Q = [I_W]^{\mathcal{C}'}_{\mathcal{C}}$.

(2) *Let* $\mathcal{B} = \{\mathbf{v}_1, \mathbf{v}_2, \ldots, \mathbf{v}_n\}, \mathcal{B}' = \{\mathbf{v}'_1, \mathbf{v}'_2 \ldots, \mathbf{v}'_n\}$ *be bases of* V. *Let* $P = [I]^{\mathcal{B}}_{\mathcal{B}'} = (p_{ij})$ *i.e.,* $\mathbf{v}'_j = \sum_i p_{ij}\mathbf{v}_i$. *If* $T : V \to V$ *is a linear operator then*

$$[T]_{\mathcal{B}'} = P^{-1}[T]_{\mathcal{B}}P \tag{6.48}$$

Proof. (1) This follows immediately from Theorem 6.4.16,

$$[I_W]^{\mathcal{C}'}_{\mathcal{C}}[T]^{\mathcal{C}}_{\mathcal{B}} = [T]^{\mathcal{C}'}_{\mathcal{B}} = [T]^{\mathcal{C}'}_{\mathcal{B}'}[I_V]^{\mathcal{B}'}_{\mathcal{B}}$$

(2) The equality (6.48) is a special case of (6.47). □

Example 6.4.27. *Let* $V = P_2(F)$. *Let*

$$\mathcal{B} = \{1, x, x^2\} \text{ and } \mathcal{B}' = \left\{\mathbf{v}_1 = 1 + x, \mathbf{v}_2 = x - x^2, \mathbf{v}_3 = x^2\right\}$$

be bases of V. *Let* $T : P_2(F) \to P_2(F)$ *be the linear map defined by* $T(f(x)) = f(x - 3)$. *Then*

$$[T]_{\mathcal{B}} = \begin{bmatrix} 1 & -3 & 9 \\ 0 & 1 & -6 \\ 0 & 0 & 1 \end{bmatrix}$$

To find $[T]_{\mathcal{B}'}$ *we need to write* $T(\mathbf{v}_i)$ *as a linear combination of* $\mathcal{B}'$. *For this, it is immediate that*

$$[I]^{\mathcal{B}}_{\mathcal{B}'} = \begin{bmatrix} 1 & 0 & 0 \\ 1 & 1 & 0 \\ 0 & -1 & 1 \end{bmatrix} = P^{-1}$$

Hence we have the change of coordinate matrix

$$[I]_{\mathcal{B}}^{\mathcal{B}'} = P = \begin{bmatrix} 1 & 0 & 0 \\ -1 & 1 & 0 \\ -1 & 1 & 1 \end{bmatrix}$$

Thus we have $[T(\mathbf{v})]_{\mathcal{B}'} = [I]_{\mathcal{B}}^{\mathcal{B}'}[T(\mathbf{v})]_{\mathcal{B}} = P[T(\mathbf{v})]_{\mathcal{B}}$. *Hence*

$$[T(\mathbf{v}_1)]_{\mathcal{B}'} = P[T(\mathbf{v}_1)]_{\mathcal{B}} = \begin{bmatrix} 1 & 0 & 0 \\ -1 & 1 & 0 \\ -1 & 1 & 1 \end{bmatrix} \begin{bmatrix} -2 \\ 1 \\ 0 \end{bmatrix} = \begin{bmatrix} -2 \\ 3 \\ 3 \end{bmatrix}$$

and hence we have $T(\mathbf{v}_1) = -2 + x = -2\mathbf{v}_1 + 3\mathbf{v}_2 + 3\mathbf{v}_3$. *Similarly, we have* $T(\mathbf{v}_2) = -12\mathbf{v}_1 + 19\mathbf{v}_2 + 18\mathbf{v}_3$ *and* $T(\mathbf{v}_3) = 9\mathbf{v}_1 - 15\mathbf{v}_2 - 14\mathbf{v}_3$. *Therefore, we have*

$$[T]_{\mathcal{B}'} = \begin{bmatrix} -2 & -12 & 9 \\ 3 & 19 & -15 \\ 3 & 18 & -14 \end{bmatrix}$$

Also we have $P^{-1}[T]_{\mathcal{B}}P = [T]_{\mathcal{B}'}$.

We restate Theorem 4.3.1 for the matrices over F instead of the real field $\mathbb{R}$.

Corollary 6.4.28. *Let* $V = F^n$, $A \in M(n, F)$ *and* $\mathcal{C} = \{\mathbf{v}_1, \mathbf{v}_2, \ldots, \mathbf{v}_n\} \subseteq F^n$ *be a basis of* V. *Let* $T_A : F^n \to F^n$ *be the matrix linear map defined by* $T_A(\mathbf{v}) = A\mathbf{v}$. *Then*

$$[T_A]_{\mathcal{C}} = P^{-1}AP$$

where the j*-th column* $P^{(j)}$ *of* P *is the coordinate* $[\mathbf{v}_j]_{\mathcal{E}}$ $(\mathbf{v}_j \in \mathcal{C})$ *and* $\mathcal{E}$ *is the standard basis of* F^n.

Proof. We have $T_A(\mathbf{v}) = A\mathbf{v}$ where $\mathbf{v} = \begin{bmatrix} x_1 \\ \vdots \\ x_n \end{bmatrix}$. We saw $[T_A]_{\mathcal{E}} = A$ where $\mathcal{E} = \{\mathbf{e}^1, \mathbf{e}^2, \ldots, \mathbf{e}^n\}$ is the standard basis. Let $P = [I]_{\mathcal{C}}^{\mathcal{E}}$ then we have $[T_A]_{\mathcal{C}} = P^{-1}[T_A]_{\mathcal{E}}P = P^{-1}AP$. Now the j-th column of $P = (p_{ij})$ is given by $I(\mathbf{v}_j) = \sum_i p_{ij}\mathbf{e}^i$ i.e., the j-th column of P is nothing but the column vector $[\mathbf{v}_j]_{\mathcal{E}}$. □

Recall (Definition 4.3.8) that $A, B \in M(n, F)$ are **similar** if there is an invertible matrix $P \in M(n, F)$ such that $B = P^{-1}AP$ in which case we write $A \sim B$. Hence we can say that the matrix representation of a matrix linear map of $A \in M(n, F)$ in different basis is similar to A.

For $A, B, C \in M(n, F)$ we will write $A \sim B$ if they are similar i.e., there is $P \in M(n)$ such that $B = P^{-1}AP$. We want to show $\sim$ on $M(n, F)$ satisfies the following conditions

(i) $A \sim A$

(ii) If $A \sim B$ then $B \sim A$.

(iii) If $A \sim B$ and $B \sim C$ then $A \sim C$.

for all $A, B, C \in M(n, F)$. That is $\sim$ is an equivalence relation (Definition A.2.1).

Theorem 6.4.29. *Similarity is an equivalence relation.*

Proof. For (i), $A = I^{-1}AI$. Hence $A \sim A$. For (ii), suppose $A \sim B$ say $B = P^{-1}AP$. Then $A = (P^{-1})^{-1}B(P^{-1})$. Hence $B \sim A$. Now, for (iii), suppose $A \sim B$ and $B \sim C$ say $B = P^{-1}AP$ and $C = Q^{-1}BQ$. Then $C = (PQ)^{-1}A(PQ)$. Hence $A \sim C$. □

Next we show that the determinant as well as the trace is a well defined function on the linear operators. Recall if $A = [a_{ij}] \in M(n, F)$ then we defined (Definition 3.3.19) the trace of A to be the sum of diagonal entries of A i.e.,

$$\mathrm{Tr}\,(A) = \sum_{i=n}^{n} a_{ii}$$

Over the field $\mathbb{R}$ of reals we showed in Corollary 3.3.12 and Corollary 3.3.21 that similar matrices have the same determinant and the same trace. The proofs for the corollaries work for the matrices over arbitrary fields. We record the result omitting the proofs.

Theorem 6.4.30. *Similar matrices have the same determinant and the same trace: If P is an invertible matrix then*

$$\det(P^{-1}AP) = \det(A), \;\; \mathrm{Tr}\,(P^{-1}AP) = \mathrm{Tr}\,(A)$$

In particular, if T is a linear operator and $\mathcal{B}, \mathcal{C}$ are bases of V: Then

$$\det([T]_{\mathcal{B}}) = \det([T]_{\mathcal{C}}), \;\; \mathrm{Tr}\,([T]_{\mathcal{B}}) = \mathrm{Tr}\,([T]_{\mathcal{C}})$$

Theorem 6.4.30 justifies the following definition.

Definition 6.4.31. *Let V be a vector space over F with a basis $\mathcal{B}$ and let $T \in L(V)$. We define the* **determinant** *and* **trace** *of T by*

$$\det(T) = \det([T]_{\mathcal{B}}) \quad \text{and} \quad \mathrm{Tr}\,(T) = \mathrm{Tr}\,([T]_{\mathcal{B}})$$

Corollary 6.4.32. *Let T be a linear operator on a vector space V. Then T is invertible if and only if* $\det(T) \neq 0$.

Proof. By Theorem 6.4.20, T is invertible if and only if $[T]_{\mathcal{B}}$ is invertible. By Theorem 3.3.7, $[T]_{\mathcal{B}}$ is invertible if and only if $\det[T]_{\mathcal{B}} \neq 0$. □

Example 6.4.33. *In* Example 6.4.27 *we have*

$$[T]_{\mathcal{B}} = \begin{bmatrix} 1 & -3 & 9 \\ 0 & 1 & -6 \\ 0 & 0 & 1 \end{bmatrix} \text{ and } [T]_{\mathcal{B}'} = \begin{bmatrix} -2 & -12 & 9 \\ 3 & 19 & -15 \\ 3 & 18 & -14 \end{bmatrix}$$

Now $\det([T]_{\mathcal{B}}) = 1$ *which is easy to find. But* $\det([T]_{\mathcal{B}'})$ *looks so complicated. By* Theorem 6.4.30 *we also have* $\det([T]_{\mathcal{B}'}) = 1$. *The readers should check this out. Also observe that* $\mathrm{Tr}\,([T]_{\mathcal{B}}) = \mathrm{Tr}\,([T]_{\mathcal{B}'}) = 3$.

Also by Corollary 6.4.32, *T in* Example 6.4.27 *is invertible. The inverse of T is $T^{-1}(f(x)) = f(x+3)$ which follows easily from the definition of T.*

Answers for Practice Problems 6.4

(6.4.1) We compute $[T(1)]_{\mathcal{E}} = \begin{bmatrix} 1 \\ 1 \end{bmatrix}, [T(x)]_{\mathcal{E}} = \begin{bmatrix} 1 \\ 2 \end{bmatrix}, [T(x^2)]_{\mathcal{E}} = \begin{bmatrix} 1 \\ 4 \end{bmatrix}$. Hence we have $[T]^{\mathcal{E}}_{\mathcal{B}} = \begin{bmatrix} 1 & 1 & 1 \\ 1 & 2 & 4 \end{bmatrix}$.

(6.4.2) Choose a basis for $M(2, F)$:

$$\mathcal{B} = \left\{ E_{11} = \begin{bmatrix} 1 & 0 \\ 0 & 0 \end{bmatrix}, E_{12} = \begin{bmatrix} 0 & 1 \\ 0 & 0 \end{bmatrix}, E_{21} = \begin{bmatrix} 0 & 0 \\ 1 & 0 \end{bmatrix}, E_{22} = \begin{bmatrix} 0 & 0 \\ 0 & 1 \end{bmatrix} \right\}$$

Then $T(E_{11}) = E_{11},\ T(E_{12}) = E_{12},\ T(E_{21}) = E_{12} + E_{21},\ T(E_{22}) = E_{21} + E_{22}$. Hence

$$[T]_{\mathcal{B}} = \begin{bmatrix} 1 & 0 & 0 & 0 \\ 0 & 1 & 1 & 0 \\ 0 & 0 & 1 & 1 \\ 0 & 0 & 0 & 1 \end{bmatrix} \quad \text{and} \quad [T]_{\mathcal{B}}^{-1} = \begin{bmatrix} 1 & 0 & 0 & 0 \\ 0 & 1 & -1 & 1 \\ 0 & 0 & 1 & -1 \\ 0 & 0 & 0 & 1 \end{bmatrix}$$

As $\det[T]_{\mathcal{B}} = 1$ we see that T is an isomorphism.

Now $\mathbf{v} = \begin{bmatrix} 1 & 1 \\ 0 & 1 \end{bmatrix} = E_{11} + E_{12} + E_{22}$. Hence we have $[\mathbf{v}]_{\mathcal{B}} = \begin{bmatrix} 1 \\ 1 \\ 0 \\ 1 \end{bmatrix}$. Thus

$$[T^{-1}(\mathbf{v})]_{\mathcal{B}} = [T^{-1}]_{\mathcal{B}}[\mathbf{v}]_{\mathcal{B}} = [T]_{\mathcal{B}}^{-1}[\mathbf{v}]_{\mathcal{B}} = \begin{bmatrix} 1 & 0 & 0 & 0 \\ 0 & 1 & -1 & 1 \\ 0 & 0 & 1 & -1 \\ 0 & 0 & 0 & 1 \end{bmatrix} \begin{bmatrix} 1 \\ 1 \\ 0 \\ 1 \end{bmatrix} = \begin{bmatrix} 1 \\ 2 \\ -1 \\ 1 \end{bmatrix}$$

Therefore

$$T^{-1}(\mathbf{v}) = E_{11} + 2E_{12} - E_{21} + E_{22} = \begin{bmatrix} 1 & 2 \\ -1 & 1 \end{bmatrix}$$

6.4 Exercises

1. Let $S, T \in L(V)$ and $\mathcal{B}$ be a basis of V. Let $A = [S]_{\mathcal{B}}, B = [T]_{\mathcal{B}}$.
   (a) Prove that $ST = TS$ if and only if $AB = BA$.
   (b) Prove that if $ST = I_V$ then $TS = I_V$.
2. Prove Theorem 6.4.9.
3. Complete the proof of Theorem 6.4.12.
4. For $f, g \in F[t]$ and $T \in L(V)$ prove the equalities (6.42)
$$(f + g)(T) = f(T) + g(T), \quad (fg)(T) = f(T)g(T)$$
5. Prove Corollary 6.4.22.
6. Let $A, B \in M(n, \mathbb{R})$. Prove that if A be symmetric and B is skew symmetric then $\mathrm{Tr}\,(AB) = 0$.
7. Let $A \in M(n, F)$ and $f(t) \in F[t]$. Prove the following equalities.
   (a) $f(T_A) = T_{f(A)}$

(b) If $B = P^{-1}AP$ for some invertible matrix P then $f(B) = P^{-1}f(A)P$.

8. Let $A, B \in M(n)$. Prove that A and B are similar if and only if A^t and B^t are similar.

9. Prove that $A = \begin{bmatrix} 1 & 4 & 7 \\ 2 & 5 & 8 \\ 3 & 6 & 9 \end{bmatrix}, B = \begin{bmatrix} 9 & 6 & 3 \\ 8 & 5 & 2 \\ 7 & 4 & 1 \end{bmatrix}$ are similar. Find P such that $P^{-1}AP = B$.

10. Let $A = [a_{ij}] \in M(n, F)$ be an upper triangular matrix. Prove that $\text{Tr}\,(A^2) = \sum_{i=1}^{n} a_{ii}^2$.

11. Prove the true statement(s) and find a counter example for the false statement(s).

    (a) An invertible matrix is not nilpotent.

    (b) Let $A, B \in M(n, F)$ be invertible matrices. If A and B are similar then so are A^{-1} and B^{-1}.

    (c) If $A \in M(n)$ is nilpotent then A is upper or lower triangular with zeros on the diagonal.

    (d) If an upper triangular matrix $A \in M(n, F)$ is nilpotent then all the diagonal entries of A are zero.

    (e) If T is a linear operator on V and $\mathcal{B}$ is a basis for V then $\dim R(T) = \rho[T]_{\mathcal{B}}$,

    (f) Let $\mathcal{B} = \{\mathbf{v}_i\}, \mathcal{B}' = \{\mathbf{v}'_i\}$ be bases for a vector space. Let $\mathbf{v}'_j = \sum p_{ij}\mathbf{v}_i$ and $\mathbf{v}_j = \sum q_{ij}\mathbf{v}'_i$ and $P = (p_{ij}), Q = (q_{ij})$ then $PQ = I_n$.

    (g) For any $A \in M(n)$ we have $\text{Tr}\,(A^2) = \text{Tr}\,(A)^2$.

    (h) For any $A \in M(n)$ we have $\text{Tr}\,(A^2) = \sum A_{ii}^2$.

    (i) If $T \in L(V)$ is nilpotent then $R(T) \neq V$.

    (j) IF $T \in L(V)$ and $R(T) \neq V$ then T is nilpotent.

    (k) If $A, B \in M(n)$ are similar then A is nilpotent if and only if so is B.

    (l) If N_1, N_2 are nilpotent matrices then $N_1 + N_2$ is also nilpotent.

    (m) If D is a diagonal matrix of size n then $AD = DA$ for all $A \in M(n)$.

12. Let V, W be vector spaces over F with $\dim W = 1$. If $T : V \to W$ is a nonzero linear map then T is surjective. Deduce that the map $\phi : L(F, V) \to V$ defined by $\phi(T) = T(1)$ is an isomorphism.

13. Let U, V, W be vector spaces. Prove that there are natural isomorphisms

    (a) $\phi : L(U \times V, W) \xrightarrow{\cong} L(U, W) \times L(V, W)$

    (b) $\psi : L(U, V \times W) \xrightarrow{\cong} L(U, V) \times L(U, W)$

    (Hint: For $T : U \times V \to W$ define $T_U(\mathbf{u}) = T(\mathbf{u}, \mathbf{0}), T_V(\mathbf{v}) = T(\mathbf{0}, \mathbf{v})$. Then show that the association $T \mapsto (T_U, T_V)$ is the required isomorphism.)

14. Let $T \in L(V)$. Prove the following inclusions.

    (a) $N(T) \subseteq N(T^2) \subseteq N(T^3) \subseteq \cdots$

    (b) $R(T) \supseteq R(T^2) \supseteq R(T^3) \supseteq \cdots$

15. Let T be a projection (Definition 6.3.15) on a vector space V. Prove that there is a basis $\mathcal{B}$ for which $[T]_{\mathcal{B}} = \begin{bmatrix} I_r & 0 \\ 0 & 0 \end{bmatrix}$ where I_r is the $r \times r$ identity matrix.

16. Prove that a linear operator $T \in L(V)$ on V is nilpotent if and only if $[T]_{\mathcal{B}}$ is a nilpotent matrix for any basis $\mathcal{B}$ of V.

17. Let T be a nonzero linear operator on a finite dimensional vector space. Prove that the following conditions are equivalent.

    (a) $N(T^i) = N(T^{i+1})$ for all $i \geq 1$

    (b) $N(T^i) = N(T^{i+1})$ for some $i \geq 1$

    (c) $N(T) = N(T^2)$

18. Let T be a nonzero linear operator on a finite dimensional vector space. Prove that the following conditions are equivalent.

    (a) The linear operator T is nilpotent.

    (b) $N(T) \subsetneq N(T^2) \subsetneq \cdots \subsetneq N(T^{i-1}) \subsetneq N(T^i) = N(T^{i+1}) = \cdots = V$ for some i.

    (c) $\nu(T) < \nu(T^2)$

19. (a) Prove that $T \in L(V)$ is nilpotent if and only if there is a basis $\mathcal{B}$ for which $[T]_{\mathcal{B}}$ is upper triangular with zeros on the diagonal. (Hint: Use Exercise 18 above.)

    (b) Prove that $A \in M(n, F)$ nilpotent if and only if A is similar to an upper triangular matrix with zeros on the diagonal.

20. Prove that if T is a nilpotent linear operator then $\det(T) = 0$ and $\mathrm{Tr}\,(T) = 0$.

21. Let N_1, N_2 be nilpotent linear operators on V. Prove that $N_1 + N_2$ is also nilpotent if $N_1 N_2 = N_2 N_1$.

22. Let $N \in L(V)$ be a nilpotent linear operators on a vector space V over F.

    (a) Prove that $I_V + N$ is an invertible linear operator.

    (b) Let $f(t) = a_0 + a_1 t + a_2 t^2 + \cdots + a_n t^n \in F[t]$. Prove that $f(N)$ is nilpotent if $a_0 = 0$. Also prove that $f(N)$ is invertible if $a_0 \neq 0$.

    (c) Let $f(t) = a_i t^i + a_{i+1} t^{i+1} + \cdots + a_n t^n$ with $a_i \neq 0$. Prove that $\rho(f(N)) = \rho(N^i)$.

23. Let $M_L(n, F)$ be the $n \times n$ lower triangular matrices over F. (We may replace lower triangular by upper triangular in this exercise.)

    (a) Prove that a matrix $L \in M_L(n, F)$ is nilpotent if and only if the diagonals of L are zero.

    (b) Prove that a matrix $L \in M_L(n, F)$ is invertible if and only if the diagonals of L are all nonzero.

    (c) Prove that the set of all nilpotent matrices in $M_L(n, F)$ form a subspace.

24. Let $D(n, F)$ be the set of all diagonal matrices in $M(n, F)$ over F.

(a) Prove that $D(n, F)$ is a subspace of $M(n, F)$ of dimension n. And if $D = \text{diag}[d_1, d_2, ..., d_n]$, $D' = \text{diag}[d'_1, d'_2, ..., d'_n] \in D(n, F)$ then $DD' = D'D$ and they are equal to $\text{diag}[d_1 d'_1, d_2 d'_2, ..., d_n d'_n]$.

(b) Let $D = \text{diag}[d_1, d_2, ..., d_n]$ and $f(t) \in F[t]$, Prove that

$$f(D) = \text{diag}[f(d_1), f(d_2), ..., f(d_n)]$$

(c) Let $A \in M(n, F)$. Prove that if $AB = BA$ for all $B \in M(n, F)$ then $A = cI_n$ for some $c \in F$. (Hint: Let $E_{ii} = \text{diag}[0, ..., 1, ..., 0]$ with 1 on the i-th place. Deduce that A is a diagonal matrix from the fact that $E_{ii}A = AE_{ii}$. Now let I_{ij} be the matrix obtained by interchanging the i-th and j-th row of the identity matrix I_n (Definition 2.5.10). If A is a diagonal matrix conclude that all the diagonal entries are the same from the fact that $I_{1i}A = AI_{1i}$.)

6.5 Linear Functionals and Dual Spaces

6.5.1 Dual Spaces and Dual Basis

Recall that if V, W are vector spaces over F then

$$L(V, W) = \{\text{ all } F\text{-linear maps } T : V \to W\}$$

is a vector space of dimension $\dim(V)\dim(W)$ over F (Corollary 6.4.11) . When $W = F$, a linear map $f : V \to F$ is called a **linear functional** (Definition 6.3.1).

Definition 6.5.1. *Let V be a vector space over F. Let*

$$V^{\#} = \{\text{all linear functionals } f : V \to F\}^{\dagger} \tag{6.49}$$

which we call the **dual space** *of V.*

Note that the dual $V^{\#}$ of V is a vector space over F of the same dimension:

$$\dim(V^{\#}) = \dim(V) \tag{6.50}$$

In particular, V and $V^{\#}$ are isomorphic (Corollary 6.3.30) but there is no preferred isomorphism. Namely every choices of bases for V and $V^{\#}$ determines an isomorphism. If the dimension of V is infinite then V is never isomorphic to $V^{\#}$ (See Appendix D, Corollary D.3.2).

If V has an extra structure, say the dot product on $\mathbb{R}^n$, then there is a natural isomorphism between V and $V^{\#}$ as in the next example.

Example 6.5.2. *Let $V = \mathbb{R}^n$. For $\mathbf{v} \in V$ define $\Delta_{\mathbf{v}} : V \to \mathbb{R}$ by $\Delta_{\mathbf{v}}(\mathbf{x}) = \mathbf{v} \cdot \mathbf{x}$* (Definition 5.2.1). *Then we have $\Delta_{\mathbf{v}} \in V^{\#}$ by the properties of dot product* (Theorem 5.2.2).

Now consider the map $\phi : V \to V^{\#}$ defined by $\phi(\mathbf{v}) = \Delta_{\mathbf{v}}$. We claim that ϕ is an isomorphism. It is easy to show that ϕ is linear. Now if $\Delta_{\mathbf{v}} = 0$ i.e., $\mathbf{v} \cdot \mathbf{x} = 0$ for all $\mathbf{x} \in \mathbb{R}^n$ then $\mathbf{v} = \mathbf{0}$ by the property of the dot, product (Theorem 5.2.2). *Hence ϕ is injective. Since* $\dim V = \dim V^{\#}$ *we conclude that $\phi : V \to V^{\#}$ is an isomorphism* (Theorem 6.3.26). *In*

†The usual notation for the dual space of V is V^*. But we reserve it for something else in Chapter 8.

particular, if $f \in V^{\#}$ *and* $\mathbf{x} = \begin{bmatrix} x_1 \\ \vdots \\ x_n \end{bmatrix}$ *then* $f(\mathbf{x}) = a_1x_1 + \cdots + a_nx_n$ *for some uniquely determined* $\mathbf{v} = \begin{bmatrix} a_1 \\ \vdots \\ a_n \end{bmatrix} \in \mathbb{R}^n$ *i.e.,* $f(\mathbf{x}) = \Delta_{\mathbf{v}}(\mathbf{x})$ (Theorem 2.2.8).

We will study such vector spaces with an extra structure in Chapter 8.

Example 6.5.3. *Let* $V = P_n(\mathbb{R})$ *and let* $a, b \in \mathbb{R}$. *For* $p \in V$, *define*

$$e_a(p) = p(a),\ I_{ab}(p) = \int_a^b p(x)\,dx,\ D_a(p) = p'(a)$$

Then e_a, I_{ab}, D_a *are linear functionals on* V.

Example 6.5.4. *Let* $V = M(n, F)$. *Then the trace map* $\mathrm{Tr} : M(n, F) \to F$ *is a linear functional.*

For $A = [a_{ij}] \in V$ *define* $p_{ij}(A) = a_{ij}$. *Then* $p_{ij} \in V^{\#}$ *and* $\mathrm{Tr} = \sum p_{ii}$.

Next we give a criterion for a set of linear functionals to be a basis.

Theorem 6.5.5. *Let* $\{\mathbf{v}_1, \mathbf{v}_2, ..., \mathbf{v}_n\}$ *be a basis of a vector space* V *over* F. *Let* $\{f_1, f_2, ..., f_n\}$ *be a set of linear functionals. Then* $\{f_1, f_2, ..., f_n\}$ *is a basis of* $V^{\#}$ *if and only if the matrix* $A := (f_i(\mathbf{v}_j)) \in M(n, F)$ *is invertible.*

In this case, let $\mathbf{v} = \sum c_i\mathbf{v}_i \in V$ *and* $f_i(\mathbf{v}) = b_i$. *Let* $\mathbf{b} = \begin{bmatrix} b_1 \\ \vdots \\ b_n \end{bmatrix}$, $\mathbf{c} = \begin{bmatrix} c_1 \\ \vdots \\ c_n \end{bmatrix}$. *Then we have* $\mathbf{c} = A^{-1}\mathbf{b}$.

Let $f = \sum_i e_if_i \in V^{\#}$ *and* $f(\mathbf{v}_i) = d_i$. *Let* $\mathbf{e} = \begin{bmatrix} e_1 \\ \vdots \\ e_n \end{bmatrix}$ *and* $\mathbf{d} = \begin{bmatrix} d_1 \\ \vdots \\ d_n \end{bmatrix}$. *Then* $\mathbf{e} = (A^t)^{-1}\mathbf{d}$.

Proof. Assume $\{f_1, f_2, ..., f_n\}$ is a basis of $V^{\#}$. Since A is invertible if and only if $B := A^t$ is invertible (6.39), it suffices to show that the columns of B are linearly independent (Theorem 2.5.14). Suppose $a_1B^{(1)} + a_2B^{(2)} + \cdots a_nB^{(n)} = \mathbf{0}$ i.e.,

$$a_1 \begin{bmatrix} f_1(\mathbf{v}_1) \\ \vdots \\ f_1(\mathbf{v}_n) \end{bmatrix} + a_2 \begin{bmatrix} f_2(\mathbf{v}_1) \\ \vdots \\ f_2(\mathbf{v}_n) \end{bmatrix} + \cdots + a_n \begin{bmatrix} f_n(\mathbf{v}_1) \\ \vdots \\ f_n(\mathbf{v}_n) \end{bmatrix} = \begin{bmatrix} 0 \\ \vdots \\ 0 \end{bmatrix}$$

Thus $(\sum_i a_if_i)(\mathbf{v}_j) = 0$ for all j. Therefore the linear functional $\sum a_if_i$ is identically zero on V (Theorem 6.3.9) i.e., $\sum a_if_i$ is zero as an element of $V^{\#}$. Since $\{f_i\}$ is a basis we conclude that $a_i = 0$ for all i; the columns of B are linearly independent; B is invertible.

Now suppose $A = (f_i(\mathbf{v}_j))$ is invertible and $\sum a_if_i = 0$. Hence we have $(\sum_i a_if_i)(\mathbf{v}_j) = \sum_i a_if_i(\mathbf{v}_j) = \sum_i f_i(\mathbf{v}_j)a_i = 0$ for all j; $B_{(j)} \cdot \mathbf{v} = \mathbf{0}$ where $\mathbf{v} = \begin{bmatrix} a_1 \\ \vdots \\ a_n \end{bmatrix}$. Since $B = A^t$ is invertible we conclude that $\mathbf{v} = \mathbf{0}$. Hence $\{f_1, f_2, ..., f_n\}$ is linearly independent and hence it is a basis of $V^{\#}$ since $\dim V^{\#} = n$ (Corollary 6.2.16).

To find $\mathbf{c}$, we have $b_i = f_i(\mathbf{v}) = \sum c_j f_i(\mathbf{v}_j) = A_{(i)} \begin{bmatrix} c_1 \\ \vdots \\ c_n \end{bmatrix}$. Hence $\mathbf{b} = A\mathbf{c}$.

The proof of the last statement is Exercise 9. $\square$

Example 6.5.6. *Let $V = P_2(\mathbb{R})$ and let $\mathcal{B}' = \{e_0, e_1, e_2\}$ be the evaluation maps at $0, 1, 2$ i.e., $e_i(p) = p(i)$ $(i = 0, 1, 2)$ for $p \in V$. We want to show that $\mathcal{B}'$ is a basis of $V^\#$. We choose a basis $\mathcal{B} = \{1, x, x^2\}$ for V. By* Theorem 6.5.5, *we need to show the matrix $A = (e_i(x^j))$ is invertible. Now we have $A = \begin{bmatrix} 1 & 0 & 0 \\ 1 & 1 & 1 \\ 1 & 2 & 4 \end{bmatrix}$ and $|A| = 2$. Hence A is invertible and therefore $\mathcal{B}'$ is a basis for $V^\#$.*

Now we look for $p(x) \in V$ such that $e_i(p) = i + 1$ $(i = 1, 2, 3)$; $\mathbf{b} = \begin{bmatrix} e_0(p) \\ e_1(p) \\ e_2(p) \end{bmatrix} = \begin{bmatrix} 1 \\ 2 \\ 3 \end{bmatrix}$. Let $p(x) = c_0 + c_1 x + c_2 x^2$ and $\mathbf{c} = \begin{bmatrix} c_0 \\ c_1 \\ c_2 \end{bmatrix}$. Then by Theorem 6.5.5, *we have $\mathbf{c} = A^{-1}\mathbf{b}$. Now $A^{-1} = \frac{1}{2} \begin{bmatrix} 2 & 0 & 0 \\ -3 & 4 & -1 \\ 1 & -2 & 1 \end{bmatrix}$ and hence $\mathbf{c} = A^{-1}\mathbf{b} = \begin{bmatrix} 1 \\ 1 \\ 0 \end{bmatrix}$. Therefore $p(x) = 1 + x$.*

Practice Problem 6.5.1. *Referring to* Example 6.5.6, *express the linear functional $e_{\frac{1}{2}}$, the evaluation map at $\frac{1}{2}$, as a linear combination of $\mathcal{B}'$.*

Let $\mathcal{B} = \{\mathbf{v}_1, \mathbf{v}_2, \ldots, \mathbf{v}_n\}$ be a basis of V. Define a linear map $f_i : V \to F$ by requiring (Theorem 6.3.9)

$$f_i(\mathbf{v}_j) = \delta_{ij} = \begin{cases} 1 & (\text{ if } i = j) \\ 0 & (\text{ if } i \neq j) \end{cases}$$

and let

$$\mathcal{B}^\# = \{f_1, f_2, \ldots, f_n\} \tag{6.51}$$

Corollary 6.5.7. *The linear functionals $\mathcal{B}^\#$ is a basis for $V^\#$.*

Proof. Follows from Theorem 6.5.5 since $(f_i(\mathbf{v}_j)) = I_n$. $\square$

Definition 6.5.8. *The basis (6.51) of $V^\#$ is called the* **dual basis** *of the basis $\mathcal{B} = \{\mathbf{v}_1, \mathbf{v}_2, \ldots, \mathbf{v}_n\}$.*

Example 6.5.9. *Let $V = \mathbb{R}^2$ and let $\mathcal{B} = \left\{ \mathbf{v}_1 = \begin{bmatrix} 2 \\ 1 \end{bmatrix}, \mathbf{v}_2 = \begin{bmatrix} 3 \\ 1 \end{bmatrix} \right\}$ be a basis of V. We want to find the dual basis $\mathcal{B}^\# = \{f_1, f_2\}$. Let $f_1(x, y) = ax + by$. Then $f_1(\mathbf{v}_1) = f_1(2, 1) = 2a + b = 1$ and $f_1(\mathbf{v}_2) = f_1(3, 1) = 3a + b = 0$. Solving this we get $a = -1, b = 3$. Hence $f_1(x, y) = -x + 3y$. Similarly, $f_2(x, y) = x - 2y$.*

Example 6.5.10. *Consider the vector space $V = P_n(F)$ of polynomials of degree $\leq n$ with* $\dim(P_n) = n + 1$ *over F. Let $x_0, x_1, ..., x_n \in F$. Recall that the Lagrange polynomials* (3.33)

$$p_i(x) = \frac{(x - x_0) \cdots (x - x_{i-1})(x - x_{i+1}) \cdots (x - x_n)}{(x_i - x_0) \cdots (x_i - x_{i-1})(x_i - x_{i+1}) \cdots (x_i - x_n)} \quad (i = 0, 2, ..., n) \tag{6.52}$$

are elements of V and they are linearly independent. In fact, if $\sum c_i p_i(x) = 0$ for some c_i's. Then evaluating both sides at $x = x_j$ we get $c_j = 0$ for all j. Since $\mathcal{B} = \{p_0, p_1, ..., p_n\}$ is a linearly independent vectors in $n+1$ dimensional vector space $\mathcal{B}$ is a basis of V.

Let $e_i : V \to F$ $(i = 0, 1, ..., n)$ be the evaluation map at x_i defined by $e_i(f) = f(x_i)$. Then it is easy to check that $e_i \in V^{\#}$ and

$$e_i(p_j(x)) = \delta_{ij}$$

Hence $\mathcal{B}^{\#} = \{e_0, e_1, ..., e_n\}$ is the dual basis of $\mathcal{B} = \{p_0, p_1, ..., p_n\}$.

Example 6.5.11. *Let $V = P_n(\mathbb{R})$ be the vector space of real polynomials of degree $\le n$ and $\mathcal{B} = \{1, x, x^2, ..., x^n\}$ be the standard basis for V. Let $D_k = \left.\frac{1}{k!}\frac{d^k}{dx^k}\right|_{x=0}$ where $\left.\frac{d^k}{dx^k}\right|_{x=0}$ denotes the k-th derivative at $x = 0$ with the convention $0! = 1$ and $\left.\frac{d^0}{dx^0}\right|_{x=0}(p) = p(0)$ for all $p \in V$. Then it is easy to show that $D_i \in V^{\#}$ and $D_i(x^j) = \delta_{ij}$. Hence $\{D_0, D_1, ..., D_n\}$ is the dual basis for $\mathcal{B}$.*

Practice Problem 6.5.2. (1) *Referring to* Example 6.5.6, *we saw $\mathcal{B}' = \{e_0, e_1, e_2\}$ is a basis for $V^{\#}$. Express the dual basis $\mathcal{B}^{\#} = \{f_1, f_2, f_3\}$ of $\mathcal{B} = \{1, x, x^2\}$ as linear combinations of $\mathcal{B}'$.*

(2) *Express the evaluation $e_{\frac{1}{2}}$ at $\frac{1}{2}$ as a linear combination of D_0, D_1, D_2.*

Following result is an immediate consequence of Theorem 6.5.5.

Corollary 6.5.12. *Let $\mathcal{B} = \{\mathbf{v}_1, \mathbf{v}_2, \ldots, \mathbf{v}_n\}$ be a basis of V and $\mathcal{B}^{\#} = \{f_1, f_2, \ldots, f_n\}$ be the dual basis of $\mathcal{B}$. Then for any $\mathbf{v} \in V$ and $f \in V^{\#}$ we have*

$$\mathbf{v} = \sum f_i(\mathbf{v})\mathbf{v}_i \quad \text{and} \quad f = \sum_{i=1}^{n} f(\mathbf{v}_i) f_i \tag{6.53}$$

In particular, $\mathbf{v} = \mathbf{0}$ if and only if $f_i(\mathbf{v}) = 0$ for all i. And $f = 0$ if and only if $f(\mathbf{v}_i) = 0$ for all i.

Proof. We use Theorem 6.5.5: The matrix $A = (f_i(\mathbf{v}_j)) = I_n$ since $\mathcal{B}$ and $\mathcal{B}^{\#}$ are dual basis. Since $b_i = f_i(\mathbf{v})$ we have $\mathbf{c} = A^{-1}\mathbf{b} = \mathbf{b}$ and hence $\mathbf{v} = \sum_i f_i(\mathbf{v})$.

For the second equality, we have $d_i = f(\mathbf{v}_i)$. By the theorem we see that $\mathbf{e} = (A^t)^{-1}\mathbf{d} = \mathbf{d}$ and hence $f = \sum_i f(\mathbf{v}_i) f_i$.

The last statements are obvious from (6.53). □

Example 6.5.13. *In* Example 6.5.10, *we have the Lagrange interpolation formula* (3.34)*: For $p \in V$*

$$p = \sum p(x_i) p_i$$

by (1) *of* Corollary 6.5.12 *since f_i is the evaluation e_i at x_i.*

6.5.2 Linear Maps on Dual Spaces and Double Dual

Let $T : V \to W$ be a linear map. Let $g \in W^{\#}$ i.e., $g : W \to F$ is a linear functional. Then the composition $g \circ T$ is a linear functional on V. Hence $g \circ T \in V^{\#}$. We write

$$T^{\#}(g) = g \circ T \tag{6.54}$$

In this way, we get a map

$$T^\# : W^\# \to V^\# \tag{6.55}$$

Now we claim that $T^\#$ is a linear map. In fact, let $g_1, g_2 \in W^\#$. Then (Theorem 6.4.13)

$$T^\#(g_1 + g_2) = (g_1 + g_2) \circ T = g_1 \circ T + g_2 \circ T = T^\#(g_1) + T^\#(g_2)$$

Similarly, $T^\#(cg) = cT^\#(g)$. We will call $T^\#$ the **dual** of T.

If $S : U \to V$ is also a linear map then

$$(T \circ S)^\# = S^\# \circ T^\# \tag{6.56}$$

In fact, for $g \in W^\#$ we have $(T \circ S)^\#(g) = T \circ S \circ g = T^\#(S \circ g) = T^\# S^\#(g)$. Also it is obvious that

$$(I_V)^\# = I_{V^\#}$$

We summarize our observations.

Theorem 6.5.14. *Let $T : V \to W$ be a linear map of vector spaces. Then*

$$T^\# : W^\# \to V^\#$$

defined by $T^\#(g) = g \circ T$ $(g \in W^\#)$ is linear. If $S : U \to V$ is a linear map then we have

$$(T \circ S)^\# = S^\# \circ T^\# \tag{6.57}$$

Now we show that the matrix representation of $T^\#$ with respect to the dual bases is the transpose of the matrix representation of T.

Theorem 6.5.15. *Let $T : V \to W$ be a linear map and $\mathcal{B}, \mathcal{C}$ be bases of V, W. Then*

$$[T^\#]_{\mathcal{C}^\#}^{\mathcal{B}^\#} = \left([T]_{\mathcal{B}}^{\mathcal{C}}\right)^t \tag{6.58}$$

In particular, if $V = W$, $\mathcal{B} = \mathcal{C}$ then

$$[T^\#]_{\mathcal{B}^\#} = \left([T]_{\mathcal{B}}\right)^t \tag{6.59}$$

If T is invertible then

$$(T^\#)^{-1} = (T^{-1})^\# \tag{6.60}$$

Proof. Let $\mathcal{B} = \{\mathbf{v}_1, \mathbf{v}_2, \ldots, \mathbf{v}_n\}, \mathcal{C} = \{\mathbf{w}_1, \mathbf{w}_2, \ldots, \mathbf{w}_m\}$. Let $\mathcal{B}^\# = \{f_1, f_2, \ldots, f_n\}$, $\mathcal{C}^\# = \{g_1, g_2, \ldots, g_m\}$ be the dual bases of $\mathcal{B}, \mathcal{C}$. And write $A = [T]_{\mathcal{B}}^{\mathcal{C}}$, $B = [T^\#]_{\mathcal{C}^\#}^{\mathcal{B}^\#}$. Then

$$T^\#(g_j) = g_j \circ T = \sum_k B_{kj} f_k \tag{6.61}$$

Now, evaluate the second and third terms of (6.61) at $\mathbf{v}_i$:

$$g_j T(\mathbf{v}_i) = g_j\Big(\sum_l A_{li}\mathbf{w}_l\Big) = A_{ji} \text{ and } \Big(\sum_k B_{kj} f_k\Big)(\mathbf{v}_i) = B_{ij}$$

Therefore $B_{ij} = A_{ji} = (A^t)_{ij}$. Hence $B = A^t$.

To see (6.60) we compute: $(T^{-1})^\# T^\# = (TT^{-1})^\# = (I_W)^\# = I_{W^\#}$. Similarly, we have $T^\#(T^{-1})^\# = I_{V^\#}$. ☐

We can express (6.58) in the commutative diagram

$$\begin{array}{ccc} W^{\#} & \xrightarrow{T^{\#}} & V^{\#} \\ \phi_{\mathcal{C}^{\#}} \downarrow & & \downarrow \phi_{\mathcal{B}^{\#}} \\ F^m & \xrightarrow[T_{A^t}]{} & F^n \end{array}$$

If $T = T_A : F^n \to F^n$ and $\mathcal{E}$ is the standard basis then (6.59) becomes

$$[T_A^{\#}]_{\mathcal{E}^{\#}} = T_{A^t} \tag{6.62}$$

Hence we can view $T^{\#}$ as the linear map corresponding to the transpose of matrices. The formula (6.60) is the formula for the linear maps corresponding to the fact $(A^t)^{-1} = (A^{-1})^t$ (6.39) which is about the matrices.

Corollary 6.5.16. *Let $T : V \to W$ be a linear map. Then T is surjective* (resp. *injective, isomorphic*) *if and only if $T^{\#}$ is injective* (resp. *surjective, isomorphic*).

Proof. We know this is true when $T = T_A$ by Corollary 2.4.3. Let $\mathcal{B}, \mathcal{C}$ be bases for V, W and $A = [T]_{\mathcal{B}}^{\mathcal{C}}$. Then $A^t = [T^{\#}]_{\mathcal{C}^{\#}}^{\mathcal{B}^{\#}}$ by Theorem 6.5.15. By Corollary 6.4.5, T is surjective if and only if so is T_A. By Corollary 2.4.3, this is equivalent to that T_{A^t} is injective and again, this is equivalent to that $T^{\#}$ is injective.

The other cases are similarly proved. □

Example 6.5.17. (1) *Let $V = P_2(\mathbb{R})$ and let $T : V \to V$ be defined by the derivation: $T(f) = f'$. Let $\mathcal{B} = \{1, x, x^2\}$ be the standard basis for V. In* Example 6.5.11 *we showed that $\mathcal{B}^{\#} = \{D_0, D_1, D_2\}$ is the dual basis of $\mathcal{B}$. To find $T^{\#}(D_i)$, let $f(x) = a + bx + cx^2 \in V$. (Or we can use* Corollary 6.5.12.) *Then we have $T^{\#}(D_0)(f) = D_0 \circ T(f) = D_0(f') = b$. Hence we have $T^{\#}(D_0) = D_1$. Now $T^{\#}(D_1)(f) = D_1 \circ T(f) = D_1(f') = 2c$. Therefore we have $T^{\#}(D_1) = 2D_2$. Similarly, we have $T^{\#}(D_2) = 0$. Hence we have*

$$[T^{\#}]_{\mathcal{B}^{\#}} = \begin{bmatrix} 0 & 0 & 0 \\ 1 & 0 & 0 \\ 0 & 2 & 0 \end{bmatrix} = [T]_{\mathcal{B}}^t$$

(2) *Let $T : P_2(\mathbb{R}) \to \mathbb{R}^3$ be defined by $T(p(x)) = \begin{bmatrix} p(-1) \\ p(0) \\ p(1) \end{bmatrix}$. And let $\mathcal{B} = \{1, x, x^2\}, \mathcal{C} = \{\mathbf{e}^1, \mathbf{e}^2, \mathbf{e}^3\}$. Let's find $[T^{\#}]_{\mathcal{C}^{\#}}^{\mathcal{B}^{\#}}$ and $[T]_{\mathcal{B}}^{\mathcal{C}}$ to illustrate* Theorem 6.5.15.

Let $\mathbf{v}_1 = 1, \mathbf{v}_2 = x, \mathbf{v}_3 = x^2$. Then

$$T(\mathbf{v}_1) = \begin{bmatrix} 1 \\ 1 \\ 1 \end{bmatrix}, T(\mathbf{v}_2) = \begin{bmatrix} -1 \\ 0 \\ 1 \end{bmatrix}, T(\mathbf{v}_3) = \begin{bmatrix} 1 \\ 0 \\ 1 \end{bmatrix}$$

Hence

$$[T]_{\mathcal{B}}^{\mathcal{C}} = \begin{bmatrix} 1 & -1 & 1 \\ 1 & 0 & 0 \\ 1 & 1 & 1 \end{bmatrix}$$

On the other hand, let $\mathcal{B}^\# = \{f_1, f_2, f_3\}, \mathcal{C}^\# = \{g_1, g_2, g_3\}$. *We have* $f_i(\mathbf{v}_j) = \delta_{ij} = g_i(\mathbf{e}^j)$. *For* $T^\#(g_1)$ *we have,* $T^\#(g_1) = \sum_{i=0}^{2} T^\#(g_1)(x^i)f_i = \sum_{i=0}^{2} g_1 \circ T(x^i)f_i$, *by* Corollary 6.5.12. *Hence*

$$T^\#(g_1) = g_1 \begin{bmatrix} 1 \\ 1 \\ 1 \end{bmatrix} f_1 + g_1 \begin{bmatrix} -1 \\ 0 \\ 1 \end{bmatrix} f_2 + g_1 \begin{bmatrix} 1 \\ 0 \\ 1 \end{bmatrix} f_3 = f_1 - f_2 + f_3$$

Similarly, $T^\#(g_2) = f_1$ *and* $T^\#(g_3) = f_1 + f_2 + f_3$. *Hence*

$$[T^\#]_{\mathcal{C}^\#}^{\mathcal{B}^\#} = \begin{bmatrix} 1 & 1 & 1 \\ -1 & 0 & 1 \\ 1 & 0 & 1 \end{bmatrix}$$

which is the transpose of $[T]_{\mathcal{B}}^{\mathcal{C}}$ *as expected.*

Let V be a finite dimensional vector space over F. Consider the **double dual**

$$V^{\#\#} = (V^\#)^\# = L(V^\#, F) = \{\text{all linear functionals } \phi : V^\# \to F\}$$

For $\mathbf{v} \in V$, define

$$\hat{\mathbf{v}} : V^\# \to F; \quad \hat{\mathbf{v}}(f) = f(\mathbf{v}) \text{ for } f \in V^\# \tag{6.63}$$

One checks that $\hat{\mathbf{v}} \in V^{\#\#}$. Now we define

$$\psi = \psi_V : V \to V^{\#\#}; \quad \psi(\mathbf{v}) = \hat{\mathbf{v}} \tag{6.64}$$

Then ψ is a linear map (Exercise 14).

We know that V and $V^{\#\#}$ are isomorphic because their dimensions are the same. Now we will show that ψ is a natural isomorphism between V and $V^{\#\#}$ which does not depend on the choice of a basis in contrast to the case between V and $V^\#$ (See the remarks following Definition 6.5.1.)

Theorem 6.5.18. *Let V be a finite dimensional vector space.*

(1) *The map ψ defined by* (6.64) *is a natural isomorphism.*

(2) *If $T : V \to W$ is a linear map and $T^{\#\#} : V^{\#\#} \to W^{\#\#}$ is its double dual then* $T^{\#\#}(\hat{\mathbf{v}}) = \widehat{T(\mathbf{v})}$. *That is the diagram*

$$\begin{array}{ccc} V & \xrightarrow{T} & W \\ \psi_V \downarrow & & \downarrow \psi_W \\ V^{\#\#} & \xrightarrow[T^{\#\#}]{} & W^{\#\#} \end{array} \tag{6.65}$$

is commutative i.e., $\psi_W \circ T = T^{\#\#} \circ \psi_V$.

Proof. (1) Let $\mathcal{B} = \{\mathbf{v}_1, \mathbf{v}_2, ..., \mathbf{v}_n\}$ be a basis of V and $\mathcal{B}^\# = \{f_1, f_2, ..., f_n\}$ be its dual basis. Then $\{\hat{\mathbf{v}}_1, \hat{\mathbf{v}}_2, ..., \hat{\mathbf{v}}_n\}$ is a dual basis of $\mathcal{B}^\#$ since $\hat{\mathbf{v}}_i(f_j) = f_j(\mathbf{v}_i) = \delta_{ij}$. Hence ψ maps a basis of V to a basis of of $V^{\#\#}$. Therefore ψ is an isomorphism by Theorem 6.3.29 .

(2) We need to show $T^{\#\#}(\hat{\mathbf{v}}) = \widehat{T(\mathbf{v})}$. Let $g \in W^\#$ i.e., $g : W \to F$ is linear. Then $\widehat{T(\mathbf{v})}(g) = g(T(\mathbf{v}))$. On the other hand,

$$\left(T^{\#\#}(\hat{\mathbf{v}})\right)(g) = (\hat{\mathbf{v}} \circ T^\#)(g) = \hat{\mathbf{v}}(T^\#(g)) = \hat{\mathbf{v}}(g \circ T) = g \circ T(\mathbf{v})$$

Hence $T^{\#\#}(\hat{\mathbf{v}}) = \widehat{T(\mathbf{v})}$ i.e., $\psi_W \circ T = T^{\#\#} \circ \psi_V$ □

If we identify V, W with $V^{\#\#}, W^{\#\#}$ through the isomorphisms ψ_V, ψ_W then (6.65) says

$$T^{\#\#} = T$$

which corresponds to the equality $A^{tt} = A$ for the square matrices.

Example 6.5.19. *Let $V = P_2(\mathbb{R})$ and $\mathcal{B} = \{1, x, x^2\}$ be the standard basis of V. Let $I_i(p) = \int_0^i p(x)dx$ $(i = 1, 2, 3)$ for $p \in V$. To see if $\mathcal{B}' = \{I_1, I_2, I_3\}$ is a basis for $V^{\#}$ consider the matrix* $A = I_i(x^{j-1}) = \begin{bmatrix} 1 & \frac{1}{2} & \frac{1}{3} \\ 2 & 2 & \frac{8}{3} \\ 3 & \frac{9}{2} & 9 \end{bmatrix}$. *The determinant $|A| = 2$ and therefore by* Theorem 6.5.5 *we see that $\mathcal{B}'$ is a basis for $V^{\#}$.*

Now we look for the dual basis $\mathcal{B}'^{\#} = \{f_1, f_2, f_3\}$ of $\mathcal{B}'$ by identifying $V^{\#\#}$ with V by Theorem 6.5.18. *Hence we need to find f_1, f_2, f_3 satisfying $I_i(f_j) = \delta_{ij}$. To find f_1 we choose* $\mathbf{b} = \begin{bmatrix} I_1(f_1) \\ I_2(f_1) \\ I_3(f_1) \end{bmatrix} = \begin{bmatrix} 1 \\ 0 \\ 0 \end{bmatrix}$. *To use* Theorem 6.5.5 *we need* $A^{-1} = \begin{bmatrix} 3 & -\frac{3}{2} & \frac{1}{3} \\ -5 & 4 & -1 \\ \frac{3}{2} & -\frac{3}{2} & \frac{1}{2} \end{bmatrix}$. *Hence* $\mathbf{c} = A^{-1}\mathbf{b} = \begin{bmatrix} 3 \\ -5 \\ \frac{3}{2} \end{bmatrix}$ *which corresponds to $f_1(x) = 3 - 5x + \frac{3}{2}x^2$. Similarly, we find that $f_2(x) = -\frac{3}{2} + 4x - \frac{3}{2}x^2$ and $f_3(x) = \frac{1}{3} - x + \frac{1}{2}x^2$.*

Practice Problem 6.5.3. *Let $V = P_2(\mathbb{R})$ and $\mathcal{B}' = \{e_0, e_1, e_2\}$ be the basis of $V^{\#}$ as in* Example 6.5.6. *Identifying $V^{\#\#}$ with V by ψ by* Theorem 6.5.18 *find the dual basis of $\mathcal{B}'$ as linear combinations of $\{1, x, x^2\}$.*

Now we introduce another view point of duality of the vector spaces. At the first reading, one can skip the rest of this section.

Definition 6.5.20. *Let V, W be vector spaces over F. A map $\langle\ ,\ \rangle : V \times W \to F$ is called a* **bilinear form** *or* **bilinear pairing** *if it is linear in the first and the second variable – the conditions* (1), (2) *of* Theorem 3.1.2 *for the determinant of 2×2 matrices (See also* §8.6). *We say the pairing $\langle\ ,\ \rangle$ is* **nondegenerate** *if $\{\mathbf{v} \mid \langle \mathbf{v}, \mathbf{w}\rangle = 0 \text{ for all } \mathbf{w} \in W\} = \mathbf{0}$ and $\{\mathbf{w} \mid \langle \mathbf{v}, \mathbf{w}\rangle = 0 \text{ for all } \mathbf{v} \in V\} = \mathbf{0}$* (Cf. Theorem 8.5.10).

Let V be a finite dimensional vector space. Define a bilinear pairing

$$\langle\ ,\ \rangle : V \times V^{\#} \to F; \quad \langle \mathbf{v}, f\rangle = f(\mathbf{v}) \tag{6.66}$$

Corollary 6.5.12 says the pairing (6.66) is nondegenerate (Exercise 15).

Corollary 6.5.21. *Let $T : V \to V$ be a linear map. Then $T^{\#} : V^{\#} \to V^{\#}$ is a unique linear operator on $V^{\#}$ satisfying*

$$\langle T(\mathbf{v}), f\rangle = \langle \mathbf{v}, T^{\#}(f)\rangle \tag{6.67}$$

Proof. For uniqueness suppose $\langle T(\mathbf{v}), f\rangle = \langle \mathbf{v}, S(f)\rangle$ for some linear operator S on $V^{\#}$. Then

$$0 = \langle \mathbf{v}, T^{\#}(f)\rangle - \langle \mathbf{v}, S(f)\rangle = \langle \mathbf{v}, T^{\#}(f) - S(f)\rangle$$

for all $\mathbf{v}$. Therefore $T^{\#}(f) = S(f)$ by the nondegeneracy mentioned above. e And we have the equality by the definition of $T^{\#}$ (6.54). □

Suppose $\langle\ ,\ \rangle : V\times W \to F$ is a bilinear pairing. For $\mathbf{w} \in W$ the map $\phi_{\mathbf{w}}$ defined by $\phi_{\mathbf{w}}(\mathbf{v}) = \langle \mathbf{v}, \mathbf{w}\rangle$ is a linear functional. Hence we have a map $\phi : W \to V^{\#}$ defined by $\phi(\mathbf{w}) = \phi_{\mathbf{w}}$. Similarly, if $\mathbf{v} \in V$ then we define $\psi_{\mathbf{v}} \in W^{\#}$ by $\psi_{\mathbf{v}}(\mathbf{w}) = \langle \mathbf{v}, \mathbf{w}\rangle$. And we get a map $\psi : V \to W^{\#}$ defined by $\psi(\mathbf{v}) = \psi_{\mathbf{v}}$.

$$\phi : W \to V^{\#}, \quad \psi : V \to W^{\#} \tag{6.68}$$

We will show that ϕ and ψ are isomorphisms if the pairing is nondegenerate.

Corollary 6.5.22. *Let $\langle\ ,\ \rangle : V\times W \to F$ be a nondegenerate bilinear pairing.*

(1) The maps $\phi : W \to V^{\#}$, $\psi : V \to W^{\#}$ of (6.68) are isomorphisms. In particular, $\dim V = \dim W$.

(2) If $T : V \to V$ is a linear map then there is a unique linear map $T^* : W \to W$ which we call the **adjoint** *of T satisfying*

$$\langle T(\mathbf{v}), \mathbf{w}\rangle = \langle \mathbf{v}, T^*(\mathbf{w})\rangle \tag{6.69}$$

Similarly, if $S : W \to W$ is a linear map then there is a unique linear map $S^* : V \to V$ such that

$$\langle S^*(\mathbf{v}), \mathbf{w}\rangle = \langle \mathbf{v}, S(\mathbf{w})\rangle \tag{6.70}$$

Proof. (1) Linearity of $\phi_{\mathbf{w}} \in V^{\#}$ follows from the properties of bilinear pairing. Since the pairing is nondegenerate ϕ is injective. Hence $\dim W \le \dim V^{\#} = \dim V$. Now, by considering ψ, we see that $\dim V \le \dim W$ and hence $\dim V = \dim W$. Since ϕ, ψ are injective maps between the vector spaces of the same dimension we conclude that they are isomorphisms (Theorem 6.3.26).

(2) Let $\mathbf{w} \in W$. The association $\mathbf{v} \mapsto \langle T(\mathbf{v}), \mathbf{w}\rangle$ is a linear functional. By (1), there is a unique $\mathbf{w}_0 \in W$ such that $\langle T(\mathbf{v}), \mathbf{w}\rangle = \langle \mathbf{v}, \mathbf{w}_0\rangle$. Define $T^*(\mathbf{w}) = \mathbf{w}_0$. Then, obviously, T^* satisfies (6.69). To check linearity of T^* we need to prove $T^*(\mathbf{w}_1+\mathbf{w}_2) = T^*(\mathbf{w}_1)+T^*(\mathbf{w}_2)$. For this it suffices to show $\langle \mathbf{v}, T^*(\mathbf{w}_1+\mathbf{w}_2)\rangle = \langle \mathbf{v}, T^*(\mathbf{w}_1)+T^*(\mathbf{w}_2)\rangle$ for all $\mathbf{v} \in V$ by nondegeneracy of the pairing. Now we have $\langle \mathbf{v}, T^*(\mathbf{w}_1 + \mathbf{w}_2)\rangle = \langle T(\mathbf{v}), \mathbf{w}_1 + \mathbf{w}_2\rangle = \langle T(\mathbf{v}), \mathbf{w}_1\rangle + \langle T(\mathbf{v}), \mathbf{w}_2\rangle$. On the other hand, $\langle \mathbf{v}, T^*(\mathbf{w}_1) + T^*(\mathbf{w}_2)\rangle = \langle \mathbf{v}, T^*(\mathbf{w}_1)\rangle + \langle \mathbf{v}, T^*(\mathbf{w}_2)\rangle = \langle T(\mathbf{v}), \mathbf{w}_1\rangle + \langle T(\mathbf{v}), \mathbf{w}_2\rangle$. Hence we have the desired equality. The equality $T^*(c\mathbf{w}) = cT^*(\mathbf{w})$ is similarly proved.

Proof for the statement about S is an exercise. □

Let V, W be vector spaces with bases $\mathcal{B} = \{\mathbf{v}_i\}, \mathcal{C} = \{\mathbf{w}_j\}$. Let $\mathcal{B}^{\#} = \{f_i\}, \mathcal{C}^{\#} = \{g_j\}$ be their dual bases. Let $\langle -, -\rangle$ be a nondegenerate bilinear form on $V\times W$.

If $f_i \in V^{\#}$ then, since $\phi : W \to V^{\#}$ is an isomorphism, there is $\hat{\mathbf{w}}_i \in W$ such that $\langle \mathbf{v}, \hat{\mathbf{w}}_i\rangle = f_i(\mathbf{v})$ for all $\mathbf{v} \in V$. Similarly there is $\mathbf{v}_i^* \in V$ such that $\langle \mathbf{v}_i^*, \mathbf{w}\rangle = g_i(\mathbf{w})$ for all $\mathbf{w} \in W$. Hence we have

$$\langle \mathbf{v}_j, \hat{\mathbf{w}}_i\rangle = \delta_{ij}, \ \langle \mathbf{v}_i^*, \mathbf{w}_j\rangle = \delta_{ij}$$

We will call $\hat{\mathcal{C}} = \{\hat{\mathbf{w}}_i\}$ (resp. $\mathcal{B}^* = \{\mathbf{v}_i^*\}$) the **dual basis of** $\mathcal{B}$ (resp. $\mathcal{C}$) with respect to the nondegenerate pairing $\langle\ ,\ \rangle$.

Let T be a linear operator on V and let $T^* : W \to W$ be the adjoint of T. We want to find $[T^*]_{\hat{\mathcal{C}}}$. Let $T^*(\hat{\mathbf{w}}_j) = \sum c_{ij}\hat{\mathbf{w}}_i$. Then

$$c_{ij} = \left\langle \mathbf{v}_i, \sum c_{kj}\hat{\mathbf{w}}_k \right\rangle = \langle \mathbf{v}_i, T^*(\hat{\mathbf{w}}_j)\rangle = \langle T(\mathbf{v}_i), \hat{\mathbf{w}}_j\rangle = \left\langle \sum a_{ki}\mathbf{v}_k, \hat{\mathbf{w}}_j \right\rangle = a_{ji}$$

Hence we have $[T^*]_{\hat{\mathcal{C}}} = [T]^t_{\mathcal{B}}$. Similarly, if $S : W \to W$ then $[S^*]_{\mathcal{B}^*} = [S]^t_{\mathcal{C}}$.

$$[T^*]_{\hat{\mathcal{C}}} = [T]^t_{\mathcal{B}},\ [S^*]_{\mathcal{B}^*} = [S]^t_{\mathcal{C}} \tag{6.71}$$

Answers for Practice Problems 6.5

(6.5.1) Let $e_{\frac{1}{2}} = a_0e_0 + a_1e_1 + a_2e_2$. We have to determine $a_0.a_1, a_2$. Evaluate both sides on the basis vectors $\{1, x, x^2\}$ of V to get $\begin{cases} 1 &= a_0 + a_1 + a_2 \\ \frac{1}{2} &= \quad a_1 + 2a_2 \\ \frac{1}{4} &= \quad a_1 + 4a_2 \end{cases}$. If we let $A = \begin{bmatrix} 1 & 1 & 1 \\ 0 & 1 & 2 \\ 0 & 1 & 4 \end{bmatrix}$ and $\mathbf{a} = \begin{bmatrix} a_0 \\ a_1 \\ a_2 \end{bmatrix}$, $\mathbf{c} = \frac{1}{4}\begin{bmatrix} 4 \\ 2 \\ 1 \end{bmatrix}$ then the equality above becomes $A\mathbf{a} = \mathbf{c}$. Now we have $A^{-1} = \frac{1}{2}\begin{bmatrix} 2 & -1 & 1 \\ 0 & 4 & -2 \\ 0 & -1 & 1 \end{bmatrix}$. Hence $\mathbf{a} = \frac{1}{8}\begin{bmatrix} 3 \\ 6 \\ -1 \end{bmatrix}$. Therefore, $e_{\frac{1}{2}} = \frac{3}{8}e_0 + \frac{3}{4}e_1 - \frac{1}{8}e_2$.

(6.5.2) (1) Let $f_1 = a_0e_0 + a_1e_1 + a_2e_2$. We have to determine a_0, a_1, a_2. Evaluate both sides on the basis vectors $\{1, x, x^2\}$ of V to get $\begin{cases} 1 &= a_0 + a_1 + a_2 \\ 0 &= \quad a_1 + 2a_2 \\ 0 &= \quad a_1 + 4a_2 \end{cases}$. If we let $A = \begin{bmatrix} 1 & 1 & 1 \\ 0 & 1 & 2 \\ 0 & 1 & 4 \end{bmatrix}$ and $\mathbf{a} = \begin{bmatrix} a_0 \\ a_1 \\ a_2 \end{bmatrix}$, $\mathbf{c} = \begin{bmatrix} 1 \\ 0 \\ 0 \end{bmatrix}$ then the equality above becomes $A\mathbf{a} = \mathbf{c}$. Now we have $A^{-1} = \frac{1}{2}\begin{bmatrix} 2 & -1 & 1 \\ 0 & 4 & -2 \\ 0 & -1 & 1 \end{bmatrix}$. Hence $\mathbf{a} = \begin{bmatrix} 1 \\ 0 \\ 0 \end{bmatrix}$. Therefore $f_1 = e_0$. Similarly, we have $f_2 = -\frac{3}{2}e_0 + 2e_1 - \frac{1}{2}e_2$ and $f_3 = \frac{1}{2}e_0 - e_1 + \frac{1}{2}e_2$.

(2) Let $e_{\frac{1}{2}} = aD_0 + bD_1 + cD_2$. Evaluating at $1, x, x^2$ we get $e_{\frac{1}{2}} = D_0 + \frac{1}{2}D_1 + \frac{1}{4}D_2$.

(6.5.3) Let $\{\mathbf{v}_1, \mathbf{v}_2, \mathbf{v}_3\}$ be the dual basis of $\mathcal{B}'$ and let $\mathbf{v}_1 = a + bx + cx^2$. To find a, b, c we consider $\begin{cases} 1 = e_0(\mathbf{v}_1) = a \\ 0 = e_1(\mathbf{v}_1) = a + b + c \\ 0 = e_2(\mathbf{v}_1) = a + 2b + 4c \end{cases}$. Hence if we let $A = \begin{bmatrix} 1 & 0 & 0 \\ 1 & 1 & 1 \\ 1 & 2 & 4 \end{bmatrix}$ and $\mathbf{a} = \begin{bmatrix} a \\ b \\ c \end{bmatrix}$ then we have $A\mathbf{a} = \mathbf{e}^1$. Now $A^{-1} = \frac{1}{2}\begin{bmatrix} 2 & 0 & 0 \\ -3 & 4 & -1 \\ 1 & -2 & 1 \end{bmatrix}$. Hence $\mathbf{v}_1 = 1 - \frac{3}{2}x + \frac{1}{2}x^2$. Similarly, $\mathbf{v}_2 = 2x - x^2$ and $\mathbf{v}_3 = -\frac{1}{2}x + x^2$.

6.5 Exercises

Assume all vector spaces are finite dimensional unless otherwise stated.

1. Prove the true statement(s) and find a counter example for the false statement(s).

   (a) Let V be a vector space over F and W be a subspace of V. Let $f : W \to F$ be a linear map. If we define $\bar{f}(\mathbf{v}) = \begin{cases} f(\mathbf{v}) & (\mathbf{v} \in W) \\ 0 & (\mathbf{v} \notin W) \end{cases}$ then $\bar{f} \in V^{\#}$.

(b) Let $V = P_2(\mathbb{R})$ and x_0, x_1, x_2, x_3 be four distinct real numbers. Then the evaluation maps then the set of evaluations $\{e_{x_0}, e_{x_1}, e_{x_2}, e_{x_3}\}$ is a linearly independent subset of $V^{\#}$.

(c) If $T^{\#} : V^{\#} \to V^{\#}$ is injective then $T : V \to V$ is surjective.

(d) If V is finite dimensional then $\dim V = \dim V^{\#}$.

(e) If $T : V \to W$ is a linear map then $T^{\#} : V^{\#} \to W^{\#}$ is also linear.

(f) The map $\det : M(n, F) \to F$ is a linear functional.

(g) The map $\mathrm{Tr} : M(n, F) \to F$ is a linear functional.

(h) The map $\psi : V \to V^{\#\#}$ of (6.64) is an isomorphism for the vector spaces of infinite dimension.

(i) If $T : V \to W$ is a linear map and $\mathcal{B}, \mathcal{C}$ are bases of V, W Then $([T]_{\mathcal{B}}^{\mathcal{C}})^t = [T]_{\mathcal{B}^{\#}}^{\mathcal{C}^{\#}}$.

(j) If V is a finite dimensional vector space then the association $T \mapsto T^{\#}$ defines an isomorphism $L(V) \to L(V^{\#})$ of vector spaces.

2. Let $V = P_2(\mathbb{R})$ and $W = \mathbb{R}^2$. Define $T : V \to W$ by $T(p) = \begin{bmatrix} p(1) \\ p'(2) \end{bmatrix}$. Let $\mathcal{B} = \{1, x, x^2\}$, $\mathcal{C} = \{\mathbf{e}^1, \mathbf{e}^2\}$ be the standard bases for V and W.

   (a) Let $f(a, b) = 2a - b$ be an element of $W^{\#}$. Find $T^{\#}(f)(1 + 2x^2)$.

   (b) Find $[T^{\#}]_{\mathcal{C}^{\#}}^{\mathcal{B}^{\#}}$ without using Theorem 6.5.15.

   (c) Find $[T]_{\mathcal{B}}^{\mathcal{C}}$ and show $[T^{\#}]_{\mathcal{C}^{\#}}^{\mathcal{B}^{\#}} = ([T]_{\mathcal{B}}^{\mathcal{C}})^t$.

3. Let $V = \mathbb{R}^n$ and $\mathbf{v} \in V$. Define $\Delta_{\mathbf{v}} : V \to \mathbb{R}$ by $\Delta_{\mathbf{v}}(\mathbf{x}) = \mathbf{x} \cdot \mathbf{v}$ (Definition 5.2.1). Prove that $\Delta_{\mathbf{v}_1}, \Delta_{\mathbf{v}_2}, ..., \Delta_{\mathbf{v}_k}$ is linearly independent if and only if $\{\mathbf{v}_1, \mathbf{v}_2, ..., \mathbf{v}_k\}$ is linearly independent.

4. Let V be a vector space. Let $\{f_1, f_2, ..., f_n\}$ be a basis of the dual space $V^{\#}$. Let $\{\mathbf{v}_1, \mathbf{v}_2, ..., \mathbf{v}_n\}$ be a set of vectors. Prove that the matrix $(f_i(\mathbf{v}_j)) \in M(n, F)$ is invertible if and only if $\{\mathbf{v}_1, \mathbf{v}_2, ..., \mathbf{v}_n\}$ is a basis of V.

5. Let $V = P_n(\mathbb{R})$ and $x_0, x_1, ..., x_k$ $(k \le n)$ be distinct real numbers. Prove that the set of evaluation maps $\{e_{x_1}, e_{x_2}, ..., e_{x_k}\}$ is linearly independent.

6. Let $V = P_2(\mathbb{R})$. For $k \in \mathbb{R}$ and $p \in V$, define $I_k(p) = \int_0^k p(x)\,dx$. Prove that $\{I_1, I_2, I_3\}$ is a basis for $V^{\#}$.

7. Let $V = P_2(\mathbb{R})$. Define $\mathcal{B} = \{f_1, f_2, f_3\} \subseteq V^{\#}$ by $f_1(p) = p(1)$, $f_2(p) = p'(1)$, $f_3(p) = \int_0^1 p(x)dx$ for $p \in V$.

   (a) Prove that $\mathcal{B} = \{f_1, f_2, f_3\}$ is a basis for $V^{\#}$.

   (b) Find the dual basis for V of $\mathcal{B}$ when we identify V with $V^{\#\#}$ by Theorem 6.5.18.

8. Let U, V be vector spaces and let $S : U \to V$ be a linear map. Let $\{\mathbf{u}_1, \mathbf{u}_2, ..., \mathbf{u}_n\}$ be a basis of U with its dual basis $\{f_1, f_2, ..., f_n\}$. Prove that $S(\mathbf{u}) = \sum_i f_i(\mathbf{u}) S(\mathbf{u}_i)$ for $\mathbf{u} \in U$.

9. Let $\mathcal{B} = \{\mathbf{v}_1, \mathbf{v}_2, ..., \mathbf{v}_n\}$ be a basis of V and let $\{f_1, f_2, ..., f_n\}$ be a basis of $V^\#$. If we let $A = (f_i(\mathbf{v}_j))$ then A is an invertible matrix by Theorem 6.5.5. Let $f = \sum_i e_i f_i \in V^\#$ and $f(\mathbf{v}_i) = d_i$. Let $\mathbf{e} = \begin{bmatrix} e_1 \\ \vdots \\ e_n \end{bmatrix}$ and $\mathbf{d} = \begin{bmatrix} d_1 \\ \vdots \\ d_n \end{bmatrix}$. Then $\mathbf{e} = (A^t)^{-1}\mathbf{d}$.

10. Let $V = P_2(\mathbb{R})$ be the vector space of real polynomials of degree ≤ 2.
    (a) Let e_i $(i = 1, 2)$ be the evaluations at the points $1, 2$. Let $D(3)$ be the derivation $\frac{d}{dx}\big|_{x=3}$ at $x = 3$. Use Theorem 6.5.5 to prove that $\mathcal{C} = \{e_1, e_2, D(3)\}$ is a basis for the dual space $V^\#$ of V.
    (b) Find $f \in V$ such that $f(1) = -1, f(2) = 3$ and $D(3)(f) = -2$. (Hint: Exercise 9)

11. The evaluation map e_x at $x \in \mathbb{R}$ is a linear functional on $V = P_2(\mathbb{R})$. Show that there is a linear combination of $\{e_0, e_1, e_2\}$ which is not of the form e_x at some $x \in \mathbb{R}$.

12. Let $\mathcal{B} = \{\mathbf{v}_1, \mathbf{v}_2, ..., \mathbf{v}_n\}$ be a basis for a vector space V. Let $S = \{f_1, f_2, ..., f_r\} \subseteq V^\#$ with $r \leq n$. Prove that S is linearly independent if and only if the rank of the $n \times r$ matrix $(f_j(\mathbf{v}_i))$ is r.

13. Let $a, a_0, a_1, ..., a_n \in \mathbb{Q}$ be distinct rational numbers. Prove that there are uniquely determined rational numbers $c_0, c_1, ..., c_n \in \mathbb{Q}$ such that

$$\begin{cases} c_0 + c_1 + \cdots + c_n = 1 \\ c_0 a_0 + c_1 a_1 + \cdots + c_n a_n = a \\ c_0 a_0^2 + c_1 a_1^2 + \cdots + c_n a_n^2 = a^2 \\ \cdots\cdots\cdots \\ c_0 a_0^n + c_1 a_1^n + \cdots + c_n a_n^n = a^n \end{cases}$$

(Hint: Use the fact that $\{e_{a_0}, e_{a_1}, ..., e_{a_n}\}$ is a basis for the dual space of $P_n(\mathbb{Q})$.)

14. Prove that ψ of (6.64) is linear.

15. Let V be a vector space over F. Define

$$\beta : V \times V^\# \to F$$

by $\beta(\mathbf{v}, f) = f(\mathbf{v})$. Prove the following statements.
    (a) Show that if $V = \mathbb{R}^n$ then β can be identified with the dot product $V \times V \to \mathbb{R}; (\mathbf{v}, \mathbf{w}) \mapsto \mathbf{v} \cdot \mathbf{w}$ (Definition 5.2.1) under the isomorphism ϕ of Example 6.5.2.
    (b) Prove that if we fix f then β is linear in the first variable.
    (c) Prove that if we fix $\mathbf{v}$ then β is linear in the second variable.
    (d) Prove that if $\beta(\mathbf{v}, f) = 0$ for all $f \in V^\#$ then $\mathbf{v} = \mathbf{0}$.
    (e) Prove that if $\beta(\mathbf{v}, f) = 0$ for all $\mathbf{v} \in V$ then $f = \mathbf{0}$.

16. Let $\langle \ , \ \rangle : V \times W \to F$ be a nondegenerate pairing and T_1, T_2 be linear operators on V and $c \in F$. As in Corollary 6.5.22, write T^* for the adjoint of T with respect to the pairing $\langle \ , \ \rangle$. Prove the formulas

$$(T_1 + T_2)^* = T_1^* + T_2^*, \quad (T_1 T_2)^* = T_2^* T_1^*, \ (cT_1)^* = cT_1^*$$

17. Let V be a finite dimensional vector space over F and S be a subset of V. We define the **annihilator** S^0 of S by

$$S^0 = \{f \in V^\# \mid \beta(\mathbf{v}, f) = 0 \text{ for all } \mathbf{v} \in S\}$$

   (a) Let $V = \mathbb{R}^n$ and W be a subspace of V. Prove that W^0 is naturally isomorphic to $W^\perp$ i.e., the isomorphism of Example 6.5.2 induces an isomorphism between W^0 and $W^\perp$.

   (b) Prove that if S is a subset and $W = \operatorname{Span}(S)$ then $S^0 = W^0$.

   (c) Let W be a subspace. Prove that W^0 is a subspace of $V^\#$ and $\dim W^0 = \dim V - \dim W$.

   (d) Prove that $W^{00} = \psi(W)$ and $(S^0)^0 = \operatorname{Span} \psi(S)$ where ψ is defined by (6.64).

   (e) Let W_1, W_2 be subspaces of V. Prove that $W_1 \subseteq W_2$ if and only if $W_1^0 \supseteq W_2^0$. In particular, $W_1 = W_2$ if and only if $W_1^0 = W_2^0$.

   (f) Let W_1, W_2 be subspaces of V. Prove that $(W_1 + W_2)^0 = W_1^0 \cap W_2^0$.

   (g) Let W_1, W_2 be subspaces of V. Prove that $(W_1 \cap W_2)^0 = W_1^0 + W_2^0$.

   (h) Let W be a subspace of V and let $\{f_1, f_2, ..., f_r\}$ be a basis of W^0. Prove that $W = \bigcap_{i=1}^r N(f_i)$.

   (i) Let W be a subspace of V and $\mathbf{v} \in V \setminus W$. Prove that there is $f \in W^0$ such that $f(\mathbf{v}) \neq 0$.

18. Let $T : V \to W$ be a linear map. Prove that $N(T^\#) = R(T)^0$.

19. Let $A \in M(m \times n, F)$. Use Exercise 17c and Exercise 18 to deduce $\rho(A) = \rho(A^t)$ which is Theorem 2.3.12.

20. Let $T : V \to W$ be a linear map and U be a subspace of W. Prove that $T^\#(U^0) = (T^{-1}(U))^0$.

21. Let $T : V \to W$ be a linear map and W be a subspace of V. Prove that W is T-invariant if and only if W^0 is $T^\#$-invariant. (A subspace W of V is defined to be **invariant** (or **T-invariant**) if $T(W) \subseteq W$ (Definition 6.8.1).)

22. Let $V = M(n, F)$. For $\mathbf{u}, \mathbf{v} \in F^n$ we define $\phi_{\mathbf{u},\mathbf{v}} : V \to F$ by

$$\phi_{\mathbf{u},\mathbf{v}}(A) = \mathbf{u}^t A \mathbf{v}$$

where $\mathbf{u}^t \in F_n$ is the row vector obtained by taking the transpose of the column vector $\mathbf{u} \in F^n$.

   (a) Prove that $\phi_{\mathbf{u},\mathbf{v}}$ is a linear functional on V.

   (b) Prove that $\phi_{\mathbf{u},\mathbf{v}}$ is bilinear in $\mathbf{u}, \mathbf{v}$ i.e.,

$$\phi_{\mathbf{u}_1+\mathbf{u}_2,\mathbf{v}} = \phi_{\mathbf{u}_1,\mathbf{v}} + \phi_{\mathbf{u}_2,\mathbf{v}}, \qquad \phi_{c\mathbf{u},\mathbf{v}} = c\phi_{\mathbf{u},\mathbf{v}}$$
$$\phi_{\mathbf{u},\mathbf{v}_1+\mathbf{v}_2} = \phi_{\mathbf{u},\mathbf{v}_1} + \phi_{\mathbf{u},\mathbf{v}_2}, \qquad \phi_{\mathbf{u},c\mathbf{v}} = c\phi_{\mathbf{u},\mathbf{v}}$$

   (c) Let $\mathcal{E} = \{E_{ij} \mid 1 \leq i, j \leq n\}$ be the standard basis of V (Definition 6.2.8). Let $\phi_{ij} = \phi_{\mathbf{e}^{(i)},\mathbf{e}^{(j)}}$. Prove that $\mathcal{E}^\# = \{\phi_{ij} \mid 1 \leq i, j \leq n\}$ is the dual basis of $\mathcal{E}$.

(d) Express the trace $\mathrm{Tr} : V \to F$ as a linear combination of $\{\phi_{ij} = \phi_{\mathbf{e}^{(i)},\mathbf{e}^{(j)}}\}$.

(e) We say that a linear functional ϕ on V is **invariant under conjugation** if $\phi(P^{-1}AP) = \phi(A)$ for all invertible matrices P. Prove that the linear functionals invariant under conjugation form a subspace of $V^{\#}$ containing the trace map Tr.

(f) Prove that if $\phi \in V^{\#}$ is invariant under conjugation then $\phi = c\,\mathrm{Tr}$ for some $c \in F$. Hence the dimension of the subspace of linear functionals invariant under conjugation is 1. (Hint: Write $\phi = \sum_{i,j} c_{ij}\phi_{ij}$ and evaluate at E_{ij}'s and $P^{-1}E_{ij}P$ for various elementary matrices P's using invariance under conjugation. Also see Exercise 7.3.11.)

6.6 Tensor Product and Exterior Product

This section is a brief introduction to multilinear algebra. Contents of this section will not be used in later chapters except in some exercises.

6.6.1 Tensor Product

Let U, V be vector spaces over a field F not necessarily finite dimensional. Let $\mathcal{F}(U \times V)$ be the set of all functions $f : U \times V \to F$ such that $f(\mathbf{u}, \mathbf{v}) \neq 0$ for all but finitely many $(\mathbf{u}, \mathbf{v})$'s in $U \times V$. We define addition and scalar multiplication on $\mathcal{F}(U \times V)$ in an obvious way: For $f, g \in \mathcal{F}(U \times V)$ and $c \in F$ we define

$$(f+g)(\mathbf{u}, \mathbf{v}) = f(\mathbf{u}, \mathbf{v}) + g(\mathbf{u}, \mathbf{v}), \ \ (cf)(\mathbf{u}, \mathbf{v}) = c(f(\mathbf{u}, \mathbf{v}))$$

It is easy to show that $\mathcal{F}(U \times V)$ is an F-vector space.

For $(\mathbf{u}, \mathbf{v}) \in U \times V$, define $[\mathbf{u}, \mathbf{v}] \in \mathcal{F}(U \times V)$ by

$$[\mathbf{u}, \mathbf{v}](\mathbf{x}, \mathbf{y}) = \begin{cases} 1 & \text{if } (\mathbf{x}, \mathbf{y}) = (\mathbf{u}, \mathbf{v}) \\ 0 & \text{otherwise} \end{cases}$$

then $[\mathbf{u}, \mathbf{v}] \in \mathcal{F}(U \times V)$. Hence $\sum c_{ij}[\mathbf{u}_i, \mathbf{v}_j]$ is the function f on $U \times V$ with $f(\mathbf{u}_i, \mathbf{v}_i) = c_{ij}$ and $f(\mathbf{x}, \mathbf{y}) = 0$ for $(\mathbf{x}, \mathbf{y}) \neq (\mathbf{u}_i, \mathbf{v}_j)$ for any i, j. Since any $f \in \mathcal{F}(U \times V)$ is nonzero at only finitely many elements of $U \times V$ we see that f is of the form $\sum c_{ij}[\mathbf{u}_i, \mathbf{v}_j]$. Thus we can write

$$\mathcal{F}(U \times V) = \left\{ \sum c_{ij}[\mathbf{u}_i, \mathbf{v}_j] \;\middle|\; \mathbf{u}_i \in U,\ \mathbf{v}_j \in V, c_{ij} \in F, 1 \le i \le n, 1 \le j \le m \right\}$$

In this notation, the addition and scalar multiplication becomes

$$\sum c_{ij}[\mathbf{u}_i, \mathbf{v}_j] + \sum d_{ij}[\mathbf{u}_i, \mathbf{v}_j] = \sum (c_{ij} + d_{ij})[\mathbf{u}_i, \mathbf{v}_j]$$
$$c\left(\sum c_{ij}[\mathbf{u}_i, \mathbf{v}_j]\right) = \sum cc_{ij}[\mathbf{u}_i, \mathbf{v}_j]$$

Then the zero function $0 : U \times V \to F$ assigning 0 to every $(\mathbf{u}, \mathbf{v})$ serves as the zero element of $\mathcal{F}(U \times V)$ under the operation.

There is a natural map $U \times V \to F(U \times V)$ assigning $[\mathbf{u}, \mathbf{v}]$ to each $(\mathbf{u}, \mathbf{v}) \in U \times V$. Also notice that $[\mathbf{u}, \mathbf{v}] + [\mathbf{u}', \mathbf{v}'] \neq [\mathbf{u} + \mathbf{u}', \mathbf{v} + \mathbf{v}']$ in $F(U \times V)$. If $\{(\mathbf{u}_i, \mathbf{v}_i)\}$ are distinct pairs then $\sum c_i[\mathbf{u}_i, \mathbf{v}_i] = \mathbf{0}$ if and only if $c_i = 0$ for all i. Hence the set $\{[\mathbf{u}_i, \mathbf{v}_i]\}$ is linearly independent in $\mathcal{F}(U \times V)$ whenever $\{(\mathbf{u}_i, \mathbf{v}_i)\}$ are distinct pairs. In general, $\mathcal{F}(U \times V)$ is an infinite dimensional vector space.

Definition 6.6.1. *Let $B(U\times V)$ be the subspace of $\mathcal{F}(U\times V)$ generated by the elements of the forms*

(1) $[a\mathbf{u}+b\mathbf{u}',\mathbf{v}]-a[\mathbf{u},\mathbf{v}]-b[\mathbf{u}',\mathbf{v}]$

(2) $[\mathbf{u},c\mathbf{v}+d\mathbf{v}']-c[\mathbf{u},\mathbf{v}]-d[\mathbf{u},\mathbf{v}']$

where $\mathbf{u},\mathbf{u}'\in U$, $\mathbf{v},\mathbf{v}'\in V$ and $a,b,c,d\in F$. We define the **tensor product** *$U\otimes_F V$ or $U\otimes V$ of U and V by the quotient space* (Exercise 6.1.9)

$$U\otimes_F V=\mathcal{F}(U\times V)/B(U\times V) \tag{6.72}$$

The composition $U\times V\to\mathcal{F}(U\times V)\to U\otimes V$ gives a natural map

$$\phi: U\times V\longrightarrow U\otimes V \tag{6.73}$$

and we write $\mathbf{u}\otimes\mathbf{v}$ for $\phi(\mathbf{u},\mathbf{v})$.

With this notation we can describe $U\otimes V$ more succinctly as follows: $U\otimes V$ is generated by the elements of the form $\mathbf{u}\otimes\mathbf{v}$ over F

$$U\otimes V=\left\{\sum\mathbf{u}_i\otimes\mathbf{v}_i\ \middle|\ \mathbf{u}_i\in V,\ \mathbf{v}_i\in W\right\}$$

with the following identifications

(1′) $(a\mathbf{u}+b\mathbf{u}')\otimes\mathbf{v}=a(\mathbf{u}\otimes\mathbf{v})+b(\mathbf{u}'\otimes\mathbf{v})$

(2′) $\mathbf{u}\otimes(c\mathbf{v}+d\mathbf{v}')=c(\mathbf{u}\otimes\mathbf{v})+d(\mathbf{u}\otimes\mathbf{v}')$

By taking $b=0$ in (1′) and taking $c=a, d=0$ in (2′) we have

$$a(\mathbf{u}\otimes\mathbf{v})=(a\mathbf{u})\otimes\mathbf{v}=\mathbf{u}\otimes(a\mathbf{v}) \tag{6.74}$$

We define addition and the scalar multiplication by

$$\left(\sum\mathbf{u}_i\otimes\mathbf{v}_i\right)+\left(\sum\mathbf{u}'_j\otimes\mathbf{v}'_j\right)=\sum\mathbf{u}_i\otimes\mathbf{v}_i+\sum\mathbf{u}'_j\otimes\mathbf{v}'_j$$
$$c\left(\sum\mathbf{u}_i\otimes\mathbf{v}_i\right)=\sum c\mathbf{u}_i\otimes\mathbf{v}_i$$

Observe that not every element of $U\otimes V$ is of the form $\mathbf{u}\otimes\mathbf{v}$. We can only say that every element of $U\otimes V$ is a finite sum $\sum\mathbf{u}_i\otimes\mathbf{v}_i$.

We notice that

$$\mathbf{0}\otimes\mathbf{0}+\mathbf{u}\otimes\mathbf{v}=\mathbf{u}\otimes\mathbf{v} \tag{6.75}$$

In fact, we have

$$\mathbf{0}\otimes\mathbf{0}+\mathbf{u}\otimes\mathbf{v}=0\cdot\mathbf{u}\otimes\mathbf{0}+\mathbf{u}\otimes\mathbf{v}=\mathbf{u}\otimes 0\cdot\mathbf{0}+\mathbf{u}\otimes\mathbf{v}=\mathbf{u}\otimes(\mathbf{0}+\mathbf{v})=\mathbf{u}\otimes\mathbf{v}$$

Hence $\mathbf{0}\otimes\mathbf{0}$ serves as a zero vector in $U\otimes V$. Further

$$\mathbf{0}\otimes\mathbf{v}=\mathbf{u}\otimes\mathbf{0}=\mathbf{0}\otimes\mathbf{0} \tag{6.76}$$

Because $\mathbf{0}\otimes\mathbf{v}=0\cdot\mathbf{0}\otimes\mathbf{v}=\mathbf{0}\otimes 0\cdot\mathbf{v}=\mathbf{0}\otimes\mathbf{0}$. Similarly, $\mathbf{u}\otimes\mathbf{0}=\mathbf{0}\otimes\mathbf{0}$.

We also have

$$-(\mathbf{u}\otimes\mathbf{v})=(-\mathbf{u})\otimes\mathbf{v}=\mathbf{u}\otimes(-\mathbf{v}) \tag{6.77}$$

Theorem 6.6.2. *Let U, V be vector spaces over F. Suppose $S = \{\mathbf{u}_1, \mathbf{u}_2, ..., \mathbf{u}_n\}$ and $T = \{\mathbf{v}_1, \mathbf{v}_2, ..., \mathbf{v}_m\}$ span U and V respectively. Then $U \otimes V$ is spanned by the vectors*

$$S \otimes T := \{\mathbf{u}_i \otimes \mathbf{v}_j \mid i = 1, 2, ..., n; j = 1, 2, ..., m\}$$

Proof. To see this we need to show that $\mathbf{u} \otimes \mathbf{v}$ $(\mathbf{u} \in U, \mathbf{v} \in V)$ is a linear combination of $S \otimes T$. Write $\mathbf{u} = \sum \alpha_i \mathbf{u}_i, \mathbf{v} = \sum b_j \mathbf{v}_j$. Then $\mathbf{u} \otimes \mathbf{v} = (\sum a_i \mathbf{u}_i) \otimes (\sum b_j \mathbf{v}_j) = \sum_{i,j} a_i b_j \mathbf{u}_i \otimes \mathbf{v}_j$ by the properties $(1'), (2')$. □

Definition 6.6.3. *Let $V_1, V_2, ..., V_k$ and W be vector spaces. A map*

$$f : V_1 \times V_2 \times \cdots \times V_k \to W$$

is **multilinear**(*or k*-**multilinear**) *if*

$$f(\mathbf{v}_1, ..., a\mathbf{v}_i + b\mathbf{v}'_i, ..., \mathbf{v}_k) = af(\mathbf{v}_1, ..., \mathbf{v}_i, ..., \mathbf{v}_k) + bf(\mathbf{v}_1, ..., \mathbf{v}'_i, ..., \mathbf{v}_k)$$

for each $i = 1, 2, ..., k$. If $k = 2$ we say f is **bilinear**.

Notice that if $f : V_1 \times V_2 \times \cdots \times V_k \to W$ is multilinear then

$$f(\mathbf{v}_1, ..., \mathbf{0}, ..., \mathbf{v}_k) = \mathbf{0} \tag{6.78}$$

In fact,

$$\begin{aligned} f(\mathbf{v}_1, ..., \mathbf{0}, ..., \mathbf{v}_k) &= f(\mathbf{v}_1, ..., \mathbf{0} + \mathbf{0}, ..., \mathbf{v}_k) \\ &= f(\mathbf{v}_1, ..., \mathbf{0}, ..., \mathbf{v}_k) + f(\mathbf{v}_1, ..., \mathbf{0}, ..., \mathbf{v}_k) \end{aligned}$$

which proves (6.78).

By the conditions $(1'), (2')$ the map $\phi : U \times V \to U \otimes V$ of (6.73) is bilinear.

Theorem 6.6.4. *Let U, V, W be a vector space over F. Let $f : U \times V \to W$ be a bilinear map. Then there is a unique linear map $\bar{f} : U \otimes V \to W$ such that $\bar{f} \circ \phi = f$ i.e., the diagram*

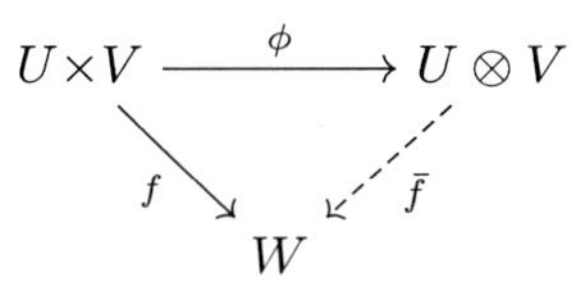

(6.79)

is commutative.

Proof. For uniqueness, suppose $\bar{f}_1$ and $\bar{f}_2$ satisfies the condition of the theorem. Then $\bar{f}_1(\mathbf{u} \otimes \mathbf{v}) = \bar{f}_2(\mathbf{u} \otimes \mathbf{v})$. Since $\mathbf{u} \otimes \mathbf{v}$ generate $U \otimes V$ and $\bar{f}_1, \bar{f}_2$ coincide on the set of generators we conclude that $\bar{f}_1 = \bar{f}_2$ on $U \otimes V$ (Theorem 6.3.9 and the remark following it.)

For existence, we define $\bar{f}(\mathbf{u} \otimes \mathbf{v}) = f(\mathbf{u}, \mathbf{v})$ on the generators of $U \otimes V$ and extend this to $U \otimes V$ by requiring linearity. We need to show that $\bar{f}$ vanishes on $B(U \times V)$ (Exercise 6.3.20). In fact, $\bar{f}((a\mathbf{u} + b\mathbf{u}') \otimes \mathbf{v}) = a\bar{f}(\mathbf{u} \otimes \mathbf{v}) + b\bar{f}(\mathbf{u}' \otimes \mathbf{v})$. Now $\bar{f}((a\mathbf{u} + b\mathbf{u}') \otimes \mathbf{v}) = f(a\mathbf{u} + b\mathbf{u}', \mathbf{v}) = af(\mathbf{u}, \mathbf{v}) + bf(\mathbf{u}', \mathbf{v}) = a\bar{f}(\mathbf{u} \otimes \mathbf{v}) + b\bar{f}(\mathbf{u}' \otimes \mathbf{v})$ since f is bilinear. Similarly, the value of $\bar{f}$ for the other types of elements in $B(U \times V)$ are $\mathbf{0}$. Finally, $\bar{f}$ is linear since we extended $\bar{f}(\mathbf{u} \otimes \mathbf{v}) = f(\mathbf{u}, \mathbf{v})$ by requiring linearity. □

By Theorem 6.6.4, there is a surjection from the set of bilinear maps $f : U\times V \to W$ to the set of all linear maps $\bar{f} : U \otimes V \to W$. On the other hand, if $\bar{f} : U \otimes V \to W$ is a linear map then $\bar{f} \circ \phi$ is bilinear. Hence there is a bijection

$$\left\{\begin{matrix} \text{bilinear map} \\ f : U\times V \to W \end{matrix}\right\} \xrightarrow{\text{bijection}} \left\{\begin{matrix} \text{linear map} \\ \bar{f} : U \otimes V \to W \end{matrix}\right\} \tag{6.80}$$

Therefore defining a linear map $\bar{f} : U \otimes V \to W$ is the same as defining a bilinear map $f : U\times V \to W$.

Now suppose we have linear maps $T : U \to V$ and $T' : U' \to V'$. Then we have a bilinear map $U\times U' \to U' \otimes V'$ assigning $T(\mathbf{u}) \otimes T'(\mathbf{v})$ to each $(\mathbf{u}, \mathbf{u}')$. Hence we have a linear map

$$T \otimes T' : U \otimes U' \to V \otimes V' \tag{6.81}$$

by Theorem 6.6.4.

Theorem 6.6.5. *Let $S_i : U_i \to V_i$ $(i = 1, 2)$ and $T_i : V_i \to W_i$ $(i = 1, 2)$ be linear maps. Then*

$$(T_1 \otimes T_2) \circ (S_1 \otimes S_2) = (T_1 \circ S_1) \otimes (T_2 \circ S_2) \tag{6.82}$$

If $T_i : V_i \to W_i$ $(i = 1, 2)$ are isomorphisms then $T_1 \otimes T_2$ is also an isomorphism and

$$(T_1 \otimes T_2)^{-1} = T_1^{-1} \otimes T_2^{-1} \tag{6.83}$$

Proof. Exercise 7 □

Next we prove various isomorphisms between tensor product of vector spaces.

Theorem 6.6.6. *Let U, U_1, U_2 and V be vector spaces over a field F. There are natural isomorphisms*

$$U \otimes V \xrightarrow{\cong} V \otimes U \tag{6.84}$$

$$U_1 \otimes (U_2 \otimes U_3) \xrightarrow{\cong} (U_1 \otimes U_2) \otimes U_3 \tag{6.85}$$

$$(U_1 \oplus U_2) \otimes V \xrightarrow{\cong} (U_1 \otimes V) \oplus (U_2 \otimes V) \tag{6.86}$$

$$F \otimes U \xrightarrow{\cong} U \tag{6.87}$$

Proof. To see (6.84), we have a bilinear map $\phi : U\times V \to V \otimes U$ defined by $\phi(\mathbf{u}, \mathbf{v}) = \mathbf{v} \otimes \mathbf{u}$ which induces $\bar{\phi} : U \otimes V \to V \otimes U$ given by $\bar{\phi}(\mathbf{u} \otimes \mathbf{v}) = \mathbf{v} \otimes \mathbf{u}$. Now the bilinear map $\psi : V\times U \to U \otimes V$ induces $\bar{\psi} : V \otimes U \to U \otimes V$ given by $\bar{\psi}(\mathbf{v} \otimes \mathbf{u}) = \mathbf{u} \otimes \mathbf{v}$. Obviously $\bar{\phi}$ and $\bar{\psi}$ are inverses each other. Hence $\bar{\phi}$ is the required isomorphism.

To prove (6.85), first construct a linear map $\phi : U_1 \otimes (U_2 \otimes U_3) \to (U_1 \otimes U_2) \otimes U_3$. Let $\mathbf{u} \in U_1$ and define $\phi_{\mathbf{u}_1} : U_2\times U_3 \to (U_1 \otimes U_2) \otimes U_3$ by $\phi_{\mathbf{u}_1}(\mathbf{u}_2, \mathbf{u}_3) = (\mathbf{u}_1 \otimes \mathbf{u}_2) \otimes \mathbf{u}_3$. Hence by Theorem 6.6.4 we get a linear map $\bar{\phi}_{\mathbf{u}_1} : U_2 \otimes U_3 \to (U_1 \otimes U_2) \otimes U_3$. Now we define $U_1\times(U_2 \otimes U_3) \to (U_1 \otimes U_2) \otimes U_3$ by $(\mathbf{u}_1, \alpha) \mapsto \phi_{\mathbf{u}_1}(\alpha)$ for $\alpha \in U_2 \otimes U_3$ which is also bilinear. Hence this map induces a linear map $\phi : U_1 \otimes (U_2 \otimes U_3) \to (U_1 \otimes U_2) \otimes U_3$ with $\phi(\mathbf{u}_1 \otimes (\mathbf{u}_2 \otimes \mathbf{u}_3)) = (\mathbf{u}_1 \otimes \mathbf{u}_2) \otimes \mathbf{u}_3$ for $\mathbf{u}_i \in U_i$ $(i = 1, 2, 3)$. Similar construction gives us a map $\psi : (U_1 \otimes U_2) \otimes U_3 \to U_1 \otimes (U_2 \otimes U_3)$ with $\psi((\mathbf{u}_1 \otimes \mathbf{u}_2) \otimes \mathbf{u}_3) = \mathbf{u}_1 \otimes (\mathbf{u}_2 \otimes \mathbf{u}_3)$. Obviously, ϕ and ψ are inverses to each other.

To prove (6.86), we identify the direct sum with direct product by Theorem 6.3.14. Define a bilinear map $\phi : (U_1 \oplus U_2)\times V \to (U_1 \otimes V) \oplus (U_2 \otimes V)$ by $\phi(\mathbf{u}_1 + \mathbf{u}_2, \mathbf{v}) = \mathbf{u}_1 \otimes \mathbf{v} + \mathbf{u}_2 \otimes \mathbf{v}$.

By Theorem 6.6.4, there is a linear map $\bar{\phi} : (U_1 \oplus U_2) \otimes V \to (U_1 \otimes V) \oplus (U_2 \otimes V)$ given by $\bar{\phi}((\mathbf{u}_1 + \mathbf{u}_2) \otimes \mathbf{v}) = \mathbf{u}_1 \otimes \mathbf{v} + \mathbf{u}_2 \otimes \mathbf{v}$. To find the inverse of $\bar{\phi}$ define $\psi_i : U_i \times V \to (U_1 \oplus U_2) \otimes V$ $(i = 1, 2)$ by $\psi_i(\mathbf{u}_i, \mathbf{v}) = \mathbf{u}_i \otimes \mathbf{v}$. Since ψ_i are bilinear we get a linear maps $\bar{\psi}_i : U_i \otimes V \to (U_1 \oplus U_2) \otimes V$ $(i = 1, 2)$. Now define $\bar{\psi} := \bar{\psi}_1 + \bar{\psi}_2 : (U_1 \otimes V) \oplus (U_2 \otimes V) \to (U_1 \oplus U_2) \otimes V$ which is given by $\bar{\psi}(\mathbf{u}_1 \otimes \mathbf{v} + \mathbf{u}_2 \otimes \mathbf{v}') = \mathbf{u}_1 \otimes \mathbf{v} + \mathbf{u}_2 \otimes \mathbf{v}'$. One checks easily that $\bar{\psi} \circ \bar{\phi} = I$ and $\bar{\phi} \circ \bar{\psi} = I$. Hence $\bar{\phi}$ is the required isomorphism.

For the isomorphism (6.87), let $\bar{\phi} : F \otimes U \to U$ be the linear map induced by the bilinear map $\phi : F \times U \to U$ defined by $\phi(c, \mathbf{u}) = c\mathbf{u}$. Now define $\psi : U \to F \otimes U$ by $\psi(\mathbf{u}) = 1 \otimes \mathbf{u}$. Now it is easy to show that ϕ and ψ are inverses to each other. □

Theorem 6.6.7. *Let U and V be vector spaces over F with the bases $\mathcal{B} = \{\mathbf{u}_1, \mathbf{u}_2, ..., \mathbf{u}_n\}$ and $\mathcal{C} = \{\mathbf{v}_1, \mathbf{v}_2, ..., \mathbf{v}_m\}$ respectively. Then*

$$\mathcal{B} \otimes \mathcal{C} := \{\mathbf{u}_i \otimes \mathbf{v}_j \mid i = 1, 2, ..., n; j = 1, 2, ..., m\} \tag{6.88}$$

is a basis for $U \otimes V$. In particular,

$$\dim(U \otimes V) = \dim(U) \dim(V)$$

Proof. There are isomorphisms $S : U \overset{\cong}{\to} F^n$ and $T : V \overset{\cong}{\to} F^m$ determined by $S(\mathbf{u}_i) = \mathbf{e}^i$ and $T(\mathbf{v}_j) = \bar{\mathbf{e}}^j$ where $\mathcal{E}^n = \{\mathbf{e}^i\}, \mathcal{E}^m = \{\bar{\mathbf{e}}^j\}$ are the canonical bases for F^n, F^m respectively. Now $U \otimes V \to F^n \otimes F^m$ is an isomorphism by (6.83). Hence to show $\mathcal{B} \otimes \mathcal{C}$ is a basis we need to show $\mathcal{E}^n \otimes \mathcal{E}^m = \{\mathbf{e}^i \otimes \bar{\mathbf{e}}^j\}$ is a basis for $F^n \otimes F^m$ (Theorem 4.2.24). By Theorem 6.6.2, $\mathcal{E}^n \otimes \mathcal{E}^m$ span $F^n \otimes F^m$ and consists of nm vectors. We need to show $\dim F^n \otimes F^m = nm$ by Corollary 6.2.16. Now we have

$$\begin{aligned} F^n \otimes F^m &\overset{\cong}{\to} (\overbrace{F \oplus F \oplus \cdots \oplus F}^{n\text{-times}}) \otimes F^m && \text{by (6.86)} \\ &\overset{\cong}{\to} (F \otimes F^m) \oplus (F \otimes F^m) \oplus \cdots \oplus (F \otimes F^m) && \text{by (6.87)} \\ &\overset{\cong}{\to} \overbrace{F^m \oplus F^m \oplus \cdots \oplus F^m}^{n\text{-times}} \overset{\cong}{\to} F^{mn} \end{aligned}$$

which is the n-times direct sum of F^m; $\dim F^n \otimes F^m = nm$. Hence $\mathcal{E}^n \otimes \mathcal{E}^m$ is a basis of $F^n \otimes F^m$. □

Corollary 6.6.8. *Let $T_i : U_i \to V_i$ $(i = 1, 2)$ be linear maps. Then $\rho(T_1 \otimes T_2) = \rho(T_1)\rho(T_2)$.*

Proof. By Theorem 6.6.7, it suffices to show that $R(T_1 \otimes T_2) = R(T_1) \otimes R(T_2)$. In fact, since $T_1 \otimes T_2(\mathbf{u} \otimes \mathbf{u}') = T_1(\mathbf{u}) \otimes T_2(\mathbf{u}')$ we see that $R(T_1 \otimes T_2) \subseteq R(T_1) \otimes R(T_2)$. On the other hand, we have $\left(\sum_i T_1(\mathbf{u}_i)\right) \otimes \left(\sum_j T_2(\mathbf{u}'_j)\right) = \sum_{i,j} T_1(\mathbf{u}_i) \otimes T_2(\mathbf{u}'_j) = \sum_{i,j} T_1 \otimes T_2(\mathbf{u}_i \otimes \mathbf{u}'_j) = T_1 \otimes T_2(\sum_{i,j} \mathbf{u}_i \otimes \mathbf{u}'_j)$ we conclude that $R(T_1 \otimes T_2) = R(T_1) \otimes R(T_2)$. Hence $\rho(T_1 \otimes T_2) = \rho(T_1)\rho(T_2)$. □

Corollary 6.6.9. *Let $T_i : U_i \to V_i$ $(i = 1, 2)$ be linear maps. If T_1, T_2 are injective (resp. surjective, resp. isomorphism) then $T_1 \otimes T_2$ is injective (resp. surjective, resp. isomorphism).*

Proof. Since T_1, T_2 are injective we have $\rho(T_1) = \dim U_1, \rho(T_2) = \dim U_2$ (Corollary 4.2.31). By the theorem $\rho(T_1 \otimes T_2) = \rho(T_1)\rho(T_2) = \dim U_1 \dim U_2 = \dim(U_1 \otimes U_2)$. By Corollary 4.2.31 again we conclude that $T_1 \otimes T_2$ is injective. (Observe that proving this by the definition of injectivity will not be easy.)

Surjectivity can be easily checked. □

Corollary 6.6.10. *Let U, V be vector spaces and U_1, V_2 be their subspaces. Then the null space of the natural map $\phi : U \otimes V \to U/U_1 \otimes V/V_1$ given by $\phi(\mathbf{u} \otimes \mathbf{v}) = \bar{\mathbf{u}} \otimes \bar{\mathbf{v}}$ is $U_1 \otimes V + U \otimes V_1$. And there is an isomorphism*

$$\bar{\phi} : U \otimes V/(U_1 \otimes V + U \otimes V_1) \to U/U_1 \otimes V/V_1 \tag{6.89}$$

Proof. The bilinear map $U \times V \to U/U_1 \otimes V/V_1$ induces $\phi : U \otimes V \to U/U_1 \otimes V/V_1$ which is surjective by Corollary 6.6.9. By Exercise 6.3.20, we need to show that $N(\phi) = U_1 \otimes V + U \otimes V_1$. By the properties of tensor product it is obvious that $N(\phi) \supseteq U_1 \otimes V + U \otimes V_1$.

To prove that the inclusion is in fact equal, let $\{\mathbf{u}_1, ..., \mathbf{u}_k\}$ be a basis for U_1 and extend this to a basis $\{\mathbf{u}_1, ..., \mathbf{u}_k, \mathbf{u}_{k+1}, ..., \mathbf{u}_n\}$ for U. And let $\{\mathbf{v}_1, ..., \mathbf{v}_h\}$ be a basis for V_1 and extend this to a basis $\{\mathbf{v}_1, ..., \mathbf{v}_h, \mathbf{v}_{h+1}, ..., \mathbf{v}_m\}$ for V. By the rank theorem we have $\rho(\phi)+\nu(\phi) = mn$. Now $\rho(\phi) = (n-k)(m-h)$ (Corollary 6.6.8) and hence $\nu(\phi) = mn - (n-k)(m-h) = mk + nh - hk$. Let

$$S = \{\mathbf{u}_i \otimes \mathbf{v}_j \mid 1 \le i \le k, 1 \le j \le m\} \cup \{\mathbf{u}_i \otimes \mathbf{v}_j \mid k+1 \le i \le n, 1 \le j \le h\}$$

which consists of $mk + nh - hk$ vectors. Obviously, $S \subseteq U_1 \otimes V + U \otimes V_1$ and S being a part of a basis (Theorem 6.6.7), is also linearly independent. Hence we have

$$mk + (n-k)m \le \dim(U_1 \otimes V + U \otimes V_1) \le \dim N(\phi) = mk + nh - kh$$

Hence $U_1 \otimes V + U \otimes V_1 = N(\phi)$.

The isomorphism (6.89) follows form Exercise 6.3.20 (a). □

We will need the following result in the next subsection.

Corollary 6.6.11. *Let U, V be vector spaces and U_1, V_1 be subspaces of U, V respectively. Let W be a vector space and $f : U \times V \to W$ be a bilinear map. Then f induces a bilinear map $f' : U/U_1 \times V/V_1 \to W$ if and only if f vanishes on $U_1 \times V$ and $U \times V_1$.*

Proof. By Corollary 6.6.10, the null space of $U \otimes V \to U/U_1 \otimes V/V_1$ is $U_1 \otimes V + U \otimes V_1$. Hence the linear maps $U \otimes V \to W$ vanishing on $U_1 \otimes V + U \otimes V_1$ is in one-to-one correspondence with the linear maps $U/U_1 \otimes V/V_1 \to W$ (Exercise 6.3.20) which is, in turn, in one-to-one correspondence with the bilinear maps $U/U_1 \times V/V_1 \to W$.

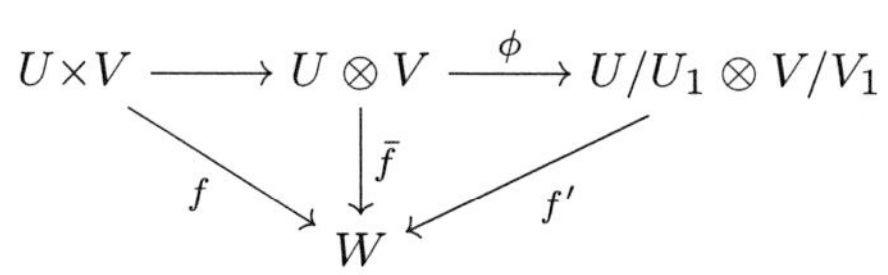

Then $f : U \times V \to W$ is the bilinear map corresponding to $\bar{f}$ vanishing on $U_1 \otimes V + U \otimes V_1$ if and only if f vanishes on $U_1 \times V$ and $U \times V_1$. As contended. □

Now we define tensor products of matrices so called Kronecker product. And we show that the matrix representation of tensor product of linear maps is the Kronecker product.

Definition 6.6.12. *If A is an $m \times n$ matrix and B is a $p \times q$ matrix, then the* **Kronecker product** *$A \otimes B$ is the $mp \times nq$ matrix:*

$$A \otimes B = \begin{bmatrix} a_{11}B & a_{12}B & \cdots & a_{1n}B \\ a_{21}B & a_{22}B & \cdots & a_{2n}B \\ \vdots & \vdots & \ddots & \vdots \\ a_{m1}B & a_{m2}B & \cdots & a_{mn}B \end{bmatrix} \tag{6.90}$$

More explicitly, $(A \otimes B)^{((i-1)q+j)} = \begin{bmatrix} a_{1i}B^{(j)} \\ a_{2i}B^{(j)} \\ \vdots \\ a_{mi}B^{(j)} \end{bmatrix}$ *where* $1 \le i \le n$ *and* $1 \le j \le q$.

Example 6.6.13. *If* $A = \begin{bmatrix} a & b \\ c & d \end{bmatrix}$ *and* $B = \begin{bmatrix} 1 & 2 & 3 \\ 4 & 5 & 6 \\ 7 & 8 & 9 \end{bmatrix}$ *then we have*

$$A \otimes B = \begin{bmatrix} a\begin{bmatrix} 1 & 2 & 3 \\ 4 & 5 & 6 \\ 7 & 8 & 9 \end{bmatrix} & b\begin{bmatrix} 1 & 2 & 3 \\ 4 & 5 & 6 \\ 7 & 8 & 9 \end{bmatrix} \\ c\begin{bmatrix} 1 & 2 & 3 \\ 4 & 5 & 6 \\ 7 & 8 & 9 \end{bmatrix} & d\begin{bmatrix} 1 & 2 & 3 \\ 4 & 5 & 6 \\ 7 & 8 & 9 \end{bmatrix} \end{bmatrix} = \begin{bmatrix} a & 2a & 3a & b & 2b & 3b \\ 4a & 5a & 6a & 4b & 5b & 6b \\ 7a & 8a & 9a & 7b & 8b & 9b \\ c & 2c & 3c & d & 2d & 3d \\ 4c & 5c & 6c & 4d & 5d & 6d \\ 7c & 8c & 9c & 7d & 8d & 9d \end{bmatrix}$$

Theorem 6.6.14. *Let* U, V *be vector spaces over* F *and* $S : U \to U$, $T : V \to V$ *be linear maps. Let* $\mathcal{B} = \{\mathbf{u}_1, \mathbf{u}_2, ..., \mathbf{u}_n\}, \mathcal{C} = \{\mathbf{v}_1, \mathbf{v}_2, ..., \mathbf{v}_m\}$ *be bases for* U, V. *Let* $\mathcal{B} \otimes \mathcal{C} := \{\mathbf{u}_i \otimes \mathbf{v}_j \mid i = 1, 2, ..., n; j = 1, 2, ..., m\}$ *be the basis for* $U \otimes V$ *with the lexicographic ordering. Then*

$$[S \otimes T]_{\mathcal{B}\otimes\mathcal{C}} = [S]_{\mathcal{B}} \otimes [T]_{\mathcal{C}}$$

Proof. Let $A = [S]_{\mathcal{B}}, B = [T]_{\mathcal{C}}$. The coefficients of $\mathbf{u}_k \otimes \mathbf{v}_l$ in $(S \otimes T)(\mathbf{u}_i \otimes \mathbf{v}_j) = (\sum_{k=1}^{n} a_{ki}\mathbf{u}_k) \otimes (\sum_{l=1}^{m} b_{lj}\mathbf{v}_l) = \sum_{k,l} a_{ki}b_{lj}\mathbf{u}_k \otimes \mathbf{v}_l$ listed lexicographically is the $((i-1)m+j)$-th column of $[S \otimes T]_{\mathcal{B}\otimes\mathcal{C}}$. Now that is the same as the $((i-1)m+j)$-th column of $[S]_{\mathcal{B}} \otimes [T]_{\mathcal{C}}$. □

Let V be a vector space over a field F. For each integer $r \ge 0$ we let $T^0(V) = F$ and

$$T^r(V) = V^{\otimes r} = \overbrace{V \otimes V \otimes \cdots \otimes V}^{r\text{-times}} \quad (r \ge 1)$$

which is unambiguously defined by (6.85).

Writing V^r for the r-times direct product of V we have a natural multilinear map

$$\phi^r : V^r \longrightarrow T^r(V) \tag{6.91}$$

which maps $(\mathbf{v}_1, \mathbf{v}_2, \ldots, \mathbf{v}_r)$ to $\mathbf{v}_1 \otimes \mathbf{v}_2 \otimes \cdots \otimes \mathbf{v}_r$.

Theorem 6.6.15. *Let* V, W *be vector spaces over* F *and let* $g : V^r \to W$ *be a multilinear map. Then there is a unique linear map* $\bar{g} : T^r(V) \to W$ *making the diagram commutative:*

$$\begin{array}{ccc} V^r & \xrightarrow{\phi^r} & T^r(V) \\ & {\scriptstyle g}\searrow & \swarrow{\scriptstyle \bar{g}} \\ & W & \end{array} \tag{6.92}$$

Proof. For $\mathbf{v} \in V$ define $g_{\mathbf{v}} : V^{r-1} \to W$ by $g_{\mathbf{v}}(\mathbf{v}_1, ..., \mathbf{v}_{r-1}) = g(\mathbf{v}_1, ..., \mathbf{v}_{r-1}, \mathbf{v})$. Then $g_{\mathbf{v}}$ is $(r-1)$-multilinear. By induction, there is a linear map $\bar{g}_{\mathbf{v}} : T^{r-1}(V) \to W$ such that $\bar{g}_{\mathbf{v}} \circ \phi^{r-1} = g_{\mathbf{v}}$. Now define $g' : T^{r-1}(V) \times V \to W$ by $g'(\mathbf{v}_1 \otimes \cdots \otimes \mathbf{v}_{r-1}, \mathbf{v}) = g_{\mathbf{v}}(\mathbf{v}_1, ..., \mathbf{v}_{r-1})$.

It is easy to check g' is bilinear. Hence, by Theorem 6.6.4, there is a linear map $\bar{g} : T^r(V) \to W$ such that $g' = \bar{g} \circ \phi^r$. Then

$$\begin{aligned} g(\mathbf{v}_1, ..., \mathbf{v}_{r-1}, \mathbf{v}) &= g_{\mathbf{v}}(\mathbf{v}_1, ..., \mathbf{v}_{r-1}) \\ &= \bar{g}_{\mathbf{v}} \circ \phi^{r-1}(\mathbf{v}_1, ..., \mathbf{v}_{r-1}) \\ &= \bar{g}_{\mathbf{v}}(\mathbf{v}_1 \otimes \cdots \otimes \mathbf{v}_{r-1}) \\ &= g'(\mathbf{v}_1 \otimes \cdots \otimes \mathbf{v}_{r-1}, \mathbf{v}) \\ &= \bar{g}(\mathbf{v}_1 \otimes \cdots \otimes \mathbf{v}_{r-1} \otimes \mathbf{v}) \end{aligned}$$

Therefore $\bar{g} \circ \phi^r = g$. □

Hence there is a bijection between the set of all multilinear maps $g : V^r \to W$ and the set of all linear maps $\bar{g} : T^r(V) \to W$.

Further we have a bilinear map,

$$\otimes : T^r(V) \times T^s(V) \longrightarrow T^{r+s}(V) \tag{6.93}$$

which is given by $(\mathbf{v}_1 \otimes \cdots \otimes \mathbf{v}_r, \mathbf{v}_{r+1} \otimes \cdots \otimes \mathbf{v}_{r+s}) \mapsto \mathbf{v}_1 \otimes \cdots \otimes \mathbf{v}_{r+s}$.

6.6.2 Exterior Product

Let V be a vector space over F and let I_r be the subspace of $T^r(V)$ generated by the elements of the form

$$\mathbf{v}_1 \otimes \mathbf{v}_2 \otimes \cdots \otimes \mathbf{v}_r, \quad \mathbf{v}_i = \mathbf{v}_j \text{ for some } i \neq j \ (1 \leq i, j \leq r).$$

We define

$$\bigwedge^r(V) = \begin{cases} F & (r = 0) \\ T^r(V)/I_r & (r > 0) \end{cases} \tag{6.94}$$

which we call the r-th **exterior product** of V. We have a canonical multilinear map

$$\psi^r : V^r \longrightarrow \bigwedge^r(V) \tag{6.95}$$

given by the composition $V^r \to T^r(V) \to \bigwedge^r(V)$. We denote the image of $(\mathbf{v}_1, \mathbf{v}_2, \ldots, \mathbf{v}_r)$ under ψ^r by $\mathbf{v}_1 \wedge \mathbf{v}_2 \wedge \cdots \wedge \mathbf{v}_r$:

$$\psi^r(\mathbf{v}_1, \mathbf{v}_2, ..., \mathbf{v}_r) = \mathbf{v}_1 \wedge \mathbf{v}_2 \wedge \cdots \wedge \mathbf{v}_r$$

Since ψ^r is surjective and V^r is spanned spanned by vectors of the form $(\mathbf{v}_1, \mathbf{v}_2, ..., \mathbf{v}_r)$ with $\mathbf{v}_i \in V$ we see that $\bigwedge^r(V)$ is a vector space spanned by $\mathbf{v}_1 \wedge \mathbf{v}_2 \wedge \cdots \wedge \mathbf{v}_r$. Note that not every elements of $\bigwedge^r(V)$ is of the form $\mathbf{v}_1 \wedge \mathbf{v}_2 \wedge \cdots \wedge \mathbf{v}_r$ but a linear combination of those forms.

Note that the map ψ^r is multilinear and **alternating** (i.e., $\mathbf{v}_1 \wedge \cdots \wedge \mathbf{v}_r = 0$ if $\mathbf{v}_i = \mathbf{v}_j$ $(i \neq j)$) by the definition of I_r. Hence, if $\text{char}(F) \neq 2$ then

$$\mathbf{v}_1 \wedge \cdots \wedge \mathbf{v}_i \wedge \mathbf{v}_{i+1} \wedge \cdots \wedge \mathbf{v}_r = -\mathbf{v}_1 \wedge \cdots \wedge \mathbf{v}_{i+1} \wedge \mathbf{v}_i \wedge \cdots \wedge \mathbf{v}_r \tag{6.96}$$

In fact, by using multilinearlity we have

$$\begin{aligned}
\mathbf{0} &= \mathbf{v}_1 \wedge \cdots \wedge (\mathbf{v}_i + \mathbf{v}_{i+1}) \wedge (\mathbf{v}_i + \mathbf{v}_{i+1}) \wedge \cdots \wedge \mathbf{v}_r \\
&= \mathbf{v}_1 \wedge \cdots \wedge \mathbf{v}_i \wedge \mathbf{v}_i \wedge \cdots \wedge \mathbf{v}_r + \mathbf{v}_1 \wedge \cdots \wedge \mathbf{v}_i \wedge \mathbf{v}_{i+1} \wedge \cdots \wedge \mathbf{v}_r \\
&\quad + \mathbf{v}_1 \wedge \cdots \wedge \mathbf{v}_{i+1} \wedge \mathbf{v}_i \wedge \cdots \wedge \mathbf{v}_r + \mathbf{v}_1 \wedge \cdots \wedge \mathbf{v}_{i+1} \wedge \mathbf{v}_{i+1} \wedge \cdots \wedge \mathbf{v}_r
\end{aligned}$$

and the first and the last term vanishes by alternating property. Hence we get (6.96).

Suppose $\{\mathbf{v}_1, \mathbf{v}_2, ..., \mathbf{v}_n\}$ span V. Then $\{\mathbf{v}_{i_1} \otimes \cdots \otimes \mathbf{v}_{i_k} | 1 \leq i_s \leq n\}$ span $T^k(V)$ by Theorem 6.6.2. By using the property (6.96) we see that $\bigwedge^k(V)$ is generated by

$$\{\mathbf{v}_{i_1} \wedge \mathbf{v}_{i_2} \wedge \cdots \wedge \mathbf{v}_{i_k} | i_1 < i_2 < \cdots < i_k\} \tag{6.97}$$

Theorem 6.6.16. *Let V, W be vector spaces over F and let $f : V^r \to W$ be multilinear and alternating (i.e., $f(\mathbf{v}_1, \mathbf{v}_2, \ldots, \mathbf{v}_r) = 0$ whenever $\mathbf{v}_i = \mathbf{v}_j$ for $i \neq j$). Then there is a unique linear map $\bar{f} : \bigwedge^r(V) \to W$ such that $f = \bar{f} \circ \psi^r$.*

$$\begin{array}{ccc}
V^r & \xrightarrow{\psi^r} & \bigwedge^r(V) \\
 & {\scriptstyle f}\searrow \quad \swarrow{\scriptstyle \bar{f}} & \\
 & W &
\end{array} \tag{6.98}$$

Proof. Since f is multilinear there is a unique linear map $\bar{f}' : T^r(V) \to W$ such that $f = \bar{f}' \circ \phi^r$ (Theorem 6.6.15). Since f is alternating we see that $\bar{f}'$ vanish on I_r. Denote the natural map $T^r(V) \to \bigwedge^r(V)$ by ψ so that $\phi^r \circ \psi = \psi^r$. Since $\bar{f}'$ vanish on I_r there is a unique linear map $\bar{f} : \bigwedge^r(V) \to W$ such that $\bar{f}' = \bar{f} \circ \psi$ (Exercise 6.3.20). Hence we have $f = \bar{f}' \circ \phi^r = \bar{f} \circ \psi \circ \phi^r = \bar{f} \circ \psi^r$. □

Hence there is a bijection between the set of all alternating multilinear maps $g : V^r \to W$ and the set of all linear maps $\bar{g} : \bigwedge^r(V) \to W$.

The composition $T^i(V) \times T^j(V) \to T^{i+j}(V) \to \bigwedge^{i+j}(V)$ vanishes on $I_i \times T^j(V)$ and $T^i(V) \times I_j$ (both are contained in I_{i+j}). Hence, by Corollary 6.6.11, the multiplication of tensor product (6.93) induces a multiplication which is bilinear on the exterior products

$$\bigwedge^i(V) \times \bigwedge^j(V) \longrightarrow \bigwedge^{i+j}(V) \tag{6.99}$$

satisfying, for $\mathbf{x} \in \bigwedge^i(V)$, $\mathbf{y}$, $\mathbf{y}_1$, $\mathbf{y}_2 \in \bigwedge^j(V)$ and $\mathbf{z} \in \bigwedge^k(V)$ (Exercise 15)

$$\begin{aligned}
\mathbf{x} \wedge (\mathbf{y}_1 + \mathbf{y}_2) &= \mathbf{x} \wedge \mathbf{y}_1 + \mathbf{x} \wedge \mathbf{y}_2 \\
(\mathbf{x} \wedge \mathbf{y}) \wedge \mathbf{z} &= \mathbf{x} \wedge (\mathbf{y} \wedge \mathbf{z})
\end{aligned} \tag{6.100}$$

The exterior product is skew symmetric i.e.,

$$\mathbf{x} \wedge \mathbf{y} = (-1)^{ij} \mathbf{y} \wedge \mathbf{x} \tag{6.101}$$

where $\mathbf{x} \in \bigwedge^i(V)$, $\mathbf{y} \in \bigwedge^j(V)$ which can be easily seen by using (6.96).

A linear map $T : V \to W$ induces a linear map

$$\bigwedge^n(T) : \bigwedge^n(V) \to \bigwedge^n(W) \tag{6.102}$$

If T is injective (resp. surjective, resp. isomorphism) then so is $\bigwedge^n(T)$ respectively (Exercise 20).

Theorem 6.6.17. *Let $V = F^n$ and let $\{\mathbf{e}^1, \mathbf{e}^2, \ldots, \mathbf{e}^n\}$ be the standard basis of V. If $r > n$, then $\bigwedge^r V = \mathbf{0}$. If $1 \le r \le n$, then $\bigwedge^r V$ is a vector space of dimension $\binom{n}{r}$ and*

$$\{\, \mathbf{e}^{i_1} \wedge \cdots \wedge \mathbf{e}^{i_r} \mid 1 \le i_1 < \cdots < i_r \le n\} \tag{6.103}$$

form a basis of $\bigwedge^r V$.

Proof. We first prove this for $r = n$. Since every element of V is of the form $\sum c_i \mathbf{e}^i$ $(c_i \in F)$, we see that $\mathbf{e}^1 \wedge \cdots \wedge \mathbf{e}^n$ generate $\bigwedge^n(V)$ by (6.97); $\dim \bigwedge^n(V) \le 1$. On the other hand, since the determinant is alternating multilinear map (Theorem 3.3.14) we see that, by Theorem 6.6.16, there is a unique linear map D making the diagram

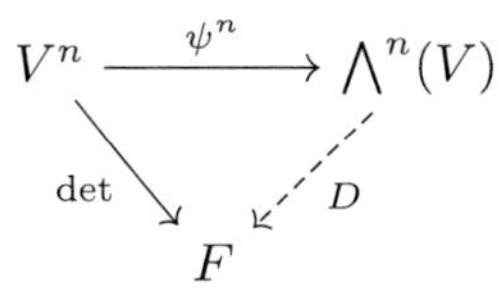

(6.104)

commutative. Since the determinant is surjective, D is also surjective. If $c\,\mathbf{e}^1 \wedge \cdots \wedge \mathbf{e}^n$ maps to 0 in F under D, then $0 = \det[c\,\mathbf{e}^1 \ \cdots \ \mathbf{e}^n] = c$. Hence the map D is also injective; D is an isomorphism with $D(\mathbf{e}^1 \wedge \cdots \wedge \mathbf{e}^n) = 1$. Therefore $\bigwedge^n(V)$ is one dimensional vector space with a basis $\mathbf{e}^1 \wedge \cdots \wedge \mathbf{e}^n$.

Now suppose $0 \le r < n$. By (6.97), $\bigwedge^r(V)$ is generated by

$$\mathbf{e}^{i_1} \wedge \cdots \wedge \mathbf{e}^{i_r} \qquad (1 \le i_1 < \cdots < i_r \le n) \tag{6.105}$$

For a multi-index $I = (i_1, i_2, ..., i_r)$ with $i_1 < \cdots < i_r$ write $\mathbf{e}^I = \mathbf{e}^{i_1} \wedge \cdots \wedge \mathbf{e}^{i_r}$. And write $I' = (i_{r+1}, ..., i_n\}$ where $i_{r+1} < \cdots < i_n$ and $\{i_1, i_2, ..., i_r\} \cup \{i_{r+1}, ..., i_n\} = \{1, 2, ..., n\}$. We only need to prove the vectors of (6.105) are linearly independent. Suppose we have a relation

$$\sum a_J \mathbf{e}^J = \mathbf{0}, \quad (J) = (j_1, \ldots, j_r) \tag{6.106}$$

Take $\wedge$-product (6.106) with $\mathbf{e}^{I'}$. If $J \ne I$ then there is $i \in J \cap I'$; hence both $\mathbf{e}^J$ and $\mathbf{e}^{I'}$ contains $\mathbf{e}^i$. Therefore, $\mathbf{e}^J \wedge \mathbf{e}^{I'} = \mathbf{0}$ and hence

$$a_I \mathbf{e}^I \wedge \mathbf{e}^{I'} = \pm a_I \mathbf{e}^1 \wedge \cdots \wedge \mathbf{e}^n = \mathbf{0}$$

By the case for $r = n$, we see that $a_I = 0$ for all I. □

Corollary 6.6.18. *Let $\mathbf{v}_1, \mathbf{v}_2, ..., \mathbf{v}_n \in F^n$ and $A := [\mathbf{v}_1 \ \mathbf{v}_2 \ \cdots \ \mathbf{v}_n]$, the matrix consisting of the columns $\mathbf{v}_i$. Then*

$$\mathbf{v}_1 \wedge \mathbf{v}_2 \wedge \cdots \wedge \mathbf{v}_n = \det(A)\mathbf{e}^1 \wedge \mathbf{e}^2 \wedge \cdots \wedge \mathbf{e}^n$$

Proof. Suppose $\mathbf{v}_1 \wedge \cdots \wedge \mathbf{v}_n = c\,\mathbf{e}^1 \wedge \cdots \wedge \mathbf{e}^n$. Then, with the notations of the proof of Theorem 6.6.17, we have $\det[\mathbf{v}_1 \ \mathbf{v}_2 \ \cdots \ \mathbf{v}_n] = D(c\,\mathbf{e}^1 \wedge \cdots \wedge \mathbf{e}^n) = \det[c\,\mathbf{e}^1 \cdots \mathbf{e}^n] = c$ by commutativity of the diagram (6.104). Hence we have the desired equality. □

Corollary 6.6.19. *Let V be a vector space over F with a basis $\{\mathbf{v}_1, \mathbf{v}_2, \cdots, \mathbf{v}_n\}$. Then*

$$\{\mathbf{v}_{i_1} \wedge \cdots \wedge \mathbf{v}_{i_r} \mid 0 < i_1 < \cdots < i_r \le n\}$$

is a basis for $\bigwedge^r(V)$.

Proof. Let $T : V \to F^n$ be the isomorphism given by the coordinate map wit respect to the given basis. Then $\bigwedge^r(T) : \bigwedge^r(V) \to \bigwedge^r(F^n)$ is an isomorphism. Since $\{\mathbf{e}^{i_1} \wedge \cdots \wedge \mathbf{e}^{i_r} \mid 0 < i_1 < \cdots < i_r \leq n\}$ is a basis for $\bigwedge^r(F^n)$ the corresponding vectors $\{\mathbf{v}_{i_1} \wedge \cdots \wedge \mathbf{v}_{i_r} \mid 0 < i_1 < \cdots < i_r \leq n\}$ is a basis for $\bigwedge^r(V)$. □

We prove $\det A \det B = \det AB$ (Theorem 3.3.11) by using the exterior product.

Corollary 6.6.20. *Let $A, B \in M(n)$. Then $\det(AB) = \det(A)\det(B)$.*

Proof. Let $\{\mathbf{e}^1, \mathbf{e}^2, ..., \mathbf{e}^n\}$ be the standard basis of $V = F^n$. Then we have

$$\begin{aligned}
&\textstyle\bigwedge^n(T_{BA})(\mathbf{e}^1 \wedge \mathbf{e}^2 \wedge \cdots \wedge \mathbf{e}^n) \\
&\quad = \textstyle\bigwedge^n(T_B \circ T_A)(\mathbf{e}^1 \wedge \mathbf{e}^2 \wedge \cdots \wedge \mathbf{e}^n) \\
&\quad = \textstyle\bigwedge^n T_B \circ \bigwedge^n T_A(\mathbf{e}^1 \wedge \mathbf{e}^2 \wedge \cdots \wedge \mathbf{e}^n) \\
&\quad = \textstyle\bigwedge^n T_B(A\mathbf{e}^1 \wedge A\mathbf{e}^2 \wedge \cdots \wedge A\mathbf{e}^n) && \text{(Corollary 6.6.18)} \\
&\quad = \textstyle\bigwedge^n T_B(\det[A^{(1)}\ A^{(2)} \cdots A^{(n)}]\mathbf{e}^1 \wedge \cdots \wedge \mathbf{e}^n) \\
&\quad = \textstyle\bigwedge^n T_B(\det(A)\mathbf{e}^1 \wedge \mathbf{e}^2 \wedge \cdots \wedge \mathbf{e}^n) && \text{(since } \textstyle\bigwedge^n T_B \text{ is linear)} \\
&\quad = \det(A)\textstyle\bigwedge^n T_B(\mathbf{e}^1 \wedge \mathbf{e}^2 \wedge \cdots \wedge \mathbf{e}^n) \\
&\quad = \det(A)\det(B)\mathbf{e}^1 \wedge \mathbf{e}^2 \wedge \cdots \wedge \mathbf{e}^n
\end{aligned}$$

On the other hand,

$$\begin{aligned}
&\textstyle\bigwedge^n(T_{BA})(\mathbf{e}^1 \wedge \mathbf{e}^2 \wedge \cdots \wedge \mathbf{e}^n) \\
&\quad = BA\mathbf{e}^1 \wedge BA\mathbf{e}^2 \wedge \cdots \wedge BA\mathbf{e}^n && \text{(Corollary 6.6.18)} \\
&\quad = \det[(BA)^{(1)} \cdots (BA)^{(n)}]\mathbf{e}^1 \wedge \cdots \wedge \mathbf{e}^n \\
&\quad = \det(BA)\mathbf{e}^1 \wedge \mathbf{e}^2 \wedge \cdots \wedge \mathbf{e}^n
\end{aligned}$$

Hence $\det(AB) = \det(A)\det(B)$. □

Let $V = F^n$ and $A \in M(n, F)$. Let $\eta \in \bigwedge^n V$. Then we have

$$\textstyle\bigwedge^n T_A(\eta) = \det(A)\eta \tag{6.107}$$

In fact, if $\eta = c\mathbf{e}^1 \wedge \mathbf{e}^2 \wedge \cdots \wedge \mathbf{e}^n$ then we have $\bigwedge^n T_A(\eta) = \bigwedge^n T_A(c\mathbf{e}^1 \wedge \mathbf{e}^2 \wedge \cdots \wedge \mathbf{e}^n) = c\bigwedge^n T_A(\mathbf{e}^1 \wedge \mathbf{e}^2 \wedge \cdots \wedge \mathbf{e}^n) = c\det(A)\mathbf{e}^1 \wedge \mathbf{e}^2 \wedge \cdots \wedge \mathbf{e}^n = \det(A)c\mathbf{e}^1 \wedge \mathbf{e}^2 \wedge \cdots \wedge \mathbf{e}^n = \det(A)\eta$.

Next we give a criteria for a set of vectors to be linearly dependent.

Theorem 6.6.21. *Let V be a vector space over F and $S = \{\mathbf{v}_1, \mathbf{v}_2, ..., \mathbf{v}_k\} \subseteq V$. Then S is linearly dependent if and only if $\mathbf{v}_1 \wedge \mathbf{v}_2 \wedge \cdots \wedge \mathbf{v}_k = \mathbf{0}$.*

Proof. If $\{\mathbf{v}_1, \mathbf{v}_2, ..., \mathbf{v}_k\}$ is linearly dependent then one of $\mathbf{v}_i$ is a linear combination of the other vectors say, $\mathbf{v}_1 = \sum_{i>1} c_i\mathbf{v}_i$. Then $\mathbf{v}_1 \wedge \mathbf{v}_2 \wedge \cdots \wedge \mathbf{v}_k = (\sum_{i>1} c_i\mathbf{v}_i) \wedge \mathbf{v}_2 \wedge \cdots \wedge \mathbf{v}_k) = \sum_{i>1}(c_i\mathbf{v}_i) \wedge \mathbf{v}_2 \wedge \cdots \wedge \mathbf{v}_k) = \mathbf{0}$ since $\mathbf{v}_i \wedge \mathbf{v}_2 \wedge \cdots \wedge \mathbf{v}_k = \mathbf{0}$ for $i > 1$.

Conversely, suppose $\mathbf{v}_1 \wedge \mathbf{v}_2 \wedge \cdots \wedge \mathbf{v}_k = \mathbf{0}$. Assume S is linearly independent. Then we can extend S to a basis $\{\mathbf{v}_1, \mathbf{v}_2, ..., \mathbf{v}_k, \mathbf{v}_{k+1}, ..., \mathbf{v}_n\}$ of V. Since $\mathbf{v}_1 \wedge \mathbf{v}_2 \wedge \cdots \wedge \mathbf{v}_k = \mathbf{0}$ we have $\mathbf{v}_1 \wedge \mathbf{v}_2 \cdots \wedge \mathbf{v}_n = \mathbf{0}$. On the other hand, if we let $A = [\mathbf{v}_1\ \mathbf{v}_2 \cdots \mathbf{v}_n]$ then, by Corollary 6.6.18, $\mathbf{v}_1 \wedge \mathbf{v}_2 \wedge \cdots \wedge \mathbf{v}_n = \det(A)\mathbf{e}^1 \wedge \mathbf{e}^2 \wedge \cdots \wedge \mathbf{e}^n$ which is nonzero since $\det(A) \neq 0$. A contradiction. Hence S is linearly independent. □

Let $V = F^n$ and $\{\mathbf{e}^1, \mathbf{e}^2, ..., \mathbf{e}^n\}$ be the standard basis of V. We have the product

$$\bigwedge^k(V)\times\bigwedge^{n-k}(V) \to \bigwedge^n(V)$$

given by (6.99). Now $\bigwedge^n(V)$ is 1-dimensional with a basis $\mathbf{e}^1\wedge\mathbf{e}^2\wedge\cdots\wedge\mathbf{e}^n$. If $\mathbf{x} \in \bigwedge^k(V)$, $\mathbf{y} \in \bigwedge^{n-k}(V)$ then $\mathbf{x}\wedge\mathbf{y} = c\,\mathbf{e}^1 \wedge \mathbf{e}^2 \wedge \cdots \wedge \mathbf{e}^n$. In this case, we write $\langle \mathbf{x}, \mathbf{y}\rangle = c$. Hence we have a bilinear pairing

$$\bigwedge^k(V)\times\bigwedge^{n-k}(V) \to F, \quad (\mathbf{x}, \mathbf{y}) \mapsto c \tag{6.108}$$

assigning c to $\langle \mathbf{x}, \mathbf{y}\rangle$. We want to show that the pairing is nondegenerate (Definition 6.5.20).

Theorem 6.6.22 (Hodge Duality). *The bilinear pairing* (6.108) *is nondegenerate.*

Proof. Let $\mathcal{I}_k$ be set of all subsets of $\{1, 2, ..., n\}$ consisting of k elements, $I = \{i_1, i_2, ..., i_k\}$ with $i_1 < \cdots < i_k$. For $I \in \mathcal{I}_k$ write $\mathbf{e}^I = \mathbf{e}^{i_1}\wedge\mathbf{e}^{i_2}\wedge\cdots\wedge\mathbf{e}^{i_k}$ and $I' = \{1, 2, ..., n\}\setminus I \in \mathcal{I}_{n-k}$. Let $\mathbf{x} \in \bigwedge^k V$ and write $\mathbf{x} = \sum_{I\in\mathcal{I}_k} c_I\mathbf{e}^I$. Suppose $\mathbf{x}\wedge\mathbf{y} = \mathbf{0}$ for all $\mathbf{y} \in \bigwedge^{n-k}V$. We need to show that $\mathbf{x} = \mathbf{0}$. Since $\mathbf{x}\wedge\mathbf{e}^{I'} = \mathbf{0}$ for all $I \in \mathcal{I}_k$ we have $\pm c_I\mathbf{e}^1 \wedge \cdots \wedge \mathbf{e}^n = 0$; $c_I = 0$. Therefore $\mathbf{x} = \mathbf{0}$. Similarly if $\mathbf{x}\wedge\mathbf{y} = \mathbf{0}$ for all $\mathbf{x} \in \bigwedge^k V$ then $\mathbf{y} = \mathbf{0}$. □

Now we give an interpretation of the adjoint matrix as a matrix representation of a linear map. Consider the nondegenerate pairing (6.108)

$$V\times\bigwedge^{n-1}V \to F$$

Let $\operatorname{adj}(T) : V \to V$ be the adjoint of $\bigwedge^{n-1}T$ with respect to this pairing (Corollary 6.5.22). That is

$$\langle \mathbf{v}, \bigwedge^{n-1}T(\eta)\rangle = \langle \operatorname{adj}(T)(\mathbf{v}), \eta\rangle \tag{6.109}$$

Theorem 6.6.23. *Let $A \in M(n, F)$ and $T = T_A$. Let $\mathcal{E} = \{\mathbf{e}^1, \mathbf{e}^2, ..., \mathbf{e}^n\}$ be the standard basis of $V = F^n$. With the notations above, the matrix representation of the linear map $\operatorname{adj}(T) : V \to V$ with respect to the standard basis $\mathcal{E}$ is the adjoint matrix* $\operatorname{adj}(A)$.

Proof. Let $\mathbf{f}_i = (-1)^{i-1}\mathbf{e}^1 \wedge \cdots \wedge \mathbf{e}^{i-1} \wedge \mathbf{e}^{i+1} \wedge \cdots \wedge \mathbf{e}^n$. Let $\operatorname{adj}(T)\mathbf{e}^j = \sum c_{kj}\mathbf{e}^k$. Then by (6.109), we have $\langle \operatorname{adj}T(\mathbf{e}^j), \mathbf{f}_i\rangle = \sum c_{kj}\langle \mathbf{e}^k, \mathbf{f}_i\rangle = c_{ij}$. On the other hand,

$$\begin{aligned}\langle \mathbf{e}^j, \bigwedge^{n-1}T(\mathbf{f}_i)\rangle &= \langle \mathbf{e}^j, (-1)^{i-1}A^{(1)} \wedge \cdots \wedge A^{(i-1)} \wedge A^{(i+1)} \wedge \cdots \wedge A^{(n)}\rangle \\ &= (-1)^{i-1}(-1)^{i-1}A^{(1)} \wedge \cdots \wedge A^{(i-1)} \wedge \mathbf{e}^j \wedge A^{(i+1)} \wedge \cdots \wedge A^{(n)} \\ &= (-1)^{i+j}|\tilde{A_{ji}}|\end{aligned}$$

Hence we have $[\operatorname{adj}(T)]_{\mathcal{E}} = \operatorname{adj}(A)$. □

Corollary 6.6.24. *Let $A \in M(n)$. We have*

$$\rho(\operatorname{adj}(A)) = \begin{cases} n & \text{(if } A \text{ is invertible)} \\ 1 & \text{(if } \rho(A) = n-1) \\ 0 & \text{(if } \rho(A) < n-1) \end{cases}$$

Proof. Let $T = T_A$. Since $\operatorname{adj}(A)$ is a matrix representation of the adjoint of $\bigwedge^{n-1}T$ we see, by (6.71), $\rho(\operatorname{adj}(A)) = \rho(\bigwedge^{n-1}T_A) = \dim\bigwedge^{n-1}R(T_A)$. If A is invertible then $R(T_A) = F^n$ and $\dim\bigwedge^{n-1}R(T_A) = n$. If $\rho(A) = n-1$ i.e., $\dim R(T_A) = n-1$ then $\rho(\operatorname{adj}(A)) = \rho(\bigwedge^{n-1}(T)) = \dim\bigwedge^{n-1}R(T_A) = 1$. If $\rho(A) < n-1$ then $\dim R(T_A) < n-1$ and hence $\dim\bigwedge^{n-1}R(T_A) = 0$. Therefore $\rho(\operatorname{adj}(A)) = 0$ (Theorem 6.6.17). □

The following corollary givers another proof of Exercise 3.4.6.

Corollary 6.6.25. *Let $A \in M(n, F)$. Then we have*

$$\mathrm{adj}(A)A = \det(A)I_n$$

Proof. Let $T = T_A$ and $V = F^n$. For $\mathbf{v} \in V$ and $\eta \in \bigwedge^{n-1} V$ we compute

$$\begin{aligned}
\langle \mathrm{adj}(T) \circ T(\mathbf{v}), \eta \rangle &= \langle T(\mathbf{v}), \textstyle\bigwedge^{n-1} T(\eta) \rangle \\
&= T(\mathbf{v}) \wedge \textstyle\bigwedge^{n-1} T(\eta) \\
&= \textstyle\bigwedge^{n} T(\mathbf{v} \wedge \eta) \\
&= \det(A)\mathbf{v} \wedge \eta & \text{by (6.107)} \\
&= \langle \det(A)\mathbf{v}, \eta \rangle
\end{aligned}$$

Therefore we conclude that $\mathrm{adj}(T) \circ T = \det(A)I_V$. Representing this into the matrices with respect to the standard basis we get $\mathrm{adj}(A)A = \det(A)I_n$. □

Example 6.6.26. *Let* $A = \begin{bmatrix} 1 & -1 & 0 & 1 \\ 1 & 2 & -1 & 0 \\ -1 & 0 & 2 & 0 \\ 1 & 1 & 1 & 1 \end{bmatrix}$. *Then* $\mathrm{adj}(A) = \begin{bmatrix} 4 & 4 & 4 & -4 \\ -1 & -1 & -1 & 1 \\ 2 & 2 & 2 & -2 \\ -5 & -5 & -5 & 5 \end{bmatrix}$ *and hence* $\rho(A) = 3,\ \rho(\mathrm{adj}(A)) = 1$.

6.6 Exercises

1. Prove the true statement(s) and find a counter example for the false statement(s).

   (a) If $f : V_1 \times V_2 \to W$ is bilinear then f is linear as well.

   (b) The map $f : \mathbb{R} \times \mathbb{R} \to \mathbb{R}$ defined by $f(x, y) = 2xy$ is a bilinear map.

   (c) The map $f : \mathbb{R} \times \mathbb{R} \to \mathbb{R}$ defined by $f(x, y) = 2x + y$ is a bilinear map.

   (d) If $f_i : V_i \to W$ $(i = 1, 2)$ are linear then $f_1 + f_2 : V_1 \times V_2 \to W$ defined by $(f_1 + f_2)(\mathbf{v}_1, \mathbf{v}_2) = f_1(\mathbf{v}_1) + f_2(\mathbf{v}_2)$ is bilinear.

   (e) Every element of $U \otimes V$ is of the form $\mathbf{u} \otimes \mathbf{v}$ with $\mathbf{u} \in U, \mathbf{v} \in V$.

   (f) If $\mathbf{v}_1, \mathbf{v}_2 \in V$ then $\mathbf{v}_1 \otimes \mathbf{v}_2 = \mathbf{v}_2 \otimes \mathbf{v}_1$ in $V \otimes V$.

   (g) If U, V are vector spaces of dimension m, n then $U \otimes V$ is a vector space of dimension $m + n$.

   (h) If $f : V_1 \times V_2 \to W$ is bilinear and U_1, U_2 are subspaces of $V_1 \times V_2$ on which $f = 0$. Then $f(U_1 + U_2) = 0$.

   (i) If $\mathbf{u} \in U, \mathbf{v} \in V$ are nonzero vectors then $\mathbf{u} \otimes \mathbf{v} \in U \otimes V$ is also a nonzero vector.

   (j) If $\mathbf{u} \in U$ is nonzero and $\{\mathbf{v}_i\}$ is a set of linearly independent vectors of V then $\{\mathbf{u} \otimes \mathbf{v}_i\}$ is also linearly independent in $U \otimes V$.

   (k) The map $\phi : U \times V \to U \otimes V$ defined by $\phi(\mathbf{u}, \mathbf{v}) = \mathbf{u} \otimes \mathbf{v}$ is a linear map.

   (l) If U_1, U_2 are subspaces of U and V_1, V_2 are subspaces of V then $U_1 \otimes V_1 + U_2 \otimes V_2 = (U_1 + U_2) \otimes (V_1 + V_2)$.

(m) If $\mathbf{v}_1, \mathbf{v}_2$ are nonzero vectors in V then $\mathbf{v}_1 \wedge \mathbf{v}_2 \neq \mathbf{0}$ in $\bigwedge^2 V$.

(n) The map $\phi : V \times V \to \bigwedge^2 V$ defined by $\phi(\mathbf{v}_1, \mathbf{v}_2) = \mathbf{v}_1 \wedge \mathbf{v}_2$ is a bilinear map.

(o) If $\mathbf{v}_1, \mathbf{v}_2$ are nonzero vectors in V then $\mathbf{v}_1 \wedge \mathbf{v}_2 = \mathbf{v}_2 \wedge \mathbf{v}_1$ in $\bigwedge^2 V$.

(p) If $\dim V = 1$ then $\bigwedge^2 V = \mathbf{0}$.

(q) If $A \in M(2, F)$ then $\rho(\operatorname{adj}(A)) = \rho(A)$.

(r) If $\dim V = n$ then $\dim \bigwedge^{n-k} V = \dim \bigwedge^k V$.

(s) If $\dim V = n$ then and $\mathbf{v}_1, ..., \mathbf{v}_{n+1} \in V$ are $n+1$ vectors. Then $\mathbf{v}_1 \wedge \cdots \wedge \mathbf{v}_{n+1} = \mathbf{0}$.

2. Prove that if $f : V_1 \times V_2 \to W$ is bilinear then $f_{\mathbf{v}_1, \mathbf{v}_2} : V_1 \times V_2 \to W$ defined by $f_{\mathbf{v}_1, \mathbf{v}_2}(\mathbf{x}_1, \mathbf{x}_2) = f(\mathbf{v}_1, \mathbf{x}_2) + f(\mathbf{x}_1, \mathbf{v}_2)$ is a linear map on $V_1 \times V_2$.

3. Prove that if $f : \mathbb{R}^{n_1} \times \mathbb{R}^{n_2} \to \mathbb{R}^m$ is bilinear then f is a polynomial as a function. (See Remark (2.1.2).)

4. Let $S : U \to U'$ and $T : V \to V'$ be isomorphisms. Prove that $S \otimes T : U \otimes V \to U' \otimes V'$ is also an isomorphism and prove that $(S \otimes T)^{-1} = S^{-1} \otimes T^{-1}$.

5. Let U', U'' and V', V'' be subspaces of U and V respectively.

   (a) Prove that $(U' \otimes V) \cap (U \otimes V') = U' \otimes V'$ in $U \otimes V$.

   (b) More generally, is $(U' \cap U'') \otimes (V' \cap V'') = (U' \otimes V') \cap (U'' \otimes V'')$ true?

6. Let $S : U \to U'$ and $T : V \to V'$ be linear maps.

   (a) Prove that $R(S \otimes T) = R(S) \otimes R(T)$ and $\rho(S \otimes T) = \rho(S)\rho(T)$.

   (b) Prove that

$$
\begin{aligned}
N(S \otimes T) &= N(S) \otimes V + U \otimes N(T) \\
\nu(S \otimes T) &= \nu(S)\nu(T) + \nu(S)\rho(T) + \nu(T)\rho(S)
\end{aligned}
$$

   (c) Prove that $R(\bigwedge^k(T)) = \bigwedge^k(R(T))$.

7. Prove Theorem 6.6.5.

8. Let V be a vector space over $\mathbb{R}$. As $\mathbb{C}$ is a vector space over $\mathbb{R}$ we can form the tensor product $V \otimes_{\mathbb{R}} \mathbb{C}$ as $\mathbb{R}$ vector spaces. Define a scalar multiplication by $\mathbb{C}$ on $V \otimes_{\mathbb{R}} \mathbb{C}$ by

$$
\mathbb{C} \times (V \otimes_{\mathbb{R}} \mathbb{C}) \to V \otimes_{\mathbb{R}} \mathbb{C} \qquad (\alpha, \mathbf{v} \otimes \beta) \mapsto \mathbf{v} \otimes (\alpha\beta)
$$

   (a) Prove that $V \otimes_{\mathbb{R}} \mathbb{C}$ is a vector space over $\mathbb{C}$.

   (b) Prove that if $\mathcal{B} = \{\mathbf{v}_i\}_{i=1}^n$ is a basis for V then $\bar{\mathcal{B}} = \{\mathbf{v}_i \otimes 1\}_{i=1}^n$ is a basis for $V \otimes_{\mathbb{R}} \mathbb{C}$ over $\mathbb{C}$. (Hint: $\dim_{\mathbb{R}} V \otimes_{\mathbb{R}} \mathbb{C} = 2 \dim V$.)

   (c) Prove that $V \otimes_{\mathbb{R}} \mathbb{C}$ is isomorphic to the complexification $V^{\mathbb{C}}$ (Exercise 6.3.16). (Hint: For $\mathbf{v} \in V$ and $\alpha = x + iy \in \mathbb{C}$ define $f(\mathbf{v}, \alpha) = x\mathbf{v} + iy\mathbf{v} \in V^{\mathbb{C}}$. Show that $f : V \times \mathbb{C} \to V^{\mathbb{C}}$ is bilinear. By Theorem 6.6.4, we get a $\mathbb{R}$-linear $\bar{f} : V \otimes_{\mathbb{R}} \mathbb{C} \to V^{\mathbb{C}}$. Now show $\bar{f}$ is a $\mathbb{C}$-isomorphism.)

9. Let U, U', V, V' be vector space over F.

(a) Prove that the linear map $L(U,U') \otimes L(V,V') \to L(U \otimes U', V \otimes V')$ induced by the bilinear map $(S,T) \mapsto S \otimes T$ is an isomorphism. (Hint: Show the map is injective by representing S,T as matrices and check that the dimensions of the vector spaces are the same.)

(b) Prove that the map $\phi : U^{\#} \otimes V \overset{\cong}{\to} L(U,V)$ defined by $\phi(f \otimes \mathbf{v})(\mathbf{u}) = f(\mathbf{u})\mathbf{v}$ is an isomorphism. (Hint: Let $U' = F, V = F, V' = V$ in (a).)

(c) Let $\{\mathbf{u}_1, \mathbf{u}_2, ..., \mathbf{u}_n\}$ be a basis of U with it dual $\{f_1, f_2, ..., f_n\}$. For $S \in L(U,V)$ define $\psi(S) = \sum_i f_i \otimes S(\mathbf{u}_i)$. Prove that ψ is the the inverse of ϕ in (b). (Hint:Exercise 6.5.8.)

10. Prove that the map $\phi : L(U \otimes V, W) \overset{\cong}{\to} L(U, L(V,W))$ defined by $\phi(f)(\mathbf{u})(\mathbf{v}) = f(\mathbf{u} \otimes \mathbf{v})$ is an isomorphism.

11. (a) Let $A = \begin{bmatrix} a & b \\ c & d \end{bmatrix}, B = \begin{bmatrix} 1 & 2 & 3 \\ 4 & 5 & 6 \\ 7 & 8 & 9 \end{bmatrix}$. Find $A \otimes B$ and $B \otimes A$ and show that there are permutation matrices (Definition 3.3.15) P, Q such that $A \otimes B = Q(B \otimes A)P$.

(b) Let $A \in M(m \times n), B \in M(p \times q)$. Prove that there are permutation matrices P, Q (Definition 3.3.15) such that $A \otimes B = Q(B \otimes A)P$. If $A \in M(n), B \in M(q)$ then $Q = P^{-1}$; $A \otimes B$ and $B \otimes A$ are similar. (Hint: Use the commutative diagram

$$\begin{array}{ccc} F^n \otimes F^q & \xrightarrow{T_{A \otimes B}} & F^m \otimes F^p \\ \sigma \downarrow & & \downarrow \bar{\sigma} \\ F^q \otimes F^n & \xrightarrow[T_{B \otimes A}]{} & F^p \otimes F^m \end{array}$$

where the vertical arrows are the isomorphisms of (6.84) of Theorem 6.6.6.)

12. Let $A \in M(n,F)$ and $B \in M(m,F)$. Prove that $\mathrm{Tr}\,(A \otimes B) = \mathrm{Tr}\,(A)\mathrm{Tr}\,(B)$. Can you find $\det(A \otimes B)$? (See Exercise 7.3.22.)

13. Let $A, A' \in M(n,F)$ and $B, B' \in M(m,F)$. Prove that if $A \sim A', B \sim B'$ then $A \otimes B \sim A' \otimes B'$ (Definition 4.3.8).

14. Let $A \in M(m \times n), B \in M(p \times q)$.

(a) Prove that $T_A \otimes T_B = T_{A \otimes B}$. That is $A \otimes B$ is the matrix representation of $T_A \otimes T_B$ with respect to the basis $\{\mathbf{e}^i \otimes \bar{\mathbf{e}}^j \mid 1 \le i \le n, 1 \le j \le q\}$ of $F^n \otimes F^q$ in lexicographic ordering where $\mathcal{E} = \{\mathbf{e}^1, \mathbf{e}^2, ..., \mathbf{e}^n\}$, $\bar{\mathcal{E}} = \{\bar{\mathbf{e}}^1, \bar{\mathbf{e}}^2, ..., \bar{\mathbf{e}}^q\}$ are the standard bases for F^n, F^q.

(b) Prove that $\rho(A \otimes B) = \rho(A)\rho(B)$ for $A \in M(m \times n), B \in M(p \times q)$.

15. Prove the equalities (6.100) and (6.101).

16. Let $V = F^n$ and $\{\mathbf{v}_1, \ldots, \mathbf{v}_n\} \subseteq V$ be a subset. Let $\{\mathbf{e}^1, \ldots, \mathbf{e}^n\}$ be the standard basis of V. Prove that $\{\mathbf{v}_1, \ldots, \mathbf{v}_n\}$ form a basis of V if and only if there is a nonzero $c \in F$ such that $\mathbf{v}_1 \wedge \cdots \wedge \mathbf{v}_n = c\mathbf{e}^1 \wedge \cdots \wedge \mathbf{e}^n$.

17. Let V be a vector space and $f : V^n \to W$ be multilinear and alternating. Prove that if $\mathbf{v}_1 \wedge \mathbf{v}_2 \wedge \cdots \wedge \mathbf{v}_n = \mathbf{0}$ then $f(\mathbf{v}_1, \mathbf{v}_2, ..., \mathbf{v}_n) = \mathbf{0}$. (Hint: Use Theorem 6.6.16.)

18. Let $f : V^{n-1} \to F$ be an alternating multilinear map. Prove that if $\mathbf{v}_1 \wedge \cdots \wedge \mathbf{v}_n = \mathbf{0}$ then $\sum(-1)^i f(\mathbf{v}_1, \ldots, \widehat{\mathbf{v}}_i, \ldots, \mathbf{v}_n)\mathbf{v}_i = \mathbf{0}$ where $\widehat{\mathbf{v}}_i$ denote deletion. (Hint: Show that $g(\mathbf{v}_1, \mathbf{v}_2, ..., \mathbf{v}_n) = \sum(-1)^i f(\mathbf{v}_1, ..., \hat{\mathbf{v}}_i, ..., \mathbf{v}_n)\mathbf{v}_i$ is alternating. And use Exercise 17.)

19. Prove that $\mathbf{v}_1 \wedge \cdots \wedge \mathbf{v}_n \neq \mathbf{0}$ if and only if there is an alternating multilinear map $f : V^n \to F$ such that $f(\mathbf{v}_1, \ldots, \mathbf{v}_n) \neq 0$.

20. Prove that if $T : V \to W$ is an injective (resp. surjective, resp. isomorphism) linear map then $\bigwedge^r(T) : \bigwedge^r(V) \to \bigwedge^r(W)$ is injective (resp. surjective, resp. isomorphism). (Hint: Use Corollary 4.2.31.)

21. Let $V = F^n$. Prove that there is a natural isomorphism $\bigwedge^k V^\# \overset{\cong}{\to} (\bigwedge^k V)^\#$. (Hint: Use Corollary 6.5.22.)

22. Let $A, B \in M(n, F)$. Prove that $\text{adj}(AB) = \text{adj}(B)\text{adj}(A)$.

23. Prove or disprove: Let $f : U_1 \times U_2 \to W$ be a bilinear map and $\bar{f} : U_1 \otimes U_2 \to W$ be its induced map. If $U_i^0 = \{\mathbf{u}_i \mid f(\mathbf{u}_1, \mathbf{u}_2) = \mathbf{0} \text{ for all } \mathbf{u}_j \in U_j \ (j \neq i)\}$ then $N(\bar{f}) = U_1 \otimes U_2^0 + U_1^0 \otimes U_2$.

24. Let U, V be subspaces of a vector space W and let $W = U \oplus V$. Prove that there is an isomorphism
$$\bigoplus_{p+q=k} \bigwedge^p(U) \otimes \bigwedge^q(V) \overset{\cong}{\to} \bigwedge^k(W)$$

For the following exercises one needs some knowledge from advanced calculus.

25. Let $f : \mathbb{R}^n \to \mathbb{R}^m$ be a function. We say f is **differentiable** at $\mathbf{v} \in \mathbb{R}^n$ if there is a linear map $L : \mathbb{R}^n \to \mathbb{R}^m$ such that
$$\lim_{\mathbf{h} \to \mathbf{0}} \frac{\|f(\mathbf{v} + \mathbf{h}) - f(\mathbf{v}) - L(\mathbf{h})\|}{\|\mathbf{h}\|} = 0$$
In this case, we write $Df(\mathbf{v})$ for L. Here $\|\mathbf{v}\| = \sqrt{v_1^2 + \cdots + v_n^2}$ for $\mathbf{v} = \begin{bmatrix} v_1 \\ \vdots \\ v_n \end{bmatrix}$. Prove the following statements.
    (a) If $m = n = 1$ and $f : \mathbb{R} \to \mathbb{R}$ is differentiable at $a \in \mathbb{R}$ then $L(\mathbf{x}) = f'(a)\mathbf{x}$.
    (b) If f is a linear map then $Df(\mathbf{v}) = f$ for all $\mathbf{v} \in \mathbb{R}^n$.
    (c) If $f : \mathbb{R}^n \to \mathbb{R}^m$ is linear then there is $c > 0$ such that $\dfrac{\|f(\mathbf{v})\|}{\|\mathbf{v}\|} \leq c$ for all $\mathbf{v}$.
    (d) If $f : \mathbb{R}^{n_1} \times \mathbb{R}^{n_2} \to \mathbb{R}^m$ is bilinear then there is $c > 0$ such that $\|f(\mathbf{v}_1, \mathbf{v}_2)\| \leq c\|\mathbf{v}_1\| \cdot \|\mathbf{v}_2\|$ for all $(\mathbf{v}_1, \mathbf{v}_2)$ and $\displaystyle\lim_{(\mathbf{v}_1, \mathbf{v}_2) \to \mathbf{0}} \frac{\|f(\mathbf{v}_1, \mathbf{v}_2)\|}{\|(\mathbf{v}_1, \mathbf{v}_2)\|} = \mathbf{0}$.
    (e) If $f : \mathbb{R}^{n_1} \times \mathbb{R}^{n_2} \to \mathbb{R}^m$ is bilinear then $Df(\mathbf{v}_1, \mathbf{v}_2) = f_{\mathbf{v}_1, \mathbf{v}_2}$ where $f_{\mathbf{v}_1, \mathbf{v}_2} : \mathbb{R}^{n_1} \times \mathbb{R}^{n_2} \to \mathbb{R}^m$ is defined by $f_{\mathbf{v}_1, \mathbf{v}_2}(\mathbf{x}_1, \mathbf{x}_2) = f(\mathbf{v}_1, \mathbf{x}_2) + f(\mathbf{x}_1, \mathbf{v}_2)$ as in Exercise 2.

(f) The derivative of the multilinear map $\det : \mathbb{R}^n \times \cdots \times \mathbb{R}^n \to \mathbb{R}$ is given by

$$D\det(\mathbf{v}_1, \mathbf{v}_2, ..., \mathbf{v}_n)(\mathbf{x}_1, \mathbf{x}_2, ..., \mathbf{x}_n) = \sum_{i=1}^{n} \det(\mathbf{v}_1, ..., \mathbf{v}_{i-1}, \mathbf{x}_i, \mathbf{v}_{i+1}, ..., \mathbf{v}_n)$$

26. Let $\mathbf{v}_1, \mathbf{v}_2, ..., \mathbf{v}_{n-1} \in \mathbb{R}^n$. Then $\phi(\mathbf{v}) = \det[\mathbf{v}_1\ \mathbf{v}_2 \cdots \mathbf{v}_{n-1}\ \mathbf{v}]$ is a linear functional. Hence there is a unique $\mathbf{w}$ such that $\mathbf{w} \cdot \mathbf{v} = \phi(\mathbf{v})$ for all $\mathbf{v}$ (Example 6.5.2). We denote $\mathbf{w}$ by

$$\mathbf{w} = \mathbf{v}_1 \times \mathbf{v}_2 \times \cdots \times \mathbf{v}_{n-1}$$

which we call the **cross product** of $\mathbf{v}_1, \mathbf{v}_2, ..., \mathbf{v}_{n-1}$ i.e.,

$$(\mathbf{v}_1 \times \mathbf{v}_2 \times \cdots \times \mathbf{v}_{n-1}) \cdot \mathbf{v} = \det[\mathbf{v}_1\ \mathbf{v}_2 \cdots \mathbf{v}_{n-1}\ \mathbf{v}]$$

Prove the following statements.

(a) $\mathbf{v}_{\sigma(1)} \times \mathbf{v}_{\sigma(2)} \times \cdots \times \mathbf{v}_{\sigma(n-1)} = \mathrm{sign}(\sigma)\mathbf{v}_1 \times \mathbf{v}_2 \times \cdots \times \mathbf{v}_{n-1}$ (See Definition 3.3.15 for $\mathrm{sign}(\sigma)$.)

(b) $\mathbf{v}_1 \times \cdots \times a\mathbf{v}_i \times \cdots \times \mathbf{v}_{n-1} = a\mathbf{v}_1 \times \cdots \times \mathbf{v}_i \times \cdots \times \mathbf{v}_{n-1}$

(c) $\mathbf{v}_1 \times \cdots \times (\mathbf{v}_i + \mathbf{v}_i') \times \cdots \times \mathbf{v}_{n-1} = \mathbf{v}_1 \times \cdots \times \mathbf{v}_i \times \cdots \times \mathbf{v}_{n-1} + \mathbf{v}_1 \times \cdots \times \mathbf{v}_i' \times \cdots \times \mathbf{v}_{n-1}$

(d) If $\{\mathbf{e}^1, \mathbf{e}^2, \mathbf{e}^3\}$ is the standard basis of $\mathbb{R}^3$ then we have $\mathbf{e}^1 \times \mathbf{e}^2 = \mathbf{e}^3$, $\mathbf{e}^2 \times \mathbf{e}^3 = \mathbf{e}^1$, $\mathbf{e}^3 \times \mathbf{e}^1 = \mathbf{e}^2$.

(e) If $\mathbf{u} = \begin{bmatrix} a_1 \\ a_2 \\ a_3 \end{bmatrix}, \mathbf{v} = \begin{bmatrix} b_1 \\ b_2 \\ b_3 \end{bmatrix}$ then $\mathbf{u} \times \mathbf{v} = \begin{bmatrix} a_2b_3 - a_3b_2 \\ a_3b_1 - a_1b_3 \\ a_2b_3 - a_3b_2 \end{bmatrix}$ which can be written as $\det \begin{bmatrix} \mathbf{e}^1 & a_1 & b_1 \\ \mathbf{e}^2 & a_2 & b_2 \\ \mathbf{e}^3 & a_3 & b_3 \end{bmatrix}$. This is the familiar cross product defined in calculus.

6.7 Diagonalization of Linear Maps

In §5.1, we studied diagonalization of the real square matrices. In this section, we investigate diagonalization of the linear operators on a vector space over a field. We will define diagonalizability of a linear operator in the same way as we did for the matrices (Definition 5.1.16).

Definition 6.7.1. *We say that $T \in L(V)$ is* **diagonalizable** *if there is a basis $\mathcal{B} = \{\mathbf{v}_1, \mathbf{v}_2, ..., \mathbf{v}_n\}$ for which $[T]_{\mathcal{B}}$ is diagonal. We say $A \in M(n, F)$ is* **diagonalizable** *if $T_A : F^n \to F^n$ is diagonalizable.*

As before, we denote the $n \times n$ diagonal matrix with $\lambda_1, \lambda_2, ..., \lambda_n$ along the diagonal by $\mathrm{diag}[\lambda_1, \lambda_2, ..., \lambda_n]$.

$$\mathrm{diag}[\lambda_1, \lambda_2, ..., \lambda_n] = \begin{bmatrix} \lambda_1 & & & \\ & \lambda_2 & & \\ & & \ddots & \\ & & & \lambda_n \end{bmatrix} \tag{6.110}$$

Suppose there is a basis $\mathcal{B} = \{\mathbf{v}_1, \mathbf{v}_2, \ldots, \mathbf{v}_n\}$ of V for which $[T]_{\mathcal{B}}$ is diagonal i.e.,

$$[T]_{\mathcal{B}} = \mathrm{diag}[\lambda_1, \lambda_2, ..., \lambda_n] \tag{6.111}$$

Then $T(\mathbf{v}_i) = \lambda_i \mathbf{v}_i$. Conversely, if there is a basis $\mathcal{B} = \{\mathbf{v}_1, \mathbf{v}_2, \ldots, \mathbf{v}_n\}$ such that $T(\mathbf{v}_i) = \lambda_i \mathbf{v}_i$ then we have the matrix representation (6.111) in the basis of $\mathcal{B}$.

Definition 6.7.2. *Let $T \in L(V)$. A nonzero $\mathbf{v} \in V$ is an* **eigenvector** *for T if $T(\mathbf{v}) = \lambda \mathbf{v}$ for some $\lambda \in F$ i.e., $\mathbf{v} \in N(T - \lambda I)$. The scalar $\lambda \in F$ is called the* **eigenvalue** *with an eigenvector $\mathbf{v}$.*

If $A \in M(n, F)$ then an eigenvector $\mathbf{v} \in F^n$ for T_A is called an **eigenvector** *for A and an eigenvalue $\lambda \in F$ for T_A is called an* **eigenvalue** *for A with an eigenvector $\mathbf{v}$.*

We sum up our discussion in

Theorem 6.7.3. *Let $T \in L(V)$. Then T is diagonalizable if and only if there is a basis $\mathcal{B} = \{\mathbf{v}_1, \mathbf{v}_2, \ldots, \mathbf{v}_n\}$ consisting of eigenvectors. If $T(\mathbf{v}_i) = \lambda_i \mathbf{v}_i$ then*

$$[T]_{\mathcal{B}} = \operatorname{diag}[\lambda_1, \lambda_2, \ldots, \lambda_n]$$

Example 6.7.4. *Let $V = M(n, \mathbb{R})$ and $T : V \to V$ be the linear map defined by $T(A) = A^t$. If $A \in V$ is symmetric then A is an eigenvector with eigenvalue 1. If A is skew symmetric then A is an eigenvector of T with the eigenvalue -1. When $n = 3$ the matrix of T with respect to the standard basis $\mathcal{B} = \{E_{ij}\}$* (Example 6.2.8 (4)) *is*

$$[T]_{\mathcal{B}} = \begin{bmatrix} 1 & 0 & 0 & 0 & 0 & 0 & 0 & 0 & 0 \\ 0 & 0 & 0 & 1 & 0 & 0 & 0 & 0 & 0 \\ 0 & 0 & 0 & 0 & 0 & 0 & 1 & 0 & 0 \\ 0 & 1 & 0 & 0 & 0 & 0 & 0 & 0 & 0 \\ 0 & 0 & 0 & 0 & 1 & 0 & 0 & 0 & 0 \\ 0 & 0 & 0 & 0 & 0 & 0 & 0 & 1 & 0 \\ 0 & 0 & 1 & 0 & 0 & 0 & 0 & 0 & 0 \\ 0 & 0 & 0 & 0 & 0 & 1 & 0 & 0 & 0 \\ 0 & 0 & 0 & 0 & 0 & 0 & 0 & 0 & 1 \end{bmatrix}$$

In fact, T is diagonalizable with the eigenvalues ± 1 (Exercise 18).

Note that λ is an eigenvalue of T if and only if there is nonzero $\mathbf{v}$ such that $T(\mathbf{v}) = \lambda \mathbf{v}$ i.e., $(T - \lambda I)(\mathbf{v}) = \mathbf{0}$. That is $N(T - \lambda I) \neq \mathbf{0}$ which is equivalent to the fact that $T - \lambda I$ is not an isomorphism by Theorem 6.3.26. We state this fact formally in:

Theorem 6.7.5. *Let $T \in L(V)$ (resp. $A \in M(n, F)$) with a basis $\mathcal{B}$ of V. Then λ is an eigenvalue of T* (resp. A) *if and only if* $\det([T]_{\mathcal{B}} - \lambda I) = 0$ *(resp.* $\det(A - \lambda I) = 0$*).*

And $\mathbf{v} \in V$ (resp. $\mathbf{v} \in F^n$) is an eigenvector for the eigenvalue λ if and only if $(T - \lambda I)\mathbf{v} = \mathbf{0}$ (resp. $(A - \lambda I_n)\mathbf{v} = \mathbf{0}$).

Proof. Since $\mathbf{v} \in N(T - \lambda I_V)$ if and only if $[\mathbf{v}]_{\mathcal{B}} \in N([T - \lambda I]_{\mathcal{B}}) = N([T]_{\mathcal{B}} - I_n)$ (Theorem 6.4.2) we only need to prove the theorem for matrices.

Now the theorem for the matrices follows from:

$$\begin{aligned} \det(A - \lambda I_n) = 0 &\Leftrightarrow T_A - \lambda I_n \text{ not injective (Theorem 3.3.7)} \\ &\Leftrightarrow N(T_A - \lambda I_n) \neq 0 \\ &\Leftrightarrow \text{ there is nonzero } \mathbf{v} \text{ such that } (A - \lambda I_n)(\mathbf{v}) = 0 \\ &\Leftrightarrow \text{ there is nonzero } \mathbf{v} \text{ such that } A\mathbf{v} = \lambda \mathbf{v} \end{aligned}$$

The statement for the eigenvectors are obvious. □

We defined the characteristic polynomial for the real matrices in Definition 5.1.9. Now we define the characteristic polynomials for the linear operators of a finite dimensional vector spaces over a field F.

Definition 6.7.6. *Let $T \in L(V)$ and choose a basis $\mathcal{B}$ of V. The* **characteristic polynomial** $f_T(t) \in F[t]$ *of T is defined to be the polynomial*

$$f_T(t) = \det([T]_{\mathcal{B}} - tI_n)$$

If $A \in M(n, F)$ then the **characteristic polynomial** $f_A(t)$ *of A is defined to be the characteristic polynomial for $T_A \in L(F^n)$ i.e., $f_A(t) = \det(A - tI_n)$. When the reference to T or A is clear we will simply write $f(t)$ for $f_T(t)$ or $f_A(t)$.*

The characteristic polynomial of a linear operator is independent of the choice of a basis as we have seen for the real case in Corollary 5.1.14. Here we recall its proof. Suppose $\mathcal{C}$ another basis of V. Then there is an invertible matrix P such that $[T]_{\mathcal{C}} = P^{-1}[T]_{\mathcal{B}}P$ (Theorem 6.4.26). Hence, by Theorem 3.3.12 and Theorem 3.3.11, we have

$$\det([T]_{\mathcal{C}} - tI_n) = \det(P^{-1}[T]_{\mathcal{B}}P - tI_n) = \det\left(P^{-1}([T]_{\mathcal{B}} - tI_n)P\right) = \det([T]_{\mathcal{B}} - tI_n)$$

Example 6.7.7. *In* Example 6.4.3, *we let $V = P_3(\mathbb{R})$ and $T : V \to V$ be given by $T(f(x)) = (x+1)f'(x)$. Let $\mathcal{B} = \{1, x, x^2, x^3\}$ be the standard basis of V. To find $[T]_B$ let's compute*

$$T(1) = 0,\ T(x) = 1 + x,\ T(x^2) = 2x + 2x^2,\ T(x^3) = 3x^2 + 3x^3$$

Hence

$$[T]_B = \begin{bmatrix} 0 & 1 & 0 & 0 \\ 0 & 1 & 2 & 0 \\ 0 & 0 & 2 & 3 \\ 0 & 0 & 0 & 3 \end{bmatrix}$$

The characteristic polynomial of T is then $f_T(t) = t(t-1)(t-2)(t-3)$. Hence the eigenvalues are $\lambda = 0, 1, 2, 3$.

Since the characteristic polynomial of a linear map is the same as the characteristic polynomial of its matrix representation with respect to a basis we may consider only the characteristic polynomial of a matrix as we do in the following theorem.

Theorem 6.7.8. *Let $A \in M(n, F)$.*

(1) Then the characteristic polynomials $f_A(t)$ is of degree n with the leading coefficient $(-1)^n$. There are at most n distinct eigenvalues of A.

(2) If $f_A(t) = (-1)^n t^n + a_{n-1}t^{n-1} + \cdots + a_1 t + a_0$ then

$$\det(A) = a_0 \quad \text{and} \quad \mathrm{Tr}\,(A) = (-1)^{n-1}a_{n-1} \tag{6.112}$$

Proof. Since

$$f_A(t) = \det \begin{bmatrix} a_{11} - t & a_{12} & \cdots & \\ a_{21} & a_{22} - t & & \\ & & \ddots & \\ & & & a_{nn} - t \end{bmatrix} \tag{6.113}$$

the highest degree term in t in the expansion of the determinant is obtained from the multiplication of the daigonal:

$$\begin{aligned}&(a_{11}-t)(a_{22}-t)\cdots(a_{nn}-t)\\&=(-1)^n t^n+(-1)^{n-1}(a_{11}+a_{22}+\cdots+a_{nn})t^{n-1}+\cdots\end{aligned}\tag{6.114}$$

We already observed that $a_0=\det(A)$ in Theorem 5.1.10.

To see $\mathrm{Tr}\,(A)=(-1)^{n-1}a_{n-1}$ we only have to show that the terms containing t^{n-1} in the determinant expansion of (6.113) comes only from (6.114). In fact, let $B=A-tI_n$. Then by Theorem 3.3.16,

$$\det B=\sum \mathrm{sign}(\sigma)B_{1\sigma(1)}B_{2\sigma(2)}\cdots B_{n\sigma(n)}\tag{6.115}$$

In the term $B_{1\sigma(1)}B_{2\sigma(2)}\cdots B_{n\sigma(n)}$, $B_{i\sigma(i)}$ contains t only if $i=\sigma(i)$. If i is the first index such that $i\neq\sigma(i)$ then there is $j>i$ such that $j\neq\sigma(j)$ since σ is a permutation of $\{1,2...,n\}$ i.e., if there is a factor $B_{i\sigma(i)}$ not containing t then there is another factor not containing t. Hence the terms in the determinant expansion other than the term of (6.114) contains the diagonals of B at most $n-2$ times. And hence the degree of such terms in t is $\leq n-2$ as required.

The fact that the number of distinct eigenvalues does not exceed n follows from the general fact that a polynomial of degree n has at most n distinct roots (Theorem C.1.4). □

If $A\in M(n,F)$ then the characteristic polynomial $f_A(t)$ has its coefficients in F. By Theorem C.1.5, we can always find a bigger field K containing F such that $f(t)$ splits. For example, if $f_A(t)\in\mathbb{Q}[t]$ then f_A has all the roots in $\mathbb{C}$ (Theorem C.2.1). Next we show that similar matrices have the same characteristic polynomial.

Corollary 6.7.9. *Let $F\subseteq K$ be fields and $A,B\in M(n,F)$.*

(1) If there is an invertible $P\in M(n,K)$ such that $B=P^{-1}AP\in M(n,K)$ i.e., A and B are similar over K then $f_B(t)=f_A(t)$. In particular, $\det(A)=\det(B)$ and $\mathrm{Tr}\,(A)=\mathrm{Tr}\,(B)$.

(2) If $f(t)=(-1)^n\prod_{i=1}^n(t-\lambda_i)$ is the characteristic polynomial of A with $\lambda_i\in K$ then

$$\mathrm{Tr}\,(A)=\sum_{i=1}^n\lambda_i \quad\text{and}\quad \det(A)=\prod_{i=1}^n\lambda_i\tag{6.116}$$

Proof. (1) Since $B-tI=P^{-1}AP-tI=P^{-1}(A-tI)P$ we have $f_B(t)=\det(B-tI)=\det P^{-1}(A-tI)P=\det P^{-1}\det(A-tI)\det P=\det(A-tI)=f_A(t)$. Hence $\det(A)=\det(B)$ and $\mathrm{Tr}\,(A)=\mathrm{Tr}\,(B)$ by Theorem 6.7.8.

(2) Let $f(t)=(-1)^n(t^n+a_{n-1}t^{n-1}+\cdots+a_1t+a_0)$ be the characteristic polynomial of A. Since $f(t)=(-1)^n\prod_{i=1}^n(t-\lambda_i)$ also we get $-\mathrm{Tr}\,(A)=a_{n-1}=-(\sum\lambda_i)$ and $\det(A)=a_0=\prod\lambda_i$ by Theorem 6.7.8. □

Example 6.7.10. *Let $A=\begin{bmatrix}1&2\\-1&1\end{bmatrix}\in M(2,\mathbb{R})$ and let $P=\begin{bmatrix}i&-2\\1&-i\end{bmatrix}\in M(2,\mathbb{C})$. Then $B:=P^{-1}AP=\frac{1}{3}\begin{bmatrix}3-4i&2\\-1&3+4i\end{bmatrix}$. Then $f_A(t)=t^2-2t+3$ and $f_B(t)=(t-(1+\sqrt{2}i))(t-(1-\sqrt{2}i))=t^2-2t+3$. If we let $\lambda_1=1+\sqrt{2}i$, $\lambda_2=1-\sqrt{2}i$ then $\det(A)=3$ and $\det(B)=\lambda_1\lambda_2=3$. Also $\mathrm{Tr}\,A=2$ and $\mathrm{Tr}\,B=-(\lambda_1+\lambda_2)=2$.*

Example 6.7.11. *In* Example 5.1.19, *we considered* $T_\theta : \mathbb{R}^2 \to \mathbb{R}^2$ *which is the matrix linear map for the matrix* $A_\theta = \begin{bmatrix} \cos\theta & -\sin\theta \\ \sin\theta & \cos\theta \end{bmatrix}$. *The characteristic polynomial* $f(t) = t^2 - (2\cos\theta)t+1$ *of* A_θ *has roots* $t = \cos\theta \pm \sqrt{\cos^2\theta - 1} = \cos\theta \pm i\sin\theta = e^{\pm i\theta}$ (See Appendix B3). *Hence* $f(t) = (t - e^{i\theta})(t - e^{-i\theta})$. *Therefore* T_θ *has complex eigenvalues* $\lambda_1 = e^{i\theta}, \lambda_2 = e^{-i\theta}$. *Hence* $\operatorname{Tr} A_\theta = \lambda_1 + \lambda_2 = e^{i\theta} + e^{-i\theta} = 2\cos\theta$ *and* $\det A_\theta = \lambda_1\lambda_2 = e^{i\theta}e^{-i\theta} = 1$.

To find an eigenvector of a linear map we first find eigenvectors of its matrix representation and translate it in terms of eigenvectors of the linear map. Here is an example:

Example 6.7.12. *In* Example 6.7.7, *we let* $V = P_3(\mathbb{R})$ *and* $T : V \to V$ *be given by* $T(f(x)) = (x+1)f'(x)$. *Let* $\mathcal{B} = \{1, x, x^2, x^3\}$ *be the standard basis of* V. *We got*

$$[T]_B = \begin{bmatrix} 0 & 1 & 0 & 0 \\ 0 & 1 & 2 & 0 \\ 0 & 0 & 2 & 3 \\ 0 & 0 & 0 & 3 \end{bmatrix}$$

The characteristic polynomial of T *is then* $f_T(t) = t(t-1)(t-2)(t-3)$ *and hence the eigenvalues are* $\lambda = 0, 1, 2, 3$.

Let's find an eigenvector for the eigenvalue $\lambda = 2$. *First we find an eigenvector for the matrix* $A = [T]_{\mathcal{B}}$: *We need to find the null space of* $N(A - 2I)$:

$$A - 2I = \begin{bmatrix} -1 & 1 & 0 & 0 \\ 0 & -1 & 2 & 0 \\ 0 & 0 & 0 & 3 \\ 0 & 0 & 0 & 1 \end{bmatrix}$$

One finds that $\mathbf{a} = \begin{bmatrix} 2 \\ 2 \\ 1 \\ 0 \end{bmatrix}$ *is a solution for* $(A - 2I)\mathbf{x} = \mathbf{0}$. *That is the* $\mathcal{B}$*-coordinate of an eigenvector for* $\lambda = 2$ *is* $\mathbf{a}$. *Hence the corresponding eigenvector for* $\lambda = 2$ *is* $\mathbf{v} = 2 + 2x + x^2$.

Example 6.7.13. *Let* $V = \mathcal{C}^\infty(\mathbb{R})$ *be the vector space of infinitely differentiable functions on* $\mathbb{R}$. *E.g, polynomials,* $\sin x$, $\cos x$, e^x *belong to* V. *Let* $T \in L(V)$ *be defined by* $T(f) = f'$. *What are the eigenvectors for* T? *We know that if* $f_\lambda(t) = e^{\lambda t}$ *then* $T(f_\lambda(t)) = \lambda f_\lambda(t)$ *for each* $\lambda \in \mathbb{R}$. *Hence* $f_\lambda(t)$ *is an eigenvector of* T *for the eigenvalue* $\lambda \in \mathbb{R}$.

Practice Problem 6.7.1. *Let* V *be the dual of* $P_2(\mathbb{R})$. *Choose a basis* $\mathcal{B}' = \{e_0, e_1, e_2\}$, *the evaluations at* $0, 1, 2$, *of* V (Example 6.5.6). *Define a linear map* $T : V \to V$ *by* $T(e_i) = e_{i+1}$ $(i = 0, 1, 2)$. *Find* $[T]_{\mathcal{B}'}$ *and the characteristic polynomial of the linear operator* T.

The following theorem is the same as Theorem 5.1.6 which was stated in terms of real square matrices whereas the following theorem is stated in terms of linear operators of a vector space. Since the proof of the following theorem is the same as the proof of Theorem 5.1.6 we omit the proof.

Theorem 6.7.14. *Let* T *be a linear operator on a vector space* V. *Suppose* T *has* k *distinct eigenvalues* $\lambda_1, \lambda_2, ..., \lambda_k$ *with the eigenvectors* $\mathbf{v}_1, \mathbf{v}_2, ..., \mathbf{v}_k$. *Then* $\mathbf{v}_1, \mathbf{v}_2, \ldots, \mathbf{v}_k$ *are linearly independent.*

Corollary 6.7.15. *Let T be a linear operator on a vector space V of dimension n. If T has n distinct eigenvalues then T is diagonalizable.*

Proof. Any n linear independent vectors in n dimensional vector space is a basis. □

Example 6.7.16. *Consider $T_\theta : \mathbb{R}^2 \to \mathbb{R}^2$ of* Example 6.7.11 *which is the matrix linear map for $A_\theta = \begin{bmatrix} \cos\theta & -\sin\theta \\ \sin\theta & \cos\theta \end{bmatrix}$. The characteristic polynomial $f(t) = t^2 - (2\cos\theta)t + 1$ of A_θ has real roots ± 1 when $\cos\theta = \pm 1$ i.e., $\theta = 0$ or π. In these cases, $T_\theta = \pm I_2$. Hence other than the values $\theta = 0, \pi$ the linear map T_θ is not diagonalizable over $\mathbb{R}$.*

If we view $T_\theta^{\mathbb{C}} : \mathbb{C}^2 \to \mathbb{C}^2$ as a linear operator on $V = \mathbb{C}^2$ over $F = \mathbb{C}$ then $f(t)$ $f(t) = (t - e^{i\theta})(t - e^{-i\theta})$ (Example 6.7.11). *Hence T_θ has eigenvalues $\lambda_1 = e^{i\theta}, \lambda_2 = e^{-i\theta}$. The corresponding eigenvectors are $\mathbf{v}_1 = \begin{bmatrix} 1 \\ -i \end{bmatrix}, \mathbf{v}_2 = \begin{bmatrix} 1 \\ i \end{bmatrix}$ which form a basis $\mathcal{B} = \{\mathbf{v}_1, \mathbf{v}_2\}$ of $V = \mathbb{C}^2$ and $[T_\theta]_{\mathcal{B}} = \begin{bmatrix} e^{i\theta} & 0 \\ 0 & e^{-i\theta} \end{bmatrix}$* (Exercise 6). *Thus T_θ is diagonalizable for all θ over $\mathbb{C}$.*

We defined the eigenspace for the real matrices (Definition 5.1.5). Now we define the eigenspaces for the linear operators in the same manner.

Definition 6.7.17. *Let $T \in L(V)$ and λ be an eigenvalue of T. The* **eigenspace** *for λ is*

$$E_\lambda = \{\mathbf{v} \in V \mid T(\mathbf{v}) = \lambda\mathbf{v}\}$$

Observe that $E_\lambda = N(T - \lambda I)$ is a subspace since it is a null space of the linear map $T - \lambda I$ and $E_\lambda = \{\text{eigenvectors for } \lambda\} \cup \{\mathbf{0}\}$. As before we have

$$E_\lambda \cap E_\mu = (0) \qquad \text{if } \lambda \neq \mu \tag{6.117}$$

We recall its proof: If $\mathbf{v} \in E_\lambda \cap E_\mu$ then $\lambda\mathbf{v} = T(\mathbf{v}) = \mu\mathbf{v}$. Hence $(\lambda - \mu)\mathbf{v} = 0$; $\mathbf{v} = 0$.

The first part of the following result is Corollary 5.1.8 in terms of linear maps.

Corollary 6.7.18. *Let T be a linear operator on V and $\lambda_1, ..., \lambda_k$ are distinct eigenvalues of T and $E_{\lambda_1}, E_{\lambda_2}, ..., E_{\lambda_k}$ be the corresponding eigenspace.*

(1) *If $\mathbf{v}_i \in E_{\lambda_i}$ and $\mathbf{v}_1 + \mathbf{v}_2 + \cdots + \mathbf{v}_k = \mathbf{0}$ then $\mathbf{v}_i = \mathbf{0}$ for all i. Further we have*

$$E_{\lambda_i} \cap \sum_{j<i} E_{\lambda_j} = \{\mathbf{0}\} \tag{6.118}$$

(2) *If S_i is a linearly independent subset of E_{λ_i} then $\cup_i S_i$ is a linearly independent set.*

Proof. (1) The same proof of Corollary 5.1.8 works in this case.
(2) Exercise 4. □

We recall Definition 5.1.9: A polynomial $f(t) \in F[t]$ is said to be **split over** F if $f(t) = c\prod_{i=1}^{n}(t - a_i)$ with $a_i \in F$ i.e., all the roots of f belong F. If $f(t) = (t - \lambda)^e g(t)$ and $g(\lambda) \neq 0$ then we say that the **multiplicity of λ in $f(t)$ is** e.

Theorem 6.7.19. *Let T be a linear operator on a vector space V over F. If T is diagonalizable then the characteristic polynomial $f(t)$ of T splits over F.*

Proof. Let $\mathcal{B}$ be a basis of V for which $[T]_{\mathcal{B}} = \text{diag}(\lambda_1, \lambda_2, \ldots, \lambda_n)$. Then

$$\begin{aligned} f(t) &= \det([T]_{\mathcal{B}} - tI) \\ &= \det(\text{diag}[\lambda_1 - t, \lambda_2 - t, \ldots, \lambda_n - t]) \\ &= \prod_{i=1}^{n} (\lambda_i - t) \end{aligned}$$

Hence $f(t)$ splits. □

The converse is false as we will see shortly (Example 6.7.21). And not every linear operator is diagonalizable (Example 5.1.19). We want to find conditions for a linear map to be diagonalizable.

The following theorem is Theorem 5.1.21 in terms of linear operators instead of matrices. For a proof we simply represent T as a matrix with respect to a basis to reduce the proof to the proof of Theorem 5.1.21

Theorem 6.7.20. *Let λ be an eigenvalue of $T \in L(V)$ with multiplicity e. Then*

$$1 \le \dim E_\lambda \le e$$

Proof. Let $\{\mathbf{v}_1, \mathbf{v}_2, \ldots, \mathbf{v}_p\} \subseteq E_\lambda$ be a basis. Extends this to a basis $\mathcal{B} = \{\mathbf{v}_1, \mathbf{v}_2, \ldots, \mathbf{v}_n\}$ for V. Then

$$A := [T]_{\mathcal{B}} = \left[\begin{array}{cccc|c} \lambda & & & & \\ & \lambda & & & * \\ & & \ddots & & * \\ & & & \lambda & \\ \hline 0 & 0 & \cdots & 0 & \\ & & \cdots & & C \\ 0 & 0 & \cdots & 0 & \end{array}\right] \tag{6.119}$$

where the left upper corner is $\text{diag}[\lambda, \ldots, \lambda]$ of size p. Hence the characteristic polynomial is $\det(A - tI) = (\lambda - t)^p \det(C - tI)$ (Theorem 3.3.17). On the other hand, $f_T(t) = (t - \lambda)^e h(t)$ with $(t - \lambda) \nmid h(t)$ since λ has multiplicity e. And hence $(t - \lambda)^p | (t - \lambda)^e h(t)$ and $(t - \lambda) \nmid h(t)$. Thus $e \ge p = \dim E_\lambda$. □

Example 6.7.21. *Let $T : P_2(\mathbb{R}) \to P_2(\mathbb{R})$ be defined by $T(f(x)) = f'(x)$. Let's see if T is diagonalizable. To find the characteristic polynomial choose a basis $\mathcal{B} = \{1, x, x^2\}$ and the matrix representation of T with respect to $\mathcal{B}$ is*

$$[T]_{\mathcal{B}} = \begin{bmatrix} 0 & 1 & 0 \\ 0 & 0 & 2 \\ 0 & 0 & 0 \end{bmatrix}$$

Hence the characteristic polynomial is $f(t) = -t^3$ and the eigenvalue is $\lambda = 0$. The eigenvectors for $\lambda = 0$ are the nonzero vectors of $N([T]_{\mathcal{B}})$. There is only one linearly independent vector $\mathbf{v} = \begin{bmatrix} 1 \\ 0 \\ 0 \end{bmatrix}$ in $N([T]_{\mathcal{B}})$ which corresponds to the constant $1 \in P_2(\mathbb{R})$. Hence there is no basis consisting of eigenvectors of T and therefore T is not diagonalizable.

Example 6.7.22. *Let $A \in M(n, \mathbb{R})$ and $T = T_A$ be a linear operator on $\mathbb{R}^n$. Then we can also view T as a linear map on $V = \mathbb{C}^n$, as in* Example 6.7.16 *in which case we write*

$$T^{\mathbb{C}} : \mathbb{C}^n \to \mathbb{C}^n$$

For an eigenvalue $\lambda \in \mathbb{R}$, we write

$$E_\lambda = N(T - \lambda) \subseteq \mathbb{R}^n, \ E_{\lambda,\mathbb{C}} = N(T^{\mathbb{C}} - \lambda) \subseteq \mathbb{C}^n$$

By Example 6.4.7, *we have*

$$\dim_{\mathbb{R}} E_\lambda = \dim_{\mathbb{C}} E_{\lambda,\mathbb{C}} \tag{6.120}$$

If $\lambda \in \mathbb{C}$ is a nonreal eigenvalue of A then, by Theorem C.2.3, *$\bar{\lambda}$ is also an eigenvalue. By* Theorem 6.3.28, *$\nu(A - \lambda) = \nu(\overline{A - \lambda}) = \nu(A - \bar{\lambda})$. Hence we have*

$$\dim_{\mathbb{C}} E_\lambda = \dim_{\mathbb{C}} E_{\bar{\lambda}} \tag{6.121}$$

Practice Problem 6.7.2. *Let $T : \mathbb{R}^3 \to \mathbb{R}^3$ be defined by $T(\mathbf{v}) = \begin{bmatrix} a+b+c \\ b+2c \\ 2b+4c \end{bmatrix}$ where $\mathbf{v} = \begin{bmatrix} a \\ b \\ c \end{bmatrix}$. Decide whether T is diagonalizable. If T is diagonalizable find a basis consisting of eigenvectors.*

The following result is Theorem 5.1.22 in terms of linear operators instead of matrices.

Theorem 6.7.23. *Let $T \in L(V)$ with $\dim V = n$. Assume the characteristic polynomial of T splits say, $f(t) = (-1)^n \prod_{i=1}^{k} (t - \lambda_i)^{e_i}$. Then T is diagonalizable if and only if $\dim E_{\lambda_i} = e_i$.*

Proof. The proof of Theorem 5.1.22 works in this case with necessary changes. □

Corollary 6.7.24. *Let $T \in L(V)$ and $\mathcal{B}$ be a basis for V. Let $A := [T]_{\mathcal{B}}$. Then T is diagonalizable if and only if A is diagonalizable.*

Proof. Exercise 8. □

We sum up the conditions for diagonalizability of a linear operator in the following corollary.

Corollary 6.7.25. *Let T be a linear operator on a vector space V of dimension n with the characteristic polynomial $f(t) = (-1)^n \prod_{i=1}^{k} (t - \lambda_i)^{e_i}$. The following conditions are equivalent.*

(1) *The linear operator T is diagonalizable.*

(2) *There is a basis consisting of eigenvectors of T.*

(3) *There is a basis $\mathcal{B}_i$ of E_{λ_i} such that $\mathcal{B} = \mathcal{B}_1 \cup \mathcal{B}_2 \cup \cdots \cup \mathcal{B}_k$ is a basis of V.*

(4) $\sum_{i=1}^{k} \dim E_{\lambda_i} = n$

(5) $\dim E_{\lambda_i} = e_i \ (i = 1, 2, ..., k)$

(6) $V = E_{\lambda_1} \oplus E_{\lambda_2} \oplus \cdots \oplus E_{\lambda_k}$

Proof. (1) $\Leftrightarrow$ (2) is the definition of diagonalizability.

(2) $\Rightarrow$ (3) Let $\mathcal{B}$ be a basis consisting of eigenvectors and let $\mathcal{B}_i = \mathcal{B} \cap E_{\lambda_i}$. Then $\mathcal{B} = \cup_i \mathcal{B}_i$ since each $\mathbf{v} \in \mathcal{B}$ belongs to one of E_{λ_i}. Since the union is disjoint (by (6.117)) we have $\sum \#\mathcal{B}_i = \#\mathcal{B} = n$. On the other hand, $\#\mathcal{B}_i \leq e_i$ (Theorem 6.7.20) and $\sum e_i = n$ since $f(t) = \pm \prod_{i=1}^{k}(t-\lambda_i)^{e_i}$ is of degree n. Hence $\#\mathcal{B}_i = e_i$. Since $\mathcal{B}_i$ consists of e_i linearly independent vectors in E_{λ_i} and $\dim E_{\lambda_i} \leq e_i$ we conclude that $\mathcal{B}_i$ is a basis of E_{λ_i}.

(3) $\Rightarrow$ (4) $n = \dim V \overset{(3)}{=} \sum \#\mathcal{B}_i = \sum \dim E_{\lambda_i}$

(4) $\Rightarrow$ (5) $\dim E_{\lambda_i} \leq e_i$ and both add up to n. Hence $\dim E_{\lambda_i} = e_i$.

(5) $\Rightarrow$ (1) Theorem 6.7.23.

(3) $\Leftrightarrow$ (6) Theorem 6.2.36 (5). $\square$

Answers for Practice Problems 6.7

(6.7.1) To find the matrix representation of T compute: $T(e_0) = e_1, T(e_1) = e_2$. For $T(e_2)$ let $T(e_2) = e_3 = ae_0 + be_1 + ce_2$. We need to find a, b, c. Evaluating at $1, x, x^2$ we get

$$\begin{cases} 1 &= a+b+c \\ 3 &= \quad b+2c \\ 9 &= \quad b+4c \end{cases}$$

Solving this we get $e_3 = 4e_0 - 3e_1 + 3e_2$. Hence

$$[T]_{\mathcal{B}'} = \begin{bmatrix} 0 & 0 & 4 \\ 1 & 0 & -3 \\ 0 & 1 & 3 \end{bmatrix}$$

Thus the characteristic polynomial is $f_T(t) = \det([T]_{\mathcal{B}'} - tI) = -t^3 + 3t^2 - 3t + 4$.

(6.7.2) To find the characteristic polynomial let $\mathcal{E} = \{\mathbf{e}^1, \mathbf{e}^2, \mathbf{e}^3\}$ be the standard basis. Then

$$A = [T]_{\mathcal{E}} = \begin{bmatrix} 1 & 1 & 1 \\ 0 & 1 & 2 \\ 0 & 2 & 4 \end{bmatrix}$$

Hence the characteristic polynomial is $f_T(t)) = \det([T]_{\mathcal{E}} - tI) = -t(t-1)(t-5)$. Therefore, the eigenvalues are $\lambda_1 = 0, \lambda_2 = 1, \lambda_3 = 5$ with the multiplicities 1. Since T has three distinct eigenvalues, T is diagonalizable.

The eigenvalues and the corresponding eigenvectors are given by

$\lambda_1 = 0$	$\lambda_2 = 1$	$\lambda_3 = 5$
$\mathbf{v}_1 = \begin{bmatrix} 1 \\ -2 \\ 1 \end{bmatrix}$	$\mathbf{v}_2 = \begin{bmatrix} 1 \\ 0 \\ 0 \end{bmatrix}$	$\mathbf{v}_3 = \begin{bmatrix} 3 \\ 4 \\ 8 \end{bmatrix}$

Also if we let $P = \begin{bmatrix} 1 & 1 & 3 \\ -2 & 0 & 4 \\ 1 & 0 & 8 \end{bmatrix}$ then $P^{-1}AP = \operatorname{diag}[0, 1, 5]$.

6.7 Exercises

1. Let $V = M(2,\mathbb{C})$. For $\begin{bmatrix} a & b \\ c & d \end{bmatrix} \in V$ define $T \in L(V)$ by $T(A) = \begin{bmatrix} d & a \\ b & c \end{bmatrix}$. Decide whether T is diagonalizable or not. If so find a basis consisting of eigenvectors.

2. Prove that a triangular matrix has at least one eigenvector.

3. Prove that $\begin{bmatrix} 0 & -1 \\ 1 & 0 \end{bmatrix} \in M(2,\mathbb{R})$ is not similar to any real 2×2 triangular matrix. (Hint: Use Exercise 2.)

4. Let $T \in L(V)$ and $\{\lambda_1, \lambda_2, \dots, \lambda_k\}$ be its distinct eigenvalues. Prove that if S_i be linearly independent vectors in E_{λ_i} then $S = S_1 \cup S_2 \cup \cdots \cup S_k$ is linearly independent.

5. Prove the true statement(s) and find a counter example for the false statement(s).

   (a) Every linear operator has an eigenvalue.

   (b) For each eigenvalue of a linear operator there is at least one eigenvector.

   (c) If 0 is an eigenvalue of a linear operator T then T is not an isomorphism.

   (d) If 0 is not an eigenvalue of a linear operator T then T is an isomorphism.

   (e) A symmetric upper triangular matrix is a diagonal matrix.

   (f) If $T \in L(V)$ with $\dim V = n$. If T has two distinct eigenvalues $\{\lambda_1, \lambda_2\}$ and $\dim E_{\lambda_1} = n-1$ then T is diagonalizable.

   (g) Let $A \in M(n,F)$ and $P \in M(n,F)$ be an invertible matrix. If $P^{-1}AP$ is a diagonal matrix then the columns of P are the eigenvectors of A.

   (h) Let $A, B \in M(n)$ be similar matrices. Then A is daigonalizable if and only if so is B.

   (i) If $A, B \in M(n)$ are similar then there is unique P satisfying $B = P^{-1}AP$.

6. Carry out the detailed computations to find the eigenvectors of T_θ in Example 6.7.16.

7. Let $A_\theta = \begin{bmatrix} \cos\theta & -\sin\theta \\ \sin\theta & \cos\theta \end{bmatrix}$. Prove that $A_\theta^n = A_{n\theta}$ for $n \in \mathbb{Z}$. (Hint: Use diagonalization over $\mathbb{C}$.)

8. Let T be a linear operator on a vector space V over F. Prove that T is diagonalizable if and only if the matrix $[T]_{\mathcal{B}}$ is diagonalizable for any basis $\mathcal{B}$ of V.

9. Let T be a linear operator on a vector space V over F.

   (a) Prove that if λ is an eigenvalue of T then λ^k is an eigenvalue of T^k.

   (b) Prove that if $f(t) \in F[t]$ and λ is an eigenvalue of T then $f(\lambda)$ is an eigenvalue of $f(T)$.

   (c) Prove that if T is invertible and λ is an eigenvalue then λ^{-1} is an eigenvalue of T^{-1}.

10. Suppose that the characteristic polynomial of a linear operator $T \in L(V)$ splits. Assume that there is a basis $\mathcal{B}$ of V such that $[T]_\mathcal{B}$ is lower (or upper) triangular. (We will show that this is in fact true in Corollary 7.2.14.)

    (a) Prove that the diagonal entries of $[T]_\mathcal{B}$ are the eigenvalues $\lambda_1, \lambda_2, ..., \lambda_n$ of T.

    (b) Prove that $\det(T) = \lambda_1\lambda_2\cdots\lambda_n$ and $\mathrm{Tr}\,(T) = \lambda_1 + \lambda_2 + \cdots + \lambda_n$.

    (c) Prove that $\mathrm{Tr}\,(T^k) = \lambda_1^k + \lambda_2^k + \cdots + \lambda_n^k$.

    (d) Prove that if $f(t)$ is a polynomial then the diagonal entries of the matrix representation $[f(T)]_\mathcal{B}$ are $f(\lambda_1), f(\lambda_2), ..., f(\lambda_n)$ which are the eigenvalues of $f(T)$.

11. Let T be a nilpotent linear operator on a vector space V of dimension n.

    (a) Prove that if T is diagonalizable then T is the zero linear operator.

    (b) Prove that the characteristic polynomial of T is $f(t) = (-1)^n t^n$. (Hint: Exercise 6.4.19)

    (c) Prove that $\det(T) = 0$ and $\mathrm{Tr}\,(T) = 0$ (Exercise 6.4.20).

12. Let T be a linear operator on a vector space V. Let λ^2 be an eigenvalue of T^2 for some $\lambda \in F$. Prove that one of λ or $-\lambda$ is an eigenvalues of T.

13. Let T be a linear operator on V and W be an invariant subspace of T i.e., $T(W) \subseteq W$. Let $\mathbf{v}_1, \ldots, \mathbf{v}_k$ be eigenvectors of T corresponding to the distinct eigenvalues $\lambda_1, \ldots, \lambda_k$. Prove that if $\mathbf{v}_1 + \cdots + \mathbf{v}_k \in W$ then $\mathbf{v}_i \in W$ for each i. (Hint: Use induction on k.)

14. Let T be a diagonalizable linear operator of V and let W be an invariant subspace i.e., $T(W) \subseteq W$. Prove that the restriction T_W of T on W is also diagonalizable.

15. Let $S, T \in L(V)$. We say S and T are **simultaneously diagonalizable** if there is a basis $\mathcal{B}$ of V such that both $[S]_\mathcal{B}$ and $[T]_\mathcal{B}$ are diagonal. Prove that if S and T are simultaneously diagonalizable then $ST = TS$. (For the converse of this see Exercise 6.8.18)

16. Let $A \in M(n, F)$ and let $f_A(t) \in F[t]$ be the characteristic polynomial of A. Let $c \in F$ be a nonzero constant.

    (a) Express the characteristic polynomial $f_{cA}(t)$ of cA in terms of c and $f_A(t)$. (See Exercise 5.1.9)

    (b) Prove that λ is an eigenvalue for A if and only if $c\lambda$ is an eigenvalue of cA.

    (c) Prove that $\mathbf{v}$ is an eigenvector for A with the eigenvalue λ if and only if $\mathbf{v}$ is an eigenvector for cA with the eigenvalue $c\lambda$.

17. Let $A, B \in M(n, \mathbb{C})$. Prove that if B is invertible then there are fintely many $t \in \mathbb{C}$ for which $A + tB$ is not invertible. (This result is in accordance with Remark 3.3.4.) Find an example of 2×2 matrices A, B such that $A+tB$ is nonsingular for all t. (Hint: Choose a singular B.)

18. Let $V = M(n, \mathbb{R})$ and let $T : V \to V$ be the linear map defined by $T(A) = A^t$. Use Exercise 6.2.8 to show that T has the eigenvalues $1, -1$. Show that T is diagonalizable and find a basis $\mathcal{B}$ of V such that $[T]_\mathcal{B}$ is diagonal with 1's and -1's along the diagonal. Show that the characteristic polynomial of T is $f_T(t) = (-1)^n(t-1)^{\frac{n(n+1)}{2}}(t+1)^{\frac{n(n-1)}{2}}$.

19. Prove that the $n \times n$ matrices

$$A = \begin{bmatrix} 1 & 1 & \cdots & 1 \\ 1 & 1 & \cdots & 1 \\ & \cdots & \cdots & \\ 1 & 1 & \cdots & 1 \end{bmatrix}, \quad B = \begin{bmatrix} n & 0 & \cdots & 0 \\ 0 & 0 & \cdots & 0 \\ & \cdots & \cdots & \\ 0 & 0 & \cdots & 0 \end{bmatrix}$$

are similar. Show that the characteristic polynomial of A is $f(t) = (-1)^n t^{n-1}(t-n)$.

20. Let λ, μ be eigenvalues of $A \in M(n, F)$, $B \in M(m, F)$ respectively.

(a) Prove that $\lambda\mu$ is an eigenvalue of $A \otimes B$.

(b) Prove that $\lambda + \mu$ is an eigenvalue of $A \otimes I_m + I_n \otimes B$.

6.8 Invariant Subspaces and Cayley-Hamilton Theorem.

Definition 6.8.1. *Let V be a vector space and $T \in L(V)$. A subspace W of V is said to be T-**invariant** subspace (or **invariant** under T) if $T(W) \subseteq W$. If W is a T-invariant subspace then the restriction T_W of T to W is a linear map*

$$T_W : W \to W$$

If W is T-invariant and $f(t)$ is the characteristic polynomial of T then we often write the characteristic polynomial of T_W by $f_W(t)$.

If $V = W_1 \oplus W_2$ and W_1, W_2 are T-invariant then $T_{W_i} : W_i \to W_i$ $(i = 1, 2)$ are linear maps. For $\mathbf{v} \in V$ write $\mathbf{v} = \mathbf{w}_1 + \mathbf{w}_2$ $(\mathbf{w}_i \in W_i)$. We define the **direct sum** $T_{W_1} \oplus T_{W_2}$ by

$$(T_{W_1} \oplus T_{W_2})(\mathbf{v}) = T_{W_1}(\mathbf{w}_1) + T_{W_2}(\mathbf{w}_2)$$

Hence we can write $T = T_{W_1} \oplus T_{W_2}$.

Example 6.8.2. (1) *Let $T : \mathbb{R}^2 \to \mathbb{R}^2$ be given by $T(\mathbf{x}) = \begin{bmatrix} x \\ x+y \end{bmatrix}$ where $\mathbf{x} = \begin{bmatrix} x \\ y \end{bmatrix}$. Let*

$$W_1 = \left\{ \begin{bmatrix} x \\ 0 \end{bmatrix} \middle| \, x \in \mathbb{R} \right\}, \; W_2 = \left\{ \begin{bmatrix} x \\ x \end{bmatrix} \middle| \, x \in \mathbb{R} \right\}, \; W_3 = \left\{ \begin{bmatrix} 0 \\ y \end{bmatrix} \middle| \, y \in \mathbb{R} \right\}$$

Then W_1 and W_2 are not T-invariant but W_3 is T-invariant.

(2) *For any $T \in L(V)$, the trivial subspace $\{\mathbf{0}\}$ and V itself are always T-invariant.*

(3) *Let $T_\theta : \mathbb{R}^2 \to \mathbb{R}^2$* (Example 6.7.16) *be the linear map given by rotation by the angle θ. The only T_θ invariant subspace the trivial subspace and the whole space. Because any one dimensional subspace* Span $\{\mathbf{v}\}$ *is invariant under T_θ only if $T_\theta(\mathbf{v}) = \lambda\mathbf{v}$ for some λ. But there is no such nonzero $\mathbf{v}$.*

(4) *If $T \in L(V)$ then the null space $N(T)$ and the range space $R(T)$ are T-invariant subspaces.*

(5) *If λ is an eigenvalue of $T \in L(V)$ then eigenspace E_λ is a T-invariant subspace. More generally, $N\left((T-\lambda)^k\right)$ is T-invariant for each positive integer k.* (Check this.)

Theorem 6.8.3. *Let $V = W_1 \oplus W_2$ and $T : V \to V$ be a linear operator. Let $\mathcal{B}_1, \mathcal{B}_2$ be bases for W_1, W_2.*

(1) If W_1 is T-invariant then the matrix representation of T with respect to $\mathcal{B} = \mathcal{B}_1 \cup \mathcal{B}_2$ is of the form

$$[T]_{\mathcal{B}} = \left[\begin{array}{c|c} [T_{W_1}]_{\mathcal{B}_1} & * \\ \hline 0 & * \end{array}\right] \tag{6.122}$$

(2) If W_1 and W_2 are T-invariant then $T = T_{W_1} \oplus T_{W_2}$ and the matrix representation of T with respect to $\mathcal{B}$ is of the form

$$[T]_{\mathcal{B}} = \left[\begin{array}{c|c} [T_{W_1}]_{\mathcal{B}_1} & 0 \\ \hline 0 & [T_{W_2}]_{\mathcal{B}_2} \end{array}\right] \tag{6.123}$$

Further if $f_1(t), f_2(t)$ are the characteristic polynomial of T_{W_1}, T_{W_2} then the characteristic polynomial of T is $f_1(t)f_2(t)$.

Proof. Exercise 4. □

Let A, B be square matrices. We define the **direct sum** of A and B by

$$A \oplus B = \begin{bmatrix} A & 0 \\ 0 & B \end{bmatrix} \tag{6.124}$$

so that $T_{A \oplus B} = T_A \oplus T_B$. In this notation, (6.123) becomes

$$[T]_{\mathcal{B}} = [T_{W_1}]_{\mathcal{B}_1} \oplus [T_{W_2}]_{\mathcal{B}_2}$$

Next we introduce a concept which will be important, especially, for Chapter 7.

Definition 6.8.4. *Let $T : V \to V$ be a linear map and let $\mathbf{v} \in V$ be a nonzero vector. Let*

$$C_T(\mathbf{v}) := \operatorname{Span}\{\mathbf{v}, T(\mathbf{v}), T^2(\mathbf{v}), \dots\}$$

*which we call the T-**cyclic subspace spanned (generated) by $\mathbf{v}$**. A subspace of the form $W = C_T(\mathbf{v})$ is called a T-**cyclic subspace**. If $T = T_A$ for a square matrix A then we will write $C_A(\mathbf{v})$ for $C_T(\mathbf{v})$. If the reference to T is clear then we will simply write $C(\mathbf{v})$ instead of $C_T(\mathbf{v})$. If $T = T_A$ then we will write $C_T(\mathbf{v}) = C_A(\mathbf{v})$.*

Note that $C_T(\mathbf{v})$ *is T-invariant* and we have (Exercise 9)

$$C_T(\mathbf{v}) = \{f(T)(\mathbf{v}) \mid f(t) \in F[t]\} \tag{6.125}$$

Example 6.8.5. *Define $T : \mathbb{R}^3 \to \mathbb{R}^3$ by $T(\mathbf{x}) = \begin{bmatrix} x \\ x+y \\ x+y+z \end{bmatrix}$ where $\mathbf{x} = \begin{bmatrix} x \\ y \\ z \end{bmatrix}$. Let $\mathbf{v} = \begin{bmatrix} 0 \\ 1 \\ 0 \end{bmatrix}$. Then*

$$\mathbf{v} = \begin{bmatrix} 0 \\ 1 \\ 0 \end{bmatrix}, \; T(\mathbf{v}) = \begin{bmatrix} 0 \\ 1 \\ 1 \end{bmatrix}, \; T^2(\mathbf{v}) = \begin{bmatrix} 0 \\ 1 \\ 2 \end{bmatrix}, \; T^3(\mathbf{v}) = \begin{bmatrix} 0 \\ 1 \\ 3 \end{bmatrix}, \; \cdots$$

Hence

$$C_T(\mathbf{v}) = \text{Span}\left\{\mathbf{v}, T(\mathbf{v}), T^2(\mathbf{v}), \dots\right\} = \left\{ \begin{bmatrix} 0 \\ y \\ z \end{bmatrix} \middle| \; y, z \in \mathbb{R} \right\}$$

Example 6.8.6. *Let $V = P_2(\mathbb{R})$ be the vector space of real polynomials of degree ≤ 2 and let $\mathcal{B} = \{1, x, x^2\}$ be the standard basis of V. Consider $T : V \to V$ defined by $T(f(x)) = xf(1) - f(2x+1)$ for $f(x) \in V$. Let $\mathbf{v} = 1$ and consider $C(\mathbf{v})$. Now*

$$T(1) = x - 1, \; T^2(1) = -2x, \; T^3(1) = 2x + 2, ...$$

Hence

$$C(\mathbf{v}) = \text{Span}\{1, -2x, 2x+2, ...\} = \text{Span}\{1, x\}$$

and $\dim C_T(\mathbf{v}) = 2$.

Let $f(t), g(t) \in F[t]$. We say $f(t)$ is **divisible** by $g(t)$ if $f(t) = g(t)h(t)$ for some $h(t) \in F[t]$ and, in this case, we write

$$g(t) \mid f(t)$$

Theorem 6.8.7. *Let $T : V \to V$ be a linear operator and W be a T-invariant subspace. Let T_W be the restriction of T on W then the characteristic polynomial $f(t)$ of T is divisible by the characteristic polynomial $f_W(t)$ of T_W namely*

$$f_W(t) \mid f(t)$$

Proof. Choose a basis $\mathcal{B} = \{\mathbf{v}_1, \mathbf{v}_2, \dots, \mathbf{v}_k, \mathbf{v}_{k+1}, \dots, \mathbf{v}_n\}$ of V so that $\mathcal{C} = \{\mathbf{v}_1, \mathbf{v}_2, \dots, \mathbf{v}_k\}$ is a basis of W. Let $[T_W]_\mathcal{C} = B_1 \in M(k, F)$. Then the characteristic polynomial of T_W is $f_W(t) = \det(B_1 - tI)$ and

$$[T]_\mathcal{B} = \begin{bmatrix} B_1 & B_2 \\ 0 & B_3 \end{bmatrix}$$

for some matrices B_2, B_3. Hence

$$\begin{aligned} f(t) &= \det([T]_\mathcal{B} - tI_n) \\ &= \det \begin{bmatrix} B_1 - tI & B_2 \\ 0 & B_3 - tI \end{bmatrix} \\ &= \det(B_1 - tI)\det(B_3 - tI) \qquad \text{(by Theorem 3.3.17)} \\ &= f_W(t) \cdot \det(B_3 - tI) \end{aligned}$$

Therefore $f_W(t) \mid f(t)$. □

Example 6.8.8. *Let $V = P_4(\mathbb{R})$ and $T : V \to V$ be a linear operator defined by*

$$T(f(x)) = (x^2 + 1)f''(x)$$

Let $W = \text{Span}\{1, x^2, x^4\}$. It is easy to show that W is T-invariant.

Let's find the characteristic polynomial $f_W(t)$ of T_W and $f(t)$. Choose a basis $\mathcal{B}_1 = \{1, x^2, x^4\}$ for W and extend it to a basis $\mathcal{B} = \{1, x^2, x^4, x, x^3\}$ of V. The matrix of T_W with respect to $\mathcal{B}_1$ is

$$[T_W]_{\mathcal{B}_1} = \begin{bmatrix} 0 & 2 & 0 \\ 0 & 2 & 12 \\ 0 & 0 & 12 \end{bmatrix}$$

The matrix $[T]_{\mathcal{B}}$ of T with respect to $\mathcal{B}$ is

$$[T]_{\mathcal{B}} = \begin{bmatrix} 0 & 2 & 0 & 0 & 0 \\ 0 & 2 & 12 & 0 & 0 \\ 0 & 0 & 12 & 0 & 0 \\ 0 & 0 & 0 & 0 & 6 \\ 0 & 0 & 0 & 0 & 6 \end{bmatrix}$$

The characteristic polynomial of T_W is $f_W(t) = -t(t-2)(t-12)$. And the characteristic polynomial of T is $f(t) = -t^2(t-2)(t-6)(t-12)$. Hence $f_W(t)|f(t)$.

Theorem 6.8.9. *Let $T : V \to V$ be a linear operator and W_i $(i = 1, 2, ..., k)$ be T-invariant subspaces of V such that*

$$V = W_1 \oplus W_2 \oplus \cdots \oplus W_k$$

Let T_i be the restriction of T on W_i then $T = \oplus_i T_i$. Let $\mathcal{B}$ be a basis of V and $\mathcal{B}' = \cup_i \mathcal{B}_i$ where $\mathcal{B}_i$ is a basis of W_i. Then $[T]_{\mathcal{B}}$ is similar to $[T]_{\mathcal{B}'} = \oplus_i [T]_{\mathcal{B}_i}$.

Let $f(t)$ be the characteristic polynomial of T, and $f_i(t)$ be the characteristic polynomial of T_i. Then

$$f(t) = f_1(t) f_2(t) \cdots f_k(t)$$

Proof. The equality $T = \oplus_i T_i$ is obvious. And $[T]_{\mathcal{B}}$ is similar to $[T]_{\mathcal{B}'} = \oplus_i [T]_{\mathcal{B}_i}$ because they are two matrix representations of the same linear map in two different bases.

For the characteristic polynomial it will suffice to consider the case $k = 2$. Let $\mathcal{B}_1, \mathcal{B}_2$ be bases of W_1, W_2. Then $\mathcal{B} = \mathcal{B}_1 \cup \mathcal{B}_2$ is a basis of V (Theorem 6.2.36) and

$$[T]_{\mathcal{B}} = \begin{bmatrix} [T_1]_{\mathcal{B}_1} & 0 \\ 0 & [T_2]_{\mathcal{B}_2} \end{bmatrix} \tag{6.126}$$

Then we have

$$\begin{aligned} f(t) &= \det([T]_{\mathcal{B}} - tI) \\ &= \det\left(\begin{bmatrix} [T_1]_{\mathcal{B}_1} & 0 \\ 0 & [T_2]_{\mathcal{B}_2} \end{bmatrix} - tI \right) \\ &= \det \begin{bmatrix} [T_1]_{\mathcal{B}_1} - tI & 0 \\ 0 & [T_2]_{\mathcal{B}_2} - tI \end{bmatrix} \\ &= \det([T_1]_{\mathcal{B}_1} - tI) \det([T_2]_{\mathcal{B}_2} - tI) \\ &= f_1(t) f_2(t) \end{aligned}$$

□

Theorem 6.8.10. *Let $T : V \to V$ be a linear operator and $\mathbf{v} \in V$ be a nonzero vector. Let $W = C_T(\mathbf{v})$ with $\dim W = k$ and $f_W(t)$ be the characteristic polynomial of T_W.*

The set of vectors

$$\mathcal{B}_{\mathbf{v}} := \{\mathbf{v}, T(\mathbf{v}), \ldots, T^{k-1}(\mathbf{v})\} \tag{6.127}$$

forms a basis of W.

If $a_0\mathbf{v} + a_1T(\mathbf{v}) + a_2T^2(\mathbf{v}) + \cdots + T^k(\mathbf{v}) = \mathbf{0}$ *then the matrix representation of* T_W *in the basis* $\mathcal{B}_{\mathbf{v}}$ *is the companion matrix of* $a_0 + a_1t + \cdots + a_{k-1}t^{k-1} + t^k$*:*

$$[T_W]_{\mathcal{B}_{\mathbf{v}}} = \begin{bmatrix} 0 & 0 & 0 & \cdots & 0 & -a_0 \\ 1 & 0 & 0 & & & -a_1 \\ 0 & 1 & 0 & & & -a_2 \\ \vdots & & \ddots & \ddots & & \\ 0 & & 0 & 1 & 0 & -a_{k-2} \\ 0 & \cdots & 0 & 0 & 1 & -a_{k-1} \end{bmatrix} \tag{6.128}$$

The characteristic polynomial of T_W *is given by*

$$f_W(t) = (-1)^k(a_0 + a_1t + \cdots + a_{k-1}t^{k-1} + t^k)$$

Proof. To show $\mathcal{B}_{\mathbf{v}}$ is a basis we only need to show that $S := \{\mathbf{v}, T(\mathbf{v}), \ldots, T^{k-1}(\mathbf{v})\}$ spans W since $\dim W = k$ (Corollary 6.2.16). For this we need to show that $T^{k+i}(\mathbf{v}) \in S$ for $i \geq 0$. Now we have $T^k(\mathbf{v}) = -\sum_{i=0}^{k-1} a_iT^i(\mathbf{v}) \in \text{Span}(S)$. To use induction assume $T^{k+i-1}(\mathbf{v}) = \sum_{i=0}^{k-1} c_iT^i(\mathbf{v}) \in \text{Span}(S)$. Then

$$T^{k+i}(\mathbf{v}) = T\left(T^{k+i-1}(\mathbf{v})\right) = T\left(\sum_{i=0}^{k-1} c_iT^i(\mathbf{v})\right) = \sum_{i=0}^{k-2} c_iT^{i+1}(\mathbf{v}) + c_{k-1}T^k(\mathbf{v})$$

The first sum in the last equality and the second term both belong to Span (S). Therefore we see that S span W.

It is immediate that the matrix representation of T_W in the basis $\mathcal{B}_{\mathbf{v}}$ is the companion matrix (6.128). Now by Theorem 3.4.24, we have

$$f_W(t) = \det([T_W]_{\mathcal{B}} - tI) = (-1)^k(t^k + a_{k-1}t^{k-1} + \cdots + a_1t + a_0)$$ □

Definition 6.8.11. *The basis of* $C_T(\mathbf{v})$ *of the form* (6.127) *is called a* T**-cyclic basis** *generated by* $\mathbf{v}$ *and we will denote it by* $\mathcal{B}_{\mathbf{v}}$ *:*

$$\mathcal{B}_{\mathbf{v}} = \{\mathbf{v}, T(\mathbf{v}), \ldots, T^{k-1}(\mathbf{v})\}$$

When $W = V$ and $T = T_A$ for some $A \in M(n, F)$ we obtain

Corollary 6.8.12. *Let* $A \in M(n, F)$ *with the characteristic polynomial* $f_A(t)$*. Assume there is* $\mathbf{v} \in F^n$ *such that* $C_A(\mathbf{v}) = F^n$*. Then* A *is similar to the companion matrix of* $(-1)^n f_A(t)$*.*

Proof. By Theorem 6.8.10, $V := F^n$ has a cyclic basis $\mathcal{B}_{\mathbf{v}}$. The matrix representation of T_A with respect to the cyclic basis is the companion matrix C of $(-1)^n f_A(t)$. Hence A and C are similar by Theorem 6.4.26. □

Example 6.8.13. *In* Example 6.8.5, *the cyclic basis of* $C_T(\mathbf{v})$ *is* $\mathcal{B}_{\mathbf{v}} = \{\mathbf{v}, T(\mathbf{v})\}$ *since* $\dim C_T(\mathbf{v}) = 2$*. Since* $T^2(\mathbf{v}) = -\mathbf{v} + 2T(\mathbf{v})$ *we have* $[T_W]_{\mathcal{B}_{\mathbf{v}}} = \begin{bmatrix} 0 & 1 \\ 1 & -2 \end{bmatrix}$.

The cyclic basis of $C_T(\mathbf{v})$ *in* Example 6.8.6 *is* $\mathcal{B}_{\mathbf{v}} = \{\mathbf{v}, T(\mathbf{v})\} = \{1, x - 1\}$ *since* $\dim C_T(\mathbf{v}) = 2$*. Since* $T^2(\mathbf{v}) = -2x = 2\mathbf{v} - 2T(\mathbf{v})$ *we have* $[T_W]_{\mathcal{B}_{\mathbf{v}}} = \begin{bmatrix} 0 & -2 \\ 1 & 2 \end{bmatrix}$.

Practice Problem 6.8.1. *Let $V = M(2, F)$ and $T : V \to V$ be defined by $T(A) = A^t$. Let $\mathbf{v} = \begin{bmatrix} 1 & 1 \\ 0 & 0 \end{bmatrix}$. Find $W = C_T(\mathbf{v})$ and $\mathcal{B}_{\mathbf{v}}$. Find the characteristic polynomial $f_W(t)$ of T_W and $[T_W]_{\mathcal{B}_{\mathbf{v}}}$.*

Recall (Definition 6.4.21) that if T is a linear operator on a vector space V and $f(t) = \sum a_i x^i \in F[t]$ is a polynomial then $f(T) = \sum a_i T^i$ is also a linear operator on V.

Theorem 6.8.14 (Cayley-Hamilton Theorem). *Let $T : V \to V$ be a linear operator with the characteristic polynomial $f_T(t)$. Then $f_T(T)$ is the zero linear operator on V.*

Proof. Let $\mathbf{v} \in V$. We have to prove that $f_T(T)(\mathbf{v}) = 0$. If $\mathbf{v} = 0$ then there is nothing to prove. So assume $\mathbf{v}$ is nonzero and let $W = C_T(\mathbf{v})$ with $\dim W = k$. If

$$T^k(\mathbf{v}) = -(a_{k-1}T^{k-1}(\mathbf{v}) + \cdots + a_1 T(\mathbf{v}) + a_0 \mathbf{v}) \tag{6.129}$$

then by Theorem 6.8.10

$$g(t) = (-1)^k(t^k + a_{k-1}t^{k-1} + \cdots + a_1 t + a_0) \tag{6.130}$$

is the characteristic polynomial of T_W. The equality (6.129) implies $g(T)(\mathbf{v}) = \mathbf{0}$. By Theorem 6.8.7, $f_T(t) = h(t)g(t)$ for some $h(t) \in F[t]$. Hence $f_T(T)(\mathbf{v}) = h(T)g(T)(\mathbf{v}) = h(T)(g(T)(\mathbf{v})) = \mathbf{0}$. □

Corollary 6.8.15. *Let $A \in M(n, F)$ and $f_A(t)$ be the characteristic polynomial of A. Then $f_A(A) = 0$, the zero matrix in $M(n, F)$.*

Proof. Since $f_A(t) = f_{T_A}(t)$ we have, by Theorem 6.8.14 and Exercise 6.4.7, $0 = f_A(T_A) = T_{f_A(A)}$. Hence $f_A(A) = 0$. □

If the characteristic polynomial of a matrix $A \in M(n, F)$ is $(-1)^n t^n$ then, by Cayley-Hamilton Theorem, $A^n = 0$. By Exercise 6.7.11, we see that the converse is also true.

Example 6.8.16. *Let* $A = \begin{bmatrix} 1 & 0 & 1 \\ 0 & 2 & 1 \\ 1 & 1 & 1 \end{bmatrix}$. *Let us illustrate* Corollary 6.8.15. *The characteristic polynomial of A is $f_A(t) = -t^3 + 4t^2 - 3t - 1$. We have*

$$A^2 = \begin{bmatrix} 2 & 1 & 2 \\ 1 & 5 & 3 \\ 2 & 3 & 3 \end{bmatrix}, \quad A^3 = \begin{bmatrix} 4 & 4 & 5 \\ 4 & 13 & 9 \\ 5 & 9 & 8 \end{bmatrix}$$

Now it is easy to check that $-A^3 + 4A^2 - 3A - I = 0$.

Practice Problem 6.8.2. *With* $A = \begin{bmatrix} 1 & 0 & 1 \\ 0 & 0 & 1 \\ 1 & 1 & 0 \end{bmatrix}$ *illustrate* Corollary 6.8.15.

Answers for Practice Problems 6.8

(6.8.1) We have $\mathbf{v} = \begin{bmatrix} 1 & 1 \\ 0 & 0 \end{bmatrix}$, $T(\mathbf{v}) = \begin{bmatrix} 1 & 0 \\ 1 & 0 \end{bmatrix}$, $T^2(\mathbf{v}) = \mathbf{v}$ and so on. Hence $T^2(\mathbf{v}) - \mathbf{v} = \mathbf{0}$ and hence the characteristic polynomial of T_W is $f_W(t) = t^2 - 1$.

And $C_T(\mathbf{v}) = \left\{ \begin{bmatrix} a+b & a \\ b & 0 \end{bmatrix} \middle| a, b \in F \right\}$ and we can choose $\mathcal{B}_{\mathbf{v}} = \left\{ \begin{bmatrix} 1 & 1 \\ 0 & 0 \end{bmatrix}, \begin{bmatrix} 1 & 0 \\ 1 & 0 \end{bmatrix} \right\}$ as a basis of W. Since $T^2(\mathbf{v}) = \mathbf{v}$ we have $[T_W]_{\mathcal{B}_{\mathbf{v}}} = \begin{bmatrix} 0 & 1 \\ 1 & 0 \end{bmatrix}$.

(6.8.2) The characteristic polynomial is $f(t) = -t^3 + t^2 + 2t - 1$. And $A^2 = \begin{bmatrix} 2 & 1 & 1 \\ 1 & 1 & 0 \\ 1 & 0 & 2 \end{bmatrix}$, $A^3 = \begin{bmatrix} 3 & 1 & 3 \\ 1 & 0 & 2 \\ 3 & 2 & 1 \end{bmatrix}$. Hence $-A^3 + A^2 + 2A - I_3 = 0$.

6.8 Exercises

1. Let $T : V \to V$ is a linear operator. Let W_1, W_2 be T-invariant subspaces. Prove that $W_1 + W_2$ and $W_1 \cap W_2$ are also T-invariant subspaces.

2. Let T be a linear operator on a vector space V and let $f(t) \in F[t]$. Prove that $R(f(T))$ and $N(f(T))$ are T-invariant subspaces.

3. Let $T : V \to V$ be a linear operator and W_1 be a T-invariant subspace. Is it true that there is a T-invariant subspace W_2 such that $V = W_1 \oplus W_2$? (Hint: Consider $V = \mathbb{R}^2$ and $T = T_A$ where $A = \begin{bmatrix} 1 & 1 \\ 0 & 1 \end{bmatrix}$. Then $W_1 = \text{Span}\{\mathbf{e}^1\}$ is a T-invariant subspace. Is there a T-invariant subspace W_2 with $V = W_1 \oplus W_2$?)

4. Prove Theorem 6.8.3.

5. For each of the following linear operator T on the vector space V determine whether the given subspace W is T-invariant.

   (a) $V = P_3(\mathbb{R}), T(p(t)) = tp'(t)$ and $W = P_2(\mathbb{R})$ where $p'(t)$ denotes the derivative

   (b) $V = \mathbb{R}^3, T(a, b, c) = \begin{bmatrix} a-b \\ c \\ b-c \end{bmatrix}$ and $W = \left\{ \begin{bmatrix} x \\ 0 \\ -x \end{bmatrix} \middle| x \in \mathbb{R} \right\}$

   (c) $V = M(2, \mathbb{R}), T\left(\begin{bmatrix} a & b \\ c & d \end{bmatrix}\right) = \begin{bmatrix} d & -c \\ -b & a \end{bmatrix}$ and $W = \{A \in M(2, \mathbb{R}) \mid \text{Tr}(A) = 0\}$

   (d) $V = P_3(\mathbb{R}), T(p(t)) = p'(t)$ and $W = \left\{ p(t) \;\middle|\; \int_{-1}^{1} p(t)dt = 0 \right\}$

   (e) $V = M(3, \mathbb{R}), T\left(\begin{bmatrix} a & b \\ c & d \end{bmatrix}\right) = \begin{bmatrix} a & c \\ b & d \end{bmatrix}$ and $W = \{A \mid A = A^t\}$

6. For each linear operator T on a vector space V find a basis for the T-cyclic subspace generated by $\mathbf{v}$.

   (a) $V = \mathbb{R}^3, T(a, b, c) = \begin{bmatrix} a \\ -a \\ b \end{bmatrix}$ and $\mathbf{v} = \begin{bmatrix} 1 \\ 1 \\ 0 \end{bmatrix}$

(b) $V = \mathbb{R}^3, T(a,b,c) = \begin{bmatrix} 0 \\ a-b \\ c \end{bmatrix}$ and $\mathbf{v} = \begin{bmatrix} 1 \\ 0 \\ 0 \end{bmatrix}$

(c) $V = P_3(\mathbb{R}), T(p(t)) = p'(t)$ and $\mathbf{v} = t^2$

(d) $V = P_3(\mathbb{R}), T(p(t)) = t^2 p''(t)$ and $\mathbf{v} = t^3$

(e) $V = M(2,\mathbb{R}), T\left(\begin{bmatrix} a & b \\ c & d \end{bmatrix}\right) = \begin{bmatrix} a & a+b \\ 0 & -b \end{bmatrix}$ and $\mathbf{v} = \begin{bmatrix} 1 & 0 \\ 0 & 0 \end{bmatrix}$

7. For each linear operator T and cyclic subspaces $W = C_T(\mathbf{v})$ of Exercise 6 find the characteristic polynomial of T and T_W.

8. Prove the true statement(s) and find a counter example for the false statement(s).

   (a) Every linear operator on a nontrivial vector space V has a nonzero proper T-invariant subspace.

   (b) Let T is a linear operator on V and $\mathbf{v} \in V$. If $c \in F$ is a nonzero constant then $C_T(\mathbf{v}) = C_T(c\mathbf{v})$.

   (c) Let T is a linear operator on V and $\mathbf{v} \in V$. Then $C_T(\mathbf{v}) = C_T(T(\mathbf{v}))$.

   (d) If $\mathbf{w} \in C_T(\mathbf{v})$ then $C_T(\mathbf{w}) \subseteq C_T(\mathbf{v})$.

   (e) If $S = \{\mathbf{v}, T(\mathbf{v}), ..., T^{n-1}(\mathbf{v})\}$ is a basis of $C(\mathbf{v})$ then $\{T(\mathbf{v}), T^2(\mathbf{v}), ..., T^n(\mathbf{v})\}$ is also a basis.

   (f) Let T is a linear operator on V with $\dim(V) = n$. Then there is a polynomial $f(t)$ of degree n for which $f(T)$ is the zero linear operator on V.

   (g) There is $A \in M(3,\mathbb{R})$ whose characteristic polynomial is $f_A(t) = t^3 + t + 1$.

9. Let T be a linear operator on V and $\mathbf{v} \in V$.

   (a) Prove that if $\mathbf{v}$ is an eigenvector of T then $C_T(\mathbf{v}) = \text{Span}\{\mathbf{v}\}$.

   (b) Prove that $C_T(\mathbf{v})$ is T-invariant and $C_T(\mathbf{v}) = \{f(T)(\mathbf{v}) \mid f(t) \in F[t]\}$.

   (c) Prove that if W is a T-invariant subspace and $\mathbf{v} \in W$ then $C_T(\mathbf{v}) \subseteq W$ i.e., $C_T(\mathbf{v})$ is the smallest T-invariant subspace containing $\mathbf{v}$.

10. Let X be a subset of V. Define

$$G_T(X) = \left\{ \sum f_i(T)\mathbf{x}_i \;\middle|\; f_i \in F[t], \mathbf{x}_i \in X \right\}$$

    Hence if $X = \{\mathbf{v}\}$ then $C_T(\mathbf{v}) = G_T(X)$.

   (a) Prove that $G_T(X)$ is the smallest T-invariant subspace containing X.

   (b) Let $X = \{\mathbf{v}_1, \mathbf{v}_2, ..., \mathbf{v}_k\}$ and $W = \text{Span}(X)$. Prove that $G_T(X) = G_T(W)$.

11. Let T be a linear operator on V and $\mathbf{v} \in V$. Let $\dim C_T(\mathbf{v}) = k$ and $f, g \in F[t]$ with $\deg(f) < k$ and $\deg(g) < k$. Prove that if $f(T)(\mathbf{v}) = g(T)(\mathbf{v})$ then $f = g$.

12. Let $A \in M(n,F)$ and let $S = \{I_n, A, A^2, ...\} \subseteq M(n,F)$. Prove that $\dim \text{Span}(S) \leq n$.

13. Let T be a linear operator on a vector space V. Suppose the characteristic polynomial $f_T(t)$ of T splits. Prove that if W is a T-invariant subspace then W contains an eigenvector of T.

14. Let $T : V \to V$ be a linear operator. Let $V = W_1 \oplus W_2$ with T-invariant subspaces W_1, W_2 and let $T_i = T_{W_i}$ be the restriction of T on W_i $(i = 1, 2)$. Prove that

$$\nu(T) = \nu(T_1) + \nu(T_2) \text{ and } \rho(T) = \rho(T_1) + \rho(T_2)$$

15. Let T be a linear operator on V and W a T-invariant subspace of V. Let $\mathbf{v}_1, \ldots, \mathbf{v}_k$ be eigenvectors of T corresponding to distinct eigenvalues $\lambda_1, \ldots, \lambda_k$. Prove that if $\mathbf{v}_1 + \cdots + \mathbf{v}_k \in W$ then $\mathbf{v}_i \in W$ for each i.

16. Let $T \in L(V)$ with the characteristic polynomial $f_T(t) = (-1)^n(t - \lambda)^n$. Prove that there is a basis $\mathcal{B}$ such that $[T]_\mathcal{B}$ is upper triangular with λ along the diagonal.

17. Let T be a diagonalizable linear operator of V and let W be an invariant subspace. Prove that the restriction T_W of T on W is also diagonalizable.

18. Let S, T be diagonalizable linear operators on V such that $ST = TS$. Prove that S and T are simultaneously diagonalizable (cf. Exercise 6.6.15), namely there is a basis $\mathcal{B}$ such that both $[S]_\mathcal{B}$ and $[T]_\mathcal{B}$ are diagonal. (Hint: First show that if $\lambda_1, \lambda_2, ..., \lambda_k$ are the eigenvalues of S then E_{λ_i} is T-invariant. The restriction $T_{E_{\lambda_i}}$ of T on E_{λ_i} is also diagonalizable.)

19. Let V be a vector space of dimension n. Let T be a linear operator on V with n distinct eigenvalues. Prove that V is a cyclic space i.e., there is $\mathbf{v} \in V$ such that $C_T(\mathbf{v}) = V$. (Hint: If $\mathbf{v}_1, \mathbf{v}_2, ..., \mathbf{v}_n$ are the eigenvectors corresponding to the distinct eigenvalues then try $\mathbf{v} = \mathbf{v}_1 + \mathbf{v}_2 + \cdots \mathbf{v}_n$.)

20. Let T be a linear operator on a vector space V of dimension n. Let V be T-cyclic space with the characteristic polynomial $f(t) = \sum_{i=1}^n a_i t^i$. Prove that

$$\rho(T) = \begin{cases} n-1 & (\text{if } a_0 = 0) \\ n & (\text{if } a_0 \neq 0) \end{cases}$$

Find an example for each case.

Chapter 7

Canonical Forms

Let $A, B \in M(n, F)$. If they are similar say, $B = P^{-1}AP$ for some invertible matrix P then the matrices A and B represent the same linear operator on $V = F^n$: A is the matrix representation of T_A in the standard basis and B is the matrix representation of T_A in the basis consisting of the columns of P (Corollary 6.4.28). Consider the 'similarity class' of A (Theorem 6.4.29)

$$[A] := \{B \in M(n, F) \mid B \text{ is similar to } A\}$$

Since the similarity relation is an equivalence relation two similarity classes are either identical or disjoint (Theorem A.2.6). The similarity class of A consists of all matrix representations in all possible bases of the linear operator T_A on V. We want to choose a representative of $[A]$. To decide whether two matrices A, B are similar or not we need to see if there is P such that $B = P^{-1}AP$. For this we need to consider a big linear equation which is not easy to solve. However, if we can find such a representative then two matrices are similar if and only if their similarity class have the same representative.

If T_A is diagonalizable then there are good reasons to choose the diagonalization of A as a representative or a 'canonical form' of $[A]$. But this is not always the case as we saw before. What is then the 'best possible' form of a representative in the similarity class $[A]$? In this chapter, we will show there is some standard or canonical form that can be considered to be a representative in the similarity class $[A]$.

There are two canonical forms we will consider in this chapter. The first one is the Jordan canonical form which is close to a diagonal matrix for which we assume the characteristic polynomial splits. The second one is the rational canonical form in which the characteristic polynomial need not split.

In general, Jordan canonical form is easier to find and has wider applications. Even if the characteristic polynomial does not split often we can draw conclusions about the matrix A by considering a bigger field in which the characteristic polynomial split.

We proceed with a linear operator T on a vector space V instead of a matrix A and we look for a basis $\mathcal{B}$ in which the matrix $[T]_{\mathcal{B}}$ is a canonical form. The proof of the existence of such a canonical form is a straight forward application of a general theory in algebra – classification of finitely generated modules over a principal ideal domain. We adapt the development of the general theory to our situation. The main ingredient of the proof will be the the characteristic polynomial of $f_T(t)$ and the 'minimal polynomial'.

To find the canonical form of a linear operator T on a vector space V we will consider the vector space V together with a fixed linear operator T as a single object.

Throughout this chapter all vector spaces are of finite dimensional over a field F.

7.1 Polynomials and Minimal Polynomial

7.1.1 Ring of Polynomials

We recall the definition of polynomials (Definition 6.1.5) and prove basic properties which we will use extensively in this chapter. One may skip the proofs of this subsection in the first reading. See also Appendix C.

Let F be a field. The set of polynomials with coefficients in F is called the **ring of polynomials** over F and will be denoted by $F[t]$:

$$F[t] = \{a_0 + a_1t + a_2t^2 + \cdots + a_nt^n \mid a_i \in F\}$$

We define the **degree** of the polynomial $f(t) = a_0 + a_1t + a_2t^2 + \cdots + a_nt^n$ to be n if $n \geq 0$ and $a_n \neq 0$. And we define the degree of the zero polynomial to be $-\infty$.

If $f, g \in F[t]$ then it is easy to show

$$\deg(fg) = \deg(f) + \deg(g) \tag{7.1}$$

Algebraic properties of the polynomials $F[t]$ over a field F and the integers $\mathbb{Z}$ are very similar and they share many – almost all – properties in common. One of the basic properties of the integers is that any set S of integers there is an element of S having the smallest absolute value, whereas $\mathbb{Q}$ has no such property. The absolute value for $\mathbb{Z}$ will be replaced by the degree of the polynomials in $F[t]$. And the corresponding property of the polynomials is then for any set S of polynomials there is a polynomial of the least degree in S.

The integers of absolute value 1 which are ± 1 corresponds to the polynomials of degree 0 which are the nonzero constant polynomials. Both ± 1 of $\mathbb{Z}$ and the nonzero constants of $F[t]$ are the elements having the multiplicative inverses, so called the 'units'.

From (7.1), we see that if $fg = c$ for some nonzero constant then f and g are also nonzero constants. In fact, $\deg(fg) = \deg(f) + \deg(g) = 0$. Hence $\deg(f) = \deg(g) = 0$. Whence f and g are constants.

Another useful fact is that if and $f|g$ and $\deg f = \deg g$ then $f = cg$ for some $0 \neq c \in F$. In fact, $g = fh$ for some $h \in F[t]$. Since $\deg f = \deg g$ we have $\deg h = 0$ i.e., h is a constant.

An integer p with $|p| > 1$ is called a **prime number** if p is not a product of two integers having absolute values $< |p|$. The corresponding notion for the polynomials is irreducibility. We briefly summarize the properties of the polynomials we need in the following. The reader should find the corresponding statements for the integers.

Definition 7.1.1. *A polynomial is said to be* **monic** *if the coefficient of the highest degree term is* 1. *A nonconstant polynomial* $p(t) \in F[t]$ *is said to be* **irreducible** *if* $p(t)$ *is not a product of two nonconstant polynomials i.e., if* $p = ab$ *where* $a, b \in F[t]$ *then* a *or* b *is a constant. For* $f, g \in F[t]$ *we say* f *is* **divisible by** g *if there is an* $h \in F[t]$ *such that* $f = gh$*. In this case, we will write* $g|f$*. (We use the same notation for the integers.)*

If $m, n \in \mathbb{Z}$ and $n|m$ as well as $m|n$ then we have $m = \pm n$. For the polynomials $f, g \in F[t]$ if $f|g$ and $g|f$ then we conclude that $f = cg$ for some constant $c \in F$. In fact, $f = hg$ and $g = h'f$. Hence $f = hg = hh'f$ and hence $hh' = 1$. As noted before, h and h' are constants.

A polynomial in $F[t]$ of degree 1 is irreducible for any field F. By Fundamental Theorem of Algebra (Theorem C.2.1), the only irreducible polynomials in $\mathbb{C}[t]$ are of degree 1. The irreducible polynomials in $\mathbb{R}[t]$ are of degree 1 and quadratic polynomials having no real roots (Theorem C.2.3). There are infinitely many irreducible polynomials in $\mathbb{Q}[t]$ in all degrees even though deciding whether a given polynomial is irreducible or not is 'very' difficult in general.

Sometimes we will assume a polynomial to be monic which must be nonzero for convenience.

Theorem 7.1.2. (1) **(Division Algorithm)** *Let $f, g \in F[t]$. Then there are unique polynomials $s, r \in F[t]$ such that*

$$f = sg + r \quad \text{with } \deg(r) < \deg(g)$$

We may have $r = 0$ in which case $\deg(r) = -\infty$. And $r \in F[t]$ is nonzero if and only if $\deg(r) \geq 0$.

(2) **(Unique Factorization of Polynomials)** *Every monic polynomial $f(t) \in F[t]$ can be written uniquely as $f(t) = p_1(t)^{e_1} p_2(t)^{e_2} \cdots p_k(t)^{e_k}$ where $p_i(t)$ are monic irreducible polynomials in $F[t]$.*

Proof. See Appendix for the proofs (Theorem C.1.3). □

Definition 7.1.3. *Let $f_1, f_2, ..., f_r \in F[t]$. A subset of $F[t]$ of the form*

$$\mathcal{I} := \left\{ \sum_{i=1}^{r} a_i f_i \;\middle|\; a_i \in F[t] \right\}$$

is called an **ideal** *(generated by $f_1, f_2, ..., f_r$). When we need to specify the generators we denote the ideal $\mathcal{I}$ by $(f_1, f_2, ..., f_r)$.*

A polynomial h is a **greatest common divisor**(gcd) *of $f_1, f_2, ..., f_r$ if $h|f_i$ for all i and, if h' also divides all f_i then $h'|h$. We denote a greatest common divisor by $\gcd(f_1, f_2, ..., f_r)$. If $\gcd(f, g) = 1$ (i.e., f, g has no nonconstant common factor) then we say f and g are* **relatively prime**.

If c is nonzero constant then $(c) = F[t]$. On the other hand, if $c = 0$ then (0) consists only of the zero element of $F[t]$ 2which we call the zero ideal. If $f, g \in F[t]$ are nonconstant polynomials then (f) consists of all multiples of f in $F[t]$. And the ideal (f, g) consists of the polynomials of the form $af + bg$ where $a, b \in F[t]$:

$$(f) = \{af \mid a \in F[t]\}$$
$$(f, g) = \{af + bg \mid a, b \in F[t]\}$$

Practice Problem 7.1.1. *Let $\mathcal{I}$ be an ideal.*

(1) *If $h_1, h_2 \in \mathcal{I}$ then $h_1 + h_2 \in \mathcal{I}$.*

(2) *If $h \in \mathcal{I}$ and $f \in F[t]$ then $fh \in \mathcal{I}$.*

We note that $\gcd(f_1, f_2, ..., f_r)$ is uniquely determined only up to a constant multiple. In fact, if d and d' are greatest common divisors of $f_1, f_2, ..., f_r$ then since $d|f_i$ for all i and d' is a greatest common divisor, we have $d|d'$. Similarly, $d'|d$. Hence d and d' differ by a multiple of a nonzero constant.

Let $f, g \in F[t]$. Then we have

$$(f) \subseteq (g) \text{ if and only if } g|f \tag{7.2}$$
$$(f) = (1) \text{ if and only if } f \text{ is a nonzero constant} \tag{7.3}$$
$$(f) = (g) \text{ if and only if } f = cg \text{ for some constant } c \tag{7.4}$$

Proofs of these statements are Exercise 4.

Let $f(t) = t - \lambda$ and $g(t) = t - \mu$ be the polynomials of degree 1 ($\lambda, \mu \in F$). Then f and g are relatively prime if and only if $\lambda \neq \mu$. (Exercise 6.)

Now consider the polynomials $f(t) = t^3 + 1$ and $g(t) = t^2 - 1$ in $\mathbb{Q}[t]$. Since $f(t) = (t+1)(t^2 - t + 1)$ and $g(t) = (t+1)(t-1)$ we have $\gcd(f, g) = t + 1$. Now

$$\frac{f(t)}{t+1} = t^2 - t + 1, \ \frac{g(t)}{t+1} = t - 1$$

are relatively prime and

$$\frac{f(t)}{t+1} - t\frac{g(t)}{t+1} = t^2 - t + 1 - t(t-1) = 1$$

Therefore

$$f(t) - tg(t) = t + 1 = \gcd(f, g)$$

and hence $(\gcd(f, g)) \subseteq (f, g)$. On the other hand, we have $f = a \gcd(f, g), g = b \gcd(f, g)$ for some $a, b \in F[t]$ by the definition of gcd. Hence $f, g \in (\gcd(f, g))$. And therefore we have $(f, g) \subseteq (\gcd(f, g))$ and hence they are equal.

We will show that this is always the case.

Theorem 7.1.4. *Let $f_1, f_2, ..., f_r$ be polynomials in $F[t]$.*

(1) Let h be a polynomial of the lowest degree in $(f_1, f_2, ..., f_r)$ then $(f_1, f_2, ..., f_r) = (h)$. The polynomial h is uniquely determined up to a multiple of a constant.

(2) We have $d = \gcd(f_1, f_2, ..., f_r)$ if and only if $(d) = (f_1, f_2, ..., f_r)$.

(3) Let $f_i = \prod_j p_j^{e_{ij}}$ where p_j are monic irreducible polynomials and $e_{ij} \geq 0$ and let $k_i = \min_j\{e_{ij}\}$. Then $\gcd(f_1, f_2, ..., f_r) = \prod_{i=1}^r p_i^{k_i}$.

Proof. Not to miss the point of the proof we assume $r = 2$. We leave the proofs for $r > 2$ as exercises for the reader or refer to [1] in Bibliography.

(1) Let h be a polynomial in (f_1, f_2) of the lowest degree. Since $h \in (f_1, f_2)$ it is obvious that $(h) \subseteq (f_1, f_2)$. To show $(f_1, f_2) \subseteq (h)$ let $f \in (f_1, f_2)$. We need to show that $h|f$ by (7.2). By division algorithm, write $f = sh + r$ with $\deg(r) < \deg(h)$. Since f, h are polynomial in (f_1, f_2) we can write $f = a_1 f_1 + a_2 f_2$ and $h = b_1 f_1 + b_2 f_2$. Hence $r = f - sh = (a_1 - sb_1)f_1 + (a_2 - sb_2)f_2$ is an element of (f_1, f_2) of degree $< \deg(h)$. If $r \neq 0$ then this is a contradiction since $r \in (f_1, f_2)$ and $\deg(r) < \deg(h)$; $r = 0$. Thus $f = sh$ and hence $(f_1, f_2) \subseteq (h)$.

For uniqueness suppose $h' \in F[t]$ is another polynomial in (f_1, f_2) of the lowest degree. Then $(h) = (f_1, f_2) = (h')$. As noted before in (7.4), h and h' differs by a multiple of a nonzero constant.

(2) Assume $d = \gcd(f_1, f_2)$ and let $f_1 = h_1 d, f_2 = h_2 d$. Then, for any $a_1, a_2 \in F[t]$, we have $a_1 f_1 + a_2 f_2 = (a_1 h_1 + a_2 h_2)d \in (d)$. Hence $(f_1, f_2) \subseteq (d)$. Now, there is $h \in F[t]$ such that $(f_1, f_2) = (h)$ by (1). Then since $f_1, f_2 \in (h)$ we have $h|f_1, h|f_2$. By definition of gcd, $h|d$ and hence $(h) \supseteq (d)$. Thereby we conclude $(f_1, f_2) = (d)$.

Conversely suppose $(f_1, f_2) = (d)$. Then since $f_1, f_2 \in (d)$ we have $d|f_i$. Suppose $d'|f_i$ $(i = 1, 2)$; $f_i = a_i' d'$. We need to show $d'|d$. Since $d \in (f_1, f_2)$ we can write $d = b_1 f_1 + b_2 f_2$ for some $b_1, b_2 \in F[t]$. Hence $d = b_1 f_1 + b_2 f_2 = a_1' b_1 d' + a_2' b_2 d'$ and hence d is divisible by d'. As required.

(3) Exercise 7. □

For explicit computation of gcd without factoring the polynomials f_i's see Exercise 8.

Corollary 7.1.5. *Let $f, g \in F[t]$. Then f, g are relatively prime if and only if there are $a, b \in F[t]$ such that $af + bg = 1$. In this case, for any polynomial $h \in F[t]$ there are polynomials $c, d \in F[t]$ such that $cf + dg = h$.*

Proof. This is a special case of Theorem 7.1.4 (2). For the second statement multiply both sides of the first equality by h. □

Corollary 7.1.6. *Let $f, g, h \in F[t]$. If $h|fg$ and $(f, h) = 1$ then $h|g$.*

Proof. By Corollary 7.1.5, there are polynomials a, b such that $1 = af + bh$. Multiplying by g we get $g = afg + bhg$ which is divisible by h since $h|fg$ by assumption. □

Corollary 7.1.7. *A polynomial $p \in F[t]$ is irreducible if and only if $p|gh$ implies either $p|g$ or $p|h$.*

Proof. Assume p is irreducible and let $p|gh$ $(g, h \in F[t])$. If $p \nmid g$ then p and g are relatively prime. Hence, by Corollary 7.1.5, there is $a, b \in F[t]$ such that $1 = ap + bg$. Therefore $h = aph + bgh$ is divisible by p since $p|gh$ by assumption.

For the converse, suppose $p = gh$ $(g, h \in F[t])$. We need to show that one of g or h is a constant. Since $p = gh$ we have $p|gh$ and hence $p|g$ or $p|h$ say, $p|g$. If we let $g = pf$ we have then $p = gh = pfh$. Therefore $fh = 1$ namely h is a constant. Therefore p is irreducible. □

7.1.2 Minimal Polynomial

Let V be a vector space over F with a linear operator T. We have an operation corresponding to the scalar multiplication

$$F \times V \to V : \quad (c, \mathbf{v}) \mapsto c\mathbf{v} \tag{7.5}$$

We extend this to an operation, depending on T, to $F[t]$ on V. For $f(t) \in F[t]$ we obtain a linear operator $f(T)$ by replacing t by T. Hence if $\mathbf{v} \in V$ then $f(T)(\mathbf{v}) \in V$. In this way, we obtain an $F[t]$ operation on V:

$$F[t] \times V \to V : \ (f(t), \mathbf{v}) \mapsto f(T)(\mathbf{v}) \tag{7.6}$$

If f is a constant polynomial c then this is just the scalar multiplication (7.5). In this chapter, we consider the operation of $F[t]$ on V given by (7.6).

The vector space V with the operation (7.6) depends on T. Hence we can denote such an object by (V, T) or V_T. With this understanding we will simply write V for (V, T) or V_T.

We may write $f(t)\mathbf{v}$ for $f(T)(\mathbf{v})$. For psychological reason, we will write $f(T)(\mathbf{v})$ instead of $f(t)\mathbf{v}$ since $f(T)(\mathbf{v})$ reminds us of a vector obtained by applying the operator $f(T)$ to $\mathbf{v}$.

For $f, g \in F[t]$ and a linear operator T on V we have (Exercise 6.4.4)

$$f(T)g(T) = (fg)(T) = g(T)f(T) \tag{7.7}$$

Here, $(fg)(T)$ represent the linear operator obtained by replacing t by T to the product $f(t)g(t)$, and $f(T)g(T)$ represent the linear operator obtained by composing the linear maps $g(T)$ and $f(T)$. Hence we have

$$f^n(T) = f(T)^n \tag{7.8}$$

where left hand side of (7.8) represents the linear map obtained by replacing t by T in the polynomial $f(t)^n$ and the right hand side is obtained by taking n-times composition of the linear map $f(T)$.

Let X be a subset of V. Define

$$\mathcal{I}_{X,T} = \{g(t) \in F[t] \mid g(T)(\mathbf{v}) = \mathbf{0} \text{ for all } \mathbf{v} \in X\}$$

Shortly we will show that $\mathcal{I}_X$ is an ideal (Theorem 7.1.10).

If the reference to T is clear then we simply write $\mathcal{I}_X$ for $\mathcal{I}_{X,T}$. If $X = \{\mathbf{v}\}$ consists of a single vector $\mathbf{v}$ then we write $\mathcal{I}_{\mathbf{v}}$ for $\mathcal{I}_X$. If $f_T(t)$ is the characteristic polynomial of T then $f_T \in \mathcal{I}_X$ for any nonempty subset X of V by Cayley-Hamilton Theorem. If $X \subseteq X'$ then it is obvious that

$$\mathcal{I}_{X'} \subseteq \mathcal{I}_X \tag{7.9}$$

Definition 7.1.8. *Let V be a vector space over F with a linear operator T. Let X be a subset of V. A monic polynomial $q_{X,T}(t)$ or $q_X(t)$ of the lowest degree in $\mathcal{I}_X$ is called a* T**-minimal polynomial of** X. *That is $q_X(t)$ is a polynomial of the lowest degree satisfying $q(T)(X) = \mathbf{0}$. If $X = \{\mathbf{v}\}$ then we write $q_{\mathbf{v}}(t)$ for $q_X(t)$.*

If $T = T_A$ for some $A \in M(n, F)$ and $V = F^n$ then a T_A-minimal polynomial q_X of X will be called an A**-minimal polynomial of** X.

If $X = V$ then we call a T-minimal polynomial of V a **minimal polynomial of** T*(resp.* **of** A*) and we will denote it by q_T(resp.* **of** q_A*) or simply by q. If the reference to T is clear then we will simply call q_X a* **minimal polynomial of** X.

If $X = \emptyset$ or $X = \{\mathbf{0}\}$ then we define $q_X(t) = 1$. If T is the zero linear operator then $q_X(t) = t$ for any subset X containing a nonzero vector. If X contains a nonzero vector then $\mathcal{I}_X$ contains the characteristic polynomial and hence $\deg q_X(t) \leq \dim V$. In this case, q_X cannot be 1. And hence $\deg q_X(t) \geq 1$.

Also it is easy to show that if $A, B \in M(n)$ are similar matrices then they have the same minimal polynomial (Exercise 12).

Example 7.1.9. *Let $A = \begin{bmatrix} 0 & 1 & 0 \\ 0 & 0 & 1 \\ 0 & 0 & 0 \end{bmatrix}$ and let $T = T_A, V = \mathbb{R}^3$. Let $\mathcal{E} = \{\mathbf{e}^1, \mathbf{e}^2, \mathbf{e}^3\}$ be the standard basis of $\mathbb{R}^3$. Then $A^2 = \begin{bmatrix} 0 & 0 & 1 \\ 0 & 0 & 0 \\ 0 & 0 & 0 \end{bmatrix}$, $A^3 = 0$. Hence we have*

$$T(\mathbf{e}^1) = \mathbf{0},\ T(\mathbf{e}^2) = \mathbf{e}^1,\ T(\mathbf{e}^3) = \mathbf{e}^2;\ T^2(\mathbf{e}^2) = \mathbf{0},\ T^2(\mathbf{e}^3) = \mathbf{e}^1;\ T^3(\mathbf{e}^3) = \mathbf{0}$$

Let $X = \{\mathbf{e}^1\}$. Since $T(\mathbf{e}^1) = \mathbf{0}$ and $t \in \mathcal{I}_X$ the degree of $q_{\mathbf{e}^1}(t)$ is ≤ 1. Hence it is of the form $t + \alpha$ for some constant α. If $\alpha \neq 0$ then $(T + \alpha I)(\mathbf{e}^1) \neq \mathbf{0}$. Hence $\alpha = 0$ and hence $q_{\mathbf{e}^1}(t) = t$. Similar argument will show that $q_{\mathbf{e}^2}(t) = t^2$ and $q_{\mathbf{e}^3}(t) = t^3$. If $X = \{\mathbf{e}^1, \mathbf{e}^2\}$ then $q_X(t) = t^2$. The characteristic polynomial of T is $f(t) = -t^3$.

Recall (Definition 6.8.4) that a T-cyclic subspace generated by $\mathbf{v}$ is defined by

$$C_T(\mathbf{v}) = \text{Span}\{\mathbf{v}, T(\mathbf{v}), T^2(\mathbf{v}), ...\}$$

As we noted before in §6.7 (6.125), we have

$$C_T(\mathbf{v}) = \{f(T)(\mathbf{v}) \mid f(t) \in F[t]\} \tag{7.10}$$

The name cyclic subspace comes from (7.10), *viz.* $C_T(\mathbf{v})$ is 'polynomially', not 'scalarily', generated by a single vector $\mathbf{v}$.

If $\mathbf{w} \in C_T(\mathbf{v})$ then $\mathbf{w}$ is of the form $\mathbf{w} = f(T)(\mathbf{v})$ where $f(t) \in F[t]$. However, the polynomial f is not unique since, for example, if $g(t) = f(t) + q_{\mathbf{v}}(t)$ then $f(T)(\mathbf{v}) = g(t)(\mathbf{v})$.

Now we have

$$\mathcal{I}_{\mathbf{v}} = \mathcal{I}_{C_T(\mathbf{v})} \tag{7.11}$$

In fact, since $\{\mathbf{v}\} \subseteq C_T(\mathbf{v})$ we have $\mathcal{I}_{\mathbf{v}} \supseteq \mathcal{I}_{C_T(\mathbf{v})}$. On the other hand, if $f \in \mathcal{I}_{\mathbf{v}}$ then $f(T)(\mathbf{v}) = \mathbf{0}$. To see $f \in \mathcal{I}_{C_T(\mathbf{v})}$ let $\mathbf{w} \in C_T(\mathbf{v})$ and $\mathbf{w} = g(T)(\mathbf{v})$ for some $g \in F[t]$. Then $f(T)(\mathbf{w}) = f(T)g(T)(\mathbf{v}) = g(T)f(T)(\mathbf{v}) = \mathbf{0}$ by (7.7). Hence $f \in \mathcal{I}_{C_T(\mathbf{v})}$ as required.

Let $\mathbf{v} \in V$, $f(t) \in F[t]$ and $\mathbf{w} = f(T)(\mathbf{v}) \in C_T(\mathbf{v})$. Then we have

$$C_T(\mathbf{w}) = f(T)C_T(\mathbf{v}) \subseteq C_T(\mathbf{v}) \tag{7.12}$$

In fact, any vector in $C_T(\mathbf{w})$ is of the form $gf(T)\mathbf{v}$ for some $g \in F[t]$. But $gf(T)\mathbf{v}$ is an element of $C_T(\mathbf{v})$. Since $gf(T)\mathbf{v} = f(T)(g(T)\mathbf{v})$ we have $C_T(f(T)(\mathbf{v})) \subseteq f(T)C_T(\mathbf{v})$. The inclusion $\supseteq$ is easy to check. The last inclusion is obvious since $C_T(\mathbf{v})$ is T-invariant.

We collect some basic facts about minimal polynomials in the following two theorems.

Theorem 7.1.10. *Let $T : V \to V$ be a linear operator with the characteristic polynomial $f_T(t)$ and the minimal polynomial $q_T(t)$. Let $X \subseteq X'$ be subsets of V.*

(1) *We have $\mathcal{I}_X = (q_X)$ and q_X is uniquely determined.*

(2) *If $h(T)(X) = \mathbf{0}$ ($h \in F[t]$) then $q_X|h$. Also we have $q_X|q_{X'}|q_T|f_T$.*

(3) *The minimal polynomial $q_{\mathbf{v}}$ of $\mathbf{v}$ is the same as the minimal polynomial of $C_T(\mathbf{v})$ i.e., $q_{\mathbf{v}} = q_{C_T(\mathbf{v})}$. If $\mathbf{w} \in C(\mathbf{v})$ then $q_{\mathbf{w}}|q_{\mathbf{v}}$.*

(4) *If $\deg(q_{\mathbf{v}}) = r$ then $\mathcal{B} = \{\mathbf{v}, T(\mathbf{v}), ..., T^{r-1}(\mathbf{v})\}$ is a basis of $C_T(\mathbf{v})$.*

Proof. (1) First, we show $\mathcal{I}_X \supseteq (q_X)$. If $g \in (q_X)$ then $g = hq_X$ and hence $g(T)(X) = h(T)q_X(T)(X) = \mathbf{0}$. Thus $g \in \mathcal{I}_X$. To show $\mathcal{I}_X \subseteq (q_X)$ let $g \in \mathcal{I}_X$. By division algorithm, write $g = sq_X + r$ with $\deg r < \deg q_X$. Then $\mathbf{0} = g(T)(X) = s(T)q_X(T)(X) + r(T)(X)$. Since $g(T)(X) = \mathbf{0} = q_X(T)(X)$ we have $r(T)(X) = \mathbf{0}$. If $r \neq 0$ then, since $\deg r < \deg q_X$, this contradict to the fact that q_X is the lowest degree satisfying $q_X(T)(X) = \mathbf{0}$. Uniqueness follows form (7.4) and the fact that q_X is monic.

(2) If $h(X) = \mathbf{0}$ then $h \in \mathcal{I}_X = (q_X)$ and hence $q_X|h$. Divisibility $q_X|q_{X'}$ follows form the fact $(q_{X'}) = \mathcal{I}_{X'} \subseteq \mathcal{I}_X = (q_X)$ and (7.2). We have $q_X|q_T$ for any X because $q_T = q_V$. Since $f_T \in \mathcal{I}_V = (q_T)$ we have $q_T|f_T$.

(3) The equality $q_{\mathbf{v}} = q_{C_T(\mathbf{v})}$ follows from (7.11). Divisibility $q_{\mathbf{w}}|q_{\mathbf{v}}$ follows from (7.12) and the previous statement.

(4) Checking that $\mathcal{B}$ is a basis for $C_T(\mathbf{v})$ is a practice problem which is almost the same as the proof of Theorem 6.8.10. $\square$

Since $q_T|f_T$, if $f_T(t) = \pm\prod_{i=1}^k p_i^{e_i}$ where p_i is a monic irreducible polynomial then $q_T(t) = \prod_{i=1}^k p_i^{\epsilon_i}$ for some $0 \le \epsilon_i \le e_i$. Later we will show that $\epsilon_i > 0$ (Corollary 7.2.10).

Practice Problem 7.1.2. *Prove that $\mathcal{B}$ in* Theorem 7.1.10 (4) *is a basis.*

Example 7.1.11. *Let $A = \begin{bmatrix} 1 & 0 & 1 \\ 0 & -1 & 0 \\ 1 & 1 & -1 \end{bmatrix}$ and $T = T_A, V = \mathbb{Q}^3$. The characteristic polynomial of A is $f(t) = -(t^2-2)(t+1)$. The polynomials t^2-2, $t+1$ are irreducible over $F = \mathbb{Q}$. For any $X \subseteq V$ its minimal polynomial q_X is a divisor of $f(t)$. Hence q_X is $t+1$ or t^2-2 or $(t+1)(t^2-2)$. We compute*

$$A+I=\begin{bmatrix}2&0&1\\0&0&0\\1&1&0\end{bmatrix},\quad A^2-2I=\begin{bmatrix}0&1&0\\0&-1&0\\0&-2&0\end{bmatrix}$$

From this we conclude that $q_T = (t+1)(t^2-2)$.

Since $(T+I)(\mathbf{e}^2) = \mathbf{0}$ *and* $\mathbf{e}^2 \neq \mathbf{0}$ *we have* $q_{\mathbf{e}^2}(t) = t+1$. *Similarly if* $\mathbf{v} = \mathbf{e}^1 + \mathbf{e}^3$ *then* $q_{\mathbf{v}} = t^2 - 2$. *Let* $\mathbf{v} = \begin{bmatrix}1\\1\\0\end{bmatrix}$. *To find* $q_{\mathbf{v}}$. *We have* $(T+I)(\mathbf{v}) \neq \mathbf{0}$ *and* $(T^2-2I)(\mathbf{v}) \neq \mathbf{0}$. *Hence by the observation above we see that the minimal polynomial of* $\mathbf{v}$ *is* $q_{\mathbf{v}}(t) = (t+1)(t^2-2)$.

Example 7.1.12. *Let* $A = \begin{bmatrix}0&-1&0\\1&0&0\\0&0&-1\end{bmatrix}$ *and* $T = T_A, V = \mathbb{R}^3$. *The characteristic polynomial of* A *is* $f(t) = -(t^2+1)(t+1)$. *For any* $\mathbf{v} \in V$ *its minimal polynomial* $q_{\mathbf{v}}$ *is a divisor of* $f(t)$. *Hence* $q_{\mathbf{v}}$ *is* $t+1$ *or* t^2+1 *or* $(t+1)(t^2+1)$. *We compute*

$$A+I=\begin{bmatrix}1&-1&0\\1&1&0\\0&0&0\end{bmatrix},A^2=\begin{bmatrix}-1&0&0\\0&-1&0\\0&0&1\end{bmatrix},A^2+I=\begin{bmatrix}0&0&0\\0&0&0\\0&0&1\end{bmatrix}$$

Let $\mathbf{v} = \begin{bmatrix}-1\\1\\2\end{bmatrix}$. *Then we have* $(T+I)(\mathbf{v}) = \mathbf{0}$; $q_{\mathbf{v}}(t) = t+1$.

Now since $(T^2+I)(\mathbf{e}^1) = (T^2+I)(\mathbf{e}^2) = \mathbf{0}$ *we conclude that* $q_{\mathbf{e}^1}(t) = q_{\mathbf{e}^2}(t) = t^2+1 = q_X(t)$ *where* $X = \{\mathbf{e}^1, \mathbf{e}^2\}$.

Theorem 7.1.13. *Let* $T : V \to V$ *be a linear operator with the characteristic polynomial* $f_T(t)$ *and the minimal polynomial* $q_T(t)$. *Let* $p(t)$ *be a monic irreducible polynomial.*

(1) *If* $p|q_T$ *then there is a nonzero* $\mathbf{v}$ *such that* $q_{\mathbf{v}} = p$. *If* $\deg(p) = r$ *then* $\mathcal{B} = \{\mathbf{v}, T(\mathbf{v}), ..., T^{r-1}(\mathbf{v})\}$ *is a basis for* $C_T(\mathbf{v})$.

(2) *Let* X *be a subset of* V. *If* $p^m(T)(X) = \mathbf{0}$ *and* $p^{m-1}(T)(X) \neq \mathbf{0}$ *then the minimal polynomial* $q_X(t)$ *of* X *is* $p^m(t)$.

Proof. (1) Let $pg = q_T$. Then $pg(T)(V) = \mathbf{0}$ but $g(T)(V) \neq \mathbf{0}$ by the definition of minimal polynomial. If $\mathbf{0} \neq \mathbf{v} \in g(T)(V)$ then $p(T)(\mathbf{v}) = \mathbf{0}$. Hence $q_{\mathbf{v}}|p$. But since p is irreducible we have $p = q_{\mathbf{v}}$. The fact that $\mathcal{B}$ is a basis follow from Theorem 7.1.10 (4).

(2) Since $p^m(T)(X) = \mathbf{0}$ we have $q_X(t)|p^m(t)$. Hence $q_X(t) = p^r(t)$ for some r. As $p^{m-1}(T)(X) \neq \mathbf{0}$ we conclude that $q_X(t) = p^m(t)$. For if $p^r(T)(X) = \mathbf{0}$ for $r < m-1$ then applying $p^{m-r-1}(T)$ we have $p^{m-1}(T)(X) = \mathbf{0}$ which is a contradiction. □

Next we show that the minimal polynomials of a linear operator is the same as the minimal polynomials of its matrix representation.

Theorem 7.1.14. *Let* T *be a linear operator on* V *with a basis* $\mathcal{B}$. *Let* $A = [T]_{\mathcal{B}}$ *and* $\phi : V \to F^n$ *be the coordinate map* $\mathbf{v} \mapsto [\mathbf{v}]_{\mathcal{B}}$ (Definition 6.3.6). *Let* X *be a subset of* V *and* $\phi(X) = \bar{X}$. *Then the* T*-minimal polynomial of* $q_X(t)$ *of* X *and the* A*-minimal polynomial* $q_{\bar{X}}(t)$ *of* $\bar{X}$ *are the same. In particular,* $q_T = q_A$.

Proof. Note that if $S : V \to V$ is any linear operator, and $\mathcal{B}$ is a basis of V and if $f(t)$ is any polynomial then by Corollary 6.4.22 we have $[f(S)]_{\mathcal{B}} = f([S]_{\mathcal{B}})$. Since the coordinate map ϕ is an isomorphism we have $S(X) = \mathbf{0}$ if and only if $[S]_{\mathcal{B}}(\bar{X}) = \mathbf{0}$. Hence

$$\{g(t) \in F[t] \mid g(T)(X) = \mathbf{0}\} = \{g(t) \in F[t] \mid g([T]_{\mathcal{B}})(\bar{X}) = \mathbf{0}\}$$

Therefore, the monic polynomial of the least degree in the left side is the same as the monic polynomial of the least degree in the right side. That is $q_X(t) = q_{\bar{X}}(t)$. □

We recall the definition of companion matrix (Definition 3.4.22, Theorem 6.8.10).

Definition 7.1.15. *The basis of the form* $\mathcal{B}_{\mathbf{v}} = \{\mathbf{v}, T(\mathbf{v}), ..., T^{k-1}(\mathbf{v})\}$ *will be called a* T-**cyclic basis** *of* $C_T(\mathbf{v})$. *We call the matrix (7.16) the* **companion matrix** *of the polynomial* $f(t) = a_0 + a_1 t + \cdots + t^k$ *and we will denote it by* $C(f)$.

$$C(f) = \begin{bmatrix} 0 & 0 & \cdots & 0 & -a_0 \\ 1 & 0 & \cdots & 0 & -a_1 \\ 0 & 1 & & & -a_2 \\ & & & \ddots & \vdots \\ 0 & \cdots & 0 & 1 & -a_{k-1} \end{bmatrix} \tag{7.13}$$

The following theorem strengthens Theorem 6.8.10.

Theorem 7.1.16. *Let* T *be a linear operator on a vector space* V. *Let* $\mathbf{v} \in V$ *and* $W = C_T(\mathbf{v})$. *Then following conditions are equivalent.*

(1) $\dim W = k$

(2) *The set of vectors* $\mathcal{B}_{\mathbf{v}} = \{\mathbf{v}, T(\mathbf{v}), ..., T^{k-1}(\mathbf{v})\}$ *forms a basis of* W.

(3) *The set of vectors* $\mathcal{B}_{\mathbf{v}} = \{\mathbf{v}, T(\mathbf{v}), ..., T^{k-1}(\mathbf{v})\}$ *is linearly independent and* $\mathcal{B}_{\mathbf{v}} \cup \{T^k(\mathbf{v})\}$ *is linearly dependent.*

(4) *The minimal polynomial* $q_{\mathbf{v}}(t)$ *of* $\mathbf{v}$ *is of degree* k.

In this case, if

$$a_0\mathbf{v} + a_1 T(\mathbf{v}) + \cdots + a_{k-1}T^{k-1}(\mathbf{v}) + T^k(\mathbf{v}) = \mathbf{0} \tag{7.14}$$

then the minimal polynomial $q_{\mathbf{v}}(t)$ *of* $\mathbf{v}$ *is*

$$q_{\mathbf{v}}(t) = a_0 + a_1 t + a_2 t^2 + \cdots + a_{k-1}t^{k-1} + t^k \tag{7.15}$$

and the characteristic polynomial $f_W(t)$ *of the restriction* T_W *of* T *on* W *is* $(-1)^k q_{\mathbf{v}}(t)$. *The matrix representation of* T_W *in the basis* $\mathcal{B}_{\mathbf{v}}$ *of* W *is the companion matrix of* $q_{\mathbf{v}}$.

$$[T_W]_{\mathcal{B}_{\mathbf{v}}} = C(q_{\mathbf{v}}) \tag{7.16}$$

Proof. (1)⇒(2) Theorem 6.8.10.

(2)⇒(3) Obvious.

(3)⇒(4) Under the condition of (3) we have a relation of the form (7.14). Hence $\deg q_{\mathbf{v}} \leq k$. If $\deg q_{\mathbf{v}} < k$ then $\mathcal{B}_{\mathbf{v}}$ is linearly dependent which is a contradiction.

(4)⇒(1) Since $\deg q_{\mathbf{v}} = k$ we see $\mathcal{B}_{\mathbf{v}} = \{\mathbf{v}, T(\mathbf{v}), ..., T^{k-1}(\mathbf{v})\}$ is a basis for W by Theorem 7.1.10 (4). Hence $\dim W = k$.

Now assume one of the conditions (1)–(4) and (7.14) holds. Let $g(t) = \sum_{i=0}^{k} a_i t^i$ where a_i are defined by (7.14). We need to show that $g = q_{\mathbf{v}}$. Since $g(T)\mathbf{v} = \mathbf{0}$ we see that $q_{\mathbf{v}}|g$. Both are monic of the same degree we conclude that $g(t) = q_{\mathbf{v}}(t)$.

Since $\dim C_T(\mathbf{v}) = k$ the characteristic polynomial f_W is of degree k with the leading coefficient $(-1)^k$. Further $q_{\mathbf{v}}(t)|f_W(t)$ and $q_{\mathbf{v}}$ is monic of degree k. Thus $f_W(t) = (-1)^k q_{\mathbf{v}}(t)$.

The statement about matrix is easy to show. □

In Theorem 3.4.24, we showed that the characteristic polynomial of a companion matrix is $(-1)^k f(t)$. It is interesting to observe that we have computed the characteristic polynomial of T_W without actually computing the determinant of a matrix in the last part of the proof of Theorem 7.1.16. From the proof we extract the idea for a different proof of Theorem 3.4.24 in the following corollary.

Corollary 7.1.17. *Let $f(t) = a_0 + a_1 t + \cdots + a_{k-1}t^{n-1} + t^n$ and $C(f)$ be the companion matrix of $f(t)$. Then the characteristic polynomial of $C(f)$ is $(-1)^n f(t)$ and the minimal polynomial is $f(t)$.*

Proof. Let $A = C(f)$ and $\mathbf{v} = \mathbf{e}^1$. Then $\mathbf{v} = \mathbf{e}^1, A\mathbf{v} = \mathbf{e}^2, A^2\mathbf{v} = \mathbf{e}^3, ..., A^{n-1}\mathbf{v} = \mathbf{e}^n$; $\{\mathbf{v}, A\mathbf{v}, ..., A^{n-1}\mathbf{v}\}$ is a basis of V and $A^n\mathbf{v} = -(a_0\mathbf{v} + a_1 A\mathbf{v} + \cdots + a_{n-1}A^{n-1}\mathbf{v})$. By Theorem 7.1.16, $q_{\mathbf{v}} = a_0 + a_1 t + \cdots + t^n$ and also $f_A(t) = (-1)^n q_{\mathbf{v}}$. And hence $q_A = q_{\mathbf{v}} = f$. □

Corollary 7.1.18. *Let T be a linear operator on a vector space V and let W be a T-invariant subspace of V. Let $f_W(t)$ be the characteristic polynomial of the restriction T_W of T on W. Let $\mathbf{v} \in W$. Then we have $\deg q_{\mathbf{v}} = \deg f_W(t)$ if and only if $W = C_T(\mathbf{v})$.*

Proof. Suppose $\mathbf{v} \in W$ satisfies $\deg q_{\mathbf{v}} = \deg f_W(t)$. Since W is T-invariant we see that $C_T(\mathbf{v}) \subseteq W$. By Theorem 7.1.16, $\dim C_T(\mathbf{v}) = \deg q_{\mathbf{v}}$. On the other hand, $\dim W = \deg f_W$. Hence we have $\dim C_T(\mathbf{v}) = \dim W$ and hence $C_T(\mathbf{v}) = W$.

Conversely suppose $W = C_T(\mathbf{v})$. If $\dim W = k$ then $f_W(t) = (-1)^k q_{\mathbf{v}}(t)$ by Theorem 7.1.16. Therefore $\deg q_{\mathbf{v}} = \deg f_W$. □

If $\deg q_W = \deg f_W$ then does it follow that there is $\mathbf{v} \in W$ such that $W = C_T(\mathbf{v})$? You should be able to answer this question in the next section.

Corollary 7.1.19. *Let T be a linear operator on V and $\mathbf{v} \in V$. If $S = \{\mathbf{v}, T(\mathbf{v}), ..., T^{k-1}(\mathbf{v})\}$ is linearly independent and $S \cup \{T^k(\mathbf{v})\}$ is linearly dependent then $S \cup \{T^{k+i}(\mathbf{v})\}$ is linearly dependent for any $i > 0$.*

Proof. By Theorem 7.1.16, $S = \{\mathbf{v}, T(\mathbf{v}), ..., T^{k-1}(\mathbf{v})\}$ is a basis of $C_T(\mathbf{v})$. Hence $S \cup \{T^{k+i}(\mathbf{v})\}$ is linearly dependent for any $i > 0$. □

Answers for Practice Problems 7.1

(7.1.1) (a) Let $\mathcal{I} = (f_1, f_2, ..., f_r)$. Let $h_i = \sum_j a_j^{(i)} f_j$ $(i = 1, 2)$. Then we have

$$h_1 + h_2 = \sum_j a_j^{(1)} f_j + \sum_j a_j^{(2)} f_j = \sum_j (a_j^{(1)} + a_j^{(2)}) f_j \in \mathcal{I}$$

(b) Let $h = \sum_j a_j f_j$. Then $fh = \sum_j (h a_j) f_j$. Hence $fh \in \mathcal{I}$.

(7.1.2) The proof is almost the same as the proof of Theorem 6.8.10. To show that $\{\mathbf{v}, T(\mathbf{v}), \ldots, T^{r-1}(\mathbf{v})\}$ is linearly independent assume the contrary: $a_0\mathbf{v} + a_1 T(\mathbf{v}) + \cdots + a_j T^j(\mathbf{v}) = 0$ for some $j < r$. Then the polynomial $h(t) = \sum_{i=0}^{j} a_i t^i$ of degree $< r$ with $h(T)\mathbf{v} = \mathbf{0}$. This is a contradiction.

To prove $\mathcal{B}$ span V first we have

$$T^r(\mathbf{v}) = -\left(a_0\mathbf{v} + a_1 T(\mathbf{v}) + \cdots + a_{r-1}T^{r-1}(\mathbf{v})\right) \tag{7.17}$$

Similarly, we have

$$\begin{aligned} T^{r+1}(\mathbf{v}) = TT^r(\mathbf{v}) &= -T\left(a_0\mathbf{v} + a_1 T(\mathbf{v}) + \cdots + a_{r-1}T^{r-1}(\mathbf{v})\right) && (7.18)\\ &= -\left(a_0 T(\mathbf{v}) + a_1 T^2(\mathbf{v}) + \cdots + a_{r-1}T^r(\mathbf{v})\right) && (7.19) \end{aligned}$$

If we substitute $T^r(\mathbf{v})$ in (7.18) by the expression (7.17) then we see that $T^{r+1}(\mathbf{v})$ also belongs to the subspace $\text{Span}\{\mathbf{v}, T(\mathbf{v}), \ldots, T^{r-1}(\mathbf{v})\}$. Repeating this we see that $T^{r+l}(\mathbf{v})$ belongs to $\text{Span}\{\mathbf{v}, T(\mathbf{v}), \ldots, T^{r-1}(\mathbf{v})\}$ for any $l \geq 0$. Hence $\mathcal{B}$ spans $C_T(\mathbf{v})$.

7.1 Exercises

1. Find the corresponding statements of (7.1.2) – (7.1.7) for the integers $\mathbb{Z}$ and provide a proof of your statements if necessary.

2. Let $f(t) = t^6 - 1$ and $g(t) = t^2 + 1$ polynomials in $\mathbb{Q}[t]$.

   (a) Find $s, r \in \mathbb{Q}[t]$ such that $f = sg + r$.

   (b) Write $f(t)$ as a product of monic irreducible polynomials over $\mathbb{Q}$.

3. Let F be a field of characteristic p.

   (a) Prove that if $f(t) = t + a \in F[t]$ then $f^p(t) = t^p + a^p$. (Hint: Use binomial theorem.)

   (b) Prove that if $f(t) = \sum_{i=0}^{n} a_i t^i \in F[t]$ then $f^p(t) = \sum_{i=0}^{n} a_i^p t^{ip}$.

4. Prove the following statements.

   (a) $(f) \subseteq (g)$ if and only if $g|f$.

   (b) $(f) = (1)$ if and only if f is a nonzero constant.

   (c) $(f) = (g)$ if and only if $f = cg$ for some constant c.

   (d) $(f) = (g)$ if and only if $g|f$ and $\deg(f) = \deg(g)$.

5. Let $\mathcal{I}$ be an ideal of $F[t]$. If $g_1, g_2, ..., g_k \in \mathcal{I}$ then $(g_1, g_2, ..., g_k) \subseteq \mathcal{I}$.

6. Let $f(t) = t - \lambda$ and $g(t) = t - \mu$ be the polynomials of degree 1 with $\lambda, \mu \in F$. Then f and g are relatively prime if and only if $\lambda \neq \mu$.

7. Let $\{f_1, f_2, ..., f_r\} \subseteq F[t]$. Prove that the following d_i's differ only by a nonzero constant multiple.

   (a) d_1 is a polynomial of the lowest degree in $(f_1, f_2, ..., f_r)$

   (b) $(d_2) = (f_1, f_2, ..., f_r)$

   (c) $d_3 = \gcd(f_1, f_2, ..., f_r)$

   (d) $d_4 = \prod_{i=1}^{r} p_i^{k_i}$ where $f_i = \prod_j p_j^{e_{ij}}$ $(e_{ij} \geq 0)$ and $k_i = \min_j\{e_{ij}\}$

   (e) d_5 is of highest degree polynomial in the set of all common divisors of $f_1, f_2, ..., f_r$

8. **(Euclidean Algorithm)** Let $f, g \in F[t]$. Write

$$
\begin{aligned}
f &= q_0 g + r_0 \\
g &= q_1 r_0 + r_1 \\
r_0 &= q_2 r_1 + r_2 \\
r_1 &= q_3 r_2 + r_3 \\
&\cdots \\
r_{n-2} &= q_n r_{n-1} + r_n \\
r_{n-1} &= q_{n-1} r_n
\end{aligned}
$$

Prove that $r_n = \gcd(f, g)$ and find $a, b \in F[t]$ such that $\gcd(f, g) = af + bg$. (Hint: If $r_{i-1} = q_3 r_i + r_{i+1}$ then the set of common divisors of $\{r_{i-1}, r_i\}$ is the same as the set of common divisors of $\{r_i, r_{i+1}\}$. And $\gcd(r_{n-1}, r_n) = r_n$.)

9. Use Euclidean Algorithm to find $\gcd(t^3 + 1, t^4 - 1)$ and $\gcd(t^6 - 1, t^3 - 1)$.

10. Prove the true statement(s) and find a counter example for the false statement(s).

    (a) A polynomial $f \in \mathbb{R}[t]$ of degree 5 can have two real roots and three nonreal complex roots.

    (b) Let $p, q \in \mathbb{Q}[t]$ be monic irreducible polynomials. If they are distinct then they are relatively prime.

    (c) If $\gcd(f_1, f_2, ..., f_r) = 1$ then there are $a_1, a_2, ..., a_r \in F[t]$ such that $a_1 f_1 + a_2 f_2 + \cdots + a_r f_r = 1$.

    (d) If $\mathcal{I} = (f_1, f_2, ..., f_n)$ is an ideal of $F[t]$ then $\mathcal{I}$ is also a subspace of $F[t]$.

    (e) Let T be a linear operator and $f(t) \in F[t]$. Then $f(T) = 0$ if and only if $f([T]_{\mathcal{B}}) = 0$ for any basis $\mathcal{B}$ of V.

    (f) Let $f, g \in F[t]$ and T be a linear operator on V. Let $T_1 = f(T), T_2 = g(T)$. Then $T_1 T_2 = T_2 T_1$.

    (g) If the characteristic polynomial of a linear operator T is irreducible then the T-minimal polynomial of any subset X containing a nonzero vector is the same as the characteristic polynomial up to a sign.

    (h) If $C_T(\mathbf{v}) = C_T(\mathbf{w})$ then $\mathbf{w} = c\mathbf{v}$ for some $c \in F$.

    (i) If $f, g \in F[t]$ are distinct irreducible polynomials then $\gcd(f, g) = 1$.

    (j) Let X be a subset containing a nonzero vector of V and T be a linear operator on V. If q_X is the T-minimal polynomial of X then the characteristic polynomial is divisible by q_X.

    (k) Let T be a linear operator on V and W is a T-invariant subspace of V. If $f(t) \in F[t]$ then W is $f(T)$-invariant subspace.

    (l) Let $A, B \in M(n, F)$ be similar matrices. For a subset X of F^n, the A-minimal polynomial $q_{X,A}$ of X is the same as the B-minimal polynomial $q_{X,B}$ of X.

    (m) Let T be a linear operator on V. Let W, W' be subspaces of V. If $q_W(t) = q_{W'}(t)$ then $W = W'$.

    (n) Let T be a linear operator on V and let $V = W_1 \oplus W_2$ where W_i are T-invariant. Then $q_T = q_{W_1} q_{W_2}$ where q_{W_i} is the minimal polynomial of T_{W_i}.

    (o) If $A, B \in M(n)$ then the minimal polynomials $q_{AB}(t)$ and $q_{BA}(t)$ are the same. (Cf. Exercise 5.1.10.)

11. Let $A, B \in M(n, F)$ and $f(t) \in F[t]$. Prove that if A, B are similar then $f(A)$ is similar to $f(B)$.

12. Prove that if $A, B \in M(n, F)$ are similar then they have the same minimal polynomial.

13. Let S, T be linear operators such that $ST = TS$. For $f(t), g(t) \in F[t]$, prove that $f(T)g(S) = g(S)f(T)$.

14. Let T be a linear operator on V and W be a T-invariant subspace. Prove that W is $f(T)$-invariant for any $f(t) \in F[t]$.

15. Let T be a linear operator on V.

    (a) For $\mathbf{v} \in V$, prove that if $f(t), g(t) \in F[t]$ then

    $$C_T(fg(T)(\mathbf{v})) \subseteq C_T(f(T)(\mathbf{v})) \subseteq C_T(\mathbf{v})$$

    (b) If $\mathbf{w} \in C_T(\mathbf{v}_1) \cap C_T(\mathbf{v}_2)$ then

    $$C_T(\mathbf{w}) \subseteq C_T(\mathbf{v}_1) \cap C_T(\mathbf{v}_2)$$

    for $\mathbf{v}_1, \mathbf{v}_2 \in V$.

16. Let $T \in L(V)$ and $\mathbf{v}, \mathbf{w} \in V$.

    (a) Prove that if $C_T(\mathbf{v}) = C_T(\mathbf{w})$ then $q_{\mathbf{v}} = q_{\mathbf{w}}$.

    (b) Prove that $C_T(\mathbf{v}) = C_T(\mathbf{w})$ if and only if there is $g \in F[t]$ which is relatively prime to $q_{\mathbf{v}}$ such that $\mathbf{w} = g(T)\mathbf{v}$.

17. Let $\mathcal{B} = \{\mathbf{v}, T(\mathbf{v}), ..., T^{k-1}(\mathbf{v})\}$ be a basis of $C_T(\mathbf{v})$. Can $X = \{T(\mathbf{v}), T^2(\mathbf{v}), ..., T^k(\mathbf{v})\}$ be a basis of $C_T(\mathbf{v})$? If so what is the condition for X to be a basis ? What if we remove $T^i(\mathbf{v})$ from $\mathcal{B}$ for some i $(1 \leq i \leq k)$?

18. Let $A = \begin{bmatrix} 1 & 0 & 0 \\ 0 & 0 & 1 \\ 0 & 0 & 0 \end{bmatrix}$. Find $q_{\mathbf{v}}(t)$ for each $\mathbf{v}$.

    (a) $\mathbf{v} = \begin{bmatrix} 1 \\ 1 \\ 0 \end{bmatrix}$ $\quad$ (b) $\mathbf{v} = \begin{bmatrix} 1 \\ 0 \\ 0 \end{bmatrix}$ $\quad$ (c) $\mathbf{v} = \begin{bmatrix} 1 \\ 1 \\ 1 \end{bmatrix}$

19. Let $T \in L(V)$ and let X be a subset of V. Prove that if $W = \text{Span}(X)$ then $q_X = q_W$.

20. Let X be a subset of V. Prove that $q_X(t) = q_{G_T(X)}(t)$. (See Exercise 6.8.10 for the notation.)

21. Let T be a linear operator on V and $X \subseteq V$. Let $p \in F[t]$ is an irreducible polynomial and $Y = p(T)(X)$. Prove that if p^ϵ is the minimal polynomial of X then the minimal polynomial q_Y of Y is $p^{\epsilon-1}$.

22. Let T be a linear operator on V with the minimal polynomial $q(t) = \sum_{i=0}^n a_i t^i$. Prove that T is an isomorphism if and only if $a_0 \neq 0$.

23. In Example 7.1.12, find $N(t^2 + t + 1)$. And find $q_{\mathbf{v}}(t)$ where $\mathbf{v} = \begin{bmatrix} 1 \\ 0 \\ -1 \end{bmatrix}$.

24. Let $A \in M(n, F)$ and $V = F^n$. Prove that the following conditions are equivalent.

    (a) There is $\mathbf{v} \in V$ such that $C_A(\mathbf{v}) = V$.

    (b) The matrix A is similar to the companion matrix $C((-1)^n f_A)$ where f_A is the characteristic polynomial of A.

25. Let $A \in M(n, F)$ with the minimal polynomial $q(t)$ of degree m. Let D be the set of all $n \times n$ diagonal matrices in $M(n, F)$ and let

$$P(A) = \{f(A) \mid f(t) \in F[t]\}$$
$$Z(A) = \{B \in M(n) \mid AB = BA\}$$

(a) Prove that $P(A)$ and $Z(A)$ are subspaces of $M(n)$ and $P(A) \subseteq Z(A)$. (See also Exercise 6.4.24.)

(b) Prove that if $A = cI$ for some $c \in F$ then $Z(A) = M(n, F)$. Also prove that if $A \in D$ then $D \subseteq Z(A)$.

(c) Prove that if $B = f(A)$ then we can choose $g(t) \in F[t]$ such that $B = g(t)$ and $\deg(g) \leq m$.

(d) Prove that $\dim P(A) = m \leq n$.

(e) Let $A = \begin{bmatrix} 1 & -1 \\ 1 & 1 \end{bmatrix} \in M(2, \mathbb{R})$. Prove that $Z(A) = P(A)$ and $\dim Z(A) = 2$.

7.2 Decomposition into Cyclic Subspaces

Let V be a finite dimensional vector space and $T \in L(V)$ with the spitting characteristic polynomial $f_(t) = \prod_{i=1}^{k}(t - \lambda_i)^{e_i}$. If T is diagonalizable then we saw (Corollary 6.7.25)

$$V = E_{\lambda_1} \oplus E_{\lambda_2} \oplus \cdots \oplus E_{\lambda_k}$$

which can be viewed as a direct sum of T-invariant subspaces. If T is not diagonalizable then $V \neq \oplus E_{\lambda_i}$. Since $E_{\lambda_i} = N(T - \lambda_i)$ this means $V \neq \oplus N(T - \lambda_i)$. However, we can show that $V = \oplus N(T - \lambda_i)^{e_i}$ which is a direct sum of T-invariant subspaces. Also we can show that $N(T - \lambda_i)^{e_i} = C_T(\mathbf{v}_i)$ for some $\mathbf{v}_i \in V$.

In general, the characteristic polynomial of T does not split. In this section, we will show that there are $\mathbf{v}_1, \mathbf{v}_2, ..., \mathbf{v}_k$ such that

$$V = C_T(\mathbf{v}_1) \oplus C_T(\mathbf{v}_2) \oplus \cdots \oplus C_T(\mathbf{v}_k) \tag{7.20}$$

The whole section will be devoted to prove this and the existence of the canonical forms will follow from this fact.

As the proof is rather long winded we will summarize the steps of the proof in terms of the characteristic polynomial f_T and the minimal polynomial q_T of T:

(1) When $f_T = \pm p^e, q_T = p$ where p is a monic irreducible polynomial (Theorem 7.2.3).

(2) When $f_T = \pm p^e = \pm q_T$ where p is a monic irreducible polynomial (Theorem 7.2.6).

(3) When $f_T = \pm p^e, q_T = p^\epsilon$ where p is a monic irreducible polynomial and $\epsilon \leq e$ (Theorem 7.2.16).

(4) When $f_T = \pm \prod p_i^{e_i}, q_T = \pm \prod p_i^{\epsilon_i}$ $(e_i \leq \epsilon_i)$ where p_i is an irreducible polynomial (Theorem 7.2.24).

The proof of the steps (1) and (2) are relatively easy. The key step is (3) which will be proved by induction using the step (1). The step (4) will follow easily from the steps (1), (2) and (3).

7.2.1 Minimal Polynomial and Cyclic Subspaces

First we prove a preliminary theorem.

Theorem 7.2.1. *Let T be a linear operator on a vector space V. Let $\mathbf{v}_1, \mathbf{v}_2 \in V$ and let q_1, q_2 be the minimal polynomials of $\mathbf{v}_1, \mathbf{v}_2$. Suppose q_1 and q_2 are relatively prime. Then*

(1) We have $C_T(\mathbf{v}_1) \cap C_T(\mathbf{v}_2) = \mathbf{0}$ and $C_T(\mathbf{v}_1) \oplus C_T(\mathbf{v}_2) = C_T(\mathbf{v}_1 + \mathbf{v}_2)$. And the minimal polynomial of $\mathbf{v}_1 + \mathbf{v}_2$ is $q_1 q_2$.

(2) Both the union of cyclic bases $\mathcal{B} := \mathcal{B}_{\mathbf{v}_1} \cup \mathcal{B}_{\mathbf{v}_2}$ and $\mathcal{B}_{\mathbf{v}_1+\mathbf{v}_2}$ are bases for $C_T(\mathbf{v}_1 + \mathbf{v}_2)$ and the matrix of T restricted to $C_T(\mathbf{v}_1 + \mathbf{v}_2)$ with respect to $\mathcal{B}$ (resp. $\mathcal{B}_{\mathbf{v}_1+\mathbf{v}_2}$) is $C(q_1) \oplus C(q_2)$ (resp.$C(q_1 q_2)$). In particular, the direct sum of the companion matrices $C(q_1) \oplus C(q_2)$ and the companion matrix $C(q_1 q_2)$ of $q_1 q_2$ are similar.

Proof. (1) Since q_1, q_2 are relatively prime, by Corollary 7.1.5, there are $g_1^{(i)}, g_2^{(i)} \in F[t]$ such that $g_1^{(i)} q_1 + g_2^{(i)} q_2 = t^i$ $(i = 0, 1, 2, ...)$. Hence we have

$$g_1^{(i)} q_1(T) + g_2^{(i)} q_2(T) = T^i \ (i = 0, 1, 2, ...) \tag{7.21}$$

Let $\mathbf{v} \in C_T(\mathbf{v}_1) \cap C_T(\mathbf{v}_2)$ then, with $i = 0$ in (7.21), we have $\mathbf{v} = T^0(\mathbf{v}) = g_1^{(0)}(T) q_1(T)(\mathbf{v}) + g_2^{(0)}(T) q_2(T)(\mathbf{v}) = \mathbf{0}$ since $q_1(T)(\mathbf{v}) = \mathbf{0} = q_2(T)(\mathbf{v})$. Hence $C_T(\mathbf{v}_1) \cap C_T(\mathbf{v}_2) = \mathbf{0}$.

Obviously $C_T(\mathbf{v}_1 + \mathbf{v}_2) \subseteq C_T(\mathbf{v}_1) \oplus C_T(\mathbf{v}_2)$. To prove $C_T(\mathbf{v}_1 + \mathbf{v}_2) \supseteq C_T(\mathbf{v}_1) \oplus C_T(\mathbf{v}_2)$ it suffices to show $T^i(\mathbf{v}_1)$ and $T^i(\mathbf{v}_2)$ belong to $C_T(\mathbf{v}_1 + \mathbf{v}_2)$ for any i. By (7.21) and the fact $q_i(T)(\mathbf{v}_i) = \mathbf{0}$, we have

$$g_2^{(i)} q_2(T)(\mathbf{v}_1 + \mathbf{v}_2) = g_2^{(i)} q_2(T)(\mathbf{v}_1) = (T^i - g_1^{(i)} q_1(T))(\mathbf{v}_1) = T^i(\mathbf{v}_1)$$

Similarly, $g_1^{(j)} q_1(T)(\mathbf{v}_1 + \mathbf{v}_2) = T^j(\mathbf{v}_2)$. Hence $T^i(\mathbf{v}_1)$ and $T^j(\mathbf{v}_2)$ belong to $C_T(\mathbf{v}_1 + \mathbf{v}_2)$ as contended.

To find the minimal polynomial say, q of $\mathbf{v}_1+\mathbf{v}_2$, notice first that $q_1(T)q_2(T)(\mathbf{v}_1+\mathbf{v}_2) = \mathbf{0}$; $q | q_1 q_2$. Now suppose $h(T)(\mathbf{v}_1 + \mathbf{v}_2) = \mathbf{0}$ for some $h \in F[t]$. Then since $h(T)(\mathbf{v}_i) \in C_T(\mathbf{v}_i)$ $(i = 1, 2)$ and $h(T)(\mathbf{v}_1) = -h(T)(\mathbf{v}_2) \in C_T(\mathbf{v}_1) \cap C_T(\mathbf{v}_2) = \mathbf{0}$ we must have $h(T)(\mathbf{v}_i) = \mathbf{0}$; $q_1 | h$ and $q_2 | h$. Hence, writing $h = a_1 q_1$ we have $q_2 | a_1 q_1$. Since $(q_1, q_2) = 1$, by Corollary 7.1.6, we conclude $q_2 | a_1$. Therefore $q_1 q_2 | h$ and, in particular, $q_1 q_2 | q$. Thus $q_1 q_2$ is the minimal polynomial of $\mathbf{v}_1 + \mathbf{v}_2$.

(2) Now $\mathcal{B} = \mathcal{B}_{\mathbf{v}_1} \cup \mathcal{B}_{\mathbf{v}_2}$ is a basis of $C_T(\mathbf{v}_1) \oplus C_T(\mathbf{v}_2)$ by Theorem 6.2.36. By Theorem 7.1.16, $\mathcal{B}_{\mathbf{v}}$ is also a basis. □

Example 7.2.2. *Let $A = \begin{bmatrix} 0 & 0 & -1 \\ 1 & 0 & 0 \\ 0 & 1 & 0 \end{bmatrix}$, $V = \mathbb{R}^3$, $T = T_A$. The characteristic polynomial of A is $f(t) = -(t^3 + 1) = -(t + 1)(t^2 - t + 1)$ and the minimal polynomial $q(t)$ of A is a divisor of $f(t)$. Hence the minimal polynomial must be one of $t + 1, t^2 - t + 1, t^3 + 1$. We have*

$$A + I = \begin{bmatrix} 1 & 0 & -1 \\ 1 & 1 & 0 \\ 0 & 1 & 1 \end{bmatrix}, \ A^2 - A + I = \begin{bmatrix} 1 & -1 & 1 \\ -1 & 1 & -1 \\ 1 & -1 & 1 \end{bmatrix}$$

And $A^3 + I = 0$ by Hamilton Cayley. Hence $q(t) = t^3 + 1 = -f(t)$.

If we choose $\mathbf{v}_1 = \begin{bmatrix} 1 \\ -1 \\ 1 \end{bmatrix} \in N(A+I)$ *then* $q_{\mathbf{v}_1}(t) = t+1$ *and* $C_T(\mathbf{v}_1)$ *is of dimension* 1 *with a basis* $\{\mathbf{v}_1\}$. *If we choose* $\mathbf{v}_2 = \begin{bmatrix} 1 \\ 1 \\ 0 \end{bmatrix} \in N(A^2-A+I)$ *then* $A\mathbf{v}_2 \neq \mathbf{0}$ *and* $(A^2-A+I)(\mathbf{v}_2) = \mathbf{0}$. *Hence* $\{\mathbf{v}_2, A\mathbf{v}_2\}$ *is a basis for* $C_T(\mathbf{v}_2)$ *and* $q_{\mathbf{v}_2}(t) = t^2 - t + 1$. *Thus* $C_T(\mathbf{v}_1) \oplus C_T(\mathbf{v}_2) = V$.

The minimal polynomials $q_{\mathbf{v}_1}(t)$, $q_{\mathbf{v}_2}(t)$ *are relatively prime. Hence the cyclic subspace generated by the vector* $\mathbf{v} := \mathbf{v}_1 + \mathbf{v}_2 = \begin{bmatrix} 2 \\ 0 \\ 1 \end{bmatrix}$ *is* V *i.e.,* $V = C_T(\mathbf{v})$. *In fact,*

$$\mathbf{v} = \begin{bmatrix} 2 \\ 0 \\ 1 \end{bmatrix}, \quad A\mathbf{v} = \begin{bmatrix} -1 \\ 2 \\ 0 \end{bmatrix}, \quad A^2\mathbf{v} = \begin{bmatrix} 0 \\ -1 \\ 2 \end{bmatrix}, \quad A^3\mathbf{v} = \begin{bmatrix} -2 \\ 0 \\ -1 \end{bmatrix}$$

and hence $\{\mathbf{v}, A\mathbf{v}, A^2\mathbf{v}\}$ *is a basis of* V *and the minimal polynomial of* $\mathbf{v}$ *is* $q_{\mathbf{v}}(t) = t^3 + 1$ *since* $A^3\mathbf{v} + \mathbf{v} = \mathbf{0}$.

Note that $\mathbf{v}$ *is not a unique vector spanning the cyclic space* $V = C_T(\mathbf{v})$. *In fact,*

$$\mathbf{e}^1 = \begin{bmatrix} 1 \\ 0 \\ 0 \end{bmatrix}, \quad A\mathbf{e}^1 = \begin{bmatrix} 0 \\ 1 \\ 0 \end{bmatrix}, \quad A^2\mathbf{e}^1 = \begin{bmatrix} 0 \\ 0 \\ 1 \end{bmatrix}, \quad A^3\mathbf{e}^1 = \begin{bmatrix} -1 \\ 0 \\ 0 \end{bmatrix}$$

Hence $C_T(\mathbf{e}^1) = V$ *and* $A^3\mathbf{e}^1 + \mathbf{e}^1 = \mathbf{0}$.

Now we prove that V is a direct sum of cyclic subspaces when $f_T(t)$ is a power of an irreducible polynomial $p(t)$ and $q_T(t) = p(t)$.

Theorem 7.2.3. *Let* T *be a linear operator on a vector space* V *and let* $p(t)$ *be a monic irreducible polynomial of degree* r. *Let* $f_T(t) = (-1)^{er}p(t)^e$ *be the characteristic polynomial and let* $q_T = p(t)$ *be the minimal polynomial of* T.

(1) *For any nonzero vector* $\mathbf{v}$ *we have* $q_{\mathbf{v}} = p$. *If* $\mathbf{v}_1, \mathbf{v}_2 \in V$ *are nonzero vectors then either* $C_T(\mathbf{v}_1) \cap C_T(\mathbf{v}_2) = \mathbf{0}$ *or* $C_T(\mathbf{v}_1) = C_T(\mathbf{v}_2)$.

(2) *If* $C_T(\mathbf{v}_1) \oplus C_T(\mathbf{v}_2) \oplus \cdots \oplus C_T(\mathbf{v}_s) \subseteq V$ *then there are* $\mathbf{v}_{s+1}, ..., \mathbf{v}_e$ *such that*

$$V = C_T(\mathbf{v}_1) \oplus C_T(\mathbf{v}_2) \oplus \cdots \oplus C_T(\mathbf{v}_e)$$

with $q_{\mathbf{v}_i}(t) = p(t)$.

(3) *If* $\mathcal{B}_{\mathbf{v}_i} = \{\mathbf{v}_i, T(\mathbf{v}_i), ..., T^{r-1}(\mathbf{v}_i)\}$ *is the cyclic basis of* $C_T(\mathbf{v}_i)$ *then*

$$\mathcal{B} := \mathcal{B}_{\mathbf{v}_1} \cup \mathcal{B}_{\mathbf{v}_2} \cup \cdots \cup \mathcal{B}_{\mathbf{v}_e}$$

is a basis of V. *The matrix representation of* T *with respect to* $\mathcal{B}$ *is*

$$[T]_{\mathcal{B}} = \begin{bmatrix} C(p) & & & \\ & C(p) & & \\ & & \ddots & \\ & & & C(p) \end{bmatrix}$$

where $C(p)$ *is the companion matrix of* p.

Proof. (1) We observed that $q_{\mathbf{v}}$ is a nonconstant polynomial and $q_{\mathbf{v}}|q_T = p$ (Theorem 7.1.10). Since p is irreducible and monic we see that $q_{\mathbf{v}} = p$.

Now suppose $\mathbf{0} \neq \mathbf{w} \in C_T(\mathbf{v}_1) \cap C_T(\mathbf{v}_2)$. We need to show $C_T(\mathbf{v}_1) = C_T(\mathbf{v}_2)$. Since $\mathbf{w}$ is nonzero we see that $q_{\mathbf{w}} = p = q_{\mathbf{v}_i}$ and hence $\dim C_T(\mathbf{w}) = r = \dim C_T(\mathbf{v}_i)$. Because $C_T(\mathbf{w}) \subseteq C_T(\mathbf{v}_i)$ and their dimensions are the same (Theorem 7.1.16) we have $C_T(\mathbf{w}) = C_T(\mathbf{v}_1) = C_T(\mathbf{v}_2)$.

(2) Suppose $\mathbf{v}_{s+1}$ is a nonzero vector not in $\oplus_{i=1}^{s} C_T(\mathbf{v}_i)$. We claim that $\sum_{i=1}^{s} C_T(\mathbf{v}_i) \cap C_T(\mathbf{v}_{s+1}) = \mathbf{0}$. Suppose $\mathbf{0} \neq \mathbf{w} \in C_T(\mathbf{v}_{s+1}) \cap \sum_{i=1}^{s} C_T(\mathbf{v}_i)$. Then $C_T(\mathbf{w}) \subseteq C_T(\mathbf{v}_{s+1})$ and hence $q_{\mathbf{w}} = q_{\mathbf{v}_{s+1}} = p$ as before and hence we see that $C_T(\mathbf{w}) = C_T(\mathbf{v}_{s+1})$. Write $\mathbf{v}_{s+1} = \sum b_i T^i(\mathbf{w})$. Since $\mathbf{w} \in \sum C_T(\mathbf{v}_i)$ and $\sum C_T(\mathbf{v}_i)$ is T-invariant we conclude that $\mathbf{v}_{s+1} = \sum b_i T^i(\mathbf{w}) \in \sum C_T(\mathbf{v}_i)$ which is a contradiction.

Hence we can adjoin $C_T(\mathbf{v}_{s+1})$ to $C_T(\mathbf{v}_1) \oplus C_T(\mathbf{v}_2) \oplus \cdots \oplus C_T(\mathbf{v}_s)$ to form $C_T(\mathbf{v}_1) \oplus C_T(\mathbf{v}_2) \oplus \cdots \oplus C_T(\mathbf{v}_s) \oplus C_T(\mathbf{v}_{s+1})$ in V (Definition 6.2.36) and $q_{\mathbf{v}_{s+1}} = p$ by (1). Repeat the process until we get $V = C_T(\mathbf{v}_1) \oplus C_T(\mathbf{v}_2) \oplus \cdots \oplus C_T(\mathbf{v}_e)$.

(3) Easy to show. □

Example 7.2.4. *Let* $A = \begin{bmatrix} 0 & -1 & 1 & -1 \\ 1 & 0 & -1 & -1 \\ 0 & 0 & 0 & -1 \\ 0 & 0 & 1 & 0 \end{bmatrix}$ *and* $V = \mathbb{R}^4$, $T = T_A$. *The characteristic polynomial of* T *is* $f(t) = (t^2+1)^2$. *Now* $A^2 = \begin{bmatrix} -1 & 0 & 0 & 0 \\ 0 & -1 & 0 & 0 \\ 0 & 0 & -1 & 0 \\ 0 & 0 & 0 & -1 \end{bmatrix}$. *Hence the minimal polynomial of* T *is* $q(t) = t^2 + 1$.

Now we look at the cyclic subspaces

$$C_T(\mathbf{e}^1) = \operatorname{Span}\left\{ \mathbf{e}^1 = \begin{bmatrix} 1 \\ 0 \\ 0 \\ 0 \end{bmatrix},\ A\mathbf{e}^1 = \begin{bmatrix} 0 \\ 1 \\ 0 \\ 0 \end{bmatrix},\ A^2\mathbf{e}^1 = \begin{bmatrix} -1 \\ 0 \\ 0 \\ 0 \end{bmatrix},\ \cdots \right\} = \operatorname{Span}\{\mathbf{e}^1, A\mathbf{e}^1 = \mathbf{e}^2\}$$

$$C_T(\mathbf{e}^2) = \operatorname{Span}\left\{ \mathbf{e}^2 = \begin{bmatrix} 0 \\ 1 \\ 0 \\ 0 \end{bmatrix},\ A\mathbf{e}^2 = \begin{bmatrix} -1 \\ 0 \\ 0 \\ 0 \end{bmatrix},\ A^2\mathbf{e}^2 = \begin{bmatrix} 0 \\ -1 \\ 0 \\ 0 \end{bmatrix},\ \cdots \right\} = \operatorname{Span}\{\mathbf{e}^1, \mathbf{e}^2\}$$

$$C_T(\mathbf{e}^3) = \operatorname{Span}\left\{ \mathbf{e}^3 = \begin{bmatrix} 0 \\ 0 \\ 1 \\ 0 \end{bmatrix},\ A\mathbf{e}^3 = \begin{bmatrix} 1 \\ -1 \\ 0 \\ 1 \end{bmatrix},\ A^2\mathbf{e}^3 = \begin{bmatrix} 0 \\ 0 \\ -1 \\ 0 \end{bmatrix},\ \cdots \right\} = \operatorname{Span}\{\mathbf{e}^3, A\mathbf{e}^3\}$$

Hence $C_T(\mathbf{e}^1) = C_T(\mathbf{e}^2)$ *and* $V = C_T(\mathbf{e}^1) \oplus C_T(\mathbf{e}^3)$. *And* $\mathcal{B} = \{\mathbf{e}^1, A\mathbf{e}^1\} \cup \{\mathbf{e}^3, A\mathbf{e}^3\}$ *is a basis of* V *and the matrix of* T *with respect to* $\mathcal{B}$ *is*

$$[T]_{\mathcal{B}} = \begin{bmatrix} 0 & -1 & 0 & 0 \\ 1 & 0 & 0 & 0 \\ 0 & 0 & 0 & -1 \\ 0 & 0 & 1 & 0 \end{bmatrix}$$

Also observe that A *is similar to* $[T]_{\mathcal{B}}$.

7.2.2 Decomposing into Null Spaces $N(g)$

Definition 7.2.5. *Let V be a vector space over a field F with a linear operator T on V. For $g(t) \in F[t]$ and a T-invariant subspace W we define the* **null subspace** *$N_W(g)$ of a polynomial $g \in F[t]$ in W by the null space of the operator $g(T)$*

$$N_W(g) = \{\mathbf{v} \in W \mid g(T)(\mathbf{v}) = \mathbf{0}\}$$
$$L_W(g) = \{\mathbf{v} \in W \mid g(T)^k(\mathbf{v}) = \mathbf{0} \text{ for some } k\}$$

If $W = V$ or the reference to the subspace W is clear then we will simply write $N(g)$, $L(g)$ for $N_W(g)$, $L_W(g)$.

We define the **nullity of the polynomial** *g by*

$$\nu_W(g) := \dim N_W(g), \ \nu(g) := \dim N(g)$$

Obviously, $N_W(g) \subseteq L_W(g)$. Also it is easy to check that $N_W(g)$ and $L_W(g)$ are T-invariant subspaces. And if $h|g$ then $N_W(h) \subseteq N_W(g)$ (Exercise 19).

If $g(t) = t - \lambda$ then $N(g) = E_\lambda$. Writing L_λ for $L(t-\lambda)$ we have

$$E_\lambda = N(t-\lambda)$$
$$L_\lambda = L(t-\lambda) = \{\mathbf{v} \mid (T-\lambda)^l(\mathbf{v}) = 0 \text{ for some } l\}$$

Observe that

$$L_W(g) = \bigcup_l N_W(g^l) = L_W(g^k) \text{ for any } k \tag{7.22}$$

$$N_W(g) \subseteq N_W(g^2) \subseteq \cdots \subseteq L_W(g) \tag{7.23}$$

If $g|f$ then we have (Exercise 19)

$$N_W(g) \subseteq N_W(f) \text{ and } L_W(g) \subseteq L_W(f) \tag{7.24}$$

Let $\mathbf{v} \in V$ and $T \in L(V)$. If $q_{\mathbf{v}}(t)$ is the minimal polynomial of $\mathbf{v}$ then (Exercise 16)

$$C_T(\mathbf{v}) \subseteq N(q_{\mathbf{v}}) \tag{7.25}$$

If X is a subset of V and $q_X(t)$ is the minimal polynomial of X and if $W = N(q_X)$ then (Exercise 6)

$$q_X(t) = q_W(t) \tag{7.26}$$

Next we will consider the vector space V where $f_T(t) = q_T(t) = p(t)^e$ up to a sign for some irreducible polynomial $p(t)$.

Theorem 7.2.6. *Let T be a linear operator on a vector space V of dimension n and $p(t)$ be a monic irreducible polynomial. Suppose the characteristic polynomial $f(t)$ and the minimal polynomial $q(t)$ of T are the same as $p(t)^e$ up to a sign. If $p(T)^e(\mathbf{v}) = \mathbf{0}$ but $p(T)^{e-1}(\mathbf{v}) \neq \mathbf{0}$ for some $\mathbf{v} \in V$ then $V = C_T(\mathbf{v})$ and $\mathcal{B}_{\mathbf{v}} = \{\mathbf{v}, T(\mathbf{v}), ..., T^{n-1}(\mathbf{v})\}$ is a cyclic basis.*

The matrix representation of T with respect to $\mathcal{B}_{\mathbf{v}}$ is the companion matrix of p^e

$$[T]_{\mathcal{B}_{\mathbf{v}}} = C(p(t)^e)$$

Further we have

$$N(p^i) = p^{e-i}(T)(V) = C_T(p^{e-i}(T)(\mathbf{v})) \tag{7.27}$$

Proof. Since $p(T)^{e-1} \neq 0$ there is $\mathbf{v} \in V$ such that $p(T)^{e-1}(\mathbf{v}) \neq \mathbf{0}$. To see $\mathcal{B}_{\mathbf{v}}$ is a basis notice that the minimal polynomial $q_{\mathbf{v}}$ of $\mathbf{v}$ is $q_{\mathbf{v}} = p^e$ by Theorem 7.1.13 (2). Hence by Theorem 7.1.16, $\mathcal{B}_{\mathbf{v}}$ is a basis as contended

The second equality of (7.27) follows from the fact that $V = C_T(\mathbf{v})$ and (7.12). For the first equality the inclusion $N(p^i) \supseteq p^{e-i}(T)(V)$ is obvious. To show $N(p^i) \subseteq p^{e-i}(T)(V)$ let $\mathbf{w} \in N(p^i)$. Since $V = C_T(\mathbf{v})$ we can write $\mathbf{w} = h(T)(\mathbf{v})$ for some $h(t) \in F[t]$. Then we have $\mathbf{0} = p^i(T)(\mathbf{w}) = p^i(T)h(T)(\mathbf{v})$. Since $q_{\mathbf{v}} = p^e$ we have $p^e | p^i h$; $p^{e-i} | h$. Hence $h = p^{e-i}g$ for some $g \in F[t]$. Therefore $\mathbf{w} = h(T)(\mathbf{v}) = p^{e-i}g(T)(\mathbf{v}) \in p^{e-i}(T)(V)$. □

Practice Problem 7.2.1. *Let $A \in M(n)$ and p be a monic irreducible polynomial. Let $f = (-1)^n p^e$ be the characteristic polynomial and q be the minimal polynomial. Prove that the following conditions are equivalent.*

(a) *The matrix A is similar to the companion matrix $C(p^e)$*

(b) *The characteristic and the minimal polynomial are equal up to a sign i.e., $q = p^e$*

(c) *There is $\mathbf{v} \in V$ such that $V = C_A(\mathbf{v})$.*

Example 7.2.7. *Let* $A = \begin{bmatrix} 0 & 0 & 0 & -1 \\ 1 & 0 & 0 & 0 \\ 0 & 1 & 0 & -2 \\ 0 & 0 & 1 & 0 \end{bmatrix}$ *which is the companion matrix of $(t^2+1)^2$. Let $V = \mathbb{R}^4$ and $T = T_A$. Hence the characteristic polynomial of A is $f(t) = (t^2+1)^2$ and the minimal polynomial is either $q_A(t) = (t^2+1)$ or $q_A(t) = (t^2+1)^2$. We compute*

$$A^2 = \begin{bmatrix} 0 & 0 & -1 & 0 \\ 0 & 0 & 0 & -1 \\ 1 & 0 & -2 & 0 \\ 0 & 1 & 0 & -2 \end{bmatrix}, \quad A^2 + I = \begin{bmatrix} 1 & 0 & -1 & 0 \\ 0 & 1 & 0 & -1 \\ 1 & 0 & -1 & 0 \\ 0 & 1 & 0 & -1 \end{bmatrix}$$

Since $A^2 + I \neq 0$ we conclude that $q_A(t) = (t^2+1)^2$. By Theorem 7.2.6, *there is $\mathbf{v}$ such that $V = C_T(\mathbf{v})$. Since $(A^2+I)(\mathbf{e}^1) \neq \mathbf{0}$ we see that $\mathcal{B} = \{\mathbf{e}^1, A\mathbf{e}^1, A^2\mathbf{e}^1, A^3\mathbf{e}^1\}$ is a basis of V.*

$$\mathbf{e}^1 = \begin{bmatrix} 1 \\ 0 \\ 0 \\ 0 \end{bmatrix}, \quad A\mathbf{e}^1 = \begin{bmatrix} 0 \\ 1 \\ 0 \\ 0 \end{bmatrix}, \quad A^2\mathbf{e}^1 = \begin{bmatrix} 0 \\ 0 \\ 1 \\ 0 \end{bmatrix}, \quad A^3\mathbf{e}^1 = \begin{bmatrix} 0 \\ 0 \\ 0 \\ 1 \end{bmatrix}$$

and

$$A^4\mathbf{e}^1 = \begin{bmatrix} -1 \\ 0 \\ -2 \\ 0 \end{bmatrix} = -\mathbf{e}^1 - 2A^2\mathbf{e}^1$$

Hence $C_T(\mathbf{e}^1) = V$ and $t^4 + 2t^2 + 1$ is the characteristic polynomial by Theorem 7.1.16, *as expected. The matrix of T in the basis $\mathcal{B}$ is A which is the companion matrix of $(t^2+1)^2$. Try various $\mathbf{v}$ in place of $\mathbf{e}^1$.*

By Theorem 7.2.6 *again, $N(t^2+1) = C_T((T^2+I)(\mathbf{e}^1))$;*

$$(A^2+I)(\mathbf{e}^1) = \begin{bmatrix} 1 \\ 0 \\ 1 \\ 0 \end{bmatrix}, \quad A(A^2+I)(\mathbf{e}^1) = \begin{bmatrix} 0 \\ 1 \\ 0 \\ 1 \end{bmatrix}, \quad A^2(A^2+I)(\mathbf{e}^1) = \begin{bmatrix} -1 \\ 0 \\ -1 \\ 0 \end{bmatrix}$$

Since $\mathcal{C} = \{(A^2+I)(\mathbf{e}^1), A(A^2+I)(\mathbf{e}^1)\}$ *is linearly independent and* $\mathcal{C} \cup \{A^2(A^2+I)(\mathbf{e}^1)\}$ *is linearly dependent, by* Theorem 7.1.16, *we conclude that* $\mathcal{C}$ *is a basis for* $N(t^2+1) = C_T((T^2+I)(\mathbf{e}^1))$.

Next we will consider the the case when $q_T(t)$ is a product of relatively prime polynomials.

Theorem 7.2.8. *Let T be a linear operator on a vector space V. Let $q(t) = g(t)h(t)$ be the minimal polynomial where g and h are relatively prime nonconstant monic polynomials. Then*

(1) $N(g) = L(g)$ *and* $N(h) = L(h)$

(2) $V = N(g) \oplus N(h)$ *and* $N(g), N(h)$ *are nonzero subspaces. Further, we have* $q_{N(g)} = g$, $q_{N(h)} = h$ *and* $q = q_{N(g)} q_{N(h)}$.

(3) *The restrictions* $g(T)' : N(h) \to N(h)$ *and* $h(T)' : N(g) \to N(g)$ *of* $g(T), h(T)$ *to* $N(h), N(g)$ *respectively, are isomorphisms.*

Proof. (1) The inclusion $N(g) \subseteq L(g)$ is obvious. Now let $\mathbf{v} \in L(g)$ and $g^n(T)(\mathbf{v}) = 0$. We need to show $g(T)(\mathbf{v}) = \mathbf{0}$. Since g^n and h are relatively prime there are $a, b \in F[t]$ such that $ag^n + bh = 1$. Hence $\mathbf{v} = ag^n(T)(\mathbf{v}) + bh(T)(\mathbf{v}) = bh(T)(\mathbf{v})$. Therefore, $g(T)(\mathbf{v}) = bgh(T)(\mathbf{v}) = bq(T)(\mathbf{v}) = \mathbf{0}$. Thus $\mathbf{v} \in N(g)$.

(2) Let $ag + bh = 1$ for some $a, b \in F[t]$. If $\mathbf{v} \in V$ then $\mathbf{v} = ag(T)(\mathbf{v}) + bh(T)(\mathbf{v})$. Since $hag(T)(\mathbf{v}) = a(gh)(T)(\mathbf{v}) = \mathbf{0}$ we have $ag(T)(\mathbf{v}) \in N(h)$. Similarly, $bh(T)(\mathbf{v}) \in N(g)$. Hence $V = N(g) + N(h)$. To show $N(g) \cap N(h) = \mathbf{0}$ let $\mathbf{x} \in N(g) \cap N(h)$. Then, as before, $\mathbf{x} = ag(T)(\mathbf{x}) + bh(T)(\mathbf{x})$. But $g(T)(\mathbf{x}) = \mathbf{0} = h(T)(\mathbf{x})$. Therefore $\mathbf{x} = \mathbf{0}$.

To see $N(g), N(h)$ are nontrivial let p, p' be irreducible factors of g, h respectively. Then $N(p) \subseteq N(g), N(p') \subseteq N(h)$. And by Theorem 7.1.13 (1), $N(p)$ and $N(p')$ are nonzero subspaces.

To show $q_{N(g)} = g$ first notice that $q_{N(g)} | g$ since $g(T)(N(g)) = \mathbf{0}$. If $\deg q_{N(g)} < \deg g$ then $\deg q_{N(g)} h < \deg q = \deg gh$ $q_{N(g)} h(T)(\mathbf{v} + \mathbf{v}') = \mathbf{0}$ for all $\mathbf{v} + \mathbf{v}' \in N(g) + N(h)$. This contradicts to our assumption that $q = gh$ is the minimal polynomial. Hence $q_{N(g)} | g$ and $\deg q_{N(g)} = \deg g$ and both are monic; $q_{N(g)} = g$. Proofs for the other statements are similar or immediate.

(3) We need to show that the restriction of $g(T)$ on $N(h)$ is one-to-one. Let $\mathbf{v} \in N(h)$ and suppose $g(T)(\mathbf{v}) = \mathbf{0}$. As before, $\mathbf{v} = ag(T)(\mathbf{v}) + bh(T)(\mathbf{v})$ and this is $\mathbf{0}$ under our conditions. So $\mathbf{v} = \mathbf{0}$.

The proof for $h(T)'$ is similar and try for yourself as a practice. □

Practice Problem 7.2.2. *In* Theorem 7.2.8 *prove that* $h(T)' : N(g) \to N(g)$ *is an isomorphism.*

As a first step toward the proof of the fact that V is direct sum of cyclic subspaces (7.20) we first write V as a direct sum of $L(p^e)$'s where p is an irreducible polynomial using Theorem 7.2.8.

Theorem 7.2.9. *Let T be a linear operator on V. Let $f(t) = \pm \prod_{i=1}^k p_i(t)^{e_i}$ be the characteristic and $q(t) = \prod_{i=1}^k p_i(t)^{\epsilon_i}$ be the minimal polynomial of T where p_i is monic and irreducible. Then*

$$V = L(p_1) \oplus L(p_2) \oplus \cdots \oplus L(p_k) \quad \text{and } L(p_i) = N(p_i^{\epsilon_i}) \tag{7.28}$$

If T_i is the restriction of T on $L(p_i)$ then $T = T_1 \oplus T_2 \oplus \cdots \oplus T_k$.

Proof. By Theorem 7.2.8, we have $V = \oplus N(p_i^{\epsilon_i})$. We only need to prove that $N(p_i^{\epsilon_i}) = L(p_i)$. But this follows from Theorem 7.2.8(1) that $N(p_i^{\epsilon_i}) = L(p_i^{\epsilon_i}) = L(p_i)$ $(i = 1, 2, ..., k)$ since $\prod_{j \neq i} p_j^{\epsilon_j}$ and $p_i^{\epsilon_i}$ are relatively prime. The fact that $T = \oplus_i T_i$ is obvious (Theorem 6.8.9). □

Scrutinizing T_i in Theorem 7.2.9 we get important facts.

Corollary 7.2.10. *With the notations of* Theorem 7.2.9, *the irreducible factors of $f(t)$ and $q(t)$ are the same i.e., $e_i \geq \epsilon_i > 0$. In particular, the characteristic polynomial and the minimal polynomial have the same roots.*

Proof. Since $q|f$, an irreducible factor of q is an irreducible factor of f. On the other hand, suppose an irreducible factor of f say, p_1 is not a factor of q. Then since $p_1^{e_1}$ and q are relatively prime, $q(T)$ restricted on $N(p_1^{\epsilon_1})$ is an isomorphism by Theorem 7.2.8 (3). Hence if $\mathbf{0} \neq \mathbf{x} \in L(p_1) = N(p_1^{\epsilon_1})$ then $q(T)(\mathbf{x}) \neq \mathbf{0}$. This contradicts to the fact that q is the minimal polynomial of T. Hence an irreducible factor of f is also a factor of q. □

From this it is easy to show Cayley-Hamilton Theorem saying that $f(T) = 0$. In fact, write $\mathbf{v} = \sum_{i=1}^k \mathbf{v}_i$ $(\mathbf{v}_i \in L(p_i))$. Then since $0 < \epsilon_i \leq e_i$ and $p_i^{e_i}(T)(\mathbf{v}_i) = \mathbf{0}$ we have

$$f(T)(\mathbf{v}) = f(T)\Big(\sum_{i=1}^k \mathbf{v}_i\Big) = \sum_{i=1}^k f(T)(\mathbf{v}_i) = \sum_{i=1}^k \Big(\prod_{j \neq i} p_j^{e_j}\Big) p_i^{e_i}(T)(\mathbf{v}_i) = \mathbf{0}$$

Corollary 7.2.11. *With the notations of* Theorem 7.2.9, *suppose p_i is of degree r_i. Then the characteristic polynomial of T_i is $f_i(t) = (-1)^{r_i e_i} p_i^{e_i}$ and the minimal polynomial of T_i is $q_i(t) = p_i^{\epsilon_i}$. Further*

$$\nu(p_i^l(T)) = \nu(p_i^l(T_i)) \ (l = 1, 2, ..., \epsilon_i) \tag{7.29}$$

Proof. The equality $q_i(t) = p_i^{\epsilon_i}$ follows from Theorem 7.2.9 (2) with $V = \oplus L(p_i) = N(p^{\epsilon_i})$ and $g = p_i^{\epsilon_i}, h = \prod_{j \neq i} p_j^{\epsilon_j}$.

Now we show $f_i = p_i^{e_i}$. Since the minimal polynomial of T_i is $p_i^{\epsilon_i}$ we have $f_i = (-1)^{r_i a_i} p_i^{a_i}$ for some $a_i \geq \epsilon_i$ by Corollary 7.2.10. And by Theorem 6.8.9, $f = \prod f_i = (-1)^n \prod p_i^{a_i}$. Hence $\prod p_i^{e_i} = \prod p_i^{a_i}$. By Theorem 7.1.2 (2), we have $a_i = e_i$. Therefore we have $f_i = (-1)^{r_i e_i} p_i^{e_i}$.

To prove (7.29), we have $\nu(p_i^l(T)) = \sum_j \nu(p_i^l(T)_{L_j})$ by Exercise 6.8.14. By Theorem 7.2.8, the linear map $p_i^l(T) : N(p_j^{\epsilon_j}) \to N(p_j^{\epsilon_j})$ is an isomorphism and hence $\nu(p_i^l(T)_{L_j}) = 0$ for $i \neq j$. Therefore, $\nu(p_i^l(T)) = \nu(p_i^l(T)_{L_i}) = \nu(p_i^l(T_i))$. □

From Practice Problem 7.2.1, we know the truth of the following result when the characteristic polynomial is a power of an irreducible polynomial. Now we can remove the condition.

Corollary 7.2.12. *Let $A \in M(n)$ with the characteristic polynomial f and q be the minimal polynomial. Prove that the following conditions are equivalent.*

(1) *The matrix A is similar to the companion matrix $C((-1)^n f)$*

(2) *The characteristic and the minimal polynomial are equal up to a sign i.e., $f = q$*

(3) *There is $\mathbf{v} \in V$ such that $V = C_A(\mathbf{v})$.*

Proof. Let $f = (-1)^n \prod p_i^{e_i}$ where p_i is a monic irreducible polynomial.

(1) ⇒ (2) By assumption $A \sim C(f)$ and by Theorem 6.8.9, $C(f)$ is similar to $\oplus_i C(p_i^{e_i})$. On the other hand, the minimal polynomial of $C(p_i^{e_i})$ is $p_i^{e_i}$ by Practice Problem 7.2.1. And by Theorem 7.2.8 (2), the minimal polynomial of A is $\prod_i p_i^{e_i}$ which is the same as $(-1)^n f$.

(2) $\Rightarrow$ (3) The characteristic polynomial of T_i is $\pm p_i^{e_i}$ and the minimal polynomial is of the form $p_i^{\epsilon_i}$. Since the minimal polynomial of T_A is the product $\prod p_i^{\epsilon_i}$ which, by our assumption, equals to f we conclude that $e_i = \epsilon_i$. Hence $L(p_i) = C_A(\mathbf{v}_i)$ for some $\mathbf{v}_i \in L(p_i)$. If we let $\mathbf{v} = \sum \mathbf{v}_i$ then, by Theorem 7.2.1, we have $C_A(\mathbf{v}) = \oplus_i C_A(\mathbf{v}_i) = V$.

(3) $\Rightarrow$ (1) Now $\mathcal{B}_{\mathbf{v}} = \{\mathbf{v}, T_A\mathbf{v}, ..., T_A^{n-1}\mathbf{v}\}$ is a basis for V. The matrix representation $[T_A]_{\mathcal{B}_{\mathbf{v}}}$ is $C(f)$. Hence $A \sim C(f)$. □

Next we will show that the i-th projection according to the decomposition $V = \oplus L(p_i)$ (Theorem 6.3.12) can be expressed as a polynomial in T.

Corollary 7.2.13. *With the notations of* Theorem 7.2.9 *let π_i be the i-th projection according to the decomposition $V = \oplus L_i$. Then there is $h_i \in F[t]$ such that $\pi_i = h_i(T)$.*

Proof. Let $f_i = \prod_{j \neq i} p_j^{\epsilon_j}$. Then, since f_i's are relatively prime, we can write $\sum a_j f_j = 1$ for some $a_j \in F[t]$. By Theorem 7.2.9, write $\mathbf{v} = \sum_j \mathbf{w}_j$ $(\mathbf{w}_j \in L(p_j))$. Then $a_i f_i(T)(\mathbf{v}) = a_i f_i(T)(\sum \mathbf{w}_j) = a_i f_i(T)(\mathbf{w}_i)$. On the other hand, if $\mathbf{w}_i \in L(p_i)$ then $\mathbf{w}_i = \sum_j a_j f_j(T)(\mathbf{w}_i) = a_i f_i(T)(\mathbf{w}_i)$ since $f_j(T)\mathbf{w}_i = \mathbf{0}$ for $i \neq j$. Hence $a_i f_i(T)(\mathbf{v}) = \mathbf{w}_i$. Therefore, if we let $h_i = a_i f_i$ then $\pi_i = h_i(T)$. □

The following corollary is a preliminary version of Theorem 7.3.10.

Corollary 7.2.14. *If the characteristic polynomial splits say, $f(t) = (-1)^n \prod_{i=1}^k (t - \lambda_i)^{e_i}$ then we have*

$$V = L_{\lambda_1} \oplus L_{\lambda_2} \oplus \cdots \oplus L_{\lambda_k} \tag{7.30}$$

and there is a basis $\mathcal{B}$ of V such that $[T]_{\mathcal{B}}$ is an upper (or lower) triangular matrix with λ_i repeated e_i times along the diagonal.

Proof. The equality (7.30) is (7.28) with $p_i(t) = t - \lambda_i$.

To find a basis such that $[T]_{\mathcal{B}}$ is triangular, we let $L_i = L_{\lambda_i}$. Then $L_i = N(T - \lambda_i)^{\epsilon_i}$. Notice that $N(T - \lambda_i) \subseteq N(T - \lambda_i)^2 \subseteq \cdots$ (Exercise 6.4.14). Choose a basis $\mathcal{B}_1^{(i)}$ for $N(T - \lambda_i)$. Extend $\mathcal{B}_1^{(i)}$ to a basis $\mathcal{B}_2^{(i)}$ for $N(T - \lambda_i)^2$. In this way we get bases $\mathcal{B}_1^{(i)} \subseteq \mathcal{B}_2^{(i)} \subseteq \cdots \subseteq \mathcal{B}_{\epsilon_i}^{(i)}$ of $N(T - \lambda_i) \subseteq N(T - \lambda_i)^2 \subseteq \cdots \subseteq N(T - \lambda_i)^{\epsilon_i}$ with $\mathcal{B}^{(i)} := \mathcal{B}_{\epsilon_i}^{(i)}$.

If $\mathbf{v} \in \mathcal{B}_1^{(i)}$ then $T(\mathbf{v}) = \lambda_i \mathbf{v}$. If $\mathbf{v} \in \mathcal{B}_2^{(i)}$ then $(T - \lambda_i)(\mathbf{v}) \in N(T - \lambda_i)$ and hence it is a linear combination of $\mathcal{B}_1^{(i)}$. Therefore $T(\mathbf{v}) = \lambda_i \mathbf{v} +$(a linear combination of $\mathcal{B}_1^{(i)}$). Similarly if $\mathbf{v} \in \mathcal{B}_3^{(i)}$ then $T(\mathbf{v}) = \lambda_i \mathbf{v} +$(a linear combination of $\mathcal{B}_2^{(i)}$) and so on. Hence the matrix representation of $[T_i]_{\mathcal{B}^{(i)}}$ is of the form

$$[T_i]_{\mathcal{B}^{(i)}} = \begin{bmatrix}
\lambda_i & 0 & \cdots & 0 & * & & * & & & \\
 & \lambda_i & & \vdots & \vdots & & \vdots & & & \\
 & & \ddots & 0 & & & & & & \\
 & & & \lambda_i & * & \cdots & * & & & \\
 & & & & \lambda_i & 0 & \cdots & \vdots & \vdots & \\
 & & & & & \ddots & 0 & * & * & \\
 & & & & & & \lambda_i & * & * & \\
 & & & & & & & \lambda_i & 0 & \cdots \\
 & & & & & & & & \ddots & \\
 & & & & & & & & & \lambda_i
\end{bmatrix}$$

If we choose a basis $\mathcal{B} = \cup_i \mathcal{B}^{(i)}$ of V then, by Theorem 6.8.3

$$[T]_{\mathcal{B}} = \begin{bmatrix} [T_1]_{\mathcal{B}^{(1)}} & & & \\ & [T_2]_{\mathcal{B}^{(2)}} & & \\ & & \ddots & \\ & & & [T_k]_{\mathcal{B}^{(k)}} \end{bmatrix}$$

If $\mathcal{B}'$ is the basis of V obtained by reversing the order of $\mathcal{B}$ then $[T]_{\mathcal{B}'}$ is lower triangular. □

Let $T \in L(V)$ with the characteristic polynomial $f(t) = \prod p_i^{e_i}(t)$ and the minimal polynomial $q(t) = \prod p_i^{\epsilon_i}(t)$. By Theorem 7.2.9,

$$V = \bigoplus_i L(p_i) \text{ and } L(p_i) = N(p_i^{\epsilon_i})$$

Hence we need to study $N(p^k)$ for an irreducible p dividing the minimal polynomial q for various k. On the other hand, by Theorem 7.2.6, if the characteristic polynomial and the minimal polynomial are the same as p^e up to a sign then we have

$$N(p^k) = C_T(p^{e-k}(T)(\mathbf{v})$$

To study $C_T(p^{e-k}(T)(\mathbf{v})$, we will consider, more generally, $C_T(h(T)(\mathbf{v}))$ for a polynomial h – not necessarily a power of an irreducible – dividing the minimal polynomial of $\mathbf{v}$.

First, look at an example. Let $T \in L(V)$ and $\mathbf{v} \in V$. Suppose $q_{\mathbf{v}}(t) = h(t)g(t)$ is the minimal polynomial of $\mathbf{v}$ where $h(t) = t^2 + t + 1$ and $g(t) = t^3 + t + 1$. Hence

$$\mathcal{B}_{\mathbf{v}} = \{\mathbf{v}, T(\mathbf{v}), T^2(\mathbf{v}), T^3(\mathbf{v}), T^4(\mathbf{v})\}$$

is a basis of $W := C_T(\mathbf{v})$. Let

$$\begin{aligned} h(\mathcal{B}_{\mathbf{v}}) &= \{T^i(h(T)(\mathbf{v})|0 \le i \le 2\} \\ &= \left\{ \begin{array}{c} h(T)(\mathbf{v}) = \mathbf{v} + T(\mathbf{v}) + T^2(\mathbf{v}), \\ Th(T)(\mathbf{v}) = T(\mathbf{v}) + T^2(\mathbf{v}) + T^3(\mathbf{v}), \\ T^2h(T)(\mathbf{v}) = T^2(\mathbf{v}) + T^3(\mathbf{v}) + T^4(\mathbf{v}) \end{array} \right\} \subseteq C_T(h(T)(\mathbf{v})) \end{aligned}$$

We claim that $h(\mathcal{B}_{\mathbf{v}})$ is a basis of $C_T(h(T)(\mathbf{v}))$. To see $h(\mathcal{B}_{\mathbf{v}})$ is linearly independent consider the matrix consisting of the columns of the coordinates of the vectors of $h(\mathcal{B}_{\mathbf{v}})$ in $\mathcal{B}_{\mathbf{v}}$:

$$\begin{bmatrix} 1 & 0 & 0 \\ 1 & 1 & 0 \\ 1 & 1 & 1 \\ 0 & 1 & 1 \\ 0 & 0 & 1 \end{bmatrix}$$

Now the columns of the matrix is linearly independent. Hence $h(\mathcal{B}_{\mathbf{v}})$ is linearly independent. To see $h(\mathcal{B}_{\mathbf{v}})$ is a basis it suffice to show $h(\mathcal{B}_{\mathbf{v}}) \cup \{T^3h(T)(\mathbf{v})\}$ is linearly dependent by Theorem 7.1.16. To see this observe $\mathbf{0} = gh(T)(\mathbf{v})$:

$$\mathbf{0} = gh(T)(\mathbf{v}) = g(T)h(T)(\mathbf{v}) = (I + T + T^3)h(T)(\mathbf{v}) = (\ (*)\ + T^3)h(T)(\mathbf{v}) \qquad (7.31)$$

where $(*)$ represents the terms of degree ≤ 2 in T and hence $(*)h(T)(\mathbf{v})$ is a linear combination of $h(\mathcal{B}_{\mathbf{v}})$. And therefore $h(\mathcal{B}_{\mathbf{v}}) \cup \{T^3 f(T)(\mathbf{v})\}$ is linearly dependent by (7.31). Thus $h(\mathcal{B}_{\mathbf{v}})$ is a basis of $C_T(h(T)(\mathbf{v}))$.

We formalize the example above in the following theorem. For more general statement see Exercise 13.

Theorem 7.2.15. *Let $T \in L(V)$ and $\mathbf{v} \in V$ and $W = C_T(\mathbf{v})$. Suppose the minimal polynomial of $\mathbf{v}$ is $q_{\mathbf{v}}(t) = g(t)h(t)$ where g, h are monic polynomials of degree r, s respectively. Let $k = r + s$ and let*

$$\mathcal{B}_{\mathbf{v}} = \{\mathbf{v}, T\mathbf{v}, ..., T^{k-1}\mathbf{v}\}$$

be the cyclic basis for $C_T(\mathbf{v})$. Then the minimal polynomial of $h(T)(\mathbf{v})$ is $g(t)$ i.e., $q_{h(T)\mathbf{v}}(t) = g(t)$ and

$$h(\mathcal{B}_{\mathbf{v}}) = \{T^i h(T)(\mathbf{v}) \mid 0 \le i < r = \deg g\} \tag{7.32}$$

is a basis for $C_T(h(T)\mathbf{v})$. Further

$$C_T(h(T)(\mathbf{v})) = N_W(g) \text{ and } \nu_W(g) = \deg(g) \tag{7.33}$$

Proof. First we show $h(\mathcal{B}_{\mathbf{v}})$ is a basis for $C_T(h(T)(\mathbf{v}))$. Let $h(t) = \sum_{k=0}^{s} a_k t^k$. Then the matrix whose columns are the coordinates of the vectors in $h(\mathcal{B}_{\mathbf{v}})$ with respect to $\mathcal{B}_{\mathbf{v}}$ is

$$A := \begin{bmatrix} a_0 & 0 & 0 & \cdots \\ a_1 & a_0 & 0 & \\ a_2 & a_1 & \ddots & 0 \\ a_3 & a_2 & \ddots & a_0 \\ \vdots & \vdots & & a_1 \\ a_s & & & \\ 0 & a_s & & \vdots \\ \vdots & & \ddots & \\ 0 & 0 & & a_s \end{bmatrix}$$

which is of size is $(r + s + 1) \times r$. The columns of the matrix A, which are the coordinates of vectors in $h(\mathcal{B}_{\mathbf{v}})$ with respect to $\mathcal{B}_{\mathbf{v}}$, are linearly independent since , h being monic, $a_s = 1$. Hence $h(\mathcal{B}_{\mathbf{v}})$ is linearly independent. To see $h(\mathcal{B}_{\mathbf{v}})$ is a basis we need to show $h(\mathcal{B}_{\mathbf{v}}) \cup \{T^r h(T)(\mathbf{v})\}$ is linearly dependent by Theorem 7.1.16. In fact,

$$\mathbf{0} = gh(T)(\mathbf{v}) = g(T)h(T)(\mathbf{v}) = ((*) + T^r)h(T)(\mathbf{v}) \tag{7.34}$$

where $(*)h(T)(\mathbf{v})$ is a linear combination of $h(\mathcal{B}_{\mathbf{v}})$. Now (7.34) shows that $h(\mathcal{B}_{\mathbf{v}}) \cup \{T^r h(T)(\mathbf{v})\}$ is linearly dependent.

For $C_T(h(T)(\mathbf{v})) = N_W(g)$, the inclusion $C_T(h(T)(\mathbf{v})) \subseteq N_W(g)$ is obvious. Now let $\mathbf{w} \in W$ and $g(T)(\mathbf{w}) = \mathbf{0}$. Since $\mathcal{B}_{\mathbf{v}}$ is a basis of W we can write $\mathbf{w} = \sum a_k T^k(\mathbf{v})$. Let $l(t) = \sum a_k t^k$ and write $l = sh + r$ with $\deg(r) < \deg(h)$. Then we have

$$\mathbf{0} = g(T)(\mathbf{w}) = g(T)(sh(T) + r(T))(\mathbf{v}) = g(T)r(T)(\mathbf{v})$$

since $g(T)(sh(T))(\mathbf{v}) = \mathbf{0}$. If $r \ne 0$ then $g(T)r(T)(\mathbf{v}) = \mathbf{0}$. This is a contradiction since $\deg(rg) < \deg(gh)$ and $q_{\mathbf{v}} = gh$ is the minimal polynomial of $\mathbf{v}$. Hence $r = 0$. Therefore $\mathbf{w} = (sh)(T)(\mathbf{v}) \in C_T(h(T)(\mathbf{v}))$.

Finally $\dim C_T(h(T)(\mathbf{v}))$ is the number of elements in $h(\mathcal{B}_{\mathbf{v}})$ which is $r = \deg g$. □

Most of the time we apply the theorem in the following situation: $q = f^n$ where $\deg f = r$ and $g = f^{n-i}$, $h = f^i$ in which case we have

$$C_T(f^i(T)(\mathbf{v})) = N_W(f^{n-i}) \tag{7.35}$$

$$\dim C_T(f^i(T)(\mathbf{v})) = \nu_W(f^{n-i}(T)) = (n - i)r \tag{7.36}$$

7.2.3 Decomposing into Cyclic Subspaces

Let T be a linear operator on V with the characteristic polynomial $f(t) = \pm p^e$ and the minimal polynomial $q(t) = p^\epsilon$ for some irreducible monic polynomial $p(t)$ $(1 \le \epsilon \le e)$. The following theorem which is a main step towards the theory, asserts that V is a direct sum of cyclic subspaces under these conditions. The theorem reduces to Theorem 7.2.3 if $\epsilon = 1$ and reduces to Theorem 7.2.6 if $\epsilon = e$.

As the notations for the following theorem are rather involved it is instructive to look up the diagram (7.50) or (7.55) while reading the proof.

Theorem 7.2.16 (Decomposition into Cyclic Subspaces I). *Let T be a linear operator on a vector space V over F. Let $p(t)$ be a monic irreducible polynomial of degree r over F. Let $f(t) = (-1)^{er}p(t)^e$ be the characteristic polynomial and $q(t) = p(t)^\epsilon$ $(1 \le \epsilon \le e)$ be the minimal polynomial of T.*

(1) *There are $\mathbf{v}_1, \mathbf{v}_2, ..., \mathbf{v}_k$ such that*

$$k = \nu(p)/r \tag{7.37}$$

$$V = C_T(\mathbf{v}_1) \oplus C_T(\mathbf{v}_2) \oplus \cdots \oplus C_T(\mathbf{v}_k) \tag{7.38}$$

(2) *If the minimal polynomial of $\mathbf{v}_l$ is p^{ϵ_l} with $\epsilon_1 \ge \epsilon_2 \ge \cdots \ge \epsilon_k$ then*

$$\epsilon = \epsilon_1 \text{ and } \epsilon_1 + \epsilon_2 + \cdots + \epsilon_k = e \tag{7.39}$$

$$N(p^i) = \bigoplus_{l=1}^{k} C_T(p^{\epsilon_l - i}(T)(\mathbf{v}_l)) \quad (i = 1, 2, ...) \tag{7.40}$$

where we let $C_T(p^{\epsilon_l - i}(T)(\mathbf{v}_l)) = C_T(p^{\epsilon_l}(T)(\mathbf{v}_l))$ if $\epsilon_l \le i$. And

$$\dim C_T(p^{\epsilon_l - i}(T)(\mathbf{v}_l)) = \begin{cases} ir & (\epsilon_l > i) \\ \epsilon_l r & (\epsilon_l \le i) \end{cases} \tag{7.41}$$

(3) *If $\mathcal{B}_{\mathbf{v}_i}$ be the cyclic basis for $C_T(\mathbf{v}_i)$ then $\mathcal{B} = \mathcal{B}_{\mathbf{v}_1} \cup \mathcal{B}_{\mathbf{v}_2} \cup \cdots \cup \mathcal{B}_{\mathbf{v}_k}$ is a basis for V. The matrix of T with respect to $\mathcal{B}$ is*

$$[T]_{\mathcal{B}} = \begin{bmatrix} C(p^{\epsilon_1}) & & & \\ & C(p^{\epsilon_2}) & & \\ & & \ddots & \\ & & & C(p^{\epsilon_k}) \end{bmatrix} \tag{7.42}$$

The matrix $[T]_B$ is called the **rational canonical form** *for T with the characteristic polynomial $(-1)^{er}p^e$. If $T = T_A$ then we call $[T]_{\mathcal{B}}$ a* **rational canonical form** *for A.*

The core of the theorem is (7.38) and we prove it by induction on ϵ where p^ϵ is the minimal polynomial. The lengthy proof below consists in checking the details of the following steps.

A brief sketch of proof: If $\epsilon = 1$ the theorem reduces to Theorem 7.2.3. The minimal polynomial of $p(T)(V)$ is $p^{\epsilon-1}$ and hence by induction assumption we can find $\mathbf{v}_1, \mathbf{v}_2, ..., \mathbf{v}_s$ such that

$$p(T)(V) = C_T(p(T)(\mathbf{v}_1)) \oplus C_T(p(T)(\mathbf{v}_2)) \oplus \cdots \oplus C_T(p(T)(\mathbf{v}_s))$$

Now $C_T(p^{\epsilon_1-1}(T)(\mathbf{v}_1))\oplus\cdots\oplus C_T(p^{\epsilon_s-1}(T)(\mathbf{v}_s))$ has minimal polynomial p and we can use Theorem 7.2.3 to write

$$N(p)=C_T(p^{\epsilon_1-1}(T)(\mathbf{v}_1))\oplus\cdots\oplus C_T(p^{\epsilon_s-1}(T)(\mathbf{v}_s))\oplus C_T(\mathbf{v}_{s+1})\oplus\cdots\oplus C_T(\mathbf{v}_k)$$

Finally, we show that

$$V=C_T(\mathbf{v}_1)\oplus\cdots\oplus C_T(\mathbf{v}_s)\oplus C_T(\mathbf{v}_{s+1})\oplus\cdots\oplus C_T(\mathbf{v}_k)$$

Proof. (1) We prove (7.38) by induction on ϵ where $q=p^\epsilon$ is the minimal polynomial of T. If $\epsilon=1$ then our assertion is Theorem 7.2.3.

Suppose $\epsilon>1$ and assume the result for $\epsilon-1$. If we let $W=p(T)(V)$ then $p^{\epsilon-1}(T)(W)=\mathbf{0}$. And the minimal polynomial of the restriction T_W of T is $p^{\epsilon-1}$ (Exercise 7.1.21). Our induction assumption applies to T_W and there are $p(T)(\mathbf{v}_i)\in W$ such that

$$W=C_T(p(T)(\mathbf{v}_1))\oplus C_T(p(T)(\mathbf{v}_2))\oplus\cdots\oplus C_T(p(T)(\mathbf{v}_s)) \tag{7.43}$$

Since the minimal polynomial of $\mathbf{v}_l$ is a divisor of the characteristic polynomial $\pm p^e$ we can write the minimal polynomial of $\mathbf{v}_l$ as $q_{\mathbf{v}_l}=p^{\epsilon_l}$ with $\epsilon_1\geq\epsilon_2\geq\cdots\geq\epsilon_s$. Notice that $\epsilon_s>1$ for otherwise $C_T(p(T)(\mathbf{v}_s))=\mathbf{0}$.

We claim that $\left(\sum_{j<i}C_T(\mathbf{v}_j)\right)\cap C_T(\mathbf{v}_i)=\mathbf{0}$. In fact, if $\mathbf{w}_i\in\left(\sum_{j<i}C_T(\mathbf{v}_j)\right)\cap C_T(\mathbf{v}_i)$ then we can write $\mathbf{w}_i=\sum_{j<i}\mathbf{w}_j$ with $\mathbf{w}_j\in C_T(\mathbf{v}_j)$ $(1\leq j\leq i)$. Then $\sum_{j<i}p(T)(\mathbf{w}_j)=p(T)(\mathbf{w}_i)$ or $\mathbf{0}=\sum_{j<i}p(T)(\mathbf{w}_j)-p(T)(\mathbf{w}_i)\in\oplus_{j=1}^{s}C_T((p(T)(\mathbf{v}_j))$. Hence, by Theorem 6.2.36, $p(T)(\mathbf{w}_j)=\mathbf{0}$ $(1\leq j\leq i)$ and hence $\mathbf{w}_j\in N(p)\cap C_T(\mathbf{v}_j)\subseteq C_T(p^{\epsilon_j-1}(T)(\mathbf{v}_j))$ (Theorem 7.2.6). Since $\epsilon_i>1$ we have $C_T(p^{\epsilon_j-1}(T)(\mathbf{v}_j))\subseteq C_T(p(T)(\mathbf{v}_j))$ $(1\leq j\leq i)$ (Exercise 7.1.15) and hence $\sum_{j<i}\mathbf{w}_j=\mathbf{w}_i\in\sum_{j<i}C_T(p(T)(\mathbf{v}_j))\cap C_T(p(T)(\mathbf{v}_i))=\mathbf{0}$. Therefore $\mathbf{w}_i=\mathbf{0}$.

Since $C_T(p^{\epsilon_1-1}(T)(\mathbf{v}_1))\oplus\cdots\oplus C_T(p^{\epsilon_s-1}(T)(\mathbf{v}_s))$ is a subspace of $N(p)$ and the minimal polynomial of $N(p)$ is p, by Theorem 7.2.3, there are $\mathbf{v}_{s+1},\ldots,\mathbf{v}_k$ such that

$$N(p)=C_T(p^{\epsilon_1-1}(T)(\mathbf{v}_1))\oplus\cdots\oplus C_T(p^{\epsilon_s-1}(T)(\mathbf{v}_s))\oplus C_T(\mathbf{v}_{s+1})\oplus\cdots\oplus C_T(\mathbf{v}_k) \tag{7.44}$$

We want to show that

$$V=C_T(\mathbf{v}_1)\oplus\cdots\oplus C_T(\mathbf{v}_s)\oplus C_T(\mathbf{v}_{s+1})\oplus\cdots\oplus C_T(\mathbf{v}_k) \tag{7.45}$$

First we need to show that $\left(\sum_{j=1}^{s}C_T(\mathbf{v}_j)+\sum_{j<i}C_T(\mathbf{v}_{s+j})\right)\cap C_T(\mathbf{v}_{s+i})=\mathbf{0}$ for $1\leq i\leq k-s$. Now let $\mathbf{w}_{s+i}\in\left(\sum_{j=1}^{s}C_T(\mathbf{v}_j)+\sum_{j<i}C_T(\mathbf{v}_{s+j})\right)\cap C_T(\mathbf{v}_{s+i})$ and write $\mathbf{w}_{s+i}=\sum_{j=1}^{s}\mathbf{w}_j+\sum_{j<i}\mathbf{w}_{s+j}$ with $\mathbf{w}_j\in C(\mathbf{v}_j)$. Hence $\sum_{j=1}^{s}\mathbf{w}_j+\sum_{j<i}\mathbf{w}_{s+j}-\mathbf{w}_{s+i}=\mathbf{0}$. Since $C_T(\mathbf{v}_{s+j})\subseteq N(p)$ $(1\leq j\leq i)$ we have

$$\mathbf{0}=\sum_{j=1}^{s}p(T)(\mathbf{w}_j)+\sum_{0<j<i}p(T)(\mathbf{w}_{s+j})-p(T)(\mathbf{w}_{s+i})=\sum_{j=1}^{s}p(T)(\mathbf{w}_j)\in\bigoplus_{j=1}^{s}C_T(p(T)\mathbf{v}_j)$$

Hence, by Theorem 6.2.36 and (7.44), $p(T)(\mathbf{w}_j)=\mathbf{0}$ $(1\leq j\leq s)$ and hence $\mathbf{w}_j\in N_{C_T(\mathbf{v}_j)}(p)=C_T(p^{\epsilon_j-1}(T)(\mathbf{v}_j))$ by (7.27). Therefore

$$\mathbf{w}_{s+i}=\sum_{j=1}^{s}\mathbf{w}_j+\sum_{j<i}\mathbf{w}_{s+j}\in\left(\sum_{j=1}^{s}C_T(p^{\epsilon_j-1}(T)(\mathbf{v}_j))+\sum_{j<i}C_T(\mathbf{v}_{s+j})\right)\cap C_T(\mathbf{v}_{s+i})$$

and by (7.44), we have $\mathbf{w}_{s+i}=\mathbf{0}$.

Next we need to show that $V = \sum_{i=1}^{k} C_T(\mathbf{v}_i)$. If $\mathbf{v} \in V$ then $p(T)(\mathbf{v}) \in \oplus_{j=1}^{s} C_T(p(T)(\mathbf{v}_j))$ by (7.43). Hence we can write $p(T)(\mathbf{v}) = \sum_{j=1}^{s} f_j p(T)(\mathbf{v}_j)$ for some $f_j \in F[t]$. Thus, by (7.44), we have

$$\mathbf{v} - \sum_{j=1}^{s} f_j(T)(\mathbf{v}_j) \in N(p) = \bigoplus_{j=1}^{s} C_T(p^{\epsilon_j - 1}(T)(\mathbf{v}_j)) \oplus \bigoplus_{j=1}^{s-k} C_T(\mathbf{v}_{s+j})$$

Therefore, there are $g_j, h_{s+j} \in F[t]$ such that

$$\mathbf{v} - \sum_{j=1}^{s} f_j(T)(\mathbf{v}_j) = \sum_{j=1}^{s} g_j p^{\epsilon_j - 1}(T)(\mathbf{v}_j) + \sum_{j=1}^{s-k} h_{s+j}(T)(\mathbf{v}_{s+j})$$

and hence

$$\mathbf{v} = \sum_{j=1}^{s} (f_j + g_j p^{\epsilon_j - 1})(T)(\mathbf{v}_j) + \sum_{j=1}^{s-k} h_{s+j}(T)(\mathbf{v}_{s+j}) \in \sum_{i=1}^{k} C_T(\mathbf{v}_i)$$

as required.

So far, we have shown that if (7.44) is true then (7.45) is true. (The converse is obviously true.) And they consists of the same number say, k factors. Each factor of $N(p)$ is of dimension $r = \deg p$ and there are k of them. Hence $\nu(p) = kr$ which proves (7.37).

(2) Now it is obvious that the minimal polynomial of T is p^{ϵ_1} and hence $\epsilon = \epsilon_1$ since we assumed $q = p^{\epsilon}$. To prove (7.40) first we have $N(p^i) = \oplus N_{C_T(\mathbf{v}_i)}(p^i)$ by Exercise 7.1.17. By (7.35), $N_{C_T(\mathbf{v}_l)}(p^i) = C_T(p^{\epsilon_l - 1}(T)(\mathbf{v}_l))$. The statement about dimension (7.41) follows again from Theorem 7.2.15 with ϵ_l for n and $\epsilon_l - i$ for i in (7.36).

Let f_i and q_i be the characteristic and minimal polynomial of the restriction $T_i := T_{C_T(\mathbf{v}_i)}$ on $C_T(\mathbf{v}_i)$. Then $q_i = q_{\mathbf{v}_i} = p^{\epsilon_i}$ by our choice of $\mathbf{v}_i$. By Corollary 7.1.18, we see $f_i = \pm q_i = \pm p^{\epsilon_i}$. The characteristic polynomial of T is the product $\prod f_i = \pm \prod p^{\epsilon_i} = p^{\sum \epsilon_i}$ by Theorem 6.8.9. Since the characteristic polynomial of T is p^e we conclude that $\sum \epsilon_i = e$ which proves (7.39).

(3) The assertion for the matrix of T with respect to $\mathcal{B} = \cup_i \mathcal{B}_{\mathbf{v}_i}$ is obvious now. □

The following result is a corollary of the proof of (7.38).

Corollary 7.2.17. *Under the same assumption of* Theorem 7.2.16, *the condition*

$$N(p) = C_T(p^{\epsilon_1 - 1}(T)(\mathbf{v}_1)) \oplus \cdots \oplus C_T(p^{\epsilon_s - 1}(T)(\mathbf{v}_s)) \oplus C_T(\mathbf{v}_{s+1}) \oplus \cdots \oplus C_T(\mathbf{v}_k) \quad (7.46)$$

is equivalent to the condition

$$V = C_T(\mathbf{v}_1) \oplus \cdots \oplus C_T(\mathbf{v}_s) \oplus C_T(\mathbf{v}_{s+1}) \oplus \cdots \oplus C_T(\mathbf{v}_k) \quad (7.47)$$

In particular, the number of factors k in the decomposition of V into cyclic subspaces is the same as the number of factors in the decomposition of $N(p)$ into cyclic subspaces which is given by

$$\nu(p(T)) = k \deg p \quad (7.48)$$

Proof. If (7.47) holds then it is obvious that (7.46) is true. The converse was shown during the proof of Theorem 7.2.16 (1) and (7.40). The equality (7.48) is a part of the theorem. □

In the decomposition of V in (7.38) the vectors $\mathbf{v}_i$ are not unique nor the cyclic subspaces $C_T(\mathbf{v}_i)$. What is unique is the matrix representation $[T]_{\mathcal{B}}$ of (7.42) as we see in the following example.

Example 7.2.18. *Let* $A = \begin{bmatrix} 0 & 0 & -1 & & \\ 1 & 0 & -3 & & \\ 0 & 1 & -3 & & \\ & & & 0 & -1 \\ & & & 1 & -2 \end{bmatrix}$. *Let* $V = \mathbb{R}^5$, $T = T_A$. *The characteristic polynomial is* $f(t) = -(t+1)^5$. *The minimal polynomial is a divisor of* $f(t)$. *We compute*

$$A + I = \begin{bmatrix} 1 & 0 & -1 & & \\ 1 & 1 & -3 & & \\ 0 & 1 & -2 & & \\ & & & 1 & -1 \\ & & & 1 & -1 \end{bmatrix}, \ (A+I)^2 = \begin{bmatrix} 1 & -1 & 1 & & \\ 2 & -2 & 2 & & \\ 1 & -1 & 1 & & \\ & & & 0 & 0 \\ & & & 0 & 0 \end{bmatrix}, \ (A+I)^3 = \begin{bmatrix} 0 & 0 & 0 & 0 & 0 \\ 0 & 0 & 0 & 0 & 0 \\ 0 & 0 & 0 & 0 & 0 \\ 0 & 0 & 0 & 0 & 0 \\ 0 & 0 & 0 & 0 & 0 \end{bmatrix}$$

Hence the minimal polynomial is $q(t) = (t+1)^3$. *Consider* $C_T(\mathbf{e}^1)$:

$$\mathbf{e}^1 = \begin{bmatrix} 1 \\ 0 \\ 0 \\ 0 \\ 0 \end{bmatrix}, \ A\mathbf{e}^1 = \begin{bmatrix} 0 \\ 1 \\ 0 \\ 0 \\ 0 \end{bmatrix}, \ A^2\mathbf{e}^1 = \begin{bmatrix} 0 \\ 0 \\ 1 \\ 0 \\ 0 \end{bmatrix}, \ A^3\mathbf{e}^1 = \begin{bmatrix} -1 \\ -3 \\ -3 \\ 0 \\ 0 \end{bmatrix}$$

Since $\mathcal{B}_{\mathbf{e}^1} = \{\mathbf{e}^1, A\mathbf{e}^1, A^2\mathbf{e}^1\}$ *is linearly independent and* $\{\mathbf{e}^1, A\mathbf{e}^1, A^2\mathbf{e}^1, A^3\mathbf{e}^1\}$ *is linearly dependent with the relation* $-\mathbf{e}^1 - 3A\mathbf{e}^1 - 3A^2\mathbf{e}^1 = A^3\mathbf{e}^1$ *we see that* $q_{\mathbf{e}^1}(t) = (t+1)^3$ *and* $\{\mathbf{e}^1, A\mathbf{e}^1, A^2\mathbf{e}^1\}$ *is a basis of* $C_T(\mathbf{e}^1)$.

Now consider $C_T(\mathbf{e}^2)$:

$$\mathbf{e}^2 = \begin{bmatrix} 0 \\ 1 \\ 0 \\ 0 \\ 0 \end{bmatrix}, \ A\mathbf{e}^2 = \begin{bmatrix} 0 \\ 0 \\ 1 \\ 0 \\ 0 \end{bmatrix}, \ A^2\mathbf{e}^2 = \begin{bmatrix} -1 \\ -3 \\ -3 \\ 0 \\ 0 \end{bmatrix}, \ A^3\mathbf{e}^2 = \begin{bmatrix} 3 \\ 8 \\ 6 \\ 0 \\ 0 \end{bmatrix},$$

Again, $\mathcal{B}_{\mathbf{e}^2} = \{\mathbf{e}^2, A\mathbf{e}^2, A^2\mathbf{e}^2\}$ *is linearly independent and* $\{\mathbf{e}^2, A\mathbf{e}^2, A^2\mathbf{e}^2, A^3\mathbf{e}^2\}$ *is linearly dependent. Hence we see that* $\mathcal{B}_{\mathbf{e}^2}$ *is a basis of* $C_T(\mathbf{e}^2)$ *and* $q_{\mathbf{e}^2}(t) = (t+1)^3$ *since* $A^3\mathbf{e}^2 = -\mathbf{e}^2 - 3A\mathbf{e}^2 - 3A^2\mathbf{e}^2$. *Therefore,* $C_T(\mathbf{e}^1) = C_T(\mathbf{e}^2) = \mathrm{Span}\{\mathbf{e}^1, \mathbf{e}^2, \mathbf{e}^3\}$. *Also we can check that* $C_T(\mathbf{e}^3) = \mathrm{Span}\{\mathbf{e}^1, \mathbf{e}^2, \mathbf{e}^3\}$.

Now consider $C_T(\mathbf{e}^4)$:

$$\mathbf{e}^4 = \begin{bmatrix} 0 \\ 0 \\ 0 \\ 1 \\ 0 \end{bmatrix}, \ A\mathbf{e}^4 = \begin{bmatrix} 0 \\ 0 \\ 0 \\ 0 \\ 1 \end{bmatrix}, \ A^2\mathbf{e}^4 = \begin{bmatrix} 0 \\ 0 \\ 0 \\ -1 \\ -2 \end{bmatrix}$$

Hence $\mathcal{B}_{\mathbf{e}^4} = \{\mathbf{e}^4, A\mathbf{e}^4\}$ *is linearly independent and* $\{\mathbf{e}^4, A\mathbf{e}^4, A^2\mathbf{e}^4\}$ *is linearly dependent with the relation* $A^2\mathbf{e}^4 = -\mathbf{e}^4 - 2A\mathbf{e}^4$ *and* $q_{\mathbf{e}^4}(t) = (t+1)^2$. *And*

$$V = C_T(\mathbf{e}^1) \oplus C_T(\mathbf{e}^4) = C_T(\mathbf{e}^2) \oplus C_T(\mathbf{e}^4) = C_T(\mathbf{e}^3) \oplus C_T(\mathbf{e}^4) \tag{7.49}$$

The dimension of the first factor of (7.49) *is* 3 *and the second factor is of dimension* 2. *The matrix representation of* T *with respect to* $\mathcal{B} = \mathcal{B}_{\mathbf{e}^1} \cup \mathcal{B}_{\mathbf{e}^4}$ *is* A.

If we let $p(t) = t+1$ *then the data* (7.40) *corresponding to the decomposition* $V = C_T(\mathbf{e}^1) \oplus C_T(\mathbf{e}^4)$ *is*

$$\begin{array}{ccccc} N(p(t)) & \subseteq & N(p(t)^2) & \subseteq & N(p(t)^3) \\ C_T(p^2(T)(\mathbf{e}^1)) & & C_T(p(T)(\mathbf{e}^1)) & & C_T(\mathbf{e}^1) \\ \oplus & & \oplus & & \\ C_T(p(T)(\mathbf{e}^4)) & & C_T(\mathbf{e}^4) & & \end{array} \tag{7.50}$$

For example, $N(p(t)^2) = C_T(p(T)(\mathbf{e}^1)) \oplus C_T(\mathbf{e}^4)$, $p(T)\mathbf{e}^1 = \mathbf{e}^1 + \mathbf{e}^2$ *and the dimension of* $C_T(p(T)(\mathbf{e}^1))$ *is* 2 *by* (7.41).

If we let $\mathbf{v}_1 = \begin{bmatrix} 0 \\ 0 \\ 1 \\ 1 \\ 0 \end{bmatrix}$. *Then* $A\mathbf{v}_1 = \begin{bmatrix} -1 \\ -3 \\ -3 \\ 0 \\ 1 \end{bmatrix}$, $A^2\mathbf{v}_1 = \begin{bmatrix} 3 \\ 8 \\ 6 \\ -1 \\ -2 \end{bmatrix}$, $A^3\mathbf{v}_1 = \begin{bmatrix} -6 \\ -15 \\ -10 \\ 2 \\ 3 \end{bmatrix}$ *and these vectors satisfies the relation*

$$-\mathbf{v}_1 - 3A\mathbf{v}_1 - 3A^2\mathbf{v}_1 = A^3\mathbf{v}_1$$

Hence $\mathcal{B}_{\mathbf{v}_1} = \{\mathbf{v}_1, A\mathbf{v}_1, A^2\mathbf{v}_1\}$ *is a cyclic basis for* $C_T(\mathbf{v}_1)$, *Now if we take* $\mathbf{v}_2 = \begin{bmatrix} 0 \\ 0 \\ 0 \\ 0 \\ 1 \end{bmatrix}$ *then we have*

$$\mathbf{v}_2 = \begin{bmatrix} 0 \\ 0 \\ 0 \\ 0 \\ 1 \end{bmatrix}, \quad A\mathbf{v}_2 = \begin{bmatrix} 0 \\ 0 \\ 0 \\ -1 \\ -2 \end{bmatrix}, \quad A^2\mathbf{v}_2 = \begin{bmatrix} 0 \\ 0 \\ 0 \\ 2 \\ 3 \end{bmatrix}$$

satisfies the relation

$$-\mathbf{v}_2 - 2A\mathbf{v}_2 = A^2\mathbf{v}_2$$

and $\mathcal{B}_{\mathbf{v}_2} = \{\mathbf{v}_2, A\mathbf{v}_2\}$ *is a cyclic basis for* $C_T(\mathbf{v}_2)$.

Therefore

$$V = C(\mathbf{v}_1) \oplus C_T(\mathbf{v}_2) \tag{7.51}$$

and the matrix representation of T *with respect to* $\mathcal{B} = \mathcal{B}_{\mathbf{v}_1} \cup \mathcal{B}_{\mathbf{v}_2}$ *is* A.

Obviously the decomposition (7.49) *and* (7.51) *are distinct. But the matrix representation of the restriction of* T *on the cyclic subspaces with respect to the cyclic bases are the same.*

Practice Problem 7.2.3. *In* Example 7.2.18, *let* $\mathbf{v}_1 = \begin{bmatrix} 1 \\ 1 \\ 0 \\ 0 \\ 0 \end{bmatrix}$, $\mathbf{v}_2 = \begin{bmatrix} 0 \\ 0 \\ 0 \\ 1 \\ 1 \end{bmatrix}$. *Find the minimal polynomials of* $\mathbf{v}_1, \mathbf{v}_2$ *and compute* $C_T(\mathbf{v}_1)$, $C_T(\mathbf{v}_2)$ *to see if* $V = C_T(\mathbf{v}_1) \oplus C_T(\mathbf{v}_2)$.

The following example explains why the rational canonical form of a linear operator is independent of the choice of the generators of the cyclic subspaces.

Example 7.2.19. *Let T be a linear operator on a vector space V of dimension 24 over $\mathbb{R}$ with the characteristic polynomial $f(t) = (t^2+1)^{12}$ and the minimal polynomial $q(t) = (t^2+1)^4$. Let $p(t) = t^2+1$. Suppose we have the decomposition of a vector space V into cyclic subspaces*

$$V = C_T(\mathbf{v}_1) \oplus C_T(\mathbf{v}_2) \oplus C_T(\mathbf{v}_3) \oplus C_T(\mathbf{v}_4) \oplus C_T(\mathbf{v}_5) \tag{7.52}$$

Suppose the minimal polynomials of $\mathbf{v}_i$ are given by

$$q_{\mathbf{v}_1} = p^4,\ q_{\mathbf{v}_2} = p^4,\ q_{\mathbf{v}_3} = p^2,\ q_{\mathbf{v}_4} = p,\ q_{\mathbf{v}_5} = p \tag{7.53}$$

Let $\mathcal{B}_{\mathbf{v}_i}$ be the cyclic basis for $C_T(\mathbf{v}_i)$ and let $\mathcal{B} = \cup\mathcal{B}_{\mathbf{v}_i}$ then, by (7.42), the matrix representation of T with respect to $\mathcal{B}$ is,

$$[T]_{\mathcal{B}} = \begin{bmatrix} C(p^4) & & & & \\ & C(p^4) & & & \\ & & C(p^2) & & \\ & & & C(p) & \\ & & & & C(p) \end{bmatrix} \tag{7.54}$$

We will show we can find the matrix (7.54) can be obtained without ever mentioning $\mathbf{v}_i$ but just looking at the nullities $\nu(p^i)$ $(i = 1, 2, 3, \ldots)$.

We have a chains of subspaces (here we omit $\oplus$ along each column)

$$\begin{array}{ccccccc} N(p) & \subseteq & N(p^2) & \subseteq & N(p^3) & \subseteq & N(p^4) \\ C_T(p^3(T)(\mathbf{v}_1)) & & C_T(p^2(T)(\mathbf{v}_1)) & & C_T(p(T)(\mathbf{v}_1) & & C_T(\mathbf{v}_1) \\ C_T(p^3(T)(\mathbf{v}_2)) & & C_T(p^2(T)(\mathbf{v}_2)) & & C_T(p(T)(\mathbf{v}_2) & & C_T(\mathbf{v}_2) \\ C_T(p(T)(\mathbf{v}_3)) & & C_T(\mathbf{v}_3) & & & & \\ C_T(\mathbf{v}_4) & & & & & & \\ C_T(\mathbf{v}_5) & & & & & & \end{array} \tag{7.55}$$

The lengths, 4 in this case, of the first two rows resulted from the fact that the minimal polynomials of $\mathbf{v}_1$ and $\mathbf{v}_2$ are p^4 and the length 2 of the third row reflects the fact that the minimal polynomial of $\mathbf{v}_3$ is p^2; the length of the fourth and fifth rows reflect the fact that the minimal polynomials of $\mathbf{v}_4, \mathbf{v}_5$ are p. Therefore the number of terms ϵ_i of the i-th row in (7.55) is the exponents of p in (7.53) and the sum of exponents is 12 which is the exponent in characteristic polynomial in p.

$$\epsilon_1 = \epsilon = 4, \epsilon_2 = 4, \epsilon_3 = 2, \epsilon_4 = 1, \epsilon_5 = 1; e = \sum \epsilon_i = 12 \tag{7.56}$$

The dimension of each term in (7.55) increase by $2(= \deg p)$ whenever we move one step to the right in each row (Theorem 7.2.15). *For example, the dimensions of the first row are*

$$\dim C_T(p^3(T)(\mathbf{v}_1)) = 2,\ \dim C_T(p^2(T)(\mathbf{v}_1)) = 4,\ \dim C_T(p(T)(\mathbf{v}_1) = 6,\ \dim C_T(\mathbf{v}_1)) = 8$$

by (7.36), and the number of terms in the first row is $\dim C_T(\mathbf{v}_1)/\deg p = 8/2 = 4 = \epsilon_1$.

The matrix (7.54) *is determined by the numbers* (7.56) *– the matrix on the upper left corner is* $C(p^{\epsilon_1})$ *and the next one is* $C(p^{\epsilon_2})$ *and so on. If we know* $\mathbf{v}_i$ *we could find the length of the rows of* (7.55). *However we want to find it without knowing* $\mathbf{v}_i$ *explicitly.*

Now we can compute $\nu(p^i)$ *without finding* $\mathbf{v}_i$*'s. First by* (7.40),

$$\begin{aligned} N(p) &= \bigoplus_l C_T(p^{\epsilon_l - 1}(T)(\mathbf{v}_l)) \\ &= C_T(p^3(T)\mathbf{v}_1) \oplus C_T(p^3(T)\mathbf{v}_2) \oplus C_T(p(T)\mathbf{v}_3) \oplus C_T(\mathbf{v}_4) \oplus C_T(\mathbf{v}_5) \end{aligned}$$

and $\nu(p)$ *is the dimension of the sum of the cyclic subspaces in the first column of* (7.55) *and each cyclic subspace is of dimension* 2. *Hence* $\nu(p) = 10$ *and the number of terms of the first column is* $\nu(p)/\deg p = 5$ *which is the same as the number of terms of cyclic factors in the cyclic decomposition of* V (Corollary 7.2.17 (7.48)). *By* (7.40) *again we have*

$$\begin{aligned} N(p^2) &= \bigoplus_l C_T(p^{\epsilon_l - 2}(T)(\mathbf{v}_l)) \\ &= C_T(p^2(T)\mathbf{v}_1) \oplus C_T(p^2(T)\mathbf{v}_2) \oplus C_T(\mathbf{v}_3) \oplus C_T(\mathbf{v}_4) \oplus C_T(\mathbf{v}_5) \end{aligned}$$

And the number of terms in the first two columns is $\nu(p^2)/\deg p = (4+4+4+2+2)/2 = 8$ *and the number of terms in the first three columns is* $\nu(p^3)/2 = 10$ *and so on.*

Hence if we denote the number of terms in the i-th column by ϵ^i, *then the number of terms of the columns in* (7.55) *is given by*

$$\begin{aligned} \nu(p)/2 &= \text{number of terms in the first column} = 5 = \epsilon^1 \\ \nu(p^2)/2 &= \text{number of terms in the first two columns} = 8 = \epsilon^1 + \epsilon^2 \\ \nu(p^3)/2 &= \text{number of terms in the first three columns} = 10 = \epsilon^1 + \epsilon^2 + \epsilon^3 \\ \nu(p^4)/2 &= \text{number of terms in the first four columns} = 12 = \epsilon^1 + \epsilon^2 + \epsilon^3 + \epsilon^4 \end{aligned} \tag{7.57}$$

Let's represent each term in (7.55) *by a* $\bullet$ *using the data* (7.57)*: The first column has* $\nu(p)/2$ *dots and; the first two columns have* $\nu(p^2)/2$ *dots with the second column filled in from the top most place and; the first three columns have* $\nu(p^3)/2$ *dots and; the last four columns have* $\nu(p^4))$ *dots which we will call a* **dot diagram**.

$$\begin{array}{ccccc} \bullet & \bullet & \bullet & \bullet & \epsilon_1 \\ \bullet & \bullet & \bullet & \bullet & \epsilon_2 \\ \bullet & \bullet & & & \epsilon_3 \\ \bullet & & & & \epsilon_4 \\ \bullet & & & & \epsilon_5 \\ & & & & \\ \epsilon^1 & \epsilon^2 & \epsilon^3 & \epsilon^4 & \end{array} \tag{7.58}$$

And the number ϵ_i *of dots in the i-th row is*

$$\epsilon_1 = 4, \ \epsilon_2 = 4, \ \epsilon_3 = 2, \ \epsilon_4 = 1, \ \epsilon_5 = 1 \tag{7.59}$$

which corresponds to the exponent of p *in* $q_{\mathbf{v}_i}$ *i.e.,* $q_{\mathbf{v}_i} = p^{\epsilon_i}$. *Hence*

$$q_{\mathbf{v}_1} = p^4, \ q_{\mathbf{v}_2} = p^4, \ q_{\mathbf{v}_3} = p^2, \ q_{\mathbf{v}_4} = p, \ q_{\mathbf{v}_5} = p$$

Notice that (7.57) *depends only on* T, *not on the choice of the generators* $\mathbf{v}_i$. *The data* (7.57) *determines the dot diagram* (7.58) *which in turn determines the matrix* (7.54).

From these observations we have the following uniqueness statement. We will omit the formal proof of this fact.

Corollary 7.2.20. *Let $T \in L(V)$ with the characteristic polynomial $f(t) = p^e$ and the minimal polynomial $q(t) = p^\epsilon$ where p is an irreducible polynomial. In the decomposition of V into T-cyclic subspaces of (7.38), the numbers ϵ_i $(1 \le i \le k)$ which is the exponent of p in the minimal polynomial of $\mathbf{v}_i$ are uniquely determined independent of the choice of the generators $\mathbf{v}_i$ of the cyclic subspaces. Therefore, the rational canonical form (7.42) is uniquely determined independent of the choice of the generators $\mathbf{v}_i$'s.*

Corollary 7.2.21. *Let $A, B \in M(n, F)$ and their characteristic polynomials are p^e where p is an irreducible polynomial. Then the following conditions are equivalent.*

(a) *A and B are similar*

(b) *$\nu(p^i(A)) = \nu(p^i(B))$ for all $i \le e$ i.e., A, B have the same rational canonical form.*

(c) *$\rho(p^i(A)) = \rho(p^i(B))$ for all $i \le e$.*

Proof. (a)⇒(b) Suppose $B = P^{-1}AP$. Then $p^i(B) = P^{-1}p^i(A)P$ (Exercise 7.1.11). Hence $\nu(p^i(B)) = \nu(P^{-1}p^i(A)P) = \nu(p^i(A))$. This means that the dot diagram of T_A and T_B are the same. Hence they have the same rational canonical form.

(b)⇒(a) If $\nu(p^i(B)) = \nu(P^{-1}p^i(A)P) = \nu(p^i(A))$ for all i then the dot diagram for T_A and T_B are the same. Hence they have the same rational canonical form C. Hence $P^{-1}AP = C$ and $Q^{-1}BQ = C$ for some invertible matrices P, Q. Thus $(PQ^{-1})^{-1}A(PQ^{-1}) = B$.

(b)⇔(c) Follows from the rank theorem (Theorem 6.3.21). □

Example 7.2.22. *Let T be a linear operator on $V = \mathbb{R}^{12}$. Let $f(t) = (t^2+1)^6$ be the characteristic polynomial and $q(t) = (t^2+1)^2$ be the minimal polynomial of T. We have a decomposition*

$$V = \bigoplus_i C_T(\mathbf{v}_i)$$

What are the possible rational canonical form for T? By Corollary 7.2.20, this amounts to find all possible ϵ_i's which are the exponent of p in $q_{\mathbf{v}_i}$. Since the minimal polynomial of T is p^2 we see that $\epsilon_1 = 2$ and we require $\epsilon_1 \ge \epsilon_2 \ge \cdots \ge \epsilon_k$. Since the characteristic polynomials is p^6 we should have $\sum \epsilon_i = 6$ (Theorem 7.2.16). Hence the possible cases are

$$6 = 2+2+2 = 2+2+1+1 = 2+1+1+1+1$$

The corresponding minimal polynomials of $\mathbf{v}_i$ are

$$\begin{aligned} &q_{\mathbf{v}_1}(t) = p(t)^2,\ q_{\mathbf{v}_2}(t) = p(t)^2,\ q_{\mathbf{v}_3}(t) = p(t)^2 \\ &q_{\mathbf{v}_1}(t) = p(t)^2,\ q_{\mathbf{v}_2}(t) = p(t)^2,\ q_{\mathbf{v}_3}(t) = p(t),\ q_{\mathbf{v}_3}(t) = p(t) \\ &q_{\mathbf{v}_1}(t) = p(t)^2,\ q_{\mathbf{v}_3}(t) = p(t),\ q_{\mathbf{v}_3}(t) = p(t),\ q_{\mathbf{v}_4}(t) = p(t),\ q_{\mathbf{v}_5}(t) = p(t) \end{aligned}$$

The corresponding matrix of T with respect to $\mathcal{B} = \cup_i \mathcal{B}_{\mathbf{v}_i}$ are

$$[T]_{\mathcal{B}} = \begin{bmatrix} C(p^2) & & \\ & C(p^2) & \\ & & C(p^2) \end{bmatrix},\ [T]_{\mathcal{B}} = \begin{bmatrix} C(p^2) & & & \\ & C(p^2) & & \\ & & C(p) & \\ & & & C(p) \end{bmatrix},\ [T]_{\mathcal{B}} = \begin{bmatrix} C(p^2) & & & & \\ & C(p) & & & \\ & & C(p) & & \\ & & & C(p) & \\ & & & & C(p) \end{bmatrix}$$

When written out, for example, the second matrix looks like

$$[T]_{\mathcal{B}} = \begin{bmatrix} 0 & 0 & 0 & -1 & & & & & & & & \\ 1 & 0 & 0 & 0 & & & & & & & & \\ 0 & 1 & 0 & -2 & & & & & & & & \\ 0 & 0 & 1 & 0 & & & & & & & & \\ & & & & 0 & 0 & 0 & -1 & & & & \\ & & & & 1 & 0 & 0 & 0 & & & & \\ & & & & 0 & 1 & 0 & -2 & & & & \\ & & & & 0 & 0 & 1 & 0 & & & & \\ & & & & & & & & 0 & -1 & & \\ & & & & & & & & 1 & 0 & & \\ & & & & & & & & & & 0 & -1 \\ & & & & & & & & & & 1 & 0 \end{bmatrix}$$

Example 7.2.23. *We want to decide whether the following two matrices are similar.*

$$A = \begin{bmatrix} 0 & 0 & 0 & -1 & 0 & 0 \\ 1 & 0 & 0 & 0 & 0 & 1 \\ 0 & 1 & 0 & -2 & 0 & 0 \\ 0 & 0 & 1 & 0 & 0 & 0 \\ & & & & 0 & -1 \\ & & & & 1 & 0 \end{bmatrix}, \quad B = \begin{bmatrix} 0 & 0 & 0 & -1 & 0 & 0 \\ 1 & 0 & 0 & 0 & 0 & 1 \\ 0 & 1 & 0 & -2 & 1 & 0 \\ 0 & 0 & 1 & 0 & 0 & 0 \\ & & & & 0 & -1 \\ & & & & 1 & 0 \end{bmatrix}$$

They differ only in $(3,5)$ *position. The upper left and lower right portions of* A, B *are* $C((t^2+1)^2),\ C(t^2+1)$

The characteristic polynomials for both A *and* B *are* $f(t) = (t^2+1)^3$. *Let* $p = t^2+1$. *We compute:*

$$A^2+I = \begin{bmatrix} 1 & 0 & -1 & 0 & 0 & 0 \\ 0 & 1 & 0 & -1 & 1 & 0 \\ 1 & 0 & -1 & 0 & 0 & 1 \\ 0 & 1 & 0 & -1 & 0 & 0 \\ 0 & 0 & 0 & 0 & 0 & 0 \\ 0 & 0 & 0 & 0 & 0 & 0 \end{bmatrix}, (A^2+I)^2 = \begin{bmatrix} 0 & 0 & 0 & 0 & 0 & -1 \\ 0 & 0 & 0 & 0 & 1 & 0 \\ 0 & 0 & 0 & 0 & 0 & -1 \\ 0 & 0 & 0 & 0 & 1 & 0 \\ 0 & 0 & 0 & 0 & 0 & 0 \\ 0 & 0 & 0 & 0 & 0 & 0 \end{bmatrix}, (A^2+I)^3 = \begin{bmatrix} 0 & 0 & 0 & 0 & 0 & 0 \\ 0 & 0 & 0 & 0 & 0 & 0 \\ 0 & 0 & 0 & 0 & 0 & 0 \\ 0 & 0 & 0 & 0 & 0 & 0 \\ 0 & 0 & 0 & 0 & 0 & 0 \\ 0 & 0 & 0 & 0 & 0 & 0 \end{bmatrix}$$

On the other hand,

$$B^2+I = \begin{bmatrix} 1 & 0 & -1 & 0 & 0 & 0 \\ 0 & 1 & 0 & -1 & 1 & 0 \\ 1 & 0 & -1 & 0 & 0 & 0 \\ 0 & 1 & 0 & -1 & 1 & 0 \\ 0 & 0 & 0 & 0 & 0 & 0 \\ 0 & 0 & 0 & 0 & 0 & 0 \end{bmatrix}, \quad (B^2+I)^2 = \begin{bmatrix} 0 & 0 & 0 & 0 & 0 & 0 \\ 0 & 0 & 0 & 0 & 0 & 0 \\ 0 & 0 & 0 & 0 & 0 & 0 \\ 0 & 0 & 0 & 0 & 0 & 0 \\ 0 & 0 & 0 & 0 & 0 & 0 \\ 0 & 0 & 0 & 0 & 0 & 0 \end{bmatrix}$$

Hence we have

$$\nu(p(A)) = 2,\ \nu(p^2(A)) = 4,\ \nu(p^3(A)) = 6$$
$$\nu(p(B)) = 4,\ \nu(p^2(B)) = 6$$

The corresponding • *diagrams are*

$$\bullet\text{-diagram for } A: \quad \bullet \quad \bullet \quad \bullet$$

$$\bullet\text{-diagram for } B: \quad \begin{matrix} \bullet & \bullet \\ \bullet & \end{matrix}$$

Therefore the rational canonical form R_A, R_B for A, B are

$$R_A = \begin{bmatrix} C(p^3) \end{bmatrix}, \quad R_B = \begin{bmatrix} C(p^2) & \\ & C(p) \end{bmatrix}$$

and hence A, B are not similar (Corollary 7.2.21).

So far we considered the case when the characteristic polynomial is a power of an irreducible polynomial. We obtained good enough results to piece them together to obtain a 'global' result.

Let $T \in L(V)$ and $f(t) = \pm \prod_{i=1}^{k} p_i(t)^{e_i}$ be its characteristic polynomial p_i is monic and irreducible. By Theorem 7.2.9, we can write $V = L(p_1) \oplus L(p_2) \oplus \cdots \oplus L(p_k)$. Let T_i be the restriction of T on $L(p_i)$ then the characteristic polynomial of T_i is $\pm p_i^{e_i}$. Hence we can apply Theorem 7.2.16 to each T_i and then piece them together using Theorem 7.2.8 to get:

Theorem 7.2.24 (Decomposition into Cyclic Subspaces II). *Let $T \in L(V)$ with $\dim V = n$. Let p_i be a monic irreducible polynomial with $\deg p_i = r_i$. Let $f(t) = (-1)^n \prod_{i=1}^{k} p_i(t)^{e_i}$ be the characteristic polynomial and $q(t) = \prod_{i=1}^{k} p_i(t)^{\epsilon_i}$ be the minimal polynomial. Then*

$$\begin{aligned} V =& L(p_1) \oplus L(p_2) \oplus \cdots \oplus L(p_k) \text{ and} \\ L(p_i) =& N(p_i^{\epsilon_i}(T)) \text{ with } \dim L(p_i) = r_i e_i \end{aligned}$$

and

$$L(p_i) = C_T(\mathbf{v}_{i1}) \oplus C_T(\mathbf{v}_{i2}) \cdots \oplus C_T(\mathbf{v}_{in_i})$$

where $\dim C_T(\mathbf{v}_{ij}) = r_i \epsilon_{ij}$ and $\mathbf{v}_{ij}$ has the minimal polynomial $q_{\mathbf{v}_{ij}}$ given by

$$q_{\mathbf{v}_{i1}} = p_i^{\epsilon_i}, \ \epsilon_i = \epsilon_{i1} \tag{7.60}$$

$$q_{\mathbf{v}_{ij}} = p_i^{\epsilon_{ij}}, \ \epsilon_{i1} + \epsilon_{i2} + \cdots + \epsilon_{in_i} = e_i, \ \epsilon_{i1} \geq \epsilon_{i2} \geq \cdots \geq \epsilon_{in_i} \tag{7.61}$$

Let $\mathcal{B}_{\mathbf{v}_{ij}}$ be the cyclic basis for $C_T(\mathbf{v}_{ij})$ and let $\mathcal{B}_i = \cup_j \mathcal{B}_{\mathbf{v}_{ij}}$:

$$\begin{aligned} \mathcal{B}_i = & \{\mathbf{v}_{i1}, T(\mathbf{v}_{i1}), ..., T^{r_i\epsilon_{i1}-1}(\mathbf{v}_{i1})\} \cup \{\mathbf{v}_{i2}, T(\mathbf{v}_{i2}), ..., T^{r_i\epsilon_{i2}-1}(\mathbf{v}_{i2})\} \cup \cdots \\ & \cup \{\mathbf{v}_{ik_i}, T(\mathbf{v}_{ik_i}), ..., T^{r_i\epsilon_{ik_i}-1}(\mathbf{v}_{ik_i})\} \end{aligned}$$

If T_i is the restriction of T on $L(p_i)$ then the minimal polynomial of T_i is $q_{\mathbf{v}_{i1}} = p_i^{\epsilon_i}$ in (7.60)*. And if we let $\mathcal{B} = \cup_i \mathcal{B}_i$ then*

$$[T]_{\mathcal{B}} = \begin{bmatrix} [T_1]_{\mathcal{B}_1} & & & \\ & [T_2]_{\mathcal{B}_2} & & \\ & & \ddots & \\ & & & [T_k]_{\mathcal{B}_k} \end{bmatrix} \tag{7.62}$$

where

$$[T_i]_{\mathcal{B}_i} = \begin{bmatrix} C(p_i^{\epsilon_{i1}}) & & & \\ & C(p_i^{\epsilon_{i2}}) & & \\ & & \ddots & \\ & & & C(p_i^{\epsilon_{ik_i}}) \end{bmatrix}$$

Proof. By Corollary 7.2.11, $T = \oplus_i T_i$ and the characteristic and minimal polynomial are $p_i^{e_i}$ and $p_i^{\epsilon_i}$ respectively. Now apply Theorem 7.2.16 to each T_i. And then piece them together using Theorem 7.2.8. Checking the details of the proof will be left as an exercise. $\square$

Definition 7.2.25. *The matrix of the form* (7.62) *is called a* **rational canonical form** *for $T \in L(V)$. If $T = T_A$ then we call $[T]_B$ a* **rational canonical form** *for A. It consists of companion matrices of powers of the irreducible factors of the characteristic polynomial.*

Corollary 7.2.26. *Let $A, B \in M(n, F)$. Then A and B are similar if and only if their characteristic polynomials are the same as say, $f(t) = \prod_i p_i^{e_i}$ where p_i is a monic irreducible and $\rho(p_i^j(A)) = \rho(p_i^j(B))$ for all i, j i.e., A, B have the same rational canonical forms.*

Proof. Suppose $B = P^{-1}AP$. Then $p_i^j(B) = P^{-1}p_i^j(A)P$. Hence $\rho(p_i^j(B)) = \rho(P^{-1}p_i^j(A)P)$ $= \rho(p_i^j(A))$ by Theorem 6.3.20.

Conversely, we can assume $\nu(p_i^j(B)) = \nu(P^{-1}p_i^j(A)P) = \nu(p_i^j(A))$ for all i, j by Rank Theorem. Then the dot diagram for the restrictions of T_A and T_B on $L(p_i)$ are the same by (7.29). Hence they have the same rational canonical form C. Hence $P^{-1}AP = C = Q^{-1}BQ$ for some invertible matrices P, Q by Theorem 7.2.9. Thus $(PQ^{-1})^{-1}A(PQ^{-1}) = B$. $\square$

Observe that in Corollary 7.2.26 we only need to check $\rho(p_i^j(A)) = \rho(p_i^j(B))$ for $j \le e_i$. If the minimal polynomials of the restriction on L_i is $p_i^{\epsilon_i}$ then we can save our effort further to check the equality $\rho(p_i^j(A)) = \rho(p_i^j(B))$ for $j \le \epsilon_i$.

Answers for Practice Problems 7.2

(7.2.1) (a)⇒(b) Since $A \sim C(f)$ we have $q_A = q_{C(f)}$ by Theorem 7.1.14. By Corollary 7.1.17 we have $q_{C(f)} = f$.

(b)⇒(c) By Theorem 7.2.6, there is $\mathbf{v}$ such that $V = C(\mathbf{v})$.

(c)⇒(a) By Theorem 7.1.16, $\mathcal{B}_{\mathbf{v}} = \{\mathbf{v}, A\mathbf{v}, ..., A^{n-1}\mathbf{v}\}$ form a basis and $[T_A]_{\mathcal{B}_{\mathbf{v}}} = C(f)$. Hence $A \sim C(f)$.

(7.2.2) First notice that $N(g)$ is $h(T)$-invariant. For if $\mathbf{v} \in N(g)$ then $g(T)(h(T)(\mathbf{v})) = h(T)g(T)(\mathbf{v}) = h(T)(\mathbf{0}) = \mathbf{0}; h(T)(\mathbf{v}) \in N(g)$. Now let $ag + bh = 1$ and let $\mathbf{v} \in N(g)$. If $h(T)(\mathbf{v}) = \mathbf{0}$ then $\mathbf{v} = ag(T)(\mathbf{v}) + bh(T)(\mathbf{v}) = \mathbf{0}$. Hence $h(T) : N(g) \to N(g)$ is injective and therefore $h(T)$ is an isomorphism (Theorem 6.3.26).

(7.2.3) Since $q_{\mathbf{v}_i}(t)$ is a divisor of $(t+1)^5$ and $(A+I)^2(\mathbf{v}_1) \neq \mathbf{0}$ but $(A+I)^3(\mathbf{v}_1) = \mathbf{0}$ we have $q_{\mathbf{v}_1}(t) = (t+1)^2$. For $\mathbf{v}_2$, the same argument shows $q_{\mathbf{v}_2}(t) = (t+1)^2$.

To find $C_T(\mathbf{v}_1)$ we compute

$$\mathbf{v}_1 = \begin{bmatrix} 1 \\ 1 \\ 0 \\ 0 \\ 0 \end{bmatrix}, A\mathbf{v}_1 = \begin{bmatrix} 0 \\ 1 \\ 1 \\ 0 \\ 0 \end{bmatrix}, A^2\mathbf{v}_1 = \begin{bmatrix} -1 \\ -3 \\ -2 \\ 0 \\ 0 \end{bmatrix}$$

and $-\mathbf{v}_1 - 2A\mathbf{v}_1 = A^2\mathbf{v}_1$. Hence $\{\mathbf{v}_1, A\mathbf{v}_1\}$ is a basis of $C_T(\mathbf{v}_1)$. Similarly, $C_T(\mathbf{v}_2)$ has a basis

$$\left\{ \mathbf{v}_2 = \begin{bmatrix} 0 \\ 0 \\ 0 \\ 1 \\ 1 \end{bmatrix}, A\mathbf{v}_2 = \begin{bmatrix} 0 \\ 0 \\ 0 \\ -1 \\ 0 \end{bmatrix} \right\}$$

Therefore $C_T(\mathbf{v}_1) \oplus C_T(\mathbf{v}_2) \neq V$.

7.2 Exercises

1. Let $A = \begin{bmatrix} 0 & -1 & 0 \\ 1 & 0 & 0 \\ 0 & 0 & -1 \end{bmatrix}$ and $T = T_A, V = \mathbb{R}^3$ as in Example 7.1.12.

   (a) Show that $C_T(\mathbf{e}^3) \oplus C_T(\mathbf{e}^1) = C_T(\mathbf{e}^1 + \mathbf{e}^3) = V$.

   (b) Let $\mathcal{B}_1, \mathcal{B}_2, \mathcal{C}$ be the cyclic bases of $C_T(\mathbf{e}^3)$, $C_T(\mathbf{e}^1)$, $C_T(\mathbf{e}^1 + \mathbf{e}^3)$ and let $\mathcal{B} = \mathcal{B}_1 \cup \mathcal{B}_2$. Show that $[T]_{\mathcal{B}} = C(t+1) \oplus C(t^2+1)$ and $[T]_{\mathcal{C}} = C(t^2+t+1)$.

   (c) Find an invertible matrix P such that $P^{-1}[T]_{\mathcal{B}}P = [T]_{\mathcal{C}}$.

2. Referring Example 7.1.12, prove that $V = N(t^2+1) \oplus N(t+1)$ by finding $N(t^2+1)$ and $N(t+1)$ explictly. Also show that $T^2+I : N(t+1) \to N(t+1)$ is an isomorphism.

3. Prove the true statement(s) and find a counter example for the false statement(s).

   (a) If a triangular matrix A is diagonalizable then A is already a diagonal matrix.

   (b) Let $X = \{\mathbf{u}, \mathbf{v}\} \subseteq V$. If X is linearly independent then $q_X = q_{\mathbf{u}} q_{\mathbf{v}}$.

   (c) Let $X = \{\mathbf{u}, \mathbf{v}\} \subseteq V$. If X is linearly independent then $C(\mathbf{u}) \neq C(\mathbf{v})$.

   (d) Let $X = \{\mathbf{u}, \mathbf{v}\} \subseteq V$. If $q_{\mathbf{u}}$ and $q_{\mathbf{v}}$ are relatively prime then X is linearly independent.

   (e) Let $X = \{\mathbf{u}, \mathbf{v}\} \subseteq V$ be linearly independent. If $q_{\mathbf{u}} = q_{\mathbf{v}}$ is a monic irreducible polynomial then $C_T(\mathbf{u}) \cap C_T(\mathbf{v}) = \mathbf{0}$.

   (f) Let $T \in L(V)$ and $f(t) = p(t)^e$ be the characteristic polynomial where p is an irreducible polynomial of degree r. Then the number of factors in the decomposition of V into cyclic subspaces is $\nu(p)/r$.

   (g) Let T be a linear operator on V. Let W, W' be subspaces of V. If $q_W(t) = q_{W'}(t)$ then $W = W'$.

   (h) Let T be a linear operator on V. Let $W \subseteq W'$ be subspaces of V. If $q_W(t) = q_{W'}(t)$ then $W = W'$.

   (i) If the characteristic polynomial of $T \in L(V)$ splits then there is a basis $\mathcal{B}$ such that $[T]_{\mathcal{B}}$ is upper triangular.

   (j) Let $T \in L(V)$. If the minimal polynomial $q_T = p$ is an irreducible polynomial then the characteristic polynomial is $f(t) = \pm p^k$ for some k.

   (k) Suppose the characteristic polynomial is a power of an irreducible polynomial p with degree r. The total number of dots in the dot diagram is the same as $\frac{\dim V}{r}$.

   (l) If $h | q_{\mathbf{v}}$ for some $h \in F[t]$. Then there is $\mathbf{w} \in C_T(\mathbf{v})$ such that $h = q_{\mathbf{w}}$.

   (m) Let T be a linear operator on V. Then the characteristic polynomial of T splits if and only if there is a basis $\mathcal{B}$ such that $[T]_{\mathcal{B}}$ is upper triangular.

4. Prove that if the characteristic polynomial of $A \in M(n, \mathbb{C})$ is $f(t) = (-1)^n(t^n - 1)$ then the minimal polynomial is $q(t) = t^n - 1$ and similar to the diagonal matrix $\operatorname{diag}[\lambda_0, \lambda_2, ..., \lambda_{n-1}]$ with $\lambda_k = e^{\frac{2\pi i k}{n}}$.

5. Let T be a linear operator on a vector space V. Let W be a subspace of V. Define (Exercise 6.8.10)

$$G_T(W) = \left\{ \sum_i f_i(T)\mathbf{w}_i \;\middle|\; \mathbf{w}_i \in W, f_i \in F[t] \right\}$$

Prove that $W \subseteq G_T(W) \subseteq N(q_W)$.

6. Let X be a subset of V and $q_X(t)$ be the minimal polynomial of X. Prove that if $W = N(q_X)$ then $q_X(t) = q_W(t)$.

7. By using the fact that t^2+1 and t^2+t+1 are relatively prime show that the matrices

$$A = \begin{bmatrix} 0 & -1 & 0 & 0 \\ 1 & 0 & 0 & 0 \\ 0 & 0 & 0 & -1 \\ 0 & 0 & 1 & -1 \end{bmatrix}, \quad B = \begin{bmatrix} 0 & 0 & 0 & -1 \\ 1 & 0 & 0 & -1 \\ 0 & 1 & 0 & -2 \\ 0 & 0 & 1 & -1 \end{bmatrix}$$

are similar and find a matrix P such that $B = P^{-1}AP$.

8. Let $A = \begin{bmatrix} 0 & 0 & -1 \\ 1 & 0 & 0 \\ 0 & 1 & 0 \end{bmatrix}$, $V = \mathbb{R}^3$, $T = T_A$. By Theorem 7.2.16, write $V = \bigoplus_{i=1}^{k} C_T(\mathbf{v}_i)$.

   (a) Find the characteristic polynomial $f(t)$ and the minimal polynomial $q(t)$ of T.

   (b) Find the number of factors of cyclic subspaces k and the minimal polynomial $q_{\mathbf{v}_i}$.

   (c) Find $\mathbf{v}_i$ and bases for $C_T(\mathbf{v}_i)$ explicitly.

9. Let $A = \begin{bmatrix} 0 & 0 & 0 & 0 & 0 \\ 1 & 0 & 0 & 0 & 0 \\ 0 & 1 & -1 & 0 & 0 \\ 0 & 0 & 0 & 0 & -1 \\ 0 & 0 & 0 & 1 & 0 \end{bmatrix}$ and $V = \mathbb{R}^5$, $T = T_A$.

   (a) Decompose V into a direct sum of cyclic subspaces $\oplus_i C_T(\mathbf{v}_i)$.

   (b) Find the minimal polynomials $q_{\mathbf{v}_i}$ of $\mathbf{v}_i$.

   (c) Let $W = R(T)$ be the range space of T and let $S = T_W$ be the restriction of T to W. Decompose W into a direct sum $\oplus_i C_S(\mathbf{w}_i)$ of S-cyclic subspaces.

   (d) Find the characteristic polynomial and the minimal polynomial of S.

10. Let $V = \mathbb{R}^5, T = T_A$ with

$$A = \begin{bmatrix} 0 & 0 & -1 & 0 & 0 \\ 1 & 0 & -3 & 0 & 1 \\ 0 & 1 & -3 & 0 & 0 \\ 0 & 0 & 0 & 0 & 1 \\ 0 & 0 & 0 & 1 & -2 \end{bmatrix} = \left[\begin{array}{c|c} C((t+1)^3) & \begin{matrix} 0 & 0 \\ 0 & 1 \\ 0 & 0 \end{matrix} \\ \hline \begin{matrix} 0 & 0 & 0 \\ 0 & 0 & 0 \end{matrix} & C((t+1)^2) \end{array}\right]$$

   (a) Find the decomposition $V = \oplus_i C_T(\mathbf{v}_i)$ and find $q_{\mathbf{v}_i}$.

   (b) Find $[T]_{\mathcal{B}}$ where $\mathcal{B}$ is the union of cyclic bases $\mathcal{B}_{\mathbf{v}_i}$.

11. Let $V = \mathbb{R}^{12}$ and T be a linear operator with the characteristic polynomial $f(t) = (t^2+1)^6$ and minimal polynomial $q(t) = (t^2+1)^3$. Find all possible rational canonical form for T.

12. Let T be a linear operator on $V = \mathbb{R}^{10}$ with the characteristic polynomial $f(t) = (t^2+1)^3(t^2+2)^2$. Find all possible rational canonical forms for T and find the minimal polynomial in each case.

13. Let T be a linear operator on a vector space V. Let $\mathbf{v} \in V$ and $q_{\mathbf{v}}$ be the minimal polynomial of $\mathbf{v}$ with $\deg q_{\mathbf{v}} = k$.
    (a) Let $f \in F[t]$ and $\gcd(q_{\mathbf{v}}, f) = g$ with $\deg g = l$. Prove that $C_T(f(T)(\mathbf{v})) = C_T(g(T)(\mathbf{v}))$ and $\dim C_T(f(T)(\mathbf{v})) = k - l$.
    (b) Let W be a subspace of $C_T(\mathbf{v})$. Prove that W is a T-invariant subspace if and only if $W = C_T(h(T)(\mathbf{v}))$ for some $h \in F[t]$ with $h|q_{\mathbf{v}}$.

14. Prove or disprove: Let T be a linear operator on a vector space V. Let $\mathbf{v}_1, \mathbf{v}_2 \in V$. If W is a nontrivial T-invariant subspace of $C_T(\mathbf{v}_1) \oplus C_T(\mathbf{v}_2)$ not contained in $C_T(\mathbf{v}_1)$ nor $C_T(\mathbf{v}_2)$ then W is of the form $W = C_T(\mathbf{w}_1) \oplus C_T(\mathbf{w}_2)$ for some nonzero $\mathbf{w}_i \in C_T(\mathbf{v}_i)$ $(i = 1, 2)$.

15. Prove that if T is a projection operator i.e., $T^2 = T$ then the only possible eigenvalues are $0, 1$.

16. Let $\mathbf{v} \in V$ and $T \in L(V)$. Prove that if $q_{\mathbf{v}}(t)$ is the minimal polynomial of $\mathbf{v}$ then $C_T(\mathbf{v}) \subseteq N(q_{\mathbf{v}})$. Find an example showing that the inclusion can be proper.

17. Let T be a linear operator on V and $V = W_1 \oplus W_2$ where W_1, W_2 are T-invariant subspaces. For $f \in F[t]$ prove that
$$N(f) = N_{W_1}(f) \oplus N_{W_2}(f)$$

18. Let $A = C \oplus D$ where A, B and C are square matrices. Let R_C, R_D be the rational canonical forms for C, D. Prove that the rational canonical form for A is $R_C \oplus R_D$.

19. Let W be a T-invariant subspace of V. Prove the following statements.
    (a) If $f \in F[t]$ then $N_W(f)$ and $L_W(f)$ are T-invariant subspaces.
    (b) If $g|f$ then we have $N_W(g) \subseteq N_W(f)$ and $L_W(g) \subseteq L_W(f)$.

20. Let T be a linear operator on V with the characteristic polynomial $f(t)$ and the minimal polynomial $q(t)$.
    (a) Prove that the constant term of $f(0) \neq 0$ if and only if $q(0) \neq 0$.
    (b) Prove that T is invertible if and only if the constant term of f or q is nonzero.
    (c) Prove that if $f(t) = \sum_{i=0}^{n} a_i t^i$ with $a_0 \neq 0$ then
    $$T^{-1} = -\frac{1}{a_0}(a_1 I + a_2 T + \cdots + a_n T^{n-1})$$
    We can also use $q(t)$ instead of $f(t)$.
    (d) Let $g(t) \in F[t]$. Prove that $(f, g) = 1$ if and only if $(q, g) = 1$.
    (e) Prove that if g is relatively prime to f or q then $g(T)$ is invertible.

21. Let T be a nilpotent linear operator (i.e., $T^l = 0$ for some l) on a vector space of dimension n. Prove that the characteristic polynomial of T is $f(t) = (-1)^n t^n$ and the minimal polynomial is $q(t) = t^k$ for some $k \le n$.

22. Prove or disprove: There is $A \in M(3, \mathbb{R})$ with the minimal polynomial $q(t) = t^2 + t + 1$.

23. Let $A, B \in M(3, F)$. Prove that A, B are similar if and only if they have the same characteristic and minimal polynomials.

7.3 Jordan Canonical Form

Let T be a linear operator on a vector space V. In this section, we consider the linear operators whose characteristic polynomial splits; $f(t) = \pm \prod (t - \lambda_i)^{e_i}$. Hence the minimal polynomial is of the form $q(t) = \prod (t - \lambda_i)^{\epsilon_i}$ with $e_i \ge \epsilon_i > 0$ (Corollary 7.2.10). In this case we will show that we can find a basis with which the matrix representation of T consists of the blocks of the form

$$J(\lambda, m) := \begin{bmatrix} \lambda & 0 & 0 & \cdots & 0 \\ 1 & \lambda & 0 & & \vdots \\ 0 & 1 & \lambda & \ddots & 0 \\ \vdots & \vdots & & \ddots & 0 \\ 0 & 0 & \cdots & 1 & \lambda \end{bmatrix} \in M(m, F) \tag{7.63}$$

along the diagonal. The matrix of the form (7.63) is called the **Jordan block** of size m with eigenvalue λ. A matrix with the Jordan blocks along the diagonal is called a **Jordan canonical form**. In this section, we will show that there is a basis $\mathcal{B}$ such that $[T]_{\mathcal{B}}$ is a Jordan canonical form if the characteristic polynomial splits. Most of the results will follow by straightforward applications of the theorems of previous section.

7.3.1 Jordan Basis and Jordan Canonical Form

First, we consider the case when the characteristic polynomial is $f(t) = (-1)^e (t - \lambda)^e$ namely the characteristic polynomial splits and has a single eigenvalue λ. The cyclic subspace of $T - \lambda I$ spanned by $\mathbf{v}$ will be denoted by $C_\lambda(\mathbf{v})$:

$$C_\lambda(\mathbf{v}) := \mathrm{Span}\{\mathbf{v}, (T - \lambda)(\mathbf{v}), (T - \lambda)^2(\mathbf{v}), ...\}$$

Observe that (Exercise 1)

$$C_\lambda(\mathbf{v}) = C_T(\mathbf{v}) \tag{7.64}$$

And if $q_{\mathbf{v}}(t) = (t - \lambda)^e$ then both

$$\begin{aligned} \mathcal{B}_{\mathbf{v}}^\lambda &= \{\mathbf{v}, (T - \lambda)(\mathbf{v}), ..., (T - \lambda)^{e-1}(\mathbf{v})\} \\ \mathcal{B}_{\mathbf{v}} &= \{\mathbf{v}, T(\mathbf{v}), ..., T^{e-1}(\mathbf{v})\} \end{aligned} \tag{7.65}$$

are bases for $C_T(\mathbf{v})$ (Exercise 2). We will call $\mathcal{B}_{\mathbf{v}}^\lambda$ a **Jordan cyclic basis** for λ.

Theorem 7.2.3 with $p(t) = t - \lambda$ becomes

Theorem 7.3.1. *Let T be a linear operator on a vector space V of dimension e. Let $f_T(t) = (-1)^e(t-\lambda)^e$ be the characteristic polynomial and let $\nu(T-\lambda) = e$ i.e., $q_T = t - \lambda$. Then there are $\mathbf{v}_1, \mathbf{v}_2, ..., \mathbf{v}_e$ which form a basis $\mathcal{B}$ such that $T(\mathbf{v}_i) = \lambda \mathbf{v}_i$ $(i = 1, 2, ..., e)$. The matrix of T with respect to the bases $\mathcal{B} = \{\mathbf{v}_1, \mathbf{v}_2, ..., \mathbf{v}_e\}$ is*

$$[T]_{\mathcal{B}} = \begin{bmatrix} \lambda & & & \\ & \lambda & & \\ & & \ddots & \\ & & & \lambda \end{bmatrix}$$

Proof. Observe that in Theorem 7.2.3, the minimal polynomial of each $\mathbf{v}_i$ is $p(t) = t - \lambda$ and hence $T(\mathbf{v}_i) = \lambda \mathbf{v}_i$ and $C_T(\mathbf{v}_i) = \text{Span}\{\mathbf{v}_i\}$. □

Corollary 7.3.2. *Let T be a linear operator on a vector space V of dimension n with the characteristic polynomial $f(t) = (-1)^n \prod_{i=1}^k (t - \lambda_i)^{e_i}$. Then T is diagonalizable if and only if the minimal polynomial is $q(t) = \prod_{i=1}^k (t - \lambda_i)$.*

Proof. Exercise 6. □

The following theorem is Theorem 7.2.6 with $p(t) = t - \lambda$.

Theorem 7.3.3. *Let T be a linear operator on a vector space V over F. Let $f(t) = (-1)^e(t-\lambda)^e$ where $\lambda \in F$. If $\nu(T - \lambda) = 1$ then there is $\mathbf{v} \in V$ such that $V = C_\lambda(\mathbf{v})$ and*

$$\mathcal{B}_{\mathbf{v}}^{\lambda} = \{\mathbf{v}, (T-\lambda)(\mathbf{v}), (T-\lambda)^2(\mathbf{v}), ..., (T-\lambda)^{e-1}(\mathbf{v})\} \tag{7.66}$$

is a basis of V. The minimal polynomial of T is $q(t) = (t-\lambda)^e$.

The matrix representation of T with respect to $\mathcal{B} := \mathcal{B}_{\mathbf{v}}^{\lambda}$ is the Jordan block of size e:

$$[T]_{\mathcal{B}} = J(\lambda, e) = \begin{bmatrix} \lambda & 0 & 0 & \cdots & 0 \\ 1 & \lambda & 0 & & \vdots \\ 0 & 1 & \lambda & \ddots & 0 \\ \vdots & \vdots & & \ddots & 0 \\ 0 & 0 & \cdots & 1 & \lambda \end{bmatrix} \in M(e, F) \tag{7.67}$$

Further

$$\mathcal{B}_{\mathbf{v}}^{\lambda}(l) := \{(T-\lambda)^{e-l}(\mathbf{v}), (T-\lambda)^{e-l+1}(\mathbf{v}), ..., (T-\lambda)^{e-1}(\mathbf{v})\} \tag{7.68}$$

is a basis for $N((T-\lambda)^l)$ for $l = 1, 2, ..., e-1$.

Proof. By Theorem 7.2.6, $V = C_T(\mathbf{v})$ for some $\mathbf{v} \in V$. By Exercise 2, $\mathcal{B}_{\mathbf{v}}^{\lambda}$ is a basis for $V = C_T(\mathbf{v})$. For the matrix representation (7.67) of T with respect to $\mathcal{B}_{\mathbf{v}}^{\lambda}$ use the identity

$$T(T-\lambda)^{i-1}(\mathbf{v}) = \lambda(T-\lambda)^{i-1}(\mathbf{v}) + (T-\lambda)^i(\mathbf{v})$$

Now we need to show that (7.68) is a basis for $N((T-\lambda)^l)$. Obviously $\mathcal{B}_{\mathbf{v}}^{\lambda}(l) \subseteq N((T-\lambda)^l)$. By Theorem 7.2.15, $N((T-\lambda)^l) = C_T((T-\lambda)^{e-l})$ is of dimension l and $\mathcal{B}_{\mathbf{v}}^{\lambda}(l)$, being a part of a basis $\mathcal{B}_{\mathbf{v}}^{\lambda}$, is linearly independent. Hence $\mathcal{B}_{\mathbf{v}}^{\lambda}(l)$ is a basis for $N((T-\lambda)^l)$. □

The following theorem is Theorem 7.2.16 when the characteristic polynomial splits.

Theorem 7.3.4 (Jordan Canonical Form I). *Let T be a linear operator on a vector space V over F with the characteristic polynomial $f(t) = (-1)^e(t-\lambda)^e$ where $\lambda \in F$. If $\nu(T-\lambda) = k$ then there are $\mathbf{v}_1, \mathbf{v}_2, ..., \mathbf{v}_k$ such that*

$$V = C_\lambda(\mathbf{v}_1) \oplus C_T(\mathbf{v}_2) \oplus \cdots \oplus C_T(\mathbf{v}_k) \tag{7.69}$$

and

$$\mathcal{B}^\lambda_{\mathbf{v}_i} = \{\mathbf{v}_i, (T-\lambda)(\mathbf{v}_i), ..., (T-\lambda)^{\epsilon_i - 1}(\mathbf{v}_i)\} \quad (i = 1, 2, ..., k) \tag{7.70}$$

is a basis for $C_T(\mathbf{v}_i)$ so that

$$\mathcal{B}^\lambda = \mathcal{B}^\lambda_{\mathbf{v}_1} \cup \mathcal{B}^\lambda_{\mathbf{v}_2} \cup \cdots \cup \mathcal{B}^\lambda_{\mathbf{v}_k} \tag{7.71}$$

forms a basis for V. Here we choose $\epsilon_1 \geq \epsilon_2 \geq \cdots \geq \epsilon_k$.

The matrix representation $[T]_{\mathcal{B}^\lambda}$ of T with respect to $\mathcal{B}^\lambda$ is the matrix consisting of Jordan blocks $J(\lambda, \epsilon_i)$ along the diagonal

$$[T]_{\mathcal{B}^\lambda} = \begin{bmatrix} J(\lambda,\epsilon_1) & & & \\ & J(\lambda,\epsilon_2) & & \\ & & \ddots & \\ & & & J(\lambda,\epsilon_k) \end{bmatrix} \tag{7.72}$$

The minimal polynomial of $\mathbf{v}_i$ is $q_{\mathbf{v}_i}(t) = (t-\lambda)^{\epsilon_i}$, the minimal polynomial of T is $q_T(t) = (t-\lambda)^{\epsilon_1}$ and $\sum_{i=1}^k \epsilon_i = e$. Further

$$\begin{aligned} \mathcal{B}^\lambda(l) = &\left\{(T-\lambda)^{\epsilon_1 - l}(\mathbf{v}_1), (T-\lambda)^{\epsilon_1 - l + 1}(\mathbf{v}_1), ..., (T-\lambda)^{\epsilon_1 - 1}(\mathbf{v}_1)\right\} \\ &\cup \{(T-\lambda)^{\epsilon_2 - l}(\mathbf{v}_2), (T-\lambda)^{\epsilon_2 - l + 1}(\mathbf{v}_2), ..., (T-\lambda)^{\epsilon_2 - 1}(\mathbf{v}_2)\} \cup \cdots \\ &\cdots \cup \{(T-\lambda)^{\epsilon_k - l}(\mathbf{v}_k), (T-\lambda)^{\epsilon_k - l + 1}(\mathbf{v}_k), ..., (T-\lambda)^{\epsilon_k - 1}(\mathbf{v}_k)\} \end{aligned} \tag{7.73}$$

forms a basis for $N(T-\lambda)^l$. Here, $(T-\lambda)^{\epsilon_i - j}(\mathbf{v}_m)$ is defined only when $\epsilon_i \geq j$.

Proof. By Theorem 7.2.16, V decomposes into a direct sum of cyclic subspaces as in (7.69) and the number of factors is $\nu(t-\lambda)/\deg(t-\lambda) = k$; $V = \oplus_{i=1}^k C_T(\mathbf{v}_i) = \oplus_{i=1}^k C_\lambda(\mathbf{v}_i)$. By Theorem 7.3.3, each $C_T(\mathbf{v}_i)$ has a Jordan cyclic basis $\mathcal{B}^\lambda_{\mathbf{v}_i}$.

The fact that $\mathcal{B}^\lambda(l)$ is a basis for $N(T-\lambda)^l$ follows from dimension counting (Theorem 7.2.16) and the fact that $\mathcal{B}^\lambda(l)$, being a subset of a basis $\mathcal{B}^\lambda$, is linearly independent. □

Definition 7.3.5. *We will call the basis* (7.71) *a* **Jordan basis** *for the eigenvalue λ. The matrix* (7.72) *is called a* **Jordan canonical form** *for the linear operator T on V.*

By Corollary 7.2.20, the Jordan canonical form for T is uniquely determined independent of our choice of the Jordan basis.

In the following example, we make computations similar to Example 7.2.19 with $p(t) = t - \lambda$ instead of $p(t) = t^2 + 1$.

Example 7.3.6. *Let T be a linear operator on a vector space of dimension 9. Suppose*

$$V = C_T(\mathbf{v}_1) \oplus C_T(\mathbf{v}_2) \oplus C_T(\mathbf{v}_3) \oplus C_T(\mathbf{v}_4) \tag{7.74}$$

Let $\breve{\beta}_i$ be a Jordan cyclic basis in reversed order[†]*. Consider a basis $\breve{\mathfrak{B}} = \cup_i \breve{\beta}_i$:*

$$\breve{\mathfrak{B}} = \begin{pmatrix} \{(T-\lambda)^3(\mathbf{v}_1), & (T-\lambda)^2(\mathbf{v}_1), & (T-\lambda)(\mathbf{v}_1), & \mathbf{v}_1\} \\ \cup\{(T-\lambda)^2(\mathbf{v}_2), & (T-\lambda)(\mathbf{v}_2), & \mathbf{v}_2\} & \\ \cup \quad \{\mathbf{v}_3\} & & & \\ \cup \quad \{\mathbf{v}_4\} & & & \end{pmatrix} \tag{7.75}$$

Let us list the basis vectors for $N(T-\lambda)$, $N(T-\lambda)^2$, $N(T-\lambda)^3$ and $N(T-\lambda)^4$ in (7.75)

	$N(T-\lambda)$	$N(T-\lambda)^2$	$N(T-\lambda)^3$	$N(T-\lambda)^4$
$\breve{\beta}_1$	$(T-\lambda)^3(\mathbf{v}_1)$	$(T-\lambda)^2(\mathbf{v}_1)$	$(T-\lambda)(\mathbf{v}_1)$	$\mathbf{v}_1$
$\breve{\beta}_2$	$(T-\lambda)^2(\mathbf{v}_2)$	$(T-\lambda)(\mathbf{v}_2)$	$\mathbf{v}_2$	
$\breve{\beta}_3$	$\mathbf{v}_3$			
$\breve{\beta}_4$	$\mathbf{v}_4$			

(7.76)

where rows $\breve{\beta}_i$ of (7.76) are reversed Jordan cyclic bases in (7.75). And the vectors in the first i columns form a basis for $N(T-\lambda)^i$.

Let's represent each vector of (7.76) by $\bullet$.

$$\begin{array}{lccccc} & N(T-\lambda) \subseteq & N(T-\lambda)^2 \subseteq & N(T-\lambda)^3 \subseteq & N(T-\lambda)^4 & \\ \breve{\beta}_1 & \bullet & \bullet & \bullet & \bullet & \epsilon_1 = 4 \\ \breve{\beta}_2 & \bullet & \bullet & \bullet & & \epsilon_2 = 3 \\ \breve{\beta}_3 & \bullet & & & & \epsilon_3 = 1 \\ \breve{\beta}_4 & \bullet & & & & \epsilon_4 = 1 \\ & \epsilon^1 = 4 & \epsilon^2 = 2 & \epsilon^3 = 2 & \epsilon^4 = 1 & \end{array} \tag{7.77}$$

The four dots in the first column represent the number of basis for $N(T-\lambda)$. The six dots in the first two columns represents the number of basis for $N(T-\lambda)^2$ and so on:

$$\begin{aligned} \nu(T-\lambda) &= \epsilon^1 = 4 \\ \nu(T-\lambda)^2 &= \epsilon^1 + \epsilon^2 = 6 \\ \nu(T-\lambda)^3 &= \epsilon^1 + \epsilon^2 + \epsilon^3 = 8 \\ \nu(T-\lambda)^4 &= \epsilon^1 + \epsilon^2 + \epsilon^3 + \epsilon^4 = 9 = \dim V \end{aligned} \tag{7.78}$$

Now the dot diagram (7.77) is equivalent to the data (7.78). We can recover (7.77) from (7.78) by first filling in four dots in the first column and then filling in the dots in the second column from the topmost place until the number of the first two columns reaches $\nu(T-\lambda)^2 = 6$. And so on.

The Jordan canonical form corresponding to the dot diagram (7.77) is

$$[T]_{\mathfrak{B}} = \begin{bmatrix} J(\lambda,4) & & & \\ & J(\lambda,3) & & \\ & & J(\lambda,1) & \\ & & & J(\lambda,1) \end{bmatrix} \tag{7.79}$$

[†]By reversing the order the number of vectors in the first column becomes the number of direct factors in (7.74).

If the characteristic polynomial of $T \in L(V)$ splits then we can represent T as a rational canonical form as well as a Jordan canonical form. Here is an example how we represent a rational canonical form into the Jordan canonical form.

Example 7.3.7. *We will reconsider* Example 7.2.18 *where* $A = \begin{bmatrix} 0 & 0 & -1 & & \\ 1 & 0 & -3 & & \\ 0 & 1 & -3 & & \\ & & & 0 & -1 \\ & & & 1 & -2 \end{bmatrix}$ *and* $V = \mathbb{R}^5$, $T = T_A$. *Then A is a rational canonical form since* $A = \begin{bmatrix} C((t+1)^3) & 0 \\ 0 & C((t+1)^2) \end{bmatrix}$.
Hence the characteristic polynomial and minimal polynomial of A are $f(t) = -(t+1)^5$ and $q(t) = (t+1)^3$ respectively. And $\lambda = -1$ is a unique eigenvalue.

Now we want to find the Jordan canonical form for A. To obtain the Jordan canonical form we need to compute the nullities $\nu(T+I)^i$. We have

$$\nu(A+I) = 2,\ \nu(A+I)^2 = 4,\ \nu(A+I)^3 = 5$$

by the computation in Example 7.2.18. *Now we can complete the dot diagram – two dots in the first column, four dots in the first two columns and five dots in the first three columns:*

$$\begin{matrix} \bullet & \bullet & \bullet \\ \bullet & \bullet & \end{matrix}$$

From this we can find the Jordan canonical form for A:

$$J_A = \begin{bmatrix} 1 & 0 & 0 & & \\ 1 & 1 & 0 & & \\ 0 & 1 & 1 & & \\ & & & 1 & 0 \\ & & & 1 & 1 \end{bmatrix}$$

Practice Problem 7.3.1. *Find the Jordan form of the following matrix A.*

$$A = \begin{bmatrix} 1 & 1 & 0 & 0 & 0 & 0 \\ 0 & 1 & 1 & 1 & 0 & 1 \\ 0 & 0 & 1 & 0 & 1 & 1 \\ 0 & 0 & 0 & 1 & 1 & 1 \\ 0 & 0 & 0 & 0 & 1 & 0 \\ 0 & 0 & 0 & 0 & 0 & 1 \end{bmatrix}$$

Now we carry on further the idea of the examples above in more general situation. Suppose the characteristic polynomial is $f(t) = (-1)^e(t-\lambda)^e$ as in Theorem 7.3.4. From the Jordan basis (7.71), we can find the Jordan canonical form (7.72). We look for a method to get the Jordan canonical form (7.72) without finding the Jordan basis explicitly as we did in Example 7.2.19 for the rational canonical form. We have a chain of subspaces and numbers

$$N(T-\lambda) \subseteq N(T-\lambda)^2 \subseteq \cdots \subseteq N(T-\lambda)^{\epsilon_1} = V \tag{7.80}$$

$$\nu(T-\lambda) \leq \nu(T-\lambda)^2 \leq \cdots \leq \nu(T-\lambda)^{\epsilon_1} = e = \dim(V) \tag{7.81}$$

The last vectors of each β_i of (7.70) forms a basis for $N(T-\lambda)$ and the last two vectors of each β_i forms a basis for $N(T-\lambda)^2$ and so on. We reverse the order of β_i which we denote by $\check{\beta}_i$ as before. We let $\check{\mathfrak{B}}$ to be the union of $\check{\beta}_i$ $(i = 1, 2, ..., k)$:

$$
\begin{aligned}
\check{\beta}_i &= \{(T-\lambda)^{\epsilon_i - 1}(\mathbf{v}_i), (T-\lambda)^{\epsilon_i - 2}(\mathbf{v}_i), ..., \mathbf{v}_i\} \\
\check{\mathfrak{B}} &= \check{\beta}_1 \cup \check{\beta}_2 \cup \cdots \cup \check{\beta}_k \\
&= \begin{pmatrix} (T-\lambda)^{\epsilon_1-1}(\mathbf{v}_1), \ (T-\lambda)^{\epsilon_1-2}(\mathbf{v}_1), \ \cdots\cdots \ , \mathbf{v}_1 \\ \cup(T-\lambda)^{\epsilon_2-1}(\mathbf{v}_2), \ (T-\lambda)^{\epsilon_2-2}(\mathbf{v}_2), \ \cdots \ , \mathbf{v}_2 \\ \cup \quad \cdots \\ \cup(T-\lambda)^{\epsilon_k-1}(\mathbf{v}_k), \ \cdots \ , \mathbf{v}_k \end{pmatrix}
\end{aligned}
\tag{7.82}
$$

with $\epsilon_1 \geq \epsilon_2 \geq \cdots \geq \epsilon_k$. In this notation, the first column of $\check{\mathfrak{B}}$ is a basis of $N(T-\lambda)$ and the first two columns of $\check{\mathfrak{B}}$ is a basis for $N(T-\lambda)^2$ and so on.

Observe that to get the matrix (7.72) we do not need to know the specific vectors $\mathbf{v}_i$ in (7.70). What we need to know is the eigenvalue λ and the number ϵ_i for each i. Since we don't need to know the vectors $\mathbf{v}_i$ we represent each vector simply by a $\bullet$. Put $\nu(T-\lambda)$ dots on the first column and add $\bullet$'s from the top most place to the second column so that the number of dots of the first two columns add up to $\nu(T-\lambda)^2$ – add $\nu(T-\lambda)^2 - \nu(T-\lambda)$ dots – and so on. In this way, we get a diagram corresponding to the basis $\check{\mathfrak{B}}$ of (7.82)

$$
\text{eigenvalue } \lambda \qquad
\begin{array}{cccccc}
\bullet & \bullet & \bullet & \cdots & \bullet & \qquad \epsilon_1 \\
\bullet & \bullet & \cdots & \bullet & & \qquad \epsilon_2 \\
\vdots & \bullet & & & & \qquad \vdots \\
\bullet & & & & & \qquad \epsilon_k \\
 & & & & & \\
\epsilon^1 & \epsilon^2 & \cdots & & &
\end{array}
\tag{7.83}
$$

where we denote the number dots on the i-th column by ϵ^i and the number dots on the i-th row by ϵ_i.

Definition 7.3.8. *Let T be a linear operator on a vector space V over F. Let $f(t) = (-1)^e(t-\lambda)^e$ where $\lambda \in F$. Let $V = \oplus_{i=1}^k C_T(\mathbf{v}_i)$ and β_i be the Jordan cyclic basis of $C_T(\mathbf{v}_i)$. Let ϵ_i be the number of vectors in β_i with $\epsilon_1 \geq \epsilon_2 \geq \cdots \geq \epsilon_p$. The* **dot diagram** *of T for the eigenvalue λ consists of ϵ_i dots on the i-th row.*

A **partition** *of a natural number e is a sequence of numbers $(\epsilon_1, \epsilon_2, \ldots, \epsilon_p)$ such that $\epsilon_1 + \epsilon_2 + \cdots + \epsilon_p = e$ with $\epsilon_1 \geq \epsilon_2 \geq \cdots \geq \epsilon_p$. Hence the set of partitions of $e \in \mathbb{N}$ is*

$$
\left\{ (\epsilon_1, \epsilon_2, \ldots, \epsilon_p) \,\middle|\, \begin{array}{c} \epsilon_1 + \epsilon_2 + \cdots + \epsilon_p = e \\ \epsilon_1 \geq \epsilon_2 \geq \cdots \geq \epsilon_p \end{array} \right\}
\tag{7.84}
$$

For example, the lengths $(\epsilon_1, \epsilon_2, \ldots, \epsilon_k)$ of the Jordan cyclic basis β_i of (7.70) is a partition of e. For more about the partition of a number see "http://en.wikipedia.org/wiki/Partition (number theory)".

Let $\epsilon_1, \epsilon_2, ..., \epsilon_k$ be the numbers of rows of the dot diagram (7.83) then we have a partition

$$
\epsilon_1 + \epsilon_2 + \cdots + \epsilon_k = e \tag{7.85}
$$

and the partition (7.85) determines the dot diagram (7.83).

On the other hand, dot diagram (7.83) can be characterized by the partition of e according to the numbers of columns: If ϵ^i is the number of dots in the i-th column then we have a partition

$$\epsilon^1 + \epsilon^2 + \cdots + \epsilon^l = e \tag{7.86}$$

and, as before, the partition (7.86) determines the dot diagram (7.83) and *vice versa*.

$$\left\{ \begin{array}{c} \text{partitions of } e \\ \text{according to the columns} \end{array} \right\} \leftarrow \left\{ \begin{array}{c} \text{dot diagrams} \\ \text{with } e \text{ dots} \end{array} \right\} \rightarrow \left\{ \begin{array}{c} \text{partitions of } e \\ \text{according to the rows} \end{array} \right\} \tag{7.87}$$

In our situation, ϵ_i is the length of the i-th Jordan cyclic basis and $\epsilon^1 = \nu(T - \lambda)$ and $\epsilon^2 = \nu(T-\lambda)^2 - \nu(T-\lambda)$ and so on.

$$\left\{ \begin{array}{l} \epsilon^1 = \nu(t-\lambda) \\ \epsilon^2 = \nu(t-\lambda)^2 - \nu(t-\lambda) \\ \quad \cdots \\ \epsilon^l = \nu(t-\lambda)^l - \nu(t-\lambda)^{l-1} \end{array} \right\} \leftarrow \{\text{partition of } e\} \rightarrow \left\{ \begin{array}{rl} \epsilon_1 = & \text{number of } \beta_1 \\ \epsilon_2 = & \text{number of } \beta_2 \\ & \cdots \\ \epsilon_k = & \text{number of } \beta_k \end{array} \right\} \tag{7.88}$$

where $l = \epsilon_1$ because l is the number of columns and ϵ_1 is the length of the first row. Of course, the data on the left side of (7.88) is in one-to-one correspondence with the data

$$\nu(T-\lambda),\ \nu(T-\lambda)^2,\ \ldots,\ \nu(T-\lambda)^{\epsilon_1} \tag{7.89}$$

Practically, we can compute the nullities $\nu(T-\lambda)^i$ which determines the dot diagram (7.83). The dot diagram in turn determines the length ϵ_i of the i-th Jordan cyclic basis β_i in (7.82) which again determines the Jordan canonical form (7.72).

We summarize our discussion in the following theorem. We omit the formal proof as the truth of the statements are obvious and writing out the formal proof will be tedious.

We will say two Jordan basis are **equivalent** if they give the same Jordan matrix. Equivalences of Jordan bases is obviously an equivalence relation (Definition A.2.1) and hence we can divide the Jordan bases into the equivalence classes (Theorem A.2.6).

Theorem 7.3.9. *Let T be a linear operator on a vector space V over F. Let $f(t) = (-1)^e(t-\lambda)^e$ be the characteristic polynomial of T where $\lambda \in F$. Then there is a one-to-one correspondence among the following data.*

(1) *The equivalence classes of Jordan bases* (7.71).

(2) *Dot diagrams corresponding to the Jordan basis* (7.71).

(3) *Partition* (7.85) *of e according to the lengths of the Jordan cyclic bases on the right side of* (7.88).

(4) *Partition* (7.86) *of e according to the nullity data* (7.89).

(5) *Jordan matrices in* (7.72).

The size of the i-th Jordan block in Theorem 7.3.4 *is given by the number of dots in the i-th row in the dot diagram for the eigenvalue λ. In particular, the Jordan canonical form for T is independent of choice of Jordan basis* (7.71).

Proof. Independence of Jordan canonical form from a particular choice of the Jordan cyclic bases is a special case of Corollary 7.2.20 when $p(t) = t - \lambda$. Hence (1) is in one-to-one correspondence with the other data. □

Putting the results we have so far altogether we get:

Theorem 7.3.10 (Jordan Canonical Form II). *Let T be a linear operator on a vector space V over F of dimension n with the characteristic polynomial $f(t) = (-1)^n \prod_{i=1}^k (t-\lambda_i)^{e_i}$ and the minimal polynomial $q(t) = \prod_{i=1}^k (t-\lambda_i)^{\epsilon_i}$ where $\lambda_i \in F$. Then*

$$V = L_{\lambda_1} \oplus L_{\lambda_2} \oplus \cdots \oplus L_{\lambda_k} \text{ and } L_{\lambda_i} = N((T-\lambda_i)^{\epsilon_i}) \text{ with } \dim L_{\lambda_i} = e_i$$

and

$$L_{\lambda_i} = C_{\lambda_i}(\mathbf{v}_{i1}) \oplus C_{\lambda_i}(\mathbf{v}_{i2}) \cdots \oplus C_{\lambda_i}(\mathbf{v}_{in_i}) \tag{7.90}$$

and the minimal polynomial $q_{\mathbf{v}_{ij}}(t)$ of $\mathbf{v}_{ij}$ is given by

$$\begin{aligned} &q_{\mathbf{v}_{ij}}(t) = (t-\lambda_i)^{\epsilon_{ij}}, \ \epsilon_{i1} \geq \epsilon_{i2} \geq \cdots \geq \epsilon_{in_i} \\ &\epsilon_{i1} + \epsilon_{i2} + \cdots + \epsilon_{in_i} = e_i \end{aligned} \tag{7.91}$$

The characteristic polynomial f_i and the minimal polynomial q_i of the restriction T_i of T on $L(p_i)$ are

$$f_i(t) = (-1)^{e_i}(t-\lambda)^{e_i} \text{ and } q_i(t) = (t-\lambda)^{\epsilon_{i1}} \text{ with } \epsilon_{i1} = \epsilon_i$$

Further, we can choose a basis $\mathcal{B}_i$ as a union of Jordan cyclic basis for L_{λ_i}:

$$\begin{aligned} \mathcal{B}_i = \{&\mathbf{v}_{i1}, ((T-\lambda_i)(\mathbf{v}_{i1}), ..., ((T-\lambda_i)^{\epsilon_{i1}-1}(\mathbf{v}_{i1})\} \cup \\ &\cup \{\mathbf{v}_{i2}, ((T-\lambda_i)(\mathbf{v}_{i2}), ..., ((T-\lambda_i)^{\epsilon_{i2}-1}(\mathbf{v}_{i2})\} \cup \cdots \\ &\cup \{\mathbf{v}_{ik_i}, ((T-\lambda_i)(\mathbf{v}_{ik_i}), ..., ((T-\lambda_i)^{\epsilon_{ik_i}-1}(\mathbf{v}_{ik_i})\} \end{aligned} \tag{7.92}$$

so that $\mathcal{B} = \mathcal{B}_1 \cup \cdots \cup \mathcal{B}_k$ is a basis of V. The matrix representation of T with respect to $\mathcal{B}$ is

$$[T]_{\mathcal{B}} = \begin{bmatrix} [T_1]_{\mathcal{B}_1} & & & \\ & [T_2]_{\mathcal{B}_1} & & \\ & & \ddots & \\ & & & [T_k]_{\mathcal{B}_k} \end{bmatrix} \tag{7.93}$$

and each block $[T_i]_{\mathcal{B}_i}$ is of the form

$$[T_i]_{\mathcal{B}_i} = \begin{bmatrix} J(\lambda_i, \epsilon_{1i}) & & & \\ & J(\lambda_i, \epsilon_{2i}) & & \\ & & \ddots & \\ & & & J(\lambda_i, \epsilon_{k_i i}) \end{bmatrix} \tag{7.94}$$

where $J(\lambda_i, \epsilon_{li})$ is the Jordan block of size ϵ_{li}.

Proof. By Corollary 7.2.11, the characteristic and minimal polynomial of T_i are $\pm(t-\lambda_i)^{e_i}$ and $(t-\lambda_i)^{\epsilon_i}$. Now, the first part of the statements are immediate consequence of Theorem 7.3.4 applied to each L_{λ_i}. For the last statement simply use Corollary 6.4.28. □

In terms of matrices Theorem 7.3.10 says, if the characteristic polynomial of $A \in M(n, F)$ splits then A has a Jordan canonical form i.e., there is an invertible matrix Q such that $Q^{-1}AQ = J$ is a Jordan canonical form. Further we can choose the columns of Q to be the vectors of Jordan cyclic bases. How can we find such bases?

7.3.2 Explicit Computation of Jordan Basis

In §7.3.1, we showed how to find the Jordan matrices without finding the explicit Jordan basis. Now we want to find a specific Jordan basis. By Theorem 7.3.10, it suffices to consider the case when the characteristic polynomial of a linear operator T on V is of the form $f(t) = \pm(t-\lambda)^e$ where $\lambda \in F$. By computing $\nu(T-\lambda)^l$ $(l = 1, 2, 3, ...)$ we can form a dot diagram (7.83). If $\nu(T-\lambda) = k$, say the number of dots of the first column of the dot diagram is k, then there are $\mathbf{v}_1, \mathbf{v}_2, ..., \mathbf{v}_k$ such that

$$V = C_T(\mathbf{v}_1) \oplus C_T(\mathbf{v}_2) \oplus \cdots \oplus C_T(\mathbf{v}_k) \tag{7.95}$$

If the number of vectors of the i-th Jordan basis is ϵ_i then the minimal polynomial of $\mathbf{v}_i$ is $(t-\lambda)^{\epsilon_i}$. Hence $\mathbf{v}_i$ satisfies

$$(T-\lambda)^{\epsilon_i}(\mathbf{v}_i) = \mathbf{0} \text{ and } (T-\lambda)^{\epsilon_i - 1}(\mathbf{v}_i) \neq \mathbf{0} \tag{7.96}$$

Next theorem gives conditions for $\mathbf{v}_i$ to satisfy (7.95).

Theorem 7.3.11. *Let $T \in L(V)$ with the characteristic polynomial $f(t) = (-1)^e(t-\lambda)^e$ where $\lambda \in F$. Let $\nu(T-\lambda) = k$ and ϵ_i be the number of dots in the i-th row of the dot diagram. Suppose $\mathbf{v}_1, \mathbf{v}_2, ..., \mathbf{v}_k$ satisfy* (7.96) *and suppose, the last vectors*

$$S := \{(T-\lambda)^{\epsilon_1 - 1}(\mathbf{v}_1), (T-\lambda)^{\epsilon_2 - 1}(\mathbf{v}_2), ..., (T-\lambda)^{\epsilon_k - 1}(\mathbf{v}_k)\}$$

of $\mathcal{B}^\lambda_{\mathbf{v}_i}$ of (7.70) in Theorem 7.3.4 *is linearly independent subset of $N(T-\lambda)$. Then*

$$\mathcal{B}^\lambda = \mathcal{B}^\lambda_{\mathbf{v}_1} \cup \mathcal{B}^\lambda_{\mathbf{v}_2} \cup \cdots \cup \mathcal{B}^\lambda_{\mathbf{v}_k}$$

is a basis of V where $\mathcal{B}^\lambda_{\mathbf{v}_i} = \{\mathbf{v}_i, (T-\lambda)(\mathbf{v}_i), (T-\lambda)^2(\mathbf{v}_i), ..., (T-\lambda)^{\epsilon_i - 1}(\mathbf{v}_i)\}$.

Proof. By Corollary 7.2.17 with $p(t) = t - \lambda$, the equality (7.95) is equivalent to that $N(T-\lambda) = \oplus C_T((T-\lambda)^{\epsilon_i - 1}(\mathbf{v}_i))$. But since $C_T((T-\lambda)^{\epsilon_i - 1}(\mathbf{v}_i))$ is of dimension 1 with a basis $\{(T-\lambda)^{\epsilon_i - 1}(\mathbf{v}_i)\}$ this is equivalent to that S is linearly independent (Exercise 6.2.13). □

If there are more than one eigenvalues then we need to look at $(T-\lambda_i)_{L_i}$ for each eigenvalue λ_i and apply Theorem 7.3.4. And we need to compute $\nu((T-\lambda_i)^l_{L_i})$ for various λ_i's to get the dot diagram. We do not need to worry about the restriction on L_{λ_i} since

$$\nu(T-\lambda_i)^l = \nu((T-\lambda_i)^l_{L_i}) \tag{7.97}$$

by (7.29) of Corollary 7.2.11. Hence to find $\nu((T-\lambda_i)^l_{L_i})$ we simply compute $\nu((T-\lambda_i)^l)$. Now use Theorem 7.3.10 to combine these results.

Here is a check list to find the Jordan canonical form of a linear operator T on a vector space V of dimension n.

(1) Find the characteristic polynomial $f(t) = (-1)^n \prod_{i=1}^k (t-\lambda_i)^{e_i}$.

(2) Compute $\nu(T-\lambda_i)^l$ $(l = 1, 2, ...)$ until the nullity reaches the multiplicity e_i for each $i = 1, 2, ..., k$. Then draw dot diagram. Dot diagram for λ_i has $k_i := \nu(T-\lambda_i)$ dots in the first column and $\nu(T-\lambda_i)^2$ dots in the first two columns filled in the second column from the top most place and so on.

(3) Write down the Jordan canonical form using the dot diagram obtained in (2) above.

(4) When $k_i = \nu(T-\lambda_i) = 1$ (i.e., L_{λ_i} is a cyclic subspace) to find Jordan basis $\mathcal{B}_i$, choose $\mathbf{v}_i$ satisfying $(T-\lambda_i)^{e_i}\mathbf{v}_i = \mathbf{0}$ but $(T-\lambda_i)^{e_i-1}\mathbf{v}_i \neq \mathbf{0}$. Then

$$\mathcal{B}_i = \{\mathbf{v}_i, (T-\lambda_i)\mathbf{v}_i, ..., (T-\lambda_i)^{e_i-1}\}$$

is a Jordan cyclic basis for L_{λ_i} (Theorem 7.2.6). Note since $k_i = 1$ we have $e_i = \epsilon_i$.

When $k_i > 1$ (i.e., L_{λ_i} is a direct sum of more than one cyclic subspaces) to find the Jordan basis $\mathcal{B}_i$ of (7.92), choose $\{\mathbf{v}_{i1}, \mathbf{v}_{i2}, ..., \mathbf{v}_{ik_i}\}$ so that

$$(T-\lambda_i)^{\epsilon_{il}}\mathbf{v}_{il} = \mathbf{0} \text{ and } (T-\lambda_i)^{\epsilon_{il}-1}\mathbf{v}_{il} \neq \mathbf{0} \ (l = 1, 2, ..., k_i) \tag{7.98}$$

and

$$\{(T-\lambda_i)^{\epsilon_{i1}-1}(\mathbf{v}_{i1}), (T-\lambda_i)^{\epsilon_{i2}-1}(\mathbf{v}_{i2}), ..., (T-\lambda_i)^{\epsilon_{ik_i}-1}(\mathbf{v}_{ik_i})\} \tag{7.99}$$

form a basis of $N(T-\lambda_i)$. We only need to check its linear independence (Theorem 7.3.11).

(5) If $T = T_A$ for some matrix A and Q is the matrix whose columns are the vectors of the Jordan basis then $Q^{-1}AQ$ is the Jordan canonical form for A.

Example 7.3.12. *Let* $A = \begin{bmatrix} 1 & 0 & 0 & 1 \\ 0 & 1 & 0 & 0 \\ 1 & 0 & 0 & 0 \\ 0 & 1 & 0 & 0 \end{bmatrix}$. *Let's find the Jordan canonical form of A. The characteristic polynomial of A is* $f(t) = t^2(t-1)^2$. *Hence A has the eigenvalues* $\lambda_1 = 0, \lambda_2 = 1$. *For* $\lambda_1 = 0$, *we compute*

$$A^2 = \begin{bmatrix} 1 & 1 & 0 & 1 \\ 0 & 1 & 0 & 0 \\ 1 & 0 & 0 & 1 \\ 0 & 1 & 0 & 0 \end{bmatrix}$$

Hence $\nu(A) = 1,\ \nu(A^2) = 2$ *and the corresponding* $\bullet$*-diagram is*

$$\lambda_1 = 0: \quad \bullet \quad \bullet$$

For the eigenvalue $\lambda_2 = 1$ *we compute*

$$A - I = \begin{bmatrix} 0 & 0 & 0 & 1 \\ 0 & 0 & 0 & 0 \\ 1 & 0 & -1 & 0 \\ 0 & 1 & 0 & -1 \end{bmatrix}, \quad (A-I)^2 = \begin{bmatrix} 0 & 1 & 0 & -1 \\ 0 & 0 & 0 & 0 \\ -1 & 0 & 1 & 1 \\ 0 & -1 & 0 & 1 \end{bmatrix}$$

Hence we have $\nu(A-I) = 1, \nu(A-I)^2 = 2$ *and therefore the corresponding* $\bullet$*-diagram is*

$$\lambda_2 = 1: \quad \bullet \quad \bullet$$

Hence the Jordan canonical form J_A *for A is*

$$J_A = \begin{bmatrix} 0 & 0 & & \\ 1 & 0 & & \\ & & 1 & 0 \\ & & 1 & 1 \end{bmatrix}$$

To find a Jordan basis, by (4), we need to find $\mathbf{v}_1, \mathbf{v}_2$ *such that*

(1) $A^2\mathbf{v}_1 = \mathbf{0}, A\mathbf{v}_1 \neq \mathbf{0}$

(2) $(A - I)^2\mathbf{v}_2 = \mathbf{0}, (A - I)\mathbf{v}_2 \neq \mathbf{0}$

Now we choose

$$\mathbf{v}_1 = \begin{bmatrix} 1 \\ 0 \\ 0 \\ -1 \end{bmatrix}, A\mathbf{v}_1 = \begin{bmatrix} 0 \\ 0 \\ 1 \\ 0 \end{bmatrix}; \qquad \mathbf{v}_2 = \begin{bmatrix} 0 \\ 1 \\ -1 \\ 1 \end{bmatrix}, (A - I)\mathbf{v}_2 = \begin{bmatrix} 1 \\ 0 \\ 1 \\ 0 \end{bmatrix}$$

We don't need to check (7.99). *Hence if we let*

$$Q = \begin{bmatrix} 1 & 0 & 0 & 1 \\ 0 & 0 & 1 & 0 \\ 0 & 1 & -1 & 1 \\ -1 & 0 & 1 & 0 \end{bmatrix}$$

then $Q^{-1}AQ = J_A$.

Example 7.3.13. *Let's find the Jordan canonical form and the Jordan basis for the matrix*

$$A = \begin{bmatrix} 5 & 0 & 1 & 1 & 1 \\ 0 & 5 & 1 & 0 & 1 \\ 0 & 0 & 5 & 0 & 1 \\ 0 & 0 & 0 & 5 & 0 \\ 0 & 0 & 0 & 0 & 5 \end{bmatrix}$$

The characteristic polynomial of A *is* $f_A(t) = -(t-5)^5$ *and the only eigenvalue is* $\lambda = 5$ *with the multiplicity* $m = 5$. *We need to find* $\nu(A - \lambda)^i$ *for* $\lambda = 5$.

$$A - 5I = \begin{bmatrix} 0 & 0 & 1 & 1 & 1 \\ 0 & 0 & 1 & 0 & 1 \\ 0 & 0 & 0 & 0 & 1 \\ 0 & 0 & 0 & 0 & 0 \\ 0 & 0 & 0 & 0 & 0 \end{bmatrix}, \quad (A - 5I)^2 = \begin{bmatrix} 0 & 0 & 0 & 0 & 1 \\ 0 & 0 & 0 & 0 & 1 \\ 0 & 0 & 0 & 0 & 0 \\ 0 & 0 & 0 & 0 & 0 \\ 0 & 0 & 0 & 0 & 0 \end{bmatrix}, \quad (A - 5I)^3 = \begin{bmatrix} 0 & 0 & 0 & 0 & 0 \\ 0 & 0 & 0 & 0 & 0 \\ 0 & 0 & 0 & 0 & 0 \\ 0 & 0 & 0 & 0 & 0 \\ 0 & 0 & 0 & 0 & 0 \end{bmatrix}$$

Now $\nu(A - 5I) = 5 - \rho(A - 5I) = 5 - 3 = 2$. *Similarly,* $\nu(A - 5I)^2 = 4$ *and* $\nu(A - 5I)^3 = 5$. *Therefore the dot diagram corresponding to* $\lambda = 5$ *is*

$$\lambda = 5 \qquad \begin{array}{ccc} \bullet & \bullet & \bullet \\ \bullet & \bullet & \end{array}$$

Hence the Jordan canonical form for A *is*

$$J = \begin{bmatrix} 5 & 0 & 0 & & \\ 1 & 5 & 0 & & \\ 0 & 1 & 5 & & \\ & & & 5 & 0 \\ & & & 1 & 5 \end{bmatrix}$$

The Jordan basis for A *is of the form*

$$\{\mathbf{v}_1, (A-5)(\mathbf{v}_1), (A-5)^2(\mathbf{v}_1)\} \cup \{\mathbf{v}_2, (A-5)(\mathbf{v}_2)\}$$

For $\mathbf{v}_1, \mathbf{v}_2$, *by* (7.98), (7.99), *we require the conditions*

(1) $(A-5)^3(\mathbf{v}_1) = 0, (A-5)^2(\mathbf{v}_1) \neq 0;$

(2) $(A-5)^2(\mathbf{v}_2) = \mathbf{0}, (A-5)(\mathbf{v}_2) \neq \mathbf{0}$

(3) $\{(A-5)^2(\mathbf{v}_1), (A-5)\mathbf{v}_2\}$ is linearly independent.

Let's try $\mathbf{v}_1 = \mathbf{e}^5, \mathbf{v}_2 = \mathbf{e}^3$. Then we have

$$\mathbf{v}_1 = \begin{bmatrix} 0\\0\\0\\0\\1 \end{bmatrix}, (A-5)\mathbf{v}_1 = \begin{bmatrix} 1\\1\\1\\0\\0 \end{bmatrix}, (A-5)^2\mathbf{v}_1 = \begin{bmatrix} 1\\1\\0\\0\\0 \end{bmatrix}; \quad \mathbf{v}_2 = \begin{bmatrix} 0\\0\\1\\0\\0 \end{bmatrix}, (A-5)\mathbf{v}_2 = \begin{bmatrix} 1\\1\\0\\0\\0 \end{bmatrix}$$

Now $\{(A-5)^2(\mathbf{v}_1), (A-5)\mathbf{v}_2\}$ is linearly dependent. Hence this is a wrong choice for $\mathbf{v}_1, \mathbf{v}_2$ since they violate the condition (3). We need to make a different choice.

Now we choose $\mathbf{v}_1 = \mathbf{e}^5, \mathbf{v}_2 = \mathbf{e}^4$. Then we have

$$\mathbf{v}_1 = \begin{bmatrix} 0\\0\\0\\0\\1 \end{bmatrix}, (A-5)\mathbf{v}_1 = \begin{bmatrix} 1\\1\\1\\0\\0 \end{bmatrix}, (A-5)^2\mathbf{v}_1 = \begin{bmatrix} 1\\1\\0\\0\\0 \end{bmatrix}; \quad \mathbf{v}_2 = \begin{bmatrix} 0\\0\\0\\1\\0 \end{bmatrix}, (A-5)\mathbf{v}_2 = \begin{bmatrix} 1\\0\\0\\0\\0 \end{bmatrix}$$

Hence our choice of $\mathbf{v}_1, \mathbf{v}_2$ satisfies the required conditions and hence

$$\{\mathbf{v}_1, (A-5)(\mathbf{v}_1), (A-5)^2(\mathbf{v}_1)\} \cup \{\mathbf{v}_2, (A-5)(\mathbf{v}_2)\}$$

is a Jordan basis for A. If we let Q be the matrix consisting of the columns of the Jordan basis

$$Q = \begin{bmatrix} 0 & 1 & 1 & 0 & 1\\ 0 & 1 & 1 & 0 & 0\\ 0 & 1 & 0 & 0 & 0\\ 0 & 0 & 0 & 1 & 0\\ 1 & 0 & 0 & 0 & 0 \end{bmatrix}$$

then we have $Q^{-1}AQ = J$.

Practice Problem 7.3.2. *Referring to the* **Practice Problem 7.3.1** *find the Jordan basis of* A.

$$A = \begin{bmatrix} 1 & 1 & 0 & 0 & 0 & 0\\ 0 & 1 & 1 & 1 & 0 & 1\\ 0 & 0 & 1 & 0 & 1 & 1\\ 0 & 0 & 0 & 1 & 1 & 1\\ 0 & 0 & 0 & 0 & 1 & 0\\ 0 & 0 & 0 & 0 & 0 & 1 \end{bmatrix}$$

Example 7.3.14. *Find the Jordan canonical form for the matrix:*

$$A = \begin{bmatrix} 5 & 0 & 1 & 1 & 1 & & & \\ 0 & 5 & 0 & 0 & 1 & & & \\ 0 & 0 & 5 & 0 & 1 & & & \\ 0 & 0 & 0 & 5 & 0 & & & \\ 0 & 0 & 0 & 0 & 5 & & & \\ & & & & & 7 & 0 & 1 \\ & & & & & 0 & 7 & 1 \\ & & & & & 0 & 0 & 7 \end{bmatrix}$$

The characteristic polynomial of A *is* $f_A(t) = (t-5)^5(t-7)^3$ *and the eigenvalues are* $\lambda = 5, 7$ *with the multiplicity* $m = 5, 3$. *We need to find* $\nu(A-\lambda)^i$ *for* $\lambda = 5, 7$.

$$A - 5I = \begin{bmatrix} 0 & 0 & 1 & 1 & 1 & & & \\ 0 & 0 & 0 & 0 & 1 & & & \\ 0 & 0 & 0 & 0 & 1 & & & \\ 0 & 0 & 0 & 0 & 0 & & & \\ 0 & 0 & 0 & 0 & 0 & & & \\ & & & & & 2 & 0 & 1 \\ & & & & & 0 & 2 & 1 \\ & & & & & 0 & 0 & 2 \end{bmatrix}, \quad (A-5I)^2 = \begin{bmatrix} 0 & 0 & 0 & 0 & 1 & & & \\ 0 & 0 & 0 & 0 & 0 & & & \\ 0 & 0 & 0 & 0 & 0 & & & \\ 0 & 0 & 0 & 0 & 0 & & & \\ 0 & 0 & 0 & 0 & 0 & & & \\ & & & & & * & * & * \\ & & & & & * & * & * \\ & & & & & * & * & * \end{bmatrix}$$

and

$$(A-5I)^3 = \begin{bmatrix} 0 & 0 & 0 & 0 & 0 & & & \\ 0 & 0 & 0 & 0 & 0 & & & \\ 0 & 0 & 0 & 0 & 0 & & & \\ 0 & 0 & 0 & 0 & 0 & & & \\ 0 & 0 & 0 & 0 & 0 & & & \\ & & & & & * & * & * \\ & & & & & * & * & * \\ & & & & & * & * & * \end{bmatrix}$$

where the lower right corner is an invertible matrix.

We compute $\nu(A-5I) = 8 - \rho(A-5I) = 8 - 5 = 3$ *and* $\nu(A-5I)^2 = 4$. *And we also have* $\nu(A-5I)^3 = 5$. *Therefore, the dot diagram corresponding to* $\lambda = 5$ *is*

$$\lambda = 5 \qquad \begin{array}{lll} \bullet & \bullet & \bullet \\ \bullet & & \\ \bullet & & \end{array}$$

Hence there is a basis of the form $\{\mathbf{v}_1, (A-5)(\mathbf{v}_1), (A-5)^2(\mathbf{v}_1)\} \cup \{\mathbf{v}_2\} \cup \{\mathbf{v}_3\}$. *For* $\mathbf{v}_1, \mathbf{v}_2, \mathbf{v}_3$ *we require the conditions*

(1) $(A-5)^3(\mathbf{v}_1) = 0, (A-5)^2(\mathbf{v}_1) \neq 0$;

(2) $(A-5)(\mathbf{v}_2) = 0, (A-5)(\mathbf{v}_3) = 0$ and

(3) $(A-5)^2(\mathbf{v}_1), \mathbf{v}_2, \mathbf{v}_3$ are linearly independent.

We can choose

$$\mathbf{v}_1 = \begin{bmatrix} 0\\0\\0\\0\\1\\0\\0\\0 \end{bmatrix},\ (A-5)(\mathbf{v}_1) = \begin{bmatrix} 1\\1\\1\\0\\0\\0\\0\\0 \end{bmatrix},\ (A-5)^2(\mathbf{v}_1) = \begin{bmatrix} 1\\0\\0\\0\\0\\0\\0\\0 \end{bmatrix},\ \mathbf{v}_2 = \begin{bmatrix} 0\\1\\0\\0\\0\\0\\0\\0 \end{bmatrix},\ \mathbf{v}_3 = \begin{bmatrix} 0\\0\\1\\-1\\0\\0\\0\\0 \end{bmatrix}$$

so that $\{(A-5)^2(\mathbf{v}_1), \mathbf{v}_2, \mathbf{v}_3\}$ is linearly independent.

Similarly, for $\lambda = 7$ we have

$$A - 7I = \begin{bmatrix} -2 & 0 & 1 & 1 & 1 & & & \\ 0 & -2 & 0 & 0 & 1 & & & \\ 0 & 0 & -2 & 0 & 1 & & & \\ 0 & 0 & 0 & -2 & 0 & & & \\ 0 & 0 & 0 & 0 & -2 & & & \\ & & & & & 0 & 0 & 1 \\ & & & & & 0 & 0 & 1 \\ & & & & & 0 & 0 & 0 \end{bmatrix},\quad (A-7I)^2 = \begin{bmatrix} * & & & \\ & 0 & 0 & 0 \\ & 0 & 0 & 0 \\ & 0 & 0 & 0 \end{bmatrix}$$

where $*$ is an invertible matrix. Therefore $\nu(A-7I) = 2$, $\nu(A-7I)^2 = 3$; the dot diagram for $\lambda = 7$ is

$$\lambda = 7 \qquad \begin{array}{cc} \bullet & \bullet \\ \bullet & \end{array}$$

As before, we look for $\mathbf{w}_1, \mathbf{w}_2$ such that

(1) $(A-7)^2(\mathbf{w}_1) = 0, (A-7)(\mathbf{w}_1) \neq 0$

(2) $(A-7)(\mathbf{w}_2) = 0$

(3) $\{(A-7)(\mathbf{w}_1), \mathbf{w}_2\}$ is linearly independent

Then $\{\mathbf{w}_1, (A-7)(\mathbf{w}_1);\ \mathbf{w}_2\}$ becomes a basis for L_7: In fact, we can choose

$$\mathbf{w}_1 = \begin{bmatrix} 0\\0\\0\\0\\0\\0\\0\\1 \end{bmatrix}, (A-7)\mathbf{w}_1 = \begin{bmatrix} 0\\0\\0\\0\\0\\1\\1\\0 \end{bmatrix}, \mathbf{w}_2 = \begin{bmatrix} 0\\0\\0\\0\\0\\0\\1\\0 \end{bmatrix}$$

In conclusion, the Jordan canonical form for A is

$$J = \begin{bmatrix} 5 & 0 & 0 & & & & & \\ 1 & 5 & 0 & & & & & \\ 0 & 1 & 5 & & & & & \\ & & & 5 & & & & \\ & & & & 5 & & & \\ & & & & & 7 & 0 & \\ & & & & & 1 & 7 & \\ & & & & & & & 7 \end{bmatrix}$$

Let Q be the matrix

$$Q = \begin{bmatrix} 0 & 1 & 1 & 0 & 0 & 0 & 0 & 0 \\ 0 & 1 & 0 & 1 & 0 & 0 & 0 & 0 \\ 0 & 1 & 0 & 0 & 1 & 0 & 0 & 0 \\ 0 & 0 & 0 & 0 & -1 & 0 & 0 & 0 \\ 1 & 0 & 0 & 0 & 0 & 0 & 0 & 0 \\ 0 & 0 & 0 & 0 & 0 & 0 & 1 & 0 \\ 0 & 0 & 0 & 0 & 0 & 0 & 1 & 1 \\ 0 & 0 & 0 & 0 & 0 & 1 & 0 & 0 \end{bmatrix}$$

whose columns are the vectors

$$\{\mathbf{v}_1, (A-5)(\mathbf{v}_1), (A-5)^2(\mathbf{v}_1)\} \cup \{\mathbf{v}_2\} \cup \{\mathbf{v}_3\} \cup \{\mathbf{w}_1, (A-7)(\mathbf{w}_1)\} \cup \{\mathbf{w}_2\}$$

then we have $Q^{-1}AQ = J$.

Example 7.3.15. *We want to find all possible Jordan canonical forms with the characteristic polynomial* $f(t) = -(t-2)^5(t-3)^4$ *and minimal polynomial* $q(t) = (t-2)^2(t-3)^2$.

For $\lambda_1 = 2$ *we need to find the partitions* $(\epsilon_1, \epsilon_2, \ldots, \epsilon_p)$ *such that* $\epsilon_1 + \epsilon_2 + \cdots + \epsilon_p = 5$ *and* $\epsilon_1 = 2$. *And such partitions are* $(2,2,1), (2,1,1,1)$.

For $\lambda_2 = 3$ *we need to find the partitions* $(\epsilon_1, \epsilon_2, \ldots, \epsilon_p)$ *such that* $\epsilon_1 + \epsilon_2 + \cdots + \epsilon_p = 4$ *and* $\epsilon_1 = 2$. *And such partitions are* $(2,2), (2,1,1)$.

Hence the possible Jordan matrices are the combinations of these:

$$\begin{bmatrix} 2 &&&&&&&& \\ 1 & 2 &&&&&&& \\ && 2 &&&&&& \\ && 1 & 2 &&&&& \\ &&&& 2 &&&& \\ &&&&& 3 &&& \\ &&&&& 1 & 3 && \\ &&&&&&& 3 & \\ &&&&&&& 1 & 3 \end{bmatrix}, \quad \begin{bmatrix} 2 &&&&&&&& \\ 1 & 2 &&&&&&& \\ && 2 &&&&&& \\ && 1 & 2 &&&&& \\ &&&& 2 &&&& \\ &&&&& 3 &&& \\ &&&&& 1 & 3 && \\ &&&&&&& 3 & \\ &&&&&&&& 3 \end{bmatrix},$$

$$\begin{bmatrix} 2 &&&&&&&& \\ 1 & 2 &&&&&&& \\ && 2 &&&&&& \\ &&& 2 &&&&& \\ &&&& 2 &&&& \\ &&&&& 3 &&& \\ &&&&& 1 & 3 && \\ &&&&&&& 3 & \\ &&&&&&& 1 & 3 \end{bmatrix}, \quad \begin{bmatrix} 2 &&&&&&&& \\ 1 & 2 &&&&&&& \\ && 2 &&&&&& \\ &&& 2 &&&&& \\ &&&& 2 &&&& \\ &&&&& 3 &&& \\ &&&&& 1 & 3 && \\ &&&&&&& 3 & \\ &&&&&&&& 3 \end{bmatrix}$$

7.3.3 Applications of Jordan Canonical Form

In this subsection, we illustrate that the Jordan canonical form is one of powerful tool in linear algebra. In order to find the Jordan canonical form of a linear operator T it is necessary that the characteristic polynomial $f(t)$ splits. But in many cases, such an assumption is not necessary.

Similarity of matrices are important in that two similar matrices represent the same linear map in different bases (Corollary 4.3.7). Suppose $B = P^{-1}AP$ for some invertible P. To see whether such an invertible matrix P exists by considering the equations coming from $PB = AP$ will be very difficult. If the characteristic polynomials split then we have an easy criterion for them to be similar.

Theorem 7.3.16. *Let $A, B \in M(n, F)$. If the characteristic polynomials of A, B splits then the following conditions are equivalent.*

(1) *The matrices A and B are similar.*

(2) *The matrices A and B have the same Jordan canonical form.*

(3) *The matrices A and B have the same eigenvalues $\lambda_1, \lambda_2, ..., \lambda_k$ and for each eigenvalue λ_i we have $\rho(A - \lambda_i)^j = \rho(B - \lambda_i)^j$ for all j (up to the multiplicity of λ_i).*

Proof. (1) $\Rightarrow$ (2) Suppose A, B are similar; there is Q such that $B = Q^{-1}AQ$ and they have the same eigenvalues $\lambda_1, \dots, \lambda_k$ (Corollary 6.7.9). Further, $\nu(B-\lambda_i)^l) = \nu(Q^{-1}AQ-\lambda_i)^l) = \nu(Q^{-1}(A-\lambda_i)^l Q) = \nu((A-\lambda_i)^l)$ since Q is invertible. Hence the nullity data for the matrices A, B are the same. Hence they have the same Jordan canonical form.

(2) $\Rightarrow$ (1) Suppose A, B have the same Jordan canonical form J. Then there are invertible matrices P, Q such that $P^{-1}AP = J = Q^{-1}BQ$ (Theorem 7.3.10). Hence $A = PJP^{-1} = PQ^{-1}BQP^{-1} = (PQ^{-1})B(PQ^{-1})^{-1}$.

(2) $\Leftrightarrow$ (3) The condition (3) is equivalent to $\nu(A - \lambda_i)^j = \nu(B - \lambda_i)^j$ for all j by rank theorem (Theorem 2.3.17). By Theorem 7.3.9, the conditions are equivalent to the fact that for each λ_i the Jordan block of A and B are the same. That is their Jordan matrices are the same. □

Let $F \subseteq K$ fields. We say $A, B \in M(n, F)$ are similar over K if there is $P \in M(n, K)$ such that $B = P^{-1}AP$. Next we show that two real matrices A, B are similar already over $\mathbb{R}$ if they are similar over $\mathbb{C}$. This does not say that if $B = P^{-1}AP$ for some $P \in M(n, \mathbb{C})$ then actually $P \in M(n, \mathbb{R})$. What it says is that we can find $Q \in M(n, \mathbb{R})$ such that $B = Q^{-1}AQ$. This will be true generally for any two fields $F \subseteq K$ instead of $\mathbb{R} \subseteq \mathbb{C}$. But we will need a little more knowledge of fields for its proof. See Remark below.

Corollary 7.3.17. *Let $A, B \in M(n, \mathbb{R})$. Then A and B are similar over $\mathbb{R}$ if and only if their Jordan forms, as matrices over $\mathbb{C}$, are the same.*

Proof. If A, B are similar over $\mathbb{R}$ then obviously they are similar over $\mathbb{C}$. Hence they have the same Jordan forms.

Conversely, assume A, B viewed as matrices over $\mathbb{C}$ have the same Jordan form then they are similar over $\mathbb{C}$ i.e., there is an invertible $P \in M(n, \mathbb{C})$ such that $B = P^{-1}AP$. Then they have the same characteristic polynomial say $f(t) = \prod p_i^{e_i}$ where $p_i \in \mathbb{R}[t]$ are irreducible polynomials. Since A, B are similar over $\mathbb{C}$ we see that $p_i^k(A), p_i^k(B)$ are also similar over $\mathbb{C}$ for all k. Therefore we have $\nu(p_i^k(A)^{\mathbb{C}}) = \nu(p_i^k(B)^{\mathbb{C}})$ for all k and hence $\nu(p_i^k(A)) = \nu(p_i^k(B))$ (Example 6.4.7). By Corollary 7.2.26, we conclude that A, B are already similar over $\mathbb{R}$. □

If $A \in M(n)$ then its transpose A^t is the dual of A (Theorem 6.5.15). We will use Jordan canonical form to prove that A and A^t are similar. There seems to be no easier way to prove this other than using Jordan form.

Theorem 7.3.18. *Let $A \in M(n, F)$. If the characteristic polynomial of A splits then A and its transpose A^t are similar. If $A \in M(n, \mathbb{R})$ then A and A^t are similar.*

Proof. Let J_A be the Jordan form for A. Then there is an invertible matrix P such that $P^{-1}AP = J_A$. Hence $A^t = P^t J_A^t (P^t)^{-1}$ and hence A^t is similar to J_A^t. Now if $\mathcal{B} = \cup_i \beta_i$ is a basis for $V = F^n$ which is a union of the cyclic Jordan basis β_i. If $\breve{\beta}_i$ is the same as β_i with the reversed order and $\breve{\mathcal{B}} = \cup_i \breve{\beta}_i$ as in (7.82) then $[T_A]_{\breve{\mathcal{B}}} = J^t$. Therefore A is similar to J_A^t. Thus $A \sim J_A^t \sim A^t$. Since similarity relation is an equivalence relation (Theorem 6.4.29) we conclude that A and A^t are similar.

Now suppose $A \in M(n, \mathbb{R})$ then A, A^t are similar over $\mathbb{C}$ i.e., there is $P \in M(n, \mathbb{C})$ such that $A^t = P^{-1}AP$. By Corollary 7.3.17, A and A^t are already similar over $\mathbb{R}$ i.e., there is $Q \in M(n, \mathbb{R})$ such that $A^t = Q^{-1}AQ$. □

Could we have used the rational canonical form instead of the Jordan form in the proof of Theorem 7.3.18 ? What is the difficulty in doing so ?

Remark. *In* Theorem 7.3.16, Theorem 7.3.18 *the condition that the characteristic polynomial splits can be removed by using a little bit of field theory and module theory. In fact, we can prove "if $F \subseteq K$ are two fields and $A, B \in M(n, F)$ then A, B are similar over K if and only if they are similar over F." The proof of this is a generalization of the proof of* Corollary 7.3.17. *For a proof see* Corollary 18, §12.2 of [1] in Bibliography.

There is another implication of the Jordan canonical form: Every linear operator on a vector space whose characteristic polynomial splits can be written as a sum of so called a 'semisimple' (or diagonalizable) and a nilpotent operator.

Theorem 7.3.19. *Let $T : V \to V$ be a linear operator with the splitting characteristic polynomial. Then there are nilpotent operator N and a diagonalizable operator S on V such that $T = N + S$. The operators N and S are uniquely determined.*

Proof. Let $\mathcal{B}$ be a Jordan basis for T. Then the matrix representation map $\Phi : L(V) \to M(n, F)$ with respect to $\mathcal{B}$ is an isomorphism (Corollary 6.4.22). Let J be the Jordan canonical form of T and let $J = J_S + J_N$ where J_S is the diagonal and J_N is the lower triangular matrix. If S and N be the linear operators which correspond to J_S and J_N. Then N is nilpotent since the corresponding matrix is nilpotent and S is diagonalizable by definition.

For uniqueness, let $N + S = T = N' + S'$ where N, N' are nilpotent and S, S' are diagonalizable. Then $N - N' = S' - S$ is nilpotent as well as diagonalizable which must be zero (Exercise 6.7.11). Thus $N = N'$ and $S = S'$. □

We can go further to show that S and N can be expressed as polynomials in T and $SN = NS$ (Exercise 14).

Here is another application of Jordan canonical form. Let $A \in M(n, \mathbb{C})$ and $V = \mathbb{C}^n$. Then $T = T_A$ is a $\mathbb{C}$-linear operator on V. We can view V as a real vector space $V_{\mathbb{R}}$ of dimension $2n$ and T as an $\mathbb{R}$-linear map $T_{\mathbb{R}}$ (Exercise 6.3.15). We want to find $\det(T_{\mathbb{R}})$ in terms of $\det(T)$ using the Jordan canonical form.

Let $\mathcal{E} = \{\mathbf{e}^1, \mathbf{e}^2, ..., \mathbf{e}^n\} \subseteq \mathbb{C}^n$ be the standard basis for $\mathbb{C}$-vector space V. Let $V_{\mathbb{R}}$ be the vector space obtained from V by restricting the scalars to $\mathbb{R}$ (Exercise 6.3.15). Then

$$\begin{aligned} \mathcal{E}' &= \{\mathbf{e}^1, \mathbf{e}^2, ..., \mathbf{e}^n, i\mathbf{e}^1, i\mathbf{e}^2, ..., i\mathbf{e}^n\} \\ \mathcal{E}'' &= \{\mathbf{e}^1, i\mathbf{e}^1, \mathbf{e}^2, i\mathbf{e}^2, ..., \mathbf{e}^n, i\mathbf{e}^n\} \end{aligned}$$

are bases for $V_{\mathbb{R}}$ which differ in the ordering. For $z = a + bi$ define the 2×2 matrix

$$z^{\mathbb{R}} = \begin{bmatrix} a & -b \\ b & a \end{bmatrix}$$

Then $z^{\mathbb{R}}$ is the matrix representation of the multiplication by z map on $\mathbb{C}_{\mathbb{R}}$ with respect to the basis $\{1, i\}$ and $\det(z^{\mathbb{R}}) = |z|^2$.

Let $A \in M(n, \mathbb{C}), A = [a_{ij}]$. Let $T = T_A : V \to V$. Define the real and imaginary part of A by

$$A_{re} = ((a_{ij})_{re}) \text{ and } A_{im} = ((a_{ij})_{im})$$

We define $A', A'' \in M(2n, \mathbb{R})$ by

$$A' = \begin{bmatrix} A_{re} & -A_{im} \\ A_{im} & A_{re} \end{bmatrix} \text{ and } A'' = \begin{pmatrix} 2\times 2 \text{ matrices whose} \\ (i,j)\text{-block is } a_{ij}^{\mathbb{R}} \end{pmatrix}$$

It is easy to see that

$$A' = [T_{\mathbb{R}}]_{\mathcal{E}'}, \quad A'' = [T_{\mathbb{R}}]_{\mathcal{E}''} \tag{7.100}$$

and hence $|A'| = |A''| = \det T_{\mathbb{R}}$.

Example 7.3.20. *Let* $A = \begin{bmatrix} i & 1+i \\ 1-i & 3 \end{bmatrix}$ *and* $T = T_A$. *Then*

$$A' = [T_{\mathbb{R}}]_{\mathcal{E}'} = \begin{bmatrix} A_{re} & -A_{im} \\ A_{im} & A_{re} \end{bmatrix} = \begin{bmatrix} 0 & 1 & -1 & -1 \\ 1 & 3 & 1 & 0 \\ 1 & 1 & 0 & 1 \\ -1 & 0 & 1 & 3 \end{bmatrix}$$

and

$$A'' = [T_{\mathbb{R}}]_{\mathcal{E}''} = \begin{bmatrix} (i)^{\mathbb{R}} & (1+i)^{\mathbb{R}} \\ (1-i)^{\mathbb{R}} & (3)^{\mathbb{R}} \end{bmatrix} = \begin{bmatrix} 0 & -1 & 1 & -1 \\ 1 & 0 & 1 & 1 \\ 1 & 1 & 3 & 0 \\ -1 & 1 & 0 & 3 \end{bmatrix}$$

Now we compute the determinants: $\det(A) = -2+3i$, $\det(A') = 13 = |\det(A)|^2 = \det(A'')$.

From this example, we can expect that $\det(T_{\mathbb{R}}) = |\det(T)|^2$. We will show that this is true in general.

Theorem 7.3.21. *Let V be a $\mathbb{C}$-vector space of dimension n. Let T be a linear operator on V. Let $V_{\mathbb{R}}$ be the vector space obtained from V by restricting the scalar to $\mathbb{R}$ and $T_{\mathbb{R}}$ be the linear operator viewing T as an $\mathbb{R}$ linear map. Then we have*

$$\det(T_{\mathbb{R}}) = |\det(T)|^2$$

Proof. Obviously we may assume $T = T_A$ for some matrix $A \in M(n, \mathbb{C})$ with the eigenvalues $\lambda_i \in \mathbb{C}$. Then we have $\det(A) = \prod \lambda_i$. Hence we need to prove $\det(A'') = |\det(A)|^2$.

First observe that $(AB)'' = A''B''$ by Exercise 6.3.15 and Theorem 6.4.16. Let J be the Jordan canonical form for A; $J = P^{-1}AP$ for some invertible matrix $P \in M(n, \mathbb{C})$. Hence $J'' = (P'')^{-1}A''P''$ and therefore $\det(T_{\mathbb{R}}) = \det(A'') = \det(J'')$.

On the other hand, J'' consists of the 2×2 matrices $\lambda_i^{\mathbb{R}}$ along the diagonal and I_2 or the 2×2 zero matrix below the diagonal. Hence $\det(J'') = \prod |\lambda_i|^2 = |\det(A)|^2$ (Exercise 3.3.12). □

For more applications of Jordan canonical form of matrices see Exercise 16–22 and Exercise 8.1.33.

Answers for Practice Problems 7.3

(7.3.1) The characteristic polynomial of A is $f(t) = -(t-1)^6$. We compute:

$$A - I = \begin{bmatrix} 0&1&0&0&0&0\\ 0&0&1&1&0&1\\ 0&0&0&0&1&1\\ 0&0&0&0&1&1\\ 0&0&0&0&0&0\\ 0&0&0&0&0&0 \end{bmatrix}, \quad (A-I)^2 = \begin{bmatrix} 0&0&1&1&0&1\\ 0&0&0&0&2&2\\ 0&0&0&0&0&0\\ 0&0&0&0&0&0\\ 0&0&0&0&0&0\\ 0&0&0&0&0&0 \end{bmatrix}$$

$$(A-I)^3 = \begin{bmatrix} 0&0&0&0&2&2\\ 0&0&0&0&0&0\\ 0&0&0&0&0&0\\ 0&0&0&0&0&0\\ 0&0&0&0&0&0\\ 0&0&0&0&0&0 \end{bmatrix}, \quad (A-I)^4 = \begin{bmatrix} 0&0&0&0&0&0\\ 0&0&0&0&0&0\\ 0&0&0&0&0&0\\ 0&0&0&0&0&0\\ 0&0&0&0&0&0\\ 0&0&0&0&0&0 \end{bmatrix}$$

Hence we have

$$\nu(A-I) = 3, \nu(A-I)^2 = 4, \nu(A-I)^3 = 5, \nu(A-I)^4 = 6$$

The corresponding $\bullet$ diagram is

$$\begin{array}{cccc} \bullet & \bullet & \bullet & \bullet \\ \bullet & & & \\ \bullet & & & \end{array}$$

and the corresponding Jordan form is

$$\begin{bmatrix} 1&0&0&0&&\\ 1&1&0&0&&\\ 0&1&1&0&&\\ 0&0&1&1&&\\ &&&&1&\\ &&&&&1 \end{bmatrix}$$

(7.3.2) Jordan basis for A is of the form $\{\mathbf{v}_1, (A-I)\mathbf{v}_1, (A-I)^2\mathbf{v}_1, (A-I)^3\mathbf{v}_1\} \cup \{\mathbf{v}_2\} \cup \{\mathbf{v}_3\}$. We need to find $\mathbf{v}_1, \mathbf{v}_2, \mathbf{v}_3$ satisfying

(1) $(A-I)^3\mathbf{v}_1 \neq \mathbf{0}$ but $(A-I)^4\mathbf{v}_1 = \mathbf{0}$
(2) $\mathbf{v}_2 \neq \mathbf{0}, \mathbf{v}_3 \neq \mathbf{0}$ with $(A-I)\mathbf{v}_2 = (A-I)\mathbf{v}_3 = \mathbf{0}$
(3) $\{(A-I)^3\mathbf{v}_1, \mathbf{v}_2, \mathbf{v}_3\}$ is linearly independent

If we choose $\mathbf{v}_1 = \mathbf{e}^5, \mathbf{v}_2 = \mathbf{e}^3 - \mathbf{e}^4, \mathbf{v}_3 = \mathbf{e}^3 - \mathbf{e}^4 + \mathbf{e}^5 - \mathbf{e}^6$ then

$$(A-I)^3\mathbf{v}_1 = 2\mathbf{e}^1, (A-I)\mathbf{v}_2 = \mathbf{0}, (A-I)\mathbf{v}_3 = \mathbf{0}$$

Hence $\{(A-I)^3\mathbf{v}_1, \mathbf{v}_2, \mathbf{v}_3\}$ is linearly independent. Hence

$$\begin{aligned} &\{\mathbf{v}_1 = \mathbf{e}^5, (A-I)\mathbf{v}_1 = \mathbf{e}^3 + \mathbf{e}^4, (A-I)^2\mathbf{v}_1 = \mathbf{e}^1 + 2\mathbf{e}^2, (A-I)^3\mathbf{v}_1 = 2\mathbf{e}^1\} \\ &\cup\{\mathbf{v}_2 = \mathbf{e}^3 - \mathbf{e}^4\} \\ &\cup\{\mathbf{v}_3 = \mathbf{e}^3 - \mathbf{e}^4 + \mathbf{e}^5 - \mathbf{e}^6\} \end{aligned}$$

is a Jordan basis for A. The change of coordinate matrix that changes A into its Jordan form is

$$Q = \begin{bmatrix} 0&0&1&2&0&0\\ 0&0&2&0&0&0\\ 0&1&0&0&1&1\\ 0&1&0&0&-1&-1\\ 1&0&0&0&0&1\\ 0&0&0&0&0&-1 \end{bmatrix}$$

7.3 Exercises

In the following exercises, we write $A \oplus B$ for $A = \begin{bmatrix} A & 0 \\ 0 & B \end{bmatrix}$ for the square matrices A, B and $\oplus_i A_i$ for the direct sum of $A_1 \oplus A_2 \oplus \cdots \oplus A_n$ (§6.8).

1. Prove the equality $C_\lambda(\mathbf{v}) = C_T(\mathbf{v})$ of (7.64).

2. Prove that $\mathcal{B}_\mathbf{v}^\lambda$ in Theorem 7.3.3 is a basis of $C_T(\mathbf{v})$. (Hint: To prove that $\mathcal{B}_\mathbf{v}^\lambda$ is a basis it suffices to show $\mathcal{B}_\mathbf{v}^\lambda$ span $C_T(\mathbf{v})$.)

3. Find the Jordan canonical forms of

$$A = \begin{bmatrix} 0 & 0 & 0 & 1 \\ 0 & 0 & 0 & 1 \\ 0 & 0 & 0 & 1 \\ 0 & 0 & 0 & 0 \end{bmatrix}, \quad B = \begin{bmatrix} 0 & 0 & 1 & 1 \\ 0 & 0 & 0 & 1 \\ 0 & 0 & 0 & 0 \\ 0 & 0 & 0 & 0 \end{bmatrix}, \quad C = \begin{bmatrix} 0 & 0 & 1 & 1 \\ 0 & 0 & 1 & 1 \\ 0 & 0 & 0 & 0 \\ 0 & 0 & 0 & 0 \end{bmatrix}, \quad D = \begin{bmatrix} 0 & 0 & 1 & 1 \\ 0 & 0 & 0 & 0 \\ 0 & 0 & 0 & 0 \\ 0 & 0 & 0 & 0 \end{bmatrix}$$

4. Find the Jordan canonical forms and a Jordan basis for the following matrices.

$$A = \begin{bmatrix} 2 & 0 & 1 & 0 & 0 & & & \\ 0 & 2 & 0 & 1 & 0 & & & \\ 0 & 0 & 2 & 1 & 1 & & & \\ 0 & 0 & 0 & 2 & 0 & & & \\ 0 & 0 & 0 & 0 & 2 & & & \\ & & & & & 3 & 1 & 0 \\ & & & & & 0 & 3 & 0 \\ & & & & & 0 & 0 & 3 \end{bmatrix}, \quad B = \begin{bmatrix} 2 & 0 & 1 & 0 & 0 & & & \\ 0 & 2 & 1 & 1 & 0 & & & \\ 0 & 0 & 2 & 1 & 1 & & & \\ 0 & 0 & 0 & 2 & 0 & & & \\ 0 & 0 & 0 & 0 & 2 & & & \\ & & & & & 3 & 1 & 1 \\ & & & & & 0 & 3 & 0 \\ & & & & & 0 & 0 & 3 \end{bmatrix}$$

5. Prove the true statement(s) and find a counter example for the false statement(s).

   (a) If $A \in M(n, F)$ is diagonalizable, say $D = P^{-1}AP$ for an invertible P and a diagonal matrix D then the Jordan form for A is D.

   (b) Every matrix $A \in M(n, F)$ has a Jordan canonical form $J_A \in M(n, F)$.

   (c) Let T be a linear operator on a vector space V. If the characteristic polynomial splits then V has a unique Jordan basis for T.

   (d) If $A, B \in M(n, F)$ have the same Jordan canonical form then they are similar.

   (e) If $A, B \in M(n, F)$ have the same characteristic polynomial and minimal polynomial then they are similar.

   (f) Suppose the characteristic polynomial of a linear operator T on a vector space splits $f(t) = \pm \prod (t - \lambda_i)^{e_i}$. Then the numbers $\{\nu(T - \lambda_i)^j \mid i, j\}$ determines the Jordan canonical form of T.

   (g) If two matrices A and B have the splitting characteristic polynomials and have the same eigenvalues $\lambda_1, \lambda_2, \ldots, \lambda_k$. If for each eigenvalue λ_i we have $\rho(A - \lambda_i)^j = \rho(B - \lambda_i)^j$ for all j then they have the same minimal polynomial.

   (h) Let λ be an eigenvalue of a linear operator T on V. If $\nu(T - \lambda) = 3, \nu(T - \lambda)^2 = 5, \nu(T - \lambda)^3 = 6, \nu(T - \lambda)^4 = 6$ then the multiplicity of the eigenvalue λ is 3.

   (i) Let λ be an eigenvalue of a linear operator T on V. If $\nu(T - \lambda) = 3, \nu(T - \lambda)^2 = 5, \nu(T - \lambda)^3 = 6, \nu(T - \lambda)^4 = 6$ then the multiplicity of the eigenvalue λ is 6.

(j) For the proof of Theorem 7.3.18 we can use the rational canonical form instead of the Jordan canonical form.

(k) Let $A \in M(n,F)$. If the Jordan form of A is the identity matrix I_n then $A = I_n$.

(l) The rational canonical form for the identity matrix I_n is I_n.

(m) The Jordan canonical form for the identity matrix I_n is I_n.

6. Let $T : V \to V$ be a linear operator. Prove that T is diagonalizable if and only if the characteristic polynomial $f(t) = \pm\prod_{i=1}^{k}(t-\lambda_i)^{e_i}$ splits and the minimal polynomial is $q(t) = \prod_{i=1}^{k}(t-\lambda_i)$ i.e., $\epsilon_i = 1$.

7. Let $h(t) = \prod_{i=1}^{k}(t-\lambda_i)^{e_i}$ where λ_i are distinct and let $A = C(h) \in M(n,F)$ be the companion matrix of h. Prove that the Jordan form for A is $J_A = \oplus_{i=1}^{k} J(\lambda_i, e_i)$.

8. Let $A = \oplus_{i=1}^{k} J(\lambda, e_i)$ $(e_1 \geq e_2 \geq \cdots \geq e_k)$. Prove that the rational canonical form for A is given by $R_A = \oplus_i C((t-\lambda)^{e_i})$.

9. Let C, D be square matrices and $A = C \oplus D$. Prove that if the Jordan form for C, D are J_C, J_D then the Jordan form for A is $J_C \oplus J_D$.

10. For a nonzero $a \in F$ we write

$$J_a(\lambda, m) := \begin{bmatrix} \lambda & 0 & 0 & \cdots & 0 \\ a & \lambda & 0 & & \vdots \\ 0 & a & \lambda & \ddots & 0 \\ \vdots & \vdots & & \ddots & 0 \\ 0 & 0 & \cdots & a & \lambda \end{bmatrix} \in M(m,F)$$

(a) Prove that the Jordan canonical form of $J_a(\lambda, m)$ is $J(\lambda, m)$.

(b) More generally prove that the Jordan canonical form of $\bigoplus\limits_i J_{a_i}(\lambda_i, m_i)$ is $\bigoplus\limits_i J(\lambda_i, m_i)$.

11. Let $E_{ij}(a) \in M(n,F)$ be the matrix whose i,j entry is nonzero $a \in F$ and zero elsewhere. Prove that $E_{ij}(a)$ $(i \neq j)$ are similar to $E_{21}(1)$. And prove that $E_{ii}(a)$ $(i > 1)$ is similar to $E_{11}(a)$.

12. Let $A \in M(n,F)$ and $f(t) = (-1)^n \prod\limits_i (t-\lambda_i) \in K[t]$ be the characteristic polynomial of A where K is a field containing F (Corollary 6.7.9). Let $g(t) \in F[t]$. Prove that $\det g(A) = \prod_i g(\lambda_i)$ and $\operatorname{Tr} g(A) = \sum_i g(\lambda_i)$.

13. Let $A \in M(n,F)$ and the characteristic polynomial of A is $f(t) = (-1)^n \prod_{i=1}^{k}(t-\lambda_i)^{e_i}$. Find the Jordan canonical form for A^l $(l = 1, 2, 3, \ldots)$ in the following steps.

(a) Let $J = J(\lambda, m)$ with $(\lambda \neq 0)$ be the Jordan block of size m. Prove that the Jordan canonical form for J^l is $J(\lambda^l, m)$.

(b) Let $J = J(0, m)$ be the Jordan block of size m with the eigenvalue 0. Let $m = sl + r$. Prove that the Jordan form for J^l is given by $J(0, s+1)^r \oplus J(0, s)^{l-r}$.

(c) If the Jordan form for A is $J_A = \oplus_{\lambda_j \neq 0} J(\lambda_j, m_j) \oplus J(0, m)$ then the Jordan form for A^l is given by $J_{A^l} = \oplus_{\lambda_j \neq 0} J(\lambda_j^l, m_j) \oplus J(0, s+1)^r \oplus J(0, s)^{l-r}$ where $m = sl + r$.

14. Let T be a linear operator on a vector space V with the characteristic polynomial $f(t) = \pm\prod(t-\lambda_i)^{e_i}$. Let $V = \oplus_i L_{\lambda_i}$ and $\pi_i : V \to V$ be the projection onto L_{λ_i}.
    (a) Prove that the linear operators S and N of Theorem 7.3.19 are given by $S = \sum \lambda_i \pi_i$ and $N = T - S$ respectively.
    (b) Prove that S, N are polynomials in T and $SN = NS$. (Hint: Use Corollary 7.2.13.)
    (c) Prove that if T is diagonalizable then $T = \sum \lambda_i \pi_i$.

15. Let A, B, C be square matrices of the same size.
    (a) Prove that $\mathrm{Tr}\,(ABC) = \mathrm{Tr}\,(BCA) = \mathrm{Tr}\,(CAB)$.
    (b) Prove that $\mathrm{Tr}\,(AB)^k = \mathrm{Tr}\,(BA)^k$ for all $k = 1, 2, \cdots$.
    (c) Find an example showing $\mathrm{Tr}\,(ABC) \neq \mathrm{Tr}\,(ACB)$.
    (d) Let $A_1, A_2, ..., A_k$ be square matrices of the same size. What kind of permutation σ satisfies $\mathrm{Tr}\,(A_{\sigma(1)}A_{\sigma(2)}\cdots A_{\sigma(k)}) = \mathrm{Tr}\,(A_1A_2\cdots A_k)$?

16. Let $A \in M(n, F)$ and lct char(F) — 0. Let

$$f_A(t) = (-1)^n(t^n + a_{n-1}t^{n-1} + \cdots + a_0)$$

be the characteristic polynomial of A.
    (a) Let K be a field containing F such that $f(t) = (-1)^n\prod_{i=1}^n(t-\lambda_i)$ $(\lambda_i \in K)$. Prove that $\mathrm{Tr}\,(A^k) = \sum_i \lambda_i^k$ and $\det(A^k) = \prod \lambda_i^k$. In particular, even if $\lambda_i \notin F$ we have $\mathrm{Tr}\,(A) = \sum \lambda_i$ and $\det(A) = \prod \lambda_i$.
    (b) Define the **elementary symmetric polynomials** s_i in $\lambda_1, \lambda_2, ..., \lambda_n$ by

$$\begin{aligned}
s_0 &= 1\\
s_1 &= \lambda_1 + \lambda_2 + \cdots + \lambda_n\\
s_2 &= \lambda_1\lambda_2 + \lambda_1\lambda_3 + \cdots + \lambda_2\lambda_3 + \cdots + \lambda_{n-1}\lambda_n = \textstyle\sum_{1\le i<j\le n}\lambda_i\lambda_j\\
s_3 &= \textstyle\sum_{1\le i<j<k\le n}\lambda_i\lambda_j\lambda_k\\
&\vdots\\
s_n &= \lambda_1\lambda_2\cdots\lambda_n
\end{aligned}$$

and define the **power-sum symmetric polynomials** p_i by

$$\begin{aligned}
p_1 &= \lambda_1 + \lambda_2 + \cdots + \lambda_n\\
p_2 &= \lambda_1^2 + \lambda_2^2 + \cdots + \lambda_n^2\\
&\cdots\\
p_n &= \lambda_1^n + \lambda_2^n + \cdots + \lambda_n^n
\end{aligned}$$

Prove that

$$a_{n-i} = (-1)^i s_i$$

Also prove the **Newton's identities**

$$s_1 = p_1,\ \ 2s_2 = s_1p_1 - s_0p_2,\ \ 3s_3 = p_1s_2 - p_2s_1 + p_3$$

and, in general, prove

$$ks_k = \textstyle\sum_{i=1}^k(-1)^{i-1}s_{k-i}p_i$$

(c) Assuming char$F = 0$ prove that

$$\begin{cases} a_{n-1} = -\mathrm{Tr}(A) \\ a_{n-2} = -\frac{1}{2}(a_{n-1}\mathrm{Tr}(A) + \mathrm{Tr}(A^2)) \\ a_{n-3} = -\frac{1}{3}(a_{n-2}\mathrm{Tr}(A) + a_{n-1}\mathrm{Tr}(A^2) + \mathrm{Tr}(A^3)) \\ \quad \cdots\cdots\cdots \\ a_0 = -\frac{1}{n}(a_1\mathrm{Tr}(A) + a_2\mathrm{Tr}(A^2) + \cdots + a_{n-1}\mathrm{Tr}(A^{n-1}) + \mathrm{Tr}(A^n)). \end{cases}$$

(d) Prove that if $A, B \in M(n, F)$ and char$F = 0$ then the characteristic polynomial of AB is the same as the characteristic polynomial of BA. (See Exercise 5.1.10. Use (c) this time.)

17. Let $A, B \in M(n, F)$ with char$F = 0$. Let $f_A(t), f_B(t)$ be the characteristic polynomials of A and B.

(a) Prove that $f_A(t) = f_B(t)$ if and only if $\mathrm{Tr}\, A^k = \mathrm{Tr}\, B^k$ for all k.

(b) Prove that if $\mathrm{Tr}\, A^k = \mathrm{Tr}\, B^k$ for all k then $\det(A) = \det(B)$.

(c) Prove that if $\mathrm{Tr}\,(A^i) = 0$ $(i = 1, 2, ..., n)$ then A is nilpotent. Find an example showing that this is false if char$(F) \neq 0$.

18. Let $A, B \in M(n, F)$ with char $F = 0$. Prove that if $\mathrm{Tr}\,(A^i) = \mathrm{Tr}\,(B^i)$ for $1 \leq i < n$ then $\mathrm{Tr}\,(A^j) = \mathrm{Tr}\,(B^j)$ for all $j \geq n$.

19. Let $A \in M(n, F)$ and let $f(t) = (-1)^n \prod_{i=1}^k (t - \lambda_i)^{e_i}$ be the characteristic polynomial of A. Let

$$J_A = \oplus_{i=1}^k \left(\oplus_s J(\lambda_i, n_{i_s})\right)$$

be the Jordan form of A. Prove that A is invertible if $\lambda_i \neq 0$ for all i and the Jordan canonical form for A^{-1} is

$$J_{A^{-1}} = \oplus_{i=1}^k (\oplus_s J(\lambda_i^{-1}, n_{i_s}))$$

For a polynomial $f(t) = a_0 + a_1 t + \cdots + a_n t^n \in F[t]$ *we define the* **reciprocal polynomial** $\hat{f}(t)$ *of* $f(t)$ *by*

$$\hat{f}(t) = t^n f(1/t) = a_0 t^n + a_1 t^{n-1} + \cdots + a_n$$

20. Let $A \in M(n, F)$ with the characteristic polynomial $f(t) = (-1)^n(a_0 + a_1 t + \cdots + t^n)$. Prove that if $a_0 \neq 0$ then the characteristic polynomial of A^{-1} is given by

$$f_{A^{-1}}(t) = a_0^{-1} \hat{f}(t)$$

21. Let $A \in M(3, F)$ let $f_A(t) = -(t^3 + a_2 t^2 + a_1 t + a_0)$ be the characteristic polynomial of A. Prove that the characteristic polynomial $f_{A^2}(t)$ of A^2 is

$$f_{A^2}(t) = -\left[t^3 + (2a_1 - a_2^2)t^2 + (a_1^2 - 2a_0 a_2)t - a_0^2\right]$$

22. Suppose that A and B are square matrices of size n and m respectively. Let $\lambda_1, ..., \lambda_n$ be the eigenvalues of A and $\mu_1, ..., \mu_m$ be the eigenvalues of B (listed according to the multiplicities).

(a) Prove that the eigenvalues of $A \otimes B$ are $\lambda_i \mu_j$, $(i = 1, \ldots, n,\ j = 1, \ldots, m)$. Deduce that the characteristic polynomial of $A \otimes B$ (Definition 6.6.12) is

$$f_{A\otimes B}(t) = (-1)^{mn} \prod_{i,j} (t - \lambda_i \mu_j)$$

(b) Let A and B are square matrices of size n and m respectively. Prove that

$$\operatorname{Tr}(A \otimes B) = \operatorname{Tr}(A)\operatorname{Tr}(B)$$

$$\det(A \otimes B) = (\det A)^m (\det B)^n$$

23. Prove or disprove: Let S, T be commuting linear operators (i.e., $ST = TS$) on V. If the characteristic polynomials of S and T splits then there is a basis $\mathcal{B}$ such that $[S]_{\mathcal{B}}$ and $[T]_{\mathcal{B}}$ are the Jordan form for both S and T. (Cf. Exercise 6.8.18.)

24. Let $A \in M(k, F), B \in M(l, F)$. Suppose the characteristic polynomials of A, B split. Find a method to find the Jordan canonical form for $A \otimes B$ (Definition 6.6.12) in the following steps.

   (a) Prove that if J_A, J_B are the Jordan canonical form for A, B then $A \otimes B$ is similar to $J_A \otimes J_B$. Hence to find the Jordan canonical form for $A \otimes B$ we need to find the Jordan canonical form for $J_A \otimes J_B$.

   (b) Let $J(\lambda, m)$ be the Jordan block of size m with the eigenvalue λ. Find the Jordan form for $J(2, 2) \otimes J(3, 3)$ and $J(3, 3) \otimes J(2, 2)$ observing that they are similar (Exercise 6.6.11).

   (c) From the computations of (b) deduce that the Jordan canonical form for the Kronecker product $J(\lambda, m) \otimes J(\mu, n)$ is

$$J(\lambda\mu, m + n - 1) \oplus \big(\overbrace{J(\lambda\mu, m - 1) \oplus \cdots \oplus J(\lambda\mu, m - 1)}^{(n-1) \text{ times}}\big)$$

   (d) Apply (6.86) of Theorem 6.6.6 to find the Jordan canonical form for $J_A \otimes J_B$ which will be then the Jordan canonical for $A \otimes B$.

7.4 Rational Canonical Form

Most of this section will a recollection of §7.2 where we consider the case when the characteristic polynomial of a linear operator T does not necessarily split. In this case, the standard form of the matrix representation of a linear map is called the rational canonical form (Definition 7.2.25) which consists of companion matrices of certain polynomials along the main diagonal (Theorem 7.2.16).

We rearrange the direct sum decomposition of Theorem 7.2.24 to get another rational canonical form (Theorem 7.4.4). Also we will discuss the method to find a rational canonical form and a basis in which the matrix representation of T is a rational canonical form.

7.4.1 Rational Canonical Form

We recall the main results of §7.2. The first one is Theorem 7.2.16.

Theorem 7.4.1. *Let T be a linear operator on a vector space V with the characteristic polynomial $f(t) = \pm p(t)^e$ where p is irreducible and $\deg p = r$. If $\nu(p) = kr$ then there are $\mathbf{v}_1, \mathbf{v}_2, ..., \mathbf{v}_k \in V$ such that*

$$V = C_T(\mathbf{v}_1) \oplus C_T(\mathbf{v}_2) \oplus \cdots \oplus C_T(\mathbf{v}_k)$$

and with $\epsilon_1 \geq \epsilon_2 \geq \cdots \geq \epsilon_k$ the set of vectors

$$\begin{aligned}\mathcal{B} = \{\mathbf{v}_1, T(\mathbf{v}_1), ..., T^{r\epsilon_1-1}(\mathbf{v}_1)\} \cup \{\mathbf{v}_2, T(\mathbf{v}_2), ..., T^{r\epsilon_2-1}(\mathbf{v}_2)\} \cup \cdots \\ \cup \{\mathbf{v}_k, T(\mathbf{v}_k), ..., T^{r\epsilon_k-1}(\mathbf{v}_k)\}\end{aligned}$$

is a basis for V. The matrix representation of T in the basis $\mathcal{B}$ is

$$[T]_{\mathcal{B}} = C(p^{\epsilon_1}) \oplus C(p^{\epsilon_2}) \oplus \cdots \oplus C(p^{\epsilon_k})$$

Piecing Theorem 7.4.1 for various eigenvalues λ_i together we obtained Theorem 7.2.24 which we rephrase here.

Theorem 7.4.2. *Let T be a linear operator on V of dimension n with the characteristic polynomial $f(t) = (-1)^n \prod_{i=1}^k p_i(t)^{e_i}$ and the minimal polynomial $q(t) = \prod_{i=1}^k p_i(t)^{\epsilon_i}$ where p_i is a monic irreducible polynomial with $\deg p_i = r_i$. Then*

$$V = L(p_1) \oplus L(p_2) \oplus \cdots \oplus L(p_k) \quad \text{and } L(p_i) = N(p_i^{\epsilon_i}) \tag{7.101}$$

$$L(p_i) = C_T(\mathbf{v}_{i1}) \oplus C_T(\mathbf{v}_{i2}) \cdots \oplus C_T(\mathbf{v}_{ik_i}) \text{ with } \dim L(p_i) = r_i e_i \tag{7.102}$$

where $\mathbf{v}_{ij}$ has the minimal polynomial $q_{\mathbf{v}_{ij}}$ which is given by

$$q_{\mathbf{v}_{ij}} = p_i^{\epsilon_{ij}}, \ \epsilon_{i1} + \epsilon_{i2} + \cdots + \epsilon_{in_i} = e_i, \ \epsilon_i = \epsilon_{i1} \geq \epsilon_{i2} \geq \cdots \geq \epsilon_{in_i}$$

We can choose a basis $\mathcal{B}_i$ of $L(p_i)$ as a union of cyclic basis so that $\mathcal{B} = \mathcal{B}_1 \cup \cdots \cup \mathcal{B}_k$ is a basis of V. If we let T_i be the restriction of T on L_i then

$$[T]_{\mathcal{B}} = \bigoplus_i [T_i]_{\mathcal{B}_i} \tag{7.103}$$

Each block $[T_i]_{\mathcal{B}_i}$ is of the form

$$[T_i]_{\mathcal{B}_i} = \bigoplus_j C(p_i^{\epsilon_{ij}}) \tag{7.104}$$

where $C(p_i^{\epsilon_{li}})$ is the companion matrix of $p_i^{\epsilon_{li}}$ of size $r_i\epsilon_{li}$. The minimal polynomial of T is

$$q_T(t) = p_1^{\epsilon_{11}} p_2^{\epsilon_{21}} \cdots p_k^{\epsilon_{k1}}$$

and the characteristic polynomial and the minimal polynomial of T_i are, respectively

$$f_i(t) = (-1)^{r_i e_i} p_i^{e_i}(t) \quad \text{and} \quad q_i(t) = p_i^{\epsilon_{i1}}(t)$$

In terms of matrices, if $A \in M(n, F)$ then A has the rational canonical form i.e., there is an invertible matrix Q such that $Q^{-1}AQ$ is a rational canonical form. We can choose the columns of Q to be the vectors of $\mathcal{B}$.

Now we rearrange the direct sum decomposition of Theorem 7.4.2. First, we consider an example.

Example 7.4.3. *Suppose* $V = L(p_1) \oplus L(p_2) \oplus L(p_3)$ *and* p_1, p_2, p_3 *are distinct monic irreducible polynomials. Let*

$$\begin{aligned} L(p_1) &= C_T(\mathbf{v}_{11}) \oplus C_T(\mathbf{v}_{12}) \oplus C_T(\mathbf{v}_{13}) && \text{with } q_{\mathbf{v}_{11}} = p_1^4,\ q_{\mathbf{v}_{12}} = p_1^2,\ q_{\mathbf{v}_{13}} = p_1 \\ L(p_2) &= C_T(\mathbf{v}_{21}) \oplus C_T(\mathbf{v}_{22}) && \text{with } q_{\mathbf{v}_{21}} = p_2^3,\ q_{\mathbf{v}_{22}} = p_2 \\ L(p_3) &= C_T(\mathbf{v}_{31}) \oplus C_T(\mathbf{v}_{32}) && \text{with } q_{\mathbf{v}_{31}} = p_3^2,\ q_{\mathbf{v}_{32}} = p_3 \end{aligned}$$

If $\mathcal{B} = \bigcup_{i,j} \mathcal{B}_{\mathbf{v}_{ij}}$ *where* $\mathcal{B}_{\mathbf{v}_{ij}}$ *is the cyclic basis of* $\mathbf{v}_{ij}$ *then*

$$[T]_{\mathcal{B}} = C(p_1^4) \oplus C(p_1^2) \oplus C(p_1) \oplus C(p_2^3) \oplus C(p_2) \oplus C(p_3^3) \oplus C(p_3)$$

Now if we let $\mathbf{w}_1 = \mathbf{v}_{11} + \mathbf{v}_{21} + \mathbf{v}_{31}, \mathbf{w}_2 = \mathbf{v}_{12} + \mathbf{v}_{22} + \mathbf{v}_{32}, \mathbf{w}_3 = \mathbf{v}_{13}$ *then by* Theorem 7.2.1, *we have*

$$\begin{aligned} C_T(\mathbf{v}_{11}) \oplus C_T(\mathbf{v}_{21}) \oplus C_T(\mathbf{v}_{31}) &= C_T(\mathbf{w}_1) && \text{with } q_{\mathbf{w}_1} = p_1^4 p_2^3 p_3^2 \\ C_T(\mathbf{v}_{12}) \oplus C_T(\mathbf{v}_{22}) \oplus C_T(\mathbf{v}_{32}) &= C_T(\mathbf{w}_2) && \text{with } q_{\mathbf{w}_2} = p_1^2 p_2 p_3 \\ C_T(\mathbf{v}_{13}) &= C_T(\mathbf{w}_3) && \text{with } q_{\mathbf{w}_3} = p_1 \end{aligned}$$

Hence if we let $q_{\mathbf{w}_i} = h_i$ *and* $\mathcal{B}' = \mathcal{B}_{\mathbf{w}_1} \cup \mathcal{B}_{\mathbf{w}_2} \cup \mathcal{B}_{\mathbf{w}_3}$ *then*

$$[T]_{\mathcal{B}'} = C(h_1) \oplus C(h_2) \oplus C(h_3) \quad \text{with} \quad h_3 | h_2 | h_1$$

The minimal polynomial of T *is then* $q(t) = h_1(t)$ *and the characteristic polynomial is* $f(t) = \pm h_1(t) h_2(t) h_3(t)$.

Formally we have

Theorem 7.4.4. *Let* T *be a linear operator on* V *of dimension* n *with the characteristic polynomial* $f(t) = (-1)^n p_1(t)^{e_1} p_2(t)^{e_2} \cdots p_k(t)^{e_k}$ *where* p_i *are monic irreducible polynomials. Then there is a basis* $\mathcal{B}$ *of* V *and uniquely determined monic polynomials* $h_1, h_2, \ldots, h_m$ *such that* $h_m | h_{m-1} | \cdots | h_1$ *and such that*

$$[T]_{\mathcal{B}} = \begin{bmatrix} C(h_1) & & & \\ & C(h_2) & & \\ & & \ddots & \\ & & & C(h_m) \end{bmatrix} \tag{7.105}$$

Further the characteristic polynomial is $f(t) = (-1)^n h_1 h_2 \cdots h_m$ *and the minimal polynomial is* $q(t) = h_1$.

Proof. Let C_1 be the direct sum of the first factors of each $L(p_i)$ in (7.102). Then by Theorem 7.2.1, C_1 is again a cyclic subspace say $C_1 = C_T(\mathbf{w}_1)$ where $\mathbf{w}_1 = \mathbf{v}_{11} + \mathbf{v}_{21} + \cdots + \mathbf{v}_{k1}$ with the minimal polynomial $h_1 = p_1^{\epsilon_{11}} p_2^{\epsilon_{21}} \cdots p_k^{\epsilon_{k1}}$. And let C_2 be the direct sum of the second factors from each $L(p_i)$. Then C_2 is again a cyclic subspace say $C_2 = C_T(\mathbf{w}_2)$ where $\mathbf{w}_2 = \mathbf{v}_{12} + \mathbf{v}_{22} + \cdots + \mathbf{v}_{k2}$ with the minimal polynomial $h_2 = p_1^{\epsilon_{12}} p_2^{\epsilon_{22}} \cdots p_k^{\epsilon_{k2}}$. Then $h_2 | h_1$. Inductively, we let C_i be the direct sum of i-th factors and then C_i is again a cyclic subspace $C_T(\mathbf{w}_i)$ with minimal polynomial $h_i = p_1^{\epsilon_{1i}} p_2^{\epsilon_{2i}} \cdots p_k^{\epsilon_{ki}}$. Now $h_i | h_{i-1}$ since $\epsilon_{ji} \geq \epsilon_{(j+1)i}$.

Hence V has a cyclic basis $\mathcal{B}_j$ for $C_T(\mathbf{w}_j)$ and $\mathcal{B} = \cup \mathcal{B}_j$ is a basis for V and the matrix representation $[T]_{\mathcal{B}}$ is of the required form. ∎

The matrices (7.103) and (7.105) are similar (Exercise 2).

Definition 7.4.5. *The matrix representation of T in the form* (7.103) *of* Theorem 7.4.2 *is called the* **first rational canonical form** *and the form* (7.105) *in* Theorem 7.4.4 *is called the* **second rational canonical form**. *The polynomials $h_1, h_2, \dots, h_m$ in* Theorem 7.4.4 *are called the* **invariants of** T.

The following is a restatement of Corollary 7.2.26.

Corollary 7.4.6. *Let $A, B \in M(n, F)$. Then A and B are similar if and only if they have the same (first or second) rational canonical forms.*

Proof. Let R_A, R_B be the rational canonical form of A, B. Then there are invertible matrices P, Q such that $R_A = P^{-1}AP, R_B = Q^{-1}BQ$.

If A, B have the same rational canonical form say $R_A = R_B$ then $A = (QP^{-1})^{-1}B(QP^{-1})$. Hence A and B are similar. Conversely if A and B are similar then they have the same characteristic polynomial f. For each monic irreducible factor p of f, since A and B are similar, $N(p^i(A))$ and $N(p^i(B))$ have the same dimension. Hence the dot diagrams for A and B are the same. Since the dot diagram (see (7.58) for example and we will discuss more about the dot diagram shortly) determines the rational canonical form, we conclude that $R_A = R_B$. □

7.4.2 Explicit Computation

As indicated by the examples of §7.2, we describe a method to find the rational canonical form for a linear operator. Let T be a linear operator on a vector space V with the characteristic polynomial $f(t) = \pm p_1(t)^{e_1} p_2(t)^{e_2} \cdots p_k(t)^{e_k}$ where p_i is a monic irreducible polynomial of degree r_i.

(1) For each p_i compute

$$\nu(p_i) \le \nu(p_i^2) \le \nu(p_i^3) \le \cdots$$

until it becomes stationary. The stationary number should be $\le r_i e_i$.

(2) Make a dot diagram for $p_i(t)$: The number of dots in the first column is $\nu(p_i)/r_i$ and the number of dots in the first two columns filled in from the top most place is $\nu(p_i^2))/r_i$ and so on.

$$\text{diagram for } p_i \qquad \begin{array}{lllll} \bullet & \bullet & \cdots\cdots & \bullet & \bullet \\ \bullet & \bullet & \cdots\bullet & & \\ \bullet & \bullet & & & \\ \vdots & & & & \\ \bullet & & & & \end{array} \tag{7.106}$$

(3) If the numbers of the rows of the dot diagram for p_i are

$$\epsilon_{i1} \ge \epsilon_{i2} \ge \cdots \ge \epsilon_{in_i} \text{ with } \epsilon_{i1} + \epsilon_{i2} + \cdots + \epsilon_{in_i} = e_i$$

then the i-th block (7.104) of the rational canonical form is

$$[T_{L(p_i)}]_{\mathcal{B}_i} = \begin{bmatrix} C(p_i^{\epsilon_{1i}}) & & & \\ & C(p_i^{\epsilon_{2i}}) & & \\ & & \ddots & \\ & & & C(p_i^{\epsilon_{k_i i}}) \end{bmatrix} \tag{7.107}$$

(4) If we let $\mathcal{B} = \cup_i \mathcal{B}_i$ then

$$[T]_{\mathcal{B}} = \begin{bmatrix} [T_{L(p_1)}]_{\mathcal{B}_1} & & & \\ & [T_{L(p_2)}]_{\mathcal{B}_2} & & \\ & & \ddots & \\ & & & [T_{L(p_k)}]_{\mathcal{B}_k} \end{bmatrix} \tag{7.108}$$

First we consider the case when the characteristic polynomial of T is $f = p^e$ and the minimal polynomial is $q = p^\epsilon$. We look for the minimal polynomial p^{ϵ_l} of $\mathbf{v}_l$ in Theorem 7.4.1. We have a chain of subspaces

$$\begin{array}{ccccccc} N(p) & \subseteq & N(p^2) & \subseteq & \cdots & \subseteq & N(p^\epsilon) \\ C_T(p^{\epsilon_1-1}(T)(\mathbf{v}_1)) & & C_T(p^{\epsilon_1-2}(T)(\mathbf{v}_1)) & & \cdots & & C_T(\mathbf{v}_1)) \\ C_T(p^{\epsilon_2-1}(T)(\mathbf{v}_2)) & & C_T(p^{\epsilon_2-2}(T)(\mathbf{v}_2)) & & \cdots & & \\ \vdots & & \vdots & & & & \\ C_T(p^{\epsilon_k-1}(T)(\mathbf{v}_k)) & & & & & & \end{array} \tag{7.109}$$

where we let $C_T(p^{\epsilon_l-j}(T)(\mathbf{v}_l)) = C_T(p^{\epsilon_l}(T)(\mathbf{v}_l))$ if $\epsilon_l < j$. By Theorem 7.2.15,

$$\dim C_T(p^{\epsilon_l-i}(T)(\mathbf{v}_l)) = i \deg p \ (i = 1, 2, ..., \epsilon_l)$$
$$\dim C_T(\mathbf{v}_l) = \epsilon_l \deg p$$

where the minimal polynomial of $\mathbf{v}_l$ is p^{ϵ_l}. By Theorem 7.2.6, we have

$$\dim N(p^i) = \sum_{\substack{i \le \epsilon_l \\ l}} \dim C_T(p^{\epsilon_l-i}(T)(\mathbf{v}_l))$$

Example 7.4.7. *Let p is an irreducible polynomial. Let $T \in L(V)$ and suppose the characteristic polynomial is a power of p. Further suppose we have*

$$\begin{array}{ccccccc} N(p) & \subseteq & N(p^2) & \subseteq & N(p^3) & \subseteq & N(p^4) \\ C_T(p^{4-1}(T)(\mathbf{v}_1)) & & C_T(p^{4-2}(T)(\mathbf{v}_1)) & & C_T(p^{4-3}(T)(\mathbf{v}_1)) & & C_T(\mathbf{v}_1) \\ C_T(p^{2-1}(T)(\mathbf{v}_2)) & & C_T(\mathbf{v}_2) & & & & \\ C_T(\mathbf{v}_3) & & & & & & \\ C_T(\mathbf{v}_4) & & & & & & \end{array} \tag{7.110}$$

Represent each $C_T(p^i(T)(\mathbf{v}_j)$ by $\bullet$ to get

$$\begin{array}{cccc} \bullet & \bullet & \bullet & \bullet \\ \bullet & \bullet & & \\ \bullet & & & \\ \bullet & & & \end{array} \tag{7.111}$$

Four $\bullet$'s in the first column shows that $\nu(p) = 4 \deg p$ and the six $\bullet$'s in the first two columns shows $\nu(p^2) = 6 \deg p$ and seven $\bullet$'s in the first three columns shows $\nu(p^3) = 7 \deg p$.

Similarly, $\nu(p^4) = 8\deg p$. From these data we conclude that the characteristic polynomial is $p(t)^8$ and the minimal polynomial is $p(t)^4$. Let's find the minimal polynomial $q_{\mathbf{v}_i}$ $(i = 1, ..., 4)$ for $\mathbf{v}_i$. Four dots in the first row shows that $\dim C_T(\mathbf{v}_1) = 4\deg p$. The size of the matrix of the restriction $T_{C_T(\mathbf{v}_1)}$ of T on $C_T(\mathbf{v}_1)$, which is the companion matrix of the minimal polynomial for $\mathbf{v}_1$, is the same as $\dim C_T(\mathbf{v}_1)$. Hence the minimal polynomial $q_{\mathbf{v}_1}$ for $\mathbf{v}_1$ is $p(t)^4$. In the same manner, $q_{\mathbf{v}_2} = p(t)^2, q_{\mathbf{v}_3} = p(t), q_{\mathbf{v}_4} = p(t)$. Therefore the rational canonical form for T is

$$\begin{bmatrix} C(p^4) & & & \\ & C(p^2) & & \\ & & C(p) & \\ & & & C(p) \end{bmatrix}$$

So far we showed how to find the rational canonical form without finding the explicit bases. Now let's find explicit cyclic generator for the cyclic subspace $C_T(\mathbf{v}_i)$ in the decomposition $V = \oplus C_T(\mathbf{v}_i)$. First consider the case when the characteristic polynomial of a linear operator T on V is of the form $f(t) = \pm p^e$ where $p(t)$ is an irreducible polynomial of degree r. If $\nu(p) = kr$ then, by Theorem 7.4.1, there are nonzero vectors $\mathbf{v}_1, \mathbf{v}_2, ..., \mathbf{v}_k$ such that

$$V = C_T(\mathbf{v}_1) \oplus C_T(\mathbf{v}_2) \oplus \cdots \oplus C_T(\mathbf{v}_k) \tag{7.112}$$

If the number of i-th row of the dot diagram (7.106) is ϵ_i then the minimal polynomial of $\mathbf{v}_i$ is p^{ϵ_i}. Hence $\mathbf{v}_i$ satisfies

$$p(T)^{\epsilon_i}(\mathbf{v}_i) = \mathbf{0} \quad \text{and} \quad p(T)^{\epsilon_i - 1}(\mathbf{v}_i) \neq \mathbf{0} \tag{7.113}$$

Next theorem gives conditions for $\mathbf{v}_i$ to satisfy (7.112). When $p(t) = t - \lambda$ the following theorem reduces to Theorem 7.3.11.

Theorem 7.4.8. *Let T be a linear operator on a vector space V over F. Let $f(t) = (-1)^{er}p(t)^e$ where p is an irreducible monic polynomial of degree r. Let $\nu(p) = kr$ and ϵ_i be the number of dots in the i-th row of the dot diagram (7.106). Suppose V has the decomposition (7.112) and $\mathbf{v}_1, \mathbf{v}_2, ..., \mathbf{v}_k$ satisfy (7.113) and let*

$$S_i = \{p(T)^{\epsilon_i - 1}(\mathbf{v}_i), Tp(T)^{\epsilon_i - 1}(\mathbf{v}_1), ..., T^{r-1}p(T)^{\epsilon_i - 1}(\mathbf{v}_i)\}$$
$$S = S_1 \cup S_2 \cup \cdots \cup S_k$$

If S is linearly independent subset of $N(p)$ then

$$\mathcal{B} := \mathcal{B}_{\mathbf{v}_1} \cup \mathcal{B}_{\mathbf{v}_2} \cup \cdots \cup \mathcal{B}_{\mathbf{v}_k}$$

is a basis of V where

$$\mathcal{B}_{\mathbf{v}_i} := \{\mathbf{v}_i, T(\mathbf{v}_i), T^2(\mathbf{v}_i), ..., T^{r\epsilon_i - 1}(\mathbf{v}_i)\}$$

is a basis for $C_T(\mathbf{v}_i)$. Further the matrix representation of T with respect to $\mathcal{B}$ is

$$[T]_{\mathcal{B}} = \begin{bmatrix} C(p^{\epsilon_1}) & & & \\ & C(p^{\epsilon_2}) & & \\ & & \ddots & \\ & & & C(p^{\epsilon_k}) \end{bmatrix}$$

Proof. The condition (7.113) says $p(T)^{\epsilon_i-1}(\mathbf{v}_i)$ has minimal polynomial p. Since p is irreducible S_i is a basis for $C_T(p(T)^{\epsilon_i-1}(\mathbf{v}_i))$ (Theorem 7.1.13). Hence $\dim C_T(p(T)^{\epsilon_i-1}(\mathbf{v}_i)) = r$ and hence $\sum \dim C_T(p(T)^{\epsilon_i-1}(\mathbf{v}_i)) = kr$. Since $C_T(p(T)^{\epsilon_i-1}(\mathbf{v}_i)) \subseteq N(p)$ and $\dim N(p) = kr$ we see that $N(p) = \oplus_i C_T(p(T)^{\epsilon_i-1}(\mathbf{v}_i))$ by Theorem 6.2.36. By Corollary 7.2.17 we see $\cup_i \mathcal{B}_{\mathbf{v}_i}$ is a basis of V. The other statements are easy to check. □

If there are more than one irreducible factors in the characteristic polynomial $f = \pm \prod p_i$ of T, we use direct sum decomposition (7.101)

$$V = L(p_1) \oplus L(p_2) \oplus \cdots \oplus L(p_k) \quad \text{and } L(p_i) = N(p_i^{\epsilon_i})$$

Then use Theorem 7.4.8 to get a basis $\mathcal{B}_i$ for each $L(p_i)$ then $\cup \mathcal{B}_i$ is a basis for V.

Example 7.4.9. *Let $V = \mathbb{R}^4$ and let $T = T_A : V \to V$ where $A = \begin{bmatrix} 0 & 0 & -1 & 0 \\ 0 & 0 & 0 & -1 \\ 1 & 0 & 0 & 0 \\ 0 & 1 & 0 & 0 \end{bmatrix}$. Let's find the rational canonical for T. The characteristic polynomial of A is $f(t) = (t^2+1)^2$. Let $p(t) = t^2 + 1$. Hence the minimal polynomial is either p^2 or p. Now*

$$A^2 + I = \begin{bmatrix} 0 & 0 & 0 & 0 \\ 0 & 0 & 0 & 0 \\ 0 & 0 & 0 & 0 \\ 0 & 0 & 0 & 0 \end{bmatrix}$$

Hence the minimal polynomial is p and $\nu(p)/\deg(p) = 4/2$ is the number of cyclic factors in the decomposition of V into cyclic subspaces. Hence the dot diagram is

•
•

Therefore, there are nonzero $\mathbf{v}_1, \mathbf{v}_2$ such that $V = C_T(\mathbf{v}_1) \oplus C_T(\mathbf{v}_2)$. The rational canonical form is

$$R_A := \begin{bmatrix} C(t^2+1) & 0 \\ 0 & C(t^2+1) \end{bmatrix} = \begin{bmatrix} 0 & -1 & 0 & 0 \\ 1 & 0 & 0 & 0 \\ 0 & 0 & 0 & -1 \\ 0 & 0 & 1 & 0 \end{bmatrix}$$

We need to find $\mathbf{v}_1, \mathbf{v}_2$ satisfying

$$(A^2 + I)(\mathbf{v}_1) = \mathbf{0}, \ \mathbf{v}_1 \neq \mathbf{0} \ \text{ and } \ (A^2 + I)(\mathbf{v}_2) = \mathbf{0}, \ \mathbf{v}_2 \neq \mathbf{0}$$

and

$$S := \{\mathbf{v}_1, A\mathbf{v}_1\} \cup \{\mathbf{v}_2, A\mathbf{v}_2\}$$

is linearly independent. For the vectors $\mathbf{v}_1, \mathbf{v}_2$ satisfying these conditions we choose, since $A^2 + I = 0$, $\mathbf{v}_1 = \mathbf{e}^1, \mathbf{v}_2 = \mathbf{e}^2$. Then $\{\mathbf{v}_1, A\mathbf{v}_1\} = \{\mathbf{e}^1, \mathbf{e}^3\}$ and $\{\mathbf{v}_2, A\mathbf{v}_2\} = \{\mathbf{e}^2, \mathbf{e}^4\}$ and hence S is linearly independent. Thus $V = \mathbb{C}_T(\mathbf{v}_1) \oplus C_T(\mathbf{v}_2)$ and $\mathcal{B}_{\mathbf{v}_1} = \{\mathbf{e}^1, A\mathbf{e}^1 = \mathbf{e}^3\}$ is a basis for $C_T(\mathbf{v}_1)$ and $\mathcal{B}_{\mathbf{v}_2} = \{\mathbf{e}^2, A\mathbf{e}^2 = \mathbf{e}^4\}$ is a basis for $C_T(\mathbf{v}_2)$. If we choose Q to be the matrix formed by the vectors of β_1, β_2

$$Q = \begin{bmatrix} 1 & 0 & 0 & 0 \\ 0 & 0 & 1 & 0 \\ 0 & 1 & 0 & 0 \\ 0 & 0 & 0 & 1 \end{bmatrix}$$

then we have $Q^{-1}AQ = R_A$.

Example 7.4.10. *Consider the rational canonical form of the matrix*

$$A = \begin{bmatrix} 0 & -1 & 0 & -1 & 0 & 0 & 0 & 0 \\ 1 & 0 & 1 & 0 & 0 & 0 & 0 & 0 \\ 0 & 0 & 0 & -1 & 0 & -1 & 0 & -1 \\ 0 & 0 & 1 & 0 & 1 & 0 & 1 & 0 \\ 0 & 0 & 0 & 0 & 0 & -1 & 0 & 0 \\ 0 & 0 & 0 & 0 & 1 & 0 & 0 & 0 \\ 0 & 0 & 0 & 0 & 0 & 0 & 0 & -1 \\ 0 & 0 & 0 & 0 & 0 & 0 & 1 & 0 \end{bmatrix} = \begin{bmatrix} C & C & 0 & 0 \\ 0 & C & C & C \\ 0 & 0 & C & 0 \\ 0 & 0 & 0 & C \end{bmatrix} \in M(8, \mathbb{R})$$

where C is the companion matrix of t^2+1. The characteristic polynomial of A is $(t^2+1)^4$. Hence we need to compute $\nu(A^2+I)^i$:

$$A^2 + I = \begin{bmatrix} 0 & 0 & -2 & 0 & -1 & 0 & -1 & 0 \\ 0 & 0 & 0 & -2 & 0 & -1 & 0 & -1 \\ 0 & 0 & 0 & 0 & -2 & 0 & -2 & 0 \\ 0 & 0 & 0 & 0 & 0 & -2 & 0 & -2 \\ 0 & 0 & 0 & 0 & 0 & 0 & 0 & 0 \\ 0 & 0 & 0 & 0 & 0 & 0 & 0 & 0 \\ 0 & 0 & 0 & 0 & 0 & 0 & 0 & 0 \\ 0 & 0 & 0 & 0 & 0 & 0 & 0 & 0 \end{bmatrix}$$

and

$$(A^2+I)^2 = \begin{bmatrix} 0 & 0 & 0 & 0 & 4 & 0 & 4 & 0 \\ 0 & 0 & 0 & 0 & 0 & 4 & 0 & 4 \\ 0 & 0 & 0 & 0 & 0 & 0 & 0 & 0 \\ 0 & 0 & 0 & 0 & 0 & 0 & 0 & 0 \\ 0 & 0 & 0 & 0 & 0 & 0 & 0 & 0 \\ 0 & 0 & 0 & 0 & 0 & 0 & 0 & 0 \\ 0 & 0 & 0 & 0 & 0 & 0 & 0 & 0 \\ 0 & 0 & 0 & 0 & 0 & 0 & 0 & 0 \end{bmatrix}, \quad (A^2+I)^3 = \begin{bmatrix} 0 & 0 & 0 & 0 & 0 & 0 & 0 & 0 \\ 0 & 0 & 0 & 0 & 0 & 0 & 0 & 0 \\ 0 & 0 & 0 & 0 & 0 & 0 & 0 & 0 \\ 0 & 0 & 0 & 0 & 0 & 0 & 0 & 0 \\ 0 & 0 & 0 & 0 & 0 & 0 & 0 & 0 \\ 0 & 0 & 0 & 0 & 0 & 0 & 0 & 0 \\ 0 & 0 & 0 & 0 & 0 & 0 & 0 & 0 \\ 0 & 0 & 0 & 0 & 0 & 0 & 0 & 0 \end{bmatrix}$$

Hence the minimal polynomial is $q = (t^2+1)^3$. And $\nu(A^2+I) = 4, \nu(A^2+I)^2 = 6, \nu(A^2+I) = 8$. Since $\deg(t^2+1) = 2$ the number of dots of the dot diagram for $p = t^2+1$ in the first column is 2 and 3 in the first two columns and 4 in the first three columns. The corresponding dot diagram will be

$$\begin{array}{ccc} \bullet & \bullet & \bullet \\ \bullet & & \end{array}$$

Hence the rational canonical form of A is

$$R_A = \begin{bmatrix} C((t^2+1)^3) & \\ & C(t^2+1) \end{bmatrix} = \begin{bmatrix} 0 & 0 & 0 & 0 & 0 & -1 & & \\ 1 & 0 & 0 & 0 & 0 & 0 & & \\ 0 & 1 & 0 & 0 & 0 & -3 & & \\ 0 & 0 & 1 & 0 & 0 & 0 & & \\ 0 & 0 & 0 & 1 & 0 & -3 & & \\ 0 & 0 & 0 & 0 & 1 & 0 & & \\ & & & & & & 0 & -1 \\ & & & & & & 1 & 0 \end{bmatrix}$$

and there are $\mathbf{v}_1, \mathbf{v}_2$ such that $V = C_A(\mathbf{v}_1) \oplus C_A(\mathbf{v}_2)$ with $q_{\mathbf{v}_1}(t) = (t^2+1)^3, q_{\mathbf{v}_2}(t) = t^2+1$.

Now we want to choose a basis of V which is a union of cyclic bases $C_A(\mathbf{v}_1), C_A(\mathbf{v}_2)$. By Theorem 7.4.8, *there are nonzero $\mathbf{v}_1, \mathbf{v}_2$ such that*

$$p^3(A)(\mathbf{v}_1) = \mathbf{0}, \; p^2(A)(\mathbf{v}_1) \neq \mathbf{0} \text{ and } p(A)(\mathbf{v}_2), \; \mathbf{v}_2 \neq \mathbf{0} \tag{7.114}$$

Also we require linear independence of $S_1 \cup S_2$ *where*

$$S_1 = \{p^2(A)(\mathbf{v}_1), Ap^2(A)(\mathbf{v}_1)\} \text{ and } S_2 = \{\mathbf{v}_2, A\mathbf{v}_2\}$$

If we choose $\mathbf{v}_1 = \mathbf{e}^6, \mathbf{v}_2 = \mathbf{e}^1$ *then they satisfy* (7.114) *and*

$$S_1 = \{p^2(A)(\mathbf{v}_1) = 4\mathbf{e}^2, Ap^2(A)(\mathbf{v}_1) = -4\mathbf{e}^1\},\ S_2 = \{\mathbf{v}_2 = \mathbf{e}^1, A\mathbf{v}_2 = \mathbf{e}^2\}$$

But $S_1 \cup S_2$ *is linearly dependent. Hence this is a wrong choice* (Theorem 7.4.8).

Now let's take $\mathbf{v}_1 = \mathbf{e}^6, \mathbf{v}_2' = \mathbf{e}^6 - \mathbf{e}^8$. *Then* $S_2' := \{\mathbf{v}_2' = \mathbf{e}^6 - \mathbf{e}^8, A\mathbf{v}_2' = -\mathbf{e}^5 + \mathbf{e}^7\}$ *and hence* $S_1 \cup S_2'$ *is linearly independent. Therefore,* $V = C_T(\mathbf{v}_1) \oplus C_T(\mathbf{v}_2')$ *and*

$$\mathcal{B} := S_1 \cup S_2' = \{\mathbf{v}_1, A\mathbf{v}_1, ..., A^5\mathbf{v}_1\} \cup \{\mathbf{v}_2', A\mathbf{v}_2'\}$$

is a basis of $C_T(\mathbf{v}_1) \oplus C_T(\mathbf{v}_2')$. *If we let* $Q \in M(8)$ *be the matrix consisting of the columns of vectors in* $\mathcal{B}$

$$Q = \begin{bmatrix} 0 & 0 & 0 & 3 & 0 & -10 & 0 & 0 \\ 0 & 0 & -1 & 0 & 6 & 0 & 0 & 0 \\ 0 & -1 & 0 & 3 & 0 & -5 & 0 & 0 \\ 0 & 0 & -2 & 0 & 4 & 0 & 0 & 0 \\ 0 & -1 & 0 & 1 & 0 & -1 & 0 & -1 \\ 1 & 0 & -1 & 0 & 1 & 0 & 1 & 0 \\ 0 & 0 & 0 & 0 & 0 & 0 & 0 & 1 \\ 0 & 0 & 0 & 0 & 0 & 0 & -1 & 0 \end{bmatrix}$$

then

$$Q^{-1}AQ = R_A$$

7.4 Exercises

1. Let

$$A = \begin{bmatrix} 0 & -1 & 0 & 0 & 0 & 0 & 0 & 0 \\ 1 & 0 & 1 & 0 & 0 & 0 & 0 & 0 \\ 0 & 0 & 0 & -1 & 0 & 0 & 0 & 0 \\ 0 & 0 & 1 & 0 & 0 & 0 & 0 & 0 \\ 0 & 0 & 0 & 0 & 0 & -1 & 0 & 0 \\ 0 & 0 & 0 & 0 & 1 & 0 & 0 & 0 \\ 0 & 0 & 0 & 0 & 0 & 0 & 0 & -1 \\ 0 & 0 & 0 & 0 & 0 & 0 & 1 & 0 \end{bmatrix} \in M(8, \mathbb{R})$$

   Find the rational canonical form R_A of A and find Q such that $Q^{-1}AQ = R_A$.

2. Show that the matrices (7.103) and (7.105) are similar.

3. Let p_1, p_2, p_3 be monic irreducible polynomials over a field F. Let $R = C(p_1^2) \oplus C(p_1) \oplus C(p_1) \oplus C(p_2^3) \oplus C(p_2) \oplus C(p_3^2) \oplus C(p_3)$ be a rational canonical form of a matrix. Write the matrix R into a rational canonical form of the second type (7.105).

4. Let $A, B \in M(2, \mathbb{R})$. Let $f_A(t), f_B(t)$ be the characteristic polynomial having no real root. Prove that A and B are similar if and only if $f_A(t) = f_B(t)$.

5. In regard to Theorem 7.4.8, can we replace the condition that S is linearly independent by the condition, as in Theorem 7.3.11, that $\{p^{e_i - 1}(T)\mathbf{v}_i \mid i = 1, 2, ..., k\}$ is linearly independent?

Chapter 8

Inner Products and Bilinear Forms

In this chapter, we study the vector spaces with an extra structure called an inner product by which we can define the length of a vector and the angle between two vectors etc. We will see that a finite dimensional vector space with an inner product is basically the same as the ordinary Euclidean spaces $\mathbb{R}^n$ and $\mathbb{C}^n$ (Corollary 8.1.21). We have already studied the basic properties of the inner product structure of $\mathbb{R}^n$ in §5.2.

In §5.2, we restricted our attention to $\mathbb{R}^n$ whereas we will consider the vector space over the field $\mathbb{R}$ as well as over $\mathbb{C}$. We will mainly concern about finite dimensional inner product space except a few examples from analysis leaving further development of infinite dimensional inner product spaces to the area of analysis.

In this chapter, we exclusively consider the vector space over the fields of real numbers $\mathbb{R}$ and complex numbers $\mathbb{C}$ except the last section §8.6. Unless otherwise stated a field F will be $\mathbb{R}$ or $\mathbb{C}$. From time to time we will remind the reader of this situation

8.1 Inner Products and Norms

First, we recall some basic facts of complex numbers from Appendix B3: As a set

$$\mathbb{C} = \{a + bi \mid a, b \in \mathbb{R}\}$$

where $i^2 = -1$. If $z = a+bi$ then a is called the **real part** of z and b is called the **imaginary part** of z and we write

$$z_{re} = a, \ z_{im} = b$$

If $z_{re} = 0$ then we say z is **pure imaginary**. We represent $\mathbb{C}$ as a plane with the imaginary axis and the real axis which we call the **complex plane**. The product of two complex numbers is defined by

$$(a + bi)(c + di) = ac - bd + (ad + bc)i$$

Together with the usual addition of complex numbers $\mathbb{C}$ becomes a field containing $\mathbb{R}$.

We can also consider $\mathbb{C}$ as a vector space of dimension 2 over the field $\mathbb{R}$ of real numbers with the 'standard' basis $\{1, i\}$.

The **(complex) conjugate** $\bar{z}$ of $z = a + bi$ is defined by

$$\bar{z} = a - bi$$

Then $\bar{\bar{z}} = z$. And $z = 0$ if and only if $\bar{z} = 0$. Further

$$z\bar{z} = a^2 + b^2 = |z|^2 \tag{8.1}$$

The conjugate map $T_{\bar{z}}$ defined by $T_{\bar{z}}(z) = \bar{z}$ on $\mathbb{C}$ is an isomorphism when we view $\mathbb{C}$ as a vector space over $\mathbb{R}$ not as a vector space over $\mathbb{C}$. It is also an isomorphism of fields which leaves the real numbers fixed. Also note that $T_{\bar{z}}^2 = I$.

The **absolute value** $|z| = \sqrt{a^2 + b^2}$ is the length of the complex number $z \in \mathbb{C}$. The angle θ between the real axis and the complex number z is called the **argument** of z. Then we can write (Theorem B.3.1)

$$z = |z|(\cos\theta + i\sin\theta) = |z|e^{i\theta} \tag{8.2}$$

where $\cos\theta + i\sin\theta = e^{i\theta}$ and

$$\left|e^{i\theta}\right| = 1$$

Any complex number of absolute value 1 is of the form $e^{i\theta}$. If $z = |z|e^{i\theta}$ then we have

$$\bar{z} = |z|e^{-i\theta}, \;\; z^n = (|z|e^{i\theta})^n = |z|^n e^{in\theta}$$

and the n-th **roots of unity** $\left\{e^{\frac{2\pi ki}{n}} \mid k = 0, 1, 2, ..., n-1\right\}$ are the roots of $t^n - 1 = 0$.

8.1.1 Inner Product

Definition 8.1.1. *Let V be a vector space over $F = \mathbb{R}$ or $F = \mathbb{C}$. An* **inner product** $\langle\ ,\ \rangle$ *on V is a map*

$$V \times V \to F : \quad (\mathbf{u}, \mathbf{v}) \mapsto \langle \mathbf{u}, \mathbf{v} \rangle$$

satisfying

(1) $\langle \mathbf{u}_1 + \mathbf{u}_2, \mathbf{v} \rangle = \langle \mathbf{u}_1, \mathbf{v} \rangle + \langle \mathbf{u}_2, \mathbf{v} \rangle$

(2)* $\langle c\mathbf{u}, \mathbf{v} \rangle = \bar{c}\langle \mathbf{u}, \mathbf{v} \rangle$ *for $c \in F$.*

(3) $\langle \mathbf{u}, \mathbf{v} \rangle = \overline{\langle \mathbf{v}, \mathbf{u} \rangle}$

(4) $\langle \mathbf{u}, \mathbf{u} \rangle > 0$ *if* $\mathbf{u} \neq 0$.

A vector space with an inner product is called an **inner product space**. *The* **norm** *induced by the inner product $\langle\ ,\ \rangle$ of $\mathbf{u} \in V$ is defined by*

$$\|\mathbf{u}\| := \sqrt{\langle \mathbf{u}, \mathbf{u} \rangle} \tag{8.3}$$

We say $\mathbf{u}$ is a **unit vector** *if $\|\mathbf{u}\| = 1$.*

Let V, W be inner product spaces with the inner products $\langle\ ,\ \rangle_V, \langle\ ,\ \rangle_W$. A linear map $T : V \to W$ is an **isometry** *if $\langle \mathbf{u}, \mathbf{v} \rangle_V = \langle T(\mathbf{u}), T(\mathbf{v}) \rangle_W$ for all $\mathbf{u}, \mathbf{v} \in V$. If an isometry T is also an isomorphism then we say T is an* **isometric isomorphism**.

*Some text require linearity in the first variable whereas we require conjugate linearity. This is because we keep writing a vector of F^n as a column instead of a row.

The condition (1) together with the condition (2) say that the inner product is **conjugate linear** in the first variable. The condition (3) says the inner product is **conjugate symmetric**. The inner product $\langle\ ,\ \rangle$ is **biadditive** i.e., the condition (1) together with the condition

$$\langle \mathbf{u}, \mathbf{v}_1 + \mathbf{v}_2 \rangle = \langle \mathbf{u}, \mathbf{v}_1 \rangle + \langle \mathbf{u}, \mathbf{v}_2 \rangle$$

holds. In fact, $\langle \mathbf{u}, \mathbf{v}_1 + \mathbf{v}_2 \rangle = \overline{\langle \mathbf{v}_1 + \mathbf{v}_2, \mathbf{u} \rangle} = \overline{\langle \mathbf{v}_1, \mathbf{u} \rangle} + \overline{\langle \mathbf{v}_2, \mathbf{u} \rangle} = \langle \mathbf{u}, \mathbf{v}_1 \rangle + \langle \mathbf{u}, \mathbf{v}_2 \rangle$ by the condition (3).

If $F = \mathbb{R}$ then the condition (3) becomes

$$\langle \mathbf{u}, \mathbf{v} \rangle = \langle \mathbf{v}, \mathbf{u} \rangle$$

For the second variable, the condition (2) becomes

$$\langle \mathbf{u}, c\mathbf{v} \rangle = c\langle \mathbf{u}, \mathbf{v} \rangle \tag{8.4}$$

In fact, $\langle \mathbf{u}, c\mathbf{v} \rangle = \overline{\langle c\mathbf{v}, \mathbf{u} \rangle} = \overline{\bar{c}\langle \mathbf{v}, \mathbf{u} \rangle} = \bar{\bar{c}}\,\overline{\langle \mathbf{v}, \mathbf{u} \rangle} = c\langle \mathbf{u}, \mathbf{v} \rangle$.

Now by conjugate linearity in the first variable we have

$$\left\langle \sum_{i=1}^{n} a_i \mathbf{u}_i, \mathbf{v} \right\rangle = \sum_{i=1}^{n} \bar{a}_i \langle \mathbf{u}_i, \mathbf{v} \rangle \tag{8.5}$$

The condition (4) implies

$$\|\mathbf{u}\| = 0 \text{ if and only if } \mathbf{u} = 0 \tag{8.6}$$

To check two vectors are equal often we use the inner product structure:

$$\mathbf{v} = \mathbf{w} \text{ if and only if } \langle \mathbf{u}, \mathbf{v} \rangle = \langle \mathbf{u}, \mathbf{w} \rangle \text{ for all } \mathbf{u} \tag{8.7}$$

In fact, $\langle \mathbf{u}, \mathbf{v} \rangle = \langle \mathbf{u}, \mathbf{w} \rangle$ for all $\mathbf{u}$ if and only if $\langle \mathbf{u}, \mathbf{v} - \mathbf{w} \rangle = 0$ for all $\mathbf{u}$. If we take $\mathbf{u} = \mathbf{v} - \mathbf{w}$ then this is equivalent to $\|\mathbf{v} - \mathbf{w}\| = 0$ which means $\mathbf{v} = \mathbf{w}$ by (8.6). The same will be true if we interchange the first and the second variables in (8.7).

And

$$\langle \mathbf{u}, \mathbf{0} \rangle = 0 = \langle \mathbf{0}, \mathbf{u} \rangle \text{ for all } \mathbf{u} \in V \tag{8.8}$$

since $\langle \mathbf{u}, \mathbf{0} \rangle = \langle \mathbf{u}, \mathbf{0} + \mathbf{0} \rangle = \langle \mathbf{u}, \mathbf{0} \rangle + \langle \mathbf{u}, \mathbf{0} \rangle$. Also if $\langle \mathbf{u}, \mathbf{v} \rangle = 0$ for all $\mathbf{u}$ then $\mathbf{v} = \mathbf{0}$ and *vice versa*. That is an inner product is a 'nondegenerate pairing' (Definition 6.5.20).

Further,

$$\|c\mathbf{u}\| = |c|\,\|\mathbf{u}\|$$

since $\|c\mathbf{u}\|^2 = \langle c\mathbf{u}, c\mathbf{u} \rangle = c\bar{c}\langle \mathbf{u}, \mathbf{u} \rangle = |c|^2\|\mathbf{u}\|^2$.

Also note that an isometry is necessarily injective. For if an isometry T satisfies $T(\mathbf{u}) = \mathbf{0}$ then since $\|\mathbf{u}\| = \|T(\mathbf{u})\| = 0$ we have $\|\mathbf{u}\| = 0$ and hence $\mathbf{u} = \mathbf{0}$.

Recall Definition 6.2.24: For $A = [a_{ij}] \in M(m\times n, \mathbb{C})$ we define the **conjugate** $\bar{A}$ and **transpose conjugate** or **adjoint** A^* in $M(n\times m, \mathbb{C})$ to be

$$(\bar{A})_{ij} = \bar{a}_{ij} \text{ and } (A^*)_{ij} = \bar{a}_{ji}$$

If $A \in M(m\times n, \mathbb{R})$ then $A^* = A^t$. If $m = n = 1$, namely A is a scalar then A^* is simply the complex conjugate. We recall Theorem 6.2.25 and Theorem 6.3.28:

Theorem 8.1.2. *If $A \in M(m\times n,\mathbb{C})$ then the following formulas, whenever the operations make sense, hold.*

(1) $(A \pm B)^* = A^* \pm B^*$	(2) $(cA)^* = \overline{c}A^*$
(3) $\overline{AB} = \bar{A}\bar{B}$, $(AB)^* = B^*A^*$	(4) $A^{**} = A$
(5) $\det(\bar{A}) = \det(A^*) = \overline{\det(A)}$	(6) $\mathrm{Tr}(\bar{A}) = \mathrm{Tr}(A^*) = \overline{\mathrm{Tr}(A)}$
(7) $\rho(A) = \rho(\bar{A}) = \rho(A^*)$	(8) $\nu(A) = \nu(\bar{A}) = \nu(A^*)$

Example 8.1.3. *Let $V = F^n$ ($F = \mathbb{R}$ or $\mathbb{C}$) and* $\mathbf{u} = \begin{bmatrix} x_1 \\ \vdots \\ x_n \end{bmatrix}, \mathbf{v} = \begin{bmatrix} y_1 \\ \vdots \\ y_n \end{bmatrix}$. *Define, viewing $\mathbf{u}, \mathbf{v}$ as $1\times n$ matrix*

$$\langle \mathbf{u}, \mathbf{v}\rangle = \mathbf{u}^*\mathbf{v} = \sum_{i=1}^{n} \overline{x}_i y_i$$

which we call the **standard inner product**. *The properties* (1) – (4) *of* Definition 8.1.1 *are easy to verify using* Theorem 8.1.2. *If $F = \mathbb{R}$ then $\langle \mathbf{u}, \mathbf{v}\rangle = \mathbf{u}\cdot\mathbf{v}$ of* Definition 5.2.1. *If $F = \mathbb{C}$ and $n = 1$ then $\|z\|^2 = |z|^2$.*

Example 8.1.4. *Let $V = F_n$ of the space of row vectors* (Example 6.1.3). *We define the inner product in the same way as we did for F^n: For the row vectors $\mathbf{u} = (x_1, x_2, ..., x_n), \mathbf{v} = (y_1, y_2, ..., y_n)$ we define*

$$\langle \mathbf{u}, \mathbf{v}\rangle = \bar{\mathbf{u}}\mathbf{v}^t = \sum_{i=1}^{n} \overline{x}_i y_i$$

Observe that $\langle A^*\mathbf{u}, \mathbf{v}\rangle = \mathbf{u}^*A^{**}\mathbf{v} = \mathbf{u}^*A\mathbf{v} = \langle \mathbf{u}, A\mathbf{v}\rangle$ for $\mathbf{u}, \mathbf{v} \in F^n$. Hence

$$\langle \mathbf{u}, A\mathbf{v}\rangle = \langle A^*\mathbf{u}, \mathbf{v}\rangle \tag{8.9}$$

which says 'A^* is adjoint to A'. If we replace A by A^* then, since $A^{**} = A$, we have

$$\langle \mathbf{u}, A^*\mathbf{v}\rangle = \langle A\mathbf{u}, \mathbf{v}\rangle \tag{8.10}$$

A complex valued function f on the interval $[a, b]$ can be written as $f = f_1 + if_2$ where f_1, f_2 are real valued functions. We define $\bar{f} = f_1 - if_2$ and

$$\int_a^b f dt = \int_a^b f_1 dt + i\int_a^b f_2 dt$$

Example 8.1.5. *Let $V = C_F[a, b]$ ($a < b$) be the set of all F valued continuous functions which is infinite dimensional. For $f, g \in V$, define the* **standard inner product** *on $V = C_F[0, 1]$ by*

$$\langle f, g\rangle = \frac{1}{b-a}\int_a^b \bar{f}(t)g(t)dt \tag{8.11}$$

Then (8.11) *defines an inner product on V. The properties* (1) – (3) *of* Definition 8.1.1 *are easy to verify. Only the property* (4) *may be worth to mention. For this we need to show*

$\int_a^b |f(t)|^2 dt > 0$ *if* $f \neq 0$. *This follows from a fact from analysis*†: *If* f *is continuous and* $f(x) \geqslant 0$ *on* $[a,b]$ *which is not identically zero then* $\int_a^b f\,dx > 0$.

For example, let $V = C_{\mathbb{C}}[0, 2\pi]$ *and let* $f(x) = e^x + i\sin x,\ g(x) = i\cos x$. *Then*

$$\|f\|^2 = \frac{1}{2\pi}\int_0^{2\pi} f\bar{f}dx = \int_0^1 (e^{2x} + \sin^2 x)dx = \frac{e^{4\pi}-1}{4\pi} + \frac{1}{2}$$
$$\langle f, g\rangle = \frac{1}{2\pi}\int_0^{2\pi} \bar{f}gdx = \frac{1}{2\pi}\int_0^{2\pi} (\sin x\cos x + i\cos x e^x)dx = i\left(\frac{e^{2\pi}-1}{4\pi}\right)$$

Division by $b-a$ *in* (8.11) *is not significant. We can multiply any positive constant.*

Example 8.1.6. *Let* $V = M(n, F)$. *For* $A \in V$ *define the transpose conjugate* A^* *to be* $(A^*)_{ij} = \overline{A}_{ji}$. *And we define*

$$\langle A, B\rangle = \mathrm{Tr}(A^*B) \tag{8.12}$$

Then (8.12) *defines an inner product on* V *which we call the* **standard inner product** *on* $M(n, F)$. *We leave its proof as* Exercise 3.

Example 8.1.7. *Let* $V = \mathbb{C}^3$ *and* $A = \begin{bmatrix} 1 & 0 & 0 \\ 0 & 1 & -i \\ 0 & i & 2 \end{bmatrix}$. *We define*

$$\langle \mathbf{u}, \mathbf{v}\rangle_A = \mathbf{u}^* A\mathbf{v} \tag{8.13}$$

for $\mathbf{u}, \mathbf{v} \in V$.

The conditions (1), (2) *of* Definition 8.1.1 *are easy to check. For the condition* (3) *first note that* $A^* = A$. *Hence we have* $\langle \mathbf{u}, \mathbf{v}\rangle_A = \mathbf{u}^*A\mathbf{v} = \mathbf{u}^*A^*\mathbf{v}^{**} = (\mathbf{v}^*A\mathbf{u})^* = \overline{\langle \mathbf{v}, \mathbf{u}\rangle_A}$ *since* $\mathbf{u}A\mathbf{v}^*$ *is a constant. For the condition* (4), *let* $\mathbf{u} = \begin{bmatrix} x \\ y \\ z \end{bmatrix} \in \mathbb{C}^3$ *and we compute*

$$\begin{aligned}
\mathbf{u}^*A\mathbf{u} &= (\bar{x}, \bar{y}, \bar{z})\begin{bmatrix} 1 & 0 & 0 \\ 0 & 1 & -i \\ 0 & i & 2 \end{bmatrix}\begin{bmatrix} x \\ y \\ z \end{bmatrix} \\
&= x\bar{x} + \bar{y}(y - iz) + \bar{z}(iy + 2z) \\
&= x\bar{x} + y\bar{y} + 2z\bar{z} - i\bar{y}z + iy\bar{z} \\
&= x\bar{x} + (y - iz)(\bar{y} + i\bar{z}) + z\bar{z} \\
&= |x|^2 + |y + iz|^2 + |z|^2 > 0 \text{ if } \mathbf{u} \neq 0
\end{aligned}$$

Hence (8.13) *defines an inner product on* V.

---

†Here is a quick proof. Since $f \neq 0$ and $f \geq 0$ on $[a,b]$ there is $c \in [a,b]$ such that $f(c) > 0$. Since f is continuous at $x = c$ if we choose $\epsilon = \frac{f(c)}{2}$ then there is $\delta > 0$ such that $|f(c) - f(x)| < \frac{f(c)}{2}$ whenever $|x - c| < \delta$. Define $g(x)$ on $[a,b]$ by $g(x) = \begin{cases} \frac{f(c)}{2} & (c - \delta < x < c + \delta) \\ 0 & (\text{otherwise}) \end{cases}$. Then $f \geq g$ and $\int_a^b g(x)dx = \frac{f(c)}{2}2\delta = f(c)\delta > 0$. Hence $\int_a^b f(x)dx \geq \int_a^b g(x)dx = f(c)\delta > 0$.

Example 8.1.8. *As in* Example 6.3.3, *let* $V_1 = M(2,\mathbb{R}), V_2 = P_3(\mathbb{R}), V_3 = \mathbb{R}^4$ *and* $T_1 : V_1 \to V_3$ *and* $T_2 : V_2 \to V_3$ *be the maps defined by* $T_1(A) = \begin{bmatrix} a \\ b \\ c \\ d \end{bmatrix} = T_2(f)$ *where* $A = \begin{bmatrix} a & b \\ c & d \end{bmatrix}$ *and* $f(x) = a+bx+cx^2+dx^3$. *Then* T_1, T_2 *are isomorphisms of vector spaces. Each* V_i *is an inner product space with the standard inner product. Now* T_1 *is an isometric isomorphism but* T_2 *is not an isometry.* (Check this !)

Practice Problem 8.1.1. *Let* $A = \begin{bmatrix} 2 & 1 \\ 1 & 2 \end{bmatrix}$. *Let* $V = \mathbb{R}^2$ *and for* $\mathbf{u}, \mathbf{v} \in V$ *define* $\langle \mathbf{u}, \mathbf{v} \rangle$ *by* $\langle \mathbf{u}, \mathbf{v} \rangle = \mathbf{u}^t A \mathbf{v}$. *Prove that* $\langle -, - \rangle$ *defined in this way is an inner product on* V.

8.1.2 Orthogonalities

We saw, in Chapter 5, that orthogonal vectors in $\mathbb{R}^n$ plays an important role in various ways. That is also true for the general inner product spaces.

Definition 8.1.9. *Let* V *be an inner product space and* $\mathbf{u}, \mathbf{v} \in V$. *We say* $\mathbf{u}$ *is* **perpendicular** (*or* **orthogonal**) *to* $\mathbf{v}$ *if* $\langle \mathbf{u}, \mathbf{v} \rangle = 0$. *In this case, we write*

$$\mathbf{u} \perp \mathbf{v}$$

If S *is a subset of* V *then we define the* **orthogonal complement** $S^\perp$ *by*

$$S^\perp = \{\mathbf{v} \in V \mid \mathbf{v} \perp \mathbf{x} \text{ for all } \mathbf{x} \in S\} \tag{8.14}$$

A set of vectors $S = \{\mathbf{u}_1, \mathbf{u}_2, \mathbf{u}_3 \ldots\}$ *is said to be* **orthogonal** *if* $\mathbf{u}_i \perp \mathbf{u}_j (i \neq j)$. *We say* S *is* **orthonormal** *if* S *is orthogonal unit vectors. A basis* $\mathcal{B}$ *of* V *is an* **orthogonal basis** (*resp.* **orthonormal basis**) *if* $\mathcal{B}$ *is a basis consisting of orthogonal* (*resp. orthonormal*) *vectors.*

Example 8.1.10. *If* $V = \mathbb{C}^n$ *then the standard basis* $\{\mathbf{e}^1, \mathbf{e}^2, ..., \mathbf{e}^n\}$ *is an orthonormal basis. If* $V = M(n,\mathbb{C})$ *then the standard basis* $\{E_{ij} \mid 1 \le i,j \le n\}$ (Example 6.2.8 (4)) *is an orthonormal basis with the standard inner product* (Example 8.1.6).

As in §5.2, $S^\perp$ is a subspace even if S is just a subset of V and

$$S \cap S^\perp \subseteq \{\mathbf{0}\} \tag{8.15}$$

where the inclusion becomes equality only when $\mathbf{0} \in S$. Otherwise the intersection is empty. Also note $\{\mathbf{0}\}^\perp = V$, $V^\perp = \{\mathbf{0}\}$.

If $\mathbf{u} \perp \mathbf{v}$ then we have **Pythagorean Theorem**

$$\|\mathbf{u} + \mathbf{v}\|^2 = \|\mathbf{u}\|^2 + \|\mathbf{v}\|^2 \tag{8.16}$$

whose proof is Exercise 5.

Example 8.1.11. *Let* $H = C_{\mathbb{C}}[0, 2\pi]$, *the complex valued continuous functions on the interval* $[0, 2\pi]$ *with the inner product defined by* $\langle f, g \rangle = \dfrac{1}{2\pi}\displaystyle\int_0^{2\pi} \bar{f}(t) g(t)\, dt$. *Then* $\{f_k(t) = e^{ikt} | k \in \mathbb{Z}\}$ *are orthonormal vectors.*

A set of orthogonal vectors have nice properties as the following theorems show. As in §5.2 for $\mathbb{R}^n$, we will have a method to produce a set of orthogonal vectors which we called Gram-Schmidt orthogonalization process.

Theorem 5.2.10 for abstract inner product space becomes:

Theorem 8.1.12. *If $S = \{\mathbf{u}_1, \mathbf{u}_2, \ldots, \mathbf{u}_n\}$ is a set of nonzero orthogonal vectors then S is linearly independent.*

Proof. Suppose $\sum a_i\mathbf{u}_i = \mathbf{0}$. Take the inner product with $\mathbf{u}_j$:

$$0 = \big\langle \textstyle\sum_i a_i\mathbf{u}_i, \mathbf{u}_j\big\rangle = \bar{a}_j\langle \mathbf{u}_j, \mathbf{u}_j\rangle = \bar{a}_j\|\mathbf{u}_j\|^2$$

Since $\|\mathbf{u}_j\| \neq 0$ we have $a_j = 0$. □

The following is an abstract inner product space version of Theorem 5.2.11.

Theorem 8.1.13. *Let $S = \{\mathbf{u}_1, \mathbf{u}_2, \ldots, \mathbf{u}_n\}$ be a set of orthogonal vectors in an inner product space V. If $\mathbf{v} \in \operatorname{Span}(S)$ then*

$$\mathbf{v} = \sum_{i=1}^{n} \frac{\langle \mathbf{u}_i, \mathbf{v}\rangle}{\|\mathbf{u}_i\|^2}\mathbf{u}_i \tag{8.17}$$

If S is orthonormal then

$$\mathbf{v} = \sum_{i=1}^{n} \langle \mathbf{u}_i, \mathbf{v}\rangle\mathbf{u}_i \tag{8.18}$$

Proof. Let $\mathbf{v} = \sum a_i\mathbf{u}_i$. We need to decide a_i. Now $\langle \mathbf{u}_i, \mathbf{v}\rangle = \langle \mathbf{u}_i, \sum a_j\mathbf{u}_j\rangle = a_i\|\mathbf{u}_i\|^2$. Hence $a_i = \dfrac{\langle \mathbf{u}_i, \mathbf{v}\rangle}{\|\mathbf{u}_i\|^2}$ as required. □

Theorem 8.1.14. *Let V be a finite dimensional inner product space and $\mathcal{B} = \{\mathbf{v}_1, \mathbf{v}_2, \ldots, \mathbf{v}_n\}$ be an orthonormal basis. If $T : V \to V$ is a linear map with $A = [T]_{\mathcal{B}}$ then*

$$A_{ij} = \langle \mathbf{v}_i, T(\mathbf{v}_j)\rangle \tag{8.19}$$

Proof. Since $T(\mathbf{v}_j) = \sum_l A_{lj}\mathbf{v}_l$, we have $\langle \mathbf{v}_i, T(\mathbf{v}_j)\rangle = \langle \mathbf{v}_i, \sum_l A_{lj}\mathbf{v}_l\rangle = A_{ij}$. □

With the inner products on F^n, F_n defined in Example 8.1.3 and Example 8.1.4 we have the following theorem which is Theorem 5.2.14 when $F = \mathbb{R}$.

Theorem 8.1.15. *Let $U \in M(n, F)$. Then the following conditions are equivalent.*

(1) *The columns $\{A^{(1)}, A^{(2)}, \ldots, A^{(n)}\}$ of A are orthonormal basis of F^n.*

(2) *The rows of $\{A_{(1)}, A_{(2)}, \ldots, A_{(n)}\}$ of A are orthonormal basis of F_n.*

(3) *$A^*A = AA^* = I_n$*

Proof. The condition (1) is equivalent to $A^*A = I_n$ and this is equivalent to $A^*A = I_n$ by Theorem 2.5.3. Hence (1) and (3) are equivalent. The rest of the proof is Exercise 7. □

Definition 8.1.16. *A matrix $A \in M(n, \mathbb{C})$ satisfying $A^*A = AA^* = I$ is called a* **unitary matrix**. *If $F = \mathbb{R}$ we call such a matrix A an* **orthogonal matrix**.

Hence $A \in M(n, \mathbb{C})$ (resp. $A \in M(n, \mathbb{R})$) is unitary (resp. orthogonal) if and only if the columns of A are orthonormal. Also this is equivalent to the condition that the rows of A are orthonormal. A real unitary matrix is just an orthogonal matrix.

Example 8.1.17. (1) *We saw in* Example 5.2.16 *that the rotation map* $A_\theta = \begin{bmatrix} \cos\theta & -\sin\theta \\ \sin\theta & \cos\theta \end{bmatrix}$ *is an orthogonal matrix. Later we will show that an orthogonal matrix is similar to a matrix with* ± 1*'s and* A_θ*'s along the diagonal* (Theorem 8.4.15).

(2) *If we let* $A = \frac{1}{\sqrt{2}} \begin{bmatrix} 1 & 0 & i \\ 0 & \sqrt{2}\,i & 0 \\ i & 0 & 1 \end{bmatrix}$ *then* $A^* = \frac{1}{\sqrt{2}} \begin{bmatrix} 1 & 0 & i \\ 0 & -\sqrt{2}\,i & 0 \\ -i & 0 & 1 \end{bmatrix}$. *Hence* $AA^* = I_3$ *and hence* A *is a unitary matrix. We will show later that a unitary matrix is similar to a matrix with* $e^{i\theta}$*'s along the diagonal* (Corollary 8.4.13).

8.1.3 Orthogonal Projections and Gram Schmidt Process

Orthogonal projections of §5.2 in $\mathbb{R}^n$ can be carried over to any inner product space almost without any change.

Let V be an inner product space and $\mathbf{y} \in V$ and $W = \text{Span}\{\mathbf{u}\}$. The **orthogonal projection** $\hat{\mathbf{y}}$ of $\mathbf{y}$ onto $\mathbf{u}$ is defined by

$$\hat{\mathbf{y}} \in W \text{ and } (\mathbf{y} - \hat{\mathbf{y}}) \perp \mathbf{u} \tag{8.20}$$

We let $\hat{\mathbf{y}} = c\mathbf{u}$ and we require $(\mathbf{y} - \hat{\mathbf{y}}) \perp \mathbf{u}$ i.e., $\langle \mathbf{u}, \mathbf{y} - c\mathbf{u} \rangle = 0$; $\langle \mathbf{u}, \mathbf{y} \rangle = c\langle \mathbf{u}, \mathbf{u} \rangle$. Thus we get

$$\hat{\mathbf{y}} = \frac{\langle \mathbf{u}, \mathbf{y} \rangle}{\langle \mathbf{u}, \mathbf{u} \rangle} \mathbf{u} \tag{8.21}$$

As in §5.2, $\hat{\mathbf{y}}$ is independent of the choice of a basis of W. We will call $\hat{\mathbf{y}}$ the **orthogonal projection** of $\mathbf{y}$ onto W and we will write $P_W(\mathbf{y})$ for $\hat{\mathbf{y}}$.

If $\mathbf{w} \in W$ then $\mathbf{w} = d\mathbf{u}$ for some $d \in F$ and

$$P_W(\mathbf{w}) = \frac{\langle \mathbf{u}, d\mathbf{u} \rangle}{\langle \mathbf{u}, \mathbf{u} \rangle} \mathbf{u} = d\mathbf{u} = \mathbf{w} \tag{8.22}$$

and hence

$$P_W^2(\mathbf{y}) = P_W\left(P_W(\mathbf{y})\right) = P_W(\mathbf{y}) \tag{8.23}$$

If we need to specify the basis $\{\mathbf{u}\}$ of W we write $P_{\mathbf{u}}(\mathbf{y})$ for $P_W(\mathbf{y})$.

As we saw in §5.2, another geometric interpretation of orthogonal projection: $\hat{\mathbf{y}}$ is that the closest vector to $\mathbf{y}$ among the vectors in W i.e.,

$$\|\mathbf{y} - \hat{\mathbf{y}}\| \leq \|\mathbf{y} - \mathbf{w}\| \text{ for all } \mathbf{w} \in W \tag{8.24}$$

This will follow by writing $\mathbf{y} - \mathbf{w} = (\mathbf{y} - \hat{\mathbf{y}}) + (\hat{\mathbf{y}} - \mathbf{w})$ and from the fact $(\mathbf{y} - \hat{\mathbf{y}}) \perp (\hat{\mathbf{y}} - \mathbf{w})$ since $\hat{\mathbf{y}} - \mathbf{w} \in W$.

Now we consider the orthogonal projection onto a subspace W which is not necessarily one dimensional. If $S = \{\mathbf{u}_1, \mathbf{u}_2, ..., \mathbf{u}_p\}$ is orthogonal basis for a subspace W and $\mathbf{y} \in \text{Span}(S)$ then, in terms of orthogonal projections, (8.17) says

$$\mathbf{y} = \sum_{i=1}^{p} P_{\mathbf{u}_i}(\mathbf{y}) \tag{8.25}$$

Let W be a subspace of V and $\mathbf{y} \in V$. Now we define an orthogonal projection $\hat{\mathbf{y}} = P_W(\mathbf{y})$ of $\mathbf{y}$ onto W. As in (8.20) the properties for $\hat{\mathbf{y}}$ we require are

$$\hat{\mathbf{y}} \in W \text{ and } (\mathbf{y} - \hat{\mathbf{y}}) \perp W \tag{8.26}$$

for $\dim W \geq 1$. Let $S = \{\mathbf{u}_1, \mathbf{u}_2, ..., \mathbf{u}_p\}$ be an orthogonal basis of W. As in (5.27) of §5.2, we have

$$\hat{\mathbf{y}} = \frac{\langle \mathbf{u}_1, \mathbf{y}\rangle}{\langle \mathbf{u}_1, \mathbf{u}_1\rangle}\mathbf{u}_1 + \frac{\langle \mathbf{u}_2, \mathbf{y}\rangle}{\langle \mathbf{u}_2, \mathbf{u}_2\rangle}\mathbf{u}_2 + \cdots + \frac{\langle \mathbf{u}_p, \mathbf{y}\rangle}{\langle \mathbf{u}_p, \mathbf{u}_p\rangle}\mathbf{u}_p \tag{8.27}$$

Definition 8.1.18. *Let W be a subspace of V with an orthogonal basis $S = \{\mathbf{u}_1, \mathbf{u}_2, ..., \mathbf{u}_p\}$. The* **orthogonal projection $\hat{\mathbf{y}}$ of $\mathbf{y}$ onto** *W is defined by (8.27). Often we write $P_W(\mathbf{y})$ for $\hat{\mathbf{y}}$.*

If $S = \{\mathbf{u}_1, \mathbf{u}_2, ..., \mathbf{u}_p\}$ is orthonormal then (8.27) becomes

$$\hat{\mathbf{y}} = P_W(\mathbf{y}) = \langle \mathbf{u}_1, \mathbf{y}\rangle\mathbf{u}_1 + \langle \mathbf{u}_2, \mathbf{y}\rangle\mathbf{u}_2 + \cdots + \langle \mathbf{u}_p, \mathbf{y}\rangle\mathbf{u}_p \tag{8.28}$$

For the fact that $P_W(\mathbf{y})$ is independent of our choice of the orthonormal basis see the remark following Theorem 8.1.24.

Example 8.1.19. (1) *Let $\mathbf{y} = \begin{bmatrix} 1 \\ -1 \\ 2 \end{bmatrix}$ and W = xy-plane. Choose an orthonormal basis $\{\mathbf{e}^1, \mathbf{e}^2\}$ for W. Then*

$$P_W(\mathbf{y}) = \hat{\mathbf{y}} = \langle \mathbf{e}^1, \mathbf{y}\rangle\mathbf{e}^1 + \langle \mathbf{e}^2, \mathbf{y}\rangle\mathbf{e}^2 = \mathbf{e}^1 - \mathbf{e}^2$$

as expected.

(2) *Let $H = C_{\mathbb{C}}[0, 2\pi]$ be the inner product space of* Example 8.1.11 *with the inner product $\langle f, g\rangle = \frac{1}{2\pi}\int_0^{2\pi} \bar{f}g\, dt$. Let $S = \{1, e^{it}, e^{2it}\}$ and $W = \operatorname{Span}(S)$. Then S is an orthonormal basis for W. Let $f(t) = t^2 \in H$. What is $P_W(f)$? We have*

$$\begin{aligned} P_W(f) &= \sum_{k=0}^{2} \langle e^{ikt}, t^2\rangle e^{ikt} = \sum_{k=0}^{2} \left(\int_0^{2\pi} t^2 e^{-ikt}\, dt\right) e^{ikt} \\ &= \frac{8\pi^2}{3} \cdot 1 + 4(\pi + \pi^2 i)e^{it} + (\pi + 2\pi^2 i)e^{2it} \end{aligned}$$

We saw an orthogonal set of vectors are useful. We recall the method to produce orthogonal set of vectors from §5.2 so called Gram-Schmidt orthogonalization (Theorem 5.2.26): Let $\{\mathbf{w}_1, \mathbf{w}_2, ..., \mathbf{w}_p\} \subseteq V$ be a linearly independent vectors in an inner product space V and let $W_i = \operatorname{Span}\{\mathbf{w}_1, \mathbf{w}_2, ..., \mathbf{w}_i\}$. Define

$$\begin{aligned} \mathbf{v}_1 &= \mathbf{w}_1 \in W_1 \\ \mathbf{v}_2 &= \mathbf{w}_2 - P_{W_1}\mathbf{w}_2 = \mathbf{w}_2 - \frac{\langle \mathbf{v}_1, \mathbf{w}_2\rangle}{\langle \mathbf{v}_1, \mathbf{v}_1\rangle}\mathbf{v}_1 \in W_2 \end{aligned}$$

Then $\mathbf{v}_2 \perp W_1$ by (8.26) and $\mathbf{v}_2 \in W_2$; $\operatorname{Span}\{\mathbf{v}_1, \mathbf{v}_2\} = W_2$. Now let

$$\mathbf{v}_3 = \mathbf{w}_3 - P_{W_2}\mathbf{w}_3 = \mathbf{w}_3 - \frac{\langle \mathbf{v}_1, \mathbf{w}_3\rangle}{\langle \mathbf{v}_1, \mathbf{v}_1\rangle}\mathbf{v}_1 - \frac{\langle \mathbf{v}_2, \mathbf{w}_3\rangle}{\langle \mathbf{v}_2, \mathbf{v}_2\rangle}\mathbf{v}_2 \in W_3$$

Then $\mathbf{v}_3 \perp W_2$ by (8.26) and hence $\{\mathbf{v}_1, \mathbf{v}_2, \mathbf{v}_3\}$ is mutually orthogonal. And also $\mathbf{v}_3 \in W_3$; $\text{Span}\{\mathbf{v}_1, \mathbf{v}_2, \mathbf{v}_3\} = W_3$. Repeating this we get $\{\mathbf{v}_1, \mathbf{v}_2, ..., \mathbf{v}_i\}$ such that

$$\text{Span}\{\mathbf{v}_1, \mathbf{v}_2, ..., \mathbf{v}_i\} = W_i \quad \text{and } \mathbf{v}_{i+1} \perp W_i \tag{8.29}$$

Theorem 8.1.20 (Gram-Schmidt Orthogonalization). *Let V be an inner product space and $S = \{\mathbf{w}_1, \mathbf{w}_2, ..., \mathbf{w}_p\}$ be linearly independent subset of V. Let $W_i = \text{Span}\{\mathbf{w}_1, \mathbf{w}_2, ..., \mathbf{w}_i\}$, $W = W_p$. Then there is an orthogonal subset $\{\mathbf{v}_1, \mathbf{v}_2, ..., \mathbf{v}_p\}$ of W satisfying (8.29).*

We can choose $\mathbf{v}_i$ inductively as

$$\begin{cases} \mathbf{v}_1 = \mathbf{w}_1 \\ \mathbf{v}_2 = \mathbf{w}_2 - P_{W_1}(\mathbf{w}_2) \\ \quad \cdots \\ \mathbf{v}_i = \mathbf{w}_i - P_{W_{i-1}}(\mathbf{w}_i) \end{cases} \tag{8.30}$$

Proof. We will omit a formal proof since the our assertions are obvious by the preceding remarks. □

Corollary 8.1.21. *Let V be an inner product space of dimension n. Then there is an isometric isomorphism $\phi : V \to F^n$ where the inner product of F^n is the standard inner product. Explicitly, if $\mathcal{B}$ is an orthonormal basis of V then the coordinate map*

$$\phi_{\mathcal{B}} : V \to F^n, \qquad \mathbf{v} \mapsto [\mathbf{v}]_{\mathcal{B}}$$

is an isometric isomorphism.

Proof. Let $\mathcal{B} = \{\mathbf{v}_1, \mathbf{v}_2, ..., \mathbf{v}_n\}$ be an orthonormal basis which exists by Gram-Schmidt. We know that the coordinate map is an isomorphism of vector space (Theorem 6.3.7). We need to show the coordinate map preserves the inner product. Let $\mathbf{u} = \sum a_i\mathbf{v}_i, \mathbf{w} = \sum b_i\mathbf{v}_i$. Then we have

$$\langle \mathbf{v}, \mathbf{w}\rangle_V = \left\langle \sum a_i\mathbf{v}_i, \sum b_i\mathbf{v}_i \right\rangle_V = \sum \bar{a}_i b_i = \langle [\mathbf{v}]_{\mathcal{B}}, [\mathbf{w}]_{\mathcal{B}}\rangle_{F^n}$$

where $\langle\ ,\ \rangle_V$, $\langle\ ,\ \rangle_{F^n}$ are the inner product on V, F^n respectively. □

Let $\mathcal{B}$ be an orthonormal basis of an inner product space V and T is a linear operator and let $A = [T]_{\mathcal{B}}$. Corollary 8.1.21 says that V is not only isomorphic to F^n as vector spaces but also the inner product structure of V is the same as the inner product structure of F^n. For example, if $\mathbf{v}$ is an eigenvector for T if and only if $[\mathbf{v}]_{\mathcal{B}}$ is an eigenvector of A. And $\mathbf{u} \perp \mathbf{v}$ if and only if $[\mathbf{u}]_{\mathcal{B}} \perp [\mathbf{v}]_{\mathcal{B}}$ and so on.

Example 8.1.22. *Let $V = P_2(\mathbb{R})$ and $\langle f, g\rangle = \int_0^1 f(x)g(x)dx$. We want to find an orthonormal basis of V using the basis $\{1, x, x^2\}$.*

Let $\mathbf{w}_1 = 1, \mathbf{w}_2 = x, \mathbf{w}_3 = x^2$. Then we have

$$\mathbf{v}_1 = \mathbf{w}_1 = 1;\ \|\mathbf{v}_1\|^2 = \int_0^1 1dx = 1 \text{ and } \langle \mathbf{v}_1, \mathbf{w}_2\rangle = \int_0^1 1 \cdot xdx = \frac{1}{2}$$

$$\mathbf{v}_2 = \mathbf{w}_2 - \frac{\langle \mathbf{v}_1, \mathbf{w}_2\rangle}{\|\mathbf{v}_1\|^2}\mathbf{v}_1 = -\frac{1}{2} + x;\ \ \|\mathbf{v}_2\|^2 = \int_0^1 (-\frac{1}{2} + x)^2 dx = \frac{1}{12}$$

$$\mathbf{v}_3 = \mathbf{w}_3 - \frac{\langle \mathbf{v}_1, \mathbf{w}_3\rangle}{\|\mathbf{v}_1\|^2}\mathbf{v}_1 - \frac{\langle \mathbf{v}_2, \mathbf{w}_3\rangle}{\|\mathbf{v}_2\|^2}\mathbf{v}_2;$$

Since $\langle \mathbf{v}_1, \mathbf{w}_3\rangle = \int_0^1 1\cdot x^2dx = \frac{1}{3}$; $\langle \mathbf{v}_2, \mathbf{w}_3\rangle = \int_0^1(-\frac{1}{2}+x)x^2dx = \frac{1}{12}$ *we have*

$$\mathbf{v}_3 = \frac{1}{6} - x + x^2,\ \|\mathbf{v}_3\|^2 = \int_0^1 \left(\frac{1}{6} - x + x^2\right)^2 dx = \frac{1}{180}$$

Hence

$$\left\{\frac{\mathbf{v}_1}{\|\mathbf{v}_1\|}, \frac{\mathbf{v}_2}{\|\mathbf{v}_2\|}, \frac{\mathbf{v}_3}{\|\mathbf{v}_3\|}\right\} = \left\{1,\ 2\sqrt{3}\left(-\frac{1}{2}+x\right),\ 6\sqrt{5}\left(\frac{1}{6} - x + x^2\right)\right\}$$

is an orthonormal basis of V.

Now, let $W = \operatorname{Span}\{1, x\}$ *and* $\mathbf{y} = 1 + x + x^2$. *For the orthogonal projection* $P_W(\mathbf{y})$ *of* $\mathbf{y}$ *we compute:*

$$\begin{aligned}P_W(\mathbf{y}) &= \frac{\langle \mathbf{v}_1, \mathbf{y}\rangle}{\|\mathbf{v}_1\|^2}\mathbf{v}_1 + \frac{\langle \mathbf{v}_2, \mathbf{y}\rangle}{\|\mathbf{v}_2\|^2}\mathbf{v}_2 \\ &= \int_0^1 (1+x+x^2)dx\cdot 1 + 12\int_0^1\left(-\frac{1}{2}+x\right)(1+x+x^2)dx\cdot x \\ &= \frac{1}{6}(11+x)\end{aligned}$$

Example 8.1.23. *Let* $V = \mathbb{C}^3$ *with the inner product defined by* $\langle \mathbf{u}, \mathbf{v}\rangle_A = \mathbf{u}^*A\mathbf{v}$ *where* $A = \begin{bmatrix}1 & 0 & 0\\ 0 & 1 & -i\\ 0 & i & 2\end{bmatrix}$ *in* Example 8.1.7. *Using the standard basis* $\mathcal{B} = \{\mathbf{e}^1, \mathbf{e}^2, \mathbf{e}^3\}$ *we want to find an orthonormal basis of* V *with the inner product* $\langle -, -\rangle_A$.

We apply Gram-Schmidt orthogonalization writing $\|\mathbf{v}\|_A$ *for* $\sqrt{\langle \mathbf{v}, \mathbf{v}\rangle_A}$:

$$\begin{aligned}\mathbf{v}_1 &= \mathbf{e}^1;\ \|\mathbf{v}_1\|_A = 1 \\ \mathbf{v}_2 &= \mathbf{e}^2 - \frac{\langle \mathbf{v}_1, \mathbf{e}^2\rangle_A}{\|\mathbf{v}_1\|_A^2}\mathbf{e}^1 = \mathbf{e}^2;\ \|\mathbf{v}_2\|_A = 1 \\ \mathbf{v}_3 &= \mathbf{e}^3 - \frac{\langle \mathbf{v}_1, \mathbf{e}^3\rangle_A}{\|\mathbf{v}_1\|_A^2}\mathbf{e}^1 - \frac{\langle \mathbf{v}_2, \mathbf{e}^3\rangle_A}{\|\mathbf{v}_2\|_A^2}\mathbf{e}^2 = i\mathbf{e}^2 + \mathbf{e}^3;\ \|\mathbf{v}_3\|_A = 1\end{aligned}$$

These are unit vectors. Hence $\mathcal{C} = \{\mathbf{v}_1 = \mathbf{e}^1, \mathbf{v}_2 = \mathbf{e}^2, \mathbf{v}_3 = i\mathbf{e}^2 + \mathbf{e}^3\}$ *is an orthonormal basis for* V.

Practice Problem 8.1.2. *Let* $V = P_3(\mathbb{R})$ *with basis* $\{1, x, x^2, x^3\}$ *and* $\langle f, g\rangle = \int_0^1 f(x)g(x)dx$. *Let* $W = \operatorname{span}\{1, x, x^2\}$. *Find* $\operatorname{proj}_W(x^3)$.

The following theorem is an abstract inner product version of Theorem 5.2.20.

Theorem 8.1.24. *Let* W *be a subspace of an inner product space* V. *Every* $\mathbf{y} \in V$ *can be written uniquely as*

$$\mathbf{y} = \hat{\mathbf{y}} + \mathbf{z}$$

where $\hat{\mathbf{y}} \in W$ *and* $\mathbf{z} = \mathbf{y} - \hat{\mathbf{y}} \in W^\perp$. *If* $S = \{\mathbf{u}_1, \mathbf{u}_2, ..., \mathbf{u}_p\}$ *is an orthogonal basis for* W *then the orthogonal projection* $\hat{\mathbf{y}} = P_W(\mathbf{y})$ *is given by* (8.27).

Proof. Existence is already guaranteed by (8.27). We only need to check uniqueness. Suppose $\hat{\mathbf{y}} + \mathbf{z} = \mathbf{y} = \hat{\mathbf{y}'} + \mathbf{z}'$. Then $\hat{\mathbf{y}} - \hat{\mathbf{y}'} = \mathbf{z}' - \mathbf{z} \in W \cap W^\perp = \mathbf{0}$ by (8.15). Therefore $\hat{\mathbf{y}} = \hat{\mathbf{y}'}$ and $\mathbf{z}' = \mathbf{z}$. □

By uniqueness of $\hat{\mathbf{y}}$ we see that the projection map $P_W : V \to V$ given in (8.27) does not depends on the choice of orthogonal basis $S = \{\mathbf{u}_1, \mathbf{u}_2, ..., \mathbf{u}_p\}$ of W.

Definition 8.1.25. *Let W, W' be subspaces if V. We say V is an* **orthogonal sum** *of W and W' if $W \perp W'$ and $V = W \oplus W'$. In this case, we will write*

$$V = W \textcircled{\perp} W' \tag{8.31}$$

More generally, if $W_1, W_2, ..., W_k$ are mutually orthogonal subspaces and $V = \bigoplus_{i=1}^{k} W_i$ then we write

$$V = \textcircled{\perp}_{i=1}^{k} W_i$$

In this notation, Theorem 8.1.24 says $V = W \textcircled{\perp} W^{\perp}$ and P_W is the projection according to the direct sum decomposition $V = W \oplus W^{\perp}$. We sum up our discussion in the following theorem which is an abstraction of Theorem 5.2.20 .

Theorem 8.1.26. *Let V be a finite dimensional inner product space and W be a subspace. The orthogonal projection $P_W : V \to V$ onto W* (Definition 8.1.18) *is a linear map and*

$$R(P_W) = W, \ N(P_W) = W^{\perp} \text{ and } V = W \textcircled{\perp} W^{\perp}$$
$$P_W^2 = P_W$$
$$P_W(\mathbf{w}) = \mathbf{w} \text{ for all } \mathbf{w} \in W$$

In particular,

$$\dim V = \dim W + \dim W^{\perp} \tag{8.32}$$

If $\mathcal{B} = \{\mathbf{u}_1, \mathbf{u}_2, ..., \mathbf{u}_p\}$ is an orthonormal basis for W and $\mathcal{B}^{\perp} = \{\mathbf{u}_{p+1}, \mathbf{u}_{p+2}, ..., \mathbf{u}_n\}$ is an orthonormal basis of $W^{\perp}$ then $\mathcal{C} = \mathcal{B} \cup \mathcal{B}^{\perp}$ is a basis of V and

$$[P_W]_{\mathcal{C}} = \begin{bmatrix} I_p & \\ & 0 \end{bmatrix}$$

Proof. Checking the details of the proof of the theorem is Exercise 19 and Exercises 20. □

In Example 6.2.31, we saw that if W is a subspace of a vector space V then there are many subspaces W' such that $V = W \oplus W'$. However if V is an inner product space and W is a subspace of V then Theorem 8.1.26 shows that there is a unique subspace W' such that $V = W \textcircled{\perp} W'$ (Exercise 14).

Recall that a linear operator $T : V \to V$ on a vector space V satisfying $T^2 = T$ is called a **projection** (Definition 6.3.15). By Theorem 8.1.26, we see that an orthogonal projection on an inner product space is a projection. The projections in Example 6.3.13 are not orthogonal projections. (Check these assertions.)

Next theorem gives a characterization of an orthogonal projection for the inner product spaces.

Theorem 8.1.27. *Let V be a finite dimensional inner product space. A linear operator $T : V \to V$ is an orthogonal projection onto $R(T)$ if and only if $T^2 = T$ and $R(T)^{\perp} = N(T)$.*

Proof. If T is an orthogonal projection onto $W = R(T)$ then T satisfies the required conditions by Theorem 8.1.26.

Conversely assume T satisfies the condition. Let $\mathbf{y} \in V$. Then $\mathbf{y} = T(\mathbf{y}) + (\mathbf{y} - T(\mathbf{y}))$ with $T(\mathbf{y}) \in R(T)$ and $\mathbf{y} - T(\mathbf{y}) \in N(T) = R(T)^\perp$ since $T^2 = T$. On the other hand, by Theorem 8.1.24, we can write $\mathbf{y}$ as a sum $\mathbf{y} = \hat{\mathbf{y}} + \mathbf{z}$ in a unique way where $\hat{\mathbf{y}} \in R(T)$ is the orthogonal projection to $R(T)$ and $\mathbf{z} \in R(T)^\perp$. And hence, by uniqueness, $T(\mathbf{y}) = \hat{\mathbf{y}}$. □

Example 8.1.28. *Let* $V = \mathbb{R}^2$ *and* $A_1 = \begin{bmatrix} 1 & 1 \\ 0 & 1 \end{bmatrix}, A_2 = \begin{bmatrix} 1 & 1 \\ 0 & 0 \end{bmatrix}, A_3 = \begin{bmatrix} 1 & 0 \\ 0 & 0 \end{bmatrix}$. *Let* $T_i = T_{A_i}$ $(i = 1, 2, 3)$ *be the matrix linear maps for* A_i*'s on* $\mathbb{R}^2$. *Then* T_1 *is not a projection,* T_2 *is a projection but not an orthogonal projection. Now* T_3 *is an orthogonal projection.*

The following fact is a generalization of (8.24).

Theorem 8.1.29. *Let* W *be a subspace of an inner product space* V *and let* $\mathbf{y} \in V$. *Then for all* $\mathbf{w} \in W$ *we have*

$$\|\mathbf{y} - \hat{\mathbf{y}}\| \le \|\mathbf{y} - \mathbf{w}\| \tag{8.33}$$

That is, $\hat{\mathbf{y}}$ *is an element of* W *which is closest to* $\mathbf{y}$ *among the vectors of* W. *Equality holds if and only if* $\hat{\mathbf{y}} = \mathbf{w}$.

Proof. The proof is the same as the proof of (8.24) where W is of dimension 1. Since $\hat{\mathbf{y}}, \mathbf{w} \in W$ and $(\mathbf{y} - \hat{\mathbf{y}}) \perp W$ we have $(\mathbf{y} - \hat{\mathbf{y}}) \perp (\hat{\mathbf{y}} - \mathbf{w})$. By Pythagorean,

$$\|\mathbf{y} - \mathbf{w}\|^2 = \|(\mathbf{y} - \hat{\mathbf{y}}) + (\hat{\mathbf{y}} - \mathbf{w})\|^2 = \|(\mathbf{y} - \hat{\mathbf{y}})\|^2 + \|(\hat{\mathbf{y}} - \mathbf{w})\|^2$$

Since $\|(\hat{\mathbf{y}} - \mathbf{w})\|^2 \ge 0$ we have the desired inequality. □

8.1.4 Cauchy-Schwartz Inequality and Triangle Inequality

In §5.2, we proved Cauchy Schwartz inequality (Theorem 5.2.28) for $\mathbb{R}^n$ but the proof does not apply to arbitrary inner product spaces. We will give a proof working for any inner product space. Cauchy Schwartz inequality is important not only in linear algebra but also plays an important role in analysis.

Theorem 8.1.30. *Let* V *be an inner product space (not necessarily finite dimensional) and* $\mathbf{u}, \mathbf{v} \in V$. *Then*

(1) **(Cauchy Schwartz inequality)** $|\langle \mathbf{u}, \mathbf{v} \rangle| \le \|\mathbf{u}\| \, \|\mathbf{v}\|$

(2) **(Triangle Inequality)** $\|\mathbf{u} + \mathbf{v}\| \le \|\mathbf{u}\| + \|\mathbf{v}\|$

Proof. (1) If $\mathbf{v} = \mathbf{0}$ then the inequality is obviously true. For $\mathbf{v} \ne \mathbf{0}$ the inequality follows from

$$\begin{aligned} 0 &\le \langle \mathbf{u} - \frac{\langle \mathbf{v}, \mathbf{u} \rangle}{\|\mathbf{v}\|^2}\mathbf{v}, \mathbf{u} - \frac{\langle \mathbf{v}, \mathbf{u} \rangle}{\|\mathbf{v}\|^2}\mathbf{v} \rangle \\ &= \|\mathbf{u}\|^2 - \frac{\langle \mathbf{v}, \mathbf{u} \rangle \langle \mathbf{u}, \mathbf{v} \rangle}{\|\mathbf{v}\|^2} - \frac{\langle \mathbf{u}, \mathbf{v} \rangle \langle \mathbf{v}, \mathbf{u} \rangle}{\|\mathbf{v}\|^2} + \frac{\langle \mathbf{u}, \mathbf{v} \rangle \langle \mathbf{v}, \mathbf{u} \rangle}{\|\mathbf{v}\|^2} \\ &= \|\mathbf{u}\|^2 - \frac{|\langle \mathbf{v}, \mathbf{u} \rangle|^2}{\|\mathbf{v}\|^2} \end{aligned}$$

(2) For the triangle inequality first observe that if $z \in \mathbb{C}$ then

$$z + \bar{z} \leq |z + \bar{z}| \leq 2|z| \tag{8.34}$$

Since $\langle \mathbf{u}, \mathbf{v} \rangle = \overline{\langle \mathbf{v}, \mathbf{u} \rangle}$ we have

$$\begin{aligned}
\|\mathbf{u}+\mathbf{v}\|^2 &= \langle \mathbf{u}+\mathbf{v}, \mathbf{u}+\mathbf{v} \rangle \\
&= \langle \mathbf{u}, \mathbf{u} \rangle + \langle \mathbf{u}, \mathbf{v} \rangle + \langle \mathbf{v}, \mathbf{u} \rangle + \langle \mathbf{v}, \mathbf{v} \rangle && \text{(by (8.34))} \\
&\leq \|\mathbf{u}\|^2 + 2|\langle \mathbf{u}, \mathbf{v} \rangle| + \|\mathbf{v}\|^2 && \text{(by Cauchy Schwartz)} \\
&\leq \|\mathbf{u}\|^2 + 2\|\mathbf{u}\|\,\|\mathbf{v}\| + \|\mathbf{v}\|^2 \\
&= (\|\mathbf{u}\| + \|\mathbf{v}\|)^2
\end{aligned}$$

□

The following examples are from analysis which are good illustrations of Cauchy Schwartz and triangle inequality.

Example 8.1.31. (1) *If* $V = \mathbb{R}_n$ *with the standard inner product* (Example 8.1.4). *Let* $\mathbf{a} = (a_1, a_2, ..., a_n)$ *and* $\mathbf{b} = (b_1, b_2, ..., b_n)$. *Then by Cauchy-Schwartz we have*

$$|\langle \mathbf{a}, \mathbf{b} \rangle| = \left| \sum_{i=1}^{n} a_i b_i \right| \leq \sqrt{\sum_{i=1}^{n} a_i^2} \sqrt{\sum_{i=1}^{n} b_i^2} = \|\mathbf{a}\| \|\mathbf{b}\| \tag{8.35}$$

(2) *Let* $V = C_F[0,1]$ *of* Example 8.1.5 *then Cauchy-Schwartz and triangle inequality says*

$$\left| \int_0^1 f(t)g(t)\, dt \right|^2 \leq \int_0^1 |f(t)|^2 dt \int_0^1 |g(t)|^2 dt$$

$$\int_0^1 |f(t) + g(t)|^2\, dt \leq \left(\sqrt{\int_0^1 |f(t)|^2\, dt} + \sqrt{\int_0^1 |g(t)|^2\, dt} \right)^2$$

Cauchy Schwartz inequality proves its real worth in analysis dealing with infinite dimensional vector spaces as we see in the following two examples:

Example 8.1.32. *Let*

$$\ell^2 := \left\{ \mathbf{a} = (a_1, a_2, ...) \;\middle|\; a_i \in \mathbb{R},\ \sum_{i=1}^{\infty} |a_i|^2 < \infty \right\}$$

First, we need to show that ℓ^2 *is a vector space. If* $\mathbf{a} \in \ell^2$ *and* $c \in \mathbb{R}$ *then it is immediate that* $c\mathbf{a} \in \ell^2$. *If* $\mathbf{a}, \mathbf{b} \in \ell^2$ *then we need to show* $\mathbf{a} + \mathbf{b} \in \ell^2$ *namely,* $\sum |a_i + b_i|^2 < \infty$. *For this we first show*

$$\langle \mathbf{a}, \mathbf{b} \rangle := \sum_{i=1}^{\infty} a_i b_i$$

converges absolutely where $\mathbf{b} = (b_1, b_2, ...)$. *In fact, by Cauchy-Schwartz* (8.35) *on* $\mathbb{R}_n$ *we have*

$$\sum_{i=1}^{n} |a_i||b_i| \leq \sqrt{\sum_{i=1}^{n} |a_i|^2} \sqrt{\sum_{i=1}^{n} |b_i|^2} \leq \sqrt{\sum_{i=1}^{\infty} |a_1|^2} \sqrt{\sum_{i=1}^{\infty} |b_i|^2}$$

for all n. Hence $\sum_{i=1}^{\infty} a_i b_i$ converges absolutely. Hence

$$\sum_i |a_i + b_i|^2 \le \sum_i (|a_i| + |b_i|)^2 = \sum_i \left(|a_i|^2 + 2|a_i b_i| + |b_i|^2\right)$$
$$= \sum_i |a_i|^2 + 2\sum_i |a_i b_i| + \sum_i |b_i|^2$$

and the last sum is finite. (The last equality follows from a theorem in analysis which says: If a series converges absolutely then any rearrangement of the series converges to the same sum. See W. Rudin, Principle of Mathematical Analysis, *McGraw-Hill, 1976 (p.78)). Thus $\mathbf{a} + \mathbf{b} \in \ell^2$ and it is easy now to show ℓ^2 becomes a vector space. Also it is straightforward to check that $\langle \mathbf{a}, \mathbf{b} \rangle$ makes ℓ^2 into an inner product space.*

Notice that ℓ^2 is infinite dimensional since it has an infinite number of linear independent set $\{\mathbf{e}_i \mid i = 1, 2, ..\}$. Also note that ℓ^2 contains $\mathbb{R}_n$ for all $n = 1, 2, 3, ...$

Example 8.1.33. *Let $a_k \ge 0$ and let $\sum_{k=1}^{\infty} a_k$ is a convergent series. First, we want to show that $\sum_{k=1}^{\infty} \frac{\sqrt{a_k}}{k}$ also converges. In fact, let $\mathbf{a} = (\sqrt{a_1}, \sqrt{a_2}, \sqrt{a_3}, ...)$ and $\mathbf{b} = (1, \frac{1}{2}, \frac{1}{3}, ...)$. Then $\mathbf{a}, \mathbf{b} \in \ell^2$ since $\sum \frac{1}{k^2}$ converges. In* Example 8.1.32, *we showed that $\langle \mathbf{a}, \mathbf{b} \rangle$ converges;*

$$\langle (\sqrt{a_1}, \sqrt{a_2}, ...), (1, \frac{1}{2}, ...) \rangle = \sum_{k=1}^{\infty} \frac{\sqrt{a_k}}{k}$$

*is a convergent series. Now Cauchy Schwartz inequality (*Theorem 8.1.30) *gives*

$$\sum_{k=1}^{\infty} \frac{\sqrt{a_k}}{k} = \langle (\sqrt{a_1}, \sqrt{a_2}, ...), (1, \frac{1}{2}, ...) \rangle \le \sqrt{\sum_{k=1}^{\infty} a_k} \sqrt{\sum_{k=1}^{\infty} \frac{1}{k^2}}$$

and we know that $\sum a_k < \infty$, $\sum \frac{1}{k^2} < \infty$. And in fact, it is known that $\sum_{k=1}^{\infty} \frac{1}{k^2} = \frac{\pi^2}{6}$. Therefore, we have

$$\sum_{k=1}^{\infty} \frac{\sqrt{a_k}}{k} \le \frac{\pi}{\sqrt{6}} \sqrt{\sum_{k=1}^{\infty} a_k}$$

which is also an interesting inequality.

Answers for Practice Problems 8.1

(8.1.1) The conditions (1) and (2) follows from the properties of matrix multiplication. For (3), first observe that $\mathbf{u}^t A \mathbf{v} = (\mathbf{u}^t A \mathbf{v})^t$ since $\mathbf{u}^t A \mathbf{v} \in \mathbb{R}$. Hence we have $\langle \mathbf{v}, \mathbf{u} \rangle = \mathbf{v}^t A \mathbf{u} = (\mathbf{v}^t A \mathbf{u})^t = \mathbf{u}^t A^t \mathbf{v} = \mathbf{u}^t A \mathbf{v} = \langle \mathbf{u}, \mathbf{v} \rangle$ since A is symmetric. Now we need to check the property (4): Let $\mathbf{u} = \begin{bmatrix} x \\ y \end{bmatrix}$ then $\langle \mathbf{u}, \mathbf{u} \rangle = [x\ y] \begin{bmatrix} 2 & 1 \\ 1 & 2 \end{bmatrix} \begin{bmatrix} x \\ y \end{bmatrix} = 2x^2 + 2xy + 2y^2 = 2(x + \frac{1}{2}y)^2 + \frac{3}{2}y^2 > 0$ unless $\mathbf{u} = \mathbf{0}$.

(8.1.2) We found an orthonormal basis of W in Example 8.1.22

$$\left\{\mathbf{u}_1 = 1,\ \mathbf{u}_2 = 2\sqrt{3}(-\frac{1}{2}+x),\ \mathbf{u}_3 = 6\sqrt{5}\left(\frac{1}{6}-x+x^2\right)\right\}$$

Letting $\mathbf{y} = x^3$ we have

$$\langle \mathbf{u}_1, \mathbf{y}\rangle = \int_0^1 x^3 dx = \frac{1}{4}$$
$$\langle \mathbf{u}_2, \mathbf{y}\rangle = 2\sqrt{3}\int_0^1 x^3\left(-\frac{1}{2}+x\right)dx = \frac{3\sqrt{3}}{20}$$
$$\langle \mathbf{u}_3, \mathbf{y}\rangle = 6\sqrt{5}\int_0^1 x^3\left(\frac{1}{6}-x+x^2\right)dx = \frac{\sqrt{5}}{20}$$

Hence

$$P_W(\mathbf{y}) = \langle \mathbf{u}_1, \mathbf{y}\rangle\mathbf{u}_1 + \langle \mathbf{u}_2, \mathbf{y}\rangle\mathbf{u}_2 + \langle \mathbf{u}_1, \mathbf{y}\rangle\mathbf{u}_3 = \frac{1}{20} - \frac{3}{5}x + \frac{3}{2}x^2$$

8.1 Exercises

1. Let $V = \mathbb{R}^2$ and define $\langle \mathbf{u}, \mathbf{v}\rangle_A = \mathbf{u}^t A\mathbf{v}$ for $A \in M(2)$. Determine which of the following A defines an inner product on V.

   (a) $A = \begin{bmatrix} 1 & 1 \\ 1 & 2 \end{bmatrix}$ (b) $A = \begin{bmatrix} 1 & 2 \\ 2 & 2 \end{bmatrix}$ (c) $A = \begin{bmatrix} 1 & 1 \\ 1 & 1 \end{bmatrix}$ (d) $A = \begin{bmatrix} 1 & -1 \\ 1 & 0 \end{bmatrix}$

2. Let $\mathbf{u} \in \mathbb{C}^n$. Let $\mathbf{u} = \mathbf{u}_{re} + i\mathbf{u}_{im}$ and $\bar{\mathbf{u}} = \mathbf{u}_{re} - i\mathbf{u}_{im}$ where $\mathbf{u}_{re}, \mathbf{u}_{im}, \bar{\mathbf{u}}$ the real, imaginary part and the conjugate of $\mathbf{u}$ (Definition 6.2.24). Prove that if $\langle \mathbf{u}, \bar{\mathbf{u}}\rangle = 0$ then $\langle \mathbf{u}_{re}, \mathbf{u}_{im}\rangle = 0$ and $\|\mathbf{u}_{re}\| = \|\mathbf{u}_{im}\|$.

3. Prove that (8.12) of Example 8.1.6 defines an inner product.

4. Let V be an inner product space. Prove the **parallelogram law**

$$\|\mathbf{u}+\mathbf{v}\|^2 + \|\mathbf{u}-\mathbf{v}\|^2 = 2\|\mathbf{u}\| + 2\|\mathbf{v}\|^2$$

5. Prove Pythagorean Theorem (8.16): If $\mathbf{u} \perp \mathbf{v}$ then we have

$$\|\mathbf{u}+\mathbf{v}\|^2 = \|\mathbf{u}\|^2 + \|\mathbf{v}\|^2$$

6. Prove the following statements.

   (a) Let V be an inner product space and W be a subspace of V. Then the restriction of the inner product on W makes W into an inner product space.

   (b) Let V, W be inner product space over F and $T : V \to W$ be an injective linear map. Let $\langle\ ,\ \rangle_W$ be a inner product on W. Then $\langle\ ,\ \rangle_V$ defined by $\langle \mathbf{v}_1, \mathbf{v}_2\rangle_V = \langle T(\mathbf{v}_1), T(\mathbf{v}_2)\rangle_W$ is an inner product on V.

   (c) Let $\langle -,-\rangle_1, \langle -,-\rangle_2$ be inner products on a vector space V. Then $\langle -,-\rangle := \langle -,-\rangle_1 + \langle -,-\rangle_2$ is also an inner product on V.

(d) Let $\langle -,-\rangle_1, \langle -,-\rangle_2$ be inner products on a vector space V with a basis $\mathcal{B} = \{\mathbf{v}_i\}$. Then $\langle -,-\rangle_1 = \langle -,-\rangle_2$ if $\langle \mathbf{v}_i, \mathbf{v}_j\rangle_1 = \langle \mathbf{v}_i, \mathbf{v}_j\rangle_2$.

(e) Let $\langle -,-\rangle$ be an inner product on a vector space V and c be a positive real number. Then $c\langle -,-\rangle$ is also an inner product.

7. Complete the proof of Theorem 8.1.15.

8. Prove the true statement(s) and find counter example(s) for the false statement(s). All vector spaces are inner product spaces.

(a) If $\langle -,-\rangle$ is an inner product on a vector space V over $\mathbb{C}$. Then for each $\mathbf{v} \in V$ the map $\phi_{\mathbf{v}}(\mathbf{w}) = \langle \mathbf{v}, \mathbf{w}\rangle$ is a linear functional.

(b) If $\langle -,-\rangle$ is an inner product on a vector space V over $\mathbb{C}$. Then for each $\mathbf{v} \in V$ the map $\psi_{\mathbf{v}}(\mathbf{w}) = \langle \mathbf{w}, \mathbf{v}\rangle$ is a linear functional.

(c) If $A \in M(n,\mathbb{C})$ then $A\bar{A} = \bar{A}A$.

(d) We can apply Gram Schmidt process to any set of n vectors in a finite dimensional inner product space to get n orthogonal vectors.

(e) On the vector space $\mathbb{R}^n$ there is only one inner product namely the standard inner product.

(f) If $\langle \mathbf{u}, \mathbf{v}\rangle = 0$ for all $\mathbf{v}$ in an inner product space V then $\mathbf{u} = \mathbf{0}$.

(g) If T is a linear operator on an inner product space V such that $\langle T(\mathbf{u}), \mathbf{v}\rangle = 0$ for all $\mathbf{u}, \mathbf{v} \in V$ then T is the zero linear operator.

(h) We can define an inner product on every finite dimensional vector space over $\mathbb{R}$ or $\mathbb{C}$.

(i) Every finite dimensional inner product space has an orthonormal basis.

(j) The orthogonal complement of a set of vectors in an inner product space is a subspace.

(k) If $T : V \to V$ is an isometry on an inner product space V then T is an isometric isomorphism.

(l) Let $V = M(2,F), W = F^4$ with their standard inner products. The map $T : V \to W$ defined by $T\left(\begin{bmatrix} a & b \\ c & d \end{bmatrix}\right) = \begin{bmatrix} a \\ b \\ c \\ d \end{bmatrix}$ is an isometry.

(m) A projection of an inner product space is an orthogonal projection.

(n) The set of all inner products on a vector space forms a vector space.

9. Prove that the standard basis of $V = M(n,F)$ of Example 6.2.8 (4) is an orthonormal basis of V with respect to the standard inner product (Example 8.1.6).

10. Apply Gram Schmidt orthogonalization to the linearly independent set S of the inner product space V to get an orthonormal basis for $W = \mathrm{Span}(S)$. Also find $P_W(\mathbf{v})$ and bases for $W^{\perp}$.

(a) $V = \mathbb{C}^3$ with the inner product of Example 8.1.7 and

$$S = \left\{ \begin{bmatrix} i \\ 0 \\ 0 \end{bmatrix}, \begin{bmatrix} 0 \\ 1 \\ i \end{bmatrix} \right\}, \ \mathbf{v} = \begin{bmatrix} 0 \\ 0 \\ 1 \end{bmatrix}$$

(b) $V = \mathbb{C}^4$ with the standard inner product and

$$S = \left\{ \begin{bmatrix} 0 \\ 1 \\ i \\ 0 \end{bmatrix}, \begin{bmatrix} 1 \\ i \\ 0 \\ 0 \end{bmatrix}, \begin{bmatrix} i \\ 0 \\ 0 \\ 1 \end{bmatrix} \right\}, \ \mathbf{v} = \begin{bmatrix} 1 \\ 0 \\ 0 \\ 0 \end{bmatrix}$$

(c) $V = P_3(\mathbb{R})$ with $\langle f, g\rangle = \int_0^1 fg dx$ and $S = \{1, 1+x\}, \mathbf{v} = x^3$

(d) $V = P_3(\mathbb{R})$ with $\langle f, g\rangle = \int_0^1 fg dx$ and $S = \{1, 1+x, x^2\}, \mathbf{v} = x^3$

(e) $V = M(2, \mathbb{C})$ with the standard inner product of Example 8.1.6 and

$$S = \left\{ \begin{bmatrix} i & 0 \\ 0 & 0 \end{bmatrix}, \begin{bmatrix} i & 0 \\ 0 & 1 \end{bmatrix}, \begin{bmatrix} 1 & 0 \\ 1 & 1 \end{bmatrix} \right\}, \ \mathbf{v} = \begin{bmatrix} 1 & 0 \\ 1 & 0 \end{bmatrix}$$

(f) $V = C_{\mathbb{R}}[0, 1]$ the vector space of real valued continuous functions on the interval $[0, 1]$ with $\langle f, g\rangle = \int_0^1 f(x)g(x)dx$ and $S = \{1, \sqrt{x}\}, \mathbf{v} = x^3$. (For this you don't need to find a basis for $W^{\perp}$)

11. Let T be a linear operator on an inner product space V. Prove that T is an isomorphism if $\|T(\mathbf{u})\| = \|\mathbf{u}\|$ for all $\mathbf{u} \in V$.

12. Prove that if V, W are inner product spaces of the same dimension then there is an isometric isomorphism between V and W.

13. Let V be an inner product space and W be a subspace of V. Prove that an orthogonal projection P_W onto W is completely determined by its range W in contrast to the fact that the projection is determined by the direct sum decomposition $V = W \oplus W'$ (Example 6.3.13).

14. Prove that if $V = W_1 \oplus W_2 = W_1 \oplus W_2'$ then $W_2 = W_2'$.

15. Let V be an inner product space and W_i $(i = 1, 2, ..., k)$ be subspaces such that $W_i \perp W_j$ for $i \neq j$. Prove that if $V = \sum_i W_i$ then $V = \bigoplus_{i=1}^{k} W_i$ i.e., the sum $\sum W_i$ is direct.

16. Let V be an inner product space. Prove that

$$\|\mathbf{u} \pm \mathbf{v}\|^2 = \|\mathbf{u}\|^2 \pm 2\langle \mathbf{u}, \mathbf{v}\rangle_{re} + \|\mathbf{v}^2\|$$

17. Prove the **polar identity** for the complex inner product space corresponding to the the formula $\langle \mathbf{u}, \mathbf{v}\rangle = \frac{1}{4}\|\mathbf{u}+\mathbf{v}\|^2 - \frac{1}{4}\|\mathbf{u}-\mathbf{v}\|^2$ of Exercise 5.2.6 (a) for the real case.

$$\langle \mathbf{u}, \mathbf{v}\rangle = \frac{1}{4}\sum_{k=1}^{4} i^k \|i^k \mathbf{u} + \mathbf{v}\|^2$$

18. Let V be a finite dimensional inner product space. Let $S \subseteq T$ be subsets of V and W be a subspace of V. Then

    (a) $T^\perp \subseteq S^\perp$

    (b) $S \subseteq S^{\perp\perp}$

    (c) $W = W^{\perp\perp}$

19. Prove that $P_W : V \to V$ is a linear map with the null space $W^\perp$ and the range space W. Also prove that $P_W^2 = P_W$ and $P_W(\mathbf{w}) = \mathbf{w}$ for all $\mathbf{w} \in W$.

20. Let V be an inner product space of dimension n. Let $S = \{\mathbf{v}_1, \mathbf{v}_2, \ldots, \mathbf{v}_k\}$ be an orthonormal set of vectors and $W = \text{Span}(S)$.

    (a) Prove that S can be extended to an orthonormal basis $\mathcal{B} = S \cup \{\mathbf{v}_{k+1}, \ldots, \mathbf{v}_n\}$ of V.

    (b) Prove that $S' = \{\mathbf{v}_{k+1}, \ldots, \mathbf{v}_n\}$ is an orthonormal basis of $W^\perp$.

    (c) Let $\mathbf{v} = \sum_{i=1}^n a_i \mathbf{v}_i$. Prove that $P_W(\mathbf{v}) = \sum_{i=1}^k a_i \mathbf{v}_i$ and $\|\mathbf{v}\|^2 \le \|P_W(\mathbf{v})\|^2$.

21. Let V be an inner product space of dimension n. Let $\mathcal{B} = \{\mathbf{v}_1, \mathbf{v}_2, \ldots, \mathbf{v}_n\}$ be an orthonormal basis of V.

    (a) For $\mathbf{u}, \mathbf{v} \in V$, prove **Parseval's Identity***.

    $$\langle \mathbf{u}, \mathbf{v} \rangle = \sum_{i=1}^{n} \langle \mathbf{u}, \mathbf{v}_i \rangle \langle \mathbf{v}_i, \mathbf{v} \rangle$$

    (b) Let $S = \{\mathbf{v}_1, \mathbf{v}_2, ..., \mathbf{v}_k\}$ be a subset of $\mathcal{B}$. For $\mathbf{v} \in V$ prove **Bessel's Inequality**†

    $$\sum_{i=1}^{k} |\langle \mathbf{v}, \mathbf{v}_i \rangle|^2 \le \|\mathbf{v}\|^2$$

22. Let T be a linear operator on an inner product space V. Prove that T is an isomorphism if there is a positive constant c such that $\|T(\mathbf{v})\| \ge c\,\|\mathbf{v}\|$ for all $\mathbf{v} \in V$.

23. Let V be a finite dimensional real inner product space with an inner product $\langle -, - \rangle$. Let $V^{\mathbb{C}}$ be the complexification of V (Exercise 6.3.16). Prove that

    $$\langle \mathbf{u} + \mathbf{v}i, \tilde{\mathbf{u}} + \tilde{\mathbf{v}}i \rangle^{\mathbb{C}} := \langle \mathbf{u}, \tilde{\mathbf{u}} \rangle + \langle \mathbf{v}, \tilde{\mathbf{v}} \rangle - i\langle \mathbf{v}, \tilde{\mathbf{u}} \rangle + i\langle \mathbf{u}, \tilde{\mathbf{v}} \rangle$$

    where $\mathbf{u}, \tilde{\mathbf{u}}, \mathbf{v}, \tilde{\mathbf{u}} \in V$ defines a complex inner product on $V^{\mathbb{C}}$. Writing $\mathbf{v}' = \mathbf{v} + i\mathbf{0} \in V^{\mathbb{C}}$ for $\mathbf{v} \in V$ prove that if $\mathcal{B} = \{\mathbf{v}_i\}$ is an orthonormal basis for V then $\mathcal{B}' = \{\mathbf{v}_i'\}$ is an orthonormal basis for $V^{\mathbb{C}}$.

24. Let W be an inner product space over $\mathbb{C}$ with an inner product $\langle -, - \rangle$. Let $W_{\mathbb{R}}$ be the real vector space obtained from W by restricting the scalars to $\mathbb{R}$ (Exercise 6.3.15). Define $\langle -, - \rangle_{\mathbb{R}} = \langle -, - \rangle_{re}$. Prove that $\langle -, - \rangle_{\mathbb{R}}$ defines an inner product on $W_{\mathbb{R}}$. Further, show that $\langle \mathbf{v}i, \mathbf{v} \rangle_{\mathbb{R}} = 0$ and $\langle \mathbf{v}, \mathbf{v} \rangle_{\mathbb{R}} = \langle \mathbf{v}, \mathbf{v} \rangle$ for all $\mathbf{v} \in W$.

---

*Marc-Antoine Parseval des Chênes (1755 – 1836) was a French mathematician.

†Friedrich Wilhelm Bessel (1784 – 1846) was a German astronomer, mathematician and physicist.

25. Let V be a vector space over $\mathbb{C}$. Let $\langle -,-\rangle$ be a (real) inner product on $V_{\mathbb{R}}$ such that $\langle \mathbf{u}i, \mathbf{u}\rangle = 0$ for all $\mathbf{u} \in V$. Define

$$[\mathbf{u}, \mathbf{v}] = \langle \mathbf{u}, \mathbf{v}\rangle + \langle \mathbf{u}i, \mathbf{v}\rangle i$$

Prove that $[-,-]$ is a complex valued inner product on V.

Let V be a vector space over $\mathbb{R}$ or $\mathbb{C}$ as usual. A **norm** *on V is a map $\|\cdot\| : V \to \mathbb{R}$ with*

(1) $\|\mathbf{v}\| \geq 0$ *and* $\|\mathbf{v}\| = 0$ *if and only if* $\mathbf{v} = \mathbf{0}$.

(2) $\|c\mathbf{v}\| = |c|\|\mathbf{v}\|$ *where* $c \in F$.

(3) $\|\mathbf{u} + \mathbf{v}\| \leq \|\mathbf{u}\| + \|\mathbf{v}\|$

A vector space V over $\mathbb{R}$ or $\mathbb{C}$ provided with a norm $\|\cdot\|_V$ is called a **normed vector space**. *We define $\lim_{n\to\infty} \mathbf{x}_n = \mathbf{v}$ if $\lim_{n\to\infty} \|\mathbf{v} - \mathbf{x}_n\| = 0$. Let W be a normed vector space with a norm $\|\cdot\|_W$. If $T : V \to W$ is a linear map then we define T to be* **continuous** *if $\lim_{\mathbf{x}_n\to\mathbf{v}} T(\mathbf{x}_n) = T(\mathbf{v})$. We say T is* **bounded** *if there is an $M > 0$ such that $\|T(\mathbf{x})\|_W \leq M\|\mathbf{x}\|_V$ for all $\mathbf{x} \neq \mathbf{0}$.*

26. Prove that the following $\|\cdot\|$'s are norms on V.

(a) $V = \mathbb{C}$ or $\mathbb{R}$ and $\|z\|$ is the absolute value $|z|$.

(b) $V = \mathbb{R}^2$ and $\left\|\begin{bmatrix} a \\ b \end{bmatrix}\right\|_1 = \sqrt{a^2+b^2}$, $\left\|\begin{bmatrix} a \\ b \end{bmatrix}\right\|_2 = |a| + |b|$ and $\left\|\begin{bmatrix} a \\ b \end{bmatrix}\right\|_3 = \max\{|a|, |b|\}$ with $\begin{bmatrix} a \\ b \end{bmatrix} \in V$.

(c) V is an inner product space and $\|\mathbf{v}\| = \sqrt{\langle \mathbf{v}, \mathbf{v}\rangle}$ for $\mathbf{v} \in V$.

(d) $V = M(m{\times}n, F)$ and $\|A\| = \max_{i,j} |A_{ij}|$ for $A \in V$.

(e) $V = M(m{\times}n, F)$ and $\|A\| = \sum_{i,j} |A_{ij}|$ for $A \in V$.

(f) $V = M(m{\times}n, F)$ and $\|A\| = \sup\left\{ \frac{\|A\mathbf{x}\|}{\|\mathbf{x}\|} \,\middle|\, \mathbf{x} \neq \mathbf{0}\right\} = \sup\left\{ \frac{\|A\mathbf{x}\|}{\|\mathbf{x}\|} \,\middle|\, \|\mathbf{x}\| = 1\right\}$ for $A \in V$.

(g) $V = C_F[0,1]$ and $\|f\| = \max\{|f(t)| \mid t \in [0,1]\}$ with $f \in V$.

27. Let V be a normed vector space with the norm $\|\cdot\|$. Prove that

$$\big|\|\mathbf{u}\| - \|\mathbf{v}\|\big| \leq \|\mathbf{u} - \mathbf{v}\| \tag{8.36}$$

28. Let V, W be normed vector spaces with the norms $\|\cdot\|_V, \|\cdot\|_W$. Prove the following statements.

(a) Let $\|(\mathbf{v}, \mathbf{w})\|_{V\times W} = \|\mathbf{v}\|_V + \|\mathbf{w}\|_W$. Then $\|(-,-)\|_{V\times W}$ is a norm on $V{\times}W$.

(b) The maps $V{\times}V \to V : (\mathbf{v}, \mathbf{v}') \mapsto \mathbf{v} + \mathbf{v}'$ and $F{\times}V \to V : (c, \mathbf{v}) \mapsto c\mathbf{v}$ are continuous. (Here, the norm on $V{\times}V$ is defined in (a) and the norm on F is the usual absolute value.)

(c) If V is an inner product space with the inner product $\langle -,-\rangle$ and $\sqrt{\langle \mathbf{v}, \mathbf{v}\rangle} = \|\mathbf{v}\|_V$ then the inner product map $V{\times}V \to F; (\mathbf{v}, \mathbf{v}') \mapsto \langle \mathbf{v}, \mathbf{v}'\rangle$ is continuous.

29. Let V, W be a normed vector spaces (not necessarily finite dimensional) with the norms $\|\cdot\|_V, \|\cdot\|_W$. Let $T : V \to W$ be a linear map. Prove that the following conditions are equivalent.

    (a) The linear map T is continuous on V.

    (b) The linear map T is continuous at one point $\mathbf{0} \in V$.

    (c) The linear map T is bounded.

30. Let V be a vector space over F. Let $\|\cdot\|_1, \|\cdot\|_2$ be two norms on V. We say $\|\cdot\|_1$ and $\|\cdot\|_2$ are **equivalent** if there are positive numbers c_1, c_2 such that

$$c_1\|\mathbf{v}\|_1 \leq \|\mathbf{v}\|_2 \leq c_2\|\mathbf{v}\|_1$$

    (a) Let $\|\cdot\|_1$, $\|\cdot\|_2$ be norms on a vector space V which we denote by V_1, V_2 respectively. Prove that $\|\cdot\|_1$ and $\|\cdot\|_2$ are equivalent if and only if $I_V : V_1 \to V_2$ and $I_V : V_2 \to V_1$ are continuous.

    (b) Prove that if V is finite dimensional then any two norms are equivalent. (Hint: The continuous function $f(\mathbf{v}) = \frac{\|\mathbf{v}\|_1}{\|\mathbf{v}\|_2}$ $(\mathbf{v} \neq \mathbf{0})$ has a maximum on the compact set $S_1 = \{\mathbf{v} \in V \mid \|\mathbf{v}\| = 1\}$.)

    (c) Let $\phi : V \to F^n$ be an isomorphism. Let $\|\cdot\|$ be the norm on F^n induced by the standard inner product on F^n. Define $\|\cdot\|_\phi$ by $\|\mathbf{v}\|_\phi = \|\phi(\mathbf{v})\|$ for $\mathbf{v} \in V$. Prove that $\|\cdot\|_\phi$ is a norm on V and ϕ, ϕ^{-1} are continuous functions with respect to the norms $\|\cdot\|, \|\cdot\|_\phi$.

    (d) Let V, W be finite dimensional normed vector space with the norms $\|\cdot\|_V, \|\cdot\|_W$. Prove that if $T : V \to W$ is a linear map then T is continuous. (Hint: Use (b) to show that we can identify V, W with F^n, F^m.)

31. Let V be a vector space over the field $F = \mathbb{R}$ of real numbers. Let $\|\cdot\|$ be a norm on V satisfying the parallelogram law of Exercise 4. Prove that

$$\langle \mathbf{u}, \mathbf{v} \rangle = \frac{1}{4}\left(\|\mathbf{u}+\mathbf{v}\|^2 - \|\mathbf{u}-\mathbf{v}\|^2\right) \tag{8.37}$$

defines an inner product on V and $\|\mathbf{u}\|^2 = \langle \mathbf{u}, \mathbf{u} \rangle$.
(Hint: A function f defined on V is continuous if $f(\mathbf{x})$ approaches to $f(\mathbf{v})$ whenever $\mathbf{x}$ approaches to $\mathbf{v}$. Prove that the addition, the scalar multiplication and the norm are continuous functions. Conclude that $V \times V \to \mathbb{R}; (\mathbf{u}, \mathbf{v}) \mapsto \langle \mathbf{u}, \mathbf{v} \rangle$ is a continuous function. It is easy to show that $\langle \mathbf{u}, \mathbf{v} \rangle = \langle \mathbf{v}, \mathbf{u} \rangle$ and $\|\mathbf{u}\| = \sqrt{\langle \mathbf{u}, \mathbf{u} \rangle}$. Now use the parallelogram law

$$\begin{aligned} \|\mathbf{u}+\mathbf{v}+\mathbf{w}\|^2 &= 2\|\mathbf{u}+\mathbf{w}\|^2 + 2\|\mathbf{v}\|^2 - \|\mathbf{u}-\mathbf{v}+\mathbf{w}\|^2 \\ &= 2\|\mathbf{v}+\mathbf{w}\|^2 + 2\|\mathbf{u}\|^2 - \|\mathbf{v}-\mathbf{u}+\mathbf{w}\|^2 \end{aligned}$$

to show that $\langle \mathbf{u}+\mathbf{v}, \mathbf{w} \rangle = \langle \mathbf{u}, \mathbf{w} \rangle + \langle \mathbf{v}, \mathbf{w} \rangle$. Deduce from this that $\langle r\mathbf{u}, \mathbf{v} \rangle = r\langle \mathbf{u}, \mathbf{v} \rangle$ for all $r \in \mathbb{Z}$ and then extend this to $r \in \mathbb{Q}$. Now conclude, from the continuity of $\langle -, - \rangle$ that this is true for all $r \in \mathbb{R}$.)

32. Let V be a vector space over $\mathbb{C}$ with a norm $\|\cdot\|$ satisfying the parallelogram law. Prove that there is an inner product $\langle -, - \rangle$ on V such that $\langle \mathbf{v}, \mathbf{v} \rangle = \|\mathbf{v}\|^2$ for all $\mathbf{v} \in V$. (Hint: Repeat the same argument of Exercise 31 with Exercise 17 instead of (8.37). Or use Exercises 25 and 31.)

Let $A \in M(n, F)$. We define the **exponential** *and* **logarithm** *of A by*

$$e^A = \sum_{n=0}^{\infty} \frac{A^n}{n!} = \exp(A)$$

$$\log(A) = \sum_{n=1}^{\infty} \frac{(A-I)^n}{n}$$

33. Let $A \in M(n, F)$ ($F = \mathbb{R}$ or $\mathbb{C}$). Define the norm (Exercise 26) of A by

$$\|A\| = \max_{i,j} |A_{ij}|$$

Prove that the exponential $\exp(A)$ converges for all A and the logarithm $\log(A)$ converge for A satisfying $\|A - I\| < 1$.

34. Prove the following formulas for the real or complex $n \times n$ matrices.

    (a) $e^D = \operatorname{diag}[e^{\lambda_1}, e^{\lambda_2}, ..., e^{\lambda_n}]$ where $D = \operatorname{diag}[\lambda_1, \lambda_2, ..., \lambda_n]$
    (b) $e^A e^B = e^{(A+B)}$ if $AB = BA$
    (c) $\exp(A^t) = \exp(A)^t$
    (d) $e^A e^{-A} = I$ (Hence e^A is invertible.)
    (e) $P^{-1} e^A P = e^{P^{-1}AP}$
    (f) $e^{\operatorname{Tr}(A)} = \det e^A$ (Hint: Use (e) and Jordan. Also you will need Exercise 7.3.16.)
    (g) $\exp \circ \log(A) = A$ if $\log(A)$ is defined
    (h) $\log(AB) = \log(A) + \log(B)$ if $AB = BA$

8.2 Adjoint of Linear Maps

Let $V = F^n$ with the standard inner product and $A \in M(n, F)$. By (8.9) we see that

$$\langle \mathbf{u}, T_A(\mathbf{v}) \rangle = \langle T_{A^*}(\mathbf{u}), \mathbf{v} \rangle$$

Let T be a linear operator on an inner product space V. In this section, we will define the linear map T^* corresponding to T_{A^*} and investigate its properties. When $F = \mathbb{R}$ we will show that T^* can be viewed as the dual $T^\#$ of T under certain identification.

8.2.1 Linear Functionals and Inner Product

Let V be a finite dimensional inner product space. For $\mathbf{v} \in V$ define

$$\phi_{\mathbf{v}}(\mathbf{x}) = \langle \mathbf{v}, \mathbf{x} \rangle \ (\mathbf{x} \in V) \tag{8.38}$$

Then $\phi_{\mathbf{v}}$ is a linear functional i.e., $\phi_{\mathbf{v}} \in V^{\#}$ (Definition 6.5.1). Hence we get a map

$$\phi : V \to V^{\#} \tag{8.39}$$

which maps $\mathbf{v}$ to $\phi_{\mathbf{v}}$. If $V = \mathbb{R}^n$ then we saw that ϕ is an isomorphism (Example 6.5.2). Next theorem shows that ϕ is a bijection for any finite dimensional inner product spaces.

Theorem 8.2.1. *Let V be a finite dimensional inner product space. Let $g \in V^{\#}$ then there is a unique $\mathbf{v} \in V$ such that*

$$g(\mathbf{x}) = \langle \mathbf{v}, \mathbf{x} \rangle \tag{8.40}$$

for all $\mathbf{x} \in V$. If $\mathcal{B} = \{\mathbf{v}_1, \mathbf{v}_2, \dots, \mathbf{v}_n\}$ is an orthonormal basis of V then we can choose $\mathbf{v}$ as

$$\mathbf{v} = \sum_{i=1}^{n} \overline{g(\mathbf{v}_i)}\, \mathbf{v}_i \tag{8.41}$$

Proof. Uniqueness follows from the properties of inner product: If $\langle \mathbf{v}, \mathbf{x} \rangle = \langle \mathbf{v}', \mathbf{x} \rangle$ for all $\mathbf{x}$ then $\mathbf{v} = \mathbf{v}'$ by (8.7). For existence, we need to check that $\mathbf{v}$ given by (8.41) satisfies the required property namely, $g(\mathbf{x}) = \langle \mathbf{v}, \mathbf{x} \rangle$. Since g and $\langle \mathbf{v}, - \rangle$ are both linear, we need to check for $\mathbf{x} = \mathbf{v}_j$ for $j = 1, 2, ..., n$ (Theorem 6.3.9).

$$\langle \mathbf{v}, \mathbf{v}_j \rangle = \left\langle \sum_{i=1}^{n} \overline{g(\mathbf{v}_i)}\, \mathbf{v}_i, \mathbf{v}_j \right\rangle = g(\mathbf{v}_j)\langle \mathbf{v}_j, \mathbf{v}_j \rangle = g(\mathbf{v}_j)$$

Since two linear maps agree on a basis they are equal. □

Example 8.2.2. *Let $V = \mathbb{C}^3$ with the standard inner product. Let $f(a, b, c) = a - ib + (1+i)c$ be a linear functional. We want to find $\mathbf{v} = \begin{bmatrix} x \\ y \\ z \end{bmatrix}$ such that $f(\mathbf{u}) = \langle \mathbf{v}, \mathbf{u} \rangle$ for all $\mathbf{u} = \begin{bmatrix} a \\ b \\ c \end{bmatrix}$ in V. Since $\langle \mathbf{v}, \mathbf{u} \rangle = a\bar{x} + b\bar{y} + c\bar{z}$ and $f(\mathbf{u}) = a - ib + (1+i)c$. Thus we have $\mathbf{v} = \begin{bmatrix} 1 \\ i \\ 1-i \end{bmatrix}$.*

Example 8.2.3. *Let $V = \mathbb{C}^3$ with the inner product $\langle \mathbf{u}, \mathbf{v} \rangle_A = \mathbf{u}^* A \mathbf{v}$ where $A = \begin{bmatrix} 1 & 0 & 0 \\ 0 & 1 & -i \\ 0 & i & 2 \end{bmatrix}$ is defined in* Example 8.1.7. *Let $f \in V^{\#}$ be defined by $f(a, b, c) = 2a - ib - c$. We want to find $\mathbf{v} \in V$ such that $f(\mathbf{u}) = \langle \mathbf{v}, \mathbf{u} \rangle_A$.*

We use the orthonormal basis $\mathcal{C} = \{\mathbf{e}^1, \mathbf{e}^2, i\mathbf{e}^2 + \mathbf{e}^3\}$ of Example 8.1.23. *Now*

$$f(\mathbf{e}^1) = 2,\ f(\mathbf{e}^2) = -i,\ f(i\mathbf{e}^2 + \mathbf{e}^3) = 0$$

By Theorem 8.2.1 *we have*

$$\mathbf{v} = \overline{f(\mathbf{e}^1)}\mathbf{e}^1 + \overline{f(\mathbf{e}^2)}\mathbf{e}^2 + \overline{f(i\mathbf{e}^2 + \mathbf{e}^3)}(i\mathbf{e}^2 + \mathbf{e}^3) = 2\mathbf{e}^1 + i\mathbf{e}^2$$

Practice Problem 8.2.1. *In* Example 8.2.3, *find $\mathbf{v} = \begin{bmatrix} x \\ y \\ z \end{bmatrix} \in \mathbb{C}^3$ such that $f(\mathbf{u}) = \langle \mathbf{v}, \mathbf{u} \rangle_A$ for all $\mathbf{u} = \begin{bmatrix} a \\ b \\ c \end{bmatrix}$ by solving the equations obtained from $f(a, b, c) = \mathbf{v}^* A \mathbf{u}$.*

We defined a map

$$\phi : V \to V^{\#}$$

by $\phi(\mathbf{v})(\mathbf{x}) = \langle \mathbf{v}, \mathbf{x} \rangle = \phi_{\mathbf{v}}(\mathbf{x})$. Theorem 8.2.1 asserts that the map $\phi : V \to V^{\#}$ is surjective. If $F = \mathbb{R}$ then ϕ is a linear map. However, if $F = \mathbb{C}$ then ϕ is not linear i.e., $\phi(c\mathbf{v}) = \bar{c}\phi(\mathbf{v})$ since the inner product is conjugate linear in the first variable.

Definition 8.2.4. *Let V, W be vector space over $\mathbb{C}$. A map $T : V \to W$ will be called* **conjugate linear** *(or* **anti linear***) if*

$$T(\mathbf{v}_1 + \mathbf{v}_2) = T(\mathbf{v}_1) + T(\mathbf{v}_2) \text{ and } T(c\mathbf{v}) = \bar{c}T(\mathbf{v})$$

for all $\mathbf{v}_1, \mathbf{v}_2, \mathbf{v} \in V$ and $c \in \mathbb{C}$.

If $T : V \to W$ is conjugate linear then the null space $N(T)$ and the range $R(T)$ are subspaces. Further the rank theorem

$$\nu(T) + \rho(T) = \dim V \tag{8.42}$$

and other related theorems hold (Exercise 3). Also see Exercise 2.

Theorem 8.2.5. *The map $\phi : V \to V^{\#}$ given by $\phi(\mathbf{v}) = \phi_{\mathbf{v}}$ is an isomorphism if $F = \mathbb{R}$. If $F = \mathbb{C}$ then ϕ is a conjugate linear bijection.*

Proof. Since rank theorem holds for $\mathbb{R}$ (Theorem 6.3.21) and $\mathbb{C}$ (8.42) we need to show that $N(T) = \mathbf{0}$. But this follows from (8.7): If $\phi_{\mathbf{v}} = 0$ then $\langle \mathbf{v}, \mathbf{x} \rangle = 0$ for all $\mathbf{x} \in V$ then $\mathbf{v} = \mathbf{0}$. □

8.2.2 Adjoint of Linear Maps

If $A \in M(n.\mathbb{C})$ then we saw in (8.9) that

$$\langle \mathbf{u}, A\mathbf{v} \rangle = \langle A^*\mathbf{u}, \mathbf{v} \rangle$$

The converse of this is also true: Let $A, B \in M(n, \mathbb{C})$.

$$B = A^* \text{ if and only if } \langle \mathbf{u}, A\mathbf{v} \rangle = \langle B\mathbf{u}, \mathbf{v} \rangle \text{ for all } \mathbf{u}, \mathbf{v} \in V \tag{8.43}$$

In fact, if we take $\mathbf{u} = \mathbf{e}^i, \mathbf{v} = \mathbf{e}^j$ then $\langle \mathbf{u}, A\mathbf{v} \rangle = (\mathbf{e}^i)^* A^{(j)} = \mathbf{e}_i A^{(j)} = A_{ij}$. On the other hand, $\langle B\mathbf{u}, \mathbf{v} \rangle = (B^{(i)})^* \mathbf{e}^j = \overline{B}_{ji}$. Hence $B = A^*$.

In terms of matrix linear maps, (8.43) becomes

$$B = A^* \text{ if and only if } \langle T_B(\mathbf{u}), \mathbf{v} \rangle = \langle \mathbf{u}, T_A(\mathbf{v}) \rangle \tag{8.44}$$

For a linear operator T on an inner product space V we will define a linear map T^* corresponding to T_B by requiring the equality (8.44).

Theorem 8.2.6. *Let V be a finite dimensional inner product space and T be a linear operator on V. Then there is a unique linear map $T^* : V \to V$ such that*

$$\langle \mathbf{u}, T(\mathbf{v}) \rangle = \langle T^*(\mathbf{u}), \mathbf{v} \rangle \tag{8.45}$$

for all $\mathbf{u}, \mathbf{v} \in V$.

Proof. Fix $\mathbf{u} \in V$ and let $g(\mathbf{v}) = \langle \mathbf{u}, T(\mathbf{v}) \rangle$. Then g is a linear functional i.e., $g \in V^{\#}$. By Theorem 8.2.1 there is a unique $\mathbf{u}_0 \in V$ such that $g(\mathbf{v}) = \langle \mathbf{u}_0, \mathbf{v} \rangle$ for all $\mathbf{v} \in V$. We let $\mathbf{u}_0 = T^*(\mathbf{u})$. Then $\langle \mathbf{u}, T(\mathbf{v}) \rangle = \langle T^*(\mathbf{u}), \mathbf{v} \rangle$ and hence T^* satisfies (8.45).

Now we need to show T^* is linear: To show $T^*(\mathbf{u}_1 + \mathbf{u}_2) = T^*(\mathbf{u}_1) + T^*(\mathbf{u}_2)$ it suffices to show

$$\langle T^*(\mathbf{u}_1 + \mathbf{u}_2), \mathbf{v} \rangle = \langle T^*(\mathbf{u}_1) + T^*(\mathbf{u}_2), \mathbf{v} \rangle$$

for all $\mathbf{v} \in V$ by (8.7). Now we check:

$$\begin{aligned} \langle T^*(\mathbf{u}_1 + \mathbf{u}_2), \mathbf{v} \rangle &= \langle \mathbf{u}_1 + \mathbf{u}_2, T(\mathbf{v}) \rangle \\ &= \langle \mathbf{u}_1, T(\mathbf{v}) \rangle + \langle \mathbf{u}_2, T(\mathbf{v}) \rangle \\ &= \langle T^*(\mathbf{u}_1), \mathbf{v} \rangle + \langle (T^*)(\mathbf{u}_2), \mathbf{v} \rangle \\ &= \langle T^*(\mathbf{u}_1) + T^*(\mathbf{u}_2), \mathbf{v} \rangle \end{aligned}$$

Therefore $T^*(\mathbf{u}_1 + \mathbf{u}_2) = T^*(\mathbf{u}_1) + T^*(\mathbf{u}_2)$. The proof for $T^*(c\mathbf{u}) = cT^*(\mathbf{u})$ is an exercise. □

By Theorem 8.2.6 and the equality (8.44) we have

Corollary 8.2.7. *Let $A \in M(n, \mathbb{C})$. Then we have $(T_A)^* = T_{A^*}$.*

Note that T^* is uniquely determined by the condition (8.45). Also we have

$$\langle T(\mathbf{u}), \mathbf{v} \rangle = \langle \mathbf{u}, T^*(\mathbf{v}) \rangle \tag{8.46}$$

since $\langle T(\mathbf{u}), \mathbf{v} \rangle = \overline{\langle \mathbf{v}, T(\mathbf{u}) \rangle} = \overline{\langle T^*(\mathbf{v}), \mathbf{u} \rangle} = \langle \mathbf{u}, T^*(\mathbf{v}) \rangle$.

Definition 8.2.8. *Let V be an inner product space and $T : V \to V$ be a linear operator. The linear map $T^* : V \to V$ in* Theorem 8.2.6 *is called the* **adjoint** *of T.*

More generally we can define the adjoint $T^* : W \to V$ of a linear map $T : V \to W$. Existence and the properties of T^* similar to the case of linear operators can be proved similarly (Exercise 12).

If V is an infinite dimensional inner product space then the adjoint of a linear operator can be defined by the condition (8.45). However the existence of such T^* is not guaranteed in general. We need certain extra conditions on T and V – a bounded linear operators on a 'Hilbert space' which belong to the area of analysis.

Example 8.2.9. (1) *Let $V = P_2(\mathbb{R})$ and $\langle f, g \rangle = \int_0^1 fg dx$. Let T be the linear operator defined by $T(f(x)) = f(x+1)$. We want to find $T^*(1)$. Let $T^*(1) = ax^2 + bx + c$. To find a, b, c, we let $\mathbf{v} = 1, x, x^2$ in the equality $\langle T(\mathbf{v}), 1 \rangle = \langle \mathbf{v}, T^*(1) \rangle$ to get the system of linear equations: If $\mathbf{v} = 1$ then the equality implies $1 = \langle 1, 1 \rangle = \langle 1, ax^2 + bx + c \rangle = \int_0^1 (ax^2 + bx + c) dx = \frac{1}{3}a + \frac{1}{2}b + c$. Similarly, if $\mathbf{v} = x$ we get the equation $\frac{3}{2} = \frac{1}{4}a + \frac{1}{3}b + \frac{1}{2}c$. And if $\mathbf{v} = x^2$ then we get $\frac{7}{3} = \frac{1}{5}a + \frac{1}{4}b + \frac{1}{3}c$. Thus we need to solve the equation*

$$\begin{cases} \frac{1}{3}a + \frac{1}{2}b + \ c \ = 1 \\ \frac{1}{4}a + \frac{1}{3}b + \frac{1}{2}c = \frac{3}{2} \\ \frac{1}{5}a + \frac{1}{4}b + \frac{1}{3}c = \frac{7}{3} \end{cases}$$

Solving this we get $a = 180, b = -168, c = 25$. Hence $T^*(1) = 180x^2 - 168x + 25$.

(2) *Let* $V = P_2(\mathbb{R})$ *and* T *be the linear operator given by* $T(f) = f'$ *for* $f \in V$. *Let* T^* *be the adjoint of* T. *Let* $\mathcal{B} = \{\mathbf{v}_1 = 1, \mathbf{v}_2 = x, \mathbf{v}_3 = x^2\}$ *be the standard basis for* V. *We want to find* $[T^*]_{\mathcal{B}}$. *For this, let* $T^*(\mathbf{v}_1) = ax^2 + bx + c$ *and we use*

$$\langle T^*(\mathbf{v}_1), \mathbf{u}\rangle = \langle \mathbf{v}_1, T(\mathbf{u})\rangle$$

and set $\mathbf{u} = \mathbf{v}_1, \mathbf{v}_2, \mathbf{v}_3$ *to get a system of equations*

$$\begin{cases} \frac{1}{3}a + \frac{1}{2}b + \ c \ = 0 \\ \frac{1}{4}a + \frac{1}{3}b + \frac{1}{2}c = \frac{1}{2} \\ \frac{1}{5}a + \frac{1}{4}b + \frac{1}{3}c = \frac{2}{3} \end{cases}$$

We get the solution $a = 0, b = 12, c = -6$. *Hence we get* $T^*(\mathbf{v}_1) = 12x - 6$. *Similarly, we get* $T^*(\mathbf{v}_2) = 30x^2 - 24 + 2, T^*(\mathbf{v}_3) = 30x^2 - 26x + 3$. *Therefore, we have*

$$[T^*]_{\mathcal{B}} = \begin{bmatrix} -6 & 2 & 3 \\ 12 & -24 & -26 \\ 0 & 30 & 3 \end{bmatrix}$$

Next we show that the matrix representation of the adjoint T^* in an orthonormal basis is the adjoint of the matrix representation of T in the orthonormal basis.

Theorem 8.2.10. *Let* V *be a finite dimensional inner product space and* $T : V \to V$ *be a linear operator. Let* $\mathcal{B}$ *be an orthonormal basis of* V. *Then*

$$[T^*]_{\mathcal{B}} = [T]^*_{\mathcal{B}} \tag{8.47}$$

Proof. Let $\mathcal{B} = \{\mathbf{v}_1, \mathbf{v}_2, \ldots, \mathbf{v}_n\}$ be an orthonormal basis of V. Let $A = [T]_{\mathcal{B}}$ and $B = [T^*]_{\mathcal{B}}$. By Corollary 8.1.14 we have $A_{ij} = \langle \mathbf{v}_i, T(\mathbf{v}_j)\rangle$. On the other hand, $B_{ij} = \langle \mathbf{v}_i, \sum_k B_{kj}\mathbf{v}_k\rangle = \langle \mathbf{v}_i, T^*(\mathbf{v}_j)\rangle = \langle T(\mathbf{v}_i), \mathbf{v}_j\rangle = \langle \sum_k A_{ki}\mathbf{v}_k, \mathbf{v}_j\rangle = \bar{A}_{ji}$ by (8.46). Hence $B = A^*$ as desired. □

Corollary 8.2.11. *Let* T *be a linear operator on a finite dimensional inner product space* V. *Then*

$$\rho(T) = \rho(T^*) \ \text{ and } \ \nu(T) = \nu(T^*) \tag{8.48}$$

Proof. By Theorem 8.2.10, we have $[T^*]_{\mathcal{B}} = [T]^*_{\mathcal{B}}$ and hence $\rho([T^*]_{\mathcal{B}}) = \rho([T]^*_{\mathcal{B}})$. By Theorem 6.3.28, we have $\rho([T]^*_{\mathcal{B}}) = \rho([T]_{\mathcal{B}})$. Thus, $\rho(T) = \rho(T^*)$ and $\nu(T) = \nu(T^*)$. □

Example 8.2.12. (1) *Let* W *be a subspace of an inner product space* V. *Then, by* Theorem 8.1.26, *there are orthonormal bases* $\mathcal{B}, \mathcal{B}^\perp$ *for* $W, W^\perp$ *such that* $\mathcal{C} = \mathcal{B} \cup \mathcal{B}^\perp$ *is a basis for* V *and* $[P_W]_{\mathcal{C}} = \begin{bmatrix} I_p & \\ & 0 \end{bmatrix}$. *Hence* $[P_W]^*_{\mathcal{C}} = [P_W]_{\mathcal{C}}$ *and therefore*

$$P_W^* = P_W \tag{8.49}$$

(2) *Continuing* (2) *of* Example 8.2.9, *choose the orthonormal basis* $\mathcal{C} = \{\mathbf{u}_1 = 1, \mathbf{u}_2 = \sqrt{3}(2x - 1), \mathbf{u}_3 = \sqrt{5}(6x^2 - 6x + 1)\}$ *of* Example 8.1.22. *Its easy to find*

$$[T]_{\mathcal{C}} = \begin{bmatrix} 0 & 2\sqrt{3} & 0 \\ 0 & 0 & \frac{6\sqrt{5}}{\sqrt{3}} \\ 0 & 0 & 0 \end{bmatrix}$$

In Example 8.2.9, *we computed* $T^*(\mathbf{u}_1) = 12x - 6 = 2\sqrt{3}\mathbf{u}_2$. *Hence the first column of* $[T^*]_{\mathcal{C}}$ *is* $\begin{bmatrix} 0 \\ 2\sqrt{3} \\ 0 \end{bmatrix}$. *To compute the second and third columns of* $[T^*]_{\mathcal{C}}$ *we use* (8.19) *this time:*

$$([T^*]_{\mathcal{C}})_{ij} = \langle \mathbf{u}_i, T^*(\mathbf{u}_j) \rangle = \langle T(\mathbf{u}_i), \mathbf{u}_j \rangle$$

Now we compute:

$$\begin{aligned} ([T^*]_{\mathcal{C}})_{12} &= \langle T(\mathbf{u}_1), \mathbf{u}_2 \rangle = \langle 0, \mathbf{u}_2 \rangle = 0 \\ ([T^*]_{\mathcal{C}})_{22} &= \langle T(\mathbf{u}_2), \mathbf{u}_2 \rangle = 3\,\langle 2, 2x-1 \rangle = 0 \\ ([T^*]_{\mathcal{C}})_{32} &= \langle T(\mathbf{u}_3), \mathbf{u}_2 \rangle = \sqrt{3}\sqrt{5}\langle 12x-6, 2x-1 \rangle = \tfrac{6\sqrt{5}}{\sqrt{3}} \end{aligned}$$

Similarly, we have $([T^*]_{\mathcal{C}})_{13} = 0, ([T^*]_{\mathcal{C}})_{23} = 0, ([T^*]_{\mathcal{C}})_{33} = 0$. *Therefore, we have*

$$[T^*]_{\mathcal{C}} = \begin{bmatrix} 0 & 0 & 0 \\ 2\sqrt{3} & 0 & 0 \\ 0 & \frac{6\sqrt{5}}{\sqrt{3}} & 0 \end{bmatrix} = [T]^t_{\mathcal{C}}$$

Example 8.2.13. *Let* $V = \mathbb{C}^3$ *with the inner product* $\langle \mathbf{u}, \mathbf{v} \rangle_A = \mathbf{u}^* A\mathbf{v}$ *where* $A = \begin{bmatrix} 1 & 0 & 0 \\ 0 & 1 & -i \\ 0 & i & 2 \end{bmatrix}$ *defined in* Example 8.1.7. *Let* $T : V \to V$ *be defined by* $T(a, b, c) = \begin{bmatrix} 2a \\ 2b - (1+2i)c \\ ic \end{bmatrix}$ *and* $\mathbf{w} = \begin{bmatrix} 0 \\ 0 \\ 1 \end{bmatrix}$. *We want to find* $T^*(\mathbf{w})$. *In* Example 8.1.23, *we found an orthonormal basis* $\mathcal{C} = \{\mathbf{v}_1 = \mathbf{e}^1, \mathbf{v}_2 = \mathbf{e}^2, \mathbf{v}_3 = i\mathbf{e}^2 + \mathbf{e}^3\}$. *We have*

$$[T]_{\mathcal{C}} = \begin{bmatrix} 2 & 0 & 0 \\ 0 & 2 & 0 \\ 0 & 0 & i \end{bmatrix}, \quad [\mathbf{w}]_{\mathcal{C}} = \begin{bmatrix} 0 \\ -i \\ 1 \end{bmatrix} \tag{8.50}$$

By Theorem 8.2.10 *and* Theorem 6.4.2 *we have* $[T^*(\mathbf{w})]_{\mathcal{C}} = [T]^*_{\mathcal{C}}[\mathbf{w}]_{\mathcal{C}}$ *and hence we have* $[T^*(\mathbf{w})]_{\mathcal{C}} = [T]^*_{\mathcal{C}}[\mathbf{w}]_{\mathcal{C}} = \begin{bmatrix} 0 \\ -2i \\ -i \end{bmatrix}$. *Therefore*

$$T^*(\mathbf{w}) = -2i\mathbf{v}_2 - i\mathbf{v}_3 = \begin{bmatrix} 0 \\ 1-2i \\ -i \end{bmatrix}$$

Theorem 8.1.27 shows that the property

$$T^2 = T, \ N(T) = R(T)^{\perp} \tag{8.51}$$

characterize the orthogonal projection onto $R(T)$ i.e., if T satisfies (8.51) with $R(T) = W$ then $T = P_W$. We give another characterization of orthogonal projection.

Theorem 8.2.14. *Let V be a finite dimensional inner product space and $T : V \to V$ be a linear operator. Then T is an orthogonal projection (onto $R(T)$) if and only if*

$$T^2 = T = T^* \tag{8.52}$$

Proof. If T is an orthogonal projection then by Theorem 8.1.27 we have $T^2 = T$ and by (8.49) of Example 8.2.12 we have $T = T^*$.

Conversely, suppose T satisfies the condition (8.52). By Theorem 8.1.27 we only need to show $R(T)^\perp = N(T)$. In fact, let $\mathbf{z} \in N(T)$ and $\mathbf{y} \in V$. Then we have

$$\langle \mathbf{z}, T(\mathbf{y}) \rangle = \langle T^*(\mathbf{z}), \mathbf{y} \rangle = \langle T(\mathbf{z}), \mathbf{y} \rangle = 0$$

since $T = T^*$ and $\mathbf{z} \in N(T)$. Therefore $\mathbf{z} \in R(T)^\perp$. Thus $N(T) \subseteq R(T)^\perp$. On the other hand, if $\mathbf{z} \in R(T)^\perp$ then since $T = T^2 = T^*$ we have

$$\|T(\mathbf{z})\|^2 = \langle T(\mathbf{z}), T(\mathbf{z}) \rangle = \langle T^*T(\mathbf{z}), \mathbf{z} \rangle = \langle T(\mathbf{z}), \mathbf{z} \rangle = 0$$

Hence $T(\mathbf{z}) = \mathbf{0}$ i.e., $\mathbf{z} \in N(T)$. Thus $R(T)^\perp = N(T)$ and hence T is an orthogonal projection. □

Practice Problem 8.2.2. *Find $T^*(\mathbf{w})$ of* Example 8.2.13 *by letting $T^*(\mathbf{w}) = \begin{bmatrix} x \\ y \\ z \end{bmatrix}$ and solve the equation coming from $\langle T(\mathbf{v}), \mathbf{w} \rangle_A = \langle \mathbf{v}, T^*(\mathbf{w}) \rangle_A$ for $\mathbf{v} = \mathbf{e}^1, \mathbf{e}^2, \mathbf{e}^3$.*

The adjoint of linear operators satisfies the properties the adjoint of matrices satisfy (Theorem 6.2.25).

Theorem 8.2.15. *Let V be a finite dimensional inner product space over F and $S, T \in L(V)$ be linear operators on V. Then we have the following formulas.*

(1) $(S + T)^* = S^* + T^*$

(2) $(cT)^* = \bar{c}T^* \ (c \in F)$

(3) $(ST)^* = T^*S^*$

(4) $(T^{-1})^* = (T^*)^{-1}$

(5) $T^{**} = T$

(6) $I^* = I$

Proof. To prove (3), it suffices to show $\langle \mathbf{v}, (ST)^*(\mathbf{w}) \rangle = \langle \mathbf{v}, T^*S^*(\mathbf{w}) \rangle$. In fact,

$$\langle \mathbf{v}, (ST)^*(\mathbf{w}) \rangle = \langle ST(\mathbf{v}), \mathbf{w} \rangle = \langle T(\mathbf{v}), S^*(\mathbf{w}) \rangle = \langle \mathbf{v}, T^*S^*(\mathbf{w}) \rangle$$

The proofs of the other equalities are Exercise 4. □

Theorem 8.2.16. *Let V be a finite dimensional inner product space and T be a linear operator. Then we have*

(1) $R(T^*)^\perp = N(T)$

(2) $R(T^*) = N(T)^\perp$

Proof. If we prove one of two then the other one will follow from Exercise 8.1.18 (c). For the equality (1), we have the equality $\langle \mathbf{w}, T^*(\mathbf{v})\rangle = \langle T(\mathbf{w}), \mathbf{v}\rangle$. Hence if $\mathbf{w} \in N(T)$ then $\mathbf{w} \in R(T^*)^\perp$. And if $\mathbf{w} \in R(T^*)^\perp$ then $\langle \mathbf{w}, T^*(\mathbf{v})\rangle = 0$ for all $\mathbf{v}$. Hence $\mathbf{w} \in R(T^*)^\perp$. □

Example 8.2.17. *Let $V = F^n$ and W be a subspace of V. By* Theorem 4.1.13 *we know that there is $A \in M(n)$ such that $W = N(A)$. We can apply* Theorem 8.2.16 *to find A. Choose $\{\mathbf{v}_1, \mathbf{v}_2, ..., \mathbf{v}_k\}$ be an orthonormal basis for W and extend it to an orthonormal basis $\{\mathbf{v}_1, \mathbf{v}_2, ..., \mathbf{v}_k, \mathbf{v}_{k+1}, ..., \mathbf{v}_n\}$ of V. Then $\{\mathbf{v}_{k+1}, ..., \mathbf{v}_n\}$ is an orthonormal basis for $W^\perp$. Choose a set of vectors $\{\mathbf{w}_1, \mathbf{w}_2, ..., \mathbf{w}_k\}$ in $W^\perp$. Let $A^* = [\mathbf{w}_1\ \mathbf{w}_2 \cdots \mathbf{w}_k\ \mathbf{v}_{k+1} \cdots \mathbf{v}_n]$. Then $R(T_{A^*})^\perp = (W^\perp)^\perp = W$. Now, by* Theorem 8.2.16, *we have $N(A) = W$ where $A = (A^*)^*$.*

Definition 8.2.18. *For $f(t) = \sum a_i t^i \in \mathbb{C}[t]$ and $A = [a_{ij}] \in M(n, \mathbb{C})$ we write*

$$\bar{f}(t) = \sum_{i=0}^{n} \bar{a}_i t^i \text{ and } \bar{A} = [\,\bar{a}_{ij}] \tag{8.53}$$

Note also that $\bar{f}(\bar{\lambda}) = \overline{f(\lambda)}$. Hence

$$f(\lambda) = 0 \text{ if and only if } \bar{f}(\bar{\lambda}) = 0 \tag{8.54}$$

Theorem 8.2.19. *Let V be an finite dimensional inner product space. Let $T : V \to V$ be a linear operator with the characteristic polynomial $f_T(t) \in F[t]$ ($F = \mathbb{R}$ or $\mathbb{C}$).*

(1) *The characteristic polynomial $f_{T^*}(t)$ of T^* is $\bar{f}_T(t)$.*

(2) *A scalar $\lambda \in F$ is an eigenvalue of T if and only if $\bar{\lambda}$ is an eigenvalue of T^*.*

Proof. (1) Choose an orthonormal basis $\mathcal{B}$ and let $A = [T]_{\mathcal{B}}$. Then $A^* = [T^*]_{\mathcal{B}}$ by Theorem 8.2.10. Since $\det(\bar{A}) = \overline{\det(A)}$ (Theorem 8.1.2) we have

$$\begin{aligned} f_{T^*}(t) &= \det(A^* - tI) = \det(A - tI)^* = \det \overline{(A - tI)^t} \\ &= \overline{\det(A - tI)} = \bar{f}_T(t) \end{aligned}$$

(2) By definition, λ is an eigenvalue of T if and only if $f_T(\lambda) = 0$. By (8.54), this is equivalent to $\overline{f_T(\lambda)} = \bar{f}_T(\bar{\lambda}) = f_{T^*}(\bar{\lambda}) = 0$. That is $\bar{\lambda}$ is an eigenvalue of T^*. □

8.2.3 Adjoint and Dual of Linear Map.

In this subsection, we show that the adjoint T^* of a linear operator T can be realized as the dual $T^\#$ of §6.5 if $F = \mathbb{R}$. For $F = \mathbb{C}$ we can find a relationship between $T^\#$ and T^*.

Let V be a finite dimensional inner product space. Consider the diagram

$$\begin{array}{ccc} V & \xrightarrow{T^*} & V \\ {\scriptstyle\phi}\downarrow & & \downarrow{\scriptstyle\phi} \\ V^\# & \xrightarrow[T^\#]{} & V^\# \end{array} \tag{8.55}$$

where $\phi(\mathbf{v}) = \phi_{\mathbf{v}}$ in (8.38). Here ϕ is an isomorphism if $F = \mathbb{R}$ (Theorem 8.2.1) and conjugate isomorphism if $F = \mathbb{C}$ namely $\phi_{c\mathbf{v}} = \bar{c}\phi_{\mathbf{v}}$.

Now we show that the diagram (8.55) commutes. To see this let $\mathbf{v}, \mathbf{w} \in V$. Then

$$T^{\#} \circ \phi(\mathbf{v})(\mathbf{w}) = \phi_{\mathbf{v}} \circ T(\mathbf{w}) = \phi_{\mathbf{v}}(T(\mathbf{w})) = \langle \mathbf{v}, T(\mathbf{w}) \rangle$$

On the other hand,

$$\phi(T^*(\mathbf{v}))(\mathbf{w}) = \langle T^*(\mathbf{v}), \mathbf{w} \rangle = \langle \mathbf{v}, T(\mathbf{w}) \rangle$$

Hence the diagram above is commutative. Therefore, if we identify V with $V^{\#}$ *via* ϕ (Theorem 8.2.5) then T^* is the 'same' as $T^{\#}$. In fact, if $\mathcal{B} = \{\mathbf{v}_1, \mathbf{v}_2, \ldots, \mathbf{v}_n\}$ is an orthonormal basis and if we let $\phi(\mathbf{v}_i) = f_i$, *viz* $f_i = \langle \mathbf{v}_i, - \rangle$, then $\mathcal{B}^{\#} = \{f_1, f_2, \ldots, f_n\}$ is the dual basis of $\mathcal{B}$. If we identify $\mathbf{v}_i$ with f_i and $T^{\#}$ with T^* via ϕ as above then, by Theorem 6.5.15, we have

$$[T^*]_{\mathcal{B}} \overset{via\ \phi}{=} [T^{\#}]_{\mathcal{B}^{\#}} = [T]^*_{\mathcal{B}}$$

which is Theorem 8.2.10. The first identification is an isomorphism if $F = \mathbb{R}$ and conjugate isomorphism if $F = \mathbb{C}$.

Answers for Practice Problems 8.2

(8.2.1) We need to find $\mathbf{v} = \begin{bmatrix} x \\ y \\ z \end{bmatrix} \in \mathbb{C}^3$ such that $f(a, b, c) = \langle \mathbf{v}, \mathbf{u} \rangle_A = \mathbf{v}^* A\mathbf{u}$ for all $\mathbf{u} = \begin{bmatrix} a \\ b \\ c \end{bmatrix}$. Now $f(a, b, c) = 2a - ib - c$ and $\mathbf{v}^* A\mathbf{u} = a\bar{x} + (\bar{y} + i\bar{z})b + (-i\bar{y} + 2\bar{z})c$. Comparing these expressions we have $\bar{x} = 2,\ \bar{y} + i\bar{z} = -i,\ -i\bar{y} + 2\bar{z} = -1$. Solving this we get $x = 2, y = i, z = 0$. i.e., $\mathbf{v} = 2\mathbf{e}^1 + i\mathbf{e}^2$.

(8.2.2) We have $\langle T(\mathbf{v}), \mathbf{w} \rangle_A = \langle \mathbf{v}, T^*(\mathbf{w}) \rangle_A$ for all $\mathbf{v} \in V$. Let $T^*(\mathbf{w}) = \begin{bmatrix} x \\ y \\ z \end{bmatrix}$ and letting $\mathbf{v} = \mathbf{e}^1, \mathbf{e}^2, \mathbf{e}^3$ we obtain a system of equations: $\begin{cases} x & & = 0 \\ & y \ -iz & = -2i \\ & iy \ +2z & = 2 - i \end{cases}$. Solving this we get $x = 0, y = -2i + 1, z = -i$. Hence $T^*(\mathbf{w}) = \begin{bmatrix} 0 \\ 1 - 2i \\ -i \end{bmatrix}$.

8.2 Exercises

1. For each of the following linear operator $T \in L(V)$ find $T^*(\mathbf{w})$.

   (a) $V = \mathbb{C}^3$ with the standard inner product and $T(a, b, c) = \begin{bmatrix} a - ib \\ (1 - i)b \\ b + ic \end{bmatrix}$, $\mathbf{w} = \begin{bmatrix} -1 \\ -i \\ 0 \end{bmatrix}$

   (b) $V = \mathbb{C}^3$ with the inner product defined in Example 8.1.7 and the linear operator $T(a, b, c) = \begin{bmatrix} a - ib \\ (1 - i)b \\ b + ic \end{bmatrix}$, $\mathbf{w} = \begin{bmatrix} -1 \\ -i \\ 0 \end{bmatrix}$

(c) $V = M(2,\mathbb{C})$ with the standard inner product of Example 8.1.6 and the linear operator $T\left(\begin{bmatrix} a & b \\ c & d \end{bmatrix}\right) = \begin{bmatrix} a & -c \\ -ib & id \end{bmatrix}$ and $\mathbf{w} = \begin{bmatrix} 0 & 1 \\ i & 0 \end{bmatrix}$

(d) $V = M(2,\mathbb{C})$ with the standard inner product of Example 8.1.6 and the linear operator $T(A) = XA$ where $X = \begin{bmatrix} 1 & -i \\ i & 0 \end{bmatrix}$, $\mathbf{w} = \begin{bmatrix} 0 & 1 \\ i & 0 \end{bmatrix}$

(e) $V = P_1(\mathbb{C})$ with the inner product $\langle f, g\rangle = \int_0^1 \bar{f}(x)g(x)dx$ and the linear operator $T(f(x)) = f(ix + (1-i)), \mathbf{w} = ix$

(f) $V = P_1(\mathbb{C})$ with the inner product $\langle f, g\rangle = \int_{-1}^1 \bar{f}(x)g(x)dx$ and the linear operator $T(f(x)) = f(x+i), \mathbf{w} = 1 + ix$

2. Let V be a vector space over $\mathbb{C}$. We define the **conjugate vector space** $\overline{V}$ of V as follows: As a set $\overline{V} = V$. The addition of $\overline{V}$ is the same as that of V. The scalar multiplication on $\overline{V}$ is defined by

$$\mathbb{C} \times \overline{V} \to \overline{V}, \quad (c, \mathbf{v}) \mapsto \bar{c}\mathbf{v}$$

(a) Prove that $\overline{V}$ is a vector space over $\mathbb{C}$.

(b) Prove that the identity maps $V \to \overline{V}$ and $\overline{V} \to V$ are conjugate linear maps.

(c) Prove that a map $T : V \to W$ is conjugate linear if and only if $T : V \to \overline{W}$ is linear. Also show that this is equivalent to that $T : \overline{V} \to W$ is linear.

(d) Prove that $\mathcal{B} = \{\mathbf{v}_1, \mathbf{v}_2, ..., \mathbf{v}_n\}$ is a basis of V if and only if $\mathcal{B}$ is a basis for $\overline{V}$.

(e) Prove that $[\mathbf{v}]_{\mathcal{B}} = \begin{bmatrix} a_1 \\ a_2 \\ \vdots \\ a_n \end{bmatrix}$ is the $\mathcal{B}$-coordinate of $\mathbf{v} \in V$ if and only if $[\mathbf{v}]_{\mathcal{B}} = \begin{bmatrix} \bar{a}_1 \\ \bar{a}_2 \\ \vdots \\ \bar{a}_n \end{bmatrix}$ is the $\mathcal{B}$-coordinate of $\mathbf{v} \in \overline{V}$.

3. Show that if $T : V \to W$ is conjugate linear then the null space $N(T)$ and the range $R(T)$ are subspaces. And check the assertions (6.3.10) – (6.3.26) for conjugate linear maps.

4. Prove (1), (2), (4), (5) and (6) of Theorem 8.2.15.

5. Prove that $\det(T^*) = \overline{\det(T)}$ and $\mathrm{Tr}(T^*) = \overline{\mathrm{Tr}(T)}$ (See Definition 6.4.31 for the definition of $\det(T)$ and $\mathrm{Tr}(T)$.)

6. Let T be a linear operator on an inner product space. Prove that $(f(T))^* = \bar{f}(T^*)$ for $f(t) \in F[t]$.

7. Prove the true statement(s) and find counter example(s) for the false statement(s) about a finite dimensional inner product spaces V.

(a) Every linear functional on V is of the form $\mathbf{x} \mapsto \langle \mathbf{x}, \mathbf{v}\rangle$ for some $\mathbf{v} \in V$.

(b) If T is a linear operator on V then for any basis $\mathcal{B}$ of V we have $[T^*]_{\mathcal{B}} = [T]^*_{\mathcal{B}}$.

(c) If T is a linear operator on V then $(cT)^* = cT^*$ for a constant $c \in \mathbb{C}$.

(d) If T is a linear operator on V and $T^* = 0$ then $T = 0$.

(e) If T is an invertible linear operator on V then T^* is also invertible.

(f) Let T be a linear operator on V. If W is a T-invariant subspace of V then W is also T^*-invariant.

(g) If $A \in (n, \mathbb{C})$ is symmetric then $AA^* = A^*A$.

(h) If λ is an eigenvalue of a linear operator T on V then $\bar{\lambda}$ is an eigenvalue of T^*.

(i) The map $\phi^* : L(V) \to L(V)$ defined by $\phi * (T) = T^*$ is conjugate linear.

8. Let T be a linear operator. Prove that $\langle T(\mathbf{u}), T(\mathbf{v})\rangle = \langle \mathbf{u}, \mathbf{v}\rangle$ for all $\mathbf{u}, \mathbf{v} \in V$ if and only if $\|T(\mathbf{u})\| = \|\mathbf{u}\|$ for all $\mathbf{u} \in V$.

9. Let T be a linear operator. Prove that T is the zero operator if T^*T is the zero operator. Is the same conclusion holds if TT^* is the zero operator instead ?

10. Let T be a linear operator on an inner product space V.

    (a) Prove $N(T^*T) = N(T)$ and $N(TT^*) = N(T^*)$. Deduce that $\rho(T^*T) = \rho(T)$ and $\rho(TT^*) = \rho(T^*)$.

    (b) Prove $\rho(T^*) = \rho(T)$ and deduce that $\rho(TT^*) = \rho(T) = \rho(T^*T)$. (Cf. Theorem 4.2.35)

    (c) Deduce that if $A \in M(n, \mathbb{C})$ then $\rho(AA^*) = \rho(A^*A) = \rho(A)$. How about the nullity?

11. Let T be a linear operator on a real inner product space V. Let $T^{\mathbb{C}} : V^{\mathbb{C}} \to V^{\mathbb{C}}$ be the complexification of T (Exercise 6.3.16, Exercise 8.1.23). Prove that $(T^*)^{\mathbb{C}} = (T^{\mathbb{C}})^*$.

Let $T : V \to W$ be a linear map between the inner product spaces with the inner products $\langle\ ,\ \rangle_V, \langle\ ,\ \rangle_W$. A function $T^* : W \to V$ is called an **adjoint** *of T if*

$$\langle \mathbf{w}, T(\mathbf{v})\rangle_W = \langle T^*(\mathbf{w}), \mathbf{v}\rangle_V \tag{8.56}$$

12. With the notations above prove the following statements using (8.56).

    (a) Let $V = F^n, W = F^m$ and $A \in M(m{\times}n)$ and $\langle\ ,\ \rangle_V, \langle\ ,\ \rangle_W$ be the standard inner products. Let $B \in M(n{\times}m)$. Prove that $\langle \mathbf{w}, A\mathbf{v}\rangle_W = \langle B\mathbf{w}, \mathbf{v}\rangle_V$ if and only if $B = A^*$ which generalize (8.43).

    (b) There is a unique linear map T^* satisfying (8.56).

    (c) We have $T^{**} = T$. If $S : U \to V$ then $(T \circ S)^* = S^* \circ T^*$.

    (d) If $\mathcal{B}, \mathcal{C}$ are orthonormal bases of V, W respectively then $[T^*]^{\mathcal{B}}_{\mathcal{C}} = \left([T]^{\mathcal{C}}_{\mathcal{B}}\right)^*$

    (e) $\rho(T) = \rho(T^*)$

    (f) We have $\langle T(\mathbf{v}), \mathbf{w}\rangle_V = \langle \mathbf{v}, T^*(\mathbf{w})\rangle_W$ for $\mathbf{v} \in W, \mathbf{w} \in V$.

    (g) We have $T^*T(\mathbf{v}) = \mathbf{0}$ if and only if $T(\mathbf{v}) = \mathbf{0}$.

    (h) We have $(T^*T)^* = T^*T$ and $\langle \mathbf{v}, T^*T(\mathbf{v})\rangle \geq 0$ for all $\mathbf{v} \in V$.

    (i) $\rho(T^*T) = \rho(T)$

13. Prove an analog of Theorem 8.2.15 for $T : V \to W$ by making appropriate changes.

14. Prove that $R(T^*)^{\perp} = N(T)$ for a linear map $T : V \to W$.

15. Prove that a linear map $T : V \to W$ is an isometry if and only if $T^*T = I_W$ and $T^*T = I_V$.

16. Let V, W be inner product spaces. Let $T : V \to W$ be a linear map and $T^* : W \to V$ be defined by (8.56). Prove that the following diagram is commutative

$$\begin{array}{ccc} W & \xrightarrow{T^*} & V \\ {\scriptstyle \phi_W}\downarrow & & \downarrow{\scriptstyle \phi_V} \\ W^{\#} & \xrightarrow[T^{\#}]{} & V^{\#} \end{array}$$

where ϕ_V, ϕ_W are defined by (8.39).

8.3 Normal Operators and Spectral Theorem

A matrix $A \in M(n, F)$ will be called normal if $A^*A = AA^*$. In the following sections, we investigate the properties of normal matrices. It include Hermitian matrices, symmetric matrices, unitary matrices, orthogonal matrices.

8.3.1 Basic Properties of Normal Operators

If the characteristic polynomial of a linear operator $T \in L(V)$ splits then there is a basis for which $[T]_{\mathcal{B}}$ is upper triangular by Corollary 7.2.14. If V is an inner product space then we can do more; we can choose an orthonormal basis in which the matrix is upper triangular. We cannot use the proof of Corollary 7.2.14 since L_i's are not necessarily mutually orthogonal.

Theorem 8.3.1. *Let V be a finite dimensional inner product space and let $T : V \to V$ be a linear operator with the splitting characteristic polynomial $f_T(t) \in F[t]$. Then there is an orthonormal basis $\mathcal{B}$ such that $[T]_{\mathcal{B}}$ is upper triangular.*

Let $A \in M(n, F)$ with the splitting characteristic polynomial. Then there is $P \in M(n, F)$ with orthonormal columns such that $P^{-1}AP$ is upper triangular.

Proof. We use induction on $\dim V$. If $\dim V = 1$ then our assertion is trivial. Now suppose $\dim V = n > 1$ and assume our result for the vector spaces of dimension $< n$.

Since f_T splits so does f_{T^*} for if $f_T(t) = \pm\prod(t - \lambda_i)$ then $f_{T^*}(t) = \bar{f}_T(t) = \pm\prod(t - \bar{\lambda}_i)$ (Theorem 8.2.19). If $\bar{\lambda}$ is an eigenvalue of T then λ is an eigenvalue for T^*. And hence T^* has an eigenvector $\mathbf{w}$ corresponding to λ. By normalizing $\mathbf{w}$ we can assume $\mathbf{w}$ is a unit vector such that $T^*(\mathbf{w}) = \lambda\mathbf{w}$ and let $W = \text{span}\{\mathbf{w}\}$. We claim that $T(W^{\perp}) \subseteq W^{\perp}$. In fact, let $\mathbf{y} \in W^{\perp}$. Then $\langle T(\mathbf{y}), \mathbf{w}\rangle = \langle \mathbf{y}, T^*(\mathbf{w})\rangle = \langle \mathbf{y}, \lambda\mathbf{w}\rangle = \lambda\langle \mathbf{y}, \mathbf{w}\rangle = 0$. Hence $T(\mathbf{y}) \in W^{\perp}$.

Now $\dim W^{\perp} = n - 1$ by Theorem 8.1.26. By our induction hypothesis there is an orthonormal basis $\mathcal{C}$ of $W^{\perp}$ such that $[T_{W^{\perp}}]_{\mathcal{C}}$ is upper triangular. Then $\mathcal{B} = \mathcal{C} \cup \{\mathbf{w}\}$ is an orthonormal basis of V such that

$$[T]_{\mathcal{B}} = \left[\begin{array}{ccc|c} & & & * \\ & [T_{W^\perp}]_{\mathcal{C}} & & \vdots \\ & & & * \\ \hline 0 & \cdots & 0 & * \end{array}\right]$$

which is upper triangular where the last column is the $\mathcal{B}$-coordinate $[T(\mathbf{w})]_{\mathcal{B}}$ of $T(\mathbf{w})$. □

Definition 8.3.2. *Let V be an inner product space over F and T be a linear operator on V. We say T is* **normal** *if $TT^* = T^*T$. We say T is* **self adjoint** *if $T^* = T$. If $F = \mathbb{C}$ then a self adjoint operator T is also said to be* **Hermitian**.

We say that a matrix $A \in M(n, F)$ is **normal** *if $AA^* = A^*A$. We say a matrix A is* **self adjoint** *if $A^* = A$. When $F = \mathbb{C}$ we say A is* **Hermitian** *(resp.* **skew Hermitian***) if $A = A^*$ (resp. $A^* = -A$).*

A real symmetric matrix is a synonym for a real self adjoint matrix. And a complex self adjoint matrix is a synonym for a Hermitian matrix.

A Hermitian or skew Hermitian or self adjoint operators are normal. Since $T_A^* = T_{A^*}$ we see that $A \in M(n, F)$ is normal (resp. self adjoint, resp. Hermitian) if and only if T_A is normal (resp. self adjoint, resp. Hermitian) (Exercise 4). More generally,

Theorem 8.3.3. *Let V be a finite dimensional inner product space with an orthonormal basis $\mathcal{B}$. Let T be a linear operator on V. Then the linear operator T is normal if and only if $[T]_{\mathcal{B}}$ is normal. In particular, if $[T]_{\mathcal{B}}$ is diagonal then T is normal.*

Proof. Let $\mathcal{B}$ be an orthonormal basis. Then T is normal if and only if $[TT^*]_{\mathcal{B}} = [T^*T]_{\mathcal{B}}$. Since $[T^*]_{\mathcal{B}} = [T_{\mathcal{B}}]^*$ (Theorem 8.2.10) this is equivalent to $[T]_{\mathcal{B}}[T]_{\mathcal{B}}^* = [T]_{\mathcal{B}}^*[T]_{\mathcal{B}}$. And hence the latter is equivalent to that $[T]_{\mathcal{B}}$ is normal.

If $[T]_{\mathcal{B}}$ is diagonal then so is $[T^*]_{\mathcal{B}} = [T]_{\mathcal{B}}^*$ and they commute. Hence T is normal by the previous statement. □

We will prove that the converse of Theorem 8.3.3 is true when the characteristic polynomial splits. That is, T is normal if and only if there is an orthonormal basis consisting of eigenvectors of T (Theorem 8.3.9).

Example 8.3.4. *Let $A = \begin{bmatrix} \cos\theta & -\sin\theta \\ \sin\theta & \cos\theta \end{bmatrix} \in M(2, \mathbb{R})$. Then A is the rotation by the angle θ and $A^* = A^t = \begin{bmatrix} \cos\theta & \sin\theta \\ -\sin\theta & \cos\theta \end{bmatrix} = \begin{bmatrix} \cos(-\theta) & -\sin(-\theta) \\ \sin(-\theta) & \cos(-\theta) \end{bmatrix}$ is the rotation by the angle $-\theta$. Hence $AA^* = I = A^*A$. Therefore A is normal.*

More generally, an orthogonal or a symmetric or a skew symmetric real square matrices are normal.

Example 8.3.5. *Let $A = \begin{bmatrix} a & b \\ c & d \end{bmatrix} \in M(2, \mathbb{R})$ be normal. Since A is normal we have*

$$A^*A = \begin{bmatrix} a & c \\ b & d \end{bmatrix}\begin{bmatrix} a & b \\ c & d \end{bmatrix} = \begin{bmatrix} a^2 + c^2 & ab + cd \\ ab + cd & b^2 + d^2 \end{bmatrix}$$

$$AA^* = \begin{bmatrix} a & b \\ c & d \end{bmatrix}\begin{bmatrix} a & c \\ b & d \end{bmatrix} = \begin{bmatrix} a^2 + b^2 & ac + bd \\ ac + bd & c^2 + d^2 \end{bmatrix}$$

By comparing $(1,1)$ *positions we conclude that* $b^2 = c^2; b = \pm c$. *Hence the characteristic polynomial is* $f_A(t) = t^2 - (a+d)t + ad - b^2$ *with the discriminant* $D = (a-d)^2 + 4b^2$.

If $b = c$ *then* A *is symmetric*

$$A = \begin{bmatrix} a & b \\ b & d \end{bmatrix} \tag{8.57}$$

The characteristic polynomial $f_A(t)$ *has two distinct real eigenvalues if the discriminant* $D = (a-d)^2 + 4b^2$ *is positive i.e., if either* $b \neq 0$ *or* $a \neq d$. *It has one real eigenvalue only when* $b = 0$ *and* $a = d$ *in which case* $A = \mathrm{diag}[a, a]$.

If $b = -c \neq 0$, *then comparing the* $(1,2)$ *positions of* A^*A *and* AA^* *we have* $b(a-d) = (d-a)b$. *Hence* $a = d$. *Hence* A *is of the form*

$$A = \begin{bmatrix} a & b \\ -b & a \end{bmatrix} \tag{8.58}$$

The matrices of the form (8.58) *includes both rotations and skew-symmetric matrices. In this case, the characteristic polynomial* $f_A(t)$ *has no real root since* $b \neq 0$.

The canonical form for the real normal operators will be given in Theorem 8.3.20.

Theorem 8.3.6. *Let* V *be an inner product space over* F *and* $T : V \to V$ *be a normal operator. Then*

(1) $\|T(\mathbf{v})\| = \|T^*(\mathbf{v})\|$

(2) *For any* $c \in F$, *the linear operator* $(T - cI)$ *is normal.*

(3) *If* $T(\mathbf{v}) = \lambda\mathbf{v}$ *then* $T^*(\mathbf{v}) = \overline{\lambda}\mathbf{v}$.

(4) *If* λ_1, λ_2 *are distinct eigenvalues with the corresponding eigenvectors* $\mathbf{v}_1, \mathbf{v}_2$ *are orthogonal.*

Proof. (1) Follows from the computation:

$$\begin{aligned} \|T(\mathbf{v})\|^2 &= \langle T(\mathbf{v}), T(\mathbf{v})\rangle = \langle T^*T(\mathbf{v}), \mathbf{v}\rangle \\ &= \langle TT^*(\mathbf{v}), \mathbf{v}\rangle = \langle T^*(\mathbf{v}), T^*(\mathbf{v})\rangle = \|T^*(\mathbf{v})\|^2 \end{aligned}$$

(2) Since $(T - cI)(T - cI)^* = (T - cI)(T^* - \bar{c}I)$ (Theorem 8.2.15) we have

$$\begin{aligned} (T - cI)(T^* - \bar{c}I) &= TT^* - cT^* - \bar{c}T + |c|^2 I \\ &= T^*T - cT^* - \bar{c}T + |c|^2 I = (T^* - \bar{c}I)(T - cI) \end{aligned}$$

Hence $(T - cI)(T - cI)^* = (T - cI)^*(T - cI)$.

(3) Let $T(\mathbf{v}) = \lambda\mathbf{v}$ and let $S = T - \lambda I$. Then S is normal by (2) and $S^* = T^* - \overline{\lambda}I$ by Theorem 8.2.15. By (1), we have $0 = \|S(\mathbf{v})\| = \|S^*(\mathbf{v})\| = \|(T^* - \overline{\lambda}I)(\mathbf{v})\|$. Hence $(T^* - \bar{\lambda})(\mathbf{v}) = 0$; $T^*(\mathbf{v}) = \bar{\lambda}\mathbf{v}$.

(4) Let $T(\mathbf{v}_i) = \lambda_i\mathbf{v}_i (i = 1, 2)$ with $\lambda_1 \neq \lambda_2$. Then, by (3), we have

$$\bar{\lambda}_1\langle \mathbf{v}_1, \mathbf{v}_2\rangle = \langle \lambda_1\mathbf{v}_1, \mathbf{v}_2\rangle = \langle T(\mathbf{v}_1), \mathbf{v}_2\rangle = \langle \mathbf{v}_1, T^*(\mathbf{v}_2)\rangle = \langle \mathbf{v}_1, \overline{\lambda}_2\mathbf{v}_2\rangle = \bar{\lambda}_2\langle \mathbf{v}_1, \mathbf{v}_2\rangle$$

Since $\lambda_1 \neq \lambda_2$ we conclude that $\langle \mathbf{v}_1, \mathbf{v}_2\rangle = 0$. □

Example 8.3.7. *In* Example 6.7.16 *we considered* $T_\theta : \mathbb{C}^2 \to \mathbb{C}^2$ *which is the matrix linear map for* $A = \begin{bmatrix} \cos\theta & -\sin\theta \\ \sin\theta & \cos\theta \end{bmatrix}$. *For* $\mathbf{u} = \begin{bmatrix} a \\ b \end{bmatrix} \in \mathbb{C}^2$ *we have* $T(\mathbf{u}) = \begin{bmatrix} a\cos\theta - b\sin\theta \\ a\sin\theta + b\cos\theta \end{bmatrix}$. *By* Theorem 5.2.14 *or by a cumbersome but straightforward calculation we have* $\|T(\mathbf{u})\| = \|\mathbf{u}\|$.

In Example 6.7.11, *we showed that the characteristic polynomial* $f_A(t) = t^2 - (2\cos\theta)t + 1$ *of* A *has roots* $t = e^{\pm i\theta}$. *Hence the eigenvalues are* $\lambda_1 = e^{i\theta}, \lambda_2 = e^{-i\theta}$. *The corresponding eigenvectors are* $\mathbf{v}_1 = \begin{bmatrix} 1 \\ -i \end{bmatrix}, \mathbf{v}_2 = \begin{bmatrix} 1 \\ i \end{bmatrix}$ (Example 6.7.16). *By* Corollary 8.2.7, T_θ^* *is the matrix linear map of* $A^t = \begin{bmatrix} \cos\theta & \sin\theta \\ -\sin\theta & \cos\theta \end{bmatrix} = \begin{bmatrix} \cos(-\theta) & -\sin(-\theta) \\ \sin(-\theta) & \cos(-\theta) \end{bmatrix}$. *Hence* $T_\theta^* = T_{-\theta}$ *and* $T_\theta^* T_\theta = T_\theta T_\theta^* = I$. *Now we have*

$$\begin{aligned} T_\theta^*(\mathbf{v}_1) &= \begin{bmatrix} \cos(-\theta) & -\sin(-\theta) \\ \sin(-\theta) & \cos(-\theta) \end{bmatrix} \begin{bmatrix} 1 \\ -i \end{bmatrix} = \begin{bmatrix} \cos(-\theta) + i\sin(-\theta) \\ \sin(-\theta) - i\cos(-\theta) \end{bmatrix} \\ &= \begin{bmatrix} \cos(-\theta) + i\sin(-\theta) \\ -i(\cos(-\theta) + i\sin(-\theta)) \end{bmatrix} = e^{-i\theta} \begin{bmatrix} 1 \\ -i \end{bmatrix} \end{aligned}$$

This computation illustrate (3) *of* Theorem 8.3.6*: If* $T_\theta(\mathbf{v}_1) = e^{i\theta}\mathbf{v}_1$ *then*

$$T_\theta^*(\mathbf{v}_1) = \overline{(e^{i\theta})}\mathbf{v}_1 = e^{-i\theta}\mathbf{v}_1$$

Also $T_\theta^*(\mathbf{v}_2) = \overline{(e^{-i\theta})}\mathbf{v}_2 = e^{i\theta}\mathbf{v}_2$.

If $\mathbf{v}_1, \mathbf{v}_2$ *are the eigenvectors corresponding to the distinct eigenvalues* $e^{i\theta}, e^{-i\theta}$ *then we have* $\langle \mathbf{v}_1, \mathbf{v}_2 \rangle = [1 \;\; i] \begin{bmatrix} 1 \\ i \end{bmatrix} = 0$. *Hence* $\left\{ \frac{1}{\sqrt{2}}\mathbf{v}_1, \frac{1}{\sqrt{2}}\mathbf{v}_2 \right\}$ *is an orthonormal basis for* $\mathbb{C}^2$ *consisting of eigenvectors of* T_θ.

The following example illustrates that the eigenvectors corresponding to the distinct eigenvalues are not necessarily orthogonal if A is not normal.

Example 8.3.8. *Let* $A = \begin{bmatrix} 0 & i \\ -i & i \end{bmatrix}$. *Then* $A^* = \begin{bmatrix} 0 & i \\ -i & -i \end{bmatrix}$, $A^*A = \begin{bmatrix} 1 & -1 \\ -1 & 2 \end{bmatrix}$ *and* $AA^* = \begin{bmatrix} 1 & 1 \\ 1 & 2 \end{bmatrix}$. *Hence* A *is not normal. The characteristic polynomial is* $f(t) = t^2 - it - 1$ *and the eigenvalues are, by the quadratic formula,* $\lambda_+ = \frac{1}{2}(\sqrt{3} + i)$ *and* $\lambda_- = \frac{1}{2}(-\sqrt{3} + i)$ *with the eigenvectors* $\mathbf{v}_+ = \begin{bmatrix} \frac{1}{2}(1 + \sqrt{3}i) \\ 1 \end{bmatrix}$ *and* $\mathbf{v}_- = \begin{bmatrix} \frac{1}{2}(1 - \sqrt{3}i) \\ 1 \end{bmatrix}$. *Now* $\langle \mathbf{v}_+, \mathbf{v}_- \rangle = \frac{1}{2} - \frac{\sqrt{3}i}{2}$. *Hence the eigenvectors of* A *corresponding to distinct eigenvalues are not orthogonal.*

8.3.2 Diagonalization of Normal Operators

Now we prove an important fact that a linear operator T, when the characteristic polynomial splits, is normal if and only if there is an orthonormal basis $\mathcal{B}$ such that $[T]_\mathcal{B}$ is diagonal.

Theorem 8.3.9. *Let* T *be a linear operator on a finite dimensional inner product space* V *over* F *whose characteristic polynomial splits. Then* T *is normal if and only if there is an orthonormal basis* $\mathcal{B}$ *consisting of eigenvectors say, with the eigenvalues* $\lambda_1, \lambda_2, ..., \lambda_n$ *i.e.,*

$$[T]_\mathcal{B} = \mathrm{diag}[\lambda_1, \lambda_2, ..., \lambda_n]$$

Let $A \in M(n, F)$ whose characteristic polynomial splits. Then A is normal if and only if there is a unitary or an orthogonal (depending on whether $F = \mathbb{R}$ or $F = \mathbb{C}$) matrix P such that P^*AP is diagonal.

Proof. If there is an orthonormal basis $\mathcal{B}$ consisting of eigenvectors then $[T]_{\mathcal{B}}$ is diagonal. Hence T is normal by Theorem 8.3.3.

For the converse, by Theorem 8.3.1, there is an orthonormal basis $\mathcal{B} = \{\mathbf{v}_1, \mathbf{v}_2, \ldots, \mathbf{v}_n\}$ such that $[T]_{\mathcal{B}}$ is upper triangular. Note that $\mathbf{v}_1$ is an eigenvector of T. Let $\{\mathbf{v}_1, \mathbf{v}_2, \ldots, \mathbf{v}_{k-1}\}$ be the eigenvectors in $\mathcal{B}$ with the eigenvalues $\lambda_1, \ldots, \lambda_{k-1}$ so that

$$[T]_{\mathcal{B}} = \left[\begin{array}{cccc|ccc} \lambda_1 & & & & & & \\ & \lambda_2 & & & & C & \\ & & \ddots & & & & \\ & & & \lambda_{k-1} & & & \\ \hline 0 & \cdots & \cdots & 0 & a_{kk} & * & * \\ & \cdots & \cdots & & 0 & a_{k+1k+1} & * \\ & & & & \vdots & \ddots & \\ 0 & \cdots & \cdots & 0 & 0 & \cdots\ 0 & a_{nn} \end{array}\right] \tag{8.59}$$

is upper triangular.

We claim that $\mathbf{v}_k$ is also an eigenvector. (Once this is shown we are done by induction.) In fact, by Theorem 8.3.6, $\{\mathbf{v}_1, \mathbf{v}_2, ..., \mathbf{v}_{k-1}\}$ are also eigenvectors for T^* corresponding to the eigenvalues $\overline{\lambda}_1, \overline{\lambda}_2, ..., \overline{\lambda}_{k-1}$. Hence

$$[T^*]_{\mathcal{B}} = \left[\begin{array}{cccc|c} \overline{\lambda}_1 & & & & \\ & \overline{\lambda}_2 & & & * \\ & & \ddots & & * \\ & & & \overline{\lambda}_{k-1} & \\ \hline 0 & \cdots & \cdots & 0 & \\ & \cdots & \cdots & & * \\ & & & & * \\ 0 & \cdots & \cdots & 0 & \end{array}\right] \tag{8.60}$$

By (8.59), we have a lower triangular matrix

$$[T]^*_{\mathcal{B}} = \left[\begin{array}{cccc|ccc} \overline{\lambda}_1 & & & & 0 & \cdots & 0 \\ & \overline{\lambda}_2 & & & & \cdots & \\ & & \ddots & & 0 & \cdots & 0 \\ & & & \overline{\lambda}_{k-1} & 0 & \cdots & 0 \\ \hline & & & & \overline{a}_{kk} & 0 & \cdots 0 \\ & & & & * & \overline{a}_{k+1k+1} & \\ & & C^* & & \vdots & \ddots & \\ & & & & * & \cdots\ * & \overline{a}_{nn} \end{array}\right] \tag{8.61}$$

Now, by Theorem 8.2.10, the matrices (8.60) and (8.61) are the same. Hence $C^* = 0$ which in turn implies $C = 0$. Therefore, $T(\mathbf{v}_k) = a_{kk}\mathbf{v}_k$ by (8.59) i.e., $\mathbf{v}_k$ is an eigenvector. Inductively $\mathcal{B} = \{\mathbf{v}_1, \mathbf{v}_2, \ldots, \mathbf{v}_n\}$ is an orthonormal basis consisting of eigenvectors.

Hence a matrix A is normal if and only if there is an orthonormal basis $\mathcal{B} = \{\mathbf{v}_i\}$ of F^n with $\lambda_i \mathbf{v}_i = A\mathbf{v}_i$. That is $A[\mathbf{v}_1\ \mathbf{v}_2 \cdots \mathbf{v}_n] = \mathrm{diag}(\lambda_1, \lambda_2, ..., \lambda_n)[\mathbf{v}_1\ \mathbf{v}_2 \cdots \mathbf{v}_n]$. This is equivalent to

$$P^{-1}AP = \begin{bmatrix} \lambda_1 & & \\ & \ddots & \\ & & \lambda_n \end{bmatrix}$$

where $P^{(i)} = \mathbf{v}_i$. Since $\{P^{(i)}\}$ are orthonormal $P^* = P^{-1}$ by Theorem 8.1.15. Hence this is equivalent to $P^*AP = \mathrm{diag}[\lambda_1, \lambda_2, ..., \lambda_n]$ where P is unitary. □

If $F = \mathbb{C}$ then, by Fundamental Theorem of Algebra, the characteristic polynomial splits. Hence we have

Corollary 8.3.10. *Let T be a linear operator on a finite dimensional inner product space over $\mathbb{C}$. Then T is normal if and only if there is an orthonormal basis $\mathcal{B}$ consisting of eigenvectors. A matrix $A \in M(n, \mathbb{C})$ is normal if and only if there is a unitary matrix P such that P^*AP is diagonal.*

Definition 8.3.11. *Let $A, B \in M(n, F)$. When $F = \mathbb{C}$, we say A and B are* **unitarily equivalent** *if there is a unitary matrix U such that $B = U^*AU$. When $F = \mathbb{R}$ we say A and B are* **orthogonally equivalent** *if there is an orthogonal matrix P such that $B = P^tAP$.*

By Theorem 8.3.1, if the characteristic polynomial of A splits then A is orthogonally or unitarily equivalent to an upper triangular matrix. By Theorem 8.3.9, a normal matrix $A \in M(n, F)$ is orthogonally or unitarily equivalent to a diagonal matrix depending on whether $F = \mathbb{R}$ or $F = \mathbb{C}$ provided that the characteristic polynomial splits.

Example 8.3.12. *Let $V = \mathbb{C}^3$ with $\langle \mathbf{u}, \mathbf{v} \rangle = \mathbf{u}^*A\mathbf{v}$ where $A = \begin{bmatrix} 1 & 0 & 0 \\ 0 & 1 & -i \\ 0 & i & 2 \end{bmatrix}$ defined in* Example 8.1.7. *Define T by $T(a, b, c) = \begin{bmatrix} 2a \\ 2b - (1+2i)c \\ ci \end{bmatrix}$ as in* Example 8.2.13. *We want to see if T is normal. The matrix representation $[T]_{\mathcal{C}}$ of T with respect to the orthonormal basis $\mathcal{C} = \{\mathbf{v}_1 = \mathbf{e}^1, \mathbf{v}_2 = \mathbf{e}^2, \mathbf{v}_3 = i\mathbf{e}^2 + \mathbf{e}^3\}$ in* Example 8.2.13 *is $[T]_{\mathcal{C}} = \begin{bmatrix} 2 & 0 & 0 \\ 0 & 2 & 0 \\ 0 & 0 & i \end{bmatrix}$. Hence $\mathcal{C}$ is an orthonormal basis consisting of eigenvectors. Therefore, T is normal.*

For the following theorem we observe that if $V = W_1 \oplus W_2$ and W is a subspace then the direct sum $(W \cap W_1) \oplus (W \cap W_2) \neq W$ in general. For example, let

$$V = \mathbb{R}^2,\ W = \left\{ \begin{bmatrix} x \\ x \end{bmatrix} \middle| x \in \mathbb{R} \right\},\ W_1 = \left\{ \begin{bmatrix} x \\ 0 \end{bmatrix} \middle| x \in \mathbb{R} \right\},\ W_2 = \left\{ \begin{bmatrix} 0 \\ y \end{bmatrix} \middle| y \in \mathbb{R} \right\}$$

Then $(W \cap W_1) \oplus (W \cap W_2) = \{\mathbf{0}\}$ which is different from W.

By Theorem 8.3.9, if T is a normal operator having the splitting characteristic polynomial with the eigenvalues λ_i then we have $V = \oplus_i E_{\lambda_i}$. The following theorem shows that $W = W \cap (\oplus E_{\lambda_i})$ if W is T-invariant.

Theorem 8.3.13. *Let T be a normal operator on a finite dimensional inner product space V. Suppose the characteristic polynomial of T splits with distinct eigenvalues $\lambda_1, \lambda_2, ..., \lambda_k$. Let $V = \oplus E_{\lambda_i}$ and let W be a T-invariant subspace of V. Then*

$$W = (E_{\lambda_1} \cap W) \oplus (E_{\lambda_2} \cap W) \oplus \cdots \oplus (E_{\lambda_k} \cap W) \tag{8.62}$$

Further, W is T^*-invariant and the restriction T_W of T on W is also normal.

Proof. By Theorem 8.3.9, write $V = \oplus_{i=1}^k E_{\lambda_i}$. Obviously $\oplus_i(E_{\lambda_i} \cap W) \subseteq W$. To prove these are equal let $\mathbf{w} \in W$ and write $\mathbf{w} = \sum_{i=1}^k \mathbf{v}_i$ with $\mathbf{v}_i \in E_{\lambda_i}$. We need to show that $\mathbf{v}_i \in W$ for all i. We use induction on k. If $k = 1$ then our assertion is obvious. Suppose $k > 1$ and let $\mathbf{v}_1 + \mathbf{v}_2 + \cdots + \mathbf{v}_k \in W$. Then $T(\mathbf{v}_1 + \mathbf{v}_2 + \cdots + \mathbf{v}_k) = \lambda_1 \mathbf{v}_1 + \cdots + \lambda_k \mathbf{v}_k \in W$ since W is T-invariant. Since the eigenvalues are distinct we may assume $\lambda_k \neq 0$ and $\mathbf{v}_k \neq \mathbf{0}$ as well. Dividing through λ_k we have $\lambda_k^{-1}\lambda_1\mathbf{v}_1 + \cdots + \lambda_k^{-1}\lambda_{k-1}\mathbf{v}_{k-1} + \mathbf{v}_k \in W$. Subtracting $\mathbf{v}_1 + \mathbf{v}_2 + \cdots + \mathbf{v}_k$ we see that

$$(\lambda_k^{-1}\lambda_1 - 1)\mathbf{v}_1 + (\lambda_k^{-1}\lambda_2 - 1)\mathbf{v}_2 + \cdots + (\lambda_k^{-1}\lambda_{k-1} - 1)\mathbf{v}_{k-1} \in W$$

By induction we see that each $\mathbf{v}_1, \mathbf{v}_2, \ldots, \mathbf{v}_{k-1}$ is in W which in turn implies that $\mathbf{v}_k \in W$. Therefore, $\mathbf{w} = \sum \mathbf{v}_i \in \oplus_i E_{\lambda_i} \cap W$.

To show W is T^*-invariant write $\mathbf{w} = \sum \mathbf{v}_i$ with $\mathbf{v}_i \in E_{\lambda_i} \cap W$ by (8.62). Then, by Theorem 8.3.6, $T^*(\mathbf{w}) = \sum \bar{\lambda}_i \mathbf{v}_i \in \oplus(E_{\lambda_i} \cap W) \in W$.

To see T_W is normal choose an orthonormal basis $\mathcal{B}_i$ for $E_{\lambda_i} \cap W$ then they are the eigenvectors for T_W. Then $\mathcal{B}_i$ and $\mathcal{B}_j$ are orthogonal by Theorem 8.3.6. Hence $\mathcal{B} = \cup \mathcal{B}_i$ is an orthonormal basis for W consisting of eigenvectors. By Theorem 8.3.9, T_W is normal. □

Using Theorem 8.3.13 we can prove that if S, T are commuting normal operators then there is an orthonormal basis $\mathcal{B}$ for which both $[S]_{\mathcal{B}}, [T]_{\mathcal{B}}$ are diagonal (Exercise 8).

8.3.3 Spectral Theorem

In this subsection, we prove Spectral Theorem which is a reinterpretation of Theorem 8.3.9 in terms of orthogonal projections: Let T be a normal linear operator on an inner product space having splitting characteristic polynomial with the eigenvalues $\lambda_1, \lambda_2, ..., \lambda_k$. We will show that T is a linear combination of orthogonal projections P_i onto the eigenspaces E_{λ_i} with the coefficients λ_i namely

$$T = \sum \lambda_i P_i$$

We will assume that V is a finite dimensional inner product space for simplicity even though some theorems are true for the infinite dimensional inner product spaces also.

Recall our previous results and definitions. Let $W_1, W_2, ..., W_k$ be mutually orthogonal subspaces and $V = \bigoplus_{i=1}^{k} W_i$ then we write (Definition 8.1.25)

$$V = \bigoplus_{i=1}^{k} W_i$$

Let V be a finite dimensional inner product space and W be a subspace of V. Choose an orthonormal basis $\mathcal{B} = \{\mathbf{v}_1, \mathbf{v}_2, \ldots, \mathbf{v}_k\}$ for W. The orthogonal projection onto W is defined by (Definition 8.1.18)

$$P_W(\mathbf{v}) = \sum \langle \mathbf{v}_i, \mathbf{v} \rangle \mathbf{v}_i$$

We showed, in Theorem 8.1.26, that

$$R(P_W) = W,\ N(P_W) = W^{\perp},\ P_W^2 = P_W \tag{8.63}$$

We also showed, in Theorem 8.1.27, that (8.63) characterizes the orthogonal projection P_W: If T is a linear operator on V then $T = P_W$ if and only if

$$T^2 = T,\ R(T) = W,\ N(T) = W^{\perp} \tag{8.64}$$

Theorem 8.2.14 gives another characterization of the orthogonal projection: A linear operator T on V is an orthogonal projection onto the range $R(T)$ if and only if

$$T^2 = T = T^* \tag{8.65}$$

Now we will show that a normal operators are linear combinations of orthogonal projections if the characteristic polynomial split. Most of the statements of the following theorem are immediate from Theorem 8.3.9.

Theorem 8.3.14 (Spectral Theorem). *Let T be a normal operator on a finite dimensional inner product space V with the characteristic polynomial $f(t) = \pm\prod(t-\lambda_i)^{e_i}$. Let $E_1, E_2, \ldots, E_k$ be the eigenspaces corresponding to $\lambda_1, \lambda_2, \ldots, \lambda_k$. Let $P_1, P_2, \ldots, P_k$ be the orthogonal projections onto $E_1, E_2, \ldots, E_k$. Then*

$$V = E_1 \textcircled{\perp} E_2 \textcircled{\perp} \cdots \textcircled{\perp} E_k \text{ and } E_i^{\perp} = \textcircled{\perp}_{j\neq i} E_j \tag{8.66}$$

Further, we have

$$P_iP_j = \delta_{ij}P_i\ (1 \le i, j \le k) \tag{8.67}$$

$$I = P_1 + P_2 + \cdots + P_k \tag{8.68}$$

$$T = \lambda_1 P_1 + \lambda_2 P_2 + \cdots + \lambda_k P_k \tag{8.69}$$

Let $A \in M(n, F)$ be normal with the splitting characteristic polynomial say, $f(t) = \pm\prod(t-\lambda_i)^{e_i}$. Then there is a unitary or orthogonal matrix U depending on whether $F = \mathbb{R}$ or $F = \mathbb{C}$ such that

$$U^*AU = \begin{bmatrix} \lambda_1 I_{e_1} & & & \\ & \lambda_2 I_{e_2} & & \\ & & \ddots & \\ & & & \lambda_k I_{e_k} \end{bmatrix} \tag{8.70}$$

where I_{e_i} is the identity matrix of size e_i.

Proof. By Theorem 8.3.9, T is diagonalizable. Hence $V = E_1 \oplus E_2 \oplus \cdots \oplus E_k$ by Theorem 6.7.25. And they are mutually orthogonal by Theorem 8.3.6. Hence $V = \textcircled{\perp}_{i=1}^{k} E_i$.

Obviously $\bigoplus_{j\neq i} E_j \subseteq E_i^{\perp}$. But, by Theorem 8.1.26, $\dim E_i^{\perp} = \dim V - \dim E_i$. On the other hand, $\dim \bigoplus_{j\neq i} E_j = \sum_{j\neq i} \dim E_j = \dim V - \dim E_i$. Therefore $\bigoplus_{j\neq i} E_j = E_i^{\perp}$.

To show (8.67), let $\mathbf{x} \in V$. By (8.66), we can write $\mathbf{x} = \mathbf{x}_1 + \mathbf{x}_2 + \cdots + \mathbf{x}_k$ ($\mathbf{x}_i \in E_i$). Then $P_i(\mathbf{x}) = \mathbf{x}_i$ and $P_iP_i(\mathbf{x}) = \mathbf{x}_i = P_i(\mathbf{x})$ as before and $P_iP_j(\mathbf{x}) = P_i(\mathbf{x}_j) = 0$ for $i \neq j$. Hence we have $P_iP_j = \delta_{ij}P_i$.

For (8.68), let $\mathbf{x} = \mathbf{x}_1 + \mathbf{x}_2 + \cdots + \mathbf{x}_k$ as above. Then $P_i(\mathbf{x}) = \mathbf{x}_i$. Hence

$$\begin{aligned}(P_1 + P_2 + \cdots + P_k)(\mathbf{x}) &= P_1(\mathbf{x}) + P_2(\mathbf{x}) + \cdots + P_k(\mathbf{x}) \\ &= \mathbf{x}_1 + \mathbf{x}_2 + \cdots + \mathbf{x}_k = \mathbf{x}\end{aligned}$$

Therefore $P_1 + P_2 + \cdots + P_k = I$.

To see (8.69), let $\mathbf{x} = \mathbf{x}_1 + \mathbf{x}_2 + \cdots + \mathbf{x}_k$ as above. Then

$$\begin{aligned}T(\mathbf{x}) &= T(\mathbf{x}_1) + T(\mathbf{x}_2) + \cdots + T(\mathbf{x}_k) \\ &= \lambda_1\mathbf{x}_1 + \lambda_2\mathbf{x}_2 + \cdots + \lambda_k\mathbf{x}_k \\ &= \lambda_1P_1(\mathbf{x}) + \lambda_2P_2(\mathbf{x}) + \cdots + \lambda_kP_k(\mathbf{x})\end{aligned}$$

Thus $\lambda_1P_1 + \lambda_2P_2 + \cdots + \lambda_kP_k = T$

The statements about the matrix follows from the previous results by letting $T = T_A$. □

Definition 8.3.15. *The set of eigenvalues $\lambda_1, \lambda_2, \ldots, \lambda_k$ of* Theorem 8.3.14 *are called the* **spectrum of T**. *The representation of the identity map as a sum of orthogonal projections* (8.68) *is called a* **resolution of the identity** *and the expression of T as a linear combinations of orthogonal projections in* (8.69) *is called a* **spectral decomposition of T**.

Let $g(t) \in \mathbb{C}[t]$ and $T = \lambda_1P_1 + \lambda_2P_2 + \cdots + \lambda_kP_k$ be a spectral decomposition of T. Then by orthogonality (8.67) of P_i we have (Exercise 17)

$$g(T) = g(\lambda_1)P_1 + g(\lambda_2)P_2 + \cdots + g(\lambda_k)P_k \tag{8.71}$$

If $A \in M(n, \mathbb{C})$ is normal then, by Exercise 8.3.12, there is a polynomial $f(t) \in \mathbb{C}[t]$ such that $A^* = f(A)$. From this we can deduce a corresponding statement for linear operators but we will give a proof of this fact by using Spectral Theorem.

Corollary 8.3.16. *Let T be a linear operator on a finite dimensional inner product space V whose characteristic polynomial splits. Then T is normal if and only if $T^* = g(T)$ for some $g \in F[t]$.*

Proof. If $T^* = g(T)$ for some polynomial $g(t) \in F[t]$ then we have $Tg(T) = g(T)T$.

Conversely assume T is normal and the characteristic polynomial splits. Let $T = \lambda_1P_1 + \lambda_2P_2 + \cdots + \lambda_kP_k$ be the spectral decomposition of T. Then since P_i is an orthogonal projection we have $P_i^* = P_i$ (Example 8.2.12). Hence, by Theorem 8.2.15, we get

$$T^* = \overline{\lambda}_1P_1^* + \overline{\lambda}_2P_2^* + \cdots + \overline{\lambda}_kP_k^* = \overline{\lambda}_1P_1 + \overline{\lambda}_2P_2 + \cdots + \overline{\lambda}_kP_k$$

By Lagrange interpolation (Theorem 3.4.16), there is $g(t) \in F[t]$ of degree $\leq k$ such that $g(\lambda_i) = \overline{\lambda}_i$. Then we have

$$g(T) = g(\lambda_1)P_1 + g(\lambda_2)P_2 + \cdots + g(\lambda_k)P_k = \overline{\lambda}_1P_1 + \overline{\lambda}_2P_2 + \cdots + \overline{\lambda}_kP_k = T^*$$

As required. □

Even if the characteristic polynomial of T does not split over the field $\mathbb{R}$ of reals Corollary 8.3.16 is still true (Exercise 18).

In the following corollary, we assume the characteristic polynomial splits. We will show, in Theorem 8.4.1, that the assumption is not necessary.

Corollary 8.3.17. *Let V be a finite dimensional inner product space and $T \in L(V)$ with the splitting characteristic polynomial. Then T is self adjoint if and only if T is normal with the real eigenvalues $\lambda_1, \lambda_2, \ldots, \lambda_k$.*

Proof. Write $T = \lambda_1 P_1 + \lambda_2 P_2 + \cdots + \lambda_k P_k$ by Spectral Theorem. Then $T^* = \overline{\lambda}_1 P_1 + \overline{\lambda}_2 P_2 + \cdots + \overline{\lambda}_k P_k$. Hence $T = T^*$ if and only if $\lambda_i = \overline{\lambda}_i$; $\lambda_i \in \mathbb{R}$. $\square$

Corollary 8.3.18. *Let $T = \lambda_1 P_1 + \lambda_2 P_2 + \cdots + \lambda_k P_k$ as in* Theorem 8.3.14. *Then $P_j = g_j(T)$ for some $g_j \in F[t]$ $(j = 1, 2, ..., k)$.*

Proof. By Lagrange interpolation (Theorem 3.4.16), choose $g_j \in F[t]$ so that $g_j(\lambda_i) = \delta_{ij}$. Then $g_j(T) = g_j(\lambda_1)P_1 + g_j(\lambda_2)P_2 + \cdots + g_j(\lambda_k)P_k = P_j$. $\square$

Corollary 7.2.13 gives a similar result in terms of projections instead of orthogonal projections.

Example 8.3.19. *Let $V = \mathbb{C}^3$ with $\langle \mathbf{u}, \mathbf{v} \rangle_A = \mathbf{u}^* A \mathbf{v}$ where $A = \begin{bmatrix} 1 & 0 & 0 \\ 0 & 1 & -i \\ 0 & i & 2 \end{bmatrix}$ as in* Example 8.1.7. *And let $T : V \to V$ be defined by $T(a, b, c) = \begin{bmatrix} 2a \\ 2b - (1+2i)c \\ ic \end{bmatrix}$ as in* Example 8.2.13. *We want to find the spectral decomposition of T.*

Let $\mathcal{B}_1 = \{\mathbf{v}_1 = \mathbf{e}^1, \mathbf{v}_2 = \mathbf{e}^2\}$ and $\mathcal{B}_2 = \{\mathbf{v}_3 = i\mathbf{e}^2 + \mathbf{e}^3\}$ be orthonormal bases for the eigenspaces E_1 and E_2 corresponding to $\lambda_1 = 2, \lambda_2 = i$ in Example 8.3.12. *Let P_i be the orthogonal projections onto E_i. Then, for $\mathbf{v} = \begin{bmatrix} a \\ b \\ c \end{bmatrix} \in V$, we have*

$$P_1(\mathbf{v}) = \langle \mathbf{v}_1, \mathbf{v} \rangle_A \mathbf{v}_1 + \langle \mathbf{v}_2, \mathbf{v} \rangle_A \mathbf{v}_2 = \begin{bmatrix} a \\ b - ic \\ 0 \end{bmatrix}$$

$$P_2(\mathbf{v}) = \langle \mathbf{v}_3, \mathbf{v} \rangle_A \mathbf{v}_3 = \begin{bmatrix} 0 \\ ic \\ c \end{bmatrix}$$

Hence $P_1 + P_2 = I$ and $T = 2P_1 + iP_2$.

Let's express T^* as a polynomial in T i.e., find $g \in \mathbb{C}[t]$ such that $T^* = g(T)$. We need to find a polynomial $g(t)$ degree ≤ 1 such that $g(\lambda_i) = \bar{\lambda}_i$. Let $g(t) = at + b$ and $g(2) = 2a + b = 2$ and $g(i) = ia + b = -i$. Solving this we have $g(t) = (\frac{3+4i}{5})t + (\frac{4-8i}{5})$. Hence

$$T^* = \left(\frac{3+4i}{5}\right) T + \left(\frac{4-8i}{5}\right) I$$

8.3.4 Canonical Form for Normal Operator

The next theorem identifies the canonical form for the normal operator over the field $\mathbb{R}$ of reals. When the characteristic polynomial splits the canonical form for normal operators is given in (8.70) of Theorem 8.3.14.

Theorem 8.3.20. *Let T be a normal operator on a finite dimensional inner product space V over $\mathbb{R}$. Suppose the characteristic polynomial $f(t)$ of T is of the form*

$$f(t) = \pm \prod_{i=1}^{l}(t-\lambda_i) \prod_{k=l+1}^{s} \left(t^2 - 2a_k t + (a_k^2 + b_k^2)\right) \tag{8.72}$$

where $\lambda_i \in \mathbb{R}$ and $a_k \pm ib_k \in \mathbb{C}$ ($b_k \neq 0$) are the roots of $f(t)$. Then there is an orthonormal basis $\mathcal{B}$ of V such that $[T]_{\mathcal{B}}$ is of the form

$$[T]_{\mathcal{B}} = \begin{bmatrix} \lambda_1 & & & & & & & & & \\ & \ddots & & & & & & & & \\ & & \lambda_l & & & & & & & \\ & & & a_1 & b_1 & & & & & \\ & & & -b_1 & a_1 & & & & & \\ & & & & & a_2 & b_2 & & & \\ & & & & & -b_2 & a_2 & & & \\ & & & & & & & \ddots & & \\ & & & & & & & & a_m & b_m \\ & & & & & & & & -b_m & a_m \end{bmatrix} \tag{8.73}$$

Let $A \in M(n,\mathbb{R})$. Then A is normal if and only if there is an orthogonal matrix Q such that Q^tAQ is of the form (8.73).

Proof. By Theorem C.2.3, a monic real polynomial having the real roots λ_i and nonreal complex roots $a_k + b_k i$ ($a_k, b_k \in \mathbb{R}, b_k \neq 0$) and its complex conjugate $a_k - b_k i$ is of the form $\prod_{\lambda_i\in\mathbb{R}}(t-\lambda_i)\prod_{b_k\neq 0}(t^2 - 2a_k t + (a_k^2+b_k^2))$ which is of the form (8.72).

To get the canonical form (8.73), represent T as a matrix with respect an orthonormal basis and we assume $T = T_A : \mathbb{R}^n \to \mathbb{R}^n$ for some $A \in M(n,\mathbb{R})$ satisfying $A^tA = AA^t$. And we need to find a matrix $Q \in M(n,\mathbb{R})$ whose columns are orthonormal such that Q^tAQ is of the form (8.73).

If A is similar to the matrix of the form (8.73) then it is easy to show that A is normal (Cf. Example 8.3.5).

For the converse, we consider the complexification $T^{\mathbb{C}} : \mathbb{C}^n \to \mathbb{C}^n$ of T (Exercise 6.3.16) i.e., we view T as a linear operator on $\mathbb{C}^n$. Then $T^{\mathbb{C}}$ is also normal and, by Theorem 8.3.6, we have $V = \oplus E_{\lambda_k}$ ($\lambda_k \in \mathbb{C}$) and E_{λ_k} are mutually orthogonal. We will consider real and nonreal complex eigenvalues.

Suppose $\lambda = \lambda_j \in \mathbb{R}$ is a real eigenvalue of A with multiplicity m. By Theorem 8.3.9, we have $\nu(T^{\mathbb{C}} - \lambda I) = m$ and by Example 6.7.22 we see that $\nu(T - \lambda I) = m$. Hence we can choose m orthonormal basis consisting of eigenvectors in $\mathbb{R}^n$ for λ. Let $\mathcal{B}_{re} = \{\mathbf{w}_k\}$ be the union of orthonormal bases for E_{λ_j} ($\lambda_j \in \mathbb{R}$). Then $\mathcal{B}_{re}$ is an orthonormal basis for the direct sum $\oplus_k\{E_{\lambda_j} \mid \lambda_j \text{ is real }\}$.

Now suppose $\lambda_k = a + bi$ is a nonreal eigenvalue of $T^{\mathbb{C}}$ then $\bar{\lambda}_k$ is also an eigenvalue of $T^{\mathbb{C}}$. Let $\mathcal{B}_{\lambda_k} = \{\mathbf{u}_p\}$ be a orthonormal basis of E_{λ_k} in $\mathbb{C}^n$. If $\mathbf{u} \in \mathcal{B}_{\lambda_k}$ then by taking the conjugate of $A\mathbf{u} = \lambda_k\mathbf{u}$ we have $A\bar{\mathbf{u}} = \bar{\lambda}_k\bar{\mathbf{u}}$ since $A \in M(n,\mathbb{R})$ (See Definition 6.2.24 for the notation.) And $\mathbf{u} \perp \bar{\mathbf{u}}$ since they are eigenvectors corresponding to the distinct eigenvalues of a normal operator.

By Example 6.7.22, we have $\dim_{\mathbb{C}} E_{\lambda_k} = \dim_{\mathbb{C}} E_{\bar{\lambda}_k}$ and $\mathcal{B}_{\bar{\lambda}_k} = \{\bar{\mathbf{u}}_p\}$ is a $\mathbb{C}$-linearly independent set of vectors (Exercise 6.2.6). And they are also orthonormal since so are the vectors $\{\mathbf{u}_p\}$. Therefore $\mathcal{B}_{\bar{\lambda}_k} = \{\bar{\mathbf{u}}_p\}$ is an orthonormal basis of $E_{\bar{\lambda}_k}$.

Let $\mathbf{u}_p \in \mathcal{B}_{\lambda_k}$ and let $W_p = \mathrm{Span}_{\mathbb{C}}\{\mathbf{u}_p, \bar{\mathbf{u}}_p\}$. Then since $\{\mathbf{u}_p, \bar{\mathbf{u}}_p\}$ is $\mathbb{C}$-linearly independent $\dim_{\mathbb{C}} W_p = 2$. If $p \neq q$ then $W_p \perp W_q$ and $E_{\lambda_k} \bigoplus E_{\bar{\lambda}_k} = \bigoplus_p W_p$. Since $\mathbf{u}_p \perp \bar{\mathbf{u}}_p$, we see that $(\mathbf{u}_p)_{re} \perp (\mathbf{u}_p)_{im}$ and $\|(\mathbf{u}_p)_{re}\| = \|(\mathbf{u}_p)_{im}\|$ by Exercise 8.1.2.

Let $W'_p = \mathrm{Span}_{\mathbb{R}}\{(\mathbf{u}_p)_{re}, (\mathbf{u}_p)_{im}\}$. Then $W'_p \subseteq W_p$ and hence $W'_p \perp W'_q$ for $p \neq q$. We have then, since $A\mathbf{u}_p = \lambda_k\mathbf{u}_p$ and $\lambda_k = a + bi$ we have

$$\begin{aligned} A\left((\mathbf{u}_p)_{re}\right) &= \frac{1}{2}A(\mathbf{u}_p - \bar{\mathbf{u}}_p) = \frac{1}{2}((a+bi)\mathbf{u}_p + (a-bi)\bar{\mathbf{u}}_p) \\ &= \frac{1}{2}(a(\mathbf{u}_p + \bar{\mathbf{u}}_p) + ib(\mathbf{u}_p - \bar{\mathbf{u}}_p)) = a(\mathbf{u}_p)_{re} - b(\mathbf{u}_p)_{im} \end{aligned} \tag{8.74}$$

Similarly $A(\mathbf{u}_p)_{im} = b(\mathbf{u}_p)_{re} + a(\mathbf{u}_p)_{re}$. Hence W'_p is T-invariant. Since $1 = \|\mathbf{u}_p\|^2 = \|(\mathbf{u}_p)_{re}\|^2 + \|(\mathbf{u}_p)_{im}\|^2$ and $\|(\mathbf{u}_p)_{re}\| = \|(\mathbf{u}_p)_{im}\|$ we have $\|(\mathbf{u}_p)_{re}\| = \|(\mathbf{u}_p)_{im}\| = \frac{1}{\sqrt{2}}$. Let $\mathbf{v}_p = \sqrt{2}(\mathbf{u}_p)_{re}, \mathbf{v}'_p = \sqrt{2}(\mathbf{u}_p)_{im}$. Then $\mathcal{B}_p := \{\mathbf{v}_p, \mathbf{v}'_p\}$ are orthonormal basis for W'_p. Now by the computation (8.74) we have

$$A\mathbf{v}_p = a\mathbf{v}_p - b\mathbf{v}'_p, \quad A\mathbf{v}'_p = b\mathbf{v}_p + a\mathbf{v}'_p$$

Let T_p be the restriction of T to W'_p. Then we have

$$[T_p]_{\mathcal{B}_p} = \begin{bmatrix} a & b \\ -b & a \end{bmatrix} := A_p \tag{8.75}$$

Now it is obvious that the union $\mathcal{B} = \mathcal{B}_{re} \cup (\cup_p \mathcal{B}_p)$ forms an orthonormal basis of $\mathbb{R}^n$ and the matrix representation of T with respect to $\mathcal{B}$ consists of real λ_i's and A_p's along the diagonal as in (8.73).

Let $Q \in M(n,\mathbb{R})$ be the matrix consisting of the columns of the vectors $\mathcal{B}_{re} = \{\mathbf{w}_k\}$ and $\mathbf{v}_p, \mathbf{v}'_p$ for all k, p then Q is an orthogonal matrix and it is clear that Q^tAQ is the desired form. □

Example 8.3.21. *Let* $A = \begin{bmatrix} 1 & -1 & 0 & -1 \\ -1 & 1 & -1 & 0 \\ -1 & 0 & 1 & -1 \\ 0 & -1 & -1 & 1 \end{bmatrix}$. *Then* $AA^t = \begin{bmatrix} 3 & -2 & 0 & 0 \\ -2 & 3 & 0 & 0 \\ 0 & 0 & 3 & -2 \\ 0 & 0 & -2 & 3 \end{bmatrix} = A^tA$. *Hence A is normal. The characteristic polynomial of A is $f(t) = (t^2-1)(t^2-4t+5)$ and the eigenvalues are $\lambda_1 = 1, \lambda_2 = -1, \lambda_3 = 2+i, \bar{\lambda}_3 = 2-i$. Then* $A\begin{bmatrix} 1 \\ 1 \\ 1 \\ 1 \end{bmatrix} = -\begin{bmatrix} 1 \\ 1 \\ 1 \\ 1 \end{bmatrix}$, $A\begin{bmatrix} 1 \\ 1 \\ -1 \\ -1 \end{bmatrix} = \begin{bmatrix} 1 \\ 1 \\ -1 \\ -1 \end{bmatrix}$. *Hence* $\mathbf{w}_1 = \frac{1}{2}\begin{bmatrix} 1 \\ 1 \\ 1 \\ 1 \end{bmatrix}, \mathbf{w}_2 = \frac{1}{2}\begin{bmatrix} 1 \\ 1 \\ -1 \\ -1 \end{bmatrix}$ *are orthonormal eigenvectors for λ_1, λ_2.*

Unit eigenvectors corresponding to $\lambda_3, \bar{\lambda}_3$ *are* $\mathbf{u}_3 = \frac{1}{2}\begin{bmatrix}1\\-1\\i\\-i\end{bmatrix}$, $\bar{\mathbf{u}}_3 = \frac{1}{2}\begin{bmatrix}1\\-1\\-i\\i\end{bmatrix}$. *Let* $W_3 =$ $\mathrm{Span}_{\mathbb{C}}\{\mathbf{u}_3, \bar{\mathbf{u}}_3\}$. *Then* $(\mathbf{u}_3)_{re} = \frac{1}{2}\begin{bmatrix}1\\-1\\0\\0\end{bmatrix}$, $(\mathbf{u}_3)_{im} = \frac{1}{2}\begin{bmatrix}0\\0\\1\\-1\end{bmatrix}$ *and* $W_3' = \mathrm{Span}\{(\mathbf{u}_3)_{re}, (\mathbf{u}_3)_{im}\}$.
Then we have $\|(\mathbf{u}_3)_{re}\| = \|(\mathbf{u}_3)_{im}\| = \frac{1}{\sqrt{2}}$.

Let $\mathbf{v}_3 = \sqrt{2}(\mathbf{u}_3)_{re} = \frac{1}{\sqrt{2}}\begin{bmatrix}1\\-1\\0\\0\end{bmatrix}$ *and* $\mathbf{v}_3' = \sqrt{2}(\mathbf{u}_3)_{im} = \frac{1}{\sqrt{2}}\begin{bmatrix}0\\0\\1\\-1\end{bmatrix}$. *Let* $\mathcal{B}_3 = \{\mathbf{v}_3, \mathbf{v}_3'\}$.
Then $\mathcal{B}_3$ *is an orthonormal basis for* W_3' *and* $A\mathbf{v}_3 = 2\mathbf{v}_3 - \mathbf{v}_3'$, $A\mathbf{v}_3' = \mathbf{v}_3 + 2\mathbf{v}_3'$. *The matrix representation of* T_A *restricted to* W_3' *is* $\begin{bmatrix}2 & 1\\-1 & 2\end{bmatrix}$.

Now $\mathcal{B} = \{\mathbf{w}_1, \mathbf{w}_2, \mathbf{v}_3, \mathbf{v}_3'\}$ *is an orthonormal basis for* $V = \mathbb{R}^4$ *and*

$$[T_A]_{\mathcal{B}} = \begin{bmatrix}-1 & 0 & 0 & 0\\0 & 1 & 0 & 0\\0 & 0 & 2 & 1\\0 & 0 & -1 & 2\end{bmatrix}$$

If we take $Q = \frac{1}{2}\begin{bmatrix}1 & 1 & \sqrt{2} & 0\\1 & 1 & -\sqrt{2} & 0\\1 & -1 & 0 & \sqrt{2}\\1 & -1 & 0 & -\sqrt{2}\end{bmatrix}$ *then we have* $Q^tAQ = [T_A]_{\mathcal{B}}$.

8.3 Exercises

1. For each of the following linear operator $T \in L(V)$ determine whether T is normal, self adjoint or neither.

   (a) $V = \mathbb{R}^2$ with the standard inner product and T is defined by $T(a,b) = \begin{bmatrix}2a-b\\-a+b\end{bmatrix}$.

   (b) $V = P_2(\mathbb{R})$ with $\langle f, g\rangle = \int_0^1 fg dx$ and T is defined by $T(f) = f'$.

   (c) $V = P_2(\mathbb{R})$ with $\langle f, g\rangle = \int_0^1 fg dx$ and T is defined by $T(f(t)) = f(2t-3)$.

   (d) $V = M(2,\mathbb{R})$ with $\langle A, B\rangle = \mathrm{Tr}(B^tA)$ and T is defined by $T(A) = A^t$.

   (e) $V = M(2,\mathbb{R})$ with $\langle A, B\rangle = \mathrm{Tr}(B^*A)$ and T is defined by $T(A) = \begin{bmatrix}d & a\\b & c\end{bmatrix}$.

2. Let $A, B \in M(n,\mathbb{C})$. Prove that if A and B are unitarily equivalent and A is normal then B is also normal. State and prove a similar statement for $\mathbb{R}$ instead of $\mathbb{C}$.

3. Prove the true statement(s) and find a counter example for the false statement(s).

(a) The matrix $\begin{bmatrix} -2 & 1 \\ 1 & 2 \end{bmatrix}$ is normal.

(b) If T is a normal operator then so is T^*.

(c) Let $A, B \in M(n, F)$. If A and B are normal then AB is also normal.

(d) Let $A, B \in M(n, F)$. If A and B are similar and A is normal then B is also normal.

(e) A linear operator $T \in L(V)$ is normal if and only if the matrix $[T]_{\mathcal{B}}$ is normal for any basis $\mathcal{B}$.

(f) A self adjoint operator is normal.

(g) If $A \in M(n, \mathbb{C})$ and $A^t = A$ then A is normal.

(h) If $A \in M(n.\mathbb{R})$ is orthogonally equivalent to $B \in M(n, \mathbb{R})$ and A is normal then so is B.

4. Prove that $A \in M(n, F)$ is normal (resp. self adjoint, resp. Hermitian) if and only if T_A is normal (resp. self adjoint, resp. Hermitian).

5. Prove that a nilpotent normal operator on a finite dimensional inner product space is the zero operator.

6. Let T be a normal operator on an inner product space V. Prove that $T(\mathbf{v}) = \mathbf{0}$ if and only if $T^*(\mathbf{v}) = \mathbf{0}$.

7. Let T be a normal operator on an inner product space V over F and $f(t) \in \mathbb{R}[t]$.

   (a) Prove that T^{-1} is also normal if T is invertible.

   (b) Prove that $f(T)$ is a normal operator on V.

   (c) Prove that the eigenvalues of $f(T)$ are $f(\lambda_1), f(\lambda_2), ..., f(\lambda_n)$ if $\lambda_1, \lambda_2, ..., \lambda_n$ are the eigenvalues of T.

8. Let V be a finite dimensional vector space over $\mathbb{C}$. Let $S, T \in L(V)$ be normal operators. Prove that $ST = TS$ if and only if there is an orthonormal basis $\mathcal{B}$ such that $[S]_{\mathcal{B}}, [T]_{\mathcal{B}}$ are both diagonal in the following steps.

   (a) Let $E_\lambda^S = \{\mathbf{v} \in V \mid S(\mathbf{v}) = \lambda \mathbf{v}\}$. Prove that E_λ^S is T-invariant.

   (b) Prove that there is an orthonormal basis $\mathcal{B}$ such that both $[S]_{\mathcal{B}}, [T]_{\mathcal{B}}$ are both diagonal. (Hint: Theorem 8.3.13)

   (c) Prove that $ST = TS$ if and only if there is an orthonormal basis $\mathcal{B}$ such that $[S]_{\mathcal{B}}, [T]_{\mathcal{B}}$ are both diagonal.

9. Let T be a normal operator on an inner product space V over $\mathbb{C}$. Prove that for a positive integer n there is a normal linear operator S on V such that $S^n = T$.

10. Let $A = \begin{bmatrix} -1 & 0 & 2 & 0 \\ 0 & -1 & 0 & 2 \\ 0 & 2 & -1 & 0 \\ 2 & 0 & 0 & -1 \end{bmatrix}$. Show that A is normal. And find an orthonormal basis $\mathcal{B}$ of $V = \mathbb{R}^4$ and the matrix representation $[T_A]_{\mathcal{B}}$ which is of the form (8.73).

11. Prove that if $A \in M(n, \mathbb{R})$ is normal and its characteristic polynomial split then A is symmetric.

12. Let $A \in M(n, \mathbb{C})$ be a normal matrix. Prove that there is a polynomial $f(t) \in \mathbb{C}[t]$ of degree $< n$ such that $f(A) = A^*$. (Hint: What $f(t)$ would be if A is actually diagonal? Use Lagrange interpolation.)

13. Let T be a normal linear operator on a finite dimensional inner product space over $\mathbb{C}$. Let W be a T invariant subspace of V.

    (a) Prove that W is also T^*-invariant.

    (b) Prove that $W^\perp$ is also T-invariant.

    (c) Prove that $(T_W)^* = (T^*)_W$ namely the adjoint of the restriction to W is the same as the restriction of the adjoint.

    (d) Prove that if $\mathcal{B}$ is an orthonormal basis of W consisting of the eigenvectors of T then $\mathcal{B}$ can be extended to an orthonormal basis $\mathcal{B}'$ of V consisting of eigenvectors of T. (Hint: Use (b).)

14. Show that the following matrices are normal and find their spectral decomposition namely find the eigenvalues and orthonormal bases of the eigenspaces. Decide whether it is just normal, Hermitian or unitary.

$$A = \begin{bmatrix} 1+i & 0 & -1+i \\ 0 & 2 & 0 \\ -1+i & 0 & 1+i \end{bmatrix}, \quad B = \begin{bmatrix} 1 & 0 & i \\ 0 & 1 & 0 \\ -i & 0 & 1 \end{bmatrix}, \quad C = \begin{bmatrix} \frac{1}{\sqrt{2}} & \frac{1}{\sqrt{2}} & 0 \\ \frac{1}{\sqrt{2}} & -\frac{1}{\sqrt{2}} & 0 \\ 0 & 0 & 1 \end{bmatrix}$$

15. Let T be a projection (i.e., $T^2 = T$) on an inner product space V. Prove that T is an orthogonal projection if and only if T is normal.

16. Let T be a linear operator on a finite dimensional inner product space V.

    (a) Prove that if T is an orthogonal projection on W then $\|T(\mathbf{x})\| \le \|\mathbf{x}\|$.

    (b) Find an example of a projection (i.e., $T^2 = T$) for which the above inequality fails.

    (c) Prove that a projection T satisfying $\|T(\mathbf{x})\| = \|\mathbf{x}\|$ must be the identity.

    (d) Prove that a projection T satisfying $\|T(\mathbf{x})\| \le \|\mathbf{x}\|$ must be an orthogonal projection.

17. Let T be a normal operator on an inner product space V with a splitting characteristic polynomial. Let $T = \lambda_1 P_1 + \lambda_2 P_2 + \cdots + \lambda_k P_k$ be the spectral decomposition of T.

    (a) For $g(t) \in F[t]$ prove that $g(T) = g(\lambda_1)P_1 + g(\lambda_2)P_2 + \cdots + g(\lambda_k)P_k$.

    (b) Let S be a linear operator on V. Prove that $ST = TS$ if and only if $SP_i = P_iS$ for all i.

    (c) Prove that T is an orthogonal projection if and only if the eigenvalues are 0 or 1.

18. Prove that Corollary 8.3.16 is true even if the characteristic polynomial does not split. Namely, show that T is normal if and only if $T^* = f(T)$ for some $f \in \mathbb{R}[t]$ in the following steps.

    (a) Let $A = \begin{bmatrix} a & -b \\ b & a \end{bmatrix}$. Prove that there is $f(x) \in \mathbb{R}[x]$ of degree 1 such that $f(A) = A^t$. (Hint: Diagonalize A over $\mathbb{C}$ and observe that the Lagrange polynomial f with $f(e^{i\theta}) = e^{-i\theta}$ and $f(e^{-i\theta}) = e^{i\theta}$ has real coefficients.)

    (b) Let $a_1, a_2, ..., a_r, a'_1, a'_2, ..., a'_r$ be real numbers and $b_1, b_2, ..., b_s$ be non real complex numbers. Then the Lagrange polynomial $f(t)$ such that

    $$f(x) = \begin{cases} a'_i & (x = a_i) \\ \bar{b}_i & (x = b_i) \\ b_i & (x = \bar{b}_i) \end{cases}$$

    has real coefficients.

    (c) Use Theorem 8.3.20 to prove that if T is a normal operator on a real inner product space then there is $f(x) \in \mathbb{R}[x]$ such that $T^* = f(T)$.

8.4 Self Adjoint and Unitary Operators

In this section we consider two kinds of normal operators – the self adjoint operators *viz* operators T satisfying $T = T^*$ and unitary operators *viz* operators T satisfying $TT^* = I$.

8.4.1 Self Adjoint and Positive Operators

Theorem 8.4.1. *Let V be a finite dimensional inner product space over F and T be a linear operator on V. Then T is self adjoint if and only if there is an orthonormal basis $\mathcal{B}$ consisting of eigenvectors with real eigenvalues namely*

$$[T]_{\mathcal{B}} = \text{diag}[\lambda_1, \lambda_2, ..., \lambda_n] \text{ with } \lambda_i \in \mathbb{R} \tag{8.76}$$

In particular, the characteristic polynomial of a self adjoint operator splits over $\mathbb{R}$.

Proof. First assume V has an orthonormal basis $\mathcal{B}$ for which $[T]_{\mathcal{B}}$ is of the form (8.76). Then, by Theorem 8.2.10, we have $[T^*]_{\mathcal{B}} = [T]^*_{\mathcal{B}} = \text{diag}[\lambda_1.\lambda_2, ..., \lambda_n]^* = \text{diag}[\lambda_1.\lambda_2, ..., \lambda_n] = [T]_{\mathcal{B}}$ since $\lambda_i \in \mathbb{R}$. Hence $T^* = T$.

Conversely assume T is self adjoint. Let $\mathcal{C}$ be an orthonormal basis of V and let $A = [T]_{\mathcal{C}} \in M(n, F)$. Now since T is self adjoint, $A^* = [T_{\mathcal{C}}]^* = [T^*]_{\mathcal{C}} = [T]_{\mathcal{C}} = A$ by Theorem 8.2.10. Hence $A = A^*$. Since the coordinate map $\phi_{\mathcal{C}}$ is an isometric isomorphism (Corollary 8.1.21) we need to prove that if A is self adjoint then T_A has a basis of orthonormal eigenvectors with real eigenvalues. Even if $A \in M(n, \mathbb{R})$ we will view A as an $n \times n$ complex matrix i.e., $T_A : \mathbb{C}^n \to \mathbb{C}^n$.

We can write the characteristic polynomial of A as $f(t) = \pm \prod(t - \lambda_i)$ $(\lambda_i \in \mathbb{C})$ by the fundamental theorem of algebra (Theorem C.2.1). Since T_A is normal there is an orthonormal basis $\mathcal{B} = \{\mathbf{u}_1, \mathbf{u}_2, ..., \mathbf{u}_n\}$ consisting of eigenvectors with the eigenvalues $\{\lambda_1, \lambda_2, ..., \lambda_n\}$; $T_A(\mathbf{u}_i) = \lambda_i \mathbf{u}_i$ $(\lambda_i \in \mathbb{C})$. By Theorem 8.3.6 (3), we have $\bar{\lambda}_i \mathbf{u}_i = (T_A)^*(\mathbf{u}_i) = T_{A^*}(\mathbf{u}_i) = T_A(\mathbf{u}_i) = A\mathbf{u}_i = \lambda_i \mathbf{u}_i$. Hence we conclude that $\bar{\lambda}_i = \lambda_i$ i.e., $\lambda_i \in \mathbb{R}$. □

In Theorem 8.4.1, the scalar F can be the field $\mathbb{R}$ of reals or the field $\mathbb{C}$ of complex numbers. In the proof of Theorem 8.4.1, when $F = \mathbb{R}$ the operator T may have a complex eigenvalue λ and we cannot talk about the eigenvector for λ in this case. To make the proof go through we viewed V as an inner product space over $\mathbb{C}$ (Exercise 6.3.16, Exercise 8.1.23). This is what we actually did and for that we translate the statements in terms of operators into the statements in terms of matrices. See Exercise 3 for a proof using complexification.

Now we restate Theorem 8.4.1 for the field $\mathbb{R}$ of reals and the field $\mathbb{C}$ of complex numbers separately in the next two corollaries.

Corollary 8.4.2. *Let V be an inner product space over $\mathbb{R}$ and T be a linear operator on V. The following conditions are equivalent.*

(1) *The linear operator T is self adjoint.*

(2) *The linear operator T is normal and the characteristic polynomial splits over $\mathbb{R}$.*

(3) *There is an orthonormal basis consisting of eigenvectors with real eigenvalues.*

Further, if $A \in M(n, \mathbb{R})$ then A is symmetric if and only if there is an orthogonal matrix $P \in M(n, \mathbb{R})$ such that $P^t AP$ is a (real) diagonal matrix. In particular, a real symmetric matrix is diagonalizable.

Proof. (1)⇒(2) If T is self adjoint then T is normal. And by Theorem 8.4.1, the linear operator T is diagonalizable over $\mathbb{R}$. Therefore, the characteristic polynomial split over $\mathbb{R}$ (Theorem 6.7.19).

(2)⇒(3) Follows from Theorem 8.3.9.

(3)⇒(1) Let $\mathcal{B}$ be an orthonormal basis consisting of eigenvectors with real eigenvalues $\lambda_1, \lambda_2, ..., \lambda_n$. Then $[T]_{\mathcal{B}} = \text{diag}[\lambda_1, \lambda_2, ..., \lambda_n]$ with $\lambda_i \in \mathbb{R}$. Hence $[T^*]_{\mathcal{B}} = [T]^*_{\mathcal{B}} = \text{diag}[\lambda_1, \lambda_2, ..., \lambda_n] = [T]_{\mathcal{B}}$. Thus $T^* = T$.

For the matrix A we choose the matrix P with the columns of the orthonormal basis consisting of the eigenvectors. Then P is orthogonal since the columns are orthonormal. □

Observe that if A is real symmetric then A is normal. Hence there is an orthogonal matrix Q such that $Q^{-1}AQ$ is of the form (8.73). Since the characteristic polynomial of A split by Corollary 8.4.2, there are no 2×2 blocks of the form (8.75) in the matrix (8.73) but only the real numbers along the diagonal.

Next corollary is Theorem 8.4.1 with $F = \mathbb{C}$ and we omit the proof.

Corollary 8.4.3. *Let V be an inner product space over $\mathbb{C}$ and T be a linear operator on V. The following conditions are equivalent.*

(1) *The linear operator T is Hermitian.*

(2) *The linear operator T is normal and the characteristic polynomial splits over $\mathbb{R}$.*

(3) *There is an orthonormal basis consisting of eigenvectors with real eigenvalues.*

Further, if $A \in M(n, \mathbb{C})$ then A is Hermitian if and only if there is a unitary matrix $Q \in M(n, \mathbb{C})$ such that $Q^* AQ$ is a real diagonal matrix. In particular, the characteristic polynomial of a Hermitian matrix split over the reals $\mathbb{R}$.

The linear operator T in Example 8.3.12 is normal but not Hermitian since it has a nonreal eigenvalue.

Corollary 8.4.4. *Let T be a self adjoint operator on a finite dimensional inner product space V. If $\langle \mathbf{v}, T(\mathbf{v})\rangle = 0$ for all $\mathbf{v} \in V$ then $T = 0$.*

Proof. Using Theorem 8.4.1 choose an orthonormal basis $\mathcal{B}$ consisting of eigenvectors. For $\mathbf{v}_i \in \mathcal{B}$ let $T(\mathbf{v}_i) = \lambda_i \mathbf{v}_i$. Then by (8.4),

$$0 = \langle \mathbf{v}_i, T(\mathbf{v}_i)\rangle = \langle \mathbf{v}_i, \lambda_i \mathbf{v}_i\rangle = \lambda_i \langle \mathbf{v}_i, \mathbf{v}_i\rangle = \lambda_i$$

Hence $\lambda_i = 0$ for all i. Hence $[T]_{\mathcal{B}} = 0; T = 0$. □

If $F = \mathbb{C}$ then the conclusion of Corollary 8.4.4 is true without T being self adjoint (Exercise 11). Hence Corollary 8.4.4 is meaningful only when $F = \mathbb{R}$.

Example 8.4.5. *Let* $A = \begin{bmatrix} 1 & -1 & 1 \\ -1 & 1 & -1 \\ 1 & -1 & 1 \end{bmatrix}$. *Then A is real symmetric. Hence by* Corollary 8.4.2, *A is diagonalizable – there is an orthonormal basis consisting of eigenvectors. Now we want to find such a basis. The characteristic polynomial is $f_A(t) = -t^3 + 3t^2$. Hence the eigenvalues are $\lambda_1 = 0$ with multiplicity 2 and $\lambda_2 = 3$ with multiplicity 1. The eigenvectors corresponding to $\lambda_1 = 0$ are $\mathcal{B}_1 = \left\{ \mathbf{w}_1 = \begin{bmatrix} 1 \\ 0 \\ -1 \end{bmatrix}, \mathbf{w}_2 = \begin{bmatrix} 0 \\ 1 \\ 1 \end{bmatrix} \right\}$ and the eigenvector for $\lambda_2 = 3$ is $\mathcal{B}_2 = \left\{ \begin{bmatrix} 1 \\ -1 \\ 1 \end{bmatrix} \right\}$. Notice that $\mathcal{B}_1$ is not orthogonal. We need to apply Gram Schmidt process to make them orthogonal: $\mathbf{v}_1 = \mathbf{w}_1, \mathbf{v}_2 = \mathbf{w}_2 - \frac{\langle \mathbf{v}_1, \mathbf{w}_2\rangle}{\langle \mathbf{v}_1, \mathbf{v}_1\rangle}\mathbf{v}_1 = \frac{1}{2}\begin{bmatrix} 1 \\ 2 \\ 1 \end{bmatrix}$. To get the orthonormal basis normalize the length of the vectors:*

$$\mathcal{B}_1' = \left\{ \frac{1}{\sqrt{2}}\begin{bmatrix} 1 \\ 0 \\ -1 \end{bmatrix}, \frac{1}{\sqrt{6}}\begin{bmatrix} 1 \\ 2 \\ 1 \end{bmatrix} \right\}; \quad \mathcal{B}_2' = \left\{ \frac{1}{\sqrt{3}}\begin{bmatrix} 1 \\ -1 \\ 1 \end{bmatrix} \right\}$$

Hence if we let $P = \frac{1}{\sqrt{6}}\begin{bmatrix} \sqrt{3} & 1 & \sqrt{2} \\ 0 & 2 & -\sqrt{2} \\ -\sqrt{3} & 1 & \sqrt{2} \end{bmatrix}$ *then we have* $P^{-1}AP = P^tAP = \begin{bmatrix} 0 & & \\ & 0 & \\ & & 3 \end{bmatrix}$.

Now consider a Hermitian matrix.

Example 8.4.6. *Let* $A = \begin{bmatrix} 1 & i & 0 & 0 \\ -i & 1 & 0 & 0 \\ 0 & 0 & 1 & -i \\ 0 & 0 & i & 1 \end{bmatrix}$. *Then A is Hermitian and the characteristic polynomial is $f(t) = t^4 - 4t^3 + 4t^2 = t^2(t-2)^2$. The eigenvalues are $\lambda_1 = 0, \lambda_2 = 2$ with* *multiplicities 2. The unit eigenvectors for* $\lambda_1 = 0$ *are* $\mathbf{v}_1 = \frac{1}{\sqrt{2}}\begin{bmatrix}1\\-i\\0\\0\end{bmatrix}, \mathbf{v}_2 = \frac{1}{\sqrt{2}}\begin{bmatrix}0\\0\\1\\i\end{bmatrix}$.

And the unit eigenvectors for $\lambda_2 = 2$ *are* $\mathbf{v}_3 = \frac{1}{\sqrt{2}}\begin{bmatrix}1\\i\\0\\0\end{bmatrix}, \mathbf{v}_4 = \frac{1}{\sqrt{2}}\begin{bmatrix}0\\0\\1\\-i\end{bmatrix}$. *Hence* $\mathcal{B} = \{\mathbf{v}_1, \mathbf{v}_2, \mathbf{v}_3, \mathbf{v}_4\}$ *is an orthonormal basis consisting of the eigenvectors of* A *such that*

$$[T_A]_{\mathcal{B}} = \begin{bmatrix}0&0&0&0\\0&0&0&0\\0&0&2&0\\0&0&0&2\end{bmatrix}$$

If we let

$$Q = \frac{1}{\sqrt{2}}\begin{bmatrix}1&0&1&0\\-i&0&i&0\\0&1&0&1\\0&i&0&-i\end{bmatrix}$$

then Q *is a unitary matrix such that* $Q^*AQ = [T_A]_{\mathcal{B}} = \mathrm{diag}[0,0,2,2]$.

Next we consider the so called positive matrices which can be used to define an inner product as in Example 8.1.7.

Definition 8.4.7. *A self adjoint linear operator* T *on a finite dimensional inner product space* V *over* F *is* **positive definite** *(resp.* **semidefinite***) if* $\langle \mathbf{x}, T(\mathbf{x})\rangle > 0$ *(resp.* ≥ 0*)for all* $\mathbf{x} \neq 0$. *We say* $A \in M(n, F)$ *is* **positive definite** *or* **positive semidefinite** *if* T_A *has the corresponding property. We will denote a positive definite (resp. semidefinite) linear operator and a matrix by* $T > 0, A > 0$ *(resp.* $T \geq 0, A \geq 0$*).*

For example, if T is a linear operator then T^*T, TT^* are positive semidefinite. In fact, $\langle \mathbf{x}, T^*T(\mathbf{x})\rangle = \langle T(\mathbf{x}), T(\mathbf{x})\rangle \geq 0$.

Let $A = \begin{bmatrix}1&0&0\\0&1&-i\\0&i&2\end{bmatrix}$. Then $A^* = A$ and we showed that $\langle \mathbf{u}, A\mathbf{u}\rangle = \mathbf{u}^*A\mathbf{u} > 0$ in Example 8.1.7. Hence $A > 0$.

Theorem 8.4.8. *Let* T *be a linear operator on an* n*-dimensional inner product space* V *with an inner product* $\langle\ ,\ \rangle$. *Let* $\mathcal{C}$ *be an orthonormal basis of* V *and* $A = [T]_{\mathcal{C}} \in M(n, F)$. *Then* $T > 0$ *(resp.* ≥ 0*) if and only if* $A > 0$ *(resp.* ≥ 0*).*

Further, the following conditions are equivalent.

(1) $T > 0$

(2) *There is an orthonormal basis of eigenvectors of* T *with positive eigenvalues.*

(3) *There is a self adjoint invertible* $S \in L(V)$ *such that* $T = S^*S$.

(4) *The equality* $\langle \mathbf{u}, \mathbf{v}\rangle_T := \langle \mathbf{u}, T(\mathbf{v})\rangle$ *defines an inner product on* V.

For the semidefinite case, the following conditions are equivalent.

(1)′ $T \geq 0$

(2)′ *There is an orthonormal basis consisting of eigenvectors of T with nonnegative eigenvalues.*

(3)′ *There is a self adjoint $S \in L(V)$ such that $T = S^*S$.*

Proof. Let $\mathcal{C} = \{\mathbf{v}_i\}$ be an orthonormal basis. By Theorem 8.2.10, we have $A^* = [T]^*_{\mathcal{C}} = [T^*]_{\mathcal{C}}$. Hence we see that $T = T^*$ if and only if $A = A^*$. We need to show $\langle \mathbf{v}, T(\mathbf{v})\rangle > 0$ if and only if $\langle \mathbf{x}, A\mathbf{x}\rangle > 0$. Let $\mathbf{v} \in V$ and $\mathbf{v} = \sum a_i \mathbf{v}_i, \mathbf{x} = [\mathbf{v}]_{\mathcal{C}}$ be the $\mathcal{C}$-coordinate of $\mathbf{v}$. Now we have

$$\begin{aligned}\langle \mathbf{v}, T(\mathbf{v})\rangle &= \left\langle \sum a_i\mathbf{v}_i, \sum a_j T(\mathbf{v}_j)\right\rangle = \sum_i \bar{a}_i \left\langle \mathbf{v}_i, \sum a_j \left(\sum A_{kj}\mathbf{v}_k\right)\right\rangle \\ &= \sum_i \bar{a}_i \left\langle \mathbf{v}_i, \sum_{i,j} a_j A_{kj}\mathbf{v}_k \right\rangle = \sum \bar{a}_i A_{ij} a_j = \langle \mathbf{x}, A\mathbf{x}\rangle\end{aligned}$$

Hence $T > 0$ or $T \geq 0$ if and only if $A > 0$ or $A \geq 0$.

Now we turn to the proof of equivalences of the conditions.

(1) $\Rightarrow$ (2) By Theorem 8.4.1, we can choose an orthonormal basis $\mathcal{B} = \{\mathbf{v}_1, \mathbf{v}_2, \ldots, \mathbf{v}_n\}$ consisting of eigenvectors corresponding to the eigenvalues $\lambda_1, \lambda_2, \ldots, \lambda_n$ which are real. Then we have

$$0 < \langle \mathbf{v}_i, T(\mathbf{v}_i)\rangle = \lambda_i \langle \mathbf{v}_i, \mathbf{v}_i\rangle = \lambda_i$$

(2) $\Rightarrow$ (3) Let $\mathcal{B}$ be an orthonormal basis such that $[T]_{\mathcal{B}} = \text{diag}[\lambda_1, \lambda_2, \ldots, \lambda_n]$ with $\lambda_i > 0$. By Corollary 6.4.22, there is an $S \in L(V)$ such that $[S]_{\mathcal{B}} = \text{diag}[\sqrt{\lambda_1}, \sqrt{\lambda_2}, \ldots, \sqrt{\lambda_n}]$ which is an isomorphism since $\sqrt{\lambda_i} \neq 0$. Then we have $[S^*]_{\mathcal{B}} = [S]^*_{\mathcal{B}} = [S]_{\mathcal{B}}$ and hence $S^* = S$. And $[S^*S]_{\mathcal{B}} = [S]^2_{\mathcal{B}} = \text{diag}[\lambda_1, \lambda_2, ..., \lambda_n] = [T]_{\mathcal{B}}$. Therefore $T = S^*S$.

(3) $\Rightarrow$ (4) Since $T = S^*S$ we have $T^* = (S^*S)^* = S^*S = T$. To show conjugate symmetry we compute: $\langle \mathbf{u}, \mathbf{v}\rangle_T = \langle \mathbf{u}, T(\mathbf{v})\rangle = \langle T^*(\mathbf{u}), \mathbf{v}\rangle = \langle T(\mathbf{u}), \mathbf{v}\rangle = \overline{\langle \mathbf{v}, T(\mathbf{u})\rangle} = \overline{\langle \mathbf{v}, \mathbf{u}\rangle_T}$. Now $\langle \mathbf{v}, \mathbf{v}\rangle_T = \langle \mathbf{v}, T(\mathbf{v})\rangle = \langle \mathbf{v}, S^*S(\mathbf{v})\rangle = \langle S(\mathbf{v}), S(\mathbf{v})\rangle > 0$ for $\mathbf{v} \neq \mathbf{0}$. The other conditions for $\langle\ ,\ \rangle_T$ to be an inner product are easy to check.

(4) $\Rightarrow$ (1) First we show that T is Hermitian. Let $\mathcal{C} = \{\mathbf{v}_i\}$ be an orthonormal basis and $A = [T]_{\mathcal{C}}$. Since $\langle\ ,\ \rangle_T$ is an inner product we have $\langle \mathbf{v}_i, \mathbf{v}_j\rangle_T = \overline{\langle \mathbf{v}_j, \mathbf{v}_i\rangle_T}$. Now $\langle \mathbf{v}_i, \mathbf{v}_j\rangle_T = \langle \mathbf{v}_i, T(\mathbf{v}_j)\rangle = \sum A_{ki}\langle \mathbf{v}_i, \mathbf{v}_k\rangle = A_{ij}$ and $\overline{\langle \mathbf{v}_j, \mathbf{v}_i\rangle_T} = \overline{\langle \mathbf{v}_j, T(\mathbf{v}_i)\rangle} = \sum \bar{A}_{ki}\overline{\langle \mathbf{v}_j, \mathbf{v}_k\rangle} = \bar{A}_{ji}$ since $\mathcal{C} = \{\mathbf{v}_i\}$ is an orthonormal basis. Hence A is Hermitian and, by Theorem 8.2.10, T is also Hermitian. It is obvious that $T > 0$.

The equivalences of (1)′ – (3)′ are easy to check by inspecting the proofs above. □

Incidentally, the proof of (2) $\Rightarrow$ (3) shows that if $T \geq 0$ then there is a unique $S \geq 0$ such that $S^2 = T$.

Example 8.4.9. *In* Example 8.1.7, *we let $V = \mathbb{C}^3$ and $\langle \mathbf{u}, \mathbf{v}\rangle_A = \mathbf{u}^*A\mathbf{v}$ for $\mathbf{u}, \mathbf{v} \in V$ where $A = \begin{bmatrix} 1 & 0 & 0 \\ 0 & 1 & -i \\ 0 & i & 2 \end{bmatrix}$. Then A is Hermitian and the characteristic polynomial is $f(t) = -(t-1)(t^2-3t+1)$ and the eigenvalues of A are $1, \frac{1}{2}(3 \pm \sqrt{5})$ which are positive. Hence $\langle \mathbf{u}, \mathbf{v}\rangle_A = \mathbf{u}^*A\mathbf{v}$ defines an inner product on $\mathbb{C}^3$ as it should be.*

Theorem 8.4.10. *Let $[\ ,\ ]$ be an inner product on $V = F^n$. Then there is a unique positive $A \in M(n, F)$ such that $[\mathbf{u}, \mathbf{v}] = \langle \mathbf{u}, A\mathbf{v}\rangle$ where $\langle\ ,\ \rangle$ is the standard inner product on F^n.*

Proof. Let $A_{ij} = [\mathbf{e}^i, \mathbf{e}^j]$. We will show that $[\mathbf{u}, \mathbf{v}] := \langle \mathbf{u}, A\mathbf{v}\rangle$ and $A > 0$. In fact, let $\mathbf{u} = \sum a_i \mathbf{e}^i, \mathbf{v} = \sum b_i \mathbf{e}^i$ then

$$[\mathbf{u}, \mathbf{v}] = \left[\sum a_i \mathbf{e}^i, \sum b_i \mathbf{e}^i\right] = \sum \bar{a}_i b_j [\mathbf{e}^i, \mathbf{e}^j] = \sum_{i,j} \bar{a}_i A_{ij} b_j = \langle \mathbf{u}, A\mathbf{v}\rangle$$

Hence $[\mathbf{u}, \mathbf{v}] = \langle \mathbf{u}, A\mathbf{v}\rangle$. On the other hand, $A_{ji} = [\mathbf{e}^j, \mathbf{e}^i] = \overline{[\mathbf{e}^i, \mathbf{e}^j]} = \bar{A}_{ij}$ and hence $A^* = A$. Now $\langle \mathbf{v}, A\mathbf{v}\rangle = [\mathbf{v}, \mathbf{v}] > 0$ since $[\ ,\ ]$ is an inner product on V. Thus A is positive. Uniqueness of A is clear. □

8.4.2 Unitary and Orthogonal Operators

Next we study the norm preserving operators T say $\|T(\mathbf{v})\| = \|\mathbf{v}\|$ which we call an isometry.

Theorem 8.4.11. *Let $T : V \to V$ be a linear operator on a finite dimensional inner product space V over F. Then the following conditions are equivalent:*

(1) $TT^* = T^*T = I$

(2) *For all $\mathbf{x}, \mathbf{y} \in V$ we have $\langle T(\mathbf{x}), T(\mathbf{y})\rangle = \langle \mathbf{x}, \mathbf{y}\rangle$.*

(3) *If $\mathcal{B}$ is is an orthonormal basis of V then so is $T(\mathcal{B}) = \{T(\mathbf{v}_1), T(\mathbf{v}_2), \ldots, T(\mathbf{v}_n)\}$.*

(4) *For all $\mathbf{x} \in V$ we have $\|T(\mathbf{x})\| = \|\mathbf{x}\|$.*

Further, if T satisfies one of the equivalence conditions then the eigenvalues have absolute value 1.

Proof. (1)⇒(2) Since $T^*T = I$ we have $\langle T(\mathbf{x}), T(\mathbf{y})\rangle = \langle \mathbf{x}, T^*T(\mathbf{y})\rangle = \langle \mathbf{x}, \mathbf{y}\rangle$.

(2)⇒(3) Let $\mathcal{B} = \{\mathbf{v}_1, \mathbf{v}_2, \ldots, \mathbf{v}_n\}$ be an orthonormal basis. Then $\langle T(\mathbf{v}_i), T(\mathbf{v}_j)\rangle = \langle \mathbf{v}_i, \mathbf{v}_j\rangle = \delta_{ij}$. Hence $T(\mathcal{B}) = \{T(\mathbf{v}_1), T(\mathbf{v}_2), \ldots, T(\mathbf{v}_n)\}$ is an orthonormal basis.

(3)⇒(4) Let $\mathcal{B} = \{\mathbf{v}_1, \mathbf{v}_2, \ldots, \mathbf{v}_n\}$ be an orthonormal basis and let $\mathbf{x} = \sum a_i \mathbf{v}_i$. Then

$$\|\mathbf{x}\|^2 = \langle \textstyle\sum a_i \mathbf{v}_i, \sum a_j \mathbf{v}_j\rangle = \sum |a_i|^2$$

On the other hand,

$$\|T(\mathbf{x})\|^2 = \langle \textstyle\sum a_i T(\mathbf{v}_i), \sum a_j T(\mathbf{v}_j)\rangle = \sum |a_i|^2$$

since $T(\mathcal{B})$ is orthonormal. Hence $\|T(\mathbf{x})\| = \|\mathbf{x}\|$.

(4)⇒(1) For any $\mathbf{x} \in V$, we have

$$\langle \mathbf{x}, \mathbf{x}\rangle = \|\mathbf{x}\|^2 = \|T(\mathbf{x})\|^2 = \langle T(\mathbf{x}), T(\mathbf{x})\rangle = \langle \mathbf{x}, T^*T(\mathbf{x})\rangle$$

Hence $\langle \mathbf{x}, (I - T^*T)(\mathbf{x})\rangle = 0$ for all $\mathbf{x} \in V$. Now since $S = I - T^*T$ is self adjoint we have $S = 0$ by Corollary 8.4.4. Hence $T^*T = I$ i.e., T^* is a left inverse of T. By Theorem 6.4.20, we also have $TT^* = I$.

Finally, let $\lambda \in F$ be an eigenvalue of T with $T(\mathbf{u}) = \lambda\mathbf{u}$. Then, by (1)

$$\langle \mathbf{u}, \mathbf{u}\rangle = \langle T^*T(\mathbf{u}), \mathbf{u}\rangle = \langle T(\mathbf{u}), T(\mathbf{u})\rangle = |\lambda|^2 \langle \mathbf{u}, \mathbf{u}\rangle$$

Hence $|\lambda| = 1$. □

Definition 8.4.12. *Let V be an inner product space over F and T be a linear operator on V satisfying anyone of the equivalent conditions of* Theorem 8.4.11. *If $F = \mathbb{C}$ then we call T a* **unitary operator**. *If $F = \mathbb{R}$ then we call T an* **orthogonal operator**. *We call $A \in M(n, \mathbb{C})$ a* **unitary matrix** *if $A^*A = I_n = AA^*$. We call $A \in M(n, \mathbb{R})$ an* **orthogonal matrix** *if $A^tA = I_n = AA^t$.*

We saw an example of a unitary matrix in Example 8.3.7.

Corollary 8.4.13. *Let V be a finite dimensional inner product space over F. Let T be a linear operator on V with the characteristic polynomial $f(t) = \pm\prod_{l=1}^{n}(t - \lambda_l)$ $(\lambda_l \in F)$. Then T is unitary (resp. orthogonal) if and only if there is an orthonormal basis $\mathcal{B}$ consisting of eigenvectors with the eigenvalues of absolute value 1 i.e., $\lambda_l = e^{i\theta_l}$ for some $\theta_l \in \mathbb{R}$ (resp. $\lambda_l = \pm 1$). In this case, we have*

$$[T]_{\mathcal{B}} = \mathrm{diag}[\lambda_1, \lambda_2, ..., \lambda_n]$$

Let $A \in M(n, F)$ with the splitting characteristic polynomial. Then A is unitary (resp. orthogonal) if and only if there is a unitary (resp. orthogonal) matrix Q such that

$$Q^*AQ = \mathrm{diag}[\lambda_1, \lambda_2, ..., \lambda_n] \tag{8.77}$$

Proof. If T is unitary then obviously T is normal. By Theorem 8.3.9, there is an orthonormal basis consisting of eigenvectors. If λ_l is an eigenvalue with eigenvector $\mathbf{v}_l$ then, by Theorem 8.4.11

$$|\lambda_l|\|\mathbf{v}_l\| = \|\lambda_l \mathbf{v}_l\| = \|T(\mathbf{v}_l)\| = \|\mathbf{v}_l\|$$

Hence $|\lambda_l| = 1$.

For the converse, write $T = \lambda_1 P_1 + \lambda_2 P_2 + \cdots + \lambda_k P_k$ (Theorem 8.3.14) where $\lambda_1, \lambda_2, ..., \lambda_k$ are distinct eigenvalues. Then since $P_l^* = P_l$ and $P_l P_m = \delta_{lm} P_l$ we have

$$\begin{aligned} TT^* &= (\lambda_1 P_1 + \lambda_2 P_2 + \cdots + \lambda_k P_k)(\overline{\lambda}_1 P_1 + \overline{\lambda}_2 P_2 + \cdots + \overline{\lambda}_k P_k) \\ &= |\lambda_1|^2 P_1 + |\lambda_2|^2 P_2 + \cdots + |\lambda_k|^2 P_k = I \end{aligned}$$

since $|\lambda_l| = 1$. Hence T is unitary.

We leave the proof for the orthogonal case as an exercise. □

Example 8.4.14. *Let* $A = \begin{bmatrix} \frac{i}{\sqrt{2}} & 0 & \frac{1}{\sqrt{2}} \\ 0 & -i & 0 \\ \frac{1}{\sqrt{2}} & 0 & \frac{i}{\sqrt{2}} \end{bmatrix}$. *Then* $A^* = \begin{bmatrix} \frac{-i}{\sqrt{2}} & 0 & \frac{1}{\sqrt{2}} \\ 0 & i & 0 \\ \frac{1}{\sqrt{2}} & 0 & \frac{-i}{\sqrt{2}} \end{bmatrix}$ *and* $AA^* = I_3$. *Hence A is unitary. The characteristic polynomial is $f(t) = -t^3 + (\sqrt{2}-1)it^2 - (\sqrt{2}-1)t + i$. The roots of f are $\lambda_1 = -i = e^{\frac{3\pi i}{2}}$, $\lambda_2 = -\frac{-1}{\sqrt{2}} + \frac{i}{\sqrt{2}} = e^{\frac{3\pi i}{4}}$ and $\lambda_3 = \frac{1}{\sqrt{2}} + \frac{i}{\sqrt{2}} = e^{\frac{\pi i}{4}}$. Hence there is an orthonormal basis $\mathcal{B}$ such that*

$$[T_A]_{\mathcal{B}} = \begin{bmatrix} e^{\frac{3\pi i}{2}} & & \\ & e^{\frac{3\pi i}{4}} & \\ & & e^{\frac{\pi i}{4}} \end{bmatrix}$$

The eigenvectors for $\lambda_1, \lambda_2, \lambda_3$ are

$$\mathbf{v}_1 = \begin{bmatrix} 0 \\ 1 \\ 0 \end{bmatrix}, \mathbf{v}_2 = \frac{1}{\sqrt{2}}\begin{bmatrix} -1 \\ 0 \\ 1 \end{bmatrix}, \mathbf{v}_3 = \frac{1}{\sqrt{2}}\begin{bmatrix} 1 \\ 0 \\ 1 \end{bmatrix}$$

which are orthonormal. Hence if we let $Q = [\mathbf{v}_1\ \mathbf{v}_2\ \mathbf{v}_3]$ *then* $Q^{-1}AQ = \mathrm{diag}[\lambda_1, \lambda_2, \lambda_3]$.

The linear map T in Example 8.3.12 is not unitary since it has an eigenvalue whose absolute value $\neq 1$. Next we consider the real orthogonal matrices.

Theorem 8.4.15. *Let T be an orthogonal operator on an inner product space V over $\mathbb{R}$. Then the characteristic polynomial $f(t)$ is of the form*

$$f(t) = \pm(t-1)^a(t+1)^b \prod_{k=1}^{r}(t^2 - 2\cos\theta_k t + 1) \tag{8.78}$$

with the real roots ± 1 and the nonreal complex roots $e^{\pm i\theta_k} \in \mathbb{C}$. And there is an orthonormal basis $\mathcal{B}$ such that $[T]_{\mathcal{B}}$ is of the form;

$$[T]_{\mathcal{B}} = \begin{bmatrix} 1 & & & & & & & & & & \\ & \ddots & & & & & & & & & \\ & & 1 & & & & & & & & \\ & & & -1 & & & & & & & \\ & & & & \ddots & & & & & & \\ & & & & & -1 & & & & & \\ & & & & & & \cos\theta_1 & \sin\theta_1 & & & \\ & & & & & & -\sin\theta_1 & \cos\theta_1 & & & \\ & & & & & & & & \ddots & & \\ & & & & & & & & & \cos\theta_r & \sin\theta_r \\ & & & & & & & & & -\sin\theta_r & \cos\theta_r \end{bmatrix} \tag{8.79}$$

The real 2×2 submatrix

$$A_k := \begin{bmatrix} \cos\theta_k & \sin\theta_k \\ -\sin\theta_k & \cos\theta_k \end{bmatrix} \tag{8.80}$$

has non-real eigenvalues $\cos\theta_k \pm i\sin\theta_k = e^{\pm i\theta_k} \in \mathbb{C}$.

Let $A \in M(n, \mathbb{R})$. Then A is a real orthogonal matrix if and only if there is an orthogonal matrix Q such that Q^tAQ is of the form (8.79).

Proof. As in the proof of Theorem 8.4.1, we may assume $T = T_A$ for some orthogonal matrix A. We view A as a unitary matrix and hence the eigenvalue λ of A has absolute value 1. Hence any eigenvalue λ is of the form $\lambda = \pm 1$ for the real eigenvalues and $\lambda = e^{i\theta}$ for the nonreal eigenvalues (Theorem C.2.3). Since the characteristic polynomial $f(t)$ of T has real coefficients if $e^{i\theta}$ is a nonreal root of f then so is its conjugate $e^{-i\theta}$. Thus the the characteristic polynomial has the factors of the form $(t - e^{i\theta})(t - e^{-i\theta}) = t^2 - 2(\cos\theta)t + 1$ and hence the characteristic polynomial $f(t)$ is of the form (8.78). Since T is normal, by Theorem 8.3.20, the matrix is of the form (8.79). □

Example 8.4.16. *Consider* $A = \dfrac{1}{2}\begin{bmatrix} 1 & 1 & 1 & 1 \\ 1 & 1 & -1 & -1 \\ 1 & -1 & -1 & 1 \\ 1 & -1 & 1 & -1 \end{bmatrix}$. *Then A is orthogonal and symmetric.*

The characteristic polynomial is $f(t) = (t-1)^2(t+1)^2$; the eigenvalues are $\lambda_1 = -1, \lambda_2 = 1$ with the multiplicities 2. Hence there is an orthonormal basis $\mathcal{B}$ such that

$$[T_A]_{\mathcal{B}} = \begin{bmatrix} -1 & & & \\ & -1 & & \\ & & 1 & \\ & & & 1 \end{bmatrix}$$

Let E_1, E_2 be the eigenspaces for λ_1, λ_2. We choose $\mathbf{w}_1 = \begin{bmatrix} -1 \\ 1 \\ 0 \\ 2 \end{bmatrix}, \mathbf{w}_2 = \begin{bmatrix} -1 \\ 1 \\ 2 \\ 0 \end{bmatrix}$ *for a basis of E_1 and choose* $\mathbf{w}_3 = \begin{bmatrix} 2 \\ 0 \\ 1 \\ 1 \end{bmatrix}, \mathbf{w}_4 = \begin{bmatrix} 1 \\ 1 \\ 0 \\ 0 \end{bmatrix}$ *for a basis of E_2. We need to find an orthonormal basis for E_1, E_2: We apply Gram Schmidt:* $\mathcal{B}_1 = \left\{ \mathbf{v}_1 = \frac{1}{\sqrt{6}} \begin{bmatrix} -1 \\ 1 \\ 0 \\ 2 \end{bmatrix}, \mathbf{v}_2 = \frac{1}{2\sqrt{3}} \begin{bmatrix} -1 \\ 1 \\ 3 \\ -1 \end{bmatrix} \right\}$ *and* $\mathcal{B}_2 = \left\{ \mathbf{v}_3 = \frac{1}{2} \begin{bmatrix} 1 \\ -1 \\ 1 \\ 1 \end{bmatrix}, \mathbf{v}_4 = \frac{1}{\sqrt{2}} \begin{bmatrix} 1 \\ 1 \\ 0 \\ 0 \end{bmatrix} \right\}$ *are orthonormal bases for E_1, E_2. If we let $Q = [\mathbf{v}_1\ \mathbf{v}_2\ \mathbf{v}_3\ \mathbf{v}_4]$ then $Q^t AQ = [T_A]_{\mathcal{B}}$.*

Example 8.4.17. *Consider* $A = \frac{1}{2} \begin{bmatrix} 1 & 1 & 1 & 1 \\ 1 & 1 & -1 & -1 \\ 1 & -1 & -1 & 1 \\ -1 & 1 & -1 & 1 \end{bmatrix}$. *Then A is orthogonal and the characteristic polynomial is $f(t) = (t^2 - 1)(t^2 - t + 1)$. The eigenvalues are $\pm 1, \frac{1 \pm \sqrt{3}i}{2} = e^{\frac{\pm \pi i}{3}}$. Hence there is an orthonormal basis $\mathcal{B}$ such that*

$$[T_A]_{\mathcal{B}} = \begin{bmatrix} -1 & & & \\ & 1 & & \\ & & \cos\frac{\pi}{3} & \sin\frac{\pi}{3} \\ & & -\sin\frac{\pi}{3} & \cos\frac{\pi}{3} \end{bmatrix} = \begin{bmatrix} -1 & & & \\ & 1 & & \\ & & \frac{1}{2} & \frac{\sqrt{3}}{2} \\ & & -\frac{\sqrt{3}}{2} & \frac{1}{2} \end{bmatrix}$$

To find an orthonormal basis we follow the process given in the proof of Theorem 8.3.20*: Eigenvectors for $\lambda = -1, \lambda = 1$ are* $\mathbf{v}_1 = \frac{1}{\sqrt{6}} \begin{bmatrix} -1 \\ 1 \\ 2 \\ 0 \end{bmatrix}, \mathbf{v}_2 = \frac{1}{\sqrt{2}} \begin{bmatrix} 1 \\ 1 \\ 0 \\ 0 \end{bmatrix}$. *If we let* $\mathbf{u} = \frac{1}{\sqrt{3}} \begin{bmatrix} -i \\ i \\ -i \\ \sqrt{3} \end{bmatrix}$ *then $\mathbf{u}, \bar{\mathbf{u}}$ are the unit eigenvectors for $\lambda = \frac{1+\sqrt{3}i}{2}$ and $\bar{\lambda} = \frac{1-\sqrt{3}i}{2}$. And the normalized real and imaginary parts of $\mathbf{u}$ are* $\mathbf{v} = \begin{bmatrix} 0 \\ 0 \\ 0 \\ 1 \end{bmatrix}, \mathbf{v}' = \frac{1}{\sqrt{3}} \begin{bmatrix} -1 \\ 1 \\ -1 \\ 0 \end{bmatrix}$. *We have $A\mathbf{v} = \frac{1}{2}\mathbf{v} - \frac{\sqrt{3}}{2}\mathbf{v}'$ and $A\mathbf{v}' = \frac{\sqrt{3}}{2}\mathbf{v} + \frac{1}{2}\mathbf{v}'$. Letting*

$$Q = [\mathbf{v}_1\ \mathbf{v}_2\ \mathbf{v}\ \mathbf{v}'] = \frac{1}{\sqrt{6}} \begin{bmatrix} -1 & \sqrt{3} & 0 & -\sqrt{2} \\ 1 & \sqrt{3} & 0 & \sqrt{2} \\ 2 & 0 & 0 & -\sqrt{2} \\ 0 & 0 & \sqrt{6} & 0 \end{bmatrix}$$

we have $Q^t AQ = [T_A]_{\mathcal{B}}$.

8.4.3 Singular and Polar Decomposition

In this subsection, we decompose $A \in M(m\times n, F)$ as a product of the form $A = PSQ^*$ where P, Q are unitary matrices of size m, n respectively and $S \in M(m\times n, F)$ is of the form

$$S_{ij} = \begin{cases} s_i \geq 0 & (i = j) \\ 0 & (i \neq j) \end{cases} \tag{8.81}$$

When A is a square matrix a slight modification of the singular decomposition leads to a decomposition of the form $A = QS$ where Q is unitary and S is positive semidefinite. This is similar to the factorization of a complex number into the form $z = |z|e^{i\theta}$ where $e^{i\theta}$ corresponds to the unitary part and $|z|$ corresponds to the positive semidefinite part.

Recall, from Exercise 8.2.12, the adjoint of a linear map $T : V \to W$ of inner product spaces V, W with the inner products $\langle\ ,\ \rangle_V, \langle\ ,\ \rangle_W$. There is a unique linear map $T^* : W \to V$ satisfying

$$\langle \mathbf{w}, T(\mathbf{v})\rangle_W = \langle T^*(\mathbf{w}), \mathbf{v}\rangle_V$$

And T^*T is self adjoint linear operators on V. By Exercise 8.2.12 (h) and Theorem 8.4.8, we see that the eigenvalues are nonnegative say, $\lambda_1, \lambda_2, ..., \lambda_r$ are positive eigenvalues.

Theorem 8.4.18 (Singular Value Theorem). *Let $T : V \to W$ be a linear map of inner product spaces V, W of dimensions n, m. If $\rho(T) = r$ then there are orthonormal bases $\mathcal{B}, \mathcal{C}$ of V, W and uniquely determined positive real numbers $s_1 \geq s_2 \geq \cdots \geq s_r > 0$ such that*

$$[T]_{\mathcal{B}}^{\mathcal{C}} = \begin{bmatrix} s_1 & & & & & \\ & s_2 & & & & \\ & & \ddots & & & \\ & & & s_r & & \\ & & & & 0 & \\ & & & & & \ddots \\ & & & & & & 0 \end{bmatrix} \in M(m\times n) \tag{8.82}$$

Further, we can take s_i to be the square root of the eigenvalues λ_i of $T^*T \geq 0$.

If $A \in M(m\times n, \mathbb{C})$ then there are unitary matrices $P \in M(n, \mathbb{C})$ and $Q \in M(m, \mathbb{C})$ such that

$$A = Q\Sigma P^* \tag{8.83}$$

where $\Sigma \in M(m\times n, \mathbb{R})$ is of the form (8.82). Here $s_i = \sqrt{\lambda_i}$ $(1 \leq i \leq r)$ where $\lambda_1, \lambda_2, ..., \lambda_r$ are positive eigenvalues for A^*A and the columns of $P = [\mathbf{v}_1\ \mathbf{v}_2 \cdots \mathbf{v}_n]$ consists of the orthonormal eigenvectors of A^*A. The first r columns of Q are $\mathbf{w}_i = \dfrac{1}{s_i}A\mathbf{v}_i$ $(i = 1, 2, ..., r)$ and the rest of the $m-r$ columns $\mathbf{w}_{r+1}, ..., \mathbf{w}_m$ are chosen so that $\{\mathbf{w}_1, ..., \mathbf{w}_r, \mathbf{w}_{r+1}, ..., \mathbf{w}_m\}$ form an orthonormal basis for $\mathbb{C}^m$.

Proof. Since T^*T is positive semidefinite and $\rho(T) = \rho(T^*T) = r$ (Exercise 8.2.12) there is an orthonormal eigenvectors $\mathcal{B} = \{\mathbf{v}_1, \mathbf{v}_2, ..., \mathbf{v}_n\}$ of T^*T say $T^*T(\mathbf{v}_i) = \lambda_i$ with $\lambda_1 \geq \lambda_2 \geq \cdots \geq \lambda_r > 0$ and $\lambda_{r+1} = \cdots = \lambda_n = 0$ by Theorem 8.4.8.

Let $s_i = \sqrt{\lambda_i}$ and $\mathbf{w}_i = \dfrac{1}{s_i}T(\mathbf{v}_i)$ $(i = 1, 2, ..., r)$. We claim that $\{\mathbf{w}_1, \mathbf{w}_2, ..., \mathbf{w}_r\}$ is orthonormal. In fact,

$$\langle \mathbf{w}_i, \mathbf{w}_j\rangle_W = \left\langle \frac{1}{s_i}T(\mathbf{v}_i), \frac{1}{s_j}T(\mathbf{v}_j)\right\rangle_W = \frac{1}{s_i s_j}\left\langle T^*T(\mathbf{v}_i), \mathbf{v}_j\right\rangle_V = \frac{s_i^2}{s_i s_j}\langle \mathbf{v}_i, \mathbf{v}_j\rangle_V = \delta_{ij}$$

Now extend $\{\mathbf{w}_1, \mathbf{w}_2, ..., \mathbf{w}_r\}$ to an orthonormal basis $\mathcal{C} = \{\mathbf{w}_1, \mathbf{w}_2, ..., \mathbf{w}_m\}$ of W. Now it is obvious that $[T]_{\mathcal{B}}^{\mathcal{C}}$ is of the form (8.82).

Now we need to check uniqueness of s_i. For this, let $\mathcal{B}' = \{\mathbf{v}_1', \mathbf{v}_2', ..., \mathbf{v}_n'\}$, $\mathcal{C}' = \{\mathbf{w}_1', \mathbf{w}_2', ..., \mathbf{w}_m'\}$ be orthonormal bases of V, W such that $T(\mathbf{v}_i') = s_i'\mathbf{w}_i'$ with $s_1' \geq \cdots \geq s_r' > 0$ and $s_{r+1}' = \cdots = s_n' = 0$. We claim that $\mathbf{v}_i'$ is an eigenvector of T^*T with the eigenvalue $(s_i')^2$. (Then we will have $s_i = s_i'$ since they are the eigenvalues of T^*T.) To see this we compute:

$$\langle \mathbf{v}_j', T^*(\mathbf{w}_i')\rangle_V = \langle T(\mathbf{v}_j'), \mathbf{w}_i'\rangle_W = \langle s_j'\mathbf{w}_j', \mathbf{w}_i'\rangle_W = s_j'\delta_{ij}$$

Hence $T^*(\mathbf{w}_i') = s_i'\mathbf{v}_i'$ by the property of inner product (8.7). Therefore, we have

$$T^*T(\mathbf{v}_i') = T^*(s_i'\mathbf{w}_i') = (s_i')^2\mathbf{v}_i'$$

Thus $(s_i')^2$ is an eigenvalue for T^*T and hence $s_i = s_i'$.

For the matrix A, we let $T = T_A$ in the first part. Then there are orthonormal bases $\mathcal{B}, \mathcal{C}$ of $\mathbb{C}^n, \mathbb{C}^m$ such that $[T_A]_{\mathcal{B}}^{\mathcal{C}}$ is of the form (8.82). Let $\mathcal{E}^n, \mathcal{E}^m$ be the standard bases for $\mathbb{C}^n, \mathbb{C}^m$. Then we have $[I_m]_{\mathcal{C}}^{\mathcal{E}^m}[T_A]_{\mathcal{B}}^{\mathcal{C}} = [T_A]_{\mathcal{B}}^{\mathcal{E}^m} = [T_A]_{\mathcal{E}^n}^{\mathcal{E}^m}[I_n]_{\mathcal{B}}^{\mathcal{E}^n}$ by Theorem 6.4.26. If we let $P = [I_n]_{\mathcal{B}}^{\mathcal{E}^n}, Q = [I_m]_{\mathcal{C}}^{\mathcal{E}^m}$ then we have the desired decomposition. □

Definition 8.4.19. *The diagonal entries $s_1, s_2, ..., s_r, 0, ..., 0$ of* (8.82) *are called the* **singular values** *of T or A when $T = T_A$. The decomposition* (8.83) *is called the* **singular value decomposition** *of A.*

Example 8.4.20. *Let $A = \begin{bmatrix} 1 & 0 & -1 \\ 0 & 1 & 1 \end{bmatrix}$. Then $A^* = \begin{bmatrix} 1 & 0 \\ 0 & 1 \\ -1 & 1 \end{bmatrix}$, $A^*A = \begin{bmatrix} 1 & 0 & -1 \\ 0 & 1 & 1 \\ -1 & 1 & 2 \end{bmatrix}$. The eigenvalues of A^*A are $\lambda_1 = 3, \lambda_2 = 1, \lambda_3 = 0$ with the corresponding orthonormal eigenvectors $\mathbf{v}_1 = \frac{1}{\sqrt{6}}\begin{bmatrix} 1 \\ -1 \\ -2 \end{bmatrix}, \mathbf{v}_2 = \frac{1}{\sqrt{2}}\begin{bmatrix} 1 \\ 1 \\ 0 \end{bmatrix}, \mathbf{v}_3 = \frac{1}{\sqrt{3}}\begin{bmatrix} 1 \\ -1 \\ 1 \end{bmatrix}$. Hence $s_1 = \sqrt{3}$, $s_2 = 1$ and $\Sigma = \begin{bmatrix} \sqrt{3} & 0 & 0 \\ 0 & 1 & 0 \end{bmatrix}$. Since $P = [\mathbf{v}_1\ \mathbf{v}_2\ \mathbf{v}_3]$ we have $P = \frac{1}{\sqrt{6}}\begin{bmatrix} 1 & \sqrt{3} & \sqrt{2} \\ -1 & \sqrt{3} & -\sqrt{2} \\ -2 & 0 & \sqrt{2} \end{bmatrix}$. To get Q we compute $\frac{1}{s_1}T(\mathbf{v}_1) = \frac{1}{3\sqrt{2}}\begin{bmatrix} 1 & 0 & -1 \\ 0 & 1 & 1 \end{bmatrix}\begin{bmatrix} 1 \\ -1 \\ -2 \end{bmatrix} = \frac{1}{\sqrt{2}}\begin{bmatrix} 1 \\ -1 \end{bmatrix}$ and $\frac{1}{s_2}T(\mathbf{v}_2) = \frac{1}{\sqrt{2}}\begin{bmatrix} 1 & 0 & -1 \\ 0 & 1 & 1 \end{bmatrix}\begin{bmatrix} 1 \\ 1 \\ 0 \end{bmatrix} = \frac{1}{\sqrt{2}}\begin{bmatrix} 1 \\ 1 \end{bmatrix}$. Hence $Q = \frac{1}{\sqrt{2}}\begin{bmatrix} 1 & 1 \\ -1 & 1 \end{bmatrix}$. Finally, we obtain the singular decomposition for A:*

$$A = \left(\frac{1}{\sqrt{2}}\begin{bmatrix} 1 & 1 \\ -1 & 1 \end{bmatrix}\right)\begin{bmatrix} \sqrt{3} & 0 & 0 \\ 0 & 1 & 0 \end{bmatrix}\left(\frac{1}{\sqrt{6}}\begin{bmatrix} 1 & -1 & -2 \\ \sqrt{3} & \sqrt{3} & 0 \\ \sqrt{2} & -\sqrt{2} & \sqrt{2} \end{bmatrix}\right)$$

Theorem 8.4.21 (Polar decomposition). *Let T be a linear operator on an inner product space V. Then there are unitary operator U and a uniquely determined positive semidefinite operator $S \geq 0$ such that $T = US$.*

If $A \in M(n, \mathbb{C})$ then there is a unitary matrix U and a uniquely determined positive semidefinite matrix S such that $A = US$. If $A = Q\Sigma P^*$ is a singular decomposition we can choose $U = QP^*$ and $S = P\Sigma P^*$.

If A or T is invertible then the unitary part U is also uniquely determined.

Proof. By representing T in a matrix with respect to an orthonormal basis it suffices to prove the theorem for a matrix $A \in M(n, \mathbb{C})$. Using Theorem 8.4.18 write $A = Q\Sigma P^*$ with unitary P, Q and positive semidefinite Σ.

Write $U = QP^*$, $S = P\Sigma P^*$. Then $A = Q\Sigma P^* = QP^*P\Sigma P^* = US$. Since $UU^* = QP^*PQ^* = I_n$ we see that U is unitary. Now $\langle \mathbf{x}, S\mathbf{x}\rangle = \langle \mathbf{x}, P\Sigma P^*\mathbf{x}\rangle = \langle P^*\mathbf{x}, \Sigma P^*\mathbf{x}\rangle \geq 0$ since $\Sigma \geq 0$. Therefore $S = P\Sigma P^*$ is positive semidefinite.

For uniqueness of S suppose $A = US = U_1S_1$. Then $A^*A = SU^*US = S^2$ and similarly $A^*A = (S_1)^2$. Hence $S^2 = S_1^2$ with $S, S_1 \geq 0$. By Exercise 8.4.9 (c), we have $S = S_1$. If A is invertible then $S > 0$ and hence S is also invertible. Hence U is also uniquely determined by $U = S^{-1}A$. □

Polar decomposition is similar to the factorization of a complex number into a product of a complex number of unit length and a nonnegative real number: $z = e^{i\theta}|z|$.

Example 8.4.22. *Here is a simple example showing how to decompose a matrix into a product of unitary and a positive semidefinite matrix. Let* $A = \begin{bmatrix} 1 & 0 \\ 0 & -2 \end{bmatrix}$. *Then* $A^*A = \begin{bmatrix} 1 & 0 \\ 0 & 4 \end{bmatrix}$. *The eigenvalues of A^*A are $\lambda_1 = 1, \lambda_2 = 4$ with the corresponding orthonormal eigenvectors* $\mathbf{v}_1 = \begin{bmatrix} 1 \\ 0 \end{bmatrix}, \mathbf{v}_2 = \begin{bmatrix} 0 \\ 1 \end{bmatrix}$. *Hence* $P = [\mathbf{v}_1\ \mathbf{v}_2] = \begin{bmatrix} 1 & 0 \\ 0 & 1 \end{bmatrix}$. *And* $s_1 = \sqrt{\lambda_1} = 1, s_2 = \sqrt{\lambda_2} = 2$. *and hence* $\Sigma = \begin{bmatrix} 1 & 0 \\ 0 & 2 \end{bmatrix}$. *To find $Q = [\mathbf{w}_1\ \mathbf{w}_2]$ we let* $\mathbf{w}_1 = \dfrac{A\mathbf{v}_1}{s_1} = \begin{bmatrix} 1 \\ 0 \end{bmatrix}$ *and* $\mathbf{w}_2 = \dfrac{A\mathbf{v}_2}{s_2} = \dfrac{-2\mathbf{v}_2}{2} = \begin{bmatrix} 0 \\ -1 \end{bmatrix}$. *Hence* $Q = [\mathbf{w}_1\ \mathbf{w}_2] = \begin{bmatrix} 1 & 0 \\ 0 & -1 \end{bmatrix}$. *Finally, let* $U = QP^* = \begin{bmatrix} 1 & 0 \\ 0 & -1 \end{bmatrix}$ *which is unitary and let* $S = P\Sigma P^* = \begin{bmatrix} 1 & 0 \\ 0 & 2 \end{bmatrix}$ *which is positive semidefinite. Then* $A = US = \begin{bmatrix} 1 & 0 \\ 0 & -1 \end{bmatrix}\begin{bmatrix} 1 & 0 \\ 0 & 2 \end{bmatrix}$ *is a polar decomposition of A.*

To get the polar decompositions of some reasonably chosen matrix the numerical values get very complex. We have chosen the following example so that the numbers involved are not that bad.

Example 8.4.23. *Let* $A = \begin{bmatrix} 3 & 3 & -3 & -3 \\ 2 & -2 & 2 & -2 \\ -1 & 1 & 1 & -1 \\ 1 & -1 & -1 & 1 \end{bmatrix}$. *Then* $A^*A = \begin{bmatrix} 15 & 3 & -7 & -11 \\ 3 & 15 & -11 & -7 \\ -7 & -11 & 15 & 3 \\ -11 & -7 & 3 & 15 \end{bmatrix}$ *has the characteristic polynomial $f(t) = t^4 - 60t^3 + 992t^2 - 4608t = t(t-8)(t-16)(t-36)$. Let $\lambda_1 = 36, \lambda_2 = 16, \lambda_3 = 8, \lambda_4 = 0$ and $s_1 = 6, s_2 = 4, s_3 = 2\sqrt{2}, s_4 = 0$. Hence*

$$\Sigma = \operatorname{diag}[s_1, s_2, s_3, s_4] = \begin{bmatrix} 6 & & & \\ & 4 & & \\ & & 2\sqrt{2} & \\ & & & 0 \end{bmatrix}$$

The orthonormal eigenvectors corresponding to λ_i are

$$\mathbf{v}_1 = \frac{1}{2}\begin{bmatrix}-1\\-1\\1\\1\end{bmatrix}, \quad \mathbf{v}_2 = \frac{1}{2}\begin{bmatrix}-1\\1\\-1\\1\end{bmatrix}, \quad \mathbf{v}_3 = \frac{1}{2}\begin{bmatrix}1\\-1\\-1\\1\end{bmatrix}, \quad \mathbf{v}_4 = \frac{1}{2}\begin{bmatrix}1\\1\\1\\1\end{bmatrix}$$

Hence

$$P = [\mathbf{v}_1\ \mathbf{v}_2\ \mathbf{v}_3\ \mathbf{v}_4] = \frac{1}{2}\begin{bmatrix}-1 & -1 & 1 & 1\\-1 & 1 & -1 & 1\\1 & -1 & -1 & 1\\1 & 1 & 1 & 1\end{bmatrix}$$

Now we compute $\mathbf{w}_i = \frac{1}{s_i}A\mathbf{v}_i$ $(i = 1, 2, 3)$ and choose $\mathbf{w}_4$ so that $\{\mathbf{w}_1, \mathbf{w}_2, \mathbf{w}_3, \mathbf{w}_4\}$ to be orthonormal

$$\mathbf{w}_1 = \begin{bmatrix}-1\\0\\0\\0\end{bmatrix}, \quad \mathbf{w}_2 = \begin{bmatrix}0\\-1\\0\\0\end{bmatrix}, \quad \mathbf{w}_3 = \begin{bmatrix}0\\0\\-\frac{1}{\sqrt{2}}\\\frac{1}{\sqrt{2}}\end{bmatrix}, \quad \mathbf{w}_4 = \begin{bmatrix}0\\0\\\frac{1}{\sqrt{2}}\\\frac{1}{\sqrt{2}}\end{bmatrix}$$

Hence we can take

$$Q = [\mathbf{w}_1\ \mathbf{w}_2\ \mathbf{w}_3\ \mathbf{w}_4] = \frac{1}{\sqrt{2}}\begin{bmatrix}-\sqrt{2} & 0 & 0 & 0\\0 & -\sqrt{2} & 0 & 0\\0 & 0 & -1 & 1\\0 & 0 & 1 & 1\end{bmatrix}$$

Then we have the singular decomposition of A:

$$A = Q\Sigma P^*$$

For polar decomposition, we let $U = QP^*$ and $S = P\Sigma P^*$

$$U = \frac{1}{2\sqrt{2}}\begin{bmatrix}\sqrt{2} & \sqrt{2} & -\sqrt{2} & -\sqrt{2}\\\sqrt{2} & -\sqrt{2} & \sqrt{2} & -\sqrt{2}\\0 & 2 & 2 & 0\\2 & 0 & 0 & 2\end{bmatrix}, \quad S = \begin{bmatrix}\frac{\sqrt{2}+20}{8} & \frac{-\sqrt{2}+4}{8} & \frac{-\sqrt{2}-4}{8} & \frac{\sqrt{2}-20}{8}\\\frac{-\sqrt{2}+4}{8} & \frac{\sqrt{2}+20}{8} & \frac{\sqrt{2}-20}{8} & \frac{-\sqrt{2}-4}{8}\\\frac{-\sqrt{2}-4}{8} & \frac{\sqrt{2}-20}{8} & \frac{\sqrt{2}+20}{8} & \frac{-\sqrt{2}+4}{8}\\\frac{\sqrt{2}-20}{8} & \frac{-\sqrt{2}-4}{8} & \frac{-\sqrt{2}+4}{8} & \frac{\sqrt{2}+20}{8}\end{bmatrix}$$

Then U is unitary and S is positive semidefinite and $A = US$.

8.4 Exercises

1. (a) Let $A \in M(n, F)$. Prove that A is unitary (resp. orthogonal) if and only if T_A is a unitary (resp. orthogonal) linear operator on F^n.

   (b) Let $A \in M(n, \mathbb{R})$. Prove that A is symmetric if and only if $T_A : \mathbb{R}^n \to \mathbb{R}^n$ is self adjoint.

2. Prove the true statement(s) and find a counter example for the false statement(s).
   (a) The set of orthogonal matrices form a subspace of $M(n, \mathbb{R})$.
   (b) If $A \in M(n, \mathbb{C})$ is unitary then $T_A : \mathbb{C}^n \to \mathbb{C}^n$ is an isomorphism.
   (c) Let $A, B \in M(n, F)$. If A and B are similar and A is symmetric then so is B.
   (d) Let $A, B \in M(n, F)$. If A and B are orthogonally equivalent and A is symmetric then so is B.
   (e) A real symmetric matrix is diagonalizable.
   (f) An orthogonal projection to a subspace is a Hermitian operator.
   (g) If $A \in M(n, \mathbb{C})$ is Hermitian then the characteristic polynomial has real coefficients.
   (h) An orthogonal projection to a subspace is a unitary operator.
   (i) The linear operator T in Example 8.2.13 is unitary.
   (j) The characteristic polynomial of a real symmetric matrix splits.
   (k) Let $A \in M(n, \mathbb{C})$. If $A^t = A$ then A is diagonalizable.
   (l) Let $A \in M(n, \mathbb{C})$. If $A^t = A$ then the eigenvalues of A are real.
   (m) If $T \geq 0$ then so is T^2.
   (n) If $B \in M(n)$ is a symmetric matrix then there is $A \in M(n)$ such that $B = AA^t$.
   (o) If $A \in M(n, \mathbb{R})$ is skew symmetric then the characteristic polynomial splits over $\mathbb{R}$ and the eigenvalues are real.
   (p) A self adjoint operator on a real inner product space is a linear combination of orthogonal projections.
   (q) An orthogonal projection is self adjoint.
   (r) An orthogonal projection is a unitary operator.

3. Prove Theorem 8.4.1 by using complexification of T (Exercise 6.3.16) not using the matrix representation of T.

4. Let

$$A = \begin{bmatrix} 1 & 0 & 0 \\ 0 & 0 & 1 \\ 0 & -1 & 0 \end{bmatrix}, \ B = \begin{bmatrix} i & 0 & 0 \\ 0 & -i & 0 \\ 0 & 0 & 1 \end{bmatrix}, \ C = \begin{bmatrix} 1 & 2 & 0 \\ 0 & 2 & 3 \\ 0 & 0 & 3 \end{bmatrix}, \ D = \begin{bmatrix} 1 & 0 & 1 \\ 0 & 2 & 0 \\ 0 & 0 & 3 \end{bmatrix}, \ E = \begin{bmatrix} 1 & 0 & 0 \\ 0 & 2 & 0 \\ 0 & 0 & 3 \end{bmatrix}$$

   (a) Prove that A and B are unitarily equivalent.
   (b) Prove that C and E are similar but not unitarily equivalent.
   (c) Prove that D and E are similar but not unitarily equivalent.

5. Let T be a linear operator on an inner product space V over F. Prove the following statements.
   (a) If T is Hermitian then $\det(T), \operatorname{Tr}(T)$ are real numbers and the characteristic polynomial has real coefficients.

(b) If T is unitary or orthogonal then $|\det(T)| = 1$.

6. Let T be a Hermitian operator. Prove that $T + iI$ is an invertible operator. (Hint: $T - cI$ is invertible if and only if c is not an eigenvalue of T.)

7. Prove that if T is Hermitian and $f(t) \in \mathbb{R}[t]$ then $f(T)$ is also Hermitian.

8. Let S, T be unitary operators. Prove that ST and S^{-1} are also unitary. Prove the same statement for the orthogonal operators.

9. Let S, T be a self-adjoint linear operator on an n-dimensional inner product space V. Let $\mathcal{B}$ be an orthonormal basis of V and let $A = [T]_{\mathcal{B}}$.

   (a) Prove that $T > 0$ if and only if $\sum\limits_{i,j} A_{ij} a_j \bar{a}_i > 0$ for all nonzero $\begin{bmatrix} a_1 \\ a_2 \\ \vdots \\ a_n \end{bmatrix} \in F^n$.

   (b) Prove that if $T > 0$ then $\det(T) > 0$ and $\mathrm{Tr}(T) > 0$.

   (c) Prove that if $S \geq 0, T \geq 0$ and $S^2 = T^2$ then $S = T$.

   (d) Let T be a positive semidefinite operator on an inner product space. Prove that there is a unique positive semidefinite operator S such that $T = S^2$.

   (e) Prove that if $T > 0, U > 0$ and $TU = UT$ then $TU > 0$.

   (f) Prove that if $T > 0$ then $T^{-1} > 0$.

   (g) Prove that if $S > 0, T > 0$ then $T + U > 0$.

10. Let $A, B \in M(n, \mathbb{C})$ be unitarily equivalent. Prove that A is positive definite (resp. positive semidefinite) if and only if B is positive definite (resp. positive semidefinite) .

11. If $F = \mathbb{C}$ then Corollary 8.4.4 is true T without being self adjoint: Let $T : V \to V$ be a linear operator (not necessarily self adjoint) on an inner product space V over $F = \mathbb{C}$. Prove that if $\langle \mathbf{v}, T(\mathbf{v}) \rangle = 0$ for all $\mathbf{v}$ then $T = 0$. For $F = \mathbb{R}$ find a counter example showing this is false. (Hint: Expand $\langle \mathbf{u} + \mathbf{v}, T(\mathbf{u} + \mathbf{v}) \rangle = 0$ to get $\langle T(\mathbf{u}), \mathbf{v} \rangle - \langle T(\mathbf{v}), \mathbf{u} \rangle = 0$. And then replace $\mathbf{v}$ by $i\mathbf{v}$. Add two expressions.)

12. Let T be a linear operator on an inner product space V over $\mathbb{C}$. Prove that T is normal if and only if $\|T(\mathbf{x})\| = \|T^*(\mathbf{x})\|$ for all $\mathbf{x} \in V$.

13. A linear operator $T \in L(V)$ on an inner product space V. Prove that T is Hermitian if and only if $\langle \mathbf{v}, T(\mathbf{v}) \rangle$ is real for all $\mathbf{v} \in V$. (Hint: Use Exercise 11.)

14. Let $A \in M(n, F)$ and suppose $A \geq 0$, $\mathbf{v} \in F^n$. Prove that $A\mathbf{v} = \mathbf{0}$ if $\langle \mathbf{v}, A\mathbf{v} \rangle = 0$.

15. Prove that if $A \in M(n, \mathbb{C})$ is skew Hermitian then the eigenvalues are pure imaginary.

16. Prove that if $A \in M(n, \mathbb{R})$ is real symmetric and $A^k = I$ then $A^2 = I_n$.

17. Prove that if $A \in M(n, \mathbb{R})$ is a symmetric matrix then there is a symmetric $B \in M(n, \mathbb{R})$ such that $B^3 = A$.

18. Prove that a 2×2 unitary matrix can be written in the form

$$U = e^{i\theta}\begin{bmatrix} a & b \\ -\bar{b} & \bar{a} \end{bmatrix}$$

with $|a|^2 + |b|^2 = 1$.

19. Let $T : V \to W$ be a linear map of rank r. For each of the following T, find orthonormal bases $\mathcal{B} = \{\mathbf{v}_1, \mathbf{v}_2, ..., \mathbf{v}_n\}$ and $\mathcal{C} = \{\mathbf{w}_1, \mathbf{w}_2, ..., \mathbf{w}_m\}$ with respect to the standard inner products for V and W and the positive singular values $s_1 \geq s_2 \geq \cdots \geq s_r > 0$ such that $T(\mathbf{v}_i) = s_i\mathbf{w}_i$ $(1 \leq i \leq r)$.

   (a) $T : \mathbb{C}^2 \to \mathbb{C}^2$ defined by $T(a,b) = \begin{bmatrix} ia+b \\ a-ib \end{bmatrix}$

   (b) $T : M(2,\mathbb{R}) \to \mathbb{R}^3$ be defined by $T(A) = \begin{bmatrix} a+b \\ b+c \\ c+d \end{bmatrix}$ where $A = \begin{bmatrix} a & b \\ c & d \end{bmatrix}$

   (c) $T : P_2(\mathbb{R}) \to P_1(\mathbb{R})$ defined by $T(f(t)) = f'(t)$ and $\langle f, g\rangle = \int_0^1 f(t)g(t)dt$

20. Find singular decompositions of the following matrices.

$$A = \begin{bmatrix} 1 & 1 & 1 \\ 1 & 1 & 1 \end{bmatrix}, \qquad B = \begin{bmatrix} 1 & 0 & 1 \\ 1 & 0 & -1 \end{bmatrix}, \qquad C = \begin{bmatrix} 1 & 1 & -1 & 1 \\ 1 & -1 & 1 & 1 \end{bmatrix}$$

21. Find polar decompositions of the following matrices.

$$A = \begin{bmatrix} 0 & 2 \\ 2 & 0 \end{bmatrix}, \qquad B = \begin{bmatrix} 1 & 0 & 1 \\ 1 & 0 & 0 \\ 0 & 0 & 1 \end{bmatrix}, \qquad C = \begin{bmatrix} 1 & 1 & 1 \\ 1 & -1 & 1 \\ 1 & 1 & -1 \end{bmatrix}$$

22. Let $A \in M(n,\mathbb{C})$ and $A \geq 0$. Show that the polar decomposition as described in the proof of Theorem 8.4.21 leads to the polar decomposition $A = US$ with $U = I_n$, $S = A$.

23. Let $A \in M(n,\mathbb{C})$ be a normal matrix with the eigenvalues $\lambda_1, \lambda_2, ..., \lambda_n$ and the corresponding orthonormal eigenvectors $\{\mathbf{v}_1, \mathbf{v}_2, ..., \mathbf{v}_n\}$. Let $P = [\mathbf{v}_1\ \mathbf{v}_2 \cdots \mathbf{v}_n] \in M(n,\mathbb{C})$.

   (a) Prove that the singular values are $|\lambda_1|, |\lambda_2|, ..., |\lambda_n|$.

   (b) Let $S = \operatorname{diag}\big[|\lambda_1|, |\lambda_2|, ..., |\lambda_n|\big]$. Prove that there is $c_i \in \mathbb{C}$ with $|c_i| = 1$ such that $A = QSP^*$ is the singular value decomposition where $Q = [c_1\mathbf{v}_1\ c_2\mathbf{v}_2 \cdots c_n\mathbf{v}_n]$.

24. Let $A \in M(n,\mathbb{C})$ be positive semidefinite. Prove that the singular values are the eigenvalues of A and we can choose $P = Q$ in the singular value decomposition $A = QSP^*$.

25. Let $A \in M(n.\mathbb{C})$ with the polar decomposition $A = US$. Prove that A is normal if and only if $US = SU$. (Hint: Use Exercise 8.3.8.)

8.5 Bilinear Forms and Quadratic Forms

So far we studied the inner product spaces over the field of real numbers $\mathbb{R}$ or the field of complex numbers $\mathbb{C}$. In this section, we study "generalized inner products" on a vector space over an arbitrary field which we call 'bilinear forms'. This section is a brief introduction to the theory of bilinear forms.

Sometimes we need the assumption that the characteristic of a field F is not equal to 2. This signifies the fact that '$1 \neq -1$' or 'we can divide by 2' which is impossible if the characteristic of F is 2 in which case $2 = 1 + 1 = 0$ (Definition B.2.4). For the first two subsections (§8.5.1–§8.5.2), the field F need not be neither $\mathbb{R}$ nor $\mathbb{C}$. In the last subsection §8.5.3, we let again F to be $\mathbb{R}$ or $\mathbb{C}$.

8.5.1 Bilinear Forms

Definition 8.5.1. *Let V be a vector space over a field F. A function*

$$B : V \times V \to F$$

is said to be **biadditive** *if*

$$B(\mathbf{u}_1 + \mathbf{u}_2, \mathbf{v}) = B(\mathbf{u}_1, \mathbf{v}) + B(\mathbf{u}_2, \mathbf{v}), \ (\mathbf{u}_i, \mathbf{v} \in V) \tag{8.84}$$

$$B(\mathbf{u}, \mathbf{v}_1 + \mathbf{v}_2) = B(\mathbf{u}, \mathbf{v}_1) + B(\mathbf{u}, \mathbf{v}_2), \ (\mathbf{u}, \mathbf{v}_i \in V) \tag{8.85}$$

If a biadditive function B further satisfies

$$B(c\mathbf{u}, d\mathbf{v}) = cdB(\mathbf{u}, \mathbf{v}) \tag{8.86}$$

then we say B is a **bilinear form** *on V. We denote the set of all bilinear forms on V by $B(V)$.*

Example 8.5.2. (1) *An inner product on an inner product space over the reals $\mathbb{R}$ is a bilinear form. But the inner product over the field of complex numbers $\mathbb{C}$ is not bilinear since it does not satisfy* (8.86).

(2) *Let F be a field and $V = F^2$. Let*

$$B\left(\begin{bmatrix} a_1 \\ b_1 \end{bmatrix}, \begin{bmatrix} a_2 \\ b_2 \end{bmatrix}\right) = [a_1 \ b_1] \begin{bmatrix} 2 & 3 \\ 4 & -1 \end{bmatrix} \begin{bmatrix} a_2 \\ b_2 \end{bmatrix} = 2a_1a_2 + 3a_1b_2 + 4a_2b_1 - b_1b_2$$

Then B is bilinear by the properties of matrix operation. More generally, if $A \in M(n, F)$ and $\mathbf{u}, \mathbf{v} \in F^n$ then

$$B(\mathbf{u}, \mathbf{v}) = \mathbf{u}^t A \mathbf{v}$$

defines a bilinear form on F^n.

(3) *Let $V = F^2$. For $\mathbf{u}, \mathbf{v} \in F^2$ define $B(\mathbf{u}, \mathbf{v}) = \det[\mathbf{u} \ \mathbf{v}]$. Then the determinant map $B : V \times V \to F$ is a bilinear form* (Theorem 3.1.2).

If B is a bilinear form on V and $\mathbf{v} \in V$ then we define $\phi_{B,\mathbf{v}}(\mathbf{w}) = B(\mathbf{v}, \mathbf{w})$ and $\psi_{B,\mathbf{w}}(\mathbf{v}) = B(\mathbf{v}, \mathbf{w})$. Then

$$\phi_{B,\mathbf{v}}, \psi_{B,\mathbf{w}} : V \to F \tag{8.87}$$

are linear functionals i .e., $\phi_{B,\mathbf{v}}, \psi_{B,\mathbf{w}} \in V^{\#}$.

And for all $\mathbf{v}, \mathbf{w} \in V$ we have (Exercise 1)

$$B(\mathbf{v}, \mathbf{0}) = B(\mathbf{0}, \mathbf{w}) = 0 \tag{8.88}$$

We will say $\mathbf{u}$ and $\mathbf{v}$ are **orthogonal with respect to** B if $B(\mathbf{u}, \mathbf{v}) = 0$. We say $S = \{\mathbf{v}_1, \mathbf{v}_2, ..., \mathbf{v}_n\}$ is **orthonormal** if $B(\mathbf{v}_i, \mathbf{v}_j) = \delta_{ij}$.

Let $B, B_1, B_2 \in B(V)$ and $a \in F$. We define

$$\begin{aligned} (B_1 + B_2)(\mathbf{u}, \mathbf{v}) &= B_1(\mathbf{u}, \mathbf{v}) + B_2(\mathbf{u}, \mathbf{v}) \\ (aB)(\mathbf{u}, \mathbf{v}) &= aB(\mathbf{u}, \mathbf{v}) \end{aligned} \tag{8.89}$$

Then $B_1 + B_2, aB \in B(V)$.

Theorem 8.5.3. *Under the operations defined above $B(V)$ is a vector space over F.*

Proof. Exercise 2 □

Let $\mathcal{B} = \{\mathbf{v}_1, \mathbf{v}_2, \ldots, \mathbf{v}_n\}$ be a basis of V. Let $\langle B \rangle_{\mathcal{B}} \in M(n, F)$ be the matrix defined by

$$(\langle B \rangle_{\mathcal{B}})_{ij} = B(\mathbf{v}_i, \mathbf{v}_j) \tag{8.90}$$

which we call the **matrix representation of** B with respect to the basis $\mathcal{B}$.

Example 8.5.4. (1) *Let B be as in* Example 8.5.2 (2) *and let* $\mathcal{B} = \left\{ \mathbf{v}_1 = \begin{bmatrix} 1 \\ 1 \end{bmatrix}, \mathbf{v}_2 = \begin{bmatrix} 1 \\ -1 \end{bmatrix} \right\}$. *We have*

$$B(\mathbf{v}_1, \mathbf{v}_1) = 8, B(\mathbf{v}_1, \mathbf{v}_2) = 4, B(\mathbf{v}_2, \mathbf{v}_1) = 2, B(\mathbf{v}_2, \mathbf{v}_2) = -6$$

Hence we have

$$\langle B \rangle_{\mathcal{B}} = \begin{bmatrix} 8 & 4 \\ 2 & -6 \end{bmatrix}$$

(2) *Let $B : F^2 \times F^2 \to F$ be defined by* $B\left(\begin{bmatrix} a_1 \\ b_1 \end{bmatrix}, \begin{bmatrix} a_2 \\ b_2 \end{bmatrix} \right) = \det \begin{bmatrix} a_1 & a_2 \\ b_1 & b_2 \end{bmatrix}$. *Let $\mathcal{B}$ be the standard basis of F^2. To find $\langle B \rangle_{\mathcal{B}}$ we compute*

$$B(\mathbf{e}^1, \mathbf{e}^1) = 0, B(\mathbf{e}^1, \mathbf{e}^2) = 1, B(\mathbf{e}^2, \mathbf{e}^1) = -1, B(\mathbf{e}^2, \mathbf{e}^2) = 0$$

Hence

$$\langle B \rangle_{\mathcal{B}} = \begin{bmatrix} 0 & 1 \\ -1 & 0 \end{bmatrix}$$

Theorem 8.5.5. *Let $\mathcal{B}$ be a basis of a vector space V. Then the map*

$$\psi : B(V) \to M(n, F), \quad B \mapsto \langle B \rangle_{\mathcal{B}}$$

is an isomorphism of vector spaces. In particular, $\dim B(V) = n^2$.

Proof. Let $\mathcal{B} = \{\mathbf{v}_1, \mathbf{v}_2, ..., \mathbf{v}_n\}$. To prove ψ is linear map let $B, B' \in B(V)$. Then $(\langle B\rangle_{\mathcal{B}})_{ij} = B(\mathbf{v}_i, \mathbf{v}_j)$ and $(\langle B'\rangle_{\mathcal{B}})_{ij} = B'(\mathbf{v}_i, \mathbf{v}_j)$. Hence we have

$$\psi(B + B')_{ij} = (B + B')(\mathbf{v}_i, \mathbf{v}_j) = B(\mathbf{v}_i, \mathbf{v}_j) + B'(\mathbf{v}_i, \mathbf{v}_j) = \psi(B)_{ij} + \psi(B')_{ij}$$

Thus $\psi(B + B') = \psi(B) + \psi(B')$. Similarly, $\psi(cB) = c\psi(B)$ for $c \in F$.

Now we need to show ψ is bijective. For this we construct the inverse of ψ. Let $A \in M(n, F)$. We need to find B such that $\psi(B) = A$. If $\mathbf{u} = \sum a_i\mathbf{v}_i, \mathbf{v} = \sum b_i\mathbf{v}_i$ we define $B \in B(V)$ by

$$B(\mathbf{u}, \mathbf{v}) = [a_1, a_2, ..., a_n]A \begin{bmatrix} b_1 \\ b_2 \\ \vdots \\ b_n \end{bmatrix} = \sum_{i,j} a_i A_{ij} b_j \tag{8.91}$$

Now it is obvious, by the properties of matrix multiplication, that B defined by (8.91) is bilinear. It is an exercise to show that $\psi(B) = A$ which says that ψ is surjective. Injectivity of ψ follows from the fact that the null space of ψ is trivial. □

Let $A \in M(n, F)$. We write the **bilinear form** (8.91) **associated to** A by B_A:

$$B_A(\mathbf{u}, \mathbf{v}) = \mathbf{u}^t A\mathbf{v} \tag{8.92}$$

Corollary 8.5.6. *Let V be a vector space over a field F of dimension n with a basis $\mathcal{B}$. Let $A \in M(n, F)$. Then $B(\mathbf{u}, \mathbf{v}) = [\mathbf{u}]_{\mathcal{B}}^t\, A\, [\mathbf{v}]_{\mathcal{B}}$ if and only if $A = \langle B\rangle_{\mathcal{B}}$.*

Proof. Simply notice that the equality (8.91) says the bilinear form B attached to the matrix A is defined by $B(\mathbf{u}, \mathbf{v}) = [\mathbf{u}]_{\mathcal{B}}^t\, A\, [\mathbf{v}]_{\mathcal{B}}$. □

Recall the change of coordinate matrix (Definition 6.4.25): Let $\mathcal{B} = \{\mathbf{u}_1, \mathbf{u}_2, \ldots, \mathbf{u}_n\}$, $\mathcal{C} = \{\mathbf{v}_1, \mathbf{v}_2 \ldots, \mathbf{v}_n\}$ be bases of V. Let

$$\mathbf{v}_j = I(\mathbf{v}_j) = \sum_{i=1}^{n} Q_{ij}\mathbf{u}_i \text{ i.e., } Q = (Q_{ij}) = [I_V]_{\mathcal{C}}^{\mathcal{B}}$$

The matrix Q is called the change of coordinate matrix from $\mathcal{C}$ to $\mathcal{B}$.

Theorem 8.5.7. *Let V be a vector space over F and $\mathcal{B}, \mathcal{C}$ be bases of V. Let $Q = [I]_{\mathcal{C}}^{\mathcal{B}}$ be the change of coordinate matrix that change $\mathcal{C}$-coordinate into $\mathcal{B}$-coordinate then*

$$\langle B\rangle_{\mathcal{C}} = Q^t\langle B\rangle_{\mathcal{B}}Q$$

Proof. Let $\mathcal{B} = \{\mathbf{u}_1, \mathbf{u}_2, \ldots, \mathbf{u}_n\}$, $\mathcal{C} = \{\mathbf{v}_1, \mathbf{v}_2, \ldots, \mathbf{v}_n\}$ and let $A = \langle B\rangle_{\mathcal{B}}$, $Q = [I]_{\mathcal{C}}^{\mathcal{B}}$. Then $A = [B(\mathbf{u}_i, \mathbf{u}_j)]$ and $\mathbf{v}_j = \sum_i Q_{ij}\mathbf{u}_i$. By Theorem 8.5.6, we need to show

$$[\mathbf{u}]_{\mathcal{C}}^t Q^t AQ[\mathbf{v}]_{\mathcal{C}} = B(\mathbf{u}, \mathbf{v})$$

For this first note that $[\mathbf{u}]_{\mathcal{B}} = Q[\mathbf{u}]_{\mathcal{C}}, [\mathbf{v}]_{\mathcal{B}} = Q[\mathbf{v}]_{\mathcal{C}}$. Hence

$$B(\mathbf{u}, \mathbf{v}) = [\mathbf{u}]_{\mathcal{B}}^t A[\mathbf{v}]_{\mathcal{C}} = (Q[\mathbf{u}]_{\mathcal{C}})^t AQ[\mathbf{v}]_{\mathcal{C}} = [\mathbf{u}]_{\mathcal{C}}^t Q^t AQ[\mathbf{v}]_{\mathcal{C}}$$ □

Corollary 8.5.8. *Let B be a bilinear form on a vector space $V = F^n$ over F. Let $\mathcal{C} = \{\mathbf{v}_1, \mathbf{v}_2, ..., \mathbf{v}_n\}$ be a basis of V and $\mathcal{E}$ be the standard basis. Let $A = \langle B\rangle_{\mathcal{E}}, A' = \langle B\rangle_{\mathcal{C}}$ and $Q \in M(n,F)$ be the matrix $Q = [\mathbf{v}_1\ \mathbf{v}_2 \cdots \mathbf{v}_n]$. Then*

$$A' = Q^t AQ$$

We say that $A, A' \in M(n,F)$ are **congruent** if there is an invertible matrix Q such that $A' = Q^t AQ$. In this terminology, Theorem 8.5.7 says that $\langle B\rangle_{\mathcal{B}}, \langle B\rangle_{\mathcal{C}}$ are congruent. And Corollary 8.4.2 says that a real symmetric matrix is congruent to a diagonal matrix. Also note that congruent matrices have the same rank (Theorem 6.3.20).

Let B be a bilinear form on V. We define the **right kernel** $V^{\perp}$ and **left kernel** ${}^{\perp}V$ of B by

$$\begin{aligned} V^{\perp} &= \{\mathbf{v} \in V | B(\mathbf{u},\mathbf{v}) = 0 \text{ for all } \mathbf{u} \in V\} \\ {}^{\perp}V &= \{\mathbf{u} \in V | B(\mathbf{u},\mathbf{v}) = 0 \text{ for all } \mathbf{v} \in V\} \end{aligned} \tag{8.93}$$

Theorem 8.5.9. *Let B be a bilinear form on a vector space V of dimension n. Let $\mathcal{B}, \mathcal{C}$ be bases of V. Then*

(1) $\rho(\langle B\rangle_{\mathcal{C}}) = \rho(\langle B\rangle_{\mathcal{B}})$

(2) $\dim V^{\perp} = \dim {}^{\perp}V$

Proof. (1) Since $\langle B\rangle_{\mathcal{C}} = Q^t\langle B\rangle_{\mathcal{B}}Q$ and Q is invertible we have $\rho\langle B\rangle_{\mathcal{C}} = \rho\langle B\rangle_{\mathcal{B}}$ by Theorem 6.3.20 as we observed above.

(2) Let $A = \langle B\rangle_{\mathcal{B}}$. Then since $B(\mathbf{u},\mathbf{v}) = [\mathbf{u}]_{\mathcal{B}}^t\, A\, [\mathbf{v}]_{\mathcal{B}}$ we have

$$\begin{aligned} V^{\perp} &= \{\mathbf{v} \mid B(-,\mathbf{v}) = 0\} = \{\mathbf{v} \mid A[\mathbf{v}]_{\mathcal{B}} = 0\} = \{\mathbf{v} \mid [\mathbf{v}]_{\mathcal{B}} \in N(A)\} \\ {}^{\perp}V &= \{\mathbf{u} \mid B(\mathbf{u},-) = 0\} = \{\mathbf{u} \mid [\mathbf{u}]_{\mathcal{B}}^t A = (A^t[\mathbf{u}]_{\mathcal{B}})^t = 0\} = \{\mathbf{u} \mid A^t[\mathbf{u}]_{\mathcal{B}} = 0\} \end{aligned}$$

Hence $\dim V^{\perp} = \nu(A)$ and $\dim {}^{\perp}V = \nu(A^t)$. Since $A \in M(n,F)$, a square matrix, we have $\nu(A) = \nu(A^t)$ by Corollary 2.3.18. Thus $\dim {}^{\perp}V = \dim V^{\perp}$. □

Let B be a bilinear form on V. For $\mathbf{u} \in V$ define $\phi_B(\mathbf{u}) \in V^{\#}$ by $\phi_B(\mathbf{u})(\mathbf{v}) = B(\mathbf{u},\mathbf{v})$. Similarly, we define $\psi_B(\mathbf{v})(\mathbf{u}) = B(\mathbf{u},\mathbf{v})$. Hence we have

$$\begin{aligned} \phi_B &: V \to V^{\#}, \quad \mathbf{u} \mapsto \phi_B(\mathbf{u}) \\ \psi_B &: V \to V^{\#}, \quad \mathbf{v} \mapsto \psi_B(\mathbf{v}) \end{aligned} \tag{8.94}$$

Theorem 8.5.10. *Let B be a bilinear form on V and $\mathcal{B}$ be a basis of V. Then the following conditions are equivalent:*

(1) $V^{\perp} = \mathbf{0}$

(2) ${}^{\perp}V = \mathbf{0}$

(3) *The map $\phi_B : V \to V^{\#}$ is an isomorphism.*

(4) *The map $\psi_B : V \to V^{\#}$ is an isomorphism.*

(5) *$\langle B\rangle_{\mathcal{B}}$ is invertible.*

Proof. Equivalence of (1) and (2) follows from Theorem 8.5.9 (2). And since $N(\phi_B) = {}^{\perp}V$ and $N(\psi_B) = V^{\perp}$ we see that (3) and (4) are equivalent. Since $\dim V = \dim V^{\#}$,vanishing of $N(\phi_B)$ is equivalent to that ϕ_B is an isomorphism. Hence (1) and (3) are equivalent.

To prove (3) and (5) are equivalent, let $\mathcal{B} = \{\mathbf{v}_1, \mathbf{v}_2, ..., \mathbf{v}_n\}$ be a basis of V and $\mathcal{B}^{\#} = \{f_1, f_2, ..., f_n\}$ be the dual basis (6.51). By Corollary 6.5.12, we have

$$\phi_B(\mathbf{v}_j) = B(\mathbf{v}_j, -) = \textstyle\sum_i B(\mathbf{v}_j, \mathbf{v}_i) f_i$$

Hence $[\phi_B]_{\mathcal{B}}^{\mathcal{B}^{\#}} = (B(\mathbf{v}_j, \mathbf{v}_i)) = (B(\mathbf{v}_i, \mathbf{v}_j))^t$. By Theorem 6.4.20, ϕ_B is invertible if and only if $[\phi_B]_{\mathcal{B}}^{\mathcal{B}^{\#}}$ is invertible. Hence $[\phi_B]_{\mathcal{B}}^{\mathcal{B}^{\#}} = (B(\mathbf{v}_i, \mathbf{v}_j))^t$ is invertible if and only if $\langle B \rangle_{\mathcal{B}} = (B(\mathbf{v}_i, \mathbf{v}_j))$ is invertible. Hence (3) and (5) are equivalent. □

Definition 8.5.11. *If a bilinear form B on a vector space V satisfies one of the conditions of* Theorem 8.5.10 *then we say that B is* **nondegenerate**.

For example, if V is a finite dimensional <u>real</u> inner product space then the inner product $B = \langle\ ,\ \rangle$ is a nondegenerate bilinear form.

8.5.2 Symmetric Bilinear Forms and Quadratic Forms

Recall that the **characteristic** of a field F is not equal to 2 ($\text{char}(F) \neq 2$) if $1 + 1 \neq 0$ or $1 \neq -1$ in F (See Definition B.2.4 and also §6.1.1.). In this case, we can divide by 2 namely, $1/2$ is a well defined element of F.

Definition 8.5.12. *A bilinear form B on V is called* **symmetric** (resp. **skew symmetric**) *if $B(\mathbf{u}, \mathbf{v}) = B(\mathbf{v}, \mathbf{u})$* (resp. *$B(\mathbf{u}, \mathbf{v}) = -B(\mathbf{v}, \mathbf{u})$*). *A bilinear form B on V is said to be* **diagonalizable** *if there is a basis $\mathcal{B} = \{\mathbf{v}_1, \mathbf{v}_2, \ldots, \mathbf{v}_n\}$ such that $\langle B \rangle_{\mathcal{B}}$ is a diagonal matrix.*

If the field F is of characteristic 2 then a skew symmetric bilinear form is the same as the symmetric bilinear form.

Example 8.5.13. (1) *Let V be a real inner product space of dimension n. Then the inner product $B(-,-) = \langle -,- \rangle$ on V is symmetric and diagonalizable since if we choose $\mathcal{B}$ to be an orthonormal basis* (Theorem 8.1.20) *then $B_{\mathcal{B}} = I_n$.*

(2) *The determinant* $\det : F^2 \times F^2 \to F$ *on $M(2, F)$ viewed as a function of columns is skew symmetric unless* $\text{char}(F) \neq 2$ (Theorem 3.1.2). *But if* $\text{char}(F) = 2$ *then* $\det : F^2 \times F^2 \to F$ *is also symmetric.*

(3) *If $A \in M(n)$ is a symmetric matrix then the bilinear form B_A on $V = F^n$ associated to A is symmetric.*

(4) *Let F be a field. The trace map* $\text{Tr} : M(n, F) \times M(n, F) \to F$ *defined by $(A, B) \mapsto$* $\text{Tr}(A^t B)$ *is a nondegenerate symmetric bilinear form* (Exercise 4).

Theorem 8.5.14. *Let B be a bilinear form on a vector space V. Then B is symmetric if and only if $\langle B \rangle_{\mathcal{B}}$ is a symmetric matrix for any basis $\mathcal{B}$. In particular, a diagonalizable bilinear form is symmetric.*

Proof. Let $\mathcal{B} = \{\mathbf{v}_1, \mathbf{v}_2, \ldots, \mathbf{v}_n\}$. If B is symmetric then $B(\mathbf{v}_i, \mathbf{v}_j) = B(\mathbf{v}_j, \mathbf{v}_i)$. Hence $\langle B \rangle_{\mathcal{B}}$ is symmetric. Conversely suppose $A = \langle B \rangle_{\mathcal{B}}$ is symmetric. Then

$$\begin{aligned} B(\mathbf{u}, \mathbf{v}) &= [\mathbf{u}]_{\mathcal{B}}^t A [\mathbf{v}]_{\mathcal{B}} = [\mathbf{u}]_{\mathcal{B}}^t A^t [\mathbf{v}]_{\mathcal{B}} = (A[\mathbf{u}]_{\mathcal{B}})^t ([\mathbf{v}]_{\mathcal{B}})^{tt} \\ &= ([\mathbf{v}]_{\mathcal{B}}^t A [\mathbf{u}]_{\mathcal{B}})^t = [\mathbf{v}]_{\mathcal{B}}^t A [\mathbf{u}]_{\mathcal{B}} = B(\mathbf{v}, \mathbf{u}) \end{aligned}$$

The second to the last equality holds because it is a scalar.

If B is diagonalizable then there is a basis $\mathcal{B}$ such that $\langle B\rangle_{\mathcal{B}}$ is diagonal which is symmetric. Therefore B is symmetric. □

Theorem 8.5.15. *Let V be a vector space over a field F with $\mathrm{char}(F) \neq 2$. If B is a nonzero symmetric bilinear form on V then there is $\mathbf{v} \in V$ such that $B(\mathbf{v}, \mathbf{v}) \neq 0$.*

Proof. Since B is nonzero, there are $\mathbf{u}, \mathbf{v} \in V$ such that $B(\mathbf{u}, \mathbf{v}) \neq 0$. If $B(\mathbf{u}, \mathbf{u}) \neq 0$ or $B(\mathbf{v}, \mathbf{v}) \neq 0$ then we are done. Otherwise, $B(\mathbf{u}+\mathbf{v}, \mathbf{u}+\mathbf{v}) = 2B(\mathbf{u}, \mathbf{v}) \neq 0$ since the characteristic of F is not 2. □

The following theorem reminds us the theorem saying that a real symmetric matrix is diagonalizable (Corollary 8.4.2).

Theorem 8.5.16. *Let V be a finite dimensional vector space over F with $\mathrm{char}(F) \neq 2$. If B is a nonzero symmetric bilinear form on V then B is diagonalizable.*

Proof. We will prove this by induction on $n = \dim V$. If $n = 1$ then $\langle B\rangle_{\mathcal{B}}$ is trivially diagonal for any $\mathcal{B}$. Let $n > 1$ and assume our result for the vector spaces of dimension $< n$. We will prove our result for a vector space of dimension n. If $B = 0$ then obviously it is diagonalizable. Hence we suppose $B \neq 0$ and by Theorem 8.5.15 there is $\mathbf{v} \in V$ such that $B(\mathbf{v}, \mathbf{v}) \neq 0$. Then $\phi_B(\mathbf{v})$ defined by $\phi_B(\mathbf{v})(\mathbf{w}) = B(\mathbf{v}, \mathbf{w})$ is a nonzero linear functional $\phi_B(\mathbf{v}) : V \to F$ (hence $\phi_B(\mathbf{v})$ is onto). By Rank Theorem (Theorem 6.3.21), $\dim N(\phi_B(\mathbf{v})) = n - 1$. The restriction B on $N(\phi_B(\mathbf{v}))$ is a symmetric bilinear form and since $\dim N(\phi_B(\mathbf{v})) = n - 1$ we see there is a basis $\mathcal{B}' = \{\mathbf{v}_1, \mathbf{v}_2, \ldots, \mathbf{v}_{n-1}\}$ for which $\langle B\rangle_{\mathcal{B}'}$ is diagonal. Now if we let $\mathcal{B} = \{\mathbf{v}_1, \mathbf{v}_2, \ldots, \mathbf{v}_n = \mathbf{v}\}$ then $\langle B\rangle_{\mathcal{B}}$ is diagonal. □

Corollary 8.5.17. *Let F be a field with $\mathrm{char}(F) \neq 2$. If $A \in M(n, F)$ is symmetric then A is congruent to a diagonal matrix.*

Proof. Exercise 5. □

Theorem 8.5.18. *Let V be a finite dimensional inner product space over $\mathbb{R}$ and let B be a symmetric bilinear form on V. Then there is an orthonormal basis $\mathcal{B}$ of V, with respect to the inner product on V, for which $\langle B\rangle_{\mathcal{B}}$ is diagonal.*

Proof. By Gram Schmidt, choose an orthonormal, with respect to the inner product on V, basis $\mathcal{C} = \{\mathbf{w}_1, \mathbf{w}_2, \ldots, \mathbf{w}_n\}$ for V. Then $A := \langle B\rangle_{\mathcal{C}}$ is symmetric since B is symmetric. By Corollary 8.4.2, there is an orthogonal matrix Q such that $D := Q^tAQ = \mathrm{diag}[d_1, d_2, ..., d_n]$ is diagonal. Let $\mathbf{v}_j = \sum Q_{ij}\mathbf{w}_i$ and $\mathcal{B} = \{\mathbf{v}_1, \mathbf{v}_2, \ldots, \mathbf{v}_n\}$. Then, by Theorem 8.4.11, $\mathcal{B}$ is orthonormal since Q is an orthogonal matrix and $\mathcal{C}$ is orthonormal (Exercise 6), and $\langle B\rangle_{\mathcal{B}} = D$. In fact, $B(\mathbf{v}_i, \mathbf{v}_j) = B(\sum_k Q_{ki}\mathbf{w}_k, \sum_l Q_{lj}\mathbf{w}_l) = \sum_{k,l} Q_{ki}A_{kl}Q_{lj} = (Q^tAQ)_{il} = d_i\delta_{ij}$. □

Next theorem identifies the bilinear forms on a real vector space V which define inner products on V.

Theorem 8.5.19. *Let $V = \mathbb{R}^n$ with a basis $\mathcal{B}$. Let $B : V \times V \to \mathbb{R}$ be a nondegenerate symmetric bilinear form on V. Then B is an inner product if and only if the matrix $\langle B\rangle_{\mathcal{B}}$ is positive definite.*

Proof. Let $A = \langle B\rangle_{\mathcal{B}}$ and $\langle\ ,\ \rangle$ be the standard inner product on $V = \mathbb{R}^n$. Then

$$B(\mathbf{u},\mathbf{v}) = [\mathbf{u}]_{\mathcal{B}}^t A[\mathbf{v}]_{\mathcal{B}} = \langle [\mathbf{u}]_{\mathcal{B}}, T_A([\mathbf{v}]_{\mathcal{B}})\rangle$$

By Theorem 8.4.8, the last expression become an inner product if and only if $T_A > 0$ which is equivalent to $A > 0$. □

Let A be a real symmetric matrix. We know then A is congruent to a diagonal matrix say, $D = P^tAP$ for some invertible matrix P, by Corollary 8.5.17. For explicit diagonalization write $P = E_1E_2\cdots E_n$ where E_i's are elementary matrices. Hence $D = P^tAP = E_n^t\cdots E_2^tE_1^tAE_1E_2\cdots E_n$ is diagonal. For actual computation, suppose the elementary row operations $E_k^t\cdots E_1^tA$ reduces A to a matrix whose first column vanishes except possibly the $(1,1)$ position. Then additional columns operations – postmultiply the elementary matrices – $E_k^t\cdots E_1^tAE_1\cdots E_k$ reduces $E_k^t\cdots E_1^tA$ into a matrix, since A is symmetric, whose first row vanishes except possibly the $(1,1)$. And the resulting matrix is symmetric. Repeating this we end up with a diagonal matrix.

The following example shows that the diagonalization of a bilinear form is not unique. However, we will prove that there is a uniquely determined integer, so called the 'signature' of a symmetric bilinear form which is independent of the diagonalization (Theorem 8.5.21).

Example 8.5.20. *Let B be a bilinear form on $V = \mathbb{R}^3$ with the standard inner product. Let*

$$A = \begin{bmatrix} 1 & 2 & 3\\ 2 & 1 & 0\\ 3 & 0 & 1\end{bmatrix}$$

be the matrix representation of B with respect to the standard basis $\mathcal{E}$ say $A = \langle B\rangle_{\mathcal{E}}$. To diagonalize A we first perform the elementary row operations

$$A = \begin{bmatrix} 1 & 2 & 3\\ 2 & 1 & 0\\ 3 & 0 & 1\end{bmatrix} \xrightarrow{-2R_1+R_2} \begin{bmatrix} 1 & 2 & 3\\ 0 & -3 & -6\\ 3 & 0 & 1\end{bmatrix} \xrightarrow{-3R_1+R_3} \begin{bmatrix} 1 & 2 & 3\\ 0 & -3 & -6\\ 0 & -6 & -8\end{bmatrix} \xrightarrow{-2R_2+R_3} \begin{bmatrix} 1 & 2 & 3\\ 0 & -3 & -6\\ 0 & 0 & 4\end{bmatrix}$$

Now corresponding column operations $-2C_1+C_2, -3C_1+C_3, -2C_2+C_3$ reduce the last matrix to

$$\begin{bmatrix} 1 & 2 & 3\\ 0 & -3 & -6\\ 0 & 0 & 4\end{bmatrix} \xrightarrow{-2C_1+C_2} \begin{bmatrix} 1 & 0 & 3\\ 0 & -3 & -6\\ 0 & 0 & 4\end{bmatrix} \xrightarrow{-3C_1+C_3} \begin{bmatrix} 1 & 0 & 0\\ 0 & -3 & -6\\ 0 & 0 & 4\end{bmatrix} \xrightarrow{-2C_2+C_3} \begin{bmatrix} 1 & 0 & 0\\ 0 & -3 & 0\\ 0 & 0 & 4\end{bmatrix}$$

The matrices corresponding to the row operations are $E_{12}(-2), E_{13}(-3), E_{23}(-2)$. Hence

$$P^t = E_{23}(-2)E_{13}(-3)E_{12}(-2) = \begin{bmatrix} 1 & 0 & 0\\ 0 & 1 & 0\\ 0 & -2 & 1\end{bmatrix}\begin{bmatrix} 1 & 0 & 0\\ 0 & 1 & 0\\ -3 & 0 & 1\end{bmatrix}\begin{bmatrix} 1 & 0 & 0\\ -2 & 1 & 0\\ 0 & 0 & 1\end{bmatrix} = \begin{bmatrix} 1 & 0 & 0\\ -2 & 1 & 0\\ 1 & -2 & 1\end{bmatrix}$$

And if we let $P = \begin{bmatrix} 1 & -2 & 1\\ 0 & 1 & -2\\ 0 & 0 & 1\end{bmatrix}$ *and* $\mathcal{B} = \left\{\begin{bmatrix}1\\0\\0\end{bmatrix}, \begin{bmatrix}-2\\1\\0\end{bmatrix}, \begin{bmatrix}1\\-2\\1\end{bmatrix}\right\}$ *then* (Corollary 8.5.17)

$$\langle B\rangle_{\mathcal{B}} = P^tAP = \begin{bmatrix} 1 & 0 & 0\\ 0 & -3 & 0\\ 0 & 0 & 4\end{bmatrix} \tag{8.95}$$

On the other hand, since A is real symmetric there is an orthogonal matrix Q such that Q^tAQ is diagonal by Theorem 8.4.1 *and the diagonal entries are the eigenvalues of A. The characteristic polynomial of A is $f(t) = -(t-1)(t^2-2t-12)$ and the eigenvalues are $\lambda_1 = 1$, $\lambda_2 = 1+\sqrt{13}$, $\lambda_2 = 1-\sqrt{13}$ and the corresponding eigenvectors are $\mathbf{w}_1 = \begin{bmatrix} 0 \\ 3 \\ -2 \end{bmatrix}$, $\mathbf{w}_2 = \begin{bmatrix} \sqrt{13} \\ 2 \\ 3 \end{bmatrix}$, $\mathbf{w}_3 = \begin{bmatrix} \sqrt{13} \\ -2 \\ -3 \end{bmatrix}$. Let $\mathbf{v}_i = \dfrac{\mathbf{w}_i}{\|\mathbf{w}_i\|}$ $(i = 1,2,3)$ and $\mathcal{C} = \{\mathbf{v}_1, \mathbf{v}_2, \mathbf{v}_3\}$. Then $\mathcal{C}$ is an orthonormal basis and we have an orthogonal matrix $Q := [\mathbf{v}_1\ \mathbf{v}_2\ \mathbf{v}_3]$. We then have*

$$\langle B\rangle_{\mathcal{C}} = \begin{bmatrix} \mathbf{v}_1^t \\ \mathbf{v}_2^t \\ \mathbf{v}_3^t \end{bmatrix} A\ [\mathbf{v}_1\ \mathbf{v}_2\ \mathbf{v}_3] = Q^tAQ = \begin{bmatrix} 1 & 0 & 0 \\ 0 & 1+\sqrt{13} & 0 \\ 0 & 0 & 1-\sqrt{13} \end{bmatrix} \tag{8.96}$$

This is a diagonalization of the matrix A as well as a diagonalization of the symmetric bilinear form B_A. Hence the diagonalization of a symmetric bilinear form is not unique. This is in contrast to the fact that the diagonal matrix of a diagonalizable matrix is uniquely determined (§6.4).

Also observe that B is not an inner product since A is not positive definite.

Let B be a symmetric bilinear form over a field F with $\mathrm{char}(F) \neq 2$. Suppose $\mathcal{B}, \mathcal{C}$ be two bases of V such that $\langle B\rangle_{\mathcal{B}}, \langle B\rangle_{\mathcal{C}}$ are diagonal. We saw in Example 8.5.20 that the diagonal matrix is not unique. However, by Theorem 8.5.9, $\langle B\rangle_{\mathcal{B}}, \langle B\rangle_{\mathcal{C}}$ have the same rank. If B defines a inner product then the rank is the same as $\dim V$ namely positive definite (Theorem 8.5.19). We show next, that there is another intrinsic integer attached to B.

Theorem 8.5.21 (Sylverster's Law of Inertia). *Let V be a finite dimensional vector space over $\mathbb{R}$ and let B be a symmetric bilinear form on V. Let $\mathcal{B}$ and $\mathcal{C}$ be bases of V such that $\langle B\rangle_{\mathcal{B}}$ and $\langle B\rangle_{\mathcal{C}}$ are diagonal. Then the number of positive* (*resp.* negative) *entries of $\langle B\rangle_{\mathcal{B}}$ is the same as the number of positive* (*resp. negative*) *entries of $\langle B\rangle_{\mathcal{C}}$.*

Proof. By Theorem 8.5.9, we see that $\rho(\langle B\rangle_{\mathcal{B}}) = \rho(\langle B\rangle_{\mathcal{C}})$. Let $r = \rho(\langle B\rangle_{\mathcal{B}}) = \rho(\langle B\rangle_{\mathcal{C}})$ and

$$\mathcal{B} = \{\mathbf{v}_1, \mathbf{v}_2, \dots, \mathbf{v}_p, \dots, \mathbf{v}_r, \dots, \mathbf{v}_n\},\ \mathcal{C} = \{\mathbf{w}_1, \mathbf{w}_2, \dots, \mathbf{w}_q, \dots, \mathbf{w}_r, \dots, \mathbf{w}_n\}$$

with

$$\begin{aligned} B(\mathbf{v}_i, \mathbf{v}_i)) > 0\ (i = 1,2,\dots,p),\ B(\mathbf{v}_j, \mathbf{v}_j) < 0\ (j = p+1,\dots,r) \\ B(\mathbf{w}_i, \mathbf{w}_i) > 0\ (i = 1,2,\dots,q),\ B(\mathbf{w}_j, \mathbf{w}_j) < 0\ (j = q+1,\dots,r) \end{aligned}$$

We will show that

$$S = \{\mathbf{v}_1, \mathbf{v}_2, \dots, \mathbf{v}_p, \mathbf{w}_{q+1}, \dots, \mathbf{w}_r, \dots, \mathbf{w}_n\}$$

is linearly independent. (Then $p + (n-q) \leq n$ and hence $p \leq q$. By symmetry, $p = q$.) Assume the contrary i.e., S is linearly dependent. Then

$$a_1\mathbf{v}_1 + \cdots + a_p\mathbf{v}_p + b_{q+1}\mathbf{w}_{q+1} + \cdots + b_n\mathbf{w}_n = 0$$

Then $a_1\mathbf{v}_1 + \cdots + a_p\mathbf{v}_p = -b_{q+1}\mathbf{w}_{q+1} - \cdots - b_n\mathbf{w}_n$ and hence

$$B(a_1\mathbf{v}_1 + \cdots + a_p\mathbf{v}_p, a_1\mathbf{v}_1 + \cdots + a_p\mathbf{v}_p) = B(-b_{q+1}\mathbf{w}_{q+1} - \cdots - b_n\mathbf{w}_n, -b_{q+1}\mathbf{w}_{q+1} - \cdots - b_n\mathbf{w}_n)$$

Hence

$$\sum_{i=1}^{p} a_i^2 B(\mathbf{v}_i, \mathbf{v}_i) = \sum_{j=q+1}^{r} b_j^2 B(\mathbf{w}_j, \mathbf{w}_j)$$

But then the left hand side is > 0 and the right hand side is < 0. This is a contradiction. □

Corollary 8.5.22. *Let V be a finite dimensional vector space over $\mathbb{R}$ and let B be a symmetric bilinear form on V. There is a basis $\mathcal{B}$ such that $\langle B\rangle_{\mathcal{B}}$ is of the form*

$$\langle B\rangle_{\mathcal{B}} = \begin{bmatrix} 1 & & & & & & & & \\ & \ddots & & & & & & & \\ & & 1 & & & & & & \\ & & & -1 & & & & & \\ & & & & \ddots & & & & \\ & & & & & -1 & & & \\ & & & & & & 0 & & \\ & & & & & & & \ddots & \\ & & & & & & & & 0 \end{bmatrix} \tag{8.97}$$

and the number of $+1$ and the number of -1 are uniquely determined.

Proof. Let $\mathcal{C} = \{\mathbf{v}_1, \mathbf{v}_2, ..., \mathbf{v}_n\}$ and $B(\mathbf{v}_i, \mathbf{v}_i) = \lambda_i$. Let $\mathbf{w}_i = \dfrac{\mathbf{v}_i}{\sqrt{|\lambda_i|}}$ then $\{\mathbf{w}_1, \mathbf{w}_2, ..., \mathbf{w}_n\}$ is the required basis. The numbers of $+1$ and -1 are uniquely determined by Theorem 8.5.21. □

Definition 8.5.23. *Let B be a symmetric bilinear form (resp. a symmetric $A \in M(n, \mathbb{R})$) on a finite dimensional real vector space V. Corresponding to the matrix (8.97) for the bilinear form B (resp. the bilinear form B_A associated to A) the numbers*

$$\begin{aligned} p = &\text{ the number of } +1\text{'s} \\ n = &\text{ the number of } -1\text{'s} \\ z = &\text{ the number of } 0\text{'s} \end{aligned}$$

are called the **signature** *of B (resp. A). We denote the signature of B (resp. A) by*

$$\sigma(B) = (p, n, z), \quad \sigma(A) = (p, n, z)$$

Often, the **signature** *of B is defined simply by $\sigma = p - n$.*

In Example 8.5.20, the matrices (8.95) and (8.96) gives the same signature independent of diagonalization namely, $\sigma(B) = (2, 1, 0)$.

Let $V = \mathbb{R}^n$ and $A \in M(n, \mathbb{R})$ be a symmetric matrix. Then

$$\langle \mathbf{u}, \mathbf{v}\rangle = \mathbf{u}^t A\mathbf{v}$$

is an inner product if and only if $\sigma(A) = (n, 0, 0)$ (Exercise 9).

Let V be a vector space over F. A function $Q : V \to F$ is a **quadratic form** if there is a symmetric bilinear form B on V such that $Q(\mathbf{v}) = B(\mathbf{v}, \mathbf{v})$. If Q is a quadratic form on V and $\operatorname{char}(F) \neq 2$ then

$$B(\mathbf{u}, \mathbf{v}) = \frac{1}{2}\left[Q(\mathbf{u} + \mathbf{v}) - Q(\mathbf{u}) - Q(\mathbf{v})\right] \tag{8.98}$$

defines a bilinear form on V (Exercise 10). Hence if char $(F) \neq 2$ then there is a bijection between the set of quadratic forms and the set of symmetric bilinear forms on V.

Example 8.5.24. *Consider the quadratic polynomial*

$$f(x_1, x_2, \dots, x_n) = \sum_{i \leq j} a_{ij} x_i x_j \quad (a_{ij} \in \mathbb{R})$$

in the variables $x_1, x_2, ..., x_n$. Let $A \in M(n, \mathbb{R})$ be defined by

$$A_{ij} = \begin{cases} a_{ii} \text{ if } i = j \\ \frac{1}{2} a_{ij} \text{ if } i < j \\ \frac{1}{2} a_{ji} \text{ if } j < i \end{cases}$$

Then A as well as the bilinear form $B = B_A$ associated to A are symmetric. If $\mathbf{x} = \begin{bmatrix} x_1 \\ \vdots \\ x_n \end{bmatrix}$ then

$$f(\mathbf{x}) = B_A(\mathbf{x}, \mathbf{x})$$

is the quadratic form associated to the symmetric matrix A. Now, by Theorem 8.5.18, *there is an orthonormal basis $\mathcal{B}$ for which $\langle B \rangle_{\mathcal{B}} = \mathrm{diag}[b_1, \dots, b_n]$. For $\mathbf{z} \in \mathbb{R}^n$ let $[\mathbf{z}]_{\mathcal{B}} = \begin{bmatrix} z_1 \\ \vdots \\ z_n \end{bmatrix}$. Then*

$$f([\mathbf{z}]_{\mathcal{B}}) = [\mathbf{z}]_{\mathcal{B}}^t \langle B \rangle_{\mathcal{B}} [\mathbf{z}]_{\mathcal{B}} = \sum_{i=1}^{n} b_i z_i^2 \quad (b_i \in \mathbb{R})$$

This means that we can find a new coordinate system $z_1, z_2, \dots, z_n$ such that

$$f(z_1, z_2, \dots, z_n) = \sum_{i=1}^{n} b_i z_i^2 \quad (b_i \in \mathbb{R})$$

8.5.3 Sesqui-Linear Forms and Inner Products

In this subsection, the field F will be $\mathbb{R}$ or $\mathbb{C}$. We consider the inner product space $V = F^n$ with the standard inner product and 'forms' on V. Throughout this subsection $V = F^n$.

Definition 8.5.25. *Let $V = F^n$. A biadditive map $H : V \times V \to F$ on V is a* **sesqui-linear**† **form** *if*

$$H(c\mathbf{u}, d\mathbf{v}) = \bar{c} d H(\mathbf{u}, \mathbf{v}) \tag{8.99}$$

If $\mathcal{B} = \{\mathbf{v}_1, \mathbf{v}_2, ..., \mathbf{v}_n\}$ is a basis of V then we define $\langle H \rangle_{\mathcal{B}}$ to be an $n \times n$ matrix defined by

$$(\langle H \rangle_{\mathcal{B}})_{ij} = H(\mathbf{v}_i, \mathbf{v}_j)$$

†Sesqui is a Latin which means $1\frac{1}{2}$-times.

An inner product on F^n is a sesqui-linear form. And for $A \in M(n, F)$ we write

$$H_A(\mathbf{u}, \mathbf{v}) = \langle \mathbf{u}, \mathbf{v} \rangle_A = \mathbf{u}^* A\mathbf{v}$$

Then it is easy to show that H_A is a sesqui-linear from (Exercise 11). If $F = \mathbb{R}$ then a sesqui-linear form is simply a bilinear form.

Let $H(V)$ be the set of all sesqui-linear forms on V. We define the addition of two sesqui-linear forms and the scalar multiplication by (8.89). Then $H(V)$ becomes a vector space over F (Exercise 12).

Theorem 8.5.26. *Let $V = F^n$ and let $H(V)$ be the set of all sesqui-linear forms on V.*

(1) *Let $H \in H(V)$ and $\mathcal{E}$ be the standard basis of V. If $A = \langle H \rangle_{\mathcal{E}}$ then*

$$H(\mathbf{u}, \mathbf{v}) = \mathbf{u}^* A\mathbf{v}$$

(2) *We have an isomorphism*

$$\phi : H(V) \to M(n, F), \quad H \mapsto \langle H \rangle_{\mathcal{B}}$$

(3) *Let $\mathcal{B}, \mathcal{C}$ be bases of V and let $Q = [I]^{\mathcal{B}}_{\mathcal{C}}$ be the change of coordinate matrix that changes $\mathcal{C}$-coordinate into $\mathcal{B}$-coordinate. Then*

$$\langle H \rangle_{\mathcal{C}} = Q^* \langle H \rangle_{\mathcal{B}} Q$$

Proof. (1) Both H and H_A are sesqui-linear and they coincide on the basis $\mathcal{E}^n \times \mathcal{E}^n = \{(\mathbf{e}^i, \mathbf{e}^j) \mid 1 \le i, j \le n\}$ of $V \times V$. Hence $H = H_A$.

(2) Obviously the map ϕ is injective. Let $A \in M(n, F)$ and for $\mathbf{u}, \mathbf{v}$ let $[\mathbf{u}]_{\mathcal{B}}, [\mathbf{v}]_{\mathcal{B}}$ be their coordinates with respect to $\mathcal{B}$. We define a sesqui-linear form H by

$$H(\mathbf{u}, \mathbf{v}) = [\mathbf{u}]^*_{\mathcal{B}} A [\mathbf{v}]_{\mathcal{B}}$$

Then H is a sesqui-linear and $\langle H \rangle_{\mathcal{B}} = A$. One can check easily the map ϕ is linear.

(3) A slight change of the proof of Theorem 8.5.7 will be sufficient. Details of the proof is an exercise. □

Definition 8.5.27. *A sesqui-linear form H on V satisfying*

$$H(\mathbf{u}, \mathbf{v}) = \overline{H(\mathbf{v}, \mathbf{u})}$$

is called a **Hermitian form***. We say H is* **symmetric** *if $H(\mathbf{u}, \mathbf{v}) = H(\mathbf{v}, \mathbf{u})$.*

If $F = \mathbb{R}$ then a Hermitian sesqui-linear form is a symmetric bilinear form. An inner product on $V = \mathbb{C}^n$ is a Hermitian form.

It is clear that if H is Hermitian (or symmetric) then so is the matrix $\langle H \rangle_{\mathcal{B}}$ for any basis $\mathcal{B}$ of V. Conversely, if A is a Hermitian matrix and $H = H_A$ then

$$\overline{H(\mathbf{v}, \mathbf{u})} = \overline{\mathbf{v}^* A\mathbf{u}} = (\mathbf{v}^* A\mathbf{u})^* = \mathbf{u}^* A^* \mathbf{v} = \mathbf{u}^* A\mathbf{v} = H(\mathbf{u}, \mathbf{v})$$

since $A = A^*$. Consequently, a sesqui-linear form H is a Hermitian form if and only if $\langle H \rangle_{\mathcal{B}}$ is a Hermitian matrix.

Theorem 8.5.28. *Let H be a sesqui-linear form on $V = F^n$.*

(1) Let $H = H_A$ with $A \in M(n, F)$. Then H is Hermitian form on V if and only if A is a Hermitian matrix. In this case, there is an orthonormal basis $\mathcal{B} = \{\mathbf{v}_1, \mathbf{v}_2, ..., \mathbf{v}_n\}$ of V consisting of eigenvectors of A such that $\langle H\rangle_{\mathcal{B}}$ is real diagonal.

(2) Let $V = \mathbb{C}^n$. The sesqui-linear form H is Hermitian if and only if $H(\mathbf{v}, \mathbf{v})$ is real for all $\mathbf{v} \in V$.

Proof. The first part of (1) is proved in the remark above. For the second part, there is an orthonormal basis $\mathcal{B} = \{\mathbf{v}_1, \mathbf{v}_2, ..., \mathbf{v}_n\}$ of F^n consisting of eigenvectors of A with $A\mathbf{v}_i = \lambda_i\mathbf{v}_i$ ($\lambda_i \in \mathbb{R}$) by Theorem 8.4.1. Now we have $H(\mathbf{v}_i, \mathbf{v}_j) = \mathbf{v}_i^* A\mathbf{v}_j = \mathbf{v}_i^*\lambda_j\mathbf{v}_j = \lambda_j\delta_{ij}$ since $\mathcal{B}$ is an orthonormal basis.

(2) If H is Hermitian then $H(\mathbf{v}, \mathbf{v}) = \overline{H(\mathbf{v}, \mathbf{v})}$. Hence $H(\mathbf{v}, \mathbf{v})$ is real. Conversely, assume $H(\mathbf{v}, \mathbf{v}) \in \mathbb{R}$ for all $\mathbf{v} \in V$. Then $H(\mathbf{v}+\mathbf{u}, \mathbf{v}+\mathbf{u}) = H(\mathbf{v}, \mathbf{v}) + H(\mathbf{u}, \mathbf{v}) + H(\mathbf{v}, \mathbf{u}) + H(\mathbf{u}, \mathbf{u})$ is real. Hence $H(\mathbf{u}, \mathbf{v}) + H(\mathbf{v}, \mathbf{u}) \in \mathbb{R}$ and hence

$$H(\mathbf{u}, \mathbf{v}) + H(\mathbf{v}, \mathbf{u}) = \overline{H(\mathbf{u}, \mathbf{v})} + \overline{H(\mathbf{v}, \mathbf{u})} \tag{8.100}$$

Replacing $\mathbf{u}$ by $i\mathbf{u}$ we get

$$-iH(\mathbf{u}, \mathbf{v}) + iH(\mathbf{v}, \mathbf{u}) = i\ \overline{H(\mathbf{u}, \mathbf{v})} - i\ \overline{H(\mathbf{v}, \mathbf{u})}$$

and multiplying through i we get

$$H(\mathbf{u}, \mathbf{v}) - H(\mathbf{v}, \mathbf{u}) = -\overline{H(\mathbf{u}, \mathbf{v})} + \overline{H(\mathbf{v}, \mathbf{u})} \tag{8.101}$$

Adding (8.100) and (8.101) we get

$$2H(\mathbf{u}, \mathbf{v}) = 2\overline{H(\mathbf{v}, \mathbf{u})}$$

Hence H is a Hermitian form. □

Definition 8.5.29. *We say a Hermitian form H is* **positive definite** *($H > 0$ in symbol) if*

$$H(\mathbf{v}, \mathbf{v}) > 0 \text{ for all nonzero } \mathbf{v} \in V$$

We say a Hermitian matrix $A \in M(n, F)$ is **positive definite** ($A > 0$) (resp. **positive semidefinite** ($A \geq 0$)) (Definition 8.4.7) *if*

$$H_A(\mathbf{u}, \mathbf{u}) > 0 \text{ (resp.} H_A(\mathbf{u}, \mathbf{u}) \geq 0) \text{ for all } \mathbf{u} \neq 0$$

By Theorem 8.5.26 (1), $H = H_A$ for some $A \in M(n, F)$. Hence if $H = H_A$ then $H > 0$ if and only if $H = H_A$ and $A > 0$.

The following fact is obvious in view of Theorem 8.5.19.

Theorem 8.5.30. *The sesqui-linear form $H_A(\mathbf{u}, \mathbf{v}) = \mathbf{u}^* A\mathbf{v}$ is an inner product on $V = F^n$ if and only if A is positive definite Hermitian matrix.*

To decide whether a matrix $A \in M(n)$ is Hermitian or not is easy – just see if $A = A^*$. To test if A is positive definite we need to find the eigenvalues of A to apply Theorem 8.4.8. However often it is not easy to find the eigenvalues explicitly. The following theorem furnish a method to test positivity without finding the eigenvalues.

Recall the notation from §3.4: If $A \in M(n)$ then we write

$$A_k = \text{ upper left } k{\times}k \text{ submatrix of } A$$

Theorem 8.5.31 (Sylvester's Criterion). *Let $A \in M(n, F)$ be Hermitian. Then $A > 0$ if and only if $\det(A_s) > 0$ for $s = 1, 2, ..., n$.*

Proof. First, suppose $A > 0$. Let $\mathcal{E} = \{\mathbf{e}^1, \mathbf{e}^2, ..., \mathbf{e}^n\}$ be the standard basis of $V = F^n$. Let $V_k = \text{Span}\{\mathbf{e}^1, \mathbf{e}^2, ..., \mathbf{e}^k\}$ $(k \leq n)$ be the subspace of V with the basis $\mathcal{E}_k = \{\mathbf{e}^1, \mathbf{e}^2, ..., \mathbf{e}^k\}$. Let $H = H_A$ and H_k be the restriction of H on V_k. Then $H_k > 0$. Hence $\langle H_k \rangle_{\mathcal{E}_k} = A_k$ has positive determinant by Exercise 8.4.9 (b).

Now assume A is Hermitian $(A = A^*)$ and $|A_k| > 0$ for all k. We will use Theorem 3.4.5 which says that if $|A_k| \neq 0$ for all k then there is upper triangular matrix P with 1's on the diagonal such that $B = AP$ is a lower triangular matrix. Since P is upper triangular, P^* is lower triangular. Hence $P^*B = P^*AP$ is lower triangular. Since $A = A^*$ we conclude that P^*AP is also Hermitian. The lower triangular matrix P^*AP which is also Hermitian must be a diagonal matrix say $P^*AP = \text{diag}[d_1, d_2, ..., d_n] := D$. By Theorem 3.4.5, the determinant of the upper left $k{\times}k$ submatrix of $D = P^*B$ is the same as the determinant of the upper left $k{\times}k$ submatrix of B (Theorem 3.3.17 and Exercise 2.2.4). Hence $d_1 d_2 \cdots d_k = |D_k| = |(P^*B)_k| = |B_k|$.

Since $B = AP$ and P is upper triangular we have $B_k = A_k P_k$ which is easy to check. And since the diagonals of P are 1's we have $|B_k| = |A_k| > 0$ by our assumption for $k = 1, 2, ..., n$. Hence $d_i > 0$ for all i. Since A is Hermitian, by Theorem 8.4.1, there is an orthonormal basis consisting of eigenvectors with the eigenvalues d_i. As $d_i > 0$, by Theorem 8.4.8 (2), we conclude that $A > 0$. □

Example 8.5.32. *Let* $A = \begin{bmatrix} 5 & 2 & 3 & 4 \\ 2 & 5 & 3 & 4 \\ 3 & 3 & 5 & 4 \\ 4 & 4 & 4 & 5 \end{bmatrix}$. *Then* $\det(A) = 15$ *and*

$$\det A_1 = 5, \ \det A_2 = \begin{vmatrix} 5 & 2 \\ 2 & 5 \end{vmatrix}, \ \det A_3 = \begin{vmatrix} 5 & 2 & 3 \\ 2 & 5 & 3 \\ 3 & 3 & 5 \end{vmatrix} = 51$$

Hence $A > 0$ by Theorem 8.5.31. *The characteristic polynomial is $f(t) = t^4 - 20t^3 + 80t^2 - 92t + 15$. It is not easy to find the roots of $f(t)$.* Theorem 8.5.31 *provide us a method to test positivity of a matrix without finding the eigenvalues.*

8.6 Exercises

1. Prove the equalities (8.88).

2. Prove Theorem 8.5.3

3. Let $T : V \to W$ be a linear map and B_W be a bilinear form on W. Define B_V by

$$B_V(\mathbf{v}_1, \mathbf{v}_2) = B_W(T(\mathbf{v}_1), T(\mathbf{v}_2))$$

   (a) Prove that B_V is a bilinear form on V.

   (b) Let $\mathcal{B}, \mathcal{C}$ be bases for V, W. Prove that $\langle B_V \rangle_{\mathcal{B}} = ([T]_{\mathcal{B}}^{\mathcal{C}})^t \langle B_W \rangle_{\mathcal{C}} [T]_{\mathcal{B}}^{\mathcal{C}}$.

4. Let F be a field. Prove that the trace map $\mathrm{Tr} : M(n, F) \times M(n, F) \to F$ defined by $(A, B) \mapsto \mathrm{Tr}(A^t B)$ is a nondegenerate symmetric bilinear form. (Cf. Example 8.1.6)

5. Prove Corollary 8.5.17.

6. Let $Q \in M(n, \mathbb{R})$ be an orthogonal matrix. Let $\mathcal{B} = \{\mathbf{v}_1, \mathbf{v}_2, ..., \mathbf{v}_n\}$ be an orthonormal basis of $\mathbb{R}^n$ with respect to a symmetric bilinear form B. Prove that $\{Q\mathbf{v}_1, Q\mathbf{v}_2, ..., Q\mathbf{v}_n\}$ is an orthonormal basis of $\mathbb{R}^n$ with respect to B.

7. Suppose the following matrices are the matrix representation of a bilinear form on $V = F^n$ with respect to the standard basis. Find the matrix Q such that $Q^t A Q$ is diagonal and the basis $\mathcal{B}$ for which the matrix representation of B with respect to $\mathcal{B}$ is the diagonal matrix. Also find the signatures of the matrices when $F = \mathbb{R}$.

$$A_1 = \begin{bmatrix} 1 & 2 \\ 2 & 2 \end{bmatrix}, \quad A_2 = \begin{bmatrix} 0 & 1 \\ 1 & 0 \end{bmatrix}, \quad A_3 = \begin{bmatrix} 1 & 0 & 1 \\ 0 & 2 & 1 \\ 1 & 1 & 3 \end{bmatrix},$$

$$A_4 = \begin{bmatrix} 1 & 1 & 1 \\ 1 & 0 & -1 \\ 1 & -1 & 1 \end{bmatrix}, \quad A_5 = \begin{bmatrix} 0 & 1 & -1 \\ 1 & 0 & 1 \\ -1 & 1 & -2 \end{bmatrix}$$

8. Prove the true statement(s) and find counter example(s) for the false statement(s).

   (a) For all n the determinant map $\det : M(n) \to F$ is bilinear.

   (b) If A is a real symmetric matrix with signature $\sigma(A) = (p, n, z)$ then we may have $p < n$.

   (c) A sesqui-linear form is also bilinear.

   (d) An inner product on a complex vector space is sesqui-linear.

   (e) Let $A \in M(n, \mathbb{R})$ be a symmetric matrix. Let $\sigma(B_A) = (p, n, z)$ be the signature of the bilinear form B_A. Let $P^t A P = \mathrm{diag}[d_1, d_2, ..., d_n]$ be a diagonalization of the symmetric matrix A as in Corollary 8.4.2. Then the number of positive d's is p and the number of negative d's is n.

   (f) Let $A \in M(n, \mathbb{R})$ be a symmetric matrix. Let $P^t A P = \mathrm{diag}[d_1, d_2, ..., d_n]$ be a diagonalization of the symmetric matrix A. Then the diagonalization of symmetric bilinear form B_A (Theorem 8.5.16) is the same as the diagonalization of the matrix A (Corollary 8.4.2).

9. Let $V = \mathbb{R}^n$ and $A \in M(n, \mathbb{R})$ be a symmetric matrix. Then $\langle \mathbf{u}, \mathbf{v} \rangle = \mathbf{u}^t A \mathbf{v}$ for $\mathbf{u}, \mathbf{v} \in \mathbb{R}^n$ is an inner product if and only if $\sigma(A) = (n, 0, 0)$.

10. Prove that (8.98) defines a symmetric bilinear form on V.

11. For $A \in M(n, \mathbb{C})$ we let $H_A(\mathbf{u}, \mathbf{v}) = \langle \mathbf{u}, \mathbf{v} \rangle_A = \mathbf{u}^* A \mathbf{v}$. Prove that H_A is a sesqui-linear from.

12. Let $H(V)$ be the set of all sesqui-linear forms. Prove that $H(V)$ becomes a vector space under the operations (8.89).

13. Let $A \in M(n, \mathbb{C})$. Prove that A is a Hermitian matrix if and only if $\mathbf{u}^* A \mathbf{u} \in \mathbb{R}$ for all $\mathbf{u} \in \mathbb{C}^n$.

14. Let V be a finite dimensional vector space with a bilinear form B. Let W be a subspace of V. We define

$$W^{\perp} = \{\mathbf{v} \in V \mid B(\mathbf{v}, \mathbf{w}) = 0 \text{ for all } \mathbf{w} \in W\}$$
$$^{\perp}W = \{\mathbf{v} \in V \mid B(\mathbf{w}, \mathbf{v}) = 0 \text{ for all } \mathbf{w} \in W\}$$

   (a) Prove that if B is symmetric or skew symmetric or Hermitian then $W^{\perp} = {}^{\perp}W$.

   (b) Prove that $\dim(W^{\perp}) = \dim({}^{\perp}W)$ for any bilinear form B. (Hint: Reformulate the problem to use Corollary 4.2.29)

   (c) Let B be a bilinear form on V. Prove that if W is a subspace of V such that $V = W \oplus V^{\perp}$ then the restriction B_W of B on W is a nondegenerate bilinear form.

15. Let B be a bilinear form on a vector space V. For fixed $\mathbf{v} \in V$, the maps $B(-, \mathbf{v})$ and $B(\mathbf{v}, -)$ are linear functionals on V. Show that $V_L^{\#} := \{B(\mathbf{v}, -) \mid \mathbf{v} \in V\}$ and $V_R^{\#} := \{B(-, \mathbf{v}) \mid \mathbf{v} \in V\}$ are subspaces of $V^{\#}$. Find the dimensions of $V_L^{\#}$ and $V_R^{\#}$.

16. Let $f : U \times V \to F$ be a bilinear map with $\dim U = \dim V$. Define

$$U^0 = \{\mathbf{u} \in V \mid f(\mathbf{u}, \mathbf{v}) = 0 \text{ for all } \mathbf{v} \in V\}$$
$$V^0 = \{\mathbf{v} \in V \mid f(\mathbf{u}, \mathbf{v}) = 0 \text{ for all } \mathbf{u} \in V\}$$

   Prove that the following conditions are equivalent. (Compare with Theorem 8.5.10.)

   (a) If $\mathcal{B} = \{\mathbf{u}_i\}, \mathcal{C} = \{\mathbf{v}_i\}$ are bases for U, V then $A = (f(\mathbf{u}_i, \mathbf{v}_j))$ is invertible.

   (b) $U^0 = \mathbf{0}$

   (c) $V^0 = \mathbf{0}$

   (d) The map $\phi : V \to U^{\#}$ defined by $\phi(\mathbf{v}) = f(-, \mathbf{v})$ is an isomorphism.

   (e) The map $\psi : U \to V^{\#}$ defined by $\psi(\mathbf{u}) = f(\mathbf{u}, -)$ is an isomorphism.

Appendix A

Sets and Functions

A.1 Sets and Functions

A '**set**' is an undefined term in mathematics which can be described by saying "a set is a collection of distinguishable entities" and we should be able to decide whether an entity belongs to a set or not. Details of the materials in this chapter can be found in [3].

Definition A.1.1. Let X, Y be sets. A **function** (or **map**) f from X to Y, denoted by $f : X \to Y$, is a rule assigning an element $y = f(x)$ of Y for each $x \in X$. When $y = f(x)$ we call y the **image of** x **under** f and x is called a **preimage of** y **under** f. We call X the **domain** and Y the **codomain** of f.

The **image** or **range** $f(S)$ of a subset S of X under f is the subset of Y defined by

$$f(S) = \{y \in Y \mid y = f(x) \text{ for some } x \in S\}$$

The **inverse image** $f^{-1}(T)$ of a subset T of Y under f is defined by

$$f^{-1}(T) = \{x \in X \mid f(x) \in T\}$$

The **identity function** $I_X : X \to X$ on X is defined by $I_X(x) = x$. on X.

If $f : X \to Y$ is a function and $S \subseteq X, T \subseteq Y$ then we have

$$ff^{-1}(T) = T \cap f(X) \subseteq T \quad \text{and} \quad S \subseteq f^{-1}f(S) \tag{A.1}$$

Definition A.1.2. Let $f : X \to Y$ and $g : Y \to Z$ be functions. We define the **composition** $g \circ f : X \to Z$ of f and g by

$$g \circ f(x) = g(f(x))$$

We sometimes write gf for $g \circ f$.

If $f : X \to Y$ and $g : Y \to X$ are functions. If $g \circ f = I_X$ then we say g is a **left inverse** of f. And we say g is a **right inverse** of f if $f \circ g = I_Y$. We say g is a **two sided inverse** of f if g is a left as well as a right inverse of f.

Theorem A.1.3. *Let $f : X \to Y, g : Y \to Z$ and $h : Z \to W$ be functions. Then*

$$(h \circ g) \circ f = h \circ (g \circ f)$$

In other words, the composition of the functions is associative.

Proof. The values of both sides at $x \in X$ are equal to $h(g(f(x)))$. Hence $(h \circ g) \circ f = h \circ (g \circ f)$. □

Definition A.1.4. Let $f : X \to Y$ be a function. Then we say f is **surjective (onto)** if for any $y \in Y$ there is $x \in X$ such that $y = f(x)$. We say f is **injective (one-to-one)** if we have $x = x'$ whenever $f(x) = f(x')$. We say f is **bijective** if f injective and surjective. That is for all $y \in Y$ there is a unique $x \in X$ such that $y = f(x)$. We say f is **invertible** if f has a two sided inverse.

Theorem A.1.5. *Let $f : X \to Y$ be a function.*

(1) *The function f is injective if and only if f has a left inverse i.e., there is a function $g : Y \to X$ such that $g \circ f = I_X$.*

(2) *The function f is surjective if and only if f has a right inverse i.e., there is a function $g : Y \to X$ such that $f \circ g = I_Y$.*

(3) *The function $f : X \to Y$ is bijective if and only if f is invertible. In this case, there is a unique function $g : Y \to X$ satisfying $g \circ f = I_X$ and $f \circ g = I_Y$.*

Proof. (1) First suppose f is injective. Choose $x_0 \in X$ and define $g : Y \to X$ by

$$g(y) = \begin{cases} x & \text{if } y = f(x) \\ x_0 & \text{if } y \notin f(X) \end{cases}$$

Then it is obvious that $g \circ f = I_X$. Conversely assume f has a left inverse $g : Y \to X$ and suppose $f(x) = f(y)$. Then $x = g \circ f(x) = g \circ f(y) = y$. Hence $x = y$.

(2) Suppose $f : X \to Y$ is surjective. For each $f^{-1}(y) \subseteq X$ choose $x \in f^{-1}(y)$ and define $g(y) = x$.† Then it is easy to see that $f \circ g(y) = y$. Conversely if there is a right inverse $g : Y \to X$ of f. If $y \in Y$ then $y = f \circ g(y) = f(g(y))$. Hence if we let $x = g(y)$ then $f(x) = y$.

(3) If f is bijective then f has a left inverse g and right inverse g'. To show $g = g'$ let $y \in Y$. Then, by Theorem A.1.3, $g(y) = g(fg'(y)) = (gf)g'(y) = g'(y)$ as required. Hence f has a two sided inverse. Conversely, if f has a two sided inverse then f is injective by (1) and surjective by (2). Hence f is bijective.

For uniqueness of the inverse of f, suppose g and h be two sided inverses of f. Then $gf = I_X$ and $fh = I_Y$. Hence $g = g(fh) = (gf)h = h$. □

Observe that the proof of Theorem A.1.5 (3) shows that if $f : X \to Y$ is a bijection and if $g : Y \to X$ is a right (resp. left) inverse then g is also a left (resp. right) inverse.

Theorem A.1.6. *Let $f_1 : X \to Y$ and $f_2 : Y \to Z$ be functions.*

(1) *If f_1, f_2 are injective then so is $f_2 \circ f_1$. If $f_2 \circ f_1$ is injective then f_1 is injective.*

(2) *If f_1, f_2 are surjective then so is $f_2 \circ f_1$. If $f_2 \circ f_1$ is surjective then f_2 is surjective.*

Proof. (1) The first statement is easy to check. Suppose $f_1(x) = f_1(x')$. Then $f_2 \circ f_1(x) = f_2 \circ f_1(x')$. Since $f_2 \circ f_1$ is injective we conclude that $x = x'$.

(2) We only check the second statement. Let $z \in Z$. Since $f_2 \circ f_1 : X \to Z$ is surjective there is $x \in X$ such that $f_2 \circ f_1(x) = z$. Hence if we let $y = f_1(x)$ then $f_2(y) = z$. Thus f_2 is surjective. □

†The choice of such x depends on an axiom of the theory of sets so called Axiom of Choice.

A.2 Equivalence Relation

Definition A.2.1. Let A be a set. An **equivalence relation** on A is a subset R of $A \times A$ such that

(i) $(a, a) \in R$ for all $a \in A$ (reflexive)

(ii) If $(a, b) \in R$ then $(b, a) \in R$ (symmetric)

(iii) If $(a, b) \in R$ and $(b, c) \in R$ then $(a, c) \in R$ (transitive)

If $(a, b) \in R$ we will write $a \sim_R b$ or simply by $a \sim b$. If R is an equivalence relation then we define the **equivalence class** of a by

$$[a] := \{x \in A | x \sim a\} := \bar{a} \tag{A.2}$$

and any element of $[a]$ is called a **representative** of the equivalence class $[a]$. The equivalence class of a is also denoted by $\bar{a}$. The set of all equivalence classes is denoted by A/R

$$A/R = \{[a] \mid a \in A\}$$

Let X be a set. We say that a map $f : A \to X$ is a **well defined function** on A/R if $f(a) = f(b)$ whenever $[a] = [b]$.

Theorem A.2.2. *Let R be an equivalence relation on a set A. Then $a \sim b$ if and only if $[a] = [b]$.*

Proof. Suppose $a \sim b$ and let $c \in [a]$. Then $c \sim a$. Hence by transitivity $c \sim b$ and hence $c \in [b]$; $[a] \subseteq [b]$. Reversing the role of a and b we have $[a] \supseteq [b]$. Therefore, we get $[a] = [b]$.

Conversely, suppose $[a] = [b]$. Then $a \in [b]$. Hence $a \sim b$. □

Example A.2.3. (1) Let $A = \mathbb{Z}$ be the set of integers and $n \in \mathbb{Z}$. We define $a \sim b$ if $a - b$ is divisible by n i.e., $n|(a - b)$. The set of equivalence classes, called the equivalence classes modulo n, will be denoted by $\mathbb{Z}/n$. We will often write $\bar{a}$ for $[a]$.

(2) Let $A = \mathbb{R}$ be the set of all real numbers. For $x, y \in A$, define $x \sim y$ if $x - y \in \mathbb{Z}$. Then $\sim$ is an equivalence relation on A.

(3) Let $f : X \to Y$ be a function between the sets X, Y. Define $x \sim x'$ if $f(x) = f(x')$. Then $\sim$ is an equivalence relation on X.

Definition A.2.4. A **partition** of a set A is a family $\{A_i\}_{i \in I}$ of subsets of A such that

(i) $\bigcup_{i \in I} A_i = A$

(ii) $A_i \bigcap A_j = \emptyset$ whenever $i \neq j$

Theorem A.2.5. *Let A be a set and $\{A_i\}_{i \in I}$ be a partition of A. For a, b we define $a \sim b$ if and only if there is $i \in I$ such that $a \in A_i$ and $b \in A_i$. Then $\sim$ is an equivalence relation on A.*

Proof. Reflexivity is obvious since if $a \in A_i$ then $a \sim a$. For symmetry, if $a \sim b$ i.e., $a \in A_i$ and $b \in A_i$ then obviously $b \in A_i$ and $a \in A_i$. Hence $b \sim a$. For transitivity, let $a \sim b$ and $b \sim c$. Then there is i such that $a \in A_i$ and $b \in A_i$; there is j such that $b \in A_j$ and $c \in A_j$. Since $b \in A_i \cap A_j$ and since $\{A_i\}$ form a partition of A we conclude that $A_i = A_j$. Therefore $a \in A_i$ and $c \in A_i$. Thus $a \sim c$. □

Theorem A.2.6. *Let R be an equivalence relation on a set A.*

(1) *If $a, b \in A$ then either $[a] = [b]$ or $[a] \cap [b] = \emptyset$.*

(2) *The equivalence classes form a partition of A.*

Proof. (1) Suppose $[a] \cap [b] \neq \emptyset$. Then there is $c \in [a] \cap [b]$. Hence $c \sim b$ and $c \sim b$. Hence by transitivity $a \sim b$. By Theorem A.2.2, we conclude that $[a] = [b]$.

(2) By (1) it is obvious that the equivalence classes $\{[a] \mid a \in A\}$ form a partition of the set A. □

By the two theorems above we see that if A is a set then there is a bijection between the set of all equivalence relations on A and the set of all partitions of A.

Appendix B

Numbers

B.1 Integers $\mathbb{Z}$

In this section we collect basic properties of the integers $\mathbb{Z}$.

For $a, b \in \mathbb{Z}$ we say a is **divisible by** b if there is an $c \in \mathbb{Z}$ such that $a = bc$. In this case, we will write $b|a$ and we say b is a divisor of a. If $a, b \in \mathbb{Z}$ and $a|b$ as well as $b|a$ then we have $a = \pm b$. An integer p is called a **prime number** if the only divisors are ± 1.

Theorem B.1.1. (1) **(Division Algorithm)** *Let $a, b \in \mathbb{Z}$. Then there are unique $s, r \in \mathbb{Z}$ such that*

$$a = sb + r \quad \text{with } 0 \leq r < |b|$$

(2) **(Unique Factorization of Integers)** *Every $a \in \mathbb{Z}$ $(a > 1)$ can be written uniquely as a product $a = p_1^{e_1} p_2^{e_2} \cdots p_k^{e_k}$ where p_i's are positive prime integers.*

Proof. See [1] §0.2. □

Let $a_1, a_2, ..., a_r \in \mathbb{Z}$. We say $d \in \mathbb{Z}$ is a **greatest common divisor** (gcd) of $a_1, a_2, ..., a_r$ if $d|a_i$ for all i and if d' also divides all a_i then $d'|d$. A greatest common divisor is determined up to a sign. We write a greatest common divisor by $\gcd(a_1, a_2, ..., a_r)$. If $\gcd(a, b) = 1$ then we say a and b are **relatively prime**. Let

$$(a_1, a_2, ..., a_r) = \{n_1 a_1 + n_2 a_2 + \cdots + n_r a_r \mid n_i \in \mathbb{Z}\}$$

which we call the **ideal generated by** $a_1, a_2, ..., a_r$.

Theorem B.1.2. *Let $a_1, a_2, ..., a_r$ be positive integers.*

(1) *Let $d \in \mathbb{Z}$ be an element of $(a_1, a_2, ..., a_r)$ with the smallest absolute value then $(a_1, a_2, ..., a_r) = (d)$ and d is determined up to a sign.*

(2) *$d = \gcd(a_1, a_2, ..., a_r)$ if and only if $(d) = (a_1, a_2, ..., a_r)$.*

(3) *Let $a_i = \prod_j p_j^{e_{ij}}$ where $e_{ij} \geq 0$ and p_j's are positive primes. If $k_i = \min_j \{e_{ij}\}$ then*

$$\gcd(a_1, a_2, ..., a_r) = \prod_{i=1}^{r} p_i^{k_i}$$

Proof. Not to miss the point of the proof we assume $r = 2$. Slight changes will make the proof work for all r.

(1) Let d be an element in (a_1, a_2) with the smallest absolute value. Since $d \in (a_1, a_2)$ it is obvious that $(d) \subseteq (a_1, a_2)$. To show $(a_1, a_2) \subseteq (d)$ let $a \in (a_1, a_2)$. We need to show that $d|a$. By division algorithm, write $a = sd+r$ with $r < |d|$. Since a, d are elements in (a_1, a_2) we can write $a = a_1'a_1 + a_2'a_2$ and $d = b_1a_1 + b_2a_2$. Hence $r = a - sd = (a_1' - sb_1)a_1 + (a_2' - sb_2)a_2$ is an element of (a_1, a_2) with $r < |d|$. If $r \neq 0$ then r is an element of (a_1, a_2) smaller than $|d|$ which is a contradiction; $r = 0$. Thus $a \in (d)$; $(a_1, a_2) \subseteq (d)$.

For uniqueness suppose $d' \in \mathbb{Z}$ is another element in (a_1, a_2) with the smallest absolute value. Then $(d) = (a_1, a_2) = (d')$. As we noted before, d and d' differ by a sign.

(2) Assume $d = \gcd(a_1, a_2)$ and let $a_1 = b_1d, a_2 = b_2d$. Then $a_1'a_1 + a_2'a_2 = (a_1'b_1 + a_2'b_2)d \in (d)$ for any $a_1', a_2' \in \mathbb{Z}$. Hence $(a_1, a_2) \subseteq (d)$. Now by (1) there is $d' \in \mathbb{Z}$ such that $(a_1, a_2) = (d')$. Then since $a_1, a_2 \in (d')$ we have $d'|a_1, d'|a_2$. By the definition of gcd, $d'|d$ and hence $(d') \supseteq (d)$. Thereby we conclude $(a_1, a_2) = (d)$.

Conversely, suppose $(a_1, a_2) = (d)$. Then $a_1, a_2 \in (d)$ which means $a_i = da_i' (i = 1, 2)$ i.e., $d|a_i$. Let $d'|a_i$ $(i = 1, 2)$; $a_i = a_i'd'$. We need to show $d'|d$. Since $d \in (a_1, a_2)$ we can write $d = b_1a_1 + b_2a_2$ for some $b_i \in \mathbb{Z}$. Hence $d = b_1a_1 + b_2a_2 = a_1'b_1d' + a_2'b_2d'$ and hence d is divisible by d'. As required.

(3) Exercise. □

Corollary B.1.3. *Let $a, b \in \mathbb{Z}$. Then a, b are relatively prime if and only if there are $a', b' \in \mathbb{Z}$ such that $a'a + b'b = 1$. In this case, for any $k \in \mathbb{Z}$ there are $c, d \in \mathbb{Z}$ such that $ca + db = k$.*

Proof. This is a special case of Theorem B.1.2 (2). For the second statement multiply both sides of the equality $a'a + b'b = 1$ by k. □

Corollary B.1.4. *Let $a, b, c \in \mathbb{Z}$. Suppose $c|ab$ and $(a, c) = 1$ then $c|b$.*

Proof. By Corollary B.1.3, there are $a', b' \in Z$ such that $1 = a'a + b'c$. Multiplying through by b we get $b = a'ab + b'cb$ which is divisible by c since $c|ab$ by assumption. □

Corollary B.1.5. *An integer $p \in \mathbb{Z}$ is prime if and only if $p|bc$ implies either $p|b$ or $p|c$.*

Proof. Assume p is a prime and let $p|bc$ $(b, c \in \mathbb{Z})$. If $p \nmid b$ then p and b are relatively prime. Hence, by Corollary B.1.3, there is $a', b' \in \mathbb{Z}$ such that $1 = a'p + b'b$. Since $p|bc$ by assumption, $c = a'pc + b'bc$ is divisible by p.

Conversely, suppose $p = bc$ $(b, c \in \mathbb{Z})$. We need to show that one of b or c is ± 1. Now $p|bc$ and hence $p|b$ or $p|c$; say $p|b$. If we let $b = pa$ for some $a \in \mathbb{Z}$ we have then $p = bc = pac$. Therefore $ac = 1$ namely $c = \pm 1$. Therefore p is prime. □

The following theorem so called mathematical induction depends on the fact that a subset of the natural number $\mathbb{N}$ has the smallest number.

Theorem B.1.6 (Mathematical Induction). *Consider the statements $S(n), n = 1, 2, \ldots$. Suppose*

(1) *the statement $S(1)$ is true, and*

(2) *the statement $S(n)$ is true assuming the statement $S(n - 1)$.*

Then the statements $S(n), n = 1, 2, \ldots$ are all true.

Proof. Assume the contrary: The set $S = \{k \mid S(k) \text{ is false}\}$ is nonempty. Then S has a least member say k_0. Then $k_0 - 1 \notin S$ and hence $S(k_0 - 1)$ is true. But then the condition (2) says $S(k_0)$ is true. This is a contradiction. □

B.2 Fields and Rings

Definition B.2.1. A **field** is a set F with two operations **addition** $a + b \in F$, and **multiplication** $a \cdot b \in F$ (written simply ab) for each $a, b \in F$ satisfying the properties

(1) $a + b = b + a$ and $ab = ba$ (commutativity)

(2) $(a + b) + c = a + (b + c)$ and $a(bc) = (ab)c$ (associativity)

(3) There are two distinct elements 0 and 1 in F such that

$$a + 0 = a \quad \text{and } a \cdot 1 = a \tag{B.1}$$

(existence of identity)

(4) For each element $a \in F$ and for each nonzero element $b \in F$ there are elements $a' \in F$ and $b' \in F$ such that

$$a + a' = 0 \text{ and } bb' = 1 \tag{B.2}$$

(existence of additive inverse and multiplicative inverse)

(5) $a(b + c) = ab + ac$ (distributivity of multiplication over addition)

If we do not require the existence of multiplicative inverse in (B.2) then we call such an algebraic system a **commutative ring**. The element 0 in (B.1) is called the **zero element** and a' and b' in (B.2) are called the **additive inverse** and the **multiplicative inverse** and they are denoted by $-a$ and b^{-1} respectively.

The set $\mathbb{Q}$ of all rational numbers form a field which we call the **field of rational numbers**. The set $\mathbb{R}$ of real numbers form a field which we call the **field of real numbers**. The set $\mathbb{C}$ of all complex numbers $\{a + bi | a, b \in \mathbb{R}\}$ with $i^2 = -1$ is called the **field of complex numbers**.

On the other hand, the set $\mathbb{Z}$ of integers is not a field because, for example $2 \in \mathbb{Z}$ does not have a multiplicative inverse. The set of integers $\mathbb{Z}$ and the set of polynomials $F[t]$ with the coefficients in a field F are commutative rings.

Example B.2.2. Let $\mathbb{Q}[\sqrt{2}] = \{a + b\sqrt{2} \mid a, b \in \mathbb{Q}\}$. Then $\mathbb{Q}[\sqrt{2}]$ is a subset of $\mathbb{R}$ and closed under addition and multiplication. We need to show if $a + b\sqrt{2} \neq 0$ then it has a multiplicative inverse. In fact,

$$\frac{1}{a + b\sqrt{2}} = \frac{a - b\sqrt{2}}{(a + b\sqrt{2})(a - b\sqrt{2})} = \frac{a}{a^2 - 2b^2} - \frac{b}{a^2 - 2b^2}\sqrt{2}$$

which is again an element of $\mathbb{Q}[\sqrt{2}]$. Hence $\mathbb{Q}[\sqrt{2}]$ is a field.

Of course we can replace 2 by any square free (having no square factors) integer n and we conclude that $\mathbb{Q}[\sqrt{n}]$ is a field.

Example B.2.3. Let p be a prime number. On the set $\mathbb{Z}/p$ of equivalence classes of integers modulo p (Example A.2.3) we define the addition and multiplication by

$$[a] + [b] = [a + b] \text{ and } [a] \cdot [b] = [ab]$$

Then one checks that these are well defined operations namely if $[a] = [a']$ and $[b] = [b']$ then $[a+b] = [a'+b']$ and $[ab] = [a'b']$. Now the set $\mathbb{Z}/p$ becomes a field under these operations which we call the **finite field with** p **elements**. To check this we need to show that every nonzero element $\bar{a}$ has a multiplicative inverse. In fact, if $\bar{a}$ is nonzero then we can choose $0 < a < p$. Hence a and p are relatively prime and hence there are c, c' such that $ac + c'p = 1$ by Corollary B.1.3. Therefore $\bar{a}\bar{c} = 1$ in $\mathbb{Z}/p$. The other conditions for $\mathbb{Z}/p$ to be a field are easy to check.

Here are tables of addition and multiplication for $\mathbb{Z}/3$.

$+$	$\bar{0}$	$\bar{1}$	$\bar{2}$
$\bar{0}$	$\bar{0}$	$\bar{1}$	$\bar{2}$
$\bar{1}$	$\bar{1}$	$\bar{2}$	$\bar{0}$
$\bar{2}$	$\bar{2}$	$\bar{0}$	$\bar{1}$

$\cdot$	$\bar{0}$	$\bar{1}$	$\bar{2}$
$\bar{0}$	$\bar{0}$	$\bar{0}$	$\bar{0}$
$\bar{1}$	$\bar{0}$	$\bar{1}$	$\bar{2}$
$\bar{2}$	$\bar{0}$	$\bar{2}$	$\bar{1}$

(B.3)

Observe that for any $\bar{a} \in \mathbb{Z}/p$ we have $p\bar{a} = \bar{a} + \bar{a} + \cdots + \bar{a} = 0$, the p-times addition of $\bar{a}$, becomes 0 in $\mathbb{Z}/p$.

Definition B.2.4. Let F be a field. If there is a prime p such that $pa = 0$ for all $a \in F$ then we say the **characteristic** of F is p and we write $\text{char}(F) = p$. If there is no such p then we say F is of characteristic 0 and we write $\text{char}(F) = 0$.

The fields $\mathbb{Q}, \mathbb{R}, \mathbb{C}$ are the fields of characteristic 0 and $\mathbb{Z}/p$ is a field of characteristic p. When we deal with bilinear forms (§8.6) we sometimes need the assumption that the characteristic of a field F is not 2. This signifies that $1 \neq -1$ or $a + a \neq 0$ for all $a \neq 0$.

Theorem B.2.5. *Let p be a prime and n be a positive integer. Then there is a field with p^n elements which we denote by $\mathbb{F}_{p^n}$. The field $\mathbb{F}_{p^n}$ is uniquely determined, up to an isomorphism, and contains $\mathbb{F}_p$. Any finite field is isomorphic to one of these fields.*

Proof. See §13.5 of [1]. □

The determinant of a matrix in Chapter 3 can be defined over a commutative ring R. All the results of Chapter 3 are valid if they are stated properly. For example, a matrix A over a field F is invertible if and only if $\det(A) \neq 0$. However, if A is a square matrix over a commutative ring R then non-vanishing of $\det(A)$ is not good enough to guarantee that A is invertible. In fact,

Theorem B.2.6. *Let A be a square matrix over a commutative ring. Then A is invertible if and only if $\det(A)$ has a multiplicative inverse.*

Proof. This follows form Exercise 3.4.6 for commutative ring. See §11.4 of [1]. □

Hence if $A \in M(n, \mathbb{Z})$ is a square matrix over $\mathbb{Z}$ then A is invertible, by definition, if there is $B \in M(n, \mathbb{Z})$ such that $AB = BA = I_n$. In this case, Theorem 3.3.7 becomes: A is invertible if and only if $\det A = \pm 1$.

Instead of $\mathbb{Z}$, if the commutative ring is the polynomial ring $F[t]$ over a field F then the corresponding statement will be: $A \in M(n, F[t])$ is invertible if and only if $\det A$ is a nonzero constant.

B.3 Complex Numbers

As a set the complex numbers $\mathbb{C}$ consists of

$$\mathbb{C} = \{a + bi \mid a, b \in \mathbb{R}\}$$

where $i^2 = -1$. We view $\mathbb{C}$ as a vector space of dimension 2 over the field $\mathbb{R}$ of real numbers with the 'standard' basis $\{1, i\}$. We often represent $\mathbb{C}$ as a plane with the imaginary axis and the real axis which we call the **complex plane**. The angle θ between the real axis and the complex number z, is called the **argument** of z.

The product of two complex numbers is defined by

$$(a + bi)(c + di) = ac - bd + (ad + bc)i$$

The **(complex) conjugate** $\bar{z}$ of $z = a + bi$ is defined by

$$\bar{z} = a - bi$$

Then

$$z\bar{z} = a^2 + b^2 = |z|^2 \tag{B.4}$$

and the **absolute value** $|z| = \sqrt{a^2 + b^2}$ is the **length** of the complex number $z \in \mathbb{C}$. It is easy to prove that

$$|z_1 z_2| = |z_1||z_2| \tag{B.5}$$

Let z be a complex number with $|z| = 1$ then we can write z as

$$z = \cos\theta + i\sin\theta \tag{B.6}$$

If z is any complex number with the argument θ then we can write $z = |z|\frac{z}{|z|}$ where $\frac{z}{|z|}$ is a complex number of length 1. Hence we can write by (B.6)

$$z = |z|(\cos\theta + i\sin\theta) \tag{B.7}$$

where $|z|$ is a real number and $(\cos\theta + i\sin\theta)$ is a complex number of length 1 i.e., it is on the unit circle in the complex plane.

Theorem B.3.1. $e^{i\theta} = \cos\theta + i\sin\theta$

Proof. From calculus we have

$$\cos\theta = \sum_{n=0}^{\infty} \frac{(-1)^n \theta^{2n}}{(2n)!} = \sum_{n=0}^{\infty} \frac{(i\theta)^{2n}}{(2n)!}$$
$$i\sin\theta = i\sum_{n=0}^{\infty} \frac{(-1)^n \theta^{2n+1}}{(2n+1)!} = \sum_{n=0}^{\infty} \frac{(i\theta)^{2n+1}}{(2n+1)!}$$

Adding the last terms we get $\cos\theta + i\sin\theta = \sum_{n=0}^{\infty} \frac{(i\theta)^n}{n!} = e^{i\theta}$. □

Hence we get the **Euler's formula**

$$z = |z|(\cos\theta + i\sin\theta) = |z|e^{i\theta} \tag{B.8}$$

where θ is the argument of z. Therefore if $z_1 = |z_1|e^{\theta_1}$ and $z_2 = |z_2|e^{\theta_2}$ then

$$z_1 z_2 = |z_1||z_2|e^{i(\theta_1+\theta_2)} \tag{B.9}$$

If $z = |z|e^{\theta}$ then

$$z^n = |z|^n e^{in\theta} \ (n = 1, 2, ...) \tag{B.10}$$

Also it can be easily checked that (B.10) is valid for $n \in \mathbb{Q}$. And hence

$$\left\{ e^{\frac{2k\pi i}{n}} \middle| k = 0, 1, ..., n-1 \right\} \tag{B.11}$$

are the n-th root of unity i.e., they are n roots of the polynomial equation $t^n - 1 = 0$. More generally, the roots of the polynomial equation $t^n - z \in \mathbb{C}[t]$ is given by

$$\sqrt[n]{|z|}\ e^{\frac{2k\pi i}{n}} \ (k = 0, 1, ..., n-1) \tag{B.12}$$

where $\sqrt[n]{|z|}$ is the positive real n-th root of $|z|$.

Appendix C

Polynomials

C.1 Polynomials

Definition C.1.1. Let F be a field. The set of all **polynomials**

$$f(t) = a_0 + a_1 t + \cdots + a_{n-1}t^{n-1} + a_n t^n \quad (a_i \in F,\ n \geq 0) \tag{C.1}$$

with the coefficients in F is denoted by $F[t]$ which we call the **ring of polynomials** over F. The **degree** $\deg(f)$ of $f(t)$ in (C.1) is defined by

$$\deg(f) = \begin{cases} n & \text{if } a_n \neq 0, n \geq 0 \\ -\infty & \text{if } a_i = 0 \ (n \geq i \geq 0) \end{cases} \tag{C.2}$$

We define the addition and multiplication of the polynomials by

$$\begin{aligned} \sum_{i=0}^{n} a_i t^i + \sum_{i=0}^{n} b_i t^i &= \sum_{i=0}^{n} (a_i + b_i) t^i \\ \left(\sum_{i=0}^{n} a_i t^i\right)\left(\sum_{i=0}^{n} b_i t^i\right) &= \sum_{k \geq 0} \left(\sum_{i+j=k} a_i b_j\right) t^{i+j} \end{aligned} \tag{C.3}$$

The degree of a polynomial satisfies the formula

$$\deg(fg) = \deg(f) + \deg(g) \tag{C.4}$$

Let $X = \{-\infty, 0, 1, 2, ...\}$. For any $x \in X$ let's agree that $-\infty + x = -\infty$ and $-\infty \leq x$. Then (C.4) is valid in all cases.

The algebra of polynomials $F[t]$ with coefficients in the field F is very similar to the algebra of the integers $\mathbb{Z}$ (see (B.1)). The properties of division algorithm and uniqueness of factorization are two basic properties which are common in $\mathbb{Z}$ and $F[t]$. The division algorithm for $\mathbb{Z}$ we require the remainder r to be smaller than the "absolute value" of the quotient b whereas the division algorithm for $F[t]$ we require the "degree" of the remainder to be of lower degree than the degree of the quotient. And the "primes" of $\mathbb{Z}$ will be replaced by the "irreducible polynomials" in $F[t]$.

Definition C.1.2. A polynomial is said to be **monic** if the coefficient of the highest degree term is 1. A polynomial $p(t) \in F[t]$ is said to be **irreducible** if $p(t)$ is not a product of two nonconstant polynomials i.e., if $p = ab$ where $a, b \in F[t]$ then a or b is a constant. For $f, g \in F[t]$ we say f is **divisible by** g if there is an $h \in F[t]$ such that $f = gh$. In this case, we will write $g|f$.

In case of the integers, if $m, n \in \mathbb{Z}$ and $n|m$ as well as $m|n$ then we have $m = \pm n$. For the polynomials $f, g \in F[t]$ if $f|g$ and $g|f$ then we conclude that $f = cg$ for some constant $c \in F$. In fact, $f = hg$ and $g = h'f$. Hence $f = hg = hh'f$ and hence $hh' = 1$ and hence $\deg h + \deg h' = \deg 1 = 0$. Since $\deg h, \deg h' \geq 0$ we conclude that $\deg h = \deg h' = 0$; h and h' are constants.

Theorem C.1.3. (1) **(Division Algorithm)** *Let $f, g \in F[t]$. Then there are unique $s, r \in F[t]$ such that*

$$f = sg + r \quad \text{with } \deg(r) < \deg(g)$$

We may have $r = 0$ in which case $\deg(r) = -\infty$. *And $r \in F[t]$ is nonzero if and only if* $\deg(r) \geq 0$.

(2) **(Unique Factorization of Polynomials)** *Every monic polynomial $f(t)$ can be written uniquely as $f(t) = p_1(t)^{e_1} p_2(t)^{e_2} \cdots p_k(t)^{e_k}$ where $p_i(t)$ are monic irreducible.*

Proof. (1) Follows from long division of polynomials.

(2) Let $f(t)$ be a monic polynomial. We induct on the degree of f. If the degree of f is 1 then obviously f is irreducible. Now assume $\deg(f) > 1$. If there is no nonconstant polynomial dividing f then f is irreducible. Now suppose $f = gh$ with $\deg(g) > 0, \deg(h) > 0$. Then by (C.4), $\deg(f) > \deg(g), \deg(f) > \deg(h)$. Hence, by induction, g and h are product of irreducible monic polynomials say, $g = p_1^{e_1} p_2^{e_2} \cdots p_k^{e_k}, h = q_1^{f_1} q_2^{f_2} \cdots q_l^{f_l}$ where p_i, q_j's are monic irreducible polynomials. Hence $f = gh = p_1^{e_1} p_2^{e_2} \cdots p_k^{e_k} q_1^{f_1} q_2^{f_2} \cdots q_l^{f_l}$ as required.

For uniqueness let $p_1 p_2 \cdots p_r = f = q_1 q_2 \cdots q_s$ where p_i, q_j are irreducible monic polynomials. We need to show $r = s$ and $p_i = q_i$ with possible reindexing. Assume $r < s$. Now $p_1 | q_1 q_2 \cdots q_s$. Hence $p_1 = q_j$ for some q_j which we may assume $q_j = q_1$. Thus we have $p_2 | q_2 q_3 \cdots q_s$ and hence $p_2 = q_2$ after possible reindexing. Repeating this we arrive at $1 = q_t \cdots q_s$. Therefore $q_t = 1, \cdots, q_s = 1$ since we assumed q_i to be monic. □

Let $f(t) \in F[t]$. We say α is a **root** of $f(t)$ if $f(\alpha) = 0$.

Theorem C.1.4. *Let $f(t) \in F[t]$. Then $f(t)$ has at most n roots.*

Proof. We use induction. Let $f(t) \in F[t]$ be a polynomial of degree n. If there is no roots of f then there is nothing to prove. Suppose α is a root of f. Then $f(t) = (t - \alpha)g(t)$ where $g(t) \in F[t]$ with $\deg(g) = n - 1$. By induction assumption, $g(t)$ has $\leq n - 1$ roots. Therefore f has at most n roots. □

By Fundamental Theorem of Algebra (Theorem C.2.1), we see that any $f(t) \in \mathbb{C}[t]$ splits in $\mathbb{C}[t]$ i.e., $f(t) = \alpha \prod_i (t - \alpha_i)$. A field K is defined to be **algebraically closed** if every $f(t) \in K[t]$ splits in $K[t]$. Hence $\mathbb{C}$ is algebraical closed. Next theorem guarantees that any field F is contained in an algebraically closed field.

Theorem C.1.5. *For any field F there is an algebraically closed field K containing F. There is the 'smallest' algebraically field $\bar{F}$ containing F contained in K which we call the* **algebraic closure** *of F in K.*

Proof. See [1] §13.4 Proposition 30. See also Proposition 31. □

C.2 Real and Complex Polynomials

Theorem C.2.1 (Fundamental Theorem of Algebra). *Let $f(t) \in \mathbb{C}[t]$ be a monic polynomial. Then $f(t)$ is a product of monic polynomials of degree 1 i.e., we can write $f(t) = \prod_{i=1}^{n}(t - \alpha_i)$ with $\alpha_i \in \mathbb{C}$ which are not necessarily distinct.*

Proof. See §14.6, Theorem 35 of [1]. □

Example C.2.2. For example, to find the roots of $f(t) = t^2 + 2t - i$ we use the quadratic formula

$$\frac{-b \pm \sqrt{b^2 - 4c}}{2}$$

for $f(t) = t^2 + bt + c$. First we compute the square root part:

$$\sqrt{b^2 - 4c} = (4 + 4i)^{\frac{1}{2}} = 2\left(\frac{1}{\sqrt{2}} + \frac{i}{\sqrt{2}}\right)^{\frac{1}{2}} = 2\left(e^{\frac{i\pi}{4}}\right)^{\frac{1}{2}} = 2e^{\frac{i\pi}{8}} = 2\left(\cos\frac{\pi}{8} + i\sin\frac{\pi}{8}\right)$$

Using the half angle formula we get $\cos^2\left(\frac{\pi}{8}\right) = \frac{2+\sqrt{2}}{4}$ and $\sin^2\left(\frac{\pi}{8}\right) = \frac{2-\sqrt{2}}{4}$. Hence

$$\cos\left(\frac{\pi}{8}\right) = \frac{\sqrt{2+\sqrt{2}}}{2} \quad \text{and} \quad \sin\left(\frac{\pi}{8}\right) = \frac{\sqrt{2-\sqrt{2}}}{2}$$

since $\cos\left(\frac{\pi}{8}\right), \sin\left(\frac{\pi}{8}\right)$ are positive numbers. And therefore roots of f are

$$\frac{-2 \pm \left(\sqrt{2+\sqrt{2}} + i\sqrt{2-\sqrt{2}}\right)}{2}$$

The algebraic formulas for the cubic equation (Cardano's formula) and the formula for the quartic equation (Ferrari's formula) were found during the 16th century in Italy (§14.7 of [1]). For the polynomial equations of degree higher than 5 there is no algebraic formulas for the roots of the polynomial equations. This was proved by Abel and Galois after 300 years since the Italian mathematicians solved the cubic and quartic equations. See §14.7, Corollary 40 of [1] for its proof.

Theorem C.2.3. *Let $f(t) \in \mathbb{R}[t]$ be a monic polynomial.*

(1) *If $\alpha = a + bi$ is a nonreal root of f then its conjugate $\bar{\alpha} = a - bi$ is also a root of f. A real quadratic polynomial having the roots $\alpha, \bar{\alpha}$ is of the form $f(t) = t^2 - 2at + (a^2 + b^2)$ for some $c \in \mathbb{R}$.*

(2) *The polynomial $f(t)$ is a product of real linear or real quadratic polynomials. More precisely, suppose f has the real roots α_i $(i = 1, 2, ..., r)$ and the nonreal complex roots $\alpha_k = a_k + ib_k, \bar{\alpha}_k = a_k - ib_k$ $(k = r+1, r+2, ..., n)$ then f is of the form*

$$f(t) = \prod_{i=1}^{r}(t - \alpha_i) \prod_{k=r+1}^{n} (t^2 - 2a_k t + (a_k^2 + b_k^2))$$

Proof. (1) Let $f(\alpha) = 0$ $(\alpha \in \mathbb{C})$. Then $0 = \overline{f(\alpha)} = f(\bar{\alpha})$ since f has real coefficients. For the second statement we have $f(t) = (t - \alpha)(t - \bar{\alpha}) = t^2 - 2at + (a^2 + b^2)$.

(2) We can write $f(t) = \prod_{i=1}^{r}(t - \alpha_i)\prod_{k=r+1}^{n}(t - \alpha_k)(t - \bar{\alpha}_k)$ by (1) and Theorem C.2.1. Now $(t - \alpha_k)(t - \bar{\alpha}_k) = t^2 - 2a_k t + (a_k^2 + b_k^2)$. Hence

$$
\begin{aligned}
f(t) &= \prod_{i=1}^{r}(t-\alpha_i)\prod_{k=r+1}^{n}(t-\alpha_k)(t-\bar{\alpha}_k) \\
&= \prod_{i=1}^{r}(t-\alpha_i)\prod_{k=r+1}^{n}(t^2-2a_kt+(a_k^2+b_k^2))
\end{aligned}
$$

where the second product of the last equality is a product of real quadratics. □

Hence irreducible real polynomials are of degree 1 or 2. For the polynomials with the rational coefficients there are infinitely many irreducible polynomials in each degree. See [1] §9.4.

Appendix D

Existence of Basis and its Cardinality

D.1 Maximal Principle and Cardinal Numbers

Definition D.1.1. Let A be a set. A **partial order** on A is a relation $a \leq b$ on pairs of elements $(a, b) \in A \times A$ satisfying the following relations:

(i) $a \leq a$ for all $a \in A$

(ii) If $a \leq b$ and $b \leq a$ then $a = b$.

(iii) If $a \leq b$ and $b \leq c$ then $a \leq c$.

Let S be a subset of the partially ordered set A. An element $u \in A$ is an **upper bound** of S if $s \leq u$ for all $s \in S$. We say a and b are **comparable** if either $a \leq b$ or $b \leq a$. A subset C of A is a **chain** if any pair of elements of C are comparable.

The following theorem is known to be equivalent to one of the axioms of set theory so called the axiom of choice.

Theorem D.1.2 (Zorn's Lemma). *Let A be a partially ordered set. If every chain has an uppper bound then A has a maximal element.*

Proof. [3] §5.5 Theorem 5.20. □

Definition D.1.3. Let A, B sets. If there is a bijection $f : A \to B$ then we say that A and B have the **same cardinality**. If there is an injection $f : A \to B$ then we say A is of **less than or equal cardinality**. Among the sets having the same cardinality we choose a representative called the **cardinal number**. We will denote the cardinal numbers by $\mathbf{a}, \mathbf{b}, \mathbf{c}, \mathbf{d}, \ldots$. If there is a bijection between a set A and $\mathbf{a}$ then we will write $|A| = \mathbf{a}$. For two cardinal numbers $\mathbf{a}, \mathbf{b}$ we write $\mathbf{a} \leq \mathbf{b}$ if there is an injection $f : \mathbf{a} \to \mathbf{b}$.

By Theorem A.1.5, $\mathbf{a} \leq \mathbf{b}$ if and only if there is an surjection $f : \mathbf{b} \to \mathbf{a}$.

The cardinal number 1 represent the sets having a single element and the cardinal umber 2 represent the sets of two elements and so on. The natural number $\mathbb{N}$ is an infinite set and it is of the smallest cardinality among the infinite cardinals.

Definition D.1.4. The cardinal number of the natural numbers $\mathbb{N}$ is denoted by $\aleph_0$.*

Theorem D.1.5 (Cantor). *There is no bijection between the natural numbers* $\mathbb{N}$ *and the real numbers* $\mathbb{R}$.

Proof. See §7.1 of [3]. □

The cardinality of $\mathbb{R}$ is denoted by $\aleph_1$. And the continuum hypothesis asserts that there is no cardinality between $\aleph_0$ and $\aleph_1$. It can be shown that for any set X the cardinality of X is smaller than the cardinality of its power set $\mathcal{P}(X)$ (Theorem D.1.8). If the cardinality of X is $\aleph_\alpha$ then we denote the cardinality of $\mathcal{P}(X)$ by $\aleph_{\alpha+1}$.

Theorem D.1.6 (Schöder-Bernstein). *The relation* $\leq$ *on the cardinal numbers is a partial ordering. In particular, if* $\mathbf{c} \leq \mathbf{d}$ *and* $\mathbf{c} \geq \mathbf{d}$ *then* $\mathbf{c} = \mathbf{d}$.

Proof. See [3] §8.3 Theorem 8.14. □

Definition D.1.7. Let $\mathbf{a}, \mathbf{b}$ be cardinal numbers. Choose the disjoint sets A, B such that $|A| = \mathbf{a}$ and $|B| = \mathbf{b}$. We define the **sum** and **product** of $\mathbf{a}$ and $\mathbf{b}$ by

$$\mathbf{a} + \mathbf{b} = |A \cup B| \text{ and } \mathbf{ab} = |A \times B| \tag{D.1}$$

Let A^B be the set $\{f : B \to A\}$ of all functions from B to A. We define $\mathbf{a}^\mathbf{b}$ by

$$\mathbf{a}^\mathbf{b} = |A^B|$$

For $a \in A$ if we let $B_a = \{a\} \times B$ then $\{B_a\}_{a \in A}$ is a disjoint family of sets and by (D.1) we have

$$|\bigcup_{a \in A} B_a| = |A||B| \tag{D.2}$$

Theorem D.1.8. *If the cardinality of* X *is* $\mathbf{b}$ *then the cardinality of* $\mathcal{P}(X)$ *is* $2^\mathbf{b}$. *For any cardinal* $\mathbf{b}$ *we have* $2^\mathbf{b} > \mathbf{b}$.

Proof. See [3] Theorem 7.1. □

Theorem D.1.9. *If* $\mathbf{b}, \mathbf{c}$ *are infinite cardinal numbers then* $\mathbf{bc} = \mathbf{b} + \mathbf{c} = \max\{\mathbf{b}, \mathbf{c}\}$. *If* $\mathbf{a}$ *is a cardinal number such that* $\mathbf{a} \leq \aleph_0$ *then* $\mathbf{ac} = \mathbf{c}^\mathbf{a} = \mathbf{c}$.

Proof. See [3] §8.4 Corollary 8.27. The second statement follows from the first statement. □

Definition D.1.10. Let $\{A_i\}_{i \in I}$ be a family of pairwise disjoint sets with the cardinal numbers $|A_i| = \mathbf{a_i}$. We define the **infinite addition** and **infinite product** by

$$\sum_{i \in I} \mathbf{a}_i = \left| \bigcup_{i \in I} A_i \right| \text{ and } \prod_{i \in I} \mathbf{a}_i = \left| \prod_{i \in I} A_i \right|$$

One of the basic properties of the infinite cardinal sum and product we have:

Theorem D.1.11. *Let* $\{\mathbf{a}_i\}_{i \in I}$ *and* $\{\mathbf{b}_i\}_{i \in I}$ *be a family of cardinal numbers. If* $\mathbf{a}_i \leq \mathbf{b}_i$ *then*

$$\sum_{i \in I} \mathbf{a}_i \leq \sum_{i \in I} \mathbf{b}_i \text{ and } \prod_{i \in I} \mathbf{a}_i \leq \prod_{i \in I} \mathbf{b}_i$$

Proof. See [3] §8.5 Theorem 8.32. □

*$\aleph$ (aleph) is the first Hebrew alphabet.

Theorem D.1.12. (1) *Let A be a set and* $\mathbf{c}$ *be a cardinality. Let $\{X_\alpha\}_{\alpha\in A}$ be a family if sets such that $|X_\alpha| \le \mathbf{c}$. Then*

$$\left|\bigcup_{\alpha\in A} X_\alpha\right| \le \mathbf{c}|A|$$

(2) *Let X be an infinite set and $\mathcal{F}_X = \{S \mid S \text{ is a finite subset of } X\}$. Then $|\mathcal{F}_X| = |X|$.*

Proof. (1) Follows from Theorem D.1.11.

(2) For $n = 1, 2, \ldots$, let $X_n = \{S \subseteq X \mid \#S \le n\}$ be the family of subsets of X with $\le n$ elements and X^n be the n-times Cartesian product of X. Then there is a surjection $X^n \to X_n$ sending the ordered tuple $(x_1, x_2, \ldots, x_n)$ to the subset $\{x_1, x_2, \ldots, x_n\}$ of $\le n$ elements (because some of x_i's might be the same). Obviously there is an inclusion map $X \to X_n$. Hence $|X| \le |X_n| \le |X^n| = n|X| = |X|$ by Theorem D.1.9. Now since $\mathcal{F}_X = \cup_i X_i$ and $|X_i| = |X|$ we have $|\mathcal{F}_X| \le \aleph_0|X| = |X|$ by Theorem D.1.9 as $\aleph_0$ is the smallest infinite cardinal. Since $|X| \le |\mathcal{F}_X|$ obviously, we have $|\mathcal{F}_X| = |X|$. □

D.2 Existence and Equipotence of Bases

First we prove that every vector space has a basis.

Theorem D.2.1. *Let V be a vector space over a field F. Then V has a basis.*

Proof. Let $\mathcal{S}$ be the set of all linearly independent subsets of V partially ordered by inclusion. First $\mathcal{S} \ne \emptyset$ since for any $\mathbf{v} \ne 0$ the set $\{\mathbf{v}\}$ is linearly independent. Let $\{\mathcal{C}_\alpha\}$ be a chain in $\mathcal{S}$. To apply Zorn's lemma (Theorem D.1.2) we will show that $\{\mathcal{C}_\alpha\}$ has an upper bound. For this, let $\mathcal{C} = \bigcup_\alpha \mathcal{C}_\alpha$. We need to show $\mathcal{C}$ is an upper bound for $\mathcal{C}_\alpha$. For this we need to show $\mathcal{C}$ is linearly independent. In fact, let $a_1\mathbf{v}_1 + \cdots + a_n\mathbf{v}_n = 0$ ($\mathbf{v}_i \in \mathcal{C}$). Then there is α such that $\mathbf{v}_1, \ldots, \mathbf{v}_n \in \mathcal{C}_\alpha$ since $\{\mathbf{v}_1, \ldots, \mathbf{v}_n\}$ is contained in $\bigcup_\alpha \mathcal{C}_\alpha$. But then since $\mathcal{C}_\alpha$ is linearly independent we have $a_1 = \cdots = a_n = 0$ and hence $\mathcal{C}$ is linearly independent. Then it is obvious that $\mathcal{C}$ is an upper bound of $\{\mathcal{C}_\alpha\}$. Hence every chain in $\mathcal{S}$ has an upper bound.

By Zorn's lemma $\mathcal{S}$ has a maximal element say $\mathcal{B}$. We claim that $\mathcal{B}$ is a basis of V. Since $\mathcal{B} \in \mathcal{S}$ we see that $\mathcal{B}$ is linearly independent. Hence it suffices to show that $\mathcal{B}$ span V. Let $\mathbf{v} \in V$. Then, since $\mathcal{B}$ is maximal, $\mathcal{B} \cup \{\mathbf{v}\}$ is linearly dependent. Hence there are c's in F such that

$$c\mathbf{v} + c_1\mathbf{v}_1 + \cdots + c_n\mathbf{v}_n = \mathbf{0}$$

First note that $c \ne 0$. For otherwise, $c_1\mathbf{v}_1 + \cdots + c_n\mathbf{v}_n = 0$ which contradicts to the fact that $\mathcal{B}$ is linearly independent. Hence

$$\mathbf{v} = -\frac{1}{c}(c_1\mathbf{v}_1 + \cdots + c_n\mathbf{v}_n)$$

which proves that $\mathcal{B}$ span V. □

Let S be a linearly independent subset of V. If we take, in the proof of Theorem D.2.1, $\mathcal{S}$ to be the set of all linearly independent subset of V containing S then we obtain:

Theorem D.2.2. *Let V be a vector space over a field F and let S be a linearly independent subset of V. Then V has a basis containing S.*

Next we prove that any two bases of a vector space have the same cardinality.

Theorem D.2.3. *Any two bases of a vector space V over a field have the same cardinality.*

Proof. If V has a finite basis then any two basis have the same number of vectors by Corollary 6.2.14.

Now let $\mathcal{B}, \mathcal{B}'$ be two bases of an infinite dimensional vector space V. For $\mathbf{w} \in \mathcal{B}'$ let

$$\mathcal{B}_{\mathbf{w}} = \left\{ \mathbf{v}_i \in \mathcal{B} \,\middle|\, \mathbf{w} = \sum a_i \mathbf{v}_i \right\}$$

Then, since $\mathcal{B}$ is a basis, $\mathcal{B}_{\mathbf{w}}$ is a finite subset of $\mathcal{B}$. Next we claim that $\underset{\mathbf{w}\in\mathcal{B}'}{\cup} \mathcal{B}_{\mathbf{w}} = \mathcal{B}$. We need to show that $\underset{\mathbf{w}\in\mathcal{B}'}{\cup} \mathcal{B}_{\mathbf{w}} \supseteq \mathcal{B}$. Assume the contrary: there is $\mathbf{z} \in \mathcal{B} \setminus \underset{\mathbf{w}\in\mathcal{B}'}{\cup} \mathcal{B}_{\mathbf{w}}$. Then since $\mathcal{B}'$ is a basis of V, $\mathbf{z}$ is a linear combination of elements of $\mathcal{B}'$: $\mathbf{z} = \sum c_i \mathbf{w}_i$ ($\mathbf{w}_i \in \mathcal{B}'$). Now $\mathbf{w}_i$ is a linear combination of $\mathcal{B}_{\mathbf{w}_i}$: $\mathbf{w}_i = \sum_j d_{ij} \mathbf{v}_{ij}$ ($\mathbf{v}_{ij} \in \mathcal{B}_{\mathbf{w}_i}$)

$$\mathbf{z} = \sum c_i \mathbf{w}_i = \sum c_i \left(\sum d_{ij} \mathbf{v}_{ij} \right)$$

Hence $\mathbf{z} \in \mathcal{B}$ is a linear combination of other elements of $\mathcal{B}$. Thus $\mathcal{B}$ is linearly dependent which contradicts to the fact that $\mathcal{B}$ is a basis. This proves our claim.

Now since $|\mathcal{B}_{\mathbf{w}}| < \infty$ and $\cup_{\mathbf{w}} \mathcal{B}_{\mathbf{w}} = \mathcal{B}$ we have $\aleph_0 |\mathcal{B}'| \geqslant \left| \bigcup_{\mathbf{w}\in\mathcal{B}'} \mathcal{B}_{\mathbf{w}} \right| = |\mathcal{B}|$ (Theorem D.1.12). On the other hand, $\aleph_0 |\mathcal{B}'| = |\mathcal{B}'|$ (Theorem D.1.9). Hence $|\mathcal{B}'| \geqslant |\mathcal{B}|$. By reversing the role of $\mathcal{B}, \mathcal{B}'$ we have $|\mathcal{B}'| \geqslant |\mathcal{B}|$. Hence $|\mathcal{B}| = |\mathcal{B}'|$ by Theorem D.1.6. □

D.3 Cardinality of Basis of Dual Space

If V is a finite dimensional vector space then we proved that the bases for V and the dual $V^{\#}$of V have the same number of elements (6.50). Now we prove that the cardinality of a basis of the dual space $V^{\#}$ is bigger than the cardinality of a basis of V if $\dim(V) = \infty$. Hence, in this case, $V^{\#}$ is never isomorphic to V unlike the finite dimensional case.

The proof of the following theorem is based on Jacobson's book [2].

Theorem D.3.1. *Let V be an infinite dimensional vector space over a field F with a basis $\mathcal{B}$. Suppose the cardinality of F is $\mathbf{f}$ and the cardinality of $\mathcal{B}$ is $\mathbf{b}$. Then the cardinality of a basis of its dual space $V^{\#}$ is $\mathbf{f}^{\mathbf{b}}$.*

Proof. First we show that $|V| = \mathbf{bf}$. For each $\mathbf{v} \in V$ can be written uniquely as $\mathbf{v} = \sum_{i=1}^{n} a_i \mathbf{v}_i$ ($\mathbf{v}_i \in \mathcal{B}$). Hence for each $\mathbf{v}$ is in one-to-one correspondence with its $\mathcal{B}$-coordinate row $(a_1, a_2, ..., a_n) \in \prod_{i=1}^{n} F_{\mathbf{v}_i}$ where $F_{\mathbf{v}_i} = F$ indexed by $\mathcal{B}$. Under this identification we can write

$$V = \bigoplus_{\mathbf{v}\in\mathcal{B}} F_{\mathbf{v}} = \left\{ (a_{\mathbf{v}}) \in \prod\nolimits_{\mathbf{v}\in\mathcal{B}} \,\middle|\, \begin{array}{c} a_{\mathbf{v}} = 0 \text{ for all but} \\ \text{finitely many } \mathbf{v} \in \mathcal{B} \end{array} \right\}$$

The map $F \times \mathcal{B} \to \bigoplus_{\mathbf{v}\in\mathcal{B}} F_{\mathbf{v}}$ sending $(a, \mathbf{v})$ to $a_{\mathbf{v}} \in \bigoplus_{\mathbf{v}\in\mathcal{B}} F_{\mathbf{v}}$ is injective. Hence $|V| \geq \mathbf{bf}$.

Let

$$\mathcal{F}_{\mathcal{B}} = \{ S \mid S \text{ a finite subset of } \mathcal{B} \} \quad \text{and} \quad F_S = \prod_{\mathbf{v}\in S} F_{\mathbf{v}} \ (S \in \mathcal{F}_{\mathcal{B}})$$

Then $|\mathcal{F}_{\mathcal{B}}| = |\mathcal{B}|$ by Theorem D.1.12. And there is a surjection

$$\bigcup_{S\in\mathcal{F}_{\mathcal{B}}} F_S \to V; \ (a_{\mathbf{v}_1}, a_{\mathbf{v}_2}, ..., a_{\mathbf{v}_n}) \mapsto \sum_i a_{\mathbf{v}_i} \mathbf{v}_i$$

Since $|\mathcal{F}_{\mathcal{B}}| = |\mathcal{B}|$ we see that $|\bigcup_{S \in \mathcal{F}_{\mathcal{B}}} F_S| \leq \mathbf{bf}$. Hence $\mathbf{bf} \geq |\cup_{\mathcal{F}_{\mathcal{B}}} F_S| \geq |V|$. Therefore

$$|V| = \mathbf{bf} \tag{D.3}$$

Next we show $|V^{\#}| = \mathbf{f^b}$. If $f \in V^{\#}$ then the linear functional $f : V \to F$ is determined by its value $f(\mathbf{v})$ on each $\mathbf{v} \in \mathcal{B}$ (Theorem 6.3.9). On the other hand, if $(a_{\mathbf{v}}) \in \prod_{\mathbf{v} \in \mathcal{B}} F_{\mathbf{v}}$ then we can define a linear map $f : V \to F$ by requiring $f(\mathbf{v}) = a_{\mathbf{v}}$. Hence we can identify $V^{\#}$ with $\prod_{\mathbf{v} \in \mathcal{B}} F_{\mathbf{v}}$;

$$V^{\#} = \prod_{\mathbf{v} \in B} F_{\mathbf{v}} = F^B$$

Therefore $|V^{\#}| = \mathbf{f^b}$.

If we let $\mathbf{c}$ be the cardinality of a basis of $V^{\#}$ then we have the equality $|V^{\#}| = \mathbf{cf}$ by (D.3).

Now we claim that $\mathbf{c} \geq \mathbf{f}$. For this we construct an injection $\phi : F \to V^{\#}$ as follows: Choose $S = \{\mathbf{v}_i\}_{i \in \mathbb{N}} \subseteq \mathcal{B}$. For $a \in F$ we define the $\mathbf{v}$-th coordinate $\phi(a)_{\mathbf{v}}$ by

$$\phi(a)_{\mathbf{v}} = \begin{cases} a^i & \text{if } \mathbf{v} = \mathbf{v}_i \in S \\ 0 & \text{if } \mathbf{v} \notin S \end{cases}$$

Hence ϕ looks like a row vector

$$\phi(a) = (1, a, a^2, a^3, ...) \in \prod_{\mathbf{v} \in B} F_{\mathbf{v}} = V^{\#}$$

with infinite coordinate. Then obviously $\phi : F \to V^{\#}$ is an injection. Now the image of ϕ is linearly independent. In fact, by Theorem 3.4.14, it follows that the determinant of any n columns of $\phi(a_1), \phi(a_2), ..., \phi(a_n)$ is nonzero if a_i's are nonzero distinct elements of F. Hence the image of ϕ is linearly independent. Therefore $\mathbf{c} \geq \mathbf{f}$.

By Theorem D.1.9, we have $|V^{\#}| = \mathbf{cf} = \mathbf{c}$. Therefore $\mathbf{c} = |V^{\#}| = \mathbf{f^b}$. □

Corollary D.3.2. *If* $\dim(V) = \infty$ *then* V *is never isomorphic to its dual* $V^{\#}$.

Proof. With the same notations of Theorem D.3.1, the cardinality of basis of $V^{\#}$ is $\mathbf{f^b}$. Since $\mathbf{f} \geq 2$ (the field contains at least two elements $0, 1$), by Theorem D.1.8, we have $\mathbf{f^b} > \mathbf{b}$. □

Corollary D.3.3. *Let* V *be a vector space over* $\mathbb{Q}$ *or* $\mathbb{R}$ *or* $\mathbb{C}$ *with a basis of cardinality* $\mathbf{b} \geq \aleph_0$. *Then the cardinality of a basis of* $V^{\#}$ *is* $2^{\mathbf{b}}$.

Proof. We have $2 \leq \aleph_0 \leq \mathbf{f} \leq \aleph_1 = 2^{\aleph_0}$. Since $\mathbf{f} \leq 2^{\aleph_0}$ and since $\mathbf{b} \geq \aleph_0$ by Theorem D.1.9 we have

$$2^{\mathbf{b}} \leq \mathbf{f^b} \leq 2^{\aleph_0 \mathbf{b}} = 2^{\mathbf{b}}$$

Therefore $f^{\mathbf{b}} = 2^{\mathbf{b}}$. □

Index

Bibliography

[1] David S Dummit, Richard M. Foote, *Abstract Algebra* 3rd ed., John Wiley & Sons, Inc., 2004.

[2] Nathan Jacobson, *Lectures in Abstract Algebra* Vol II, Van Nostrand Company Inc., 1953.

[3] Charles C. Pinter, *Set Theory*, Addison-Weseley Publishing Company, Inc., 1971.

References for linear algebra texts:

[4] Stephen H. Friedberg, Arnorld J. Insel, Lawlence E. Spence, *Linear Algebra* 4th ed. Prentice Hall, 2003.

[5] Kenneth Hoffman, Ray Kunze, *Linear Algebra* 2nd ed. Prentice Hall, 1971.